K. Syrjänen L. Gissmann L. G. Koss (Eds.)

Papillomaviruses and Human Disease

With 184 Figures

Springer-Verlag Berlin Heidelberg New York
London Paris Tokyo

Kari J. Syrjänen, M. D., Director
Finnish Cancer Society
Laboratory of Pathology & Cancer Research
Kiekkotie 2
SF-70200 Kuopio, Finland

Lutz Gissmann, Prof. Dr. rer. nat.
Institut für Virusforschung
Deutsches Krebsforschungszentrum
Im Neuenheimer Feld 280
D-6900 Heidelberg, West Germany

Leopold G. Koss, M. D., Professor and Chairman
Department of Pathology, Montefiore Medical Center,
Albert Einstein College of Medicine
111 East 210th Street
Bronx, New York, 10467, USA

Cover photograph: Electron micrograph of papillomavirus particles. Courtesy of Dr. H. Zentgraf.

ISBN-13: 978-3-642-71099-5 e-ISBN-13:978-3-642-71097-1
DOI: 10.1007/978-3-642-71097-1

Library of Congress Cataloging-in-Publication Data.
Papillomaviruses and human disease.
Includes bibliographies and index. 1. Papillomavirus diseases. 2. Papillomaviruses. I. Syrjänen, Kari J. II. Gissmann, L. (Lutz), 1949- III. Koss, Leopold G. [DNLM: 1. Papillomaviruses. 2. Tumor Virus Infections. QW 165.5.P2 P2186]. RC168.P15P365 1987 616.9'25 87-26388

2123/3145-543210

Preface

In recent years, papillomaviruses in general and human papillomaviruses in particular have been recognized as possible agents of important diseases, including some forms of human cancer. The purpose of this book is to present a concise panorama of the present status of knowledge of this topic. This knowledge is as important to molecular biologists and virologists as it is to clinicians and pathologists. To bridge the gap among these diverse groups of investigators, we conceived of a book covering a broad spectrum of the basic scientific, clinical, and pathological aspects of diseases associated with papillomaviruses. Although the principal thrust of this book is directed at human papillomaviruses, fundamental knowledge of animal viruses is essential to the current understanding of the molecular mechanisms of cell transformation. For this reason, a chapter on animal viruses has also been included. Some of the experimental work having to do with the elucidation of transformation and other aspects of interaction between the virus and the cell cannot be based on human papillomaviruses because of a lack of suitable experimental models. Hence, some of the chapters dealing with fundamental aspects of viral molecular biology are based on animal models.

We were very fortunate in having persuaded a number of distinguished colleagues to contribute to this work. We hope that the unusual format of the book will have equal appeal to those engaged in basic sciences and in clinical work, as the two groups must increasingly share knowledge and collaborate for the benefit of humankind.

The book has few precedents, and the choice of topics reflects our preferences. Therefore, comments and criticism from the readers will be greatly appreciated.

K. Syrjänen
L. Gissmann
L. G. Koss

Table of Contents

List of Contributors

BREITBURD, FRANÇOISE V., Ph. D.
Unité des Papillomavirus, Unité INSERM 190, Institut Pasteur,
28, Rue du Dr. Roux
F-75725 Paris Cédex 15, France

DÜRST, MATTHIAS, Dr. rer. nat.
Institut für Virusforschung, Deutsches Krebsforschungszentrum,
Im Neuenheimer Feld 280
D-6900 Heidelberg, West Germany

FUCHS, PAWEL G., Dr. rer. nat.
Institut für Klinische Virologie, Universität Erlangen-Nürnberg,
Loschgestraße 7
D-8520 Erlangen, West Germany

GROSS, GERD, Prof. Dr. med.
Universitäts-Hautklinik und -Poliklinik, Martinistraße 52
D-2000 Hamburg 20, West Germany

GRUSSENDORF-CONEN, ELKE-INGRID, Prof. Dr. med.
Abteilung Dermatologie, Rheinisch-Westfälische TH Aachen,
Pauwelstraße
D-5100 Aachen, West Germany

HAUGEN, THOMAS H., Ph. D.
Department of Pathology, Veterans Administration Medical
Center, The University of Iowa College of Medicine
Iowa City, Iowa 52242, USA

KASHIMA, HASKINS, M. D.
The Johns Hopkins Medical Institutions, Department of
Otolaryngology, Head and Neck Surgery, 600 North Wolfe Street
Baltimore, Maryland 21205, USA

KOSS, LEOPOLD G., M. D., Professor and Chairman
Department of Pathology, Montefiore Medical Center,
Albert Einstein College of Medizin
111 East 210th Street
Bronx, New York 10467, USA

MOUNTS, PHOEBE, Ph. D.
Department of Immunology and Infectious Diseases, School of Public Health, The Johns Hopkins University
Baltimore, Maryland 21205, USA

ORIEL, J. DAVID, M. D.
University College Hospital, Department of Genito-Urinary Medicine, Gower Street
London WC1 6AU, England

PFISTER, HERBERT, Prof. Dr. rer. nat.
Institut für Klinische Virologie, Universität Erlangen-Nürnberg
D-8520 Erlangen, West Germany

PIXLEY, ELLIS C., M. D.
St. Anne's Hospital, Mt. Lawley
Western Australia 6050, Australia

RYLANDER, EVA, M. D.
Department of Gynecology, University Hospital of Umeå
S-90185 Umeå, Sweden

SCHNEIDER, ACHIM, Dr. med.
Universitätsklinikum Ulm, Zytologisch-Histologisches Labor der Frauenklinik, Prittwitzstraße 43
D-7900 Ulm/Donau, West Germany

SCHWARZ, ELISABETH, Dr. rer. nat.
Institut für Virusforschung, Deutsches Krebsforschungszentrum, Im Neuenheimer Feld 280
D-6900 Heidelberg, West Germany

SPRADBROW, PETER B., D. V. M.
Department of Veterinary Pathology and Public Health, University of Queensland, St. Lucia
Brisbane 4067, Australia

SUNDBERG, JOHN P., D. V. M.
The Jackson Laboratory
Bar Harbor, 04609, USA

SYRJÄNEN, KARI J., M. D.
Finnish Cancer Society, Laboratory of Pathology & Cancer Research, Kiekkotie 2
SF-70200 Kuopio, Finland

SYRJÄNEN, STINA M., D.D.S.
Department of Oral Pathology and Radiology, Institute of Dentistry, University of Kuopio, POB 6
SF-70211 Kuopio, Finland

TUREK, LUBOMIR P., M.D.
Department of Pathology, Veterans Administration Medical Center, The University of Iowa College of Medicine
Iowa City, Iowa 52242, USA

VON KROGH, GEO, M.D.
Department of Dermatology, Karolinska Institute at Södersjukhuset
S-10064 Stockholm 38, Sweden

Papillomaviruses: Particles, Genome Organisation and Proteins

H. Pfister and P. G. Fuchs

1 Introduction

Papillomaviruses are classified as genus *Papillomavirus* of the Papovaviridae family on the basis of their capsid structure and biochemical composition (Matthews 1982). Polyomalike viruses form the second genus and are distinguished by a smaller capsid and a shorter DNA. Both genera reveal group-specific antigens (Jenson et al. 1980; Shah et al. 1977) and the nucleic acids of individual members hybridise under conditions of reduced stringency (Law et al. 1979; Howley et al. 1979). However, there ist no cross-reactivity between papillomaviruses and polyomaviruses. DNA sequence analysis has disclosed a fundamentally different genome organisation. There is only one coding DNA strand in papillomaviruses (Engel et al. 1983) whereas separate strands code for vegetative functions and structural proteins in polyomaviruses (Griffin 1980). This indicates that papillomaviruses represent a distinct group.

2 The Virions

2.1 Capsid Structure and Physical Properties

Papillomaviruses have icosahedral capsids about 55 nm in diameter consisting only of protein (Fig. 1a). Analysis of electron micrographs revealed 12 five-coordinated and 60 six-coordinated capsomers arranged on a $T=7$ surface lattice (Finch and Klug 1965; Klug and Finch 1965). After negative staining the capsomers appear as hollow cylinders of equal height and width, which are connected at their base by fibrous structures (Yabe et al. 1979).

It is generally assumed that six-coordinated capsomers consist of six protomers. For polyomavirus, however, it was recently shown that the hexavalent capsomer is a pentamer, implying a total number of 360 protomers instead of 420 as in the case of hexamers (Rayment et al. 1982). The number of protomers can be calculated from the relative molecular mass (M_r) of the "empty" capsid (without DNA) and of the major structural protein. On the basis of a sedimentation coefficient ($S_{20,w}$) of 168 (Crawford and Crawford 1963) and a density in cesium chloride of 1.29 g/ml (Breedis et al. 1962) the molecular mass of empty capsids would be 2.35×10^7. M_r of the major capsid protein of human papillomavirus (HPV) type 1 is 57000 as deduced from the nucleotide sequence (Danos et al. 1982) in accordance with experimental data (Gissmann et al. 1977). From these results the number of protomers can be calculated to be $2.35 \times 10^7 \div 57000 = 413$, which is clearly closer to 420 than to 360, suggesting that the six-coordinated capsomers of papillomaviruses are hexamers.

DNA-containing particles have a sedimentation coefficient ($S_{20,w}$) of 300 and a density in cesium chloride of 1.34 g/ml (Breedis et al. 1962; Crawford and Crawford 1963). Owing to their lack of lipids, they are resistant to ether and other lipid solvents.

2.2 Protein Composition

The papillomavirus capsid consists of at least two structural proteins. One protein with a M_r in the range of 53000–59000 (Favre et al. 1975a; Gissmann et al. 1977; Orth et al. 1977; Lancaster and Olson 1978; Müller and Gissmann 1978; Pfister et al. 1979) represents 80% of the total viral protein (Favre et al. 1975a). A minor component has an average M_r of 70000 (Fig. 1c). Both proteins are coded for by two open reading frames of the viral DNA (see below). In HPV1, additional proteins with M_r of 53000 and 43000 were described, which seem to result from conversion of the major polypeptide (Pfister et al. 1977).

The DNA of HPV1 and bovine papillomavirus (BPV) type 1 is associated with cellular histones to form a chromatin-like complex (Favre et al. 1977; Pfister and zur Hausen 1978). As with polyoma-like viruses, the H1 histone is absent. Histones H3 and H4 are modified, possibly acetylated, which affects their electrophoretic mobility in acidic gel systems (Pfister and zur Hausen 1978). Histones were not found in empty particles or in DNA-containing particles of HPV4 or BPV3 (Gissmann et al. 1977; Pfister et al. 1979).

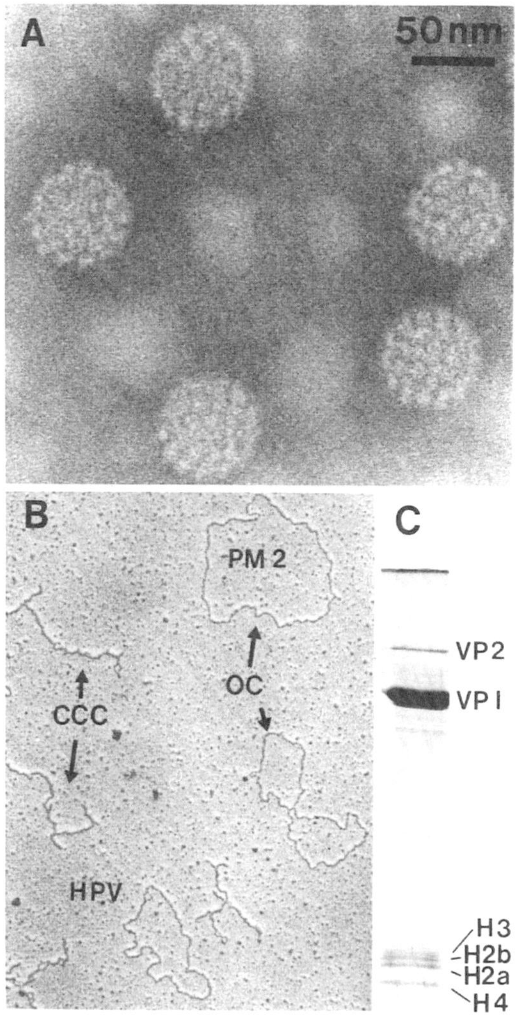

Fig. 1 A-C. Structure **(A)**, genomic DNA **(B)** and protein pattern **(C)** of papillomavirus particles. Capsids were stained with phosphotungstic acid. DNA was extracted by phenol treatment. DNA of phage PM2 was included as a size standard (9.7 kb). The molecules appear as supertwisted covalently closed circles *(CCC)* or open circles *(OC)*. Proteins were separated by polyacrylamide gel electrophoresis after disruption of viral particles by SDS. *VP1* and *VP2* represent structural proteins of the capsid shell. Histones *H3*, *H2b*, *H2a* and *H4* are associated with the viral DNA and appear in preparations of DNA-containing capsids

2.3 Nucleic Acid

Mature papillomavirus particles contain the viral genome in the form of double-stranded circular DNA molecules (Fig. 1b). For the great majority of known isolates the size of the genome is about 7.9 kb (Orth et al. 1977, 1978; Chen et al. 1982; Danos et al. 1982; Schwarz et al. 1983). Some variations have been observed in the range of ±10%, mostly among animal papillomaviruses (Pfister et al. 1979; Campo et al. 1980; Pfister and Meszaros 1980; Jarrett et al. 1984; Kremsdorf et al. 1984). The GC content of the papillomavirus genome is rather low. The average value for sequenced viruses is 42.6%, the extreme cases being HPV16 (36.5%; Seedorf et al. 1985) and deer papillomavirus (DPV) (47.5%; Groff and Lancaster 1985). The GC content of *Mastomys natalensis* papillomavirus DNA amounts to 50% (Müller and Gissmann 1978). In general, animal viruses show a higher GC content than the human ones. There is no clear GC distribution pattern throughout the genome with the exception of the noncoding regions which show slightly lower GC values as a rule.

3 Genome Organisation

The nucleotide sequence data accumulating during the past 4 years allow a more detailed insight into the genomic organisation of papillomaviruses. The complete sequences have been determined for HPV1a, 6b, 8, 11, 16, 33, BPV1, cottontail rabbit papillomavirus (CRPV) and DPV (Chen et al. 1982; Danos et al. 1982; Schwarz et al. 1983; Giri et al. 1985; Seedorf et al. 1985; Groff and Lancaster 1985; Cole and Streeck 1986; Dartmann et al. 1986; Fuchs et al. 1986). A comparison reveals a well-conserved general organisation (Fig. 2). All putative protein-coding sequences (open reading frames, ORFs) are restricted to one DNA strand. The second, presumably noncoding strand contains only short ORFs, which are conserved neither with regard to localisation nor composition. This is in line with all known data on viral transcription, which uniformly point to a single sense strand (see Schwarz, this volume).

Most major ORFs occupy similar positions relative to each other and there is no substantial difference in the length of homologous ORFs, which can be used to align the genomes. The individual frames are classified as "early" (E) or "late" (L) in analogy with other DNA viruses where genes are turned on according to a specific time schedule in the course of a productive infection. The so-called early genes are expressed shortly after infection and prior to the onset of DNA replication. Products of these genes mediate specific functions controlling replication and expression of viral DNA. In the case of tumour viruses, early gene products are also involved in transformation of the host cell. The late genes code for structural porteins of viral particles and are activated during the final stages of the viral cycle.

With papillomaviruses, there is so far really no experimental basis for discriminating between early and late, because replication in vivo is highly dependent on the differentiation of keratinocytes, and replication in vitro has not yet been analysed due to the lack of a productive cell culture system. For convenience, however, we call those ORFs early which are expressed in the nonproductive, basal part of a wart, in BPV1-transformed cells or in CRPV-induced carcinomas. ORFs, which are specifically expressed in differentiated, productive cells of a wart, are regarded as late. One should be aware of the fact that this designation is preliminary. It should also be noted that ORFs are not strict correlates of proteins. Tran-

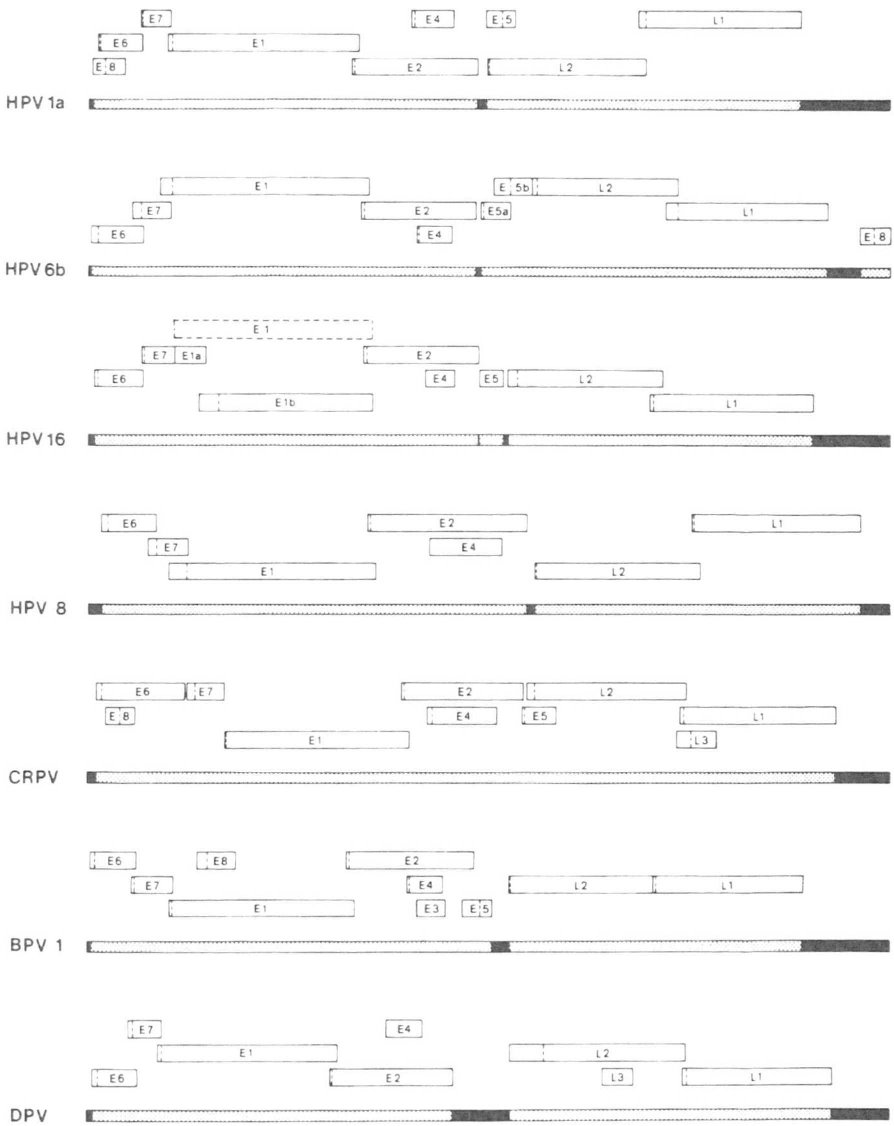

Fig. 2. Genome organisation of human papillomaviruses *(HPV) 1a, 6b, 8, 16,* bovine papillomavirus *(BPV) 1,* cottontail rabbit papillomavirus *(CRPV)* and deer papilloma virus *(DPV).* Open reading frames were displayed by means of the computer programme "FRAMES" (Devereux et al. 1984) and are indicated by *open bars. Dotted lines* within the frames represent the first methionine codon, which could serve as a start point of translation. In *HPV16* the E1 frame appeared to be split in the originally published sequence (Seedorf et al. 1985). A continuous *E1,* which was found in four new isolates (see text) is indicated by *dotted lines. Stippled areas* of the genome bar represent coding sequences and *black regions* stand for so-called noncoding regions

scripts are subject to splicing and proteins may therefore represent hybrids, originating from more than one ORF. Vice versa, one ORF may contain information for more than one portein. Therefore, it is possible that so-called early sequences partially contribute to late proteins. According to present nomenclature early and late ORFs form two blocks, which are frequently separated by noncoding sequences on both sides (Fig. 2).

3.1 The Early Region

Up to eight ORFs showing considerable overlaps can be detected in several virus types within the roughly 4.5 kb early segment. According to the proposal of Danos et al. (1983) they are referred to as E1 through E8. The ORFs E1, E2, E4, E6 and E7 are shared by all genomes analysed so far. E3 was only found in BPV1. The precise localisation of E5 and E8 varies to some extent and the putative amino acid sequences reveal no or only limited homology between different virus types.

3.1.1 Open Reading Frames E6 and E7

Two ORFs, denoted E6 and E7, can be identified at the 5'flank of the early region. The coding capacity of E6 and E7 from different viruses varies between 137-158 and 93-127 amino acid residues, respectively, assuming the first ATG codon to be the translational start signal. The only exception is the large E6 of CRPV, able to code for a polypeptide with 273 residues.

The predicted E6 and E7 proteins from different origins show a moderate degree of rather uniformly distributed homology and can easily be aligned. The conserved elements of E6 proteins appear in the N-terminal moiety of the larger CRPV E6 polypeptide. Most of the cysteine residues of E6/E7 products are arranged in the repetitive motif C-X-X-C. At least 4 repeats can be found in all putative E6 proteins with intervals of 29, 35/36, and 29 amino acid residues. E7 proteins show two cysteine repeats separated by 29 amino acids. The exceptionally large E6 protein of CRPV contains additional cysteine units in its unique part. The exact spacing of cysteine residues suggests that the tertiary structure is well conserved.

The sequence motif C-X-X-C has been identified in a number of proteins, e.g. SV40 T/t antigens, t antigens of polyoma viruses, several early proteins of adenoviruses, a family of glycoprotein hormones, some iron-containing proteins, etc. A homology between the putative E6 product and a family of ATP synthases from mitochondria, chloroplasts and bacteria has been reported in the case of CRPV. The similarities include conserved amino acid residues involved in nucleotide binding.

A functional analysis of these and other ORFs was achieved by genetic studies of BPV1. E7 seems to be important to establish a high copy number of viral DNA in persistently infected cells (Lusky and Botchan 1985). The E6 protein of BPV1 mediates morphological transformation of fibroblasts (Schiller et al. 1984).

An E6-specific protein was recently identified in BPV1 transformed cells (Androphy et al. 1985). To this end the ORF E6 was expressed in bacteria by means of a prokaryotic vector system. Antibodies were raised against the bacterial fusion protein and used for immunoprecipitation of extracts from mouse fibroblasts transformed by BPV1 E6 under control of a retroviral large terminal repeat. The alien promotor was added to enhance gene expression. A 15.5 K protein was precipitated, which fits nicely the predicted molecular weight of the E6 product. Its unchanged electrophoretic mobility in nonreducing gels suggests that there is no disulfide-linked complex of E6 and other proteins. The E6 protein has been shown to be subcellularly distributed in almost equal amounts between the nucleus and the fraction of non-nuclear membranes. As the carboxy terminus of E6 is rather rich in basic amino acids, E6 could well be a DNA-binding protein in agreement with its nuclear localisation. In view of the lack of a series of hydrophobic amino acids, E6 is probably not a transmembrane protein. In virus-transformed cells the 15.5 K protein is synthesised only in very small amounts and is hardly detectable.

An E7-specific protein was detected by immune precipitation with antisera against an E7-trp E fusion protein in extracts of two HPV16-containing cell lines, which were derived from cervical carcinomas. The protein was located in the cytoplasm (Smotkin and Wettstein 1986).

3.1.2 Open Reading Frame E1

The middle of the early region is occupied by a single open reading frame. El is the largest ORF in all papillomaviruses analysed. The amino acid sequences are highly homologous within 60% of the predicted El product, starting from the carboxy terminus of the protein. In contrast the amino terminal region of E1 is rather variable among different papillomaviruses.

ORF E1 of HPV16 turned out to be interrupted by a stop codon after 90 amino acids from the amino terminus (Seedorf et al. 1985). Four additional HPV16 isolates, which were subsequently sequenced, revealed intact E1 ORFs, however, so that the first sequence is likely to be derived from a mutant (M. Dürst and K. Seedorf, personal communication).

A comparison of E1 sequences with those of large T-proteins of polyomaviruses demonstrated significant homologies within more than 200 amino acid residues close to the carboxyterminus of E1 (Clertant and Seif 1984). Two blocks of homologous residues with the potential to form similar secondary structures correspond to sites involved in ATPase and nucleotide-binding activities of large T proteins. The T proteins are involved in the initiation of viral DNA replication. It is therefore interesting to note that E1 of BPV1 plays a role in episomal replication of the viral genome (Lusky and Botchan 1985) suggesting that conserved structural features of E1 and T reflect common functions.

3.1.3 Open Reading Frame E2-E4 Region

Two major open reading frames, E2 and E4, can be detected at the 3'end of the early region. The ORF E4 is the smaller one and in all cases entirely contained within E2-encoding sequences. The start of E2 consistently overlaps with the end of ORF E1 for 79 to 85 bp. Downstream from the first methionine of E2 there is an exactly 19 amino acid overlap with E1 in all sequenced papillomaviruses. A comparison of the putative E2 proteins from different papillomaviruses reveals a very characteristic distribution of homologies. The conserved residues are clearly confined to the N- and C-terminal parts, leaving a variable domain in the middle of the E2 polypeptide. In all known cases this nonhomologous, central part coincides with the position of ORF E4. The conserved C-terminal domain of the E2 protein showed statistically significant homology with one segment of the human *c-mos* proto-oncogene in the case of CRPV, BPV1, HPV1a and HPV6b (Giri et al. 1985). The highest score was obtained with CRPV and HPV1a, showing 29% of identical amino acid residues. Similarities were also found with regard to hydrophilicity pattern and predictable secondary structure. It is noteworthy that the part in question of the *mos* protein is also conserved among *mos*-sequences of human, mouse and rat origin. The area of E2-*mos* homology corresponds to the region of the phosphotyrosine acceptor site of a number of *onc* proteins. The interpretation of these findings is rather difficult. Generally, it might be seen in terms of evolutionary conservation of a locus, serving similar catalytic or binding functions. In certain viruses the *mos*-homology is only poorly expressed (e.g. HPV8).

The E2 gene product of BPV1 activates viral transcription in vitro (Spalholz et al. 1985) and in vivo (Kleiner et al. 1986). It is also able to stimulate *CAT* gene expression under the control of HPV8-specific sequences (Seeberger et al. 1987). This functional complementation is in line with the fact that E2 genes of different papillomaviruses are highly conserved. The C-terminal three-quarters of BPV1 E2 ORF were expressed as a bacterial fusion protein. This peptide was shown to bind to four specific sequences in the regulatory region of BPV1 between late and early parts of the genome (J. Schiller, personal communication). The binding sites share related nucleotide sequences. The motif A-C-C-(G)-X-X-X-Py-C-G-G-T-(G-C), which may be important for E2 recognition, occurs in the regulatory regions of all papillomaviruses sequenced up to now (Dartmann et al. 1986).

E4 ORFs show almost no homologies, which may point to some type-specific functions of E4-encoded proteins. The E4 ORFs of HPV8, HPV16 and DPV do not contain ATG codons over their entire lengths, so they have to be spliced to ATG-carrying 5'leader sequences in order to be translated. Homology was noted between E4 of HPV8 and a part of the transforming EBNA2 protein of Epstein-Barr Virus. From this it is tempting to speculate on the E4 product mediating some oncogenic properties of HPV8.

In the case of HPV1, in vivo synthesised, E4-related proteins were disclosed very recently (Doorbar et al. 1986). Antibodies, which were raised against an E4 fusion protein expressed in bacteria, reacted with four protein doublets of molecular weight 10/11K, 16/17K, 21/23K, and 32/34K, respectively. The 16/17K doublet represented up to 30% of the protein in sodium dodecyl sulphate (SDS) extracts of HPV1-induced warts. It was not observed in preparations from HPV2- and HPV4-containing warts or from normal palmar skin. The E4 specificity of the 16/17K proteins has been proven by the total amino acid composition, which is very similar to the predicted one, and by partial protein sequencing.

The 10/11K proteins may result from splicing part of E4 onto a conserved part of the HPV genome. The total amino acid composition did not fit that predicted for E4, and antisera against purified 10/11K proteins detected a 45K protein in extracts from both HPV1- and HPV2-induced warts. The 21/23K and 32/34K proteins were regarded as dimers of the smaller doublets.

E4-specific antibodies localised E4-specific antigens in cytoplasmic inclusion bodies in cells of the granular layer of HPV1-induced warts. The E4 protein was furthermore detected in a single preparation of purified HPV1 virions (J. Doorbar, personal communication). It seems possible, therefore, that ORF E4 plays a major role in the late phase of the viral replication cycle. This recalls the preliminary character of gene classification in papillomaviruses.

The abundant synthesis of HPV1 E4 seems to be exceptional. It remains to be established whether or not the E4 production of other papillomaviruses follows a similar time course.

3.1.4 Boundary Between Early and Late Regions

The sequences spanning the 3'-terminus of ORF E2 and the beginning of ORF L2 represent an area of quite diverse organisation. One rather consistent feature is the presence of non coding sequences. Their size differs extremely in different viruses, ranging from two small 13 and 36 bp gaps in HPV16 to a large, 602 bp region in DPV. The topography within this area is complicated by the occurrence of ORF E5, which can be found at considerably variable locations in 7 out of 8 papillomavirus types (see Fig. 2).

Using a BPV1 E5-specific antipeptide antiserum, a 6.5k protein could be precipitated from membrane fractions of BPV1-transformed cells (R. Schlegel et al. 1986). The subcellular localisation is in line with the highly hydrophobic character of this protein. Similar hydrophobicity profiles were noted for the putative E5 proteins of BPV2, HPV6 and HPV16. In DPV there exists a very short reading frame, coding for 45 amino acids, 23 of which are homologous to E5 sequences of BPV1, downstream from the first methionine (Groff and Lancaster 1985).

In the BPV1 system, fibroblast-transforming activity was assigned to ORF E5 (Yang et al. 1985; Schiller et al. 1986). It remains to be established whether or not E5 is also involved in oncogenesis by keratinocyte-specific papillomaviruses.

The hydrophobicity profile of the putative E5 protein of HPV1 looks different (R. Schlegel, personal communication). The E5 ORFs of HPV1 and CRPV overlap the conserved polyadenylation signal (see Schwarz, this volume) and the start codon of ORF L2. The sequences show some degree of homology. The meaning of this relationship is difficult to judge, however, because evolutionary restriction might be exerted by the conservation of the 5'-terminus of L2 (see below). The corresponding region of HPV8, for example, codes for homologous peptides parallel to L2 but there is no ORF with an ATG codon at the beginning.

3.2 The Late Region

The organisation of the late region is much simpler when compared to the early part of the genome. Two large ORFs, L2 and L1, occur in this area, overlapping each other slightly in most of the cases. The ORF L2 resides within the 5'half of the late region. Well over 70 amino acid residues at the N-terminus of the putative L2 proteins are highly conserved among different papillomaviruses as is a short motif of about 10 basic amino acids at the C-terminal end. The broad domain in the middle reveals only little, locally clustered, general similarities. A number of homologous sequence stretches can be identified, however, within certain groups of viruses, e.g. HPV1 and HPV8, HPV6 and HPV16, or BPV1 and DPV (Fuchs et al. 1986). It is interesting to note that this grouping reflects the tropism of the respective viruses. HPV1 and HPV8 infect the skin, HPV6 and HPV16 the mucosa, and BPV1 and DPV both fibroblasts and keratinocytes of the skin (Pfister 1984). The data suggests that L2 may be important for virus-target cell interaction. Finally, the C-terminal third of L2 is rather type specific even in the case of closely related viruses such as BPV1 and DPV.

In contrast to ORF L2, ORF L1 is monotonously conserved in all known cases. Comparing HPV1a, HPV6b, BPV1 and CRPV, the overall homology reaches 50% at the amino acid sequence level. In CRPV and DPV one additional small ORF each can be detected at different locations within the late region. It is not clear, however, whether these ORFs are actually functional.

The ORFs L1 and L2 are supposed to code for the major (VP1) and one minor capsid component, respectively. In the case of BPV1 the amino acid composition of VP1 and of the predicted polypeptide of ORF L1 are similar (Meinke and Meinke 1981; Chen et al. 1982). More direct evidence has recently been obtained by cloning late ORFs in bacterial expression vectors. Antisera where prepared against BPV1 L1 and L2, expressed as fusion proteins with β-galactosidase, and found to be active in an ELISA test with BPV1 virions as antigen (Pilacinski et al. 1984). Neutralising activity of these antisera was demonstrated by inhibition of virion-mediated transformation of C127 mouse fibroblasts. These results suggest that ORFs L1 and L2 code for protein sequences that are recognisable at the surface of BPV1 virions.

Similar studies have been undertaken for L1 and L2 of HPV1 (J. Doorbar, personal communication). Antifusion protein antibodies specific for either the N- or C-terminal half of the L2-encoded protein reacted with a minor capsid component of 78k. Antibodies against L1 detected the major 57k structural protein. Both L1 and L2 antibodies strongly stained the nuclei in the granular layer of HPV1-induced warts. The size of the protein detected by L2-specific antibodies is rather surprising. The protein is too large to originate from ORF L2 only. It did not react with available antibodies against fusion proteins specific for HPV1 E1, E2, E4, E7 and L1. Therefore, it remains to be established which ORFs are spliced together to give rise to the 78k protein.

Antisera raised against SDS-disrupted virions were only able to react with the L1-specific 57k protein (J. Doorbar, personal communication). This suggests that group-specific antigens are mainly located on this capsid component, which is in line with the high conservation of ORF L1.

3.3 Boundary Between Late and Early Regions

Roughly 400–900 bp separate the 3'end of L1 from the 5'end of E6 in different papillomaviruses. In most cases this region contains no ORFs of significant length and was therefore supposed to be noncoding. However, in HPV6b there is a small ORF designated E8 and in HPV33 there are a couple of partially overlapping

ORFs. Even with other papillomaviruses there may be some coding potential in this area. Recent genetic experiments in M. Botchan's laboratory point to a short exon in the so-called noncoding region of BPV1 (personal communication). This probably not quite appropriate term might be changed into "regulatory region" in view of the numerous control signals consistently found in this area. On the nucleotide level this function is reflected by a series of palindromes, AT-rich sequences and direct and inverted repeats. Transcription control elements will be extensively discussed by Schwarz (this volume). Last but not least the origin of replication resides in this region. Analysing replicative intermediates of BPV1 under the electron microscope (Waldeck et al. 1984), the origin was mapped close to the 3'-end of L1. When BPV1 chromatin was exposed to DNase I, the corresponding region turned out to be particularly sensitive (Rösl et al. 1983).

Through genetic experiments BPV1 sequences were identified in the regulatory region and within ORF E1 which are cis-essential for plasmid replication. These plasmid maintenance sequences (PMS) are highly homologous (Lusky and Botchan 1984). PMS1 coincides with the origin of replication, whereas no molecules were found which initiated replication at PMS2 (Waldeck et al. 1985).

4 Classification of Papillomaviruses

Papillomaviruses are classified according to host range and relatedness of nucleic acids (Coggin and zur Hausen 1979). Serology plays no role in the taxonomy of these viruses because many types replicate so poorly that there is not enough antigen for immunisation and/or for detection. The first problem can be solved now by the synthesis of viral proteins in bacterial expression systems and by immunisation of animals with *E. coli*-produced proteins. The sera may be used for the comparison of known types but are probably of limited value for the characterisation of new types. HPV1-5 were previously analysed with antisera raised against native virus particles and showed no serological cross-reactivity (Gissmann et al. 1977; Orth et al. 1977, 1978).

Each virus is named first after its natural host. This is usually unambiguous because papillomaviruses show a very restricted host range (Pfister 1984). However, the host range does not necessarily reflect the natural relationship. Isolates from different species but with similar biological properties such as BPV1 and DPV tend to be more closely related than viruses from the same species, infecting for example different tissues (Lancaster and Sundberg 1982; Jarrett 1985).

Papillomavirus DNA isolates from one species are classified according to sequence homology; if there is less than 50% cross-hybridisation when tested by reassociation in the liquid phase, the DNAs are assigned to different types. On this basis 42 human (Table 2) and 6 bovine papillomaviruses (Jarrett 1985) are differentiated at the moment. For most other species only one papillomavirus has been characterized so far (see Sundberg, this volume). In view of the rapidly increasing number of HPV types in particular (which most likely reflects greater research efforts than a special situation in humans), it might be worthwhile to think about the type criterion for a moment. Compared with other fields in virol-

Table 1. Examples for DNA cross-hybridisation between different human papillomavirus types

Papillomavirus types	% Cross-hybridisation	Reference
HPV 10a - 10P	73	Ostrow et al. 1983
HPV 10P - 10PW	60	Ostrow et al. 1983
HPV 2a - 2c	55	Fuchs and Pfister 1984
HPV 14 - 21	38	Kremsdorf et al. 1984
HPV 5 - 25	29	Gassenmaier et al. 1984
HPV 5 - 12	19	Kremsdorf et al. 1984
HPV 18 - 32	10	Orth, personal communication
BPV 1 - DPV	9	Lancaster and Sundberg 1982
HPV 6 - 13	4	Pfister et al. 1983
HPV 9 - 20	1	Gassenmaier et al. 1984
HPV 1 - 2	not detectable	Orth et al. 1977
HPV 13 - 18	not detectable	Boshart et al. 1984

ogy the criterion is rather strict. Many serotypes show much less DNA or RNA sequence divergence. The degree of cross-hybridisation must not be confused, however, with actual sequence homology. HPV6 and HPV11, for example, show only 25% cross-reactivity when analysed by reassociation kinetics but up to 90% nucleotide sequence identity within certain genome regions (Gissmann et al. 1982; Dartmann et al. 1986).

The 50% borderline is obviously arbitrary within a rather continuous spectrum of DNA relationships (Table 1). Some subtypes show as little as 55% cross-hybridisation and some types are almost as closely related. From the medical point of view a type differentiation is only meaningful if it correlates with clinically relevant biological properties. Among known papillomaviruses, pathogenetic differences were noted for isolates with 30% cross-hybridization or less. HPV5, for example, is distinguished from closely related viruses like HPV25 or HPV12 by its regular association with skin carcinomas of patients with epidermodysplasia verruciformis. In contrast, the related viruses have so far only been detected in benign tumours. This indicates that 30% cross-hybridisation may represent a limit where type differentiation becomes of interest for the clinician. However, not all viruses, which are more distantly related, differ in their biological properties. HPV6 and HPV11 induce indistinguishable lesions and the same holds true for HPV9 and HPV20, which are further apart than BPV1 and DPV from two species.

In Table 2 human papillomaviruses are listed according to DNA sequence homology. There are 15 groups (A to 0), the members of which cross-hybridise to various extents but show no relationship to other groups when tested under stringent conditions. It is important to note that viruses, which are related on the DNA level, generally share common pathogenic properties. Group B, for example, comprises two subgroups (HPV2/HPV27 and HPV3/HPV10/HPV28, respectively) and HPV29 represents a connecting link. All these viruses induce common or flat skin warts. Some differ considerably in pathology and cytopathic effects but there exist links with the morphology of common warts and the histology of flat warts, referred to as "intermediate" warts (S. Jablonska, personal communication). Group D viruses were originally isolated from patients with epidermodysplasia verruciformis. In rare cases they were detected in keratoacanthomas, solar kerato-

Table 2. Human papillomavirus types and associated human diseases

Group	HPV type	Clinical symptoms	References
A	1	Deep plantar warts, common warts	Favre et al. 1975b Gissmann et al. 1977
B	2	Common warts	Orth et al. 1977 Laurent et al. 1982
	3	Flat warts	Orth et al. 1978
	10	"Intermediate" warts	Kremsdorf et al. 1983
	27	Flat warts	Ostrow, personal communication
	28	"Intermediate" warts	Favre et al., personal communication
	29	Common warts	Favre et al., personal communication
C	4	Common warts	Gissmann et al. 1977 Pfister et al. 1980
D	5 8 9 12 14, 15 17 19-25 36-38	Flat warts, macules and pityriasis versicolor-like lesions of patients with epidermodysplasia verruciformis, occasionally found in keratoacanthomas, solar keratosis, and melanoma	Orth et al. 1980 Pfister et al. 1981 Kremsdorf et al. 1982, 1983, 1984 Gassenmaier et al. 1984 Favre et al., personal communication Scheuerlen et al., personal communication
E	6 11	Flat and exophytic condylomas, laryngeal papillomas, cervical precancerous (?) lesions	Gissmann and zur Hausen 1980 Gissmann et al. 1982, 1983
	13	Focal epithelial hyperplasia Heck	Pfister et al. 1983
F	7	Common warts	Orth et al. 1981 Ostrow et al. 1981
G	16	Flat condylomas, cervical precancerous lesions, bowenoid papulosis, cervical cancer	Dürst et al. 1983 Ikenberg et al. 1983
	31	Flat condylomas, cervical precancerous lesions	Lorincz et al. 1986
H	18	Flat condylomas, cervical precancerous lesions, cervical cancer	Boshart et al. 1984
	32	Focal epithelial hyperplasia Heck	Beaudenon et al., personal communication
	42	Flat condylomas, bowenoid papulosis	
I	26	"Intermediate" warts	Ostrow et al. 1984
J	30	Laryngeal carcinoma, exophytic condyloma	Kahn et al. 1986
K	33	Cervical cancer	Beaudenon et al., personal communication
L	34	Bowenoid papulosis, Bowen's disease	Kawashima et al. 1986
M	35	Cervical cancer	Lorincz et al., personal communication
N	39	Bowenoid papulosis, cervical cancer	Beaudenon et al., personal communication
O	41	Flat warts	zur Hausen et al., personal communication

sis and in one melanoma. It is not yet clear if they play an aetiological role in these tumors or appear as accidental passengers. Members of groups E, G and H all affect the mucosa. DNA sequence analysis of the prototypes HPV6, HPV16 and HPV18 reveals that they are significantly more closely related to each other than to viruses from skin tumors (Dartmann et al. 1986).

In summary, clear-cut biological differences can be observed among viruses which show less than 30% DNA cross-hybridisation. There are examples, however, of apparently identical pathogenicity in spite of a rather distant relationship, and some common features in tropism, pathology and cytopathogenicity are a rule so long as viral DNAs cross-hybridise under stringent conditions.

Some isolates are not easy to classify in the presented scheme. HPV32 induces the same clinical picture as HPV13, namely focal epithelial hyperplasia of the oral mucosa, but both viruses belong to different groups according to their preferential cross-hybridisation with HPV18 and HPV6/HPV11, respectively (Tables 1 and 2). Besides the clear reactivity with HPV18, there is also marginal cross-hybridisation between HPV32 and HPV6, 11, 13 and 16. A similar situation exists with HPV26, showing weak but significant cross-reactivity with HPV3, 6, 16, and 18 (Ostrow et al. 1984; Lang and Pfister, unpublished) and with HPV7, which hybridises to some extent with HPV8 and HPV16 (Oltersdorf et al. 1986).

These data can be explained by an uneven distribution of homologies throughout the genomes and by the fact that regions of maximal homology may map in different areas of the genome when different viruses are compared. As outlined in Sect. 3, there is a general pattern of homology between papillomavirus genomes. The DNA sequences are basically highly conserved within the 3'-moiety of ORF E1, the 5'-part of ORF E2 and within L1. Moderate homology occurs in the areas of E6, E7, and at the termini of ORF L2. The noncoding region, ORFs E4, E5 and the middle of L2, are rather type specific. As far as closer related types are concerned, this pattern may differ, however. Heteroduplex analysis of HPV6 and HPV11 revealed the greatest divergence within E5b, followed by the noncoding area and E5a, and finally by E4, the 3'-third of E1, and the 5'-half of L2 (Broker and Chow 1985).

These data indicate that information on overall DNA sequence homology is just not sufficient for a clinically meaningful interpretation. On the other hand we do not yet know which genes to focus on so long as our knowledge of the molecular biology of papillomaviruses remains rather limited. Therefore there is no other way of approaching papillomavirus plurality in the near future than by differentiating isolates on the basis of a DNA relationship. The present type definition seems reasonable, although it could probably be a bit more strict.

As a long-term goal we should characterise the biological activities of individual types. A comparison of isolates with similar or different pathogenic properties should help to assign biological functions to individual genes. For medical purposes it may then be possible to confine oneself to a subset of papillomavirus genes, and this could help to simplify the confusing systematics of this virus group.

References

Androphy EJ, Schiller JT, Lowy DR (1985) Identification of the protein encoded by the E6 transforming gene of bovine papillomavirus. Science 230: 442-445

Boshart M, Gissmann L, Ikenberg H, Kleinheinz A, Scheurlen W, zur Hausen H (1984) A new type of papillomavirus DNA, its presence in genital cancer biopsies and in cell lines derived from cervical cancer. EMBO J 3: 1151-1157

Breedis C, Berwick L, Anderson TF (1962) Fractionation of Shope papilloma virus in cesium chloride density gradients. Virology 17: 84-94

Broker TR, Chow LT (1985) Electron microscopiacal heteroduplex comparisons on DNAs from human papilloma virus types 6, 11, 16 and 18. J Cell Biochem (Suppl) 9C: 84

Campo MS, Moar MH, Jarret WFH, Laird HM (1980) A new papillomavirus associated with alimentary cancer in cattle. Nature 286: 180-182

Chen EY, Howley PM, Levinson AD, Seeburg PH (1982) The primary structure and genetic organization of bovine papillomavirus type 1 genome. Nature 299: 529-534

Clertant P, Seif I (1984) A common function for polyoma virus large-T and papillomavirus E1 proteins? Nature 311: 276-279

Coggin JR, zur Hausen H (1979) Workshop on papillomaviruses and cancer. Cancer Res 39: 545-546

Cole ST, Streeck RE (1986) Genome organization and nucleotide sequence of human papillomavirus, type 33, which is associated with cervical cancer. J Virol 58: 991-995

Crawford LV, Crawford EM (1963) A comparative study of polyoma and papilloma viruses. Virology 21: 258-263

Danos O, Katinka M, Yaniv M (1982) Human papillomavirus la complete DNA sequence: a novel type of genome organization among papovaviridae. EMBO J 1: 231-236

Danos O, Engel LW, Chen EY, Yaniv M, Howley PM (1983) Comparative analysis of the human type 1 and bovine type 1 papillomavirus genomes. J Virol 46: 557-566

Dartmann K, Schwarz E, Gissmann L, zur Hausen H (1986) The nucleotide sequence and genome organization of human papillomavirus type 11. Virology 151:124-130

Devereux J, Haeberli P, Smithies O (1984) A comprehensive set of sequence analysis programs for the VAX. Nucleic Acids Res 12:387-395

Doorbar J, Campbell D, Grand RJA, Gallimore PH (1986) Identification of the human papilloma virus-1a E4 gene products. EMBO J 5: 355-362

Dürst M, Gissmann L, Ikenberg H, zur Hausen H (1983) A papillomavirus DNA from a cervical carcinoma and its prevalence in cancer biopsy samples from different geographic regions. Proc Natl Acad Sci USA 80: 3812-3815

Engel LW, Heilman CA, Howley PM (1983) Transcriptional organization of bovine papillomavirus type 1. J Virol 47: 516-528

Favre M, Breitburd F, Croissant O, Orth G (1975a) Structural polypeptides of rabbit, bovine, and human papilloma viruses. J Virol 15: 1239-1247

Favre M, Orth G, Croissant O, Yaniv M (1975b) Human papillomavirus DNA: physical map. Proc Natl Acad Sci USA 72: 4810-4814

Favre M, Breitburd F, Croissant O, Orth G (1977) Chromatin-like structures obtained after alkaline disruption of bovine and human papillomaviruses. J Virol 21: 1205-1209

Finch JI, Klug A (1965) The structure of viruses of the papillomapolyoma type. III. Structure of rabbit papilloma virus. J Mol Biol 13: 1-12

Fuchs PG, Pfister H (1984) Cloning and characterization of papillomavirus type 2c DNA. Intervirology 22: 177-180

Fuchs PG, Iftner T, Weninger J, Pfister H (1986) Epidermodysplasia verruciformis-associated human papillomavirus 8: genomic sequence and comparative analysis. J Virol 58: 626-634

Gassenmaier A, Lammel M, Pfister H (1984) Molecular cloning and characterization of the DNAs of human papillomaviruses 19, 20 and 25 from a patient with epidermodysplasia verruciformis. J Virol 52: 1019-1023

Giri I, Danos O, Yaniv M (1985) Genomic structure of the cottontail rabbit (Shope) papillomavirus. Proc Natl Acad Sci USA 82: 1580-1584

Gissmann L, zur Hausen H (1980) Partial characterization of viral DNA from human genital warts (condylomata acuminata). Int J Cancer 25: 605-609

Gissmann L, Pfister H, zur Hausen H (1977) Human papilloma virus (HPV): characterization of 4 different isolates. Virology 76: 569-580
Gissmann L, Diehl V, Schulz-Coulon H-J, zur Hausen H (1982) Molecular cloning and characterization of human papilloma virus DNA derived from a laryngeal papilloma. J Virol 44: 393-400
Gissmann L, Wolnik L, Ikenberg H, Koldovsky U, Schnürch HG, zur Hausen H (1983) Human papillomavirus types 6 and 11 DNA sequences in genital and laryngeal papillomas and in some cervical cancers. Proc Natl Acad Sci USA 80: 560-563
Griffin BE (1980) Structure and genomic organization of SV40 and polyoma virus. In: Tooze J. (ed) DNA tumor viruses. Cold Spring Harbor Laboratory, Cold Spring Harbor, pp 61-123
Groff DE, Lancaster WD (1985) Molecular cloning and nucleotide sequence of deer papillomavirus. J Virol 56: 85-91
Howley PM, Israel MA, Law M-F, Martin MA (1979) A rapid method for detecting and mapping homology between heterologous DNAs. Evaluation of polyomavirus genomes. J Biol Chem 254: 4876-4883
Ikenberg H, Gissmann L, Gross G, Grußendorf-Conen E-I, zur Hausen H (1983) Human papillomavirus type 16-related DNA in genital Bowen's disease and in bowenoid papulosis. Int J Cancer 32: 563-565
Jarrett WFH (1985) The natural history of bovine papillomavirus infections. In: Klein G (ed) Advances in Viral Oncology, vol 5. Raven Press, New York, pp 83-101
Jarrett WFH, Campo MS, O'Neil BW, Laird HM, Coggins LW (1984) A novel bovine papillomavirus (BPV-6) causing true epithelial papillomas of the mammary gland skin: a member of a proposed new BPV subgroup. Virology 136: 255-264
Jenson AB, Rosenthal JR, Olson C, Pass F, Lancaster WD, Shah K (1980) Immunological relatedness of papillomaviruses from different species. J Natl Cancer Inst 64: 495-500
Kahn T, Schwarz E, zur Hausen H (1986) Molecular cloning and characterization of the DNA of a new human papillomavirus (HPV30) from a laryngeal carcinoma. Int J Cancer 37: 61-65
Kawashima M, Jablonska S, Favre M, Obalek S, Croissant O, Orth G (1986) Characterization of a new type of human papillomavirus found in a lesion of Bowen's disease of the skin. J Virol 57: 688-692
Kleiner E, Dietrich W, Pfister H (1986) Differential regulation of papilloma virus early gene expression in transformed fibroblasts and carcinoma cell lines. EMBO J 5: 1945-1950
Klug A, Finch JT (1965) Structure of virus of the papillomapolyoma type I. Human wart virus. J Mol Biol 11: 403-423
Kremsdorf D, Jablonska S, Favre M, Orth G (1982) Biochemical characterization of two types of human papillomaviruses associated with epidermodysplasia verruciformis. J Virol 43: 436-447
Kremsdorf D, Jablonska S, Favre M, Orth G (1983) Human papillomaviruses associated with epidermodysplasia verruciformis. II. Molecular cloning and biochemical characterization of human papillomavirus 3a, 8, 10 and 12 genomes. J Virol 48: 340-351
Kremsdorf D, Favre M, Jablonska S, Obalek S, Rueda LA, Lutzner MA, Blanchet-Bardon C, Van Voorst Vader PC, Orth G (1984) Molecular cloning and characterization of the genomes of nine newly recognized human papillomavirus types associated with epidermodysplasia verruciformis. J Virol 52: 1013-1018
Lancaster WD, Olsen C (1978) Demonstration of two distinct classes of bovine papilloma virus. Virology 89: 372-379
Lancaster WD, Sundberg JP (1982) Characterization of papillomaviruses isolated from cutaneous fibromas of white-tailed deer and mule deer. Virology 123: 212-216
Laurent R, Kienzler JL, Croissant O, Orth G (1982) Two anatomoclinical types of warts with plantar localization: specific cytopathogenic effects of papillomaviruses. Arch Dermatol Res 274: 101-111
Law M-F, Lancaster WD, Howley PM (1979) Conserved sequences among the genomes of papillomaviruses. J Virol 32: 199-207
Lorincz AT, Lancaster WD, Temple GF (1986) Cloning and characterization of the DNA of a new human papillomavirus from a woman with dysplasia of the uterine cervix. J Virol 58: 225-229
Lusky M, Botchan MR (1984) Characterization of the bovine papilloma virus plasmid maintenance sequences. Cell 36: 391-401

Lusky M, Botchan MR (1985) Genetic analysis of bovine papillomavirus type 1 trans-acting replication factors. J Virol 53: 955-965
Matthews REF (1982) Classification and nomenclature of viruses. Intervirology 17: 1-199
Meinke W, Meinke GC (1981) Isolation and characterization of the major capsid protein of bovine papilloma virus type 1. J Gen Virol 52: 15-24
Müller H, Gissmann L (1978) Mastomys natalensis papilloma virus (MnPV), the causative agent of epithelial proliferations: characterization of the virus particle. J Gen Virol 41: 315-323
Oltersdorf T, Campo MS, Favre M, Dartmann K, Gissmann L (1986) Molecular cloning and characterization of human papillomavirus type 7 DNA. Virology 149: 247-250
Orth G, Favre M, Croissant O (1977) Characterization of a new type of human papillomavirus that causes skin warts. J Virol 24: 108-120
Orth G, Jablonska S, Favre M, Croissant O, Jarzabek-Chorzelska M, Rzesa G (1978) Characterization of two types of human papillomaviruses in lesions of epidermodysplasia verruciformis. Proc Natl Acad Sci USA 75: 1537-1541
Orth G, Favre M, Breitburd F, Croissant O, Jablonska S, Obalek S, Jarzabek-Chorzelska M, Rzesa G (1980) Epidermodysplasia verruciformis: a model for the role of papilloma viruses in human cancer. Cold Spring Harbor Conf Cell Prolif 7: 259-282
Orth G, Jablonska S, Favre M, Croissant O, Obalek S, Jarzabek-Chorzelska M, Jibard N (1981) Identification of papillomaviruses in butcher's warts. J Invest Dermatol 76: 97-102
Ostrow RS, Krzyzek R, Pass F, Faras AJ (1981) Identification of a novel human papilloma virus in cutaneous warts of meat handlers. Virology 108: 21-27
Ostrow RS, Zachow KR, Watts S, Bender M, Pass F, Faras AJ (1983) Characterization of two HPV-3 related papillomaviruses from common warts that are distinct clinically from flat warts of epidermodysplasia verruciformis. J Invest Dermatol 80: 436-440
Ostrow RS, Zachow KR, Thompson O, Faras AJ (1984) Molecular cloning and characterization of a unique type of human papillomavirus from an immune deficient patient. J Invest Dermatol 82: 362-366
Pfister H (1984) Biology and biochemistry of papillomaviruses. Rev Physiol Biochem Pharmacol 99: 111-181
Pfister H, Meszaros J (1980) Partial characterization of a canine oral papillomavirus. Virology 104: 243-246
Pfister H, zur Hausen H (1978) Characterization of proteins of human papilloma viruses (HPV) and antibody response to HPV1. Med Microbiol Immunol 166: 13-19
Pfister H, Gissmann L, zur Hausen H (1977) Partial characterization of proteins of human papilloma viruses (HPV) 1-3. Virology 83: 131-137
Pfister H, Linz U, Gissmann L, Huchthausen B, Hoffmann D, zur Hausen H (1979) Partial characterization of a new type of bovine papillomaviruses. Virology 96: 1-8
Pfister H, Gissmann L, zur Hausen H, Gross G (1980) Characterization of human and bovine papilloma viruses and of the humoral immune response to papilloma virus infection. Cold Spring Harbor Conf Cell Prolif 7: 249-258
Pfister H, Nürnberger F, Gissman L, zur Hausen H (1981) Characterization of a human papillomavirus from epidermodysplasia verruciformis lesions of a patient from Upper Volta. Int J Cancer 27: 645-650
Pfister H, Hettich I, Runne U, Gissmann L, Chilf GN (1983) Characterization of human papillomavirus type 13 from lesions of focal epithelial hyperplasia Heck. J Virol 47: 363-366
Pilacinski WP, Glassman DL, Krzyzek RA, Sadowski PL, Robbins AK (1984) Cloning and expression in *Escherichia coli* of the bovine papillomavirus L1 and L2 open reading frames. Biotechnology 1: 356-360
Rayment I, Baker TS, Caspar DLD, Murakami WT (1982) Polyoma virus capsid structure at 22.5 Å resolution. Nature 295: 110-115
Rösl F, Waldeck W, Sauer G (1983) Isolation of episomal bovine papillomavirus chromatin and identification of a DNase I-hypersensitive region. J Virol 46: 567-574
Schiller JT, Vass WC, Lowy DR (1984) Identification of a second transforming region in bovine papillomavirus DNA. Proc Natl Acad Sci USA 81: 7880-7884
Schiller JT, Vass WC, Vousden KH, Lowy DR (1986) E5 open reading frame of bovine papillomavirus type 1 encodes a transforming gene. J Virol 57: 1-6
Schlegel R, Wade-Glass M, Rabson MS, Yang YC (1986) The E5 transforming gene of bovine papillomavirus encodes a small, hydrophobic polypeptide. Science 233: 464-467

Schwarz E, Dürst M, Demankowski C, Lattermann O, Zech R, Wolfsperger E, Suhai S, zur Hausen H (1983) DNA sequence and genome organization of genital human papillomavirus type 6b. EMBO J 2: 2341-2348
Seeberger R, Haugen TH, Turek L, Pfister H (1987) An enhancer of human papillomavirus 8 is trans-activated by the bovine papillomavirus 1 E2 function. In: Cancer Cells vol 5. Cold Spring Harbor Laboratory, New York (to be published)
Seedorf K, Krämmer G, Dürst M, Suhai S, Röwekamp WG (1985) Human papillomavirus type 16 DNA sequence. Virology 145: 181-185
Shah KV, Ozer HL, Ghazey HN, Kelly TJ (1977) Common structural antigen of papovaviruses of the simian virus 40 - polyoma subgroup. J Virol 21: 179-186
Smotkin D, Wettstein FO (1986) Transcription of human papillomavirus type 16 early genes in a cervical cancer and a cancer-derived cell line and identification of the E7 protein. Proc Natl Acad Sci USA 83: 4680-4684
Spalholz BA, Yang Y-C, Howley PM (1985) Transactivation of a bovine papillomavirus transcriptional regulatory element by the E2 gene product. Cell 42: 183-191
Waldeck W, Rösl F, Zentgraf H (1984) Origin of replication in episomal bovine papilloma virus type 1 DNA isolated from transformed cells. EMBO J 3: 2173-2178
Yabe Y, Sadakane H, Isono H (1979) Connection between capsomeres in human papilloma virus. Virology 96: 547-552
Yang Y-C, Spalholz BA, Rabson MS, Howley PM (1985) Dissociation of transforming and trans-activation functions for bovine papillomavirus type 1. Nature 318: 575-577

Methods of Identification of Human Papillomaviruses

A. Schneider

1 Introduction

Human papillomaviruses (HPV) interact with host tissues in a number of ways. The presence of HPV DNA is an obligatory prerequisite for any type of infection. In the event of transcription, viral RNA can be identified. Only with permissive infection are whole viral particles produced. All other types of infection are described as non-permissive. Thus, each HPV detection method depends on the biological state of the infection. As HPV DNA is present in all types of infections, its identification is the most sensitive method, followed by HPV antigen identification and detection of HPV particles. HPV infections may cause a broad spectrum of morphological changes of the target tissues. Regardless of the procedure selected, the specificity of HPV detection is rather high. The identification of specific types of HPV requires techniques of DNA or RNA hybridisation. In some lesions a type-specific cytopathic effect may be present. Antigen belonging to or induced by specific HPV types may be identified in the near future.

2 Light Microscopic Cytopathic Effects

The morphology of HPV-associated diseases of different organs of the human body is described extensively in other chapters of this book (see Gross, Grußendorf-Conen, Kashima, Koss, S. Syrjänen and K. Syrjänen). Morphological changes specific for some HPV types are discussed below.

2.1 Cutaneous Lesions

Cutaneous HPV infections are classified according to their anatomical distribution, clinical appearance and, more recently, histological patterns. Histological studies reveal a number of features common to most lesions regardless of type of HPV (see Grußendorf-Conen, this volume). Except for the absence of hyperkeratosis in epidermodysplasia verruciformis (EV), all histological features are represented in variable degree in diseases caused by HPV1,2,3,4,7, and HPV5,8, and 9 (Croissant et al. 1985; Table 1). Thus, distinct histological patterns allow the classification of HPV-associated cutaneous lesions. In all lesions the cytopathic effects listed in Table 1 are variable and usually present only in some of the epithelial cells. The remaining cells lack distinct morphological abnormalities. It is possible

Table 1. Morphology of the cytopathic effect caused by different HPVs in cutaneous lesions

		HPV1	HPV2	HPV3	HPV4	HPV5,8,9	HPV7
Stratum corneum	Hyperkeratosis[1]	+ + +	+ +	+	+ +	Ø	+ + +
	Parakeratosis[2]	+ + +	+ +	(+)	+	+	+
	Papillomatosis[3]	+ + +	+ +	(+)	+ + +	+ / + +	+ +
Stratum granulosum	Clear cells[4]	+	+ +	+ +	+ + +	+ +	+ +
	Cytoplasmic inclusions[5]	p np + + + / +	+	+	+ + +	+ +	p np Ø / +
Stratum germinativum	Acanthosis[6]	p np Ø / + + +	+ + +	+ +	+	+	+ +
	Clear cells[4]	Ø	+	+	+ +	+ + +	+

p, permissive infection; *np*, nonpermissive infection
[1] Increase in the thickness of the Keratin-forming layers of cells
[2] Retention of nuclei in the cornified layer
[3] Upward proliferation of papillae causing undulation of surface
[4] Nucleus surrounded by clear cytoplasm
[5] Keratohyalin or keratohyalin-like granules
[6] Elongation of the papillae of the squamous epithelium

that the latter cells may be infected in a non-permissive way or may not be infected at all. The pathognomonic cytoplasmic inclusions of HPV1 lesions are related to the accumulation of the E 4 gene product (F. Breitburd, personal communication; see also Pfister and Fuchs, this volume). It is interesting to note that in EV lesions with malignant potential distinct morphological features may be observed. In benign lesions associated with HPV3 and HPV10 infection, the degree of papillomatosis and hyperkeratosis is quite different when compared with lesions capable of progression to carcinoma and associated with HPV 5 and HPV8 (Croissant et al. 1985; Table 1).

2.2 Anogenital Lesions

Analysis of cervical smears has shown that specific cell types such as "koilocytes" and "dyskeratocytes" are pathognomonic for HPV infection (see Koss, this volume). The so-called atypical condyloma is distinguished cytologically from ordinary condyloma and cervical neoplasia by the occurrence of squamous cells with markedly atypical nuclei (Meisels et al. 1981). There is some preliminary evidence that based on these cytological features lesions associated with HPV6/11 and HPV16 can be separated. HPV6 and HPV11 are thought to be associated with ordinary flat condyloma, whereas HPV16 is found in "atypical condyloma" (Meisels and Morin 1986). It should be stressed, however, that the issue of type-specific histological alterations has not been settled as yet for the cervix. Analysis of lesions with known HPV type showed that "flat" condylomas with nuclear atypia confined to the surface epithelial cells were infected with HPV6 and HPV11. When nuclear atypia was present in all epithelial layers of koilocytotic atypia (cervical intraepithelial neoplasia with koilocytosis, CINK), HPV16 was detected in 15 of 18 biopsies analysed (see Fig. 9 A) (Crum et al. 1985). These findings are in contrast with another study in which no prediction of the HPV type could be made and the occurrence of atypical abnormal mitoses, multinucleation, koilocytosis and HPV antigen was analysed (Kadish et al. 1986). Regarding specific cytopathic effects in histological sections of anogenital lesions, condylomata acuminata caused by HPV6 and HPV11 show a distinctive combination of marked papillomatosis, acanthosis, koilocytosis and hyper- or parakeratosis (see Gross, this volume). Bowenoid papulosis of the penis and vulva appears as penile and vulvar intraepithelial neoplasia (PIN, VIN) grade III and is associated with HPV16 in more than 90% of cases (see Gross, this volume). Nuclear atypia involves the full thickness of the epithelium. Koilocytosis is absent in bowenoid papulosis.

2.3 Oral Lesions

In lesions of the oral cavity with focal epithelial hyperplasia (FEH), pathognomonic FEH cells have been described. These cells show "mitosis-like and ballooning-type nuclear degeneration" (Praetorius-Clausen 1969). This cytopathic phenomenon may be specific for HPV13 most commonly found in FEH (see S. Syrjänen, this volume).

3 Electron Microscopy

3.1 Cutaneous Lesions

HPV particles were first identified in skin warts by negative staining of ground suspensions (Strauss et al. 1949) and by transmission electron microscopy of tissue sections (Almeida et al. 1962; Fig. 1). The virus is characterised by icosahedral particles consisting of 72 capsomeres and measuring approximately 55 nm in diameter. The virions are arranged in crystalline arrays or scattered throughout the nucleus. After rupture of the nuclear membrane, virions may also be present in the cytoplasm. The concentration of viral particles varies considerably in the different wart types. Palmar and plantar warts show the highest number of viral particles followed by flat warts, whereas common warts contain a lower virus concentration (Almeida et al. 1962; Laurent et al. 1975). Occasionally, in the latter group of warts viral particles can be found. The concentration of particles also correlates with the age of the lesion. Warts that are 6–12 months old contain the highest number of

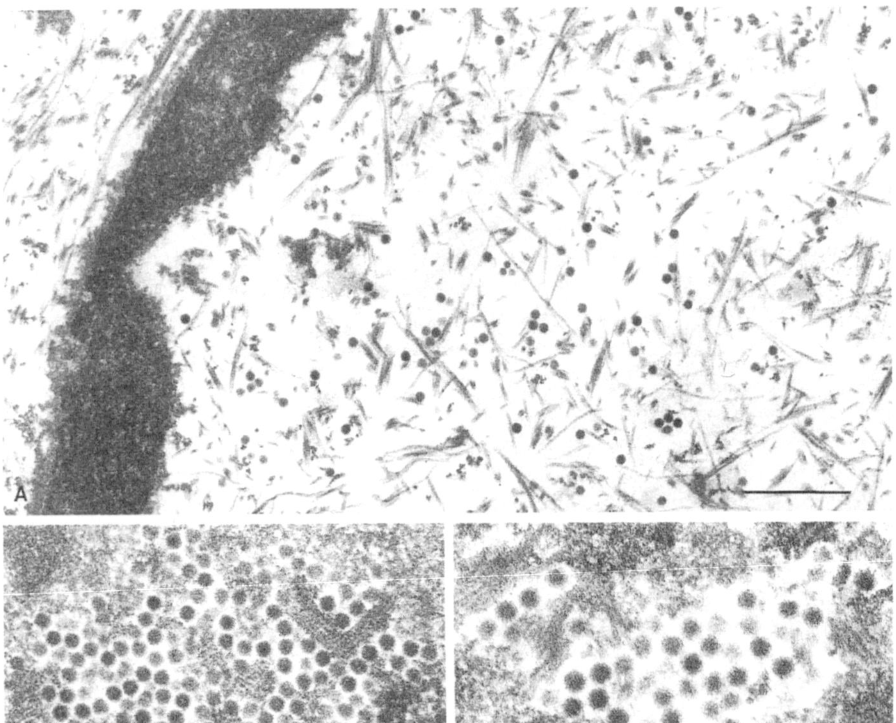

Fig. 1A–C. Electron micrographs documenting intranuclear HPV particles in squamous epithelial cells from the vagina. The *bar* represents: **A** 0.5 μm; **B, C** 0,2 μm. Courtesy of Drs. H.-W. Zentgraf and G. Heil

viral particles. Still, large differences in particle concentration can be observed in warts of the same age (Barrera-Oro et al. 1962).

3.2 Anogenital Lesions

Compared with cutaneous lesions a lower number of HPV particles was found in penile and vulvar warts (Dunn and Ogilvie 1968). HPV particles were demonstrated in about 50% of condylomata acuminata of the external genital tract (Oriel and Almeida 1970). Tissue sections and cytological smears taken from cervical condylomas were examined more recently by several investigators. HPV particles were identified in half of the cases, mainly in the nuclei of koilocytotic cells, and occasionally also in dyskeratotic cells (see Koss, this volume). Yet CIN III and invasive cervical carcinoma were found to be negative in all studies. This was also the case in the external genital tract. In four of nine cases of VIN III, only adjacent areas "showing no dysplasia" contained viral particles (Pilotti et al. 1984). Recently, HPV particles were demonstrated in cervical epithelium that showed no cytopathic effects of HPV histologically and colposcopically (Syrjänen et al. 1985).

3.3 Lesions of Other Organ Sites

HPV particles have been demonstrated repeatedly in lesions of the oral cavity such as condylomata, squamous papilloma and FEH (see S. Syrjänen, this volume).

In around 10% of laryngeal papillomas HPV particles were found by several investigators (see Kashima, this volume). There is a single case report of HPV particles present in an epithelial papilloma of the eyelid (Angevine et al. 1981).

4 Demonstration of HPV Antigen by Immunochemistry

Availability of antisera reacting against specific types of human papillomaviruses is limited (Orth et al. 1977b, 1978c; Pfister and zur Hausen 1978). Since papillomaviruses cannot be grown in tissue culture, the only source of viral particles to be used as antigen is clinical material. There is, however, a considerable variation in the rate of replication for the different HPV types in vivo. Sufficient quantities of virus particles can be harvested only in the case of HPV1,2,3,4 and EV HPV infection. From these lesions type-specific antisera have been prepared (Orth et al. 1980; Jablonska et al. 1982).

Antisera which are not type specific have been prepared by the following methods:

1. A pool of tissue from different HPV lesions is prepared, and antibodies against the external antigens of the purified intact virions are generated (Pyrhönen and Penttinen 1972)

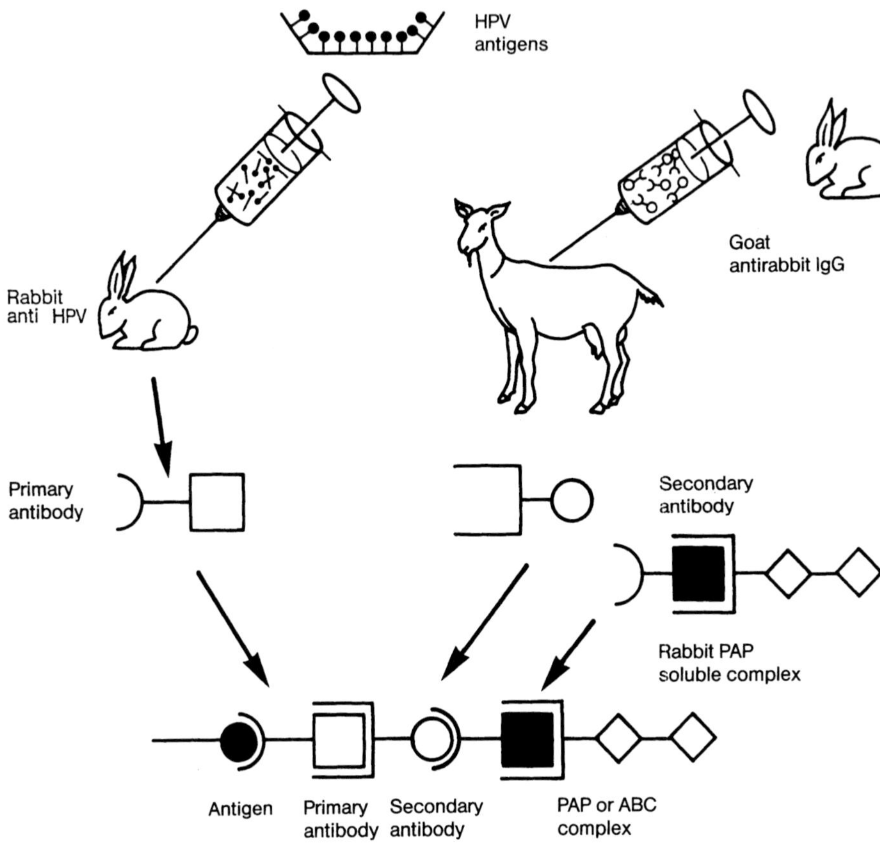

Fig. 2. Papillomavirus group-specific common antibody preparation

2. Homogenised tissue from tumours caused by the same HPV type is used as antigen. Antibodies directed against viral structural antigens and surface antigens of the virus-transformed epidermal cells have been produced (Dunn et al. 1981)
3. In addition group-specific common antibodies can be made: by using particles of HPV1, cottontail rabbit (CRPV) or bovine papillomavirus (BPV) disrupted by treatment with a detergent, internal or external antigens are exposed as targets for antibody production (Orth et al. 1978a; Jenson et al. 1980; Fig. 2)

Using antiserum against these type-specific or common antigens, frozen or routinely fixed tissues or cells can be analysed. If the primary antiserum is produced in a rabbit, it reacts with an antirabbit immunoglobulin as a secondary antibody. In order to visualise a positive reaction, a complex can be formed with a peroxidase-antiperoxidase (PAP), avidin biotin (ABC), or indirect immunofluorescence (IIF) stain (Sternberger 1974; Kawamura 1977; Lack et al. 1980; Dunn et al. 1981; Hsu et al. 1981; Gupta et al. 1983).

Table 2. HPV antigen prevalence in cervical biopsies

	Condyloma	Mild, moderate dysplasia	Severe dysplasia, carcinoma-in-situ	Invasive cancer	Total
Morin et al. 1981	12/20 (60%)	9/ 15 (60%)	0/ 0	0/ 0	21/ 35 (60%)
Kurman et al. 1983	0/ 0	77/234 (33%)	12/ 88 (14%)	0/ 0	89/322 (28%)
Syrjänen et al. 1983b	0/ 0	76/166 (46%)	45/153 (29%)	1/27 (4%)	122/346 (35%)
Guillet et al. 1983	42/51 (82%)	8/ 43 (19%)	0/ 0	0/ 0	50/ 94 (53%)
Warhol et al. 1984	0/0	11/ 28 (39%)	2/ 14 (14%)	0/ 0	13/ 42 (31%)

4.1 Cutaneous Lesions

Cutaneous papillomas from different areas were positive by immunochemistry in more than 50% (Braun et al. 1983) and common warts in almost 90% of cases (Mehregan and Nadji 1984) using the PAP technique. In contrast, premalignant lesions like Bowen's disease were negative for HPV antigen in eight and five cases, respectively (Braun et al. 1983; Gross et al. 1984).

Identical observations were made in EV lesions, where HPV antigen can only be demonstrated in benign ones (Lutzner et al. 1984).

4.2 Anogenital Lesions

Cervical lesions have been examined the most thoroughly. Cervical condylomas showed a positive reaction for HPV antigen in about 50% of cases (Woodruff et al. 1980; Ferenczy et al. 1981). Examination of CIN demonstrated a HPV-antigen detection rate inversely related to the severity of the lesion (Table 2). The presence of HPV antigen in invasive cancer of the cervix is exceptional. The same correlation was observed in external genital lesions. Between 50% and 77% of condylomata acuminata were positive for HPV antigen compared with not more than 10% of vulvar intraepithelial neoplasias (Crum et al. 1982a, b; Braun et al. 1983; Gross et al. 1984; Pilotti et al. 1984).

4.3 Lesions of Other Organ Sites

Between 18% and 48% of laryngeal papillomas examined by several investigators were positive for HPV antigen (see Kashima, K. Syrjänen, this volume).

Lesions of the oral cavity have been studied by several investigators (see S. Syrjänen, this volume). From 40% to 60% of oral cavity lesions such as condylomas and squamous cell papillomas were positive for HPV antigen. FEH is HPV antigen positive in 80% of cases. There is a high prevalence of HPV antigen-positive tissue adjacent to squamous cell carcinoma (50%), and one report describes the presence of HPV antigen in four of six carcinomas (Syrjänen et al. 1983c; Löning et al. 1985).

In squamous cell papilloma of the oesophagus, up to 31% of cases are positive for HPV antigen (see K. Syrjänen, this volume).

There are case reports on the detection of HPV antigen in squamous cell papillomas of the paranasal sinus (Syrjänen et al. 1983a) and of the conjunctiva (Völcker and Holbach 1985).

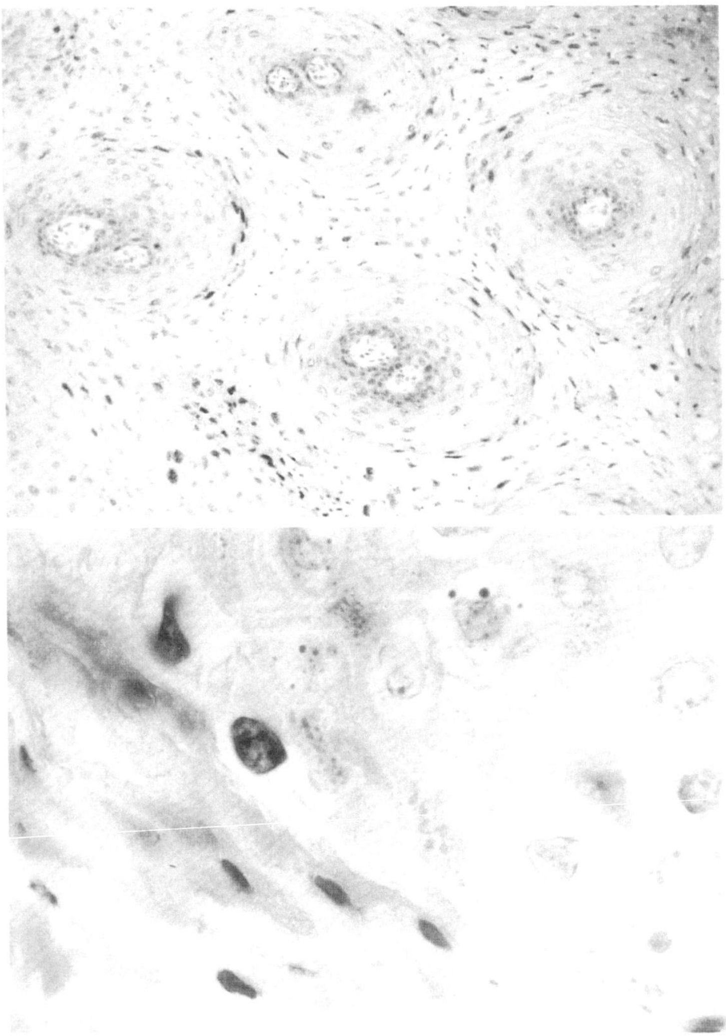

A

B

Fig. 3A, B. HPV antigen detection in a giant condyloma (Buschke-Löwenstein's tumor) using BPV1-specific antibodies and the peroxidase-antiperoxidase method. **A** ×100; **B** ×950

4.4 Antigen Distribution and Sensitivity of Techniques

The following observations were made:
(a) All tissues showed positive reactions mainly in the upper epithelial layer and specifically in the nuclei of koilocytes and dyskeratocytes
(b) Adjacent cells with identical morphological features were often negative (Fig. 3). This observation suggests that the chances of detecting positive nuclei increase with the number of biopsies. The rate of positive staining was 55% with one biopsy and 100% with four or more biopsies (Lack et al. 1980)
(c) When biopsies positive for HPV antigen were analysed ultrastructurally, HPV particles were usually identified (Jenson et al. 1980; Lack et al. 1980; Morin et al. 1981; Pilotti et al. 1984; Eng et al. 1985)

The low sensitivity and the failure in the detection of specific papillomavirus types with the group-specific common antibody technique may necessitate the development of other type-specific reagents. The use of monoclonal antibodies directed against disrupted virions is one possibility, already available for the bovine papillomavirus system (Gorra et al. 1985). As a number of HPV genomes have already been sequenced, another approach is the in vitro production of HPV peptides and proteins using suitable expression vectors (Remaut et al. 1983). Monoclonal antibodies against these in vitro-produced antigens can be prepared, as recently shown by the expression of an E6 open reading frame protein (J. Palefsky, personal communication).

5 Demonstration of HPV by Methods of Molecular Biology

5.1 Southern Blot Hybridisation

Two different methods of Southern blot hybridisation are available for HPV DNA detection.

Labelling of Viral DNA (Southern Blotting). This technique was described by Southern in 1975 and is the one most frequently used to identify HPV DNA in cell or tissue preparations. Figure 4 is an illustration of the procedure. Briefly, from biopsy material (which can be stored frozen for many years) the cellular DNA is extracted after digestion of proteins and RNA. The DNA is cleaved using different restriction endonucleases and separated by agarose gel electrophoresis. The DNA is then denatured and transferred ("blotted") onto a hybridization membrane (nitrocellulose, Gene Screen). Cloned papillomavirus DNA, labelled either with radioactive or non-radioactive compounds, is hybridised to the cellular DNA on the filter at different degrees below the melting temperature (T_m)[1].

---

[1] This temperature is by definition the point at which 50% of a given DNA exists as single-stranded molecules. It is described by the following equation (McConaughy et al. 1969):

$$T_m(°C) = 81.5 + 16.6(\log Na^+) + 0.41(\%G + C) - 0.72(\%\text{foramide}) - \frac{650}{\text{length of DNA probe}}(bp)$$

G: guanosine; *C:* cytosine, *bp:* base pair, Na^+: molar concentration of Na^+ ions

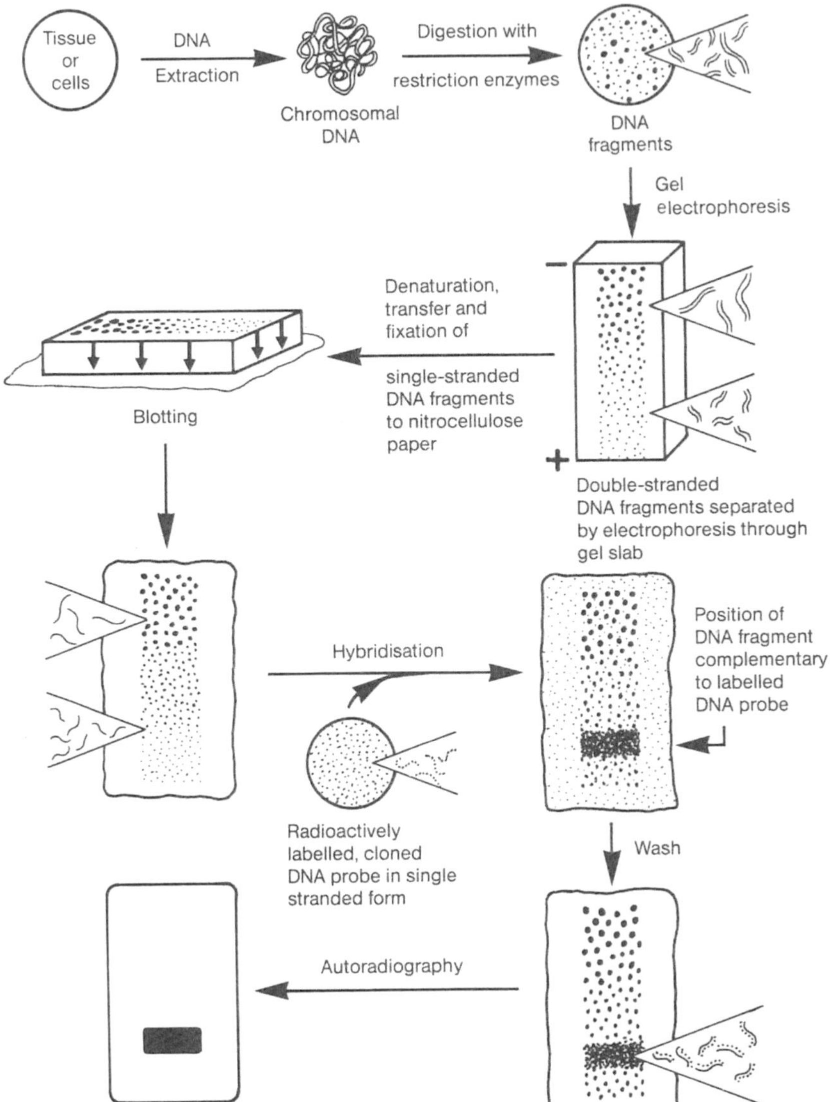

Fig. 4. Principles of Southern blot hybridisation: labelling of viral DNA

Through variation of temperature, salt and formamide concentrations, and size of the DNA probe, different levels of stringency can be applied according to the purpose of the experiment. Low-stringency conditions are suitable for screening for related HPV types ($T_m - 35\,°C$), whereas high-stringency conditions help to identify identical HPV DNAs ($T_m - 18\,°C$). After 2-4 days of hybridisation, the membrane is washed, air dried, and exposed for autoradiography for several days;

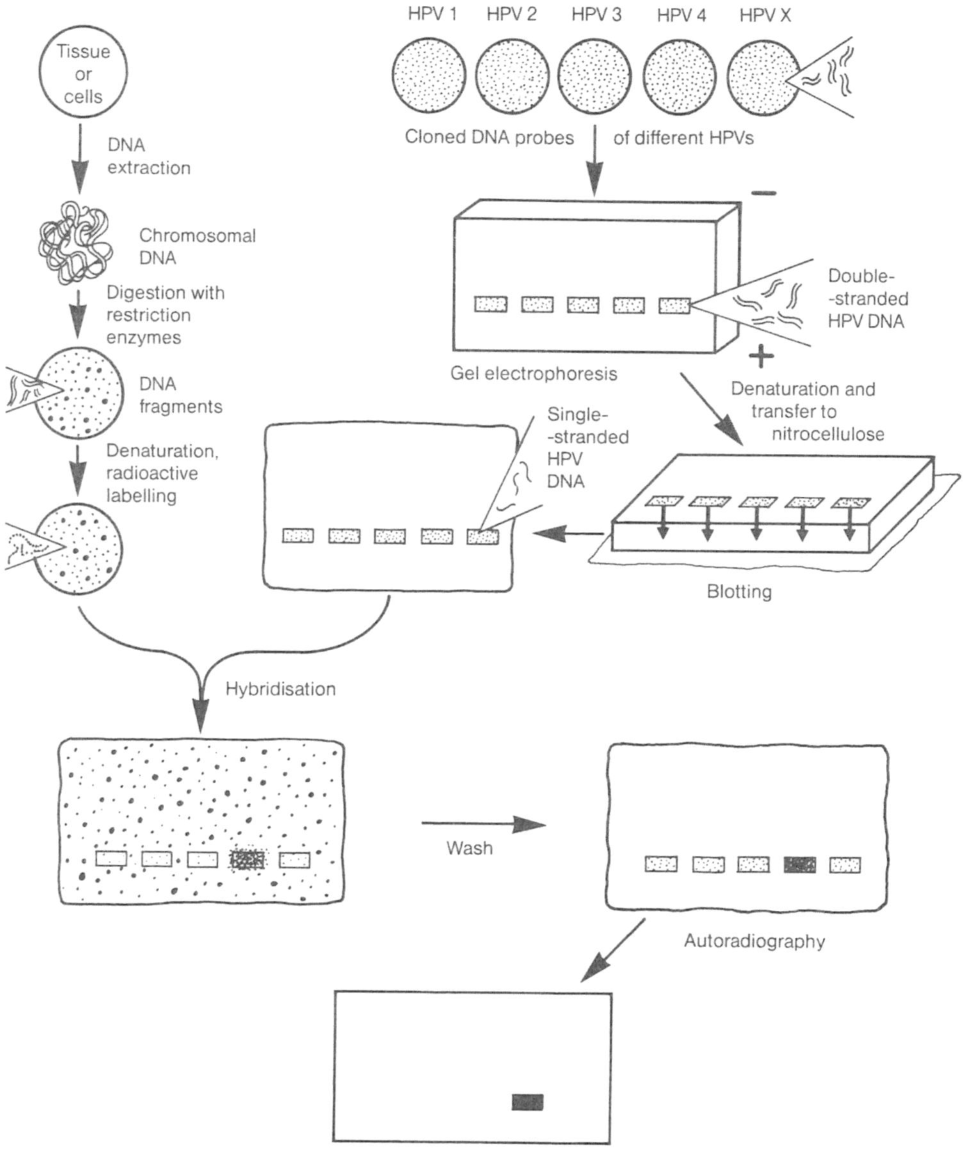

Fig. 5. Principles of Southern blot hybridisation: labelling of cellular DNA

non-radioactive markers are visualised by an appropriate reaction. From 0.1 to 0.01 HPV genome equivalents per cell can be detected.

Labelling of Nuclear DNA ("Reverse Hybridisation"). A modification of this method can be used when biopsies have to be screened for the prevalence of different HPV DNAs in one hybridisation experiment (Fig. 5). After digestion the *cel-*

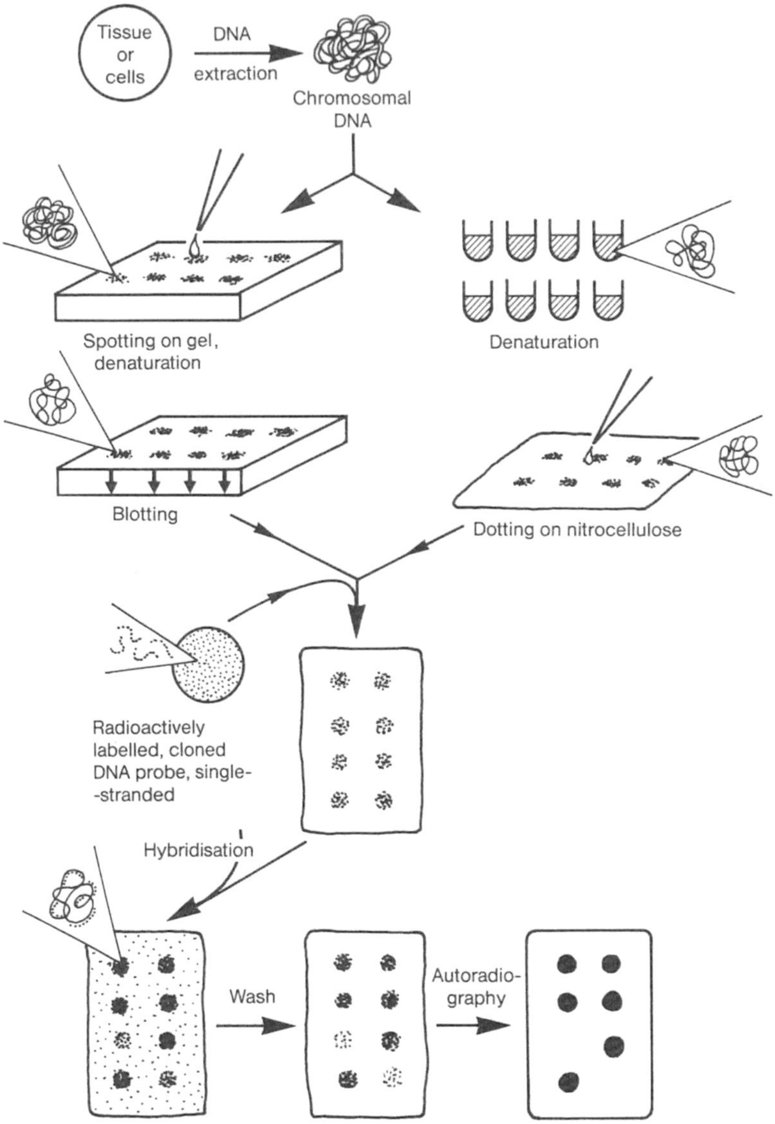

Fig. 6. Principles of dot (spot) hybridisation

lular DNA is labelled and hybridised to the cloned HPV DNAs, which have previously been blotted onto the membrane (Gissmann and Schwarz 1985; De Villiers et al. 1986). This allows the detection of about 10 HPV genome equivalents per cell, and related papillomavirus types can be distinguished by weak or strong hybridisation signals.

5.2 Northern Blot Hybridisation

This technique is used for the detection of RNA and differs from Southern blot hybridisation in the following steps (Thomas 1980; Seed 1982). Cellular RNA is extracted and separated on a denaturing gel (containing formaldehyde or methylmercury hydroxide). Thus, secondary structures of RNA are denatured, and linearised molecules are obtained. For hybridisation labelled DNA or RNA can be used, and higher conditions of stringency have to be applied as reannealing RNA molecules are more stable than DNA/DNA hybrids (Cox et al. 1984). Using this method HPV RNA was identified in tissue and cell lines of cervical carcinoma (Schwartz et al. 1985; Yee et al. 1985).

5.3 Dot (Spot) Hybridisation

In contrast to Southern blot analysis, the phenol/chloroform-extracted total cellular DNA is not digested with restriction nucleases. It is either (a) *spotted* on a gel, denatured and blotted onto a membrane (Cunningham 1983) or (b) denatured by heat and alkaline treatment and *dotted* directly onto a membrane (Kafatos et al. 1979; Thomas 1980; Fig. 6).

As the diameter of these spots can be as small as a few millimeters, it is possible to test many specimens on one filter. Screening of gynaecological smears from cytologically negative women showed between 1.3% and 11% HPV-positive specimens (Wickenden et al. 1985; Pratili et al. 1986). The sensitivity and specificity of these methods under stringent conditions give satisfactory results, whereas hybridisation under non-stringent conditions can result in false-positive reactions due to cross-hybridisation with cellular sequences. This is probably due to the high concentration of undigested cellular DNA on a small area of the filter (G. Orth, personal communication).

5.4 Filter In Situ Hybridisation

This screening method allows the examination of many different gynaecological samples in one experiment (Wagner et al. 1984). Figure 7 shows schematically the major steps for a HPV-negative and a HPV-positive example. Cells are filtered onto membranes, lysed and denatured by alkaline treatment. After neutralisation and baking they can be stored for several months. Hybridisation is carried out under stringent conditions with HPV11 (which also detects HPV6 with slightly lower efficiency) and a mixture of HPV16 and HPV18. Other HPV DNAs can be used in this hybridisation system. After washing, autoradiography is done and the result can be read after 1-5 days. Examination of cytologically positive smears with this method yields a HPV detection rate of 60%-70% (Schneider et al. 1985, 1987). More than 10000 cytologically negative women were investigated with this method in one study, and 12% were found to be positive in equal proportions for HPV6/11 and HPV16/18 (E. M. de Villiers, personal communication).

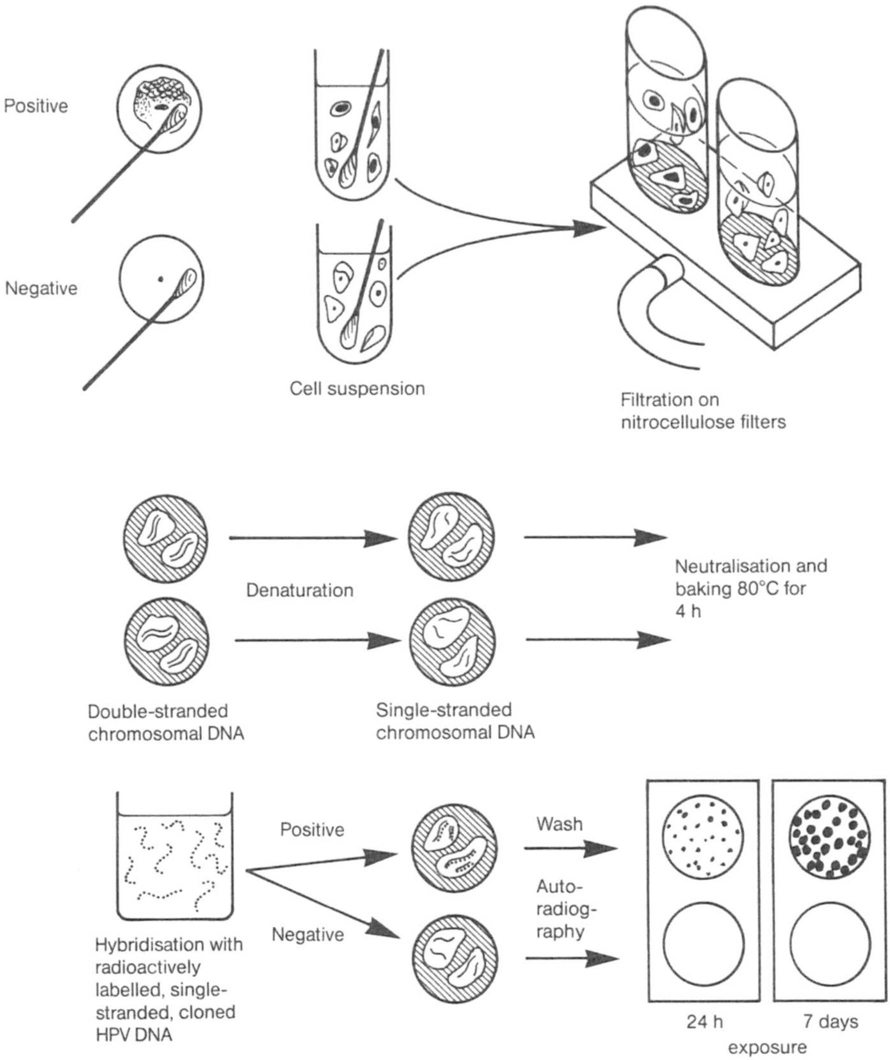

Fig. 7. Principles of filter in situ hybridisation

5.5 In Situ Hybridisation

The technique of in situ hybridisation allows the detection of DNA or RNA sequences in cytological preparations or sections of tissue (Fig. 8). Frozen or fixed material may be used, and during the procedure the morphology of the tissue must be well preserved (Fig. 9). Radioactively or chemically labelled DNA or complementary RNA are used as probes. Sections are placed on "sticky" slides. During the pretreatment the cellular DNA and RNA are exposed and denatured. A certain amount of labelled viral DNA or RNA is applied. Hybridisation is per-

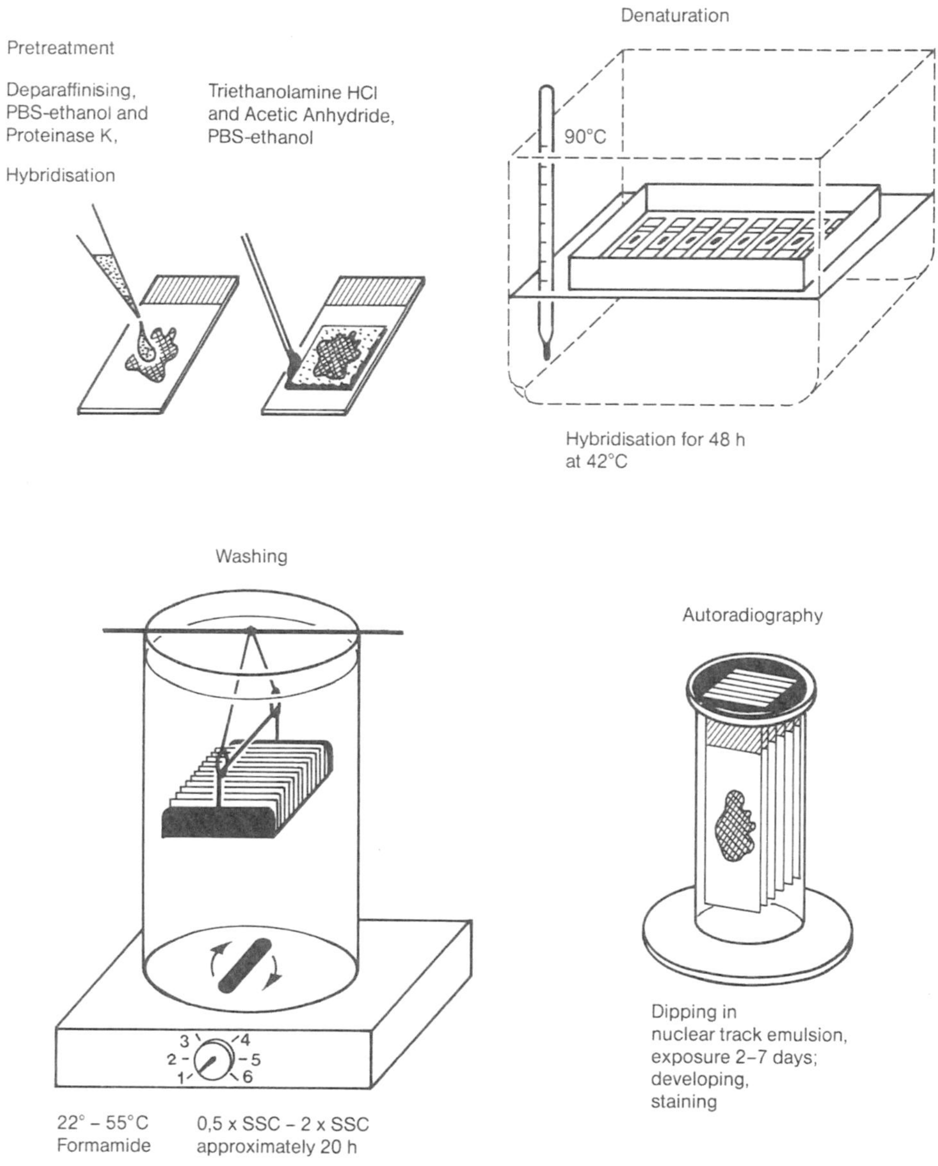

Fig. 8. Principles of in situ hybridisation on formalin-fixed sections using HPV DNA, 35 S-labelled by nick translation

formed for 12–48 h depending on the specific activity and concentration of the labelled probe. According to the stringency of the reaction required, the washing conditions are selected, and the slides are processed for autoradiography (Gall and Pardue 1971).

For papillomaviruses in situ hybridisation was first used in the CRPV model on frozen sections using tritium-labelled cRNA probes and analysed by light and

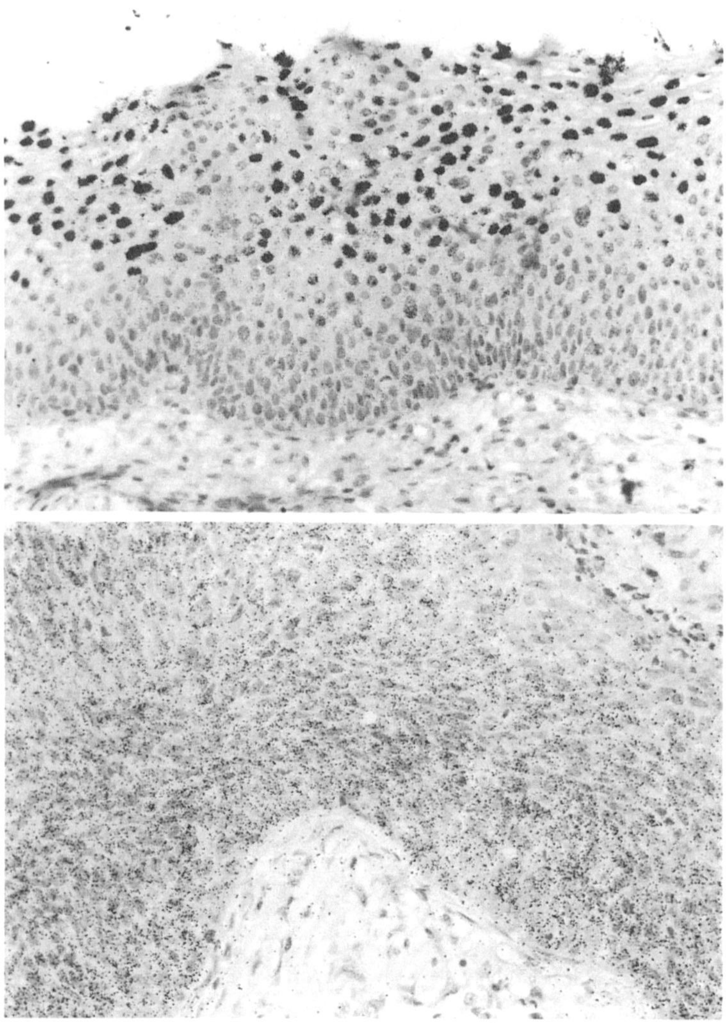

A

B

Fig. 9 A, B. Detection of HPV16 DNA by in situ hybridisation. **A** CIN II (CINK) in Carnoy-fixed cone biopsy, Paraffin embedded. Asymmetric, ^{3}H-labelled RNA was used. ×230. **B** Invasive squamous cell carcinoma of the cervix uteri fixed in formalin and embedded in paraffin. In contrast to **A** nick-translated, ^{35}S-labelled DNA was used. ×370

electron microscopy (Orth et al. 1971; Croissant et al. 1972). Studies on human material have revealed that in benign lesions vegetative viral DNA replication precedes capsidal protein production and is first detected in the suprabasal layer (Orth et al. 1977a, 1978b, 1980; Grussendorf and zur Hausen 1979; Gupta et al. 1986). Molecular in situ hybridisation also demonstrates that the beginning of viral DNA replication coincides with the appearance of the morphological cytopathic effects, showing an interference of viral DNA replication with the expression of physiological keratinisation products (Croissant et al. 1985). Using biotin-labelled DNA probes in 40% of 42 anogenital carcinomas, HPV16 or HPV18 DNA was found (Beckmann et al. 1985), and in CINK 10 of 16 lesions were positive for HPV16 (Crum et al. 1986). This method has the advantage of avoiding radioactivity, but so far only low sensitivity has been reported, as approximately 800 copies per cell can detected (Crum et al. 1986). A higher sensitivity is obtained, when asymmetric RNA probes are used, which have a greater specific activity and prevent self-reannealing of the probe (Cox et al. 1984). With this system sense and antisense probes can be produced and allow the differentiation between viral DNA and RNA during detection. In condylomas the sites of transcription of the different open reading frames of HPV6 could be localised using this technique (S. Wolynski, personal communication).

5.6 Indications for Different Hybridisation Methods

Southern Blotting. (a) Labelling of viral DNA. This method shows the highest sensitivity (0.01-0.1 copy per cell) and specificity even under non-stringent conditions. As it yields a typical restriction pattern depending on the number of cutting sites in the viral genome, different HPV subtypes can be identified. Though the procedure is time consuming, there is no alternative at the moment for the analysis of specimens with low HPV copy numbers or for the identification of new HPV types. (b) Labelling of cellular DNA. Reverse Southern blotting is a compromise between sensitivity (about 10 copies per cell) and time invested. It is the ideal screening method for material which might contain a wide range of different, already defined HPV types. The HPV DNA detection can be performed in one hybridisation experiment.

Dot (Spot) Blotting. A sensitivity of one copy per cell can be achieved with this method, and a large number of specimens can be analysed in a rather short period of time. Hybridisation is performed under stringent conditions only, so that an aliquot of each specimen has to be hybridised to each HPV type separately. This method is indicated when material has to be screened with a limited number of HPV types.

Table 3. Indications for different hybridisation methods

	Southern blot		Dot (spot)	Filter in situ	In situ
	Labelling of viral DNA	Labelling of cellular DNA			
Sensitivity	+++	++	+++	+	+
Specificity					
stringent	+++	+++	+++	+++	+++
non-stringent	++	−	−	−	+
Practicability	+	++	++	++	+
Morphology	−	−	−	−	+++

Filter in Situ Hybridisation. Compared with dot (spot) blotting, there is only a difference in sensitivity with this technique. Specificity and indication are identical. About 10^4-10^5 molecules of HPV DNA can be detected in one specimen if they are concentrated in one cell or cell cluster, resulting in a black spot on the autoradiogram. In a cervical swab containing 2×10^5 cells and only one cell infected with 10^4 HPV DNA copies, the Southern or dot (spot) blot technique wouldn't be sensitive enough for detection. On the other hand, filter in situ hybridisation cannot detect cells containing a low number of HPV DNA molecules, in spite of the fact that all cells might be infected (e.g. one copy in each of 2×10^5 cells), as the autoradiogram will show a picture wherein no clear signal-to-noise ratio can be differentiated.

In Situ Hybridisation. Depending on the method used, the sensitivity varies between 20-100 copies per cell (radioactive labelling) and 200-800 copies per cell (biotin labelling). Both methods are not as yet applicable to routine work, but they are the only choice when morphological features associated with different HPV types have to be identified. An additional advantage of the procedure is its applicability to paraffin-embedded archival tissues.

Acknowledgements. This chapter is dedicated to Professor Knörr on the occasion of his 70th birthday. I am indebted to Drs. O. Croissant, F. Breitburd and V. Schneider for their critical reading of this manuscript. Original work cited in this article was supported by the Deutsche Forschungsgemeinschaft.

References

Almeida JD, Howatson AF, Williams MG (1962) Electron microscope study of human warts: sites of virus production and nature of the inclusion bodies. J Invest Dermatol 38: 337-345

Angevine DM, Norback DH, Dortzbach RK (1981) Virus in papilloma. JAMA 246: 1087-1088

Barrera-Ora JG, Smith KO, Melnick JL (1962) Quantitation of papovavirus particles in human warts. JNCI 29: 583-595

Beckmann AM, Daling JR, McDougall JK (1985) Human papillomavirus DNA in anogenital carcinomas. J Cell Biochem (Suppl) 9c: 68

Braun L, Farmer ER, Shah KV (1983) Immunoperoxidase localization of papillomavirus antigen in cutaneous warts and bowenoid papulosis. J Med Virol 12: 187-193

Cox KH, De Leon DV, Angerer LM, Angerer RC (1984) Detection of mRNAs in sea urchin embryos by in situ hybridization using asymmetric RNA probes. Dev Biol 101: 485-502

Croissant O, Dauguet C, Jeanteur P, Orth G (1972) Use of in situ molecular hybridization technique for the detection by electron microscopy of vegetative viral DNA replication in cottontail rabbit papillomas induced by the Shope virus. CR Acad Sci (Paris) 274: 614-617

Croissant O, Breitburd F, Orth G (1985) Specificity of cytopathic effect of cutaneous human papillomaviruses. Clin Dermatol 3: 43-55

Crum CP, Braun LA, Shah KV, Fu YS, Levine RN, Fenoglio CM, Richart RM, Townsend DE (1982a) Vulvar intraepithelial neoplasia: correlation of nuclear DNA content and the presence of human papillomavirus (HPV) structural antigen. Cancer 49: 468-471

Crum CP, Fu YS, Levine RU, Richart RM, Townsend DE, Fenoglio CM (1982b) Intraepithelial squamous lesions of the vulva: biologic and histologic criteria for the distinction of condylomas from vulvar intraepithelial neoplasia. Am J Obstet Gynecol 144: 77-83

Crum CP, Mitao M, Levine RU, Silverstein S (1985) Cervical papillomaviruses segregate within morphologically distinct precancerous lesions. J Virol 54: 675-681

Crum CP, Nagai N, Levine RU, Silverstein S (1986) In situ hybridization analysis of HPV 16 DNA sequences in early cervical neoplasia. Am J Pathol 123: 174-182
Cunningham M (1983) Spot blot: a hybridization assay for specific DNA sequences in multiple samples. Anal Biochem 128: 415-421
De Villiers E-M, Schneider A, Gross G, zur Hausen H (1986) Analysis of benign and malignant urogenital tumors for human papillomavirus infection by labelling cellular DNA. Med Microbiol Immunol (Berl) 174: 281-286
Dunn AEG, Ogilvie MM (1968) Intranuclear virus particles in human genital wart tissue: observations on the ultrastructure of the epidermal layer. J Ultrastruct Res 22: 282-295
Dunn J, Weinstein L, Droegemueller W, Meinke W (1981) Immunologic detection of condylomata acuminata-specific antigens. Obstet Gynecol 57: 351-356
Eng AM, Jin YT, Matsuoka LY, Grgurich CV, Robinson J, Armin A (1985) Correlative studies of verruca vulgaris by H & E, PAP immunostaining and electron microscopy. J Cutan Pathol 12: 46-54
Ferenczy A, Braun L, Shah KV (1981) Human papillomavirus (HPV) in condylomatous lesions of cervix. Am J Surg Pathol 5: 661-670
Gall JG, Pardue ML (1971) Nucleic acid hybridization in cytological preparations. Methods Enzymol 38: 470-480
Gissmann L, Schwarz E (1985) Cloning of papillomavirus DNA. In: Becker Y (ed) Recombinant DNA research and virus. Martinus Nijhoff, Boston, pp 173-197
Gorra JB, Lancaster WD, Kurman RJ, Jenson AB (1985) Bovine papillomavirus type 1 monoclonal antibodies. JNCI 75: 121-123
Gross G, Ikenberg H, Hagedorn M, Gissmann L (1984) Nachweis von Papillomavirus Capsidantigenen und Human Papillomvirus (HPV) 16 verwandter DNS in Bowenoider Papulose und Morbus Bowen. Z Hautkr 59: 1084-1086
Grußendorf EI, zur Hausen H (1979) Localization of viral DNA-replication in sections of human warts by nucleic acid hybridization with complementary RNA of human papilloma virus type 1. Arch Dermatol Res 264: 55-63
Guillet G, Braun L, Shah K, Ferenczy A (1983) Papillomavirus in cervical condylomas with and without associated cervical intraepithelial neoplasia. J Invest Dermatol 81: 513-516
Gupta JW, Gupta PK, Shah KV, Kelly DP (1983) Distribution of human papillomavirus antigen in cervicovaginal smears and cervical tissues. Int J Gynecol Pathol 2: 160-170
Gupta J, Schneider A, Shah K (1986) Detection of papillomavirus antigen and DNA in cells and tissues. In: Peto R, zur Hausen H (eds) Viral etiology of cervical cancer. Cold Spring Harbor Laboratory, New York, pp 247-253
Hsu SM, Raine L, Fanger H (1981) Use of avidin-biotin-peroxydase complex (ABC) in immunoperoxidase technique. J Histochem Cytochem 29: 577-580
Jablonska S, Orth G, Lutzner MA (1982) Immunopathology of papillomavirus-induced tumors in different tissues. Springer Semin Immunopathol 5: 33-62
Jenson AB, Rosenthal JD, Olson C, Pass F, Lancaster WD, Shah K (1980) Immunologic relatedness of papillomaviruses from different species. JNCI 64: 495-500
Kadish AS, Burk RD, Kress Y, Calderin S, Romney SL (1986) Human papillomaviruses of different types in precancerous lesions of the uterine cervix: histologic, immunocytochemical and ultrastructural studies. Hum Pathol 17: 384-392
Kafatos FC, Jones CW, Estratiadis A (1979) Determination of nucleic acid sequence homologies and relative concentration by a dot blot hybridization procedure. Nucleic Acids Res 7: 1541-1552
Kawamura A Jr (1977) Preparation of materials: absorption with tissue powder. In: Kawamura A Jr (ed) Fluorescent antibody techniques and their application, 2nd edn. University of Tokyo Press, Tokyo and University Park Press, Baltimore, pp 65-66
Kurman RJ, Jenson AB, Lancaster WD (1983) Papillomavirus infection of the cervix: 2. Relationship to intraepithelial neoplasia based on the presence of specific viral structural proteins. Am J Surg Pathol 7: 39-52
Lack EE, Jenson AB, Smith HG, Healy GB, Pass F, Vawter GF (1980) Immunoperoxidase localization of human papillomavirus in laryngeal papillomas. Intervirology 14: 148-154
Laurent R, Agache P, Coume-Marquet J (1975) Ultrastructure of clear cells in human viral warts. J Cutan Pathol 2: 140-148
Löning T, Ikenberg H, Becker J, Gissmann L, Höpfer I, zur Hausen H (1985) Analysis of oral

papillomas, leukoplakias, and invasive carcinomas for human papillomavirus type related DNA. J Invest Dermatol 84: 417-420
Lutzner MA, Blanchet-Bardon C, Orth G (1984) Clinical observations, virologic studies, and treatment trials in patients with epidermodysplasia verruciformis, a disease induced by specific human papillomaviruses. J Invest. Dermatol 83: 18s-25s
McConaughy BL, Laird CD, McCarthy BJ (1969) Nucleic acid reassociation in formamide. Biochemistry 8: 3289-3295
Mehregan AH, Nadji M (1984) Inverted follicular keratosis and verruca vulgaris. An investigation for the papillomavirus common antigen. J Cutan Pathol 11: 99-102
Meisels A, Roy M, Fortier M, Morin C, Casas-Cordero M, Shah KV, Turgeon H (1981) Human papilloma virus infection of the cervix: the atypical condyloma. Acta Cytol 25: 7-16
Meisels A, Morin C (1986) Flat condyloma of the cervix: two variants with different prognosis. In: Peto R, zur Hausen H (eds) Viral etiology of cervical cancer. Cold Spring Harbor Laboratory, New York, pp 115-120
Morin C, Braun L, Casas-Cordero M, Shah KV, Roy M, Fortier M, Meisels A (1981) Confirmation of the papillomavirus etiology of condylomatous cervix lesions by the peroxidase-antiperoxidase technique. INCI 66: 831-835
Oriel JD, Almeida JD (1970) Demonstration of virus particles in human genital warts. Br J Vener Dis 46: 37-42
Orth G, Croissant O (1968) Characteristics of primary cultures of domestic rabbit Shope papilloma cells. CR Acad Sci [III] 266: 1084-1087
Orth G, Jeanteur P, Croissant O (1971) Evidence for and localization of vegative viral DNA replication by autographic detection of RNA-DNA hybrids in sections of tumors induced by Shope papilloma virus. Proc Natl Acad Sci USA 68: 1876-1880
Orth G, Breitburd F, Favre M (1977a) Papillomavirus: possible role in human cancer. In: Hiatt HH, Watson JD, Winsten JA (eds) Origins of human cancer. Cold Spring Harbor Laboratory, New York, pp 1043-1068
Orth G, Favre M, Croissant O (1977b) Characterization of a new type of human papillomavirus that causes skin warts. J Virol 24: 108-120
Orth G, Breitburd F, Favre M (1978a) Evidence for antigenic determinants shared by the structural polypeptides of (Shope) rabbit papillomavirus and human papillomavirus type 1. Virology 91: 243-255
Orth G, Jablonska S, Breitburd F, Favre M, Croissant O (1978b) The human papillomaviruses. Bull Cancer (Paris) 65: 151-164
Orth G, Jablonska S, Favre M, Croissant O, Jarzabek-Chorzelska M, Rzesa G (1978c) Characterization of two types of human papillomaviruses in lesions of epidermodysplasia verruciformis. Proc Natl Acad Sci USA 75: 1537-1541
Orth G, Favre M, Breitburd F, Croissant O, Jablonska S, Obalek S, Jarzabek-Chorzelska M, Rzesa G (1980) Epidermodysplasia verruciformis: a model for the role of papillomaviruses in human cancer. In: Essex M, Todaro G, zur Hausen H (eds) Viruses in naturally occurring cancers. Cold Spring Harbor Laboratory, New York, pp 259-282
Pfister H, zur Hausen H (1978) Seroepidemiological studies of human papilloma virus (HPV-1) infections. Int J Cancer 21: 161-165
Pilotti S, Rilke F, Shah KV, Delle-Torre G, De Palo G (1984) Immunohistochemical and ultrastructural evidence of papillomavirus infection associated with in situ and microinvasive squamous cell carcinoma of the vulva. Am J Surg Pathol 8: 751-761
Praetorius-Clausen F (1969) Histopathology of focal epithelial hyperplasia. Evidence of viral infection. Taendlaegebladet 73: 1013-1022
Pratili MA, LeDoussal V, Harvey P, Laval C, Bertrand F, Jibard N, Croissant O, Orth G (1986) Human papillomaviruses in the epithelial cells of the cervix uteri: frequency of types 16 and 18. Preliminary results of a clinical, cytologic and viral study. J Gynecol Obstet Biol Reprod (Paris) 15: 45-50
Pyrhönen S, Penttinen K (1972) Wart virus antibodies and the prognosis of wart disease. Lancet II: 1330-1332
Remaut E, Stanssens P, Fiers W (1983) Inducible high level synthesis of mature human fibroblast interferon in *Escherichia coli*. Nucleic Acids Res 11: 4677-4687
Schneider A, Sawada E, Gissmann L, Shah K (1987) Human papillomaviruses in women with a history of abnormal PAP smears and their male partners. Obstet Gynecol 69: 554-562

Schneider A, Kraus H, Schuhmann R, Gissmann L (1985) Papillomavirus infection of the lower genital tract: detection of viral DNA in gynecological swabs. Int J Cancer 35: 443-448

Schwarz E, Freese UK, Gissmann L, Mayer W, Roggenbuck B, Stremlau A, zur Hausen H (1985) Structure and transcription of human papillomavirus sequences in cervical carcinoma cell lines. Nature 314: 111-114

Seed B (1982) Diazotizable arylamine cellulose papers for the coupling and hybridization of nucleic acid. Nucleic Acids Res 10: 1799-1810

Southern EM (1975) Detection of specific sequences among DNA fragments separated by gel electrophoresis J Mol Biol 98: 503-517

Sternberger LA (1974) Immunocytochemistry. Prentice Hall, Englewood Cliffs, NJ, pp 110-171

Strauss MJ, Shaw EW, Bunting H, Melnick JL (1949) "Crystalline" virus-like particles from skin papillomas characterized by intranuclear inclusion bodies. Proc Soc Exp Biol Med 72: 46-50

Syrjänen K, Pyrhönen S, Syrjänen SM (1983a) Evidence suggesting human papillomavirus (HPV) etiology for the squamous cell papilloma of the paranasal sinus. Arch Geschwulstforsch 53: 77-82

Syrjänen K (1983b) Human papillomavirus lesions in association with cervical dysplasias and neoplasias. Obstet Gynecol 62: 617-624

Syrjänen K, Syrjänen S, Lamberg M, Pyrhönen S, Nuutinen J (1983c) Morphological and immunohistochemical evidence suggesting human papillomavirus (HPV) involvement in oral squamous carcinogenesis. Int J Oral Surg 12: 418-424

Syrjänen K, Väyrynen M, Hippelainen M, Castren O, Saarikoski S, Mantyjarvi R (1985) Electron microscopic assessment of cervical punch biopsies in women followed-up for human papillomavirus (HPV) lesions. Arch Geschwulstforsch 55: 131-138

Thomas PS (1980) Hybridization of denaturated RNA and small DNA fragments transferred to nitrocellulose. Proc Natl Acad Sci USA 77: 5201-5205

Völcker HE, Holbach L (1985) Pedicled papilloma of the conjunctiva with papilloma virus. Immunohistochemical detection of species specific papilloma virus antigens. Klin Monatsbl Augenheilkd 187: 212-214

Wagner D, Ikenberg H, Boehm N, Gissmann L (1984) Identification of human papillomavirus in cervical swabs by deoxyribonucleic acid in situ hybridization. Obstet Gynecol 64: 767-772

Warhol MJ, Pinkus GS, Rice RH, El-Tawil GH, Lancaster WD, Jenson AB, Kurman RJ (1984) Papillomavirus infection of the cervix: III. Relationship of the presence of viral structural proteins to the expression of involucrin. Int J Gynecol Pathol 3: 71-81

Wickenden C, Steele A, Malcolm AD, Coleman DV (1985) Screening for wart virus infection in normal and abnormal cervices by DNA hybridization of cervical scrapes. Lancet i: 65-67

Woodruff JD, Braun L, Cavalieri R, Gupta P, Pass F, Shah KV (1980) Immunologic identification of papillomavirus antigen in condyloma tissues from the female genital tract. Obstet Gynecol 56: 727-732

Yee C, Krishnan-Hewlett I, Baker CC, Schlegel R, Howley P (1985) Presence and expression of human papillomavirus sequences in human cervical carcinoma cell lines. Am J Pathol 119: 361-366

Papillomavirus Infections in Animals

J. P. Sundberg

1 Introduction

Papillomaviruses have been associated with a number of hyperplastic and neoplastic lesions in a wide variety of vertebrate species (Tables 1-3). In humans, induced lesions are usually papillomas, some of which, notably those of the lower genital tract and skin, may progress to squamous cell carcinomas. It is now known that many genetically different papillomaviruses infect human beings, that specific types have predilections for particular anatomical sites, and certain types are known to be associated with lesions that progress to carcinoma. It is not surprising to find that similar, if not identical, situations exist in other animal species.

As with human beings, most other mammalian papillomaviruses are species specific, have anatomical predilection sites, and induce benign papillomas. Notable exceptions are ruminants, in which the common papillomaviruses induce fibromas or fibropapillomas. In several species papillomas have been reported to progress to squamous cell carcinomas. Very specific cytopathology has been associated with different types of papillomaviruses which infect a single host. This chapter reviews the host range of papillomaviruses in domestic, wild, and exotic animals, the lesions, induced, and the papillomaviruses that have been isolated from the various lesions.

2 Papillomavirus Infections in Domestic Mammals

2.1 Cattle

2.1.1 Bovine Papillomas, and Fibropapillomas

The term papilloma has been used incorrectly to designate all lesions induced by papillomaviruses in domestic cattle. These viruses cause two distinct types of benign lesions. The most common are the fibropapillomas of the haired skin and lower genital tract, while cutaneous and alimentary tract papillomas make up the second major class. Since Lancaster and Olson (1978) first described two bovine papillomaviruses (BPV1 and BPV2), at least four more have been identified

Table 1. Characterized animal papillomaviruses

Host/virus	Lesion	Anatomical site	Squamous cell carcinoma in host[1]	Oncogenic in hamsters	Transforms cell lines	References
Cattle						
BPV1	Fibropapilloma	Skin	No	Yes	Yes	Lancaster and Olson 1978
BPV2	Fibropapilloma	Skin	No	Yes	Yes	Lancaster and Olson 1978
BPV3	Papilloma	Skin	No	Not Reported	Not Reported	Pfister et al. 1979
BPV4	Papilloma	Alimentary tract	Yes	Not Reported	Not Reported	Campo et al. 1980
BPV5	Fibropapilloma	Teat (skin)	No	Not Reported	Not Reported	Campo et al. 1981
BPV6	Papilloma	Teat (skin)	No	Not Reported	Not Reported	Jarrett et al. 1984b
Deer						
European elk papillomavirus	Fibropapilloma	Skin	No	Yes	Yes	Moreno-Lopez et al. 1981
White-tailed deer fibroma virus	Fibroma	Skin	No	Yes	Yes	Lancaster and Sundberg 1982; Groff et al. 1983
Mule deer fibroma virus	Fibroma	Skin	No	Not Reported	Yes	Lancaster and Sundberg 1982; Groff et al. 1983
Red deer papillomavirus	Papilloma	Skin	No	Not Reported	Not Reported	Favre, personal communication
Red deer papillomavirus	Fibropapilloma	Skin	No	Not Reported	Not Reported	Moar 1983; Moar and Jarrett 1985
Reindeer papillomavirus	Fibropapilloma	Skin	No	Not Reported	Not Reported	Moreno-Lopez et al. 1983
Horses						
Equine cutaneous papillomavirus	Papilloma	Skin	No	Not Reported	Not Reported	O'Banion et al. 1986
Equine venereal papillomavirus	Papilloma	Penis, vulva	Yes	Not Reported	Not Reported	O'Banion et al. 1986
Rabbits						
Cottontail rabbit cutaneous papillomavirus	Papilloma	Skin	Yes	Not Reported	Yes	Murphy et al. 1981; Watts et al. 1983; Nasseri and Wettstein 1984b; Giri et al. 1985
Domestic rabbit oral papillomavirus	Papilloma	Tongue	No	Yes	Not Reported	O'Banion et al. 1986
Dog						
Canine oral papillomavirus	Papilloma	Tongue, gingiva	Yes	No	Not Reported	Pfister and Meszaros 1980; Sundberg et al. 1986b
Rodent						
Mastomys natalensis papillomavirus	Kerato-acanthoma, papilloma	Skin	Yes (?)	Not Reported	Not Reported	Müller and Gissmann 1978
Birds						
Chaffinch papillomavirus	Papilloma	Skin	No	Not Reported	Not Reported	Moreno-Lopez et al. 1984
African gray parrot papillomavirus	Papilloma	Skin	No	Not Reported	Not Reported	O'Banion et al. 1986

[1] Case reports indicate some papillomas progress to squamous cell carcinomas.

Table 2. Animals with lesions in which papillomavirus antigens or virions have been detected

Host	Lesion	Anatomical site	Immuno-histo-chemistry[1]	Electron micros-copy[2]	Reference
Sheep	Fibro-papilloma	Skin	NT	+	Gibbs et al. 1975
Sheep	Papilloma, horn, hyperkeratotic scale	Ear, perineum, face	NT	+	Vanselow and Spradbrow 1982; Vanselow and Spradbrow 1983
Domestic dog	Papilloma	Skin	+	+	Watrach 1969; Davis et al. 1975; Sundberg et al. 1984
Domestic dog	Papilloma	Ocular mucous membranes	+	NT	Sundberg et al. 1984
Domestic dog	Papilloma	Penis, vulva	+	NT	Sundberg et al. 1984
Domestic cattle	Papilloma	Ocular mucous membranes	NT	+	Ford et al. 1982
Coyote	Papilloma	Tongue, gingiva	NT	+	Greig and Charlton 1973
Wolf	Papilloma	Tongue, gingiva	NT	+	Samuel et al. 1978
Opossum	Papilloma	Skin	NT	+	Koller 1972
Black-tailed deer	Fibroma	Skin	+	NT	Sundberg et al. 1984
Pronghorn antelope	Fibro-papilloma	Skin	+	NT	Sundberg et al. 1983
Beaver	Papilloma	Skin	+	+	Carlson et al. 1983
Bolivian side-necked turtle	Hyperkeratosis	Skin	NEG	+	Jacobson et al. 1982; Sundberg et al. 1984
Bear	Papilloma	Vulva	+	NT	Craft et al. 1980
Lacerta lizard	Papilloma	Skin	NT	+	Cooper et al. 1982
Colobus monkey	Papilloma	Skin	NT	+	Rangan et al. 1980
Colobus monkey	Papilloma	Penis	+	+	Sundberg 1986
Chimpanzee	Focal epithelial hyperplasia	Oral mucous membranes	NT	+	Hollander and Van Noord 1972
Giraffe	Papilloma	Skin	NT	+	Karstad and Kaminjolo 1978
Impala	Papilloma	Skin	NT	+	Karstad and Kaminjolo 1978
Indian elephant	Papilloma	Oral mucous membranes	+	NT	Sundberg et al. 1981
Sperm whale	Papilloma	Penis	NEG	+	Lambertsen et al. 1986

NT, Not tested; *NEG*, Negative

[1] Peroxidase-antiperoxidase or immunofluorescent tests using papillomavirus group-specific antiserum.

[2] Negative-stain or thin section.

Table 3. Lesions in animals which resemble papillomavirus-induced lesions, for which no etiological agent has yet been detected

Host	Lesion	Anatomical site	Reference
Domestic cat	Papilloma	Skin, oral mucous membranes	Scott 1980; Sundberg et al. 1984
Domestic cat	Fibropapilloma	Ocular mucous membranes	Sundberg et al. 1984
Laboratory rat	Keratoacanthoma	Skin	Sundberg et al. 1984
Guinea pig	Papilloma	Vulva	Sundberg et al. 1984
Gerbil	Papilloma	Skin, ocular mucous membranes	Sundberg et al. 1984
American elk	Fibroma	Skin	Sundberg et al. 1984
Yak	Fibropapilloma	Skin	Sundberg et al. 1984
Black bear	Fibroma	Skin	Sundberg et al. 1984
Dolphin	Papilloma	Skin	Sundberg et al. 1984
Parrots	Papilloma	Cloaca	Sundberg et al. 1984; Sundberg et al. 1986
Pig	Papilloma	Skin	Rieke 1980
Pig	Papilloma	Penis	Parish 1961
Goat	Papilloma	Mammary gland skin	Moulton 1961; Ficken et al. 1983; Theilen et al. 1985
Rhinoceros	Papilloma	Interdigital skin	Boever 1976
Camel	Papilloma	Skin	Sundberg 1986
Lemming	Papilloma	Stomach	Barker et al. 1982
Chamois	Papilloma	Skin	Kumer 1935
Tasmanian devil	Papilloma	Skin	Effrom et al. 1977
Skunk	Papilloma	Oral mucous membranes	Clark 1973
Badger	Papilloma	Skin	Clark 1973
Ring-tailed lemur	Papilloma	Skin	Appleby 1969
Fallow deer	Papilloma	Skin	Heidemann 1974
Caribou	Fibropapilloma	Skin	Broughton et al. 1972
Armadillo	Fibroma	Skin	Pence et al. 1983
Beaked whale	Fibroma	Vagina	Flom et al. 1980
Newt	Papilloma	Skin	Asashima and Komazaki 1980
Asian elephant	Papilloma	Vulva	Effrom et al. 1977
Red-bellied black snake	Papillomas	Skin	Effrom et al. 1977
Macacus pileatus	Papilloma	Cervix	Ratcliffe 1933

(Table 1). Some researchers have reclassified the bovine papillomaviruses into group A for those which induce fibropapillomas (BPV1, BPV2, BPV5) and group B for those which induce papillomas (BPV3, BPV4, BPV6) (Smith and Campo 1985). This division is based on similarities among the viruses of each group as well as the types of lesions induced. It is likely that additional types will be identified in the future.

The fibropapillomas are the most common skin tumors of cattle but occur on other squamous epithelial surfaces, including the rumen, omasum, vagina, vulva, penis, and anus (Head 1953; Monlux et al. 1956; Monlux and Monlux 1972; Pearson 1972; Pirie 1973; Thorsen et al. 1974; Tweddle and White 1977). The cutaneous fibropapillomas in American cattle are usually found on the haired skin of the head, neck, chin, shoulder, and dewlap; in German cattle, they are most common on the neck, ventral abdomen, legs, back, and udder. This variation in sites has been thought to be due to different forms of management (Bagdonas and Olson 1953; Monlux and Monlux 1972; Stannard and Pulley 1978), but molecular studies indicate that different virus types have predilection sites as well (Table 1). Fibropapillomas of the genitalia have been reported to be more frequent in male cattle in the United Kingdom, while they are distributed equally among male and female cattle in the United States of America (McEntee 1950; Head 1953). In the United States of America venereal lesions are rare in herds which are bred by artificial insemination (Sundberg, unpublished data). Regardless of site, lesions are most common in cattle less than 1-2 years of age (Wettimuny and Jayasekera 1970; Monlux and Monlux 1972). There does not appear to be any breed predilection for development of fibropapillomas (Monlux and Monlux 1972; Stannard and Pulley 1978).

Papillomas and fibropapillomas are frequent on the teats of cattle, particularly dairy cattle (Olson et al. 1982). In abattoirs the frequency ranges from 25%-36% (Meischke 1979a; Olson et al. 1982). Although different terms have been used, three distinct morphological types have been described: flat (atypical) papillomas,

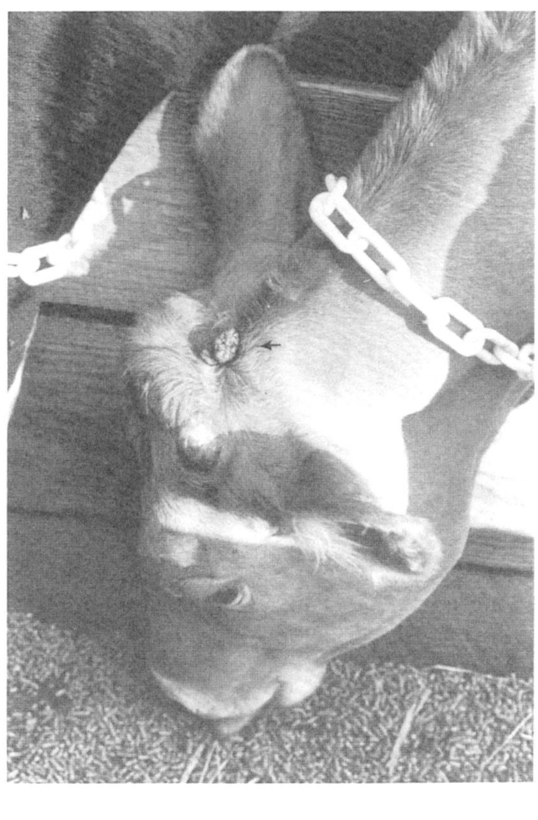

Fig. 1. A domestic cow with a cutaneous fibropapilloma between her ears and caudal to the poll *(arrow)*

filiform (atypical) papillomas, and typical fibropapillomas (Meischke 1979b; Olson et al. 1982). The term atypical refers to the first report of true papillomas in cattle (Barthold et al. 1974). These are distinct lesions and should be referred to as papillomas. Homogenates from all three types of teat lesions induced lesions on the teats of susceptible cattle, but only fibropapilloma homogenates induced fibromas in horses (Olson et al. 1982). The lesions induced were morphologically similar to the lesions of origin (Meischke 1979c).

Fibropapillomas often occur either singly (Fig. 1) or as multiple nodules. They vary in size and may reach several centimeters in diameter. The lesions may be sessile or pedunculated and lobate, fungiform, or verrucated (Moulton 1978). Microscopically, marked proliferation of the squamous epithelium, often on thin fibrovascular stalks, covers a large mass of proliferating fibroblasts in a dense matrix of collagen, hence the name fibropapilloma (Fig. 2). The venereal lesions are often ulcerated, because the prominent mesenchymal component may be only covered by a thin or moderately hyperplastic squamous epithelium. The upper layers of the stratum spinosum and stratum granulosum contain cells with clear cytoplasm occurring individually or in clusters, large, dark, basophilic, cytoplasmic, keratohyalinlike granules, and vesiculated nuclei. The nuclei of these cells stain positively for papillomavirus antigens, and viral inclusions can be easily identified within them by electron microscopy or within tumor homogenates by negative

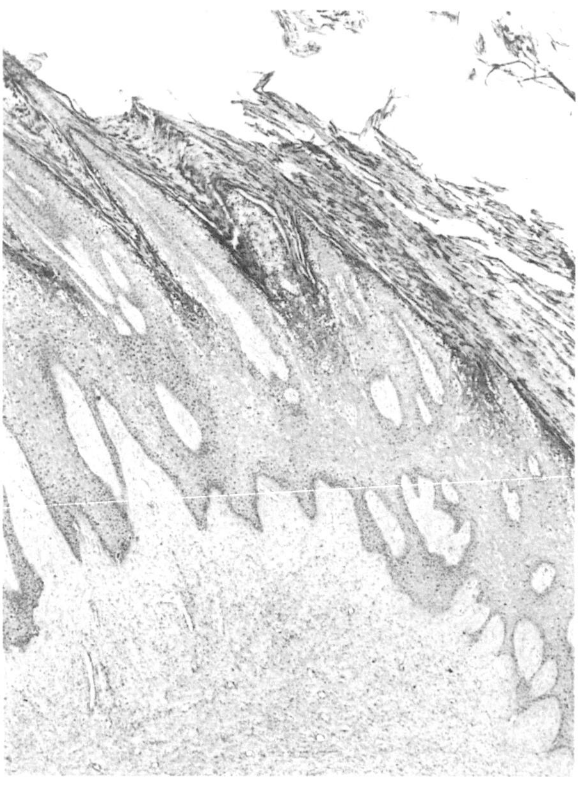

Fig. 2. A bovine cutaneous fibropapilloma with papillary projections of hyperplastic epithelium covering a homogenous mass of proliferating fibroblasts in a collagen matrix. H & E, ×65

stain electron microscopy (Fig. 3) (Brobst and Hinsman 1966; Fujimoto and Olson 1966; Tajima et al. 1968; Sundberg et al. 1984).

Bovine cutaneous fibropapillomas have been experimentally produced on scarified skin of calves by using tumor filtrates (Cheville and Olson 1964a; Fujimoto and Olson 1966; Barthold and Olson 1974a, b). Intradermal injection of the filtrates results primarily in fibroblastic proliferation (Cheville and Olson 1964a). The incubation period of experimentally produced lesions ranges from 30-59 days. The lesions regress in 1-14 months (Cheville and Olson 1964a; Barthold and Olson 1974b). In one natural outbreak animals in contact with a calf with papillomatosis developed lesions within 101-126 days (Bagdonas and Olson 1953), and the papillomas persisted for 18-173 days. Homogenates of cutaneous fibropapillomas from a bull induced typical fibropapillomas on the vulvas of 2 heifers (McEntee 1952).

Bovine papillomavirus has been observed in various layers of skin during the development of experimentally induced tumors. Particles first appear within the nuclei of cells in the stratum spinosum; later the virus is found in the stratum granulosum and stratum corneum (Brobst and Hinsman 1966; Fujimoto and Olson 1966; Tajima et al. 1968). Large intranuclear crystalline structures were present in the stratum spinosum of a naturally affected cow in which no virus particles were observed (Lepper 1967).

Neoplasms experimentally induced with homogenates of bovine fibropapillomas include meningiomas (Gordon and Olson 1968; Brobst and Dulac 1969; Robl et al. 1972) and polyps of the urinary bladder of calves (Olson et al. 1959), sarcoid-like tumors of the skin in the horse (Ragland and Spencer 1968, 1969), and fibro-

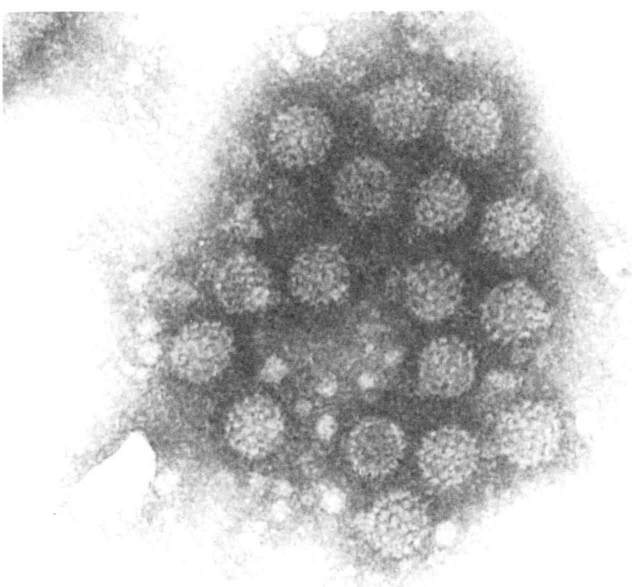

Fig. 3. Negatively stained bovine papillomavirus type 1 from a homogenized fibropapilloma. Phosphotungstic acid; *bar*, 100 nm

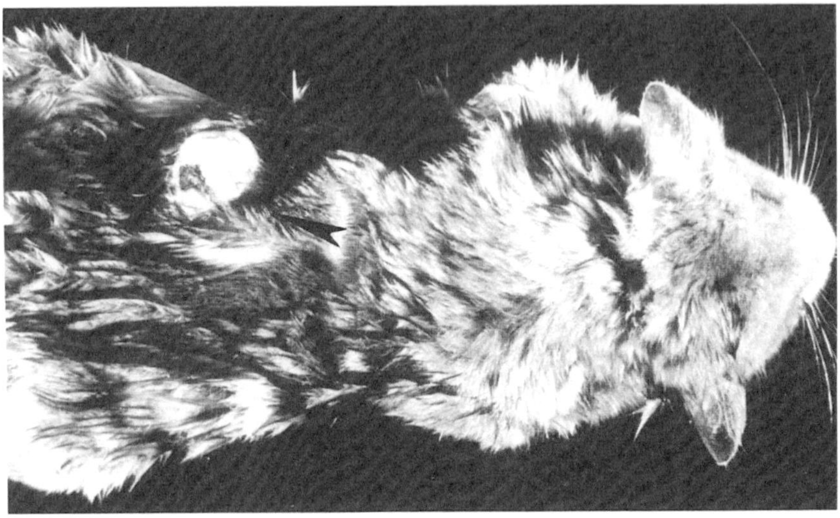

Fig. 4. Hamster with a cutaneous fibroma *(arrow head)* induced by subcutaneous injection of BPV1

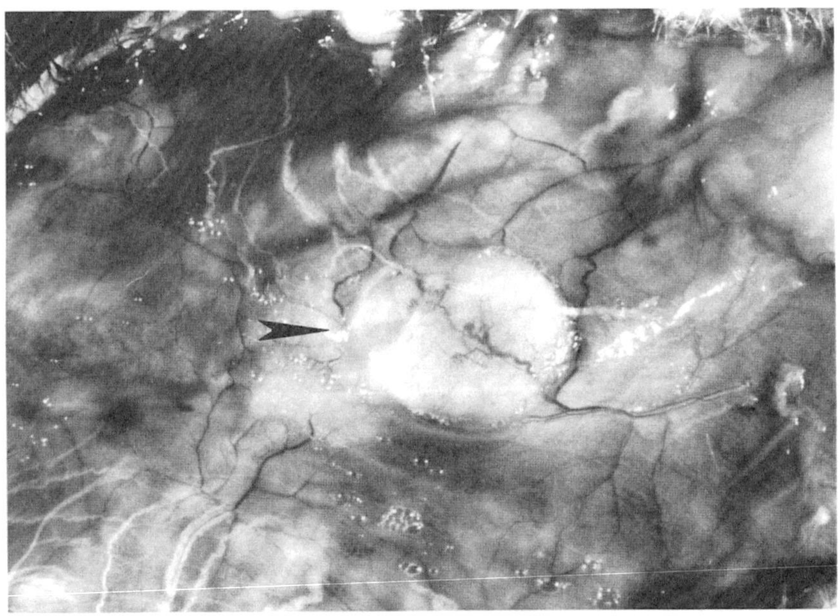

Fig. 5. BPV1-induced hamster fibroma did not invade through the hypodermis *(arrow head)*

blastomas, chondromas, and meningiomas in the hamster (Cheville 1966; Robl and Olson 1968). Both BPV1 and BPV2 were capable of tumor induction in the hamster (Fig. 4, 5) (Lancaster et al. 1979). The virus will transform cell cultures of fetal bovine conjunctiva (Black et al. 1963), skin, palate, meninges (Meischke 1979a), fetal hamster cells (Geraldes 1969, 1970; Morgan and Meinke 1980), fetal equine fibroblasts (Wood and Spradbrow 1985), fetal mouse cells (Thomas et al. 1964), and two mouse cell lines (NIH 3T3 and C127) (Dvoretzky et al. 1980).

True Papillomas, in which there is proliferation of the epithelium on thin fibrovascular stalks with no underlying mesenchymal proliferation, have been reported in cattle on haired skin (Barthold et al. 1974), interdigital skin (Rebhun et al. 1980), ocular mucous membranes (Taylor and Hanks 1972), mouth (Samuel, et al. 1985), and esophagus (Jarrett et al. 1978b). A novel bovine papillomavirus was isolated from the skin lesions and has been classified as BPV3 (Pfister et al. 1979). Since then, BPV4 has been isolated and characterized from esophageal papillomas (Campo et al. 1980), BPV6 from teat papillomas (Jarrett et al. 1984b), and BPV5 from teat fibropapillomas (Campo et al. 1981). The esophageal papillomas have been associated with squamous cell carcinomas of the esophagus in cattle, and an interaction between the papillomavirus and ingestion of bracken fern has been considered to be important in the pathogenesis of the lesions (Jarrett et al. 1978a, b). However, virus is not demonstrable in the carcinomas (Jarrett et al. 1984a). Ocular papillomas are discussed along with ocular squamous cell carcinomas.

BPV1, BPV2, and BPV5 have been isolated from fibropapillomas, and BPV3, BPV4, and BPV6 from true papillomas (Table 1). By heteroduplex mapping, it was demonstrated that BPV1 and BPV2 are closely related, but both show a high degree of sequence divergence from BPV5 (Coggins et al. 1985). The BPV3, BPV4, and BPV6 genomes exhibit moderate cross homologies with each other but minimal homology to BPV1, BPV2, and BPV5 (Coggins et al. 1985). The BPV1 genome has been completely sequenced (Chen et al. 1982), and BPV2 has been partially sequenced (Potter and Meinke 1985). Since the BPV1 genome replicates as an extrachromosomal element in transformed mouse cells, it has been used as a shuttle vector for a variety of genes (DiMaio et al. 1984; Denniston et al. 1984; Fukunaga et al. 1984; Pintel et al. 1984; Schenborn et al. 1985).

Transmission of the virus may be by direct contact to abraded areas of the skin. There is ample opportunity in most calf-raising systems for abrasion to occur at common sites of papilloma development. Natural outbreaks have occurred following rectal examinations (Tweddle and White 1977), ear tattooing (Ficarelli 1969), and dehorning (Pulley et al. 1974). Transmission of genital fibropapillomas occurs during coitus or when young bulls mount penmates. Semen from infected bulls may contain the virus (Olson et al. 1968). Bovine papillomavirus has been detected in commercially available milk from cows with teat papillomas (Meischke 1979c), and this may serve as another route of transmission.

Naturally and experimentally induced cutaneous fibropapillomas regress spontaneously due to development of immunity to the virus. Exposure to the virus during calfhood probably induces adequate immunity to account for the low incidence seen in adult cattle; however, reinfection can occur, probably due to loss of immunity or exposure to an immunologically different agent (Bagdonas and Olson 1953). Insignificant or low levels of neutralizing antibodies are produced in cattle. Serum from some unexposed horses has a slight but definite neutralizing effect. Horses with experimentally induced tumors develop neutralizing antibodies to the virus (Segre et al. 1955). Precipitin antibodies have been demonstrated in the sera of cattle with natural and experimental papillomatosis, immunized rabbits and chickens, and in cattle given commercial wart vaccine (Lee and Olson 1969a, b; Barthold et al. 1976), but their presence is not related to growth or regression of tumors (Barthold and Olson 1974b; Koller et al. 1974). An isolated case in a young bull with severe persistent fibropapillomatosis was associated with a deficient cell-mediated immunity (Duncan et al. 1975).

Vaccines have been used in the treatment of fibropapillomas, but results are variable and have only been evaluated clinically. As the disease is self-limiting, such evaluation is difficult under field conditions. Vaccination during early stages may actually prolong the disease (Olson and Skidmore 1959). New cases of fibropapillomatosis at a bull stud farm were prevented by vaccination with vaccine prepared from inactivated bovine fibropapilloma tissue or material from chicken embryos inoculated with fibropapilloma material (Olson et al. 1968; Barthold et al. 1976). Following vaccination, calves were challenged with preparations of wart tissue. The fibropapilloma tissue produced partial immunity, but no protection was produced with the chicken embryo products (Bagdonas and Olson 1954; Olson et al. 1960). When these vaccines were given to animals with relatively new fibropapillomas, the tumors grew larger (Olson and Skidmore 1959). No decrease or cessation of growth could be attributed to the vaccination. Vaccines against genital warts have been used therapeutically, as well as prophylactically, in conjunction with surgery. In 9 of 36 animals treated with autogenous vaccine, the neoplasm recurred after surgical removal; recurrence was also observed in 9 of 21 animals treated surgically without other therapy (Desmet et al. 1974).

Surgical excision of fibropapillomas has usually been unsuccessful (Amstutz 1978). In a controlled study, excision caused moderate regression in the number and size of the remaining fibropapillomas, but there was recurrence at some surgical sites (Olson and Skidmore 1959). Several surgical excisions may be required to remove penile fibropapillomas (Pearson 1972). Atypical warts (cutaneous papillomas) tend to persist and not regress (Barthold et al. 1974).

2.1.2 Bovine Squamous Cell Carcinomas

Squamous cell carcinomas are common in cattle as well as in other species. The most frequent primary sites are the unpigmented areas of the eye (Fig. 6), palpebrae, and vulva. Other sites include the horn core, esophagus, and forestomach. Ocular neoplasms have been observed worldwide, whereas squamous cell carcinomas of the vulva and horn core are limited to the tropics and the esophageal cancers to areas in Scotland and Brazil.

The incidence of squamous cell carcinoma on ocular mucous membranes has been estimated to be as high as 5% (Blodi and Ramsey 1967). Herefords, primarily females, have a significantly increased risk (Priester and Mantel 1971). In Kenya, 5% of the Ayrshire cows developed squamous cell carcinomas of either the vulva or eye if kept at an altitude of about 2200 m (Kaul and Kalra 1973). The incidence was a high as 20% in some areas. Of animals treated by veterinarians, the incidence of horn cancer has been as high as 5.20 per thousand (Kaul and Kalra 1973). The incidence was highest in castrated males; no cases were found in bulls. The neoplasm is rarely observed in animals less than 4 years of age and is most common in cattle that are 5-10 years of age (Palfi and Fabian 1969; Priester and Mantel 1971; Ivascu and Onet 1971; Kaul and Kalra 1973; Naik and Randelia 1975).

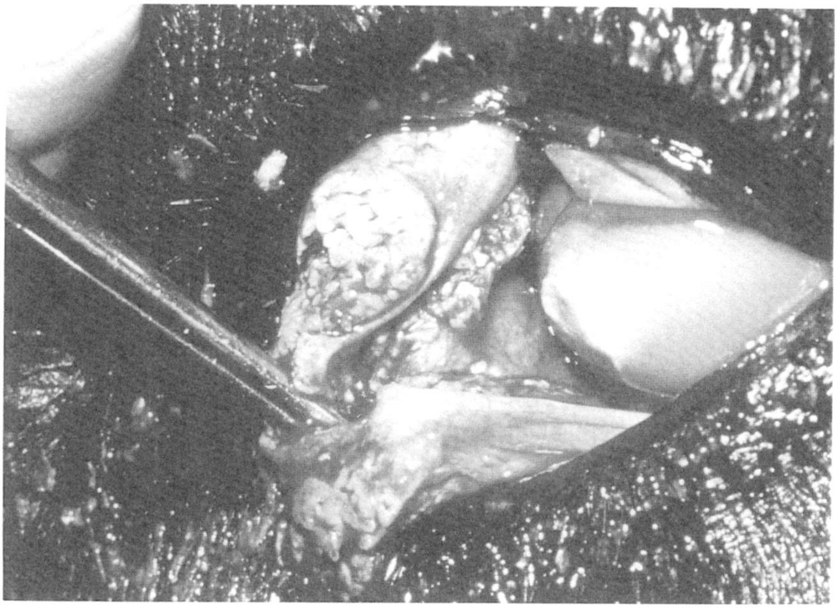

Fig. 6. Ocular squamous cell carcinoma of the unpigmented third eyelid of an Angus cow

No single cause has been identified that initiates the precancerous and cancerous changes. In virus-induced lesions ingestion of bracken fern *(Pteridium aquilinum)* has been associated with progression of esophageal papillomas to carcinomas (Jarrett et al. 1978a). A combination of genetic predisposition, lack of pigmentation, and exposure to actinic radiation has been suspected as causing ocular carcinoma (Anderson et al. 1957; Hunermund 1973; Naik and Randelia 1975). A viral etiology has been sought for the ocular neoplasms. Initially, a herpes virus, which causes infectious bovine rhinotracheitis, was suspected. This virus was isolated more often from ocular neoplasms than from normal eyes (Taylor and Hanks 1969; Epstein 1972a, b). However, transmission attempts, using cell suspensions and cell-free filtrates of ocular carcinomas, did not result in neoplastic development in animals observed for 3 years (Ivascu and Onet 1974). Therefore, this herpes virus has been considered to be an incidental finding and not the etiological agent.

Bovine ocular squamous cell carcinoma is thought to develop through a series of precancerous states of the ocular mucous membranes, which include keratotic plaques and papillomas (Fig. 7) (Monlux et al. 1957; Taylor and Hanks 1972). Papillomaviruses have also been suspected of being the etiological agent. These precancerous lesions are different from the cornified fibropapillomas of the haired skin of the palpebrae. Among 613 cattle with ocular lesions, 12% had plaques, 6.2% had papillomas, 4.9% had early squamous cell carcinomas, and 76.8% had squamous cell carcinomas (Monlux et al. 1957). Papillomaviruslike particles were observed in negatively stained homogenates of premalignant ocular lesions (Ford et al. 1982). Papillomavirus group-specific antigens were not detected in paraffin

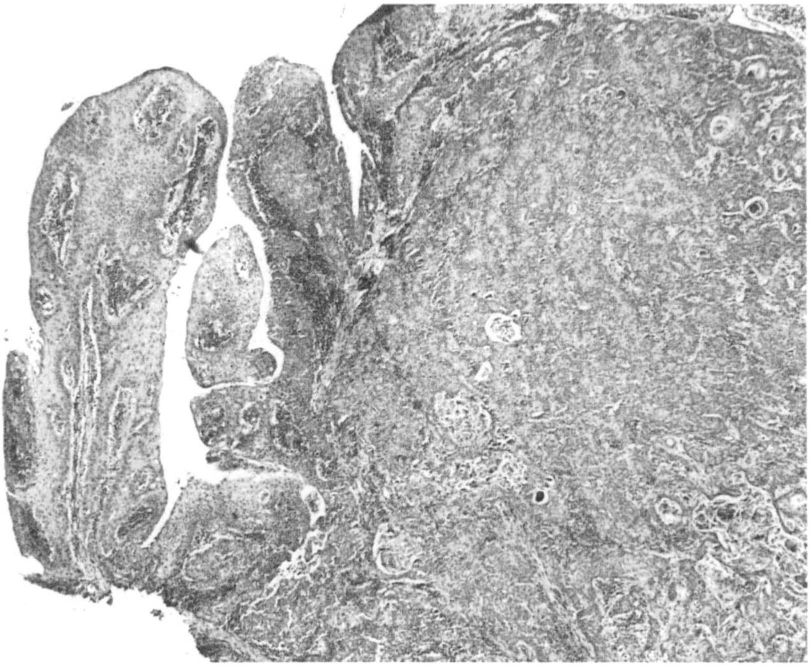

Fig. 7. Bovine ocular squamous cell carcinoma of the third eyelid of an Angus cow extending from a papilloma. H & E, ×65

sections of ocular papillomas or squamous cell carcinomas by the peroxidase-antiperoxidase technique (Sundberg et al. 1984). Southern blot hybridization of high salt-soluble DNA extracts from 20 ocular squamous cell carcinomas using a BPV1 probe under conditions of low stringency did not yield any evidence for the presence of a papillomavirus genome (O'Banion and Sundberg, unpublished data).

2.1.3 Bovine Urinary Bladder Tumors

In North America, primary neoplasms of the urinary bladder are rare in cattle (Davis et al. 1933; Plummer 1956; Brobst and Olson 1963; Migaki et al. 1971). Bladder tumors usually occur in small, well-defined geographic areas that tend to be upland wooded areas (Nandi 1969). In cattle over 2 years of age, the frequency of bladder tumors can be as high as 25% in these areas (Pamukcu 1974). Since the primary clinical presentation of affected cattle is hematuria, the syndrome has been called enzootic hematuria, bovine cystic hematuria, hematuria vesicalis bovis, and bovine vesicle fibromatosis (Migaki et al. 1971; Pamukcu 1974; Yoshikawa and Oyamada 1975). The tumors occur in cattle 4–14 years of age and are diagnosed most often in cattle 4–6 years of age (Pamukcu et al. 1970; Pamukcu 1974; Beran 1976; McKenzie 1978). No breed or sex predisposition has been reported (Pamukcu et al. 1976).

Epithelial tumors are the most common and include papillomas, adenomas, adenocarcinomas, transitional cell carcinomas, and squamous cell carcinomas. Less than 18% are mesenchymal tumors; these include fibromas, hemangiomas, and hemangiosarcomas. Lymphosarcoma may occur in the urinary bladder but is usually a multisystemic disease (Brobst and Olson 1963; Pamukcu 1974; Moulton 1978).

Ingestion of ferns over long periods of time, particularly bracken fern *(Pteridium aquilinum),* has been associated with outbreaks of bovine enzootic hematuria (Tokarnia et al. 1969; Jarrett et al. 1978a; McKenzie 1978). Bladder tumors have been induced in cattle by feeding them fresh or dried bracken. The neoplasms were histologically indistinguishable from natural tumors, and they did not develop in control animals (Pamukcu 1957; Pamukcu et al. 1967a; Price and Pamukcu 1968; Pamukcu et al. 1970; Pamukcu et al. 1976). Bracken fern may induce the tumors in certain regions alone or there may be a synergistic activity when associated with some of the bovine papillomaviruses, as has been suggested for the esophageal carcinomas in cattle (Jarrett et al. 1978a).

Numerous factors have also been suggested including dietary deficiency, ingestion of poisonous plants, deficiency of lime or excess molybdenum in the soil, infectious agents, and cystic calculi (Brobst and Olson 1963; Moulton 1978). Suspensions of cutaneous papillomas injected into the submucosa of the urinary bladder have caused the development of fibromas (Olson et al. 1959; Brobst and Olson 1965). Suspensions of spontaneous bladder tumors produced fibropapillomas of the vagina and skin as well as polypoid growths with fibromas in the urinary bladders of calves (Olson et al. 1965). Bladder tumors were not prevented by the use of a papillomavirus vaccine (Pamukcu et al. 1967b).

2.2 Horses

2.2.1 Equine Papillomas

Equine papillomas are easily diagnosed clinically and, because of the absence of alarming side effects, are seldom submitted for histological examination (Runnels and Benbrook 1942). Cutaneous papillomas represent only about 5% of all equine neoplasms submitted to diagnostic laboratories (Sastry 1959; Baker and Leyland 1975; Sundberg et al. 1977). Other surveys reveal that papillomas are the most common equine neoplasm (Head 1953; British Equine Veterinary Association 1965). Cutaneous papillomas are found in horses 1-2 years old, most often on the muzzle, but also on limbs. Papillomas also occur on the mucous membranes of the prepuce, vulva, cornea, or conjunctiva (Cook and Olson 1951; Head 1953; Blodi and Ramsey 1967; Sundberg et al. 1977; Junge et al. 1984). Congenital papillomas are not infrequent in the horse (Junge et al. 1984). Papillomas usually persist for 1-9 months and then disappear. The horse apparently develops complete immunity following infection (Cook and Olson 1951; Stannard and Pulley 1978; Sundberg et al. 1985f.).

Papillomas appear as small, elevated, circumscribed, horny masses, 2-20 mm in diameter, and range in number from 2 to more than 100 (Fig. 8). Histologically,

Fig. 8. Numerous cutaneous papillomas around the muzzle of a horse

these lesions consist of marked epithelial proliferation on a thin fibrovascular stalk in a papillary pattern. Intranuclear inclusions have been found within cells in the stratum granulosum which have abundant clear cytoplasm and numerous large, dark, basophilic, keratohyalinlike, cytoplasmic granules (Fig. 9). Cutaneous papillomas were heavily keratinized while those affecting mucous membranes, particularly the lower genital tract, had a thin stratum corneum. Papillomaviruslike particles were visualized in negatively stained preparations and viral antigens within nuclei of the stratum spinosum and upper epithelial layers (Fig. 10) (Fulton et al. 1970; Junge et al. 1984; Sundberg et al. 1984, 1985f.).

The disease appears to be transmitted by contact between horses. The infection could spread from the muzzle to the legs because horses frequently rub their legs with their nose to dislodge flies (Olson 1963).

When bacteria-free filtrates of equine papillomas were injected intradermally or subcutaneously or were applied to scarified skin, papillomas appeared 66 to 67 days after inoculation and regressed 50 days later. The virus was noninfectious when kept at 55°-65 °C for 30 min but not at 45 °C; the virus was infectious after storage for 73 days in 50% glycerol at 4 °C but not after 112 days. It was active after 185 days at −35 °C but not at 224 days. Inoculations of known active suspensions into calves, lambs, dogs, rabbits, and guinea pigs did not induce lesions (Cook and Olson 1951).

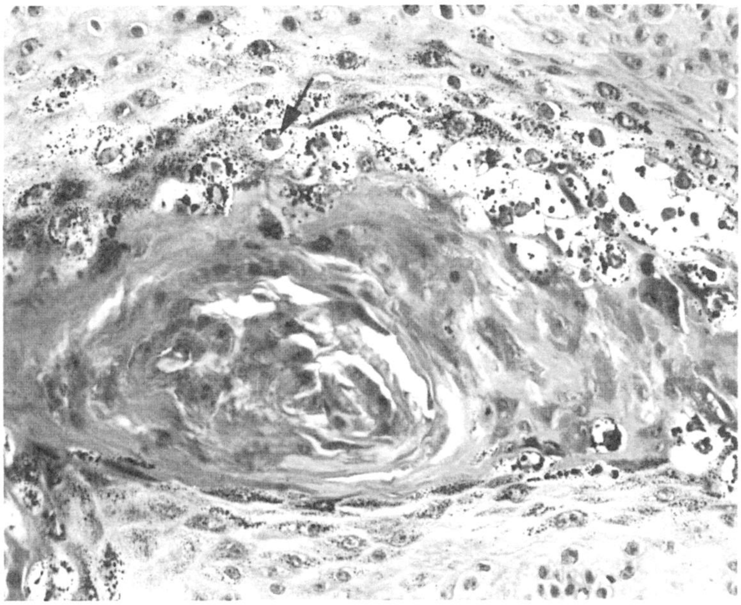

Fig. 9. Cytopathology of the equine cutaneous papillomavirus is characterized by cells of the stratum granulosum with clear cytoplasm, large, basophilic, keratohyalinlike granules, and centrally located vesicular nuclei which may contain inclusions *(arrow)*. H & E, ×400

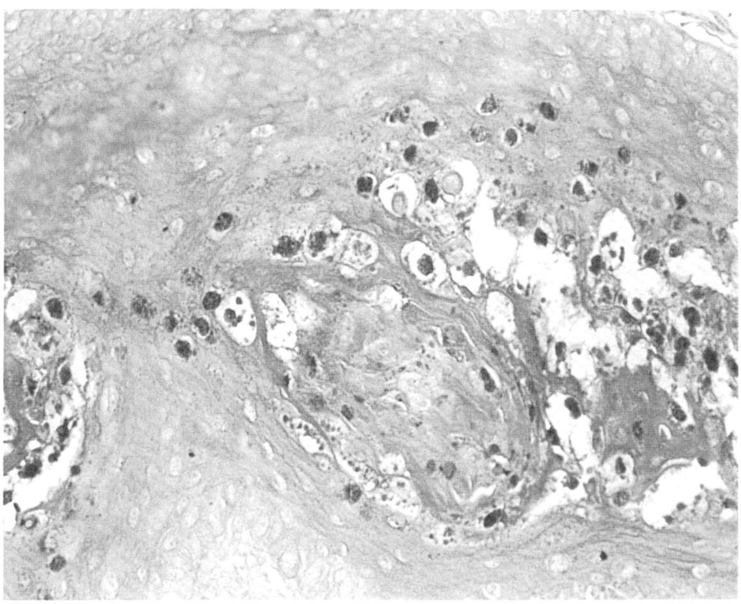

Fig. 10. Nuclei of cells in the stratum granulosum of an equine papilloma stained positively for papillomavirus antigens. Same field as Fig. 9, peroxidase-antiperoxidase technique, light green counterstain, ×400

A unique papillomavirus has recently been isolated, characterized, and its DNA cloned from cutaneous papillomas affecting the muzzle of several ponies in a herd (O'Banion et al. 1985). Identical viruses (by restriction endonuclease analysis and Southern blot hybridization) were also isolated from cutaneous papillomas affecting the muzzle and lower limbs of horses from other herds. In the same study, papillomavirus DNA was isolated from a papilloma on the shaft of the penis of a horse. This latter viral genome had a different restriction endonuclease digestion pattern and could only be detected by Southern blot hybridization under conditions of low stringency. This suggests that at least two types of papillomaviruses affect horses.

2.2.2 Equine Sarcoids

One of the most common equine neoplasms is the sarcoid (Kerr and Alden 1974; Baker and Leyland 1975; Sundberg et al. 1977). The tumors are often multiple, locally aggressive, most frequently found on the lower legs (Fig. 11), head, and prepuce, and often recur after surgical excision (Ragland et al. 1970a; Straffus et al. 1973; Hesselholt and Ingerslev 1974; Kerr and Alden 1974). They are divided into: (a) the verrucous type, which has a papillomatous appearance; (b) the fibroblastic type, which appears as firm, fairly discrete, grayish-white, slightly elevated nodules; and (c) the mixed type. The epidermis is acanthotic, and variably hyper-

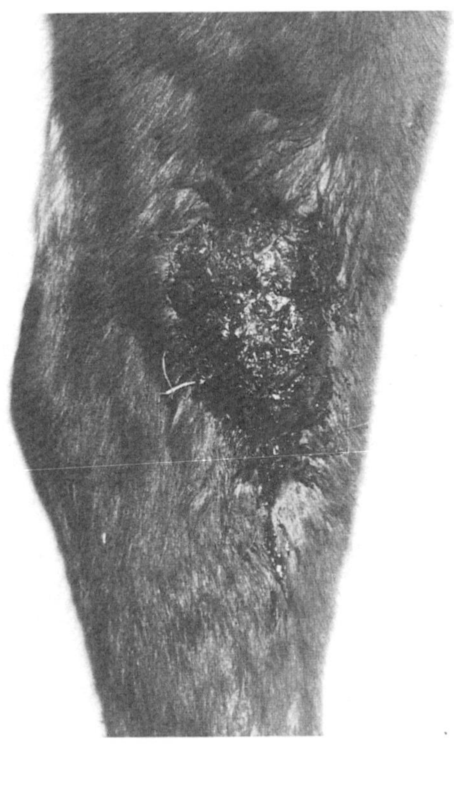

Fig. 11. An ulcerated sarcoid on the skin of the lateral hock joint of a horse

keratotic, and pseudoepitheliomatous hyperplasia is a common feature (Fig. 12). The bulk of the sarcoid consists of fibroblasts arranged in whorls and columns, containing pleomorphic nuclei and small to moderate numbers of mitotic figures in rapidly growing areas of the tumor. The fibroblasts may be oriented perpendicularly to the basement membrane at the dermoepidermal junction (Ragland et al. 1970a). The tumor resembles a fibrosarcoma or neurofibrosarcoma. There seems to be no predilection for age, breed, color, or sex (Ragland et al. 1970a; Sundberg et al. 1977).

The etiology of sarcoids in horses has yet to be definitively proven. A viral etiology has been suspected based on successful transmission using cell-free material (Voss 1964, 1969). The morphological similarity of sarcoids to bovine cutaneous fibropapillomas led to the speculation that BPV might play a role in the pathogenesis of the disease. In ponies, BPV induced tumors are histologically indistinguishable from sarcoids; but these regressed after 1 year and serial passage in horses was unsuccessful (Ragland and Spencer 1969; Ragland et al. 1970b). BPV1, BPV2, and possibly a third type, have been shown to exist as free circular episomes in fibroblasts of equine sarcoids (Lancaster et al. 1977; Lancaster and Olson 1979; Lancaster et al. 1979; Amtmann et al. 1980; Lancaster 1981). Of five sarcoids screened by Southern blot hybridization, papillomavirus DNA was identified under conditions of high stringency with a BPV2 DNA probe. The genome had restriction fragments consistent with BPV1 or BPV2 DNA. No hybridization was detected with an equine cutaneous papillomavirus DNA probe (O'Banion et al. 1985).

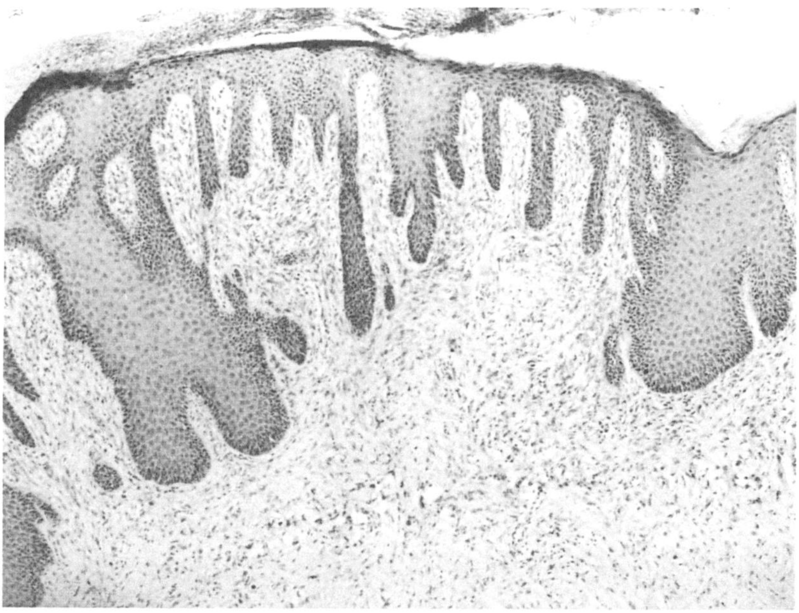

Fig. 12. An equine sarcoid with moderately hyperplastic squamous epithelium and long rete ridges extending into a large mass of proliferating fibroblasts in a dense matrix of collagen. H & E, ×65

A transformed cell line was established from a natural sarcoid case (Watson et al. 1972). Retroviruses were observed in these cells (England et al. 1973). A tumor was induced by intradermal injection of cells from this line into a immunodeficient Arabian foal and the retrovirus was isolated from explanted tumor cells (Cheevers et al. 1982). The retrovirus was isolated and characterized from cells derived from the original cell line and it was found to elicit rapid morphological transformation of primary equine dermal fibroblasts (Fatemi-Nainie et al. 1982). BPV1 has also been successfully used to transform fetal equine fibroblasts in a similar manner (Wood and Spradbrow 1985). Several cell lines were established by explant culture from naturally occurring equine sarcoids in which retroviruslike agents were never observed in ultrastructural studies (Sundberg, unpublished data). These lines were subsequently shown to contain BPV genomes (Lancaster et al. 1977). Although a number of investigators have independently demonstrated BPV DNA in sarcoids, the case containing a retrovirus may indicate a synergistic role of the two virus types, as has been observed in neoplasms of hamsters (see Sect. 2.8).

2.2.3 Equine Squamous Cell Carcinomas

Squamous cell carcinomas are also common in horses (Sundberg et al. 1977). They have an anatomical distribution similar to papillomas but tend to be diagnosed in older horses (Junge et al. 1984). These neoplasms generally appear as raised, irreg-

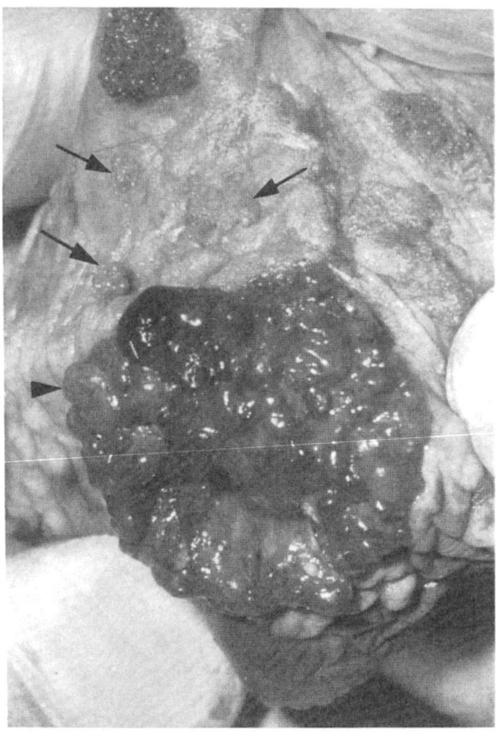

Fig. 13. Papillomas *(arrows)* and squamous cell carcinoma *(arrow head)* on the penis of a stallion

ular, and invading masses (Fig. 13). They may be heavily pigmented or totally lacking in pigmentation. Visceral metastases occur late in the disease. Microscopically, carcinomas are characterized by a proliferation of squamous epithelial cells which extensively invade the base of the lesion. Islands of cells are common in the lamina propria, and many individual cells are keratinized. Intercellular bridges and epithelial pearl formations, when present, are important diagnostic features. The mitotic index can be high, and bizarre mitotic figures are not uncommon. Papillomavirus group-specific antigens have never been detected in equine squamous cell carcinomas (Sundberg et al. 1984; Junge et al. 1984); however, since papillomaviruses have been detected in premalignant lesions of bovine ocular squamous cell carcinoma (Ford et al. 1982) and papillomavirus genomes have been detected in squamous cell carcinomas of the cervix in women (Gissmann et al. 1984), it is possible that an equine papillomaviruses might be involved with the pathogenesis of this neoplasm. A case has been observed in which papillomas that contained viral antigens and viral DNA were present on the shaft of the penis of a stallion adjacent to several locally invasive squamous cell carcinomas (O'Banion and Sundberg, unpublished data). Molecular studies of three penile, one perineal, and one ocular squamous cell carcinomas surgically removed from horses failed to demonstrate papillomavirus genomes in the tumors when an equine cutaneous papillomavirus probe was used by Southern blot hybridization under conditions of low stringency (O'Banion and Sundberg, unpublished data). Numerous cases will probably have to be screened using several different probes, including the equine venereal papillomavirus genome, before the role, if any, of these viruses can be determined in the etiology of this neoplasm.

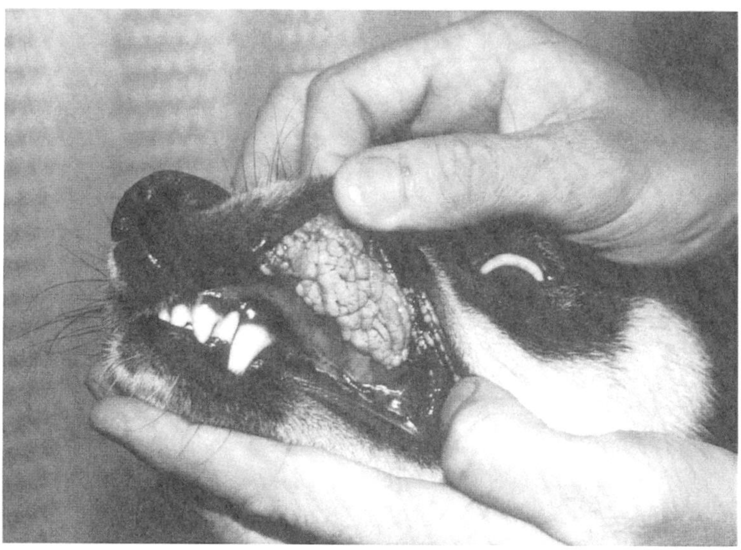

Fig. 14. Multiple papillomas in the mouth of a dog

2.3 Dogs

2.3.1 Canine Oral and Ocular Papillomas

These are infrequent tumors affecting dogs of all ages and which often regress in 3 weeks to 6 months without treatment. Papillomatosis of the conjunctiva and adnexa may be present with or without oral lesions (Belkin 1979; Bonney et al. 1980). The tumors appear as pedunculated or verrucated, unpigmented masses of various sizes and number (Fig. 14). Histologically, they consist of acanthotic and hyperkeratotic mucosa thrown into folds with vascularized cores of connective tissue which may contain microfoci of inflammatory cells (Fig. 15). Basophilic intranuclear inclusions are occasionally present in the superficial layers of the epider-

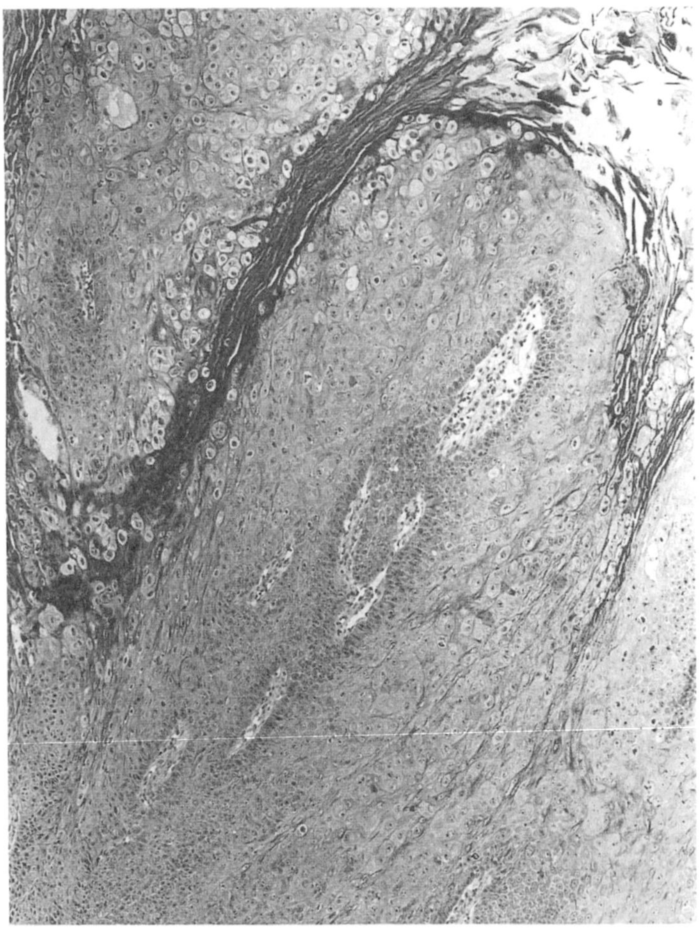

Fig. 15. Canine oral papilloma with proliferating squamous cells on a thin fibrovascular stalk. Cells of the stratum granulosum have pale staining cytoplasm and centrally located nuclei. H & E, ×65

mis (Fig. 16). The inclusions have been reported to be Feulgen positive, positive for DNA with methyl green, and fluoresce green when stained with acridine orange (Cheville and Olson 1964b; Belkin 1979). Virus particles, 40-50 nm in diameter, were visualized in negatively stained preparations and in the nuclei of cells in the stratum granulosum and stratum corneum in thin sections (Cheville and Olson 1964b). Nuclear fluorescence, corresponding to inclusion material, was shown using direct immunofluorescene with dog antiserum against canine oral papillomavirus (Cheville and Olson 1964b). Papillomavirus group-specific antigen could be detected in inclusions by the peroxidase-antiperoxidase technique (Sundberg et al. 1984).

Experimentally, the virus has been shown to have a strong affinity for the oral mucosa (DeMonbreun and Goodpasture 1932) and, to a lesser extent, to conjunctiva (Tokita and Konishi 1975; Sundberg and Olson, unpublished data) or skin around the nose and mouth (Chambers et al. 1960). Following an incubation period of 4-8 weeks, papillomas develop and then regress after an additional 4-8 weeks of growth (Chambers and Evans 1959; Sundberg and Olson, unpublished data). Watrach et al. (1970) reported progression of oral papillomas into a squamous cell carcinoma in a beagle with extensive, persistent papillomatosis. Cutaneous squamous cell carcinomas developed in several dogs at injection sites of a live canine oral papillomavirus vaccine (Sundberg et al. 1985e).

The virus causing canine oral papillomatosis lacks pathogenicity for the mouse, hamster, and guinea pig, and has not been successfully cultured in dog kidney,

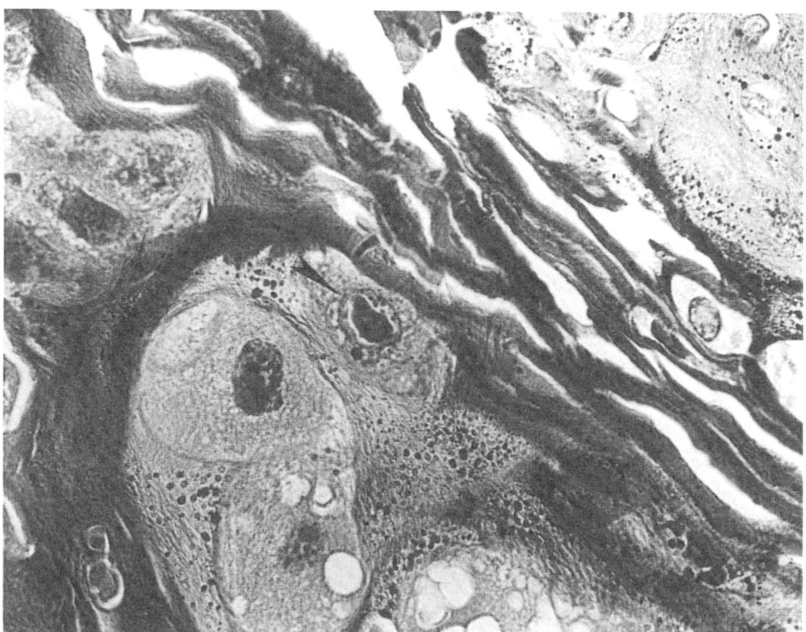

Fig. 16. Intranuclear inclusion *(arrow head)* in a cell in the stratum granulosum of a canine oral papilloma. H & E, ×650

Hela, Vero, or BHK cells. In vitro cultivation of papilloma cells following trypsinization of tumor tissue resulted in growth and keratinization of epithelial cells with detachment of cell clusters resembling a cytopathic effect. Large, swollen cells or inclusions were not observed (Tokita et al. 1977).

The etiological agents has been characterized as a papillomavirus based on morphological and molecular criteria (Cheville and Olson 1964b; Sundberg et al. 1986b). A serological relationship to HPV1 was demonstrated by indirect immunofluorescence using rabbit antiserum prepared against SDS-disrupted HPV1 (Pfister and Meszaros 1980). Antibodies against human wart virus were detected in dogs by the immunodiffusion method (Pyrhönen and Neuvonen 1978). This may be due to common antigenic determinants rather to HPV infections in dogs. The viral genome is 8.2 kb. It has been cloned and a restriction map generated (Sundberg et al. 1986b).

2.3.2 Canine Cutaneous Papillomas

Cutaneous papillomas in pet dogs are rare (Murray 1968). In racing greyhounds, papillomas often occur in young animals, 12-18 months of age, when they are first introduced to the racetrack. They are usually located on the lower limbs, distal to the carpus or hock, and are often solitary. They may be found between the foot pads, on the nail-skin junction, and along suture lines of wounds following amputation of toes or dewclaws. The papillomas are 3-5 mm in diameter with a superficial protruding surface, but the bulk of the mass lies below the line of the epidermis. Histologically, there is a hyperplastic epidermis supported by stromal papillae. Large, basophilic, intranuclear inclusions are rarely found. Virus particles, 50-55 nm in diameter, observed in negatively stained preparations, have the characteristic appearance of papovaviridae (Davis et al. 1975). Similar viruses, 45-49 nm in diameter, arranged in intranuclear crystalline arrays in cells of the stratum granulosum, have also been observed in cutaneous papillomas of pet dogs (Watrach 1969). Papillomavirus group-specific antigens have been detected in paraffin sections of routine diagnostic cases (Sundberg et al. 1984). The virus or viral genome has not been isolated. Its relationship, if any, to the canine oral papillomavirus needs to be determined.

Transmission to the skin of crossbred dogs and greyhounds was accomplished using cell-free filtrates. The incubation period ranged from 27-49 days with spontaneous regression by 59 days (Davis et al. 1975).

2.3.3 Canine Venereal Papillomas

Papillomas occur rarely on the lower genital tracts of both sexes of dogs (Fig. 17) (Sundberg et al. 1984). Papillomavirus group-specific antigens have been detected in paraffin sections in retrospective surveys (Fig. 18; Sundberg et al. 1984; Sundberg and Dunstan, unpublished data); however, the virus has not been isolated or characterized from any venereal lesions. Attempts to induce venereal papillomas with infectious material from canine oral papillomas have been unsuccessful (Tokita and Konishi 1975; Sundberg and Olson, unpublished data). It is likely that a specific canine papillomavirus type affects the lower genital tract of dogs, as is the case in human beings.

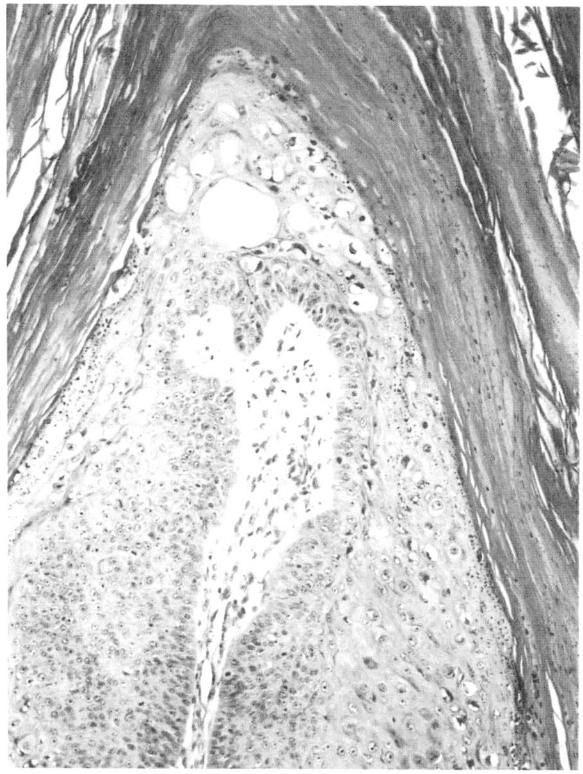

Fig. 17. Canine penile papilloma has similar microscopic features to Fig. 16. H & E, ×100

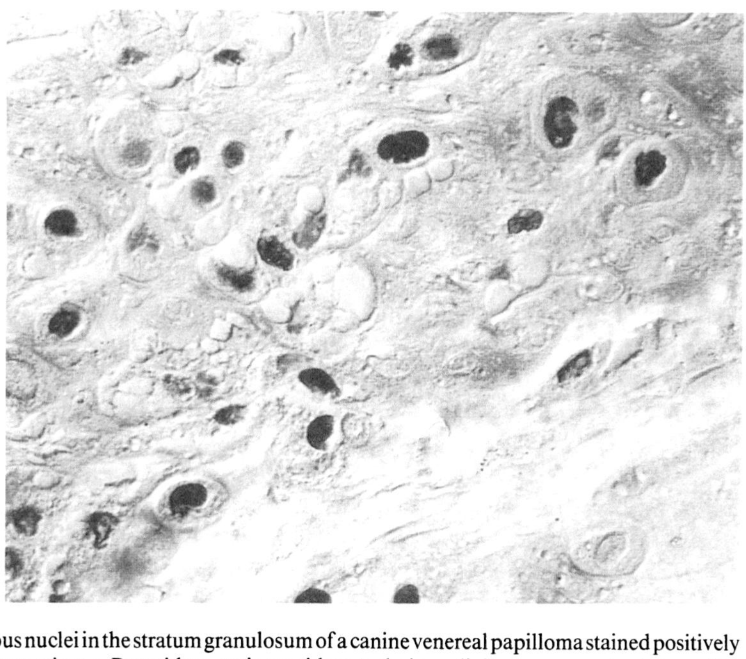

Fig. 18. Numerous nuclei in the stratum granulosum of a canine venereal papilloma stained positively for papillomavirus antigens. Peroxidase-antiperoxidase technique, light green counterstain, ×500

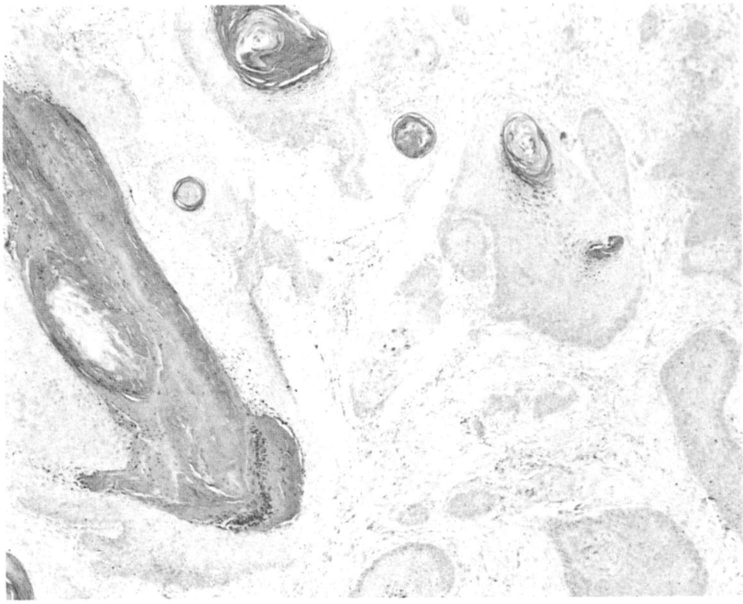

Fig. 19. Locally invasive, well-differentiated, squamous cell carcinoma of the vulva of a dog with keratin pearls. H & E, ×65

2.3.4 Canine Squamous Cell Carcinomas

Squamous cell carcinomas are not rare in dogs and affect a wide range of anatomical sites. This neoplasm is most often diagnosed on the skin or in the mouth (Strafuss et al. 1976; Sundberg and Dunstan, unpublished data). Papillomavirus group-specific antigens were detected in a spontaneous squamous cell carcinoma of the vulva (Fig. 19 and 20) and in five of six cases of cutaneous carcinoma which developed at injection sites of a live canine oral papillomavirus vaccine (Sundberg et al. 1984). Macerated tumors associated with the use of the vaccine did not induce oral papillomas on the scarified mucosa of 10- to 12-week-old puppies. Papillomas were induced on the oral mucosa of littermates when a macerated oral papilloma was applied in a similar manner. The puppies inoculated with a carcinoma were protected from challenge with infectious oral papillomavirus (Bregman et al. 1987). Experimental inoculation of five beagles with homogenates of oral papillomas used to produce the vaccine resulted in induction of a persistant, slow-growing, cystic, basosquamous carcinoma (Olson and Sundberg, unpublished data).

A canine oral papillomavirus has been isolated from oral papillomas used to prepare the vaccine associated with induction of cutaneous carcinomas. The viral genome has been cloned and characterized (Sundberg et al. 1986b). It has a restriction map similar to the limited characterization done on an oral papillomavirus isolated from a dog in Germany (Pfister and Meszaros 1980). The vaccine isolate has also been compared to two isolates from naturally occurring oral papil-

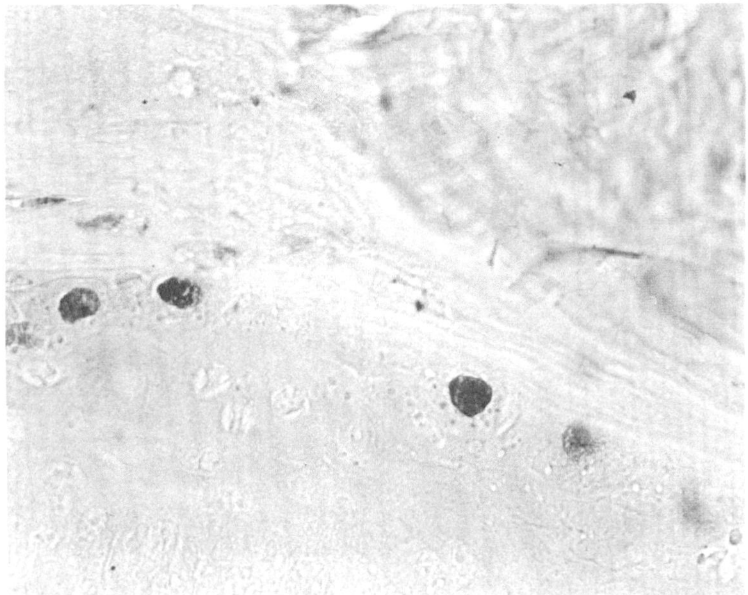

Fig. 20. Canine vulvar squamous cell carcinoma (same as Fig. 19) with nuclei stained positively for papillomavirus antigens. Peroxidase-antiperoxidase technique, light green counterstain, ×500

lomas in pet dogs by restriction endonuclease analysis and high stringency Southern blot hybridization. All three oral papillomavirus isolates appear to be similar by this method (O'Banion et al., unpublished data).

2.4 Sheep

2.4.1 Sheep Papillomas and Fibropapillomas

Papillomas and fibropapillomas are rare in sheep in most parts of the world. Cutaneous papillomas were reported on the foreleg of Southdown cross sheep in England (Gibbs et al. 1975). Although the author considered the lesion to be similar to the bovine cutaneous fibropapilloma, the description and low power photograph resemble papillomas with prominent fibrovascular stalks. The lesion lacks the large mesenchymal component typical of bovine fibropapillomas. Viruslike particles, 55 nm in diameter, resembling papillomaviruses, were found in these lesions. The lesions could be experimentally reproduced in sheep with cell-free filtrates but not in cattle or goats. Slow-growing fibromas were also induced in neonatal hamsters.

Papillomaviruslike virions were detected by negative stain electron microscopy in hyperkeratotic scales of the perineum (Vanselow and Spradbrow 1983), papillomatous areas of the face (Fig. 21) (Vanselow and Spradbrow 1982), and a highly keratinized horn on the ear (Vanselow and Spradbrow 1982). These lesions were

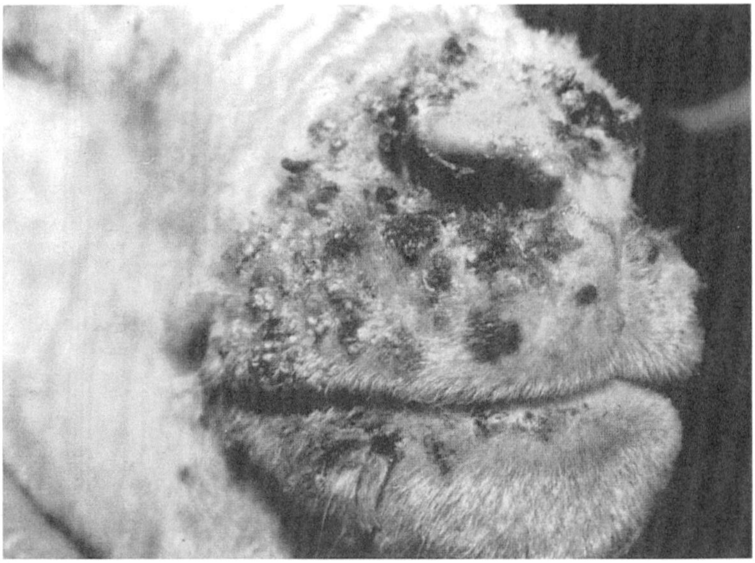

Fig. 21. Multiple cutaneous papillomas on the muzzle of a sheep (courtesy Dr. J. Samuels and P. B. Spradbrow)

observed in aged Merino sheep in Australia and were considered to be precursor lesions to squamous cell carcinomas. The viruses associated with these lesions have yet to be characterized.

2.4.2 Sheep Squamous Cell Carcinomas

Squamous cell carcinomas are rarely reported in sheep in North America because these animals are usually slaughtered at a relatively young age, often less than 1 year (Davis and Shorten 1952). Squamous cell carcinoma of the ears (Fig. 22) of aged sheep was reported to be fairly common in Australia (Dodd 1923). In Australia and South Africa, this neoplasm has been recognized with increasing frequency on or near the vulva and may be associated with the use of the radical Mules operation and short docking of tails (Vandergraaff 1976; Hawkins et al. 1981; Tustin et al. 1982; Vanselow and Spradbrow 1983). The result of this operation is that the perineum and vulvar labia are exposed to sunlight, which may be an important factor in the pathogenesis of bovine ocular squamous cell carcinoma. Squamous cell carcinomas have also been observed in wool-bearing sites which were protected from sunlight in Merino sheep. However, these neoplasms developed in the walls of cutaneous cysts in sheep that appeared to have a hereditary predisposition to the development of cysts (Carne et al. 1963).

Hyperkeratotic plaques and papillomas have been considered to be precursor lesions of squamous cell carcinomas in sheep. Viruslike particles, which resemble papillomaviruses, have been detected in these lesions by negative stain electron microscopy (Vanselow and Spradbrow 1982; Vanselow and Spradbrow 1983). Transmission attempts, using suspensions of viable cells from pooled vulvar squa-

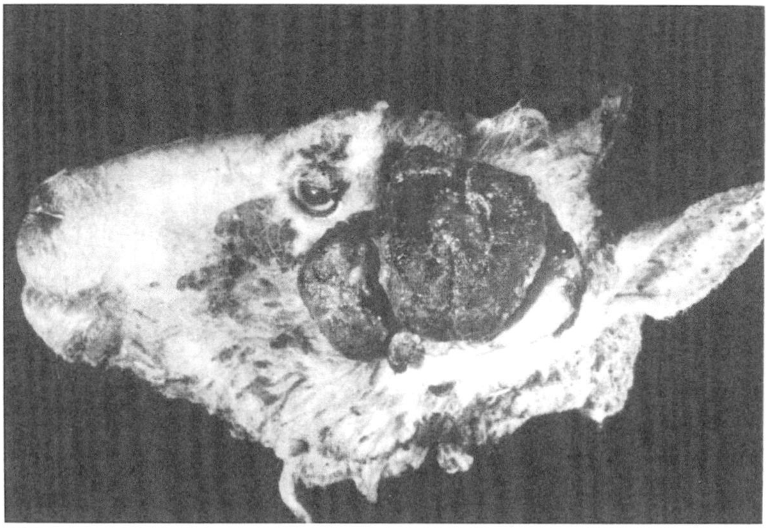

Fig. 22. Cutaneous squamous cell carcinoma ventral to the ear of a sheep (courtesy Dr. H. Gelberg)

mous cell carcinomas, were unsuccessful (Swan et al. 1983). Ovine squamous cell carcinoma has been proposed as an animal model of the human disease (Ladds and Daniels 1982). The agent has yet to be isolated and characterized.

2.5 Rabbits

2.5.1 Domestic Rabbit Oral Papillomas

Rabbit papillomatosis is usually encountered as an incidental necropsy finding (Weisbroth and Scher 1970). Typically, the lesions are found on the ventral surface of the tongues of domestic rabbits *(Oryctolagus cuniculus)* and consist of small, gray-white, sessile or pedunculated nodules measuring from 1 mm in diameter up to 4 × 5 mm. Due to the location and small size, these lesions are easily overlooked, although up to 31% of rabbits from some sources may be affected (Sundberg et al. 1985c). Filtrates are transmissible only to oral sites (not ocular or vulvar mucous membranes) of either domestic rabbits or cottontailed rabbits *(Sylvilagus floridanus)* but not to other species, and there is no cross-immunity with the cottontail rabbit cutaneous papillomavirus (Parsons and Kidd 1936, 1943; Rdzok et al. 1966; Sundberg et al. 1985c). Histologically, the oral papilloma consists of an acanthotic and hyperkeratotic epidermis on a thin fibrovascular stalk. Intranuclear inclusions are frequent in cells of the upper stratum spinosum and stratum corneum. Viruses found within these nuclei are 40 nm in diameter in thin sections and 50-52 nm in diameter in negatively stained preparations (Richter et al. 1964; Rdzok et al. 1966). Papillomavirus group-specific antigens can be detected in these cells (Sundberg et al. 1984; Sundberg et al. 1985c).

A papillomavirus was isolated and characterized from oral papillomas of New Zealand white rabbits. Analysis of restriction endonuclease digestion products revealed 3 *Bam*H1 sites which were used to clone subgenomic fragments into pBR322. All restriction fragments generated using various enzymes had patterns distinct from those published for the cottontail rabbit cutaneous papillomavirus. By Southern blot hybridization of subgenomic fragments under increasing degrees of stringency, a small region appeared to be highly conserved between the two rabbit papillomaviruses (O'Banion and Sundberg, unpublished data). The role of this region in the biology of the virus remains to be determined. It is possible that this region is related to the ability of the two viruses, which naturally affect different species, to cross species barriers in transmission trials.

2.5.2 Cottontail Rabbit Cutaneous Papilloma and Squamous Cell Carcinomas

Shope and Hurst (1933) first described a cutaneous papilloma which occurred in wild, North American cottontail rabbits *(Sylvalagus floridanus)* west of the Mississippi River. Papillomas could be induced in both wild cottontail rabbits and domestic rabbits *(Oryctolagus cunniculus)* using either clarified or filtered homogenates. The disease was transmissible in series among cottontail rabbits and could be transmitted from wild rabbits to domestic rabbits (Fig. 23) but not in series among domestic rabbits (Shope 1934). Domestic rabbits are rarely affected naturally (Hagen 1966; Sundberg, unpublished data). Rous and Kidd (1940) reported that malignant tumors developed in seven of ten domestic rabbits with papillomas persisting more than 200 days. The tumors metastasized, and transmission to other rabbits was successful. The naturally occurring papillomas occasionally became malignant in cottontail rabbits, but malignant transformation was more frequent in experimentally-induced papillomas in domestic rabbits, jackrabbits *(Lepus cali-*

Fig. 23. Cutaneous papilloma induced in a domestic rabbit by cottontail rabbit cutaneous papillomavirus (courtesy Dr. S. Maunoury and G. Orth)

fornicus), and snowshoe hares *(Lepus americanus)* Kidd and Rous 1940). In the natural host, papillomas pass through proliferative, stationary, and involutionary stages. Some rabbits kept over 6 months developed epidermoid carcinomas. Virus was recovered from 12 of 40 papillomas but not from carcinomas (Syverton 1952). Tumors were not induced by CRPV in adult rats, however, when fetal rat skin was infected with CRPV and grafted to syngeneic adult recipients, papillomas developed in 7%-10% of the grafts (Kreider et al. 1971).

The naturally occurring cutaneous papillomas in cottontail rabbits are often multiple, black or gray, firm masses with a dry, well-keratinized, fissured surface. They may be located on the shoulder, neck, back, abdomen, or inner thigh. The lesions range in size from 0.5 to 1 cm in diameter. The histological appearance consists of proliferation of cells in the basilar layer and stratum spinosum of the epidermis. The proliferative epidermis has numerous branching processes on a thin, connective tissue core (Fig. 24). Inclusions are not found, mitoses may be common, melanin is frequent, and a mild inflammatory response may be present

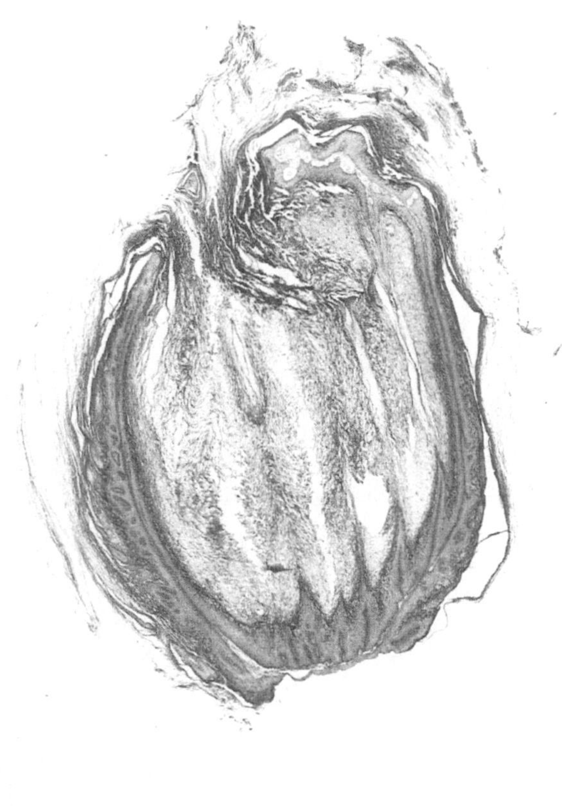

Fig. 24. Cutaneous papilloma in a cottontail rabbit (courtesy Dr. O. Croissant and G. Orth). H & E, ×15

in the dermis beneath the tumor (Moulton 1961). The papillomas in domestic rabbits have similar morphological features. The squamous cell carcinomas present as raised, fungoid, or eroding lesions. Microscopically, neoplastic squamous cells form invading islands surrounded by a marked desmoplastic response of the surrounding connective tissue. Adjacent papillomas may be undermined, infiltrated, and destroyed (Rous and Beard 1935).

Ultrastructurally, viruslike particles resembling papillomaviruses have been observed in cottontail rabbit papillomas (Stone et al. 1959; Noyes 1959, Moore 1959; Williams 1960). Immunofluorescence studies indicated that CRPV and HPV share two kinds of antigenic determinants (Orth et al. 1978). Attempts to propagate the virus in cell or organ cultures have been unsuccessful (Moulton and Lau 1964; Ishimoto et al. 1970; Ishimoto and Ito 1971). DNA extracted from papillomas or viral particles was shown to induce papillomas in several studies (Ito 1960, 1963; Kass and Knight 1965). The viral genome has been characterized (Murphy et al. 1981; Nasseri and Wettstein 1984a) and its DNA sequenced (Giri et al. 1985). The genomic organization was found to be similar to other papillomaviruses already sequenced (Danos et al. 1984). The papillomas induced in domestic rabbits do not contain infectious virions; however, extrachromasomal viral DNA in 10 to over 100 copies per cell can be found in both papillomas and carcinomas (Stevens and Wettstein 1979; Wettstein and Stevens 1980, 1982; Watts et al. 1983). In cell lines derived from transplantable carcinomas, viral DNA was integrated in head-to-tail tandem repeats (Sugawara et al. 1983; Nasseri and Wettstein 1984b; Georges et al. 1984). Morphological transformation in vitro of NIH 3T3 and C126 cells has been demonstrated using either CRPV or the viral DNA (Watts et al. 1983).

2.6 Swine Papillomas

Genital papillomas are rare, incidental necropsy findings in hogs (Parish 1961) or may be congenital in neonatal pigs (Rieke 1980). The disease has been reproduced using cell-free filtrates inoculated into genital tracts. Induced lesions underwent spontaneous regression and the animals developed immunity to reinfection (Parish 1961). No antibodies could be detected serologically in hogs resistant to reinfection until two additional injections of infectious material had been given. Antibodies were demonstrated by neutralization and conglutinating complement-absorption tests. Antigen could be detected in tissues by agar gel immunodiffusion and conglutinating complement-absorption tests (Parish 1962). No lesions resulted from inoculation of infectious material into human, pig, calf, rabbit, guinea pig, or mouse skin or into embryonated eggs (Parish 1961). The agent has yet to be visualized and characterized.

Papillomalike structures are common findings on the borders of the tongues of newborn pigs. These are normal anatomical features in pigs 15-18 days of age and are called marginal papillae (Fig. 25) (Sack 1982). Marginal papillae have been attributed to papillomavirus infection (Cheville 1983), but viral antigens or DNA could not be demonstrated in a limited study (Sundberg 1986).

Fig. 25. Papillomalike marginal papillae (normal structure) on the tongue of a young pig

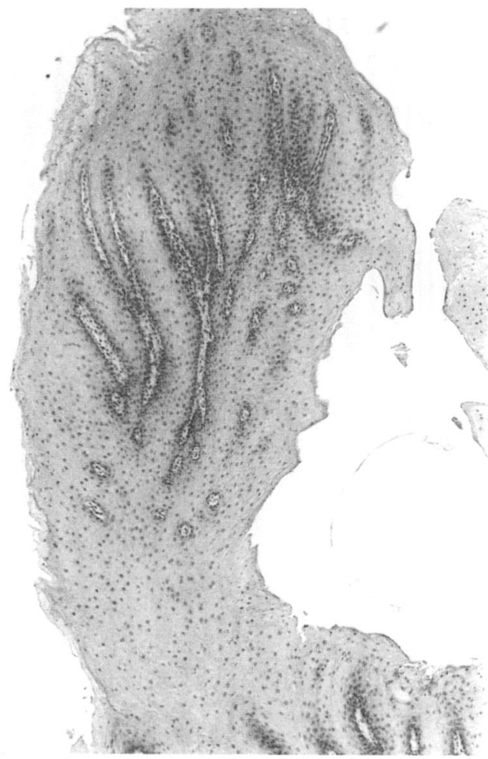

2.7 Goat Papillomas and Squamous Cell Carcinomas

Outbreaks of papillomatosis in herds of milking goats are reported rarely. Lesions may be found on the skin of the head, neck, shoulders, forelimbs, mammary glands, and the mucous membranes of the lower genital tract (Davis and Kemper 1936; Moulton 1961; Theilen et al. 1985). Some of the papillomas, particularly those on the skin of the mammary gland, may progress to squamous cell carcinoma (Moulton 1961; Ficken et al. 1983; Theilen et al. 1985). Mammary gland skin papillomas were often multiple, measured 0.25-2.0 cm in diameter, raised, hairless, and verrucated (Fig. 26). Microscopically there was exaggerated dermal papillae, hyperkeratosis, parakeratosis, acanthosis, focal erosions, and vesicle formation. The cutaneous papillomas were shallow, crusty, and white. Microscopically, these were areas of hyperkeratosis with parakeratosis and focal aggregates of inflammatory debris. Fibropapillomas of the prepuce presented as an ulcerated swelling on the ventral surface. An irregular acanthotic epidermis with hyperkeratosis and parakeratosis covered dense fibrous connective tissue with numerous fibroblasts and primitive vascular channels (Theilen et al. 1985).

Goats that lack a pigmented skin (Saanen and Saanen crosses) and are exposed to high degrees of actinic radiation are at high risk. Although mammary gland papillomatosis spreads in a herd 4-6 months after an affected goat is introduced, an infective agent has not been identified by electron microscopic or immunohis-

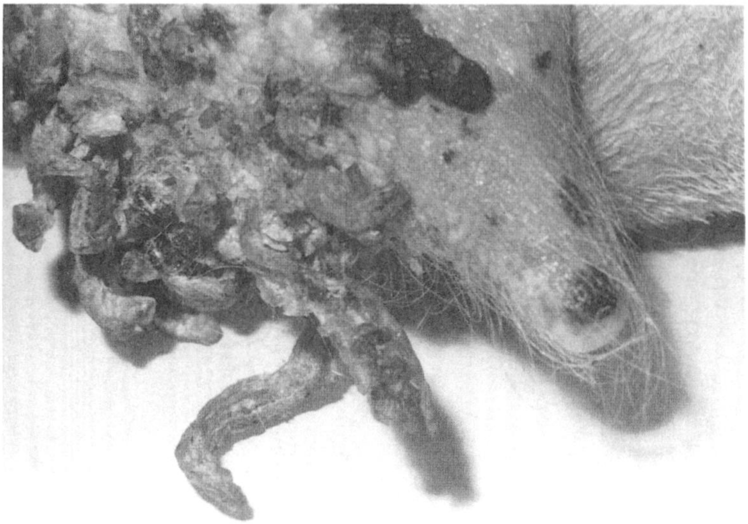

Fig. 26. Multiple papillomas on the skin over the mammary gland of a goat (courtesy Dr. J. Samuels and P. B. Spradbrow)

tochemical screening, nor has a papillomavirus genome been detected by low stringency hybridization in seven cases using a BPV1 genome as a probe (Theilen et al. 1985).

2.8 Hamster Papillomas, and Other Neoplasms

A polyomavirus was found in keratinized cells of hamster papillomas (Graffi et al. 1970). Transmission studies resulted in epidermal papillomas, lymphomas, leukemia, and subcutaneous sarcomas in hamsters; reticulum cell sarcomas in rats; and fibromas in newborn rabbits. A complication in these studies has been the isolation of C-type virus particles in the induced hamster lymphomas and sarcomas (Graffi et al. 1963, 1967, 1968). A papovavirus has also been isolated from these lesions, and the DNA showed a high degree of homology with polyoma virus (Scherneck et al. 1984). The hamster papovavirus has a length of 5.52 kb which is smaller than the papillomaviruses (Vogel et al. 1984).

A spontaneous papilloma was observed on the external nares of a Siberian hamster *(Phodopus sungorus)* (Fig. 27 and 28). Viruslike particles, 21 nm in diameter, were present in cells of the stratum corneum. Paraffin sections were negative for papillomavirus group-specific antigens using the peroxidase-antiperoxidase technique (Sundberg, unpublished data).

Papillomas have also been induced in hamsters on a pelleted diet which contained 2.0% butylated hydroxyanisole or a powdered diet containing 1.0% of the same compound for 24 weeks (Ito et al. 1983).

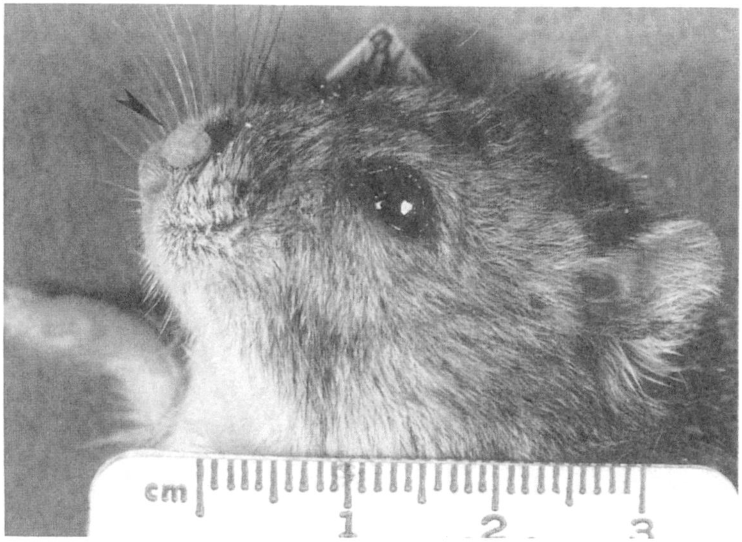

Fig. 27. Solitary papilloma adjacent to the external nares on a Siberian hamster *(arrow head)*

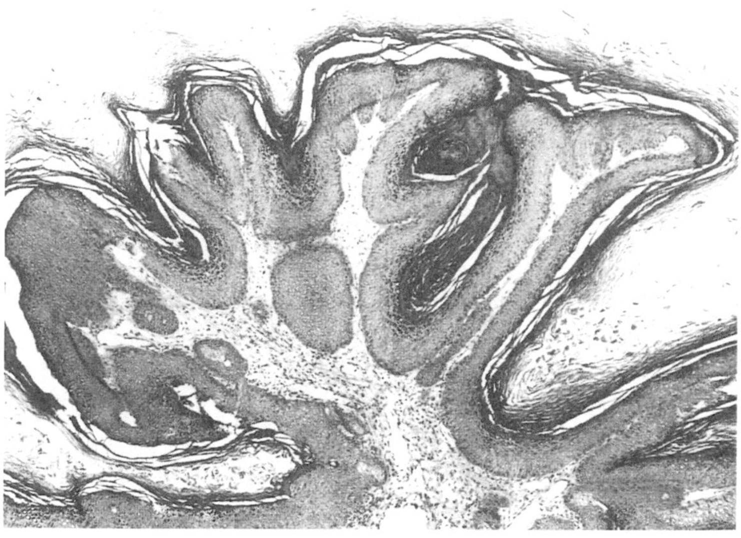

Fig. 28. Cutaneous papilloma on a Siberian hamster. H & E, ×65

2.9 Cat Papillomas, Fibropapillomas, Trichoepitheliomas, Pilomatrixomas, and Squamous Cell Carcinomas

Papillomas and cutaneous horns (hyperkeratotic papillomas) are rare in domestic cats (Scott 1980). Squamous cell carcinomas are not uncommon on the ears of old white cats (Legendre and Krahwinkle 1981). Papillomavirus group-specific antigens were not detected in a feline cutaneous or oral papilloma, one oral fibropapilloma, one trichoepithelioma, two pilomatrixomas, four cutaneous, one ocular, or one oral squamous cell carcinoma (Sundberg et al. 1984). Neither papillomaviruses nor papillomavirus DNA has been isolated from lesions in cats.

2.10 Laboratory Rodent Papillomas, Keratoacanthomas, and Squamous Cell Carcinomas

Cutaneous horns are occasionally observed on rodents used for long-term drug trials (Sundberg, unpublished data). These may be caused by the experimental compound; however, without careful workup, a viral etiology cannot be ruled out. Papillomas which progress to carcinomas have been induced in mice and rats with the use of topical carcinogens (Berenblum 1941; Iversen 1982; Reddy and Fialkow 1983). This has been the basis for the development of the initiator-promoter theory of chemical carcinogenesis. Administration of N-nitrosodimethylamine to zinc-deficient rats intragastrically resulted in the development of squamous papillomas at the junction of the fore and glandular stomach. Tumors did not develop in rats on zinc sufficient diets fed this compound (Ng et al. 1984).

Multiple cutaneous papillomas developed spontaneously in 20 of 30 mutant mice which were heterozygous for repeated epilation (Er/+) (Lutzner et al. 1985). Mice developed from one to five cutaneous papillomas and at least one cutaneous invasive squamous cell carcinoma by 2 years of age. There was no sex predilection, and animals under 6 months of age or wild type (+/+) were not affected. Attempts to identify virus either by negative stain electron microscopy, thin section electron microscopy, immunohistochemistry, or Southern blot hybridization of DNA extracts with a BPV1 probe were unsuccessful.

Nude mice have been used to test the potential oncogenicity of various papillomaviruses or the ability of transformed cell lines to be maintained in vivo (Watts et al. 1984), but productive infection has not yet been demonstrated.

Papillomaviruslike particles and papillomavirus group-specific antigens have been found in papillomas and trichoepitheliomas which occurred spontaneously in European harvest mice *(Micromys minutus)*. A supercoiled DNA molecule of about 7.6 kb was isolated, cloned, and a restriction map generated. The molecule was shown to be colinear with BPV1 DNA using well-defined subgenomic probes in Southern blot analysis (Sundberg et al. 1987).

Epithelial tumors, which morphologically resemble viral-induced papillomas of the skin as well as mucous membranes of the eye and lower genital tract, have been sporadically reported affecting gerbils and guinea pigs (Sundberg 1986). By immunohistochemical techniques papillomavirus group-specific antigens could not be detected in these lesions.

3 Papillomavirus Infection in Wild and Exotic Mammals

3.1 Deer

3.1.1 Deer Fibromas

Cutaneous fibromas are the most common neoplasms in captive and free-living deer. These tumors are most often reported in white-tailed deer *(Odocoileus virginianus)* in North America but have also been reported in mule and black-tailed deer *(Odocoileus hemionus)*, fallow deer *(Cervus dama)*, red deer *(Cervus elaphus)*, roe deer *(Capreolus capreolus)*, sika deer *(Cervus nippon)*, North American moose and European elk *(Alces alces)*, caribou *(Rangifer caribou)* (Sundberg and Nielsen 1981), American elk *(Cervus canadensis)* (Sundberg et al. 1984), and reindeer *(Rangifer tarandus)* (Sundberg 1986). The tumors described in these cervids have been given various names over the last 75 years, including fibroma, fibrosarcoma, neurofibroma, papilloma, and wart (Sundberg and Nielsen 1981). The tumor ranges in frequency in white-tailed deer from approximately 1% in surveys of hunter-killed deer at check stations (Severinghaus and Cheatum 1961; Friend 1967) to over 5% in animals submitted to diagnostic laboratories (Banasiak 1961; Reed et al. 1976). In the northeastern United States of America, 10.5% of hunter-killed, white-tailed deer were affected (Sundberg and Nielsen 1982). Fibromas occur most often in animals under 2 years of age in white-tailed deer (Leopold et al. 1951; Richards 1957; Severinghaus and Cheatum 1961; Brown 1961; Broughton et al. 1972; Sundberg and Nielsen 1982), but occasional cases are reported in aged animals (Shope et al. 1958; Severinghaus and Cheatum 1961; Roscoe et al. 1975). Males are affected more frequently than females, which may be due to the fact that more bucks are shot (either by law or hunter preference) or to increased exposure to a transmissible agent through abrasions resulting from males rubbing their antlers to shed velvet (Friend 1967; Sundberg and Nielsen 1982). In the European elk, 27 cases were found among 2200 (1.2%) animals examined (Borg 1978). The North American moose, which is the same species as the European elk, is commonly affected with cutaneous fibromas in the northeastern United States of America, and affected moose are usually over 2 years old (Sundberg et al. 1985d). Little information is available in the frequency of fibromas in other species of deer.

Deer fibromas present as firm, round, nodular skin tumors (Shope 1955; Richards 1957; Fay 1970). They range in number from 1 to 226 (Hoover 1937), and in size from 0.5 to 25 cm (Shope 1955; Roscoe et al. 1975; Sundberg and Nielsen 1981). They are often pigmented, dark brown to black, with a smooth or wrinkled surface (Fig. 29). Unpigmented, tan to white fibromas are occasionally observed (Fig. 30). The latter are usually found in areas of the body where the hair is white. Large fibromas have either a smooth or verrucated surface; they are often pedunculated and may be ulcerated. Fibromas are not locally invasive and can be easily removed surgically without recurrence (Sundberg, unpublished data). On the cut surface they are firm, white, and covered by a pigmented epidermis which is thickened to various degrees (Shope 1955; Honess and Winter 1956; Richards 1957; Fay 1970; Koller and Olson 1971). Ossifying fibromas are rare. One such

Fig. 29. Multiple pigmented fibromas on the skin of a white-tailed deer

Fig. 30. Solitary unpigmented fibroma in the white-haired area of the tail of a white-tailed deer

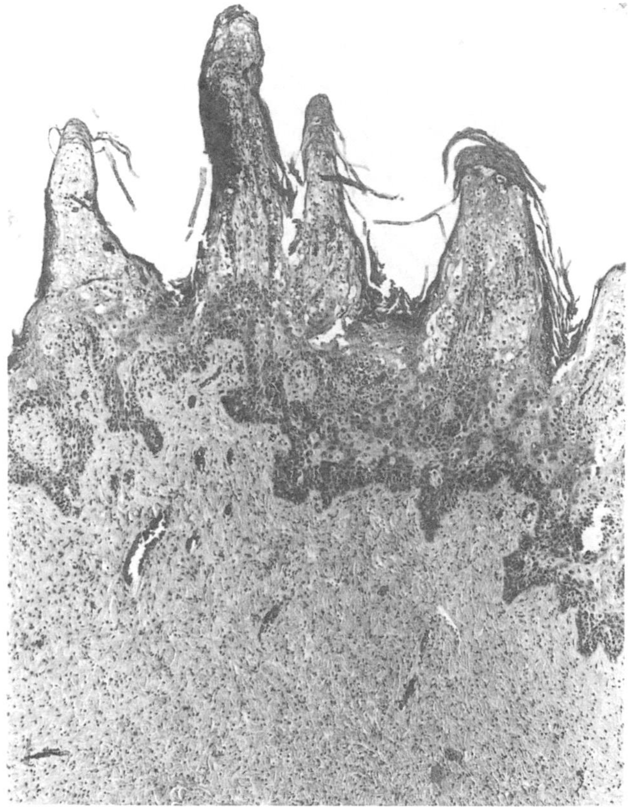

Fig. 31. Multiple short papillary projections of hyperplastic epithelium cover a large amount of fibrous connective tissue in a white-tailed deer fibroma. H & E, ×65

case had a 1-cm thick outer cortex, which resembled a typical fibroma, whereas the center was ossified, making sectioning with a knife impossible (Roscoe et al. 1975). Cutaneous fibromas may be found anywhere on the body but are most prevalent around the eyes, mouth, neck, and medial aspect of the forelimbs (Murie 1934; Honess 1939; Quortrup 1946; Wadsworth 1954; Swank 1958; Sundberg and Nielsen 1982). Pulmonary fibromatosis has been reported in white-tailed deer and European elk with cutaneous fibromatosis (Wadsworth 1954; Koller and Olson 1971; Borg 1975). The pulmonary lesions were considered to be metastases and are very rarely reported. Hunter check stations provide the opportunity to examine large numbers of free-living deer during short intervals of time. Deer are rarely presented in toto at the station; the heart and lungs are usually removed, which may explain why pulmonary lesions are considered rare.

The histological appearance of the cutaneous fibromas of deer is that of proliferating fibroblasts which produce an abundant matrix of swirling collagen bundles (Figs. 31-34). Collagen bundles tend to be arranged perpendicular to the epithelial layer, extend parallel to each other into the tumor mass, then become randomly

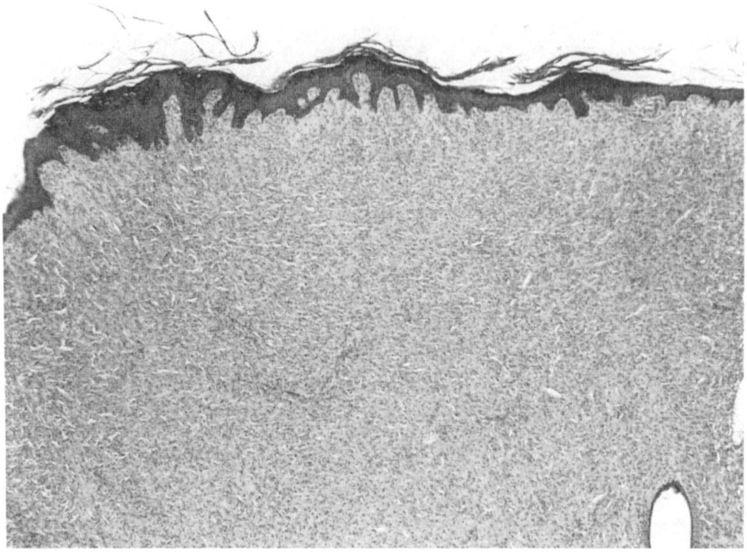

Fig. 32. Mild hyperplasia of the overlying epithelium of a white-tailed deer cutaneous fibroma. H & E, ×65

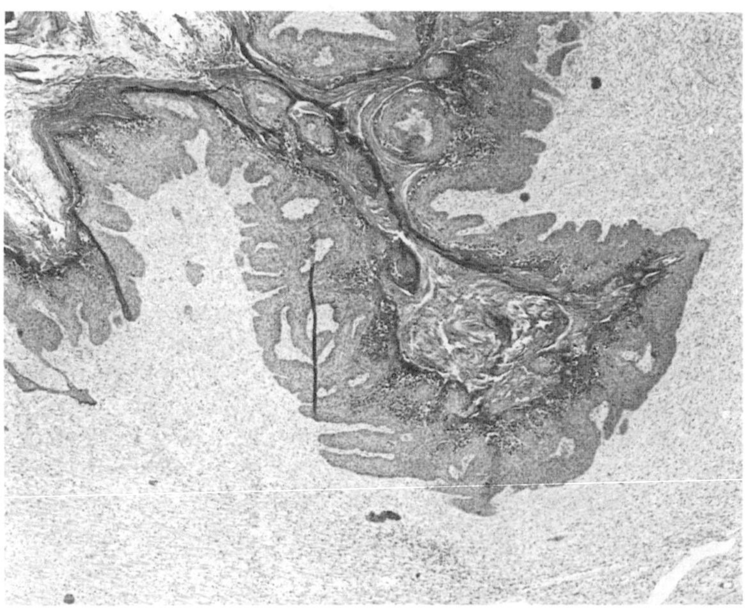

Fig. 33. Marked epithelial hyperplasia with prominent cytopathology in cells of the stratum granulosum in a mule deer fibroma. H & E, ×65

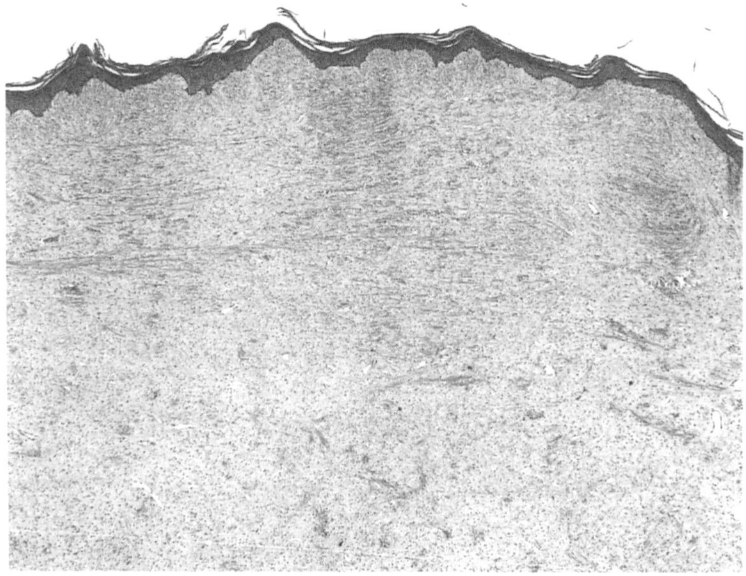

Fig. 34. Mildly hyperplastic epithelium covers abundant fibrous connective tissue in a North American moose fibroma. H & E, ×65

arranged deep in the fibroma. This forms the bulk of the pedunculated mass, and hence the morphological diagnosis of cutaneous fibroma. There is a marked difference between the densely packed connective tissue of the fibroma and the more loosely arranged connective tissue of the normal dermis. The epidermis ranges from slight acanthosis and hyperkeratosis to pseudoepitheliomatous hyperplasia, and it may form short papillary projections (fibropapilloma) (Shope et al. 1955; Fay 1970). In lesions containing papillomavirus or in which viral antigens have been confirmed, cells in the stratum granulosum exhibit specific cytopathology. The affected cells may be single or in small clusters (as in common in white-tailed deer fibromas) or may be uniformly involved (mule deer fibromas) (Sundberg et al. 1985f). The cells are swollen, have clear cytoplasm, and contain dark basophilic, cytoplasmic, keratohyalinlike granules of various sizes which are larger than normal keratohyalin granules. The morphological features of the granules and nuclei vary, depending on the host, and probably represent specific cytopathological variation induced by the different papillomaviruses. Parakeratosis is a variable feature which may be uniform, focal, or not present. This feature is often associated with the presence of large amounts of virions (Sundberg et al. 1985b).

Viral particles resembling papillomaviruses have been demonstrated in negatively stained homogenates of white-tailed deer and mule deer fibromas and within nuclei of cells in the stratum granulosum and stratum corneum in these tumors (Tajima et al. 1968; Sundberg and Nielsen 1981; Sundberg et al. 1985b). Papillomavirus group-specific antigens have also been detected in white-tailed deer and mule deer fibromas using either indirect immunofluorescence or the peroxidase-antiperoxidase technique (Sundberg et al. 1985g).

The etiology of the deer fibroma was thought to be of viral nature long before any reported transmission studies, based on morphological and epidemiological similarities to tumors in other species which had been transmitted by filterable agents (Honess 1939; Herman and Bischoff 1950). Shope (1955) successfully transmitted deer fibromas by inoculating a preparation filtered through Berkefeld N filters onto scarified skin of captive white-tailed deer. Tumor development was first observed 7 weeks following inoculation. Most of the tumors regressed 2 months after development. In white-tailed deer inoculated by rubbing a tumor homogenate containing white-tailed deer papillomavirus into puncture wounds caused by a tattoo apparatus, the initial wound healed in 2 weeks. Proliferation of fibroblasts subjacent to the epidermis occurred during the next 2–5 weeks with minimal epidermal hyperplasia. Perivascular lymphocytic infiltrates preceded regression (Sundberg et al. 1985a). Although viral antigens could not be detected in serial weekly biopsies by indirect immunofluorescence using rabbit and white-tailed deer papillomavirus serum (Sundberg et al. 1985a), episomal viral DNA could be detected in DNA extracted from the lesions by Southern blot hybridization under conditions of high stringency with a white-tailed deer papillomavirus DNA probe (O'Banion and Sundberg 1986).

Papillomaviruses which affect European elk (Moreno-Lopez et al. 1981), white-tailed deer, and mule deer (Lancaster and Sundberg 1982) have been isolated and characterized. Minor differences were noted in the restriction endonuclease patterns between white-tailed and mule deer papillomavirus genomes; however, the genomes were indistinguishable by liquid phase hybridization (Lancaster and Sundberg 1982). Both white-tailed and mule deer are members of the genus *Odocoileus*, which may explain the high degree of similarity of the viruses. Lesions collected for this study were obtained from white-tailed deer in the northeastern United States of America and from mule deer in the west, 3000 km away. The European elk papillomavirus is currently being compared to the North American deer papillomaviruses. All three viruses have little homology with BPV1 or BPV2. These viruses will transform NIH/3T3 or C127 cells (Groff et al. 1983; Stenlund et al. 1983). The European elk papillomavirus has been partially sequenced (Stenlund et al. 1983) and the white-tailed deer papillomavirus has been completely sequenced (Groff and Lancaster 1985). Preliminary characterizations of papillomaviruses isolated from red deer (Moat 1983; Moar and Jarrett 1985) and reindeer (Moreno-Lopez et al. 1983) cutaneous fibromas have been reported.

Attempts to transmit white-tailed deer fibromas to rabbits, guinea pigs, sheep, calves, monkeys, and horses have been unsuccessful (Shope et al. 1958; Koller and Olson 1972). Neoplasms developed in hamsters after 7 months when white-tailed deer fibroma suspensions were inoculated subcutaneously (Koller and Olson 1972). Lung metastases occurred in 10% of hamsters with induced subcutaneous fibromas and fibrosarcomas (Koller and Olson 1971, 1972). Similarities between bovine papillomatosis and deer fibromas have suggested a common or similar etiology. Attempts to induce fibromas by application of bovine papillomavirus to scarified skin or by intradermal injection in deer have been unsuccessful (Shope et al. 1958). The European elk papillomavirus has also been used to induce fibrosarcomas in hamsters by subcutaneous injection (Stenlund et al. 1983).

Using a suspension of bovine papillomavirus, precipitin antibodies have been demonstrated in sera from calves with either experimentally induced or natural cases of papillomatosis. Precipitin lines did not develop when these sera were tested against antigen prepared from a white-tailed deer fibroma (Lee and Olson 1969a). White-tailed deer developed hemagglutination inhibition titers to the white-tailed deer fibroma virus following experimental induction of tumors (Sundberg et al. 1985a). Hyperimmune sera prepared against BPV1, BPV2, and European elk yielded two lines by immunodiffusion against homologous antigen, but no cross-reactivity could be demonstrated between these viruses (Moreno-Lopez et al. 1981).

3.1.2 Deer Papillomas

Although many species of deer (Cervidae) have been reported to have papillomas, fibropapillomas, or fibromas, the morphological descriptions are almost uniformly of fibromas or fibropapillomas. The fibroma and fibropapilloma are probably different stages of development of the same lesion since both morphological types can be found on the same individual and the same virus was isolated from each (Sundberg, unpublished data; Moreno-Lopez, personal communication). Recently, a true papilloma was observed on the lower legs of a red deer *(Cervus elaphus)* (Favre, personal communication). The lesions consisted of epithelial proliferation on very thin fibrovascular stalks. Papillomavirus group-specific antigens were detected by the peroxidase-antiperoxidase technique, and the virus was isolated, cloned, and characterized. Comparative studies between this virus and other deer papillomaviruses are in progress.

3.2 Other Hooved Stock: Papillomas, Fibropapillomas, and Squamous Cell Carcinomas

Small papillomas were found on the feet of an impala *(Aepyceros melampus)* and on the face of a giraffe *(Giraffa camelopardalis)*. Histologically, the tumors were typical papillomas. Viral particles, characteristic of the papovavirus family, measured 38 nm and 40 nm in thin sections for the impala and giraffe tumors, respectively (Karstad and Kaminjolo 1978).

A cutaneous fibropapilloma, which contained papillomavirus group-specific antigens, was found on a pronghorn antelope *(Antelocapra americana)* (Fig. 35) (Sundberg et al. 1983). It was a solitary, black, verrucous mass. Microscopically, the fibropapilloma consisted of a large mass of proliferating fibroblasts in a dense matrix of collagen which was covered by acanthotic and hyperkeratotic epithelium thrown into papillary folds.

A yak *(Bos mutus)* had a cutaneous fibropapilloma which was morphologically similar to the cutaneous fibropapilloma of domestic cattle. Papillomavirus antigens were not detected in the lesion by the peroxidase-antiperoxidase technique (Sundberg et al. 1984). An outbreak of fibropapillomas occurred in yaks in the Toronto Zoo (Barker, personal communication).

A well-differentiated squamous cell carcinoma with a papillary pattern was

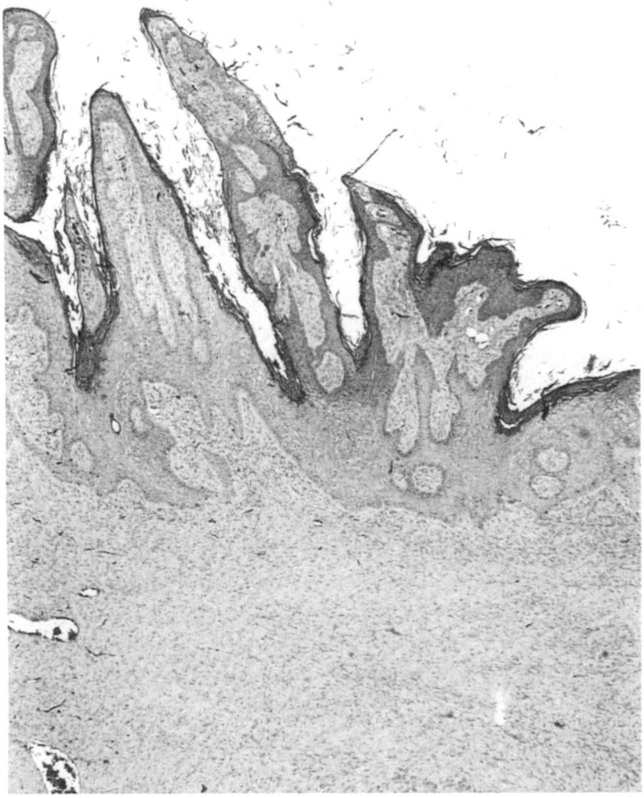

Fig. 35. Numerous papillary projections of hyperplastic epithelium on fibrous stalks cover a large mass of fibrous connective tissue in a pronghorn antelope fibropapilloma. H & E, ×65

reported on the cervix of a blackbuck *(Antelope cervicapra)* (Sundberg and McDonald 1984). Papillomavirus group-specific antigens could not be detected in the biopsy.

3.3 Beaver Papillomas

Cutaneous papillomas have been observed on the feet of beavers *(Castor canadensis)* (Fig. 36) in the eastern United States of America and Ontario, Canada (Carlson et al. 1983; Barker and Wojcinski, personal communication). The lesions were unpigmented, gray to tan in color, slightly elevated, and oval. The circumscribed masses were 2–25 mm in diameter and cornified. Microscopically, they consisted of folded, hyperplastic epidermis (Fig. 37). Viruslike particles were present in the nuclei of cells in the upper stratum granulosum and throughout the stratum corneum. Papillomavirus group-specific antigens were detected in both cases by the peroxidase-antiperoxidase technique (Sundberg 1986).

Fig. 36. Multiple sessile papillomas on the skin of the foot of a beaver (courtesy Dr. Wojcinski)

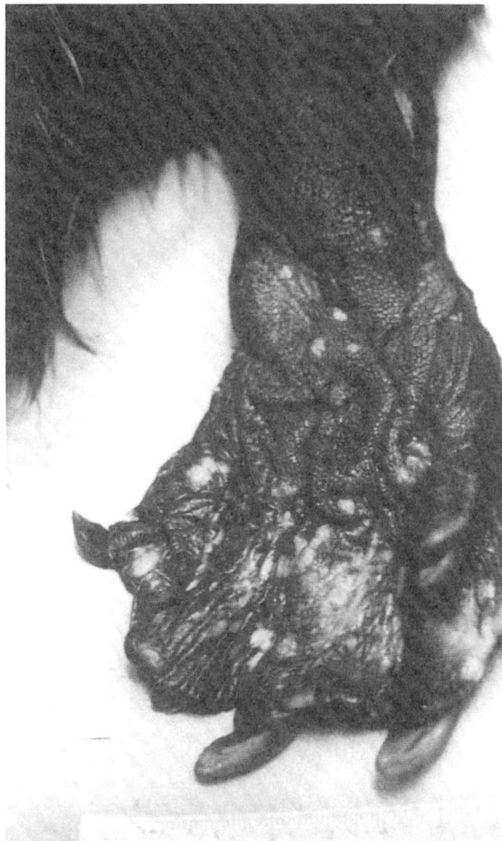

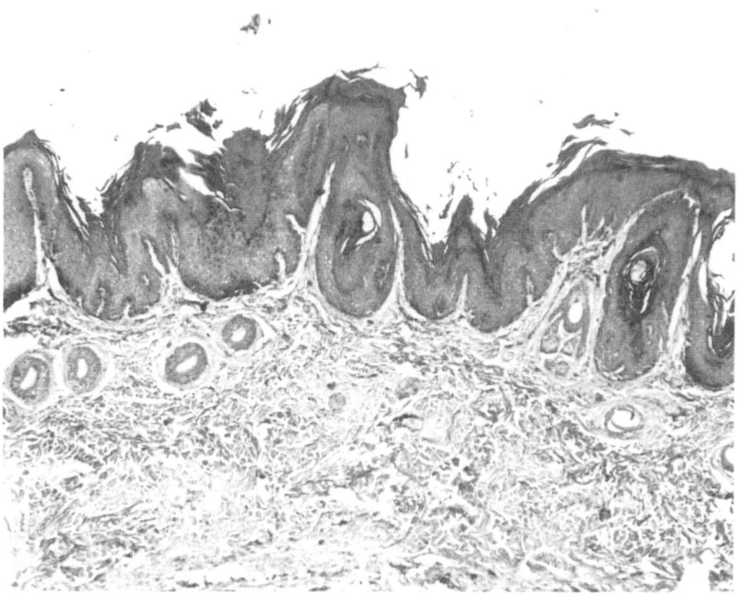

Fig. 37. Beaver cutaneous papilloma with numerous short papillary projections of hyperplastic epithelium on thin fibrovascular stalks. H & E, ×65

3.4 *Mastomys natalensis* Keratoacanthomas

A papillomavirus has been shown to be the causative agent of keratoacanthotic epithelial proliferations in *Mastomys natalensis* (Fig. 38 and 39; Müller and Gissmann 1978). Lesions resemble the keratoacanthoma of humans and dogs, but have also been classified as focal epithelial proliferations, papillomas, and squamous cell carcinomas (Burtscher et al. 1973; Rudolph and Müller 1976; Rudolph and

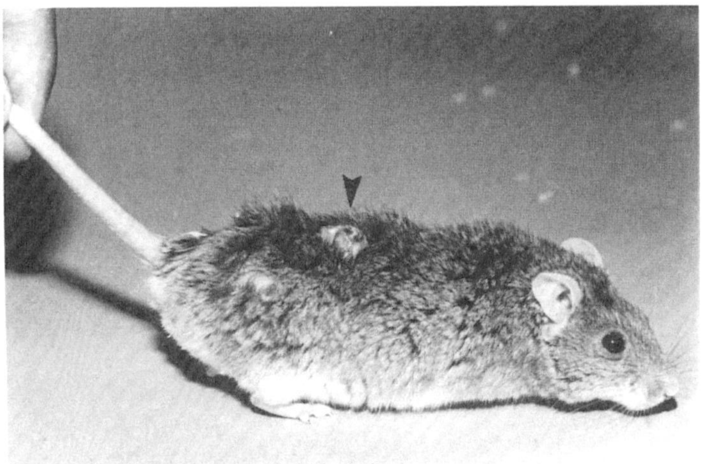

Fig. 38. *Mastomys natalensis* with a keratoacanthoma of the skin over the lumbar spine *(arrow head)*

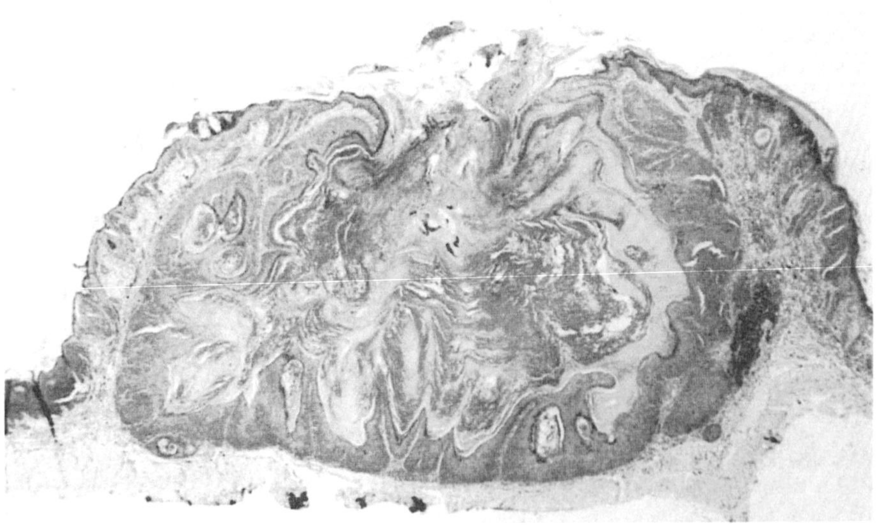

Fig. 39. *Mastomys natalensis* keratoacanthoma of the skin (courtesy Dr. E. Amtmann). H & E

Thiel 1976; Rudolph 1980). Application of tumor suspensions on depilated and scarified skin results in proliferation of the epidermis which later develops into a papilloma and keratoacanthoma. A squamous cell carcinoma may develop from the latter (Rudolph and Thiel 1976). Similar neoplasms developed when cells and virus from squamous cell carcinomas were used (Rudolph and Müller 1976).

Painting benign lesions with compounds such as dimethylbenzanthracine will induce transformation (Amtmann, personal communication). A virus with many of the characteristics of a papillomavirus has been isolated and characterized from these lesions (Müller and Gissmann 1978). The virus appears to be endogenous in many tissues of the rodent, and as the viral genome copy number increases with the age of the animal within cells of the epidermis, papillomas and other neoplasms begin to develop spontaneously (Amtmann et al. 1984). This appears to be a different phenomenon than that observed for papillomavirus infection in other mammalian species.

3.5 Wild Canid (Coyotes and Wolves) Oral Papillomas

Numerous cases of oral papillomatosis in coyotes *(Canis latrans)* and wolves *(Canis lupus)* have been reported. Nellis (1973) reported oral papillomas in 5 of 279 trapped coyotes, and 5% of nearly 500 trappers indicated they had encountered cases (Samuel et al. 1978).

Gross lesions in the two species are similar. They occur singly or in clusters ranging in diameter from 1 to 10 mm up to 300 mm. The tumors are unpigmented, have a broad base, and range in height from 1 to 8 mm. Small tumors are smooth or finely nodular, and those over 3 mm have numerous fine, pointed, papillary projections. These are present on the lips, tongue, and adjacent tissues. An acanthotic and hyperkeratotic epidermis covers thin, sometimes arborizing, connective tissue cores. The superficial layers of the epidermis sometimes contain large, amorphous, basophilic, intranuclear inclusions (Trainer et al. 1968; Broughton et al. 1970; Samuel et al. 1978). Electron microscopic examination of negatively stained extracts of coyote oral papillomas revealed a virus morphologically similar to the papovaviruses, consisting of unenveloped, round capsids of cubic symmetry, 47-57 nm in diameter (Greig and Charlton 1973). Viral particles from wolf oral papillomas were indistinguishable from those of coyotes (Samuel et al. 1978). Oral papillomatosis was produced in two beagle pups following scarification and application of a 10% suspension of oral papillomas of coyote origin (Samuel et al. 1978).

A papillomavirus has been isolated and its DNA cloned from coyote oral papillomas. By extensive restriction endonuclease analysis and high stringency Southern blot studies, this viral genome was found to be similar, if not identical to, the canine oral papillomavirus. This isolate did induce lesions in domestic dogs (Sundberg, unpublished data).

3.6 Nonhuman Primate Papillomas

Reports of papillomas in nonhuman primates are infrequent. Cutaneous papillomas were first reported in a brown cebus monkey *(Cebus apella)* (Lucke et al. 1950). Homogenates of this tumor were transmissible by scarification of the skin from New World monkeys to Old World monkeys (Lucke et al. 1950). A cutaneous horn, which microscopically was diagnosed as a keratinizing papilloma, was reported as an incidental finding in a rhesus monkey *(Macaca mulatta)* (Brown et al. 1972). Cutaneous papillomas, containing viruslike particles which were morphologically consistent with papillomaviruses, have been reported in Abyssinian colobus monkeys *(Colobus guereza)* (Rangan et al. 1980). Cutaneous papillomas have also been observed in black and white colobus monkeys *(Colobus polykomus)* (Boever and Kern 1976). A case of focal epithelial hyperplasia of the oral mucosa in chimpanzees contained intranuclear, viruslike particles which were classified as belonging to the papova group (Hollander and van Noord 1972). Oral papillomas have also been observed in baboons *(Papio papio)* (Sundberg et al. 1986c).

Papillomaviruses have been associated with benign papillomas (condyloma acuminata) as well as dysplasias, carcinoma-in-situ, and squamous cell carcinoma of the lower genital tract of human beings. Therefore, it is of interest to compare similar lesions in nonhuman primates. A keratinized papilloma was reported on the shaft of the penis of a colobus monkey *(Colobus guereza)* (Fig. 40) (Sundberg 1986). Papillomavirus structural antigens and viruslike particles were detected in the lesion. In DNA extracts of the tumor, papillomavirus sequences could be detected by Southern blot hybridization under low stringency conditions with either an HPV11 or BPV1 probe (Sundberg 1986). Dysplasia of the lower genital tract in the female crab-eating macaque *(Macaca fascicularis)* was observed in 5 of 39 experimental and 6 of 39 control monkeys (Hertig et al. 1983). Squamous cell carcinoma of the prepuce and penis with metastases has been reported in a rhesus

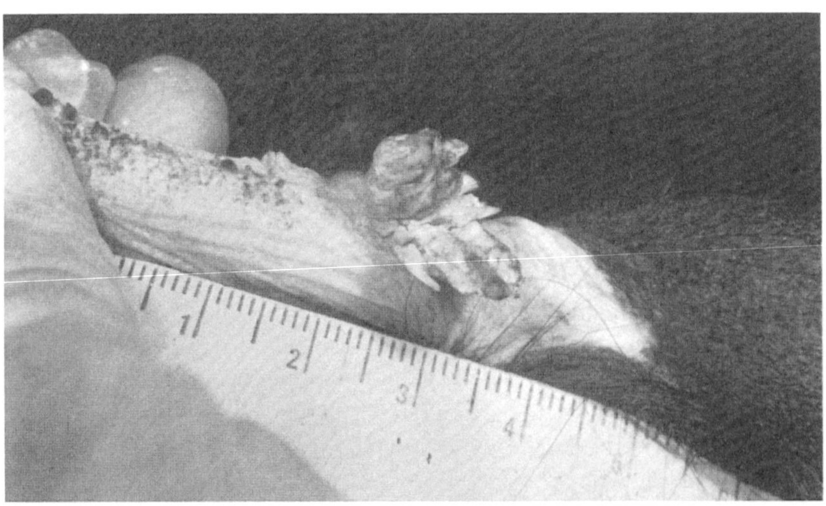

Fig. 40. Keratinized papilloma on the penis of a colobus monkey (courtesy Dr. Shima)

monkey (Hubbard et al. 1983). Immunoperoxidase screening for papillomavirus group-specific antigens was negative in a case of cervical squamous cell carcinoma and its metastases in a rhesus monkey from another colony (Sundberg 1986). The role of papillomaviruses in benign lesions of nonhuman primates appears evident in several species, but the viruses have yet to be cloned and more completely characterized.

3.7 Bear Papillomas and Fibromas

A captive zodiac bear with a history of prolonged reproductive failure had a fungating vulvar mass. Histologically this papilloma resembled human condyloma acuminatum. Intranuclear papillomavirus common antigens, similar to those seen in human condyloma acuminatum, could be demonstrated using the peroxidase-antiperoxidase technique (Craft et al. 1980).

Cutaneous fibroma was diagnosed in an American black bear *(Ursus americanus)*. The fibroma did not contain papillomavirus antigens (Sundberg et al. 1984).

3.8 Elephant Papillomas

An outbreak of papillomatosis in Indian elephants *(Elephas maximus)* occurred in the Tulsa, Oklahoma, Zoo when a 6-year-old male was placed with two older females for breeding. A rough, pale, well-circumscribed mass with a broad base developed at the mucocutaneous border of the left upper lip on the male. During the following 18 months, numerous papillomas developed on the oral (Fig. 41) and

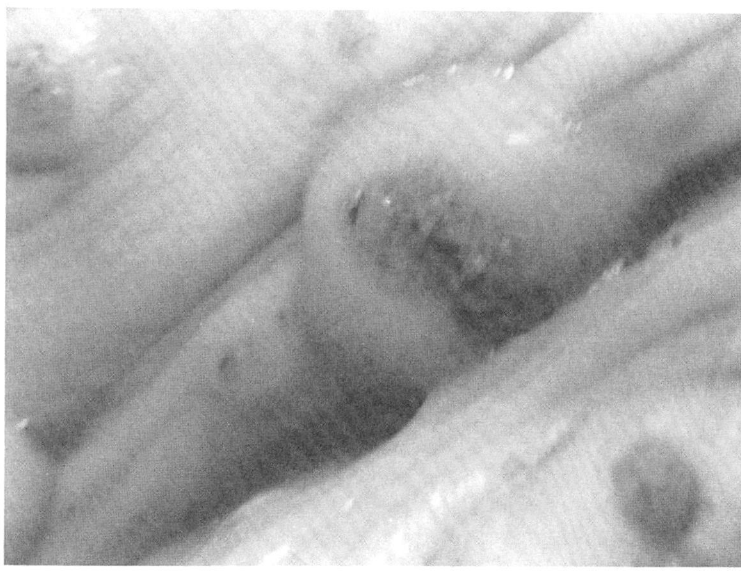

Fig. 41. Multiple oral papillomas in an elephant (courtesy Dr. W. Russell)

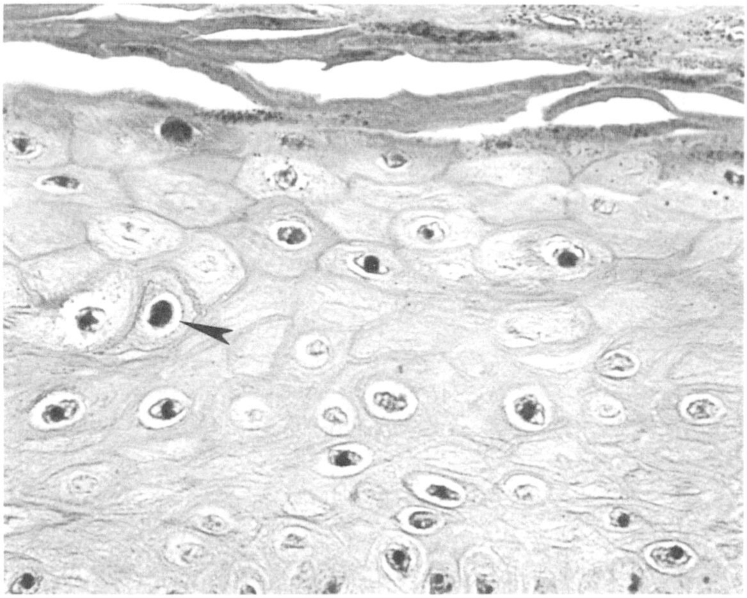

Fig. 42. Intranuclear inclusions *(arrow head)* within cells of the upper stratum spinosum and stratum granulosum in an elephant oral papilloma. H & E, ×400

nasal mucosa and dermis of the shoulder, hip, and base of the trunk. One female also developed multiple lesions of the oral labia and hard and soft palate. The histological appearance was of a proliferative epidermis thrown into numerous long, thin folds, on a thin connective tissue core. Large, basophilic, intranuclear inclusions were frequent in the stratum spinosum (Fig. 42). There was positive nuclear staining by the peroxidase-antiperoxidase technique using papillomavirus group-specific antiserum (Sundberg et al. 1981).

3.9 Rhinoceros Papillomas

A hyperkeratotic squamous papilloma was reported in the interdigital space between the second and third digits of both front feet of an adult male black rhinoceros *(Diceros bicornis)* (Boever 1976). The papillomas caused lameness. Surgical excision was curative.

3.10 Opossum Papillomas

Cutaneous papillomas were present on the foot, leg, and ear lap of a captive 1-year-old opossum. The tumors regressed and disappeared over the period of 1 year. They consisted of hyperplastic epidermis with a connective tissue base and core. Virus particles resembling papillomaviruses, 50 nm in diameter, were observed in negative-stained suspensions. Virus-containing suspensions did not produce lesions in other opossums when inoculated intradermally or when rubbed into scarified skin (Koller 1972).

3.11 Armadillo Fibromas

A cutaneous fibroma was reported between the first and second phalanges of the right forefoot of an adult male nine-banded armadillo *(Dasypus nonemcinctus)*. The lesion was morphologically similar to the papillomavirus-induced fibroma in deer, but no virions could be detected in sections examined by electron microscopy (Pence et al. 1983).

3.12 Lemming Gastric Papillomas and Squamous Cell Carcinomas

Papillomas and squamous cell carcinomas were reported in the squamous mucosa of the fermentative esophageal chamber or forestomach and the pyloric chamber of the stomach in 45 of 63 Greenland collared lemmings (Barker et al. 1982). The papillomas were 1-2 in number and 1-2 mm in size and consisted of hyperkeratotic squamous epithelium supported by a fine papillary connective tissue stroma. The carcinomas were in various stages of progression. No cause was determined for the high incidence in these animals. Two cases of gastric papillomas from lemmings in other colonies were negative for papillomavirus group-specific antigens (Sundberg 1986). Gastric hyperkeratosis and papillomatosis have been associated with parasites in the stomachs of *Microtus ochrogaster* and the muskrat *(Ondatra zibethicus)* (Cosgrove et al. 1968; Dunway et al. 1968). Candidiasis has also been associated with these lesions in lemmings (Leininger et al. 1979).

3.13 Marine Mammal Papillomas and Fibromas

Cutaneous papillomas have been reported in dolphins *(Delphinus delphis)*; however, papillomavirus antigens were not detected in lesions screened by the peroxidase-antiperoxidase technique (Sundberg et al. 1984). Penile papillomas in sperm whales *(Physeter catodon)* had intranuclear inclusions of papillomaviruslike virions detectable by electron microscopy (Lambertsen, personal communication). Two vaginal tumors adjacent to the os cervix in a beaked whale *(Mesoplodon densirostris)* were firm, gray-white, stalked, and pedunculated. They consisted of widely separated fibroblasts in an abundant collagen matrix and were not covered by an epithelium. These lesions were classified as fibromas, but no virological workup was conducted (Flom et al. 1980).

4 Papillomavirus Infections in Lower Vertebrates and Invertebrates

4.1 Bird Papillomas

The incidence of cutaneous neoplasms in wild and domestic birds is low, with the exception of squamous papillomas in the chaffinch *(Fringilla coelebs)* and brambling *(Fringilla montifringilla)* (Jennings 1968; Lina et al. 1973). In the Netherlands, 330 of 25000 chaffinches had papillomas (Lina et al. 1973). The tumors usually affect the claws, and less often, the skin at the junction of the beak and face (Jennings 1968). These tumors have a histological appearance typical of papillomas. Crystalline arrays of virus particles, with an average diameter of 37.5 nm, were found within nuclei of degenerated cells (Lina et al. 1973). In negatively stained preparations, virions measured 52 nm in diameter and had physical characteristics of papillomaviruses (Osterhaus et al. 1977). A papillomavirus has been isolated from these lesions, characterized, and partially sequenced (Moreno-Lopez et al. 1984).

Papillomas have also been reported on the skin of Amazon parrots *(Amazona ochracephala)*, African gray parrots *(Psittacus erithacus)*, budgerigars *(Melopsittacus undulatus)*, and cockatiels *(Nymphicus hollandicus)* (Petrak 1982). Jacobson et al. (1983) demonstrated papillomaviruslike particles in a cutaneous papilloma of an African gray parrot by electron microscopy and detected papillomavirus common antigens by the peroxidase-antiperoxidase technique. The papillomavirus affecting this bird has been isolated and its genome cloned and characterized (O'Banion and Sundberg, unpublished data).

Cloacal papillomas are common tumors in psittacine birds. These resemble rectal prolapses or granulation tissue. Papillomavirus antigens could not be detected in 40 cloacal papillomas or 1 carcinoma-in-situ of the cloaca (Sundberg et al. 1986a). Papillomavirus DNA could not be detected in 6 cases by Southern blot hybridization using the African gray parrot papillomavirus as a probe. Other pathogens were not detected by egg inoculation or electron microscopic studies. Lesions were not transmissible to other parrots (Sundberg et al. 1986a).

4.2 Reptile Papillomas

Typical papillomas have been reported in many species of snakes, turtles, lizards, and crocodiles (Smith et al. 1941; Schlumberger and Lucke 1948; Lunger and Clark 1978; Jacobson 1979). Electron microscopic examination of selected papillomas from a wall lizard *(Lacerta muralis)* revealed three morphologically distinct particles resembling herpesvirus, reovirus, and papovavirus (Raynaud and Adrian 1976). Virus particles were also observed in a papilloma from a green lizard *(Lacerta viridis)* (Cooper et al. 1982). The virions morphologically resembled papovaviruses but were small (25 nm). Cutaneous squamous cell carcinoma from another green lizard was negative for papillomavirus antigens (Sundberg 1986). Papillomaviruslike virus was observed by electron microscopy in hyperplastic skin lesions on the side of the head of a Bolivian side-neck turtle (Jacobson et al. 1982). This lesion did not stain with the antisera directed against papillomavirus group-specific antigens (Sundberg et al. 1984).

4.3 Amphibian Papillomas

Papillomas are among the most common tumors of amphibians. Lesions are characterized by various degrees of epidermal hyperplasia and keratinizaton. The epidermis is thrown into folds that cover the main polypoid mass which consists of fibroblasts (Reichenbach-Klinke and Eklan 1965). Numerous cases from tiger salamanders *(Ambystoma tigrinum)* have been submitted to the Registry of Tumors in Lower Animals, but reports are limited (Lucke and Schlumberger 1949). Spontaneous progressive skin papillomas have been described in newts *(Cynops pyrrhogaster)* but no etiological agents were observed (Asashima and Komazaki 1980). The frequency of these tumors in newts collected in the field is dependent upon the season (Asashima et al. 1982). Experimentally, the diameters of the papillomas in newts decreased at 4°, 25°, and 30 °C, and increased at 10° and 13 °C (Asashima et al. 1985). Amphibians are poikilotherms. Variation in body temperature changes the function of the immune system. Either immune status or tissue temperature may be important in tumor formation in these animals.

4.4 Fish Papillomas

Epidermal papillomas occur on many species of fish from both freshwater and marine habitats in widely scattered geographic areas. Epizootic outbreaks occur in Atlantic salmon *(Salmo salar)* with high mortality in hatcheries (Bylund et al. 1980), and there is a high incidence in eels *(Anguilla anguillia)* with significant economic losses (Wellings 1970). Papillomas have also been reported in flatfishes (Wellings et al. 1964; Peters and Watermann 1979) and a smooth dogfish shark *(Mustelus canis)* (Wolke and Murchelano 1976).

Papillomas vary from pink to red, depending on vascularity, to gray, brown, or black if melanocytes are present. Tumors may be flat or nodular, varied in size, single or multiple, soft to firm, and do not metastasize. The histological appearance consists of an abrupt change in epidermal thickness from normal to marked hyperplasia. Differentiation of epithelial cells in tumors varies from relatively normal to greatly enlarged cells with poorly stained or vacuolated cytoplasm to large multinucleated cells. There may be slight epidermal hyperplasia and a broad-based thickened dermis in lesions which are truly papillary with a verrucous surface consisting of multiple layers of epithelial cells supported by connective tissue (Roberts 1978).

Viruslike particles, visualized in papillomas in flathead sole *(Hippoglossoides elassodon),* measuring 44 nm in diameter, were enclosed in a single membrane 60°-70° thick, and were observed in the cytoplasm of most neoplastic cells. A granular body was also observed in these cells which was membrane bound, 160-200 nm in diameter, and composed of what appeared to be radially oriented capsomeres (Wellings and Chuinard 1964). Polyhedral viruslike particles,

52-56 nm in diameter, have been visualized in tissue cultures infected with blood from eels with epidermal papillomas. Inclusions were membrane bound. Tumor extracts and blood caused cytopathological effects on primary cultures of carp and trench gonads and permanent fish cell lines RTG-2 (rainbow trout gonads) and FHM (flathead minnow, *Pimephales promelas,* tissue posterior to anus) (Koops et al. 1970). Papillomaviruslike particles, 48-55 nm in diameter, were observed in a papilloma found on the lower lip of a brown bullhead (*Ictalurus nebulosus;* Edwards and Samsonoff 1977). Papillomas in some fish are thought to be the result of mechanical stimulation of the tissues as the fish rub against aquarium walls or may be related to the dumping of wastes containing sulfuric acid, titanium compounds, or chlorinated waste water (Peters and Watermann 1979; Grizzle et al. 1984).

4.5 Mollusc Papillomas

A conical papillomalike tumor was reported on the siphon of a horse clam (DesVoigne et al. 1970). The papillomalike mass was covered by tall columnar epithelium over smooth muscle.

References

Amstutz HE (1978) Treatment of warts in cattle. Mod Vet Pract 59: 650

Amtmann E, Muller H, Sauer G (1980) Equine connective tissue tumors contain unintegrated bovine papillomavirus DNA. J Virol 35: 962-964

Amtmann E, Volm M, Wayss K (1984) Tumour induction in the rodent *Mastomys natalensis* by activation of endogenous papilloma virus genomes. Nature 308: 291-292

Anderson DE, Lush JL, Chambers D (1957) Studies on bovine ocular squamous carcinoma ("cancer eye"). II. Relationship between eyelid pigmentation and occurrence of cancer eye lesions. J Anim Sci 16: 739-746

Appleby EC (1969) Tumours in captive wild animals: some observations and comparisons. Acta Zool Pathol Antverpiensia 48: 77-92

Asashima M, Komazaki S (1980) Spontaneous progressive skin papilloma in newts *(Cynops pyrrhogaster).* Proceedings of Japan Academy 56: 638-642

Asashima M, Komazaki S, Satou C, Oinuma T (1982) Seasonal and geographical changes of spontaneous skin papilloma in the Japanese newt *Cynops pyrrhogaster.* Cancer Res 42: 3741-3746

Asashima M, Oinuma T, Matsuyama H, Nagano M (1985) Effects of temperature on papilloma growth in the newt, *Cynops pyrrhogaster.* Cancer Res 45: 1198-1205

Bagdonas V, Olson C (1953) Observations on the epizootiology of cutaneous papillomatosis (warts) of cattle. J Am Vet Med Assoc 122: 393-397

Bagdonas V, Olson C (1954) Observations on immunity in cutaneous bovine papillomatosis. Am J Vet Res 15: 240-245

Baker JR, Leyland A (1975) Histological survey of tumours of the horse, with particular reference to those of the skin. Vet Rec 96: 419-422

Banasiak CF (1961) Deer in Maine. Maine Dept Inland Fisheries and Game Bull 6: 1-159

Barker IK, Mallory FF, Brooks RJ (1982) Spontaneous gastric squamous cell carcinomas and other neoplasms in greenland collared lemmings *(Dicrostonty groenlandicus).* Can J Comp Med 46: 307-313

Barthold SW, Olson C (1974a) Membrane antigen of bovine papillomavirus induced fibroma cells. J Natl Cancer Inst 52: 737-742

Barthold SW, Olson C (1974b) Fibroma regression in relation to antibody and challenge immunity to bovine papillomavirus. Cancer Res 34: 2436-2439

Barthold SW, Koller LD, Olson C, Studer E, Holton A (1974) Atypical warts in cattle. J Am Vet Med Assoc 165: 276-280
Barthold SW, Olson C, Larson LL (1976) Precipitin response of cattle to commercial wart vaccine. Am J Vet Res 37: 449-451
Belkin PV (1979) Ocular lesions in canine oral papillomatosis. Vet Med Small Anim Clin 74: 1520-1527
Beran GW (1976) Bovine cystic hematuria in the Philippines: a report of an enzootic area. J Am Vet Med Assoc 149: 1686-1690
Berenblum I (1941) The cocarcinogenic action of croton resin. Cancer Res 1: 44-48
Black PH, Hartley JW, Rowe WP, Huebner RJ (1963) Transformation of bovine tissue culture cells by bovine papillomavirus. Nature 199: 1016-1018
Blodi FC, Ramsey FK (1967) Ocular tumors in domestic animals. Am J Ophthalmol 64: 627-633
Boever WJ (1976) Interdigital corns in a black rhinoceros. Vet Med Small Anim Clin 71: 827-830
Boever WJ, Kern T (1976) Papillomas in black and white Colobus monkeys *(Colobus polykomus)*. J Wildl Dis 12: 180-181
Bonney CH, Koch SA, Dice PF, Confer AW (1980) Papillomatosis of conjunctiva and adnexa in dogs. J Am Vet Med Assoc 176: 48-51
Borg K (1975) Diseases in wild animals (in Swedish) *(Viltsjukdomar)*. Boktryck, Helsingborg, Sweden
Borg K (1978) Deer with fewer fibromas than expected (in Swedish) (Farre fibromalgar an vantat). Svensk Jakt 2: 83
Bregman CL, Hirth RS, Sundberg JP, Christensen EF (1986) Induction of benign and malignant cutaneous epithelial neoplasms in dogs by subcutaneous inoculation of canine oral papilloma vaccine. Vet Pathol (submitted for publication)
British Equine Veterinary Association (1965) Survey of equine disease, 1962-63. Vet Rec 77: 528-538
Brobst DF, Dulac GC (1969) Meningeal tumors induced in calves with the bovine cutaneous papillomavirus. Pathol Vet 6: 135-145
Brobst DF, Hinsman EJ (1966) Electron microscopy of the bovine cutaneous papilloma. Pathol Vet 3: 196-207
Brobst DF, Olson C (1963) Neoplastic and proliferative lesions of the bovine urinary bladder. Am J Vet Res 24: 105-111
Brobst DF, Olson C (1965) Histopathology of urinary bladder tumors induced by bovine cutaneous papilloma agent. Cancer Res 25: 12-19
Broughton E, Graesser FE, Carbyn LN, Choquette LPE (1970) Oral papillomatosis in the coyote in Western Canada. J Wildl Dis 6: 180-181
Broughton E, Miller FL, Choquette LPE (1972) Cutaneous fibropapillomas in migratory barren-ground caribou. J Wildl Dis 8: 138-140
Brown ER (1961) The black-tailed deer of Western Washington. Washington State Game Dept Biol Bull 13: 1-124
Brown RJ, Britz JL, Kupper JL, Trevathan WP (1972) Cutaneous horn in a rhesus monkey. Lab Anim Sci 22: 112-113
Burtscher H, Grunberg W, Meingassner G (1973) Infektiöse Keratoakanthome der Epidermis bei *Praomys (Mastomys) natalensis*. Naturwissenschaften 60: 209-210
Bylund G, Valtonen ET, Niemela E (1980) Observations on epidermal papillomata in wild and cultured Atlantic salmon *Salmo salar* L. in Finland. J Fish Dis 3: 525-528
Campo MS, Moar MH, Jarrett WFH, Laird HM (1980) A new papillomavirus associated with alimentary cancer in cattle. Nature 286: 180-182
Campo MS, Moar MH, Laird HM, Jarrett WFH (1981) Molecular heterogeneity and lesion site specificity of cutaneous bovine papillomaviruses. Virology 113: 323-335
Carlson BL, Hill D, Nielsen SW (1983) Cutaneous papillomatosis in a beaver. J Am Vet Med Assoc 183: 1283-1284
Carne HR, Lloyd LC, Carter HB (1963) Squamous carcinoma associated with cysts of the skin in Merino sheep. J Pathol Bact 86: 305-315
Chambers VC, Evans CA (1959) Canine oral papillomatosis. I. Virus assay and observations on the various stages of the experimental infection. Cancer Res 19: 1188-1195
Chambers VC, Evans CA, Weiser RS (1960) Canine oral papillomatosis. II. Immunological aspects of the disease. Cancer Res 20: 1083-1093

Cheevers WP, Roberson MS, Brassfield AL, Davis WC, Crawford TB (1982) Isolation of a retrovirus from cultured equine sarcoid tumor cells. Am J Vet Res 43: 804-806
Chen EY, Howley PM, Levinson AD, Seeburg PH (1982) The primary structure and genetic organization of the bovine papillomavirus type 1 genome. Nature 299: 529-534
Cheville NF (1966) Studies on connective tissue tumors in the hamster produced by bovine papillomavirus. Cancer Res 26: 2334-2339
Cheville NF (1983) Cell pathology, 2nd edn. Iowa State University Press, Ames, Iowa
Cheville NF, Olson C (1964a) Epithelial and fibroblastic proliferation in bovine cutaneous papillomatosis. Pathol Vet 1: 248-257
Cheville NF, Olson C (1964b) Cytology of the canine oral papilloma. Am J Pathol 45: 849-872
Clark KA (1973) Neoplasms of wild animals. Southwestern Vet 26: 185-188
Coggins LW, Ma JQ, Slater AA, Campo MS (1985) Sequence homologies between bovine papillomavirus genomes mapped by a novel low-stringency heteroduplex method. Virology 143: 603-611
Cook RH, Olson C (1951) Experimental transmission of cutaneous papilloma of the horse. Am J Pathol 27: 1087-1097
Cooper JE, Gschmeisser S, Holt PE (1982) Viral particles in a papilloma from a green lizard *(Lacerta viridis)*. Lab Anim 16: 12-13
Cosgrove GE, Lushbaugh WB, Humason G, Anderson MG (1968) *Synhimantus* (Nematoda) associated with gastric squamous tumors in muskrats. Bull Wild Dis Assoc 4: 54-57
Craft C, Braun L, Strandberg J (1980) The use of immunohistochemical techniques in the diagnosis of virally-induced tumors. Proc Annu Meet Am Coll Vet Pathol 31: 102
Danos O, Giri I, Thierry F, Yaniv M (1984) Papillomavirus genomes: sequences and consequences. J Invest Dermatol 83: 7s-11s
Davis CL, Kemper HE (1936) Common warts (papillomata) in goats. J Am Vet Med Assoc 88: 175-179
Davis CL, Shorten HL (1952) Carcinoma of the eye of sheep. J Am Vet Med Assoc 121: 20-24
Davis CL, Leeper RB, Shelton JE (1933) Neoplasms encountered in federally inspected establishments in Denver, Colorado. J Am Vet Med Assoc 83: 229-237
Davis PW, Huxtable CRR, Sabine M (1975) Virus papilloma in the racing greyhound. In: Jones OG (ed) The racing greyhound, vol 1. World Greyhound Racing Federation, London, pp 32-35
DeMoubreun WA, Goodpasture EW (1932) Infectious oral papillomatosis of dogs. Am J Pathol 8: 43-56
Denniston KJ, Yoneyama T, Hoyer BH, Gerin JL (1984) Expression of hepatitis B virus surface and e antigen genes cloned in bovine papillomavirus vectors. Gene 32: 357-368
Desmet P, DeMoor A, Bouters R, DeMeurichy W (1974) Treatment of papilloma of the penis in bulls. Vlaams Diergeneesk Tijdschr 43: 357-367
Des Voigne DM, Mix MC, Pauley GB (1970) A papilloma-like growth on the siphon of the horse clam, *Tresus nuttalli*. J Invert Pathol 15: 262-267
DiMaio D, Corbin V, Sibley E, Maniatis T (1984) High-level expression of a cloned HLA heavy chain gene introduced into mouse cells on a bovine papillomavirus vector. Mol Cell Biol 4: 340-350
Dodd S (1923) Cancer of the ear of sheep. J Comp Pathol 36: 231-242
Duncan JR, Corbeil LB, Davies DH, Schultz RD, Whitlock RH (1975) Persistent papillomatosis associated with immunodeficiency. Cornell Vet 65: 205-211
Dunway PB, Cosgrove GE, Story JD (1968) Capillaria and trypanosoma infestations in *Microtus ochrogaster*. Bull Wild Dis Assoc 4: 18-20
Dvoretzky I, Shober R, Chattopadhyay SK, Lowy DR (1980) A quantitative *in vitro* focus assay for bovine papillomavirus. Virology 103: 369-375
Edwards MR, Samsonoff WA (1977) Electron microscopic observations on viruslike particles of a catfish papilloma. Proc Ann Electron Microscopy Society of America Meeting 35: 394-395
Effrom M, Griner L, Benirschke K (1977) Nature and rate of neoplasia found in captive wild mammals, birds, and reptiles at necropsy. J Natl Cancer Inst 59: 185-198
England JJ, Watson RE, Larson KA (1973) Virus-like particles in an equine sarcoid cell line. Am J Vet Res 34: 1601-1603
Epstein B (1972a) Bovine herpes virus in ocular squamous cell carcinoma. Ann Soc Cient Argentina 193: 209-219
Epstein B (1972b) Isolation of bovine rhinotracheitis virus from ocular squamous cell carcinomas of cattle. Rev Vet Med 53: 105-110

Fatemi-Nainie S, Anderson LW, Cheevers WP (1982) Identification of a transforming retrovirus from cultured equine dermal fibrosarcoma. Virology 120: 490-494
Fay LD (1962) Neoplastic diseases of white-tailed deer. Proc Natl White-Tailed Deer Dis Symp (Athens, Georgia) 1: 132-137
Fay LD (1970) Skin tumors of cervidae. In: Davis JW (ed) Infectious diseases of wild mammals. Iowa State University Press, Ames, pp 385-392
Ficarelli R (1969) Observations on bovine papillomatosis transmitted by tattooing machines. Vet Ital 20: 48-57
Ficken MD, Andrews JJ, Engeltes I (1983) Papilloma-squamous cell carcinoma of the udder of a saanan goat. J Am Vet Med Assoc 183: 467
Flom JO, Brown RJ, Jones RE, Schonewald J (1980) Vaginal fibromas in a beaked whale, *Mesoplodon densirostris*. J Wildl Dis 16: 99-102
Ford JN, Jennings PA, Spradbrow PB, Francis J (1982) Evidence for papillomaviruses in ocular lesions in cattle. Res Vet Sci 32: 257-259
Friend M (1967) Skin tumors of New York deer. Bull Wildl Dis Assoc 3: 102-104
Fujimoto Y, Olson C (1966) The fine structure of the bovine wart. Pathol Vet 3: 659-684
Fukunaga R, Sokawa Y, Nagata S (1984) Constitutive production of human interferons by mouse cells with bovine papillomavirus as a vector. Proc Natl Acad Sci USA 81: 5086-5090
Fulton RE, Doane FW, MacPherson LW (1970) The fine structure of equine papillomas and the equine papilloma virus. J Ultrastruct Res 30: 328-343
Georges E, Croissant O, Bonneaud N, Orth G (1984) Physical state and transcription of the cottontail rabbit papillomavirus genome in warts and transplantable Vx2 and Vx7 carcinomas of domestic rabbits. J Virol 51: 530-538
Geraldes A (1969) Malignant transformation of hamster cells by cell-free extracts of bovine papillomas *(in vitro)*. Nature 222: 1283-1284
Geraldes A (1970) New antigens in hamster embryo cells transformed in vitro by bovine papillomavirus extracts. Nature 226: 81-82
Gibbs EPJ, Smale CJ, Lawman MJP (1975) Warts in sheep. J Comp Pathol 85: 327-334
Giri I, Danos O, Yaniv M (1985) Genomic structure of the cottontail rabbit (Shope) papillomavirus. Proc Natl Acad Sci USA 82: 1580-1584
Gissmann L, Boshart M, Dürst H, Ikenberg H, Wagner D, zur Hausen H (1984) Presence of human papillomaviruses in genital tumors. J Invest Dermatol 83: 26s-28s
Gordon DE, Olson C (1968) Meningiomas and fibroblastic neoplasia in calves induced with the bovine papillomavirus. Cancer Res 28: 2423-2431
Graffi A, Schramm T, Bender E, Horn KH, Bierwolf D (1963) Cell-free transmissible leukosis in Syrian hamsters, probably of viral aetiology. Br J Cancer 22: 577-581
Graffi A, Schramm T, Bender E, Bierwolf D, Graffi I (1967) Über einen neuen virushaltigen Hauttumor beim Goldhamster. Arch Geschwulstforsch 30: 227-283
Graffi A, Schramm T, Bender E, Graffi I, Horn KH, Bierwolf D (1968) Cell-free transmissible leukoses in syrian hamsters, probably of viral etiology. Br J Cancer 22: 577-581
Graffi A, Bender E, Schramm T, Graffi I, Bierwolf D (1970) Studies on the hamster papilloma and the hamster viral lymphoma. Comparative leukemia research 1969. Bibl Haematologica 36: 293-301
Greig AS, Charlton KM (1973) Electron microscopy of the virus of oral papillomatosis in the coyote. J Wildl Dis 9: 359-361
Grizzle JM, Melius P, Strength DR (1984) Papillomas of fish exposed to chlorinated waste water effluent. J Natl Cancer Inst 73: 1133-1142
Groff DE, Lancaster WD (1985) Molecular cloning and nucleotide sequence of deer papillomavirus. J Virol 56: 85-91
Groff DE, Sundberg JP, Lancaster WD (1983) Extrachromosomal deer fibromavirus DNA in deer fibromas and virus-transformed mouse cells. Virology 131: 546-550
Hagen KW (1966) Spontaneous papillomatosis in domestic rabbits. Bull Wildl Dis Assoc 2: 108-110
Hawkins CD, Swan RA, Chapman HM (1981) The epidemiology of squamous cell carcinoma of the perineal region of sheep. Aust Vet J 57: 455-457
Head KW (1953) Neoplastic diseases. Vet Rec 65: 926-929
Heidemann G (1974) Papillomatose bei einem Damhirsch (*Cervus dama* L. 1758). Z Jagdwiss 20: 157-158

Herman CM, Bischoff AI (1950) Papilloma, skin tumors in deer. California Fish and Game 36: 19-20

Hertig AT, Mackey JJ, Feeley G, Kampschmidt K (1983) Dysplasia of the lower genital tract in the female monkey, *Macaca fascicularis*, the crab eating macaque from southeast Asia. Am J Obstet Gynecol 145: 968-980

Hesselholt M, Ingerslev H (1974) Equine sarcoids. Danske Dyrlaegeforencing 57: 9-13

Hollander CF, Van Noord MJ (1972) Focal epithelial hyperplasia: a virus induced oral mucosal lesion in the chimpanzee. Oral Surg Oral Med Oral Pathol 33: 220-226

Honess RF (1939) A freak deer head. J Wildl Manage 3: 360-362

Honess RF, Winter KB (1956) Diseases of wildlife in Wyoming. Wyoming Game Fish Commission Bull 9: 1-279

Hoover EE (1937) Neurofibromatosis in white-tailed deer. J Mammal 18: 104-105

Hubbard GB, Wood DH, Fanton JW (1983) Squamous cell carcinoma with metastasis in a rhesus monkey *(Macaca mulatta)*. Lab Anim Sci 33: 469-472

Hunermund G (1973) Disposition of ayrshire cows to malignant tumors of the vulva and nictitating membrane with intense sunlight in the tropics. Berlin Münch Tierärztl Wochenschr 86: 414-415

Ishimoto A, Ito Y (1971) Further studies on surface antigen of Shope papilloma cells: trypsin activity. J Natl Cancer Inst 46: 353-358

Ishimoto A, Oota S, Kimura I, Miyake T, Ito Y (1970) *In vitro* cultivation and antigenicity of cottontail rabbit papilloma cells induced by the Shope papilloma virus. Cancer Res 30: 2598-2605

Ito N, Fukushima S, Imaida K, Sakata T, Masui T (1983) Induction of papilloma in the forestomach of hamsters by butylated hydroxyanisole. Gann 74: 459-461

Ito Y (1960) A tumor-producing factor extracted by phenol from papillomatous tissue (Shope) of cottontail rabbits. Virology 12: 596-601

Ito Y (1963) Studies on subviral tumorigenesis: carcinoma derived from nucleic acid-induced papillomas of rabbit skin. Acta Unio Int Contra Cancrum 19: 280-283

Ivascu I, Onet E (1971) Research into the incidence of neoplasms of the eyes in cattle. Recl Med Vet 147: 607-614

Ivascu I, Onet E (1974) Experimental and clinical observations on the etiology of ocular tumors in cattle. Schweiz Arch Tierheilkd 116: 455-459

Iversen OH (1982) Hairless mouse skin in two-stage chemical carcinogenesis. Virchows Arch [Cell Pathol] 38: 263-272

Jacobson ER (1979) Viral diseases of reptiles. Ann Proc Am Assoc Zoo Vet 13-15

Jacobson ER, Gaskin JM, Clubb S, Calderwood MB (1982) Papilloma-like virus infection in Bolivian side-neck turtles. J Am Vet Med Assoc 181: 1325-1328

Jacobson ER, Mladinich CR, Clubb S, Sundberg JP, Lancaster WD (1983) Papilloma-like virus infection in an African Gray Parrot. J Am Vet Med Assoc 183: 1307-1308

Jarrett WFH, McNeil PE, Grimshaw WTR, Selman IE, McIntyre WIM (1978a) High incidence area of cattle cancer with a possible interaction between an environmental carcinogen and a papillomavirus. Nature 274: 215-217

Jarrett WFH, Murphy J, O'Neil BW, Laird HM (1978b) Virus-induced papillomas of the alimentary tract of cattle. Int J Cancer 22: 323-328

Jarrett WFH, Campo MS, Blaxter ML, O'Neil BW, Laird NM, Moar MH, Sartirana ML (1984a) Alimentary fibropapilloma in cattle. A spontaneous tumor, nonpermissive for papillomavirus replication. J Natl Cancer Inst 73: 499-504

Jarrett WFH, Campo MS, O'Neil BW, Laird HM, Coggins LW (1984b) A novel bovine papillomavirus (BPV-6) causing true epithelial papillomas of the mammary gland skin: a member of a proposed new BPV group. Virology 136: 255-264

Jennings AR (1968) Tumors of free-living wild mammals and birds in Great Britain. Symp Zool Soc Lond 24: 273-287

Junge RE, Sundberg JP, Lancaster WD (1984) Papillomas and squamous cell carcinomas of horses. J Am Vet Med Assoc 185: 656-659

Karstad L, Kaminjolo JS (1978) Skin papillomas in an impala *(Aepyceros melampus)* and a giraffe *(Giraffa camelopardalis)*. J Wildl Dis 14: 309-313

Kass SJ, Knight CA (1965) Purification and chemical analysis of Shope papilloma virus. Virology 27: 273-281

Kaul PL, Kalra DS (1973) Incidence of horn cancer in Haryana State. Haryana Agric Univ J Res 3: 161-165

Kerr KM, Alden CL (1974) Equine neoplasia - a ten-year survey. Proc Ann Meeting Am Assoc Vet Lab Diag 17: 183-187
Kidd JG, Rous P (1940) Cancers deriving from the virus papillomas of wild rabbits under natural conditions. J Exp Med 71: 469-494
Koller LD (1972) Cutaneous papillomas on an opossum. J Natl Cancer Inst 49: 309-313
Koller LD, Olson C (1971) Pulmonary fibroblastomas in a deer with cutaneous fibromatosis. Cancer Res 31: 1371-1375
Koller LD, Olson C (1972) Attempted transmission of warts from man, cattle, and horses, and of deer fibroma to selected hosts. J Invest Dermatol 58: 366-368
Koller LD, Barthold SW, Olson C (1974) Quatitation of bovine papillomavirus and serum antibody by immunodiffusion. Am J Vet Res 35: 121-124
Koops H, Mann H, Pfitzner I, Schmid OJ, Schubert G (1970) The cauliflower disease of eels. Symp Dis Fishes and Shellfishes. Am Fisheries Soc, Washington, DC, Special Publ 5: 291-295
Kreider JW, Benjamin SA, Pruchnic WF, Strimlan CV (1971) Immunologic mechanisms in the induction and regression of shope papillomavirus-induced epidermal papillomas of rat. J Invest Dermatol 56: 102-112
Kumer L (1935) Über die Papillomatose der Gemsen. Wiener Klin Wochenschr 48: 890-891
Ladds PW, Daniels PW (1982) Animal model of human disease. Squamous cell carcinoma. Ovine squamous cell carcinoma. Am J Pathol 107: 122-123
Lambertsen RH, Kohn BA, Sundberg JP, Buergelt CD (1987) Genital papillomatosis in sperm whale bulls. Vet Pathol (submitted for publication)
Lancaster WD (1981) Apparent lack of integration of bovine papillomavirus DNA in virus-induced equine and bovine tumor cells and virus-transformed mouse cells. Virology 108: 251-255
Lancaster WD, Olson C (1978) Demonstration of two distinct classes of bovine papillomavirus. Virology 89: 372-379
Lancaster WD, Olson C (1979) Bovine papillomavirus and connective tissue tumors. In: Essex M, Todaro G, zur Hausen H (eds) Viruses in naturally occurring cancers. Cold Spring Harbor, New York, p 88
Lancaster WD, Sundberg JP (1982) Characterization of papillomaviruses isolated from cutaneous fibromas of white-tailed deer and mule deer. Virology 123: 212-216
Lancaster WD, Olson C, Meinke W (1977) Bovine papillomavirus: presence of virus-specific DNA sequences in naturally occurring equine tumors. Proc Natl Acad Sci USA 74: 524-528
Lancaster WD, Theilen GH, Olson C (1979) Hybridization of bovine papillomavirus type 1 and type 2 DNA to DNA from virus-induced hamster tumors and naturally occurring equine tumors. Intervirology 11: 227-233
Lee KP, Olson C (1969a) Precipitin response of cattle to bovine papillomavirus. Cancer Res 29: 1393-1397
Lee KP, Olson C (1969b) A gel diffusion precipitin test for bovine papillomavirus. Am J Vet Res 30: 725-731
Legendre AM, Krahwinkel DJ (1981) Feline ear tumors. J Am Anim Hosp Assoc 17: 1035-1037
Leininger JR, Folk GE, Cooper PS (1979) Gastric candidiasis in laboratory-reared brown lemmings. J Am Vet Med Assoc 175: 990-991
Leopold AS, Riney T, McCain R, Tevis L (1951) The jawbone deer herd. State of California Department of Natural Resources Division of Fish and Game. Game Bull 4: 139
Lepper AWD (1967) An intranuclear fibrillar lattice in the squamous epithelial cells of a bovine cutaneous papilloma. Vet Rec 81: 238-239
Lina PHC, Van Noord MJ, de Groot FG (1973) Detection of virus in squamous papillomas of wild bird species *Fringilla coelebs*. J Natl Cancer Inst 50: 567-571
Lucke B, Schlumberger HG (1949) Neoplasia in cold-blooded vertebrates. Physiol Rev 29: 91-126
Lucke B, Ratcliffe H, Breedis C (1950) Transmissible papilloma in monkeys. Fed Proc 9: 337
Lunger PD, Clark HF (1978) Reptilia-related viruses. Adv Virus Res 23: 159-204
Lutzner MA, Guenet JL, Breitburd F (1985) Multiple cutaneous papillomas and carcinomas that develop spontaneously in a mouse mutant, the repeated epilation heterozygote Er/+. J Natl Cancer Soc 75: 161-166
McEntee K (1950) Fibropapillomas of the external genitalia of cattle. Cornell Vet 40: 304-312
McEntee K (1952) Transmissible fibropapillomas of the external genitalia of cattle. Report NY State Vet Coll, Humphry Press, Geneve, NY, p 28

McKenzie RA (1978) Bovine enzootic hematuria in Queensland. Aust Vet J 54: 61-64
Meischke HRC (1979a) *In vitro* transformation by bovine papillomavirus. J Gen Virol 3: 473-487
Meischke HRC (1979b) A survey of bovine teat papillomas. Vet Rec 104: 28-31
Meischke HRC (1979c) Experimental transmission of bovine papillomavirus (BPV) extracted from morphologically distinct teat and cutaneous lesions and the effects of inoculation of BPV transformed fetal bovine cells. Vet Rec 104: 360-366
Migaki G, Garner FM, Carey AM (1971) Metastatic transitional cell carcinoma of the bovine urinary bladder. Cornell Vet 61: 59-70
Moar MH (1983) Characterization of a papillomavirus from red deer *(Cervus elaphus)* in Scotland. Embo workshop on papilloma viruses, 27-30 July 1983, Orenäs, Sweden, p 20
Moar MH, Jarrett WFH (1985) A cutaneous fibropapilloma from a red deer *(Cervus elaphus)* associated with a papillomavirus. Intervirology 24: 108-118
Monlux AW, Anderson WA, Davis CL (1956) A survey of tumors occurring in cattle, sheep, and swine. Am J Vet Res 17: 646-677
Monlux AW, Anderson WA, Davis CL (1957) The diagnosis of squamous cell carcinoma of the eye (cancer eye) in cattle. Am J Vet Res 18: 5-34
Monlux WS, Monlux AW (1972) Atlas of meat inspection pathology. Agricultural Handbook #367. United States Department of Agriculture, Washington, DC
Moore DH, Stone RS, Shope RS, Gelber D (1959) Ultrastructure and site of formation of rabbit papillomavirus. Proc Soc Exp Biol Med 101: 575-578
Moreno-Lopez J, Pettersson U, Dinter Z, Philipson L (1981) Characterization of a papilloma virus from the European Elk (EEPV). Virology 112: 589-595
Moreno-Lopez J, Stenlund A, Ahola H, Bergman P, Pettersson U (1983) The reindeer papillomavirus. Embo workshop on Papilloma viruses, 27-30 July 1983, Orenäs, Sweden, p 22
Moreno-Lopez J, Ahola H, Stenlund A, Osterhaus A, Pettersson U (1984) Genome of an avian papillomavirus. J Virol 51: 872-875
Morgan DM, Meinke W (1980) Isolation of clones of hamster embryo cells transformed by the bovine papillomavirus. Curr Microbiol 3: 247-251
Moulton JE (1961) Tumors in domestic animals. University of California Press, Berkeley
Moulton JE (1978) Tumors in domestic animals, 2nd edn. University of California Press, Berkeley, pp 309-345
Moulton JE, Lau D (1964) Attempts to induce neoplasia *in vitro* with Shope papillomavirus. Cornell Vet 54: 602-612
Müller H, Gissmann L (1978) *Mastomys natalensis* papillomavirus (MnPV), the causative agent of epithelial proliferations: characterization of the virus particle. J Gen Virol 41: 315-323
Murie A (1934) The moose of Isle Royale. University of Michigan, Museum of Zoology, Miscellaneous Publications #25. University of Michigan Press, Ann Arbor, MI
Murphy MF, Potter HL, Abraham JM, Morgan DM, Meinke WJ (1981) Analysis of a restriction endonuclease map for a rabbit papillomavirus DNA. Curr Microbiol 5: 349-352
Murray M (1968) Neoplasms of domestic animals in East Africa. Br Vet J 124: 514-524
Naik SN, Randelia HP (1975) Carcinoma of the eye in Indian cattle: an epidemiological aspect. Indian J Cancer 12: 310-318
Nandi SN (1969) Histopathology of enzootic bovine hematuria in the Darjeeling District of India. Br Vet J 125: 587-589
Nasseri M, Wettstein FO (1984a) Cottontail rabbit papillomavirus-specific transcripts in transplantable tumors with integrated DNA. Virology 138: 362-367
Nasseri M, Wettstein FO (1984b) Differences exist between viral transcripts in cottontail rabbit papillomavirus-induced benign and malignant tumors as well as non-virus-producing and virus-producing tumors. J Virol 51: 706-712
Nellis CH (1973) Prevalence of oral papilloma-like lesions in coyotes in Alberta. Can J Zool 51: 900
Ng WL, Fong LYY, Newberne PM (1984) Forestomach squamous papillomas in the rat: effects of dietary zinc deficiency on induction. Cancer Lett 22: 329-332
Noyes WF (1959) Studies on the Shope rabbit papillomavirus. II. The location of the infective virus in papillomas of the cottontail rabbit. J Exp Med 109: 423-428
O'Banion MK, Sundberg JP (1987) Papillomavirus genomes in experimentally-induced white-tailed deer fibromas. Am J Vet Res (in press)

O'Banion K, Sundberg JP, Reichmann ME (1985) Equine papillomavirus: partial characterization and presence in common equine skin tumors. J Cell Biochem 9c: 73
Olson C (1963) Cutaneous papillomatosis in cattle and other animals. Ann NY Acad Sci 108: 1042-1056
Olson C, Skidmore LV (1959) Therapy of experimentally produced bovine cutaneous papillomatosis with vaccines and excision. J Am Vet Med Assoc 135: 339-343
Olson C, Pamukcu AM, Brobst DF, Kowalczyk T, Satter EJ, Price JM (1959) A urinary bladder tumor induced by a bovine cutaneous papilloma agent. Cancer Res 19: 779-782
Olson C, Segre D, Skidmore LV (1960) Further observations on immunity to bovine cutaneous papillomatosis. Am J Vet Res 21: 233-242
Olson C, Pamukcu AM, Brobst DF (1965) Papilloma-like virus from bovine urinary bladder tumors. Cancer Res 25: 840-850
Olson C, Robl MG, Larson LL (1968) Cutaneous and penile bovine fibropapillomatosis and its control. J Am Vet Med Assoc 153: 1189-1194
Olson RO, Olson C, Easterday BC (1982) Papillomatosis of the bovine teat (mammary papilla). Am J Vet Res 43: 2250-2252
Orth G, Breitburd F, Favre M (1978) Evidence for antigenic determinants shared by the structural polypeptides of (Shope) rabbit papillomavirus and human papillomavirus type 1. Virology 91: 243-255
Osterhaus ADME, Ellens DJ, Horzinek MC (1977) Identification and characterization of a papillomavirus from birds (Frigillidae). Intervirology 8: 351-359
Palfi J, Fabian M (1969) Eye carcinoma in cattle. Magy Allatory Lap 24: 351-354
Pamukcu AM (1957) Tumors of the urinary bladder in cattle and water buffalo affected with bovine enzootic hematuria. Zentralb Veterinarmed 2: 185-197
Pamukcu AM (1974) Tumors of the urinary bladder. Bull WHO 50: 43-52
Pamukcu AM, Göksoy SK, Price JM (1967a) Urinary bladder neoplasms induced by feeding bracken fern *(Pteris aquilina)* to cows. Cancer Res 27: 917-924
Pamukcu AM, Olson C, Goksoy SK (1967b) Influence of a papilloma vaccine on chronic bovine enzootic hematuria. Cancer Res 27: 2197-2200
Pamukcu AM, Price JM, Bryan GT (1976) Naturally occurring and bracken fern induced bovine urinary bladder tumors. Vet Pathol 13: 110-122
Pamukcu AM, Yalciner S, Price JM, Bryan GT (1970) The effects of the coadministration of thiamine on the incidence of urinary bladder carcinomas in rats fed bracken fern. Cancer Res 30: 2671-2674
Parish WE (1961) A transmissible genital papilloma of the pig resembling condyloma acuminatum of man. J Pathol Bacteriol 81: 331-345
Parish WE (1962) An immunological study of the transmissible genital papilloma of the pig. J Pathol Bacteriol 83: 429-442
Parsons RJ, Kidd JG (1936) A virus causing oral papillomatosis in rabbits. Proc Soc Exp Biol Med 35: 441-443
Parsons RJ, Kidd JG (1943) Oral papillomatosis of rabbits; a virus disease. J Exp Med 77: 233-250
Pearson H (1972) Surgery of the male genital tract in cattle. A review of 121 cases. Vet Rec 91: 498-509
Pence DB, Tran RM, Bishop ML, Foster SH (1983) Fibroma in a nine-banded armadillo *(Dasypus novemcinctus)*. J Comp Pathol 93: 179-184
Peters N, Watermann B (1979) Three types of skin papillomas of flatfishes and their causes. Marine Ecology Progress Series 1: 269-276
Petrak ML (1982) Diseases of cage and aviary birds, 2nd ed. Lea and Febiger, Philadelphia
Pfister H, Meszaros J (1980) Partial characterization of canine oral papillomavirus. Virology 104: 243-246
Pfister H, Linz U, Gissmann L, Huchthausen B, Hoffmann D, zur Hausen H (1979) Partial characterization of a new type of bovine papillomaviruses. Virology 96: 1-8
Pintel D, Merchlinsky MJ, Ward DC (1984) Expression of minute virus of mice structural proteins in murine cell lines transformed by bovine papillomavirus-minute virus of mice plasmid chimera. J Virol 52: 320-327
Pirie HM (1973) Unusual occurrence of squamous carcinoma of the upper alimentary tract in cattle in Britain. Res Vet Sci 15: 135-138

Plummer PJG (1956) A survey of 636 tumors from domesticated animals. Can J Comp Med 20: 239-251
Potter HL, Meinke WJ (1985) Nucleotide sequence of bovine papillomavirus type 2 late region. J Gen Virology 66: 187-193
Price JM, Pamukcu AM (1968) The induction of neoplasms of the urinary bladder of the cow and the small intestine of the rat by feeding bracken fern *(Pteris aquilina)*. Cancer Res 28: 2247-2251
Priester WA, Mantel N (1971) Occurrence of tumors in domestic animals. Data from 12 US and Canadian colleges of veterinary medicine. J Natl Cancer Inst 47: 1333-1334
Pulley LT, Shively JN, Pawlicki JJ (1974) An outbreak of bovine cutaneous fibropapillomas following dehorning. Cornell Vet 64: 427-434
Pyrhönen S, Neuvonen E (1978) The occurrence of human wart-virus antibodies in dogs, pigs and cattle. Arch Virol 57: 297-305
Quortrup ER (1946) Tumors of deer. Virgina Wildlife 7: 15-18
Ragland WL, Spencer GR (1968) Attempts to relate bovine papillomavirus to the cause of equine sarcoid: Immunity to bovine papillomavirus. Am J Vet Res 27: 1363-1366
Ragland WL, Spencer GR (1969) Attempts to relate bovine papillomavirus to the cause of equine sarcoid: equidae inoculated intradermally with bovine papillomavirus. Am J Vet Res 30: 743-752
Ragland WL, Keown GH, Spencer GR (1970a) Equine sarcoid. Equine Vet J 2: 2-11
Ragland WL, McLaughlin CA, Spencer GR (1970b) Attempts to relate bovine papillomavirus to the cause of equine sarcoid: horses, donkeys, and calves inoculated with equine sarcoid extracts. Equine Vet J 2: 168-172
Rangan SRS, Gutter A, Baskin GB, Anderson D (1980) Virus associated papillomas in Colobus monkeys *(Colobus guereza)*. Lab Anim Sci 30: 885-889
Ratcliffe HL (1933) Incidence and nature of tumors in captive wild mammals and birds. Am J Cancer 17: 116-135
Raynaud MMA, Adrian M (1976) Lésions cutanées à structure papillomateuse associées à des virus chez le lezard vert *(Lacerta viridis laur)*. C R Seances Acad Sci [III] Paris 283: 847
Rdzok EJ, Shipkowitz NL, Richter WR (1966) Rabbit oral papillomatosis: Ultrastructure of experimental infection. Cancer Res 26: 160-165
Rebhun WC, Payne RM, King JM, Wolfe M, Begg SN (1980) Interdigital papillomatosis in dairy cattle. J Am Vet Med Assoc 177: 437-440
Reddy AL, Fialkow PJ (1983) Papillomas induced by initiation-promotion differ from those induced by carcinogen alone. Nature 304: 69-71
Reed DE, Shave H, Bergeland ME, Gates CE (1976) Necropsy and laboratory findings in free-living deer in South Dakota. J Am Vet Med Assoc 169: 975-979
Reichenbach-Klinke H, Elkan E (1965) The principal diseases of lower vertebrates. Academic, New York
Richards S (1957) Disease of deer and antelope. North Dakota Outdoors 17: 7, 17
Richter WR, Shipkowitz NL, Rdzok EJ (1964) Oral papillomatosis of the rabbit: an electron microscopic study. Lab Invest 13: 430-438
Rieke H (1980) Ausgedehnte kongenitale fibroepitheliale Papillome bei einem Ferkel. Dtsch Tierärztl Wochenschr 87: 412-413
Roberts RJ (1978) Fish pathology. Bailliere Tindall, London
Robl MG, Olson C (1968) Oncogenic action of bovine papillomavirus in hamsters. Cancer Res 28: 1596-1604
Robl MG, Gordon DE, Lee KP, Olson C (1972) Intracranial fibroblastic neoplasms in the hamster from bovine papillomavirus. Cancer Res 32: 2221-2225
Roscoe DE, Veikley LR, Mills M, Hinds L (1975) Debilitating ossifying fibromas of a white-tailed deer associated with ear tagging. J Wildl Dis 11: 62-65
Rous P, Beard JW (1935) The progression to carcinoma of virus-induced rabbit papillomas (Shope). J Exp Med 62: 523-548
Rous P, Kidd JG (1940) The activation, transforming and carcinogenic effects of the rabbit papillomavirus (Shope) upon implanted tar tumors. J Exp Med 71: 787-812
Rudolph R, Müller H (1976) Induktion von epidermalem Tumorwachstum in der Haut von *Mastomys natalensis* durch Übertragung virushaltigen Tumorgewebes eines Plattenepithelkarzinoms. Zbl Vet Med B23: 143-150

Rudolph R, Thiel W (1976) Pathologische Anatomie und Histologie von spontanen, epithelialen Hauttumoren bei *Mastomys natalensis*. Zbl Vet Med A 23: 429-441
Rudolph RL (1980) Neoplasien der Haut bei einem wildfarbenen Inzuchtstamm von *Mastomys natalensis* (WSA Gießen). Vet Pathol 17: 600-613
Runnells RA, Benbrook EA (1942) Epithelial tumors of horses. Am J Vet Res 3: 176-179
Sack WO (1982) Essentials of pig anatomy. Veterinary Textbooks, Ithaca, New York
Samuel JL, Spradbrow PB, Wood AL, Kelly WR (1985) Oral papillomas in cattle. Zbl Vet Med B 32: 706-714
Samuel WM, Chalmers GA, Gunson JR (1978) Oral papillomatosis in coyotes *(Canis latrans)* and wolves *(Canis lupus)* of Alberta. J Wildl Dis 14: 165-169
Sastry GA (1959) Neoplasms of animals in India. Vet Med 54: 428-430
Schenborn ET, Lund E, Mitchen JL, Dahlberg JE (1985) Expression of a human U1 RNA gene introduced into mouse cells via bovine papillomavirus DNA vectors. Mol Cel Biol 5: 1318-1326
Schlumberger HG, Lucke B (1948) Tumors of fishes, amphibians, and reptiles. Cancer Res 8: 657-754
Scott DW (1980) Feline dermatology 1900-1978: a monograph. J Am Anim Hosp Assoc 16: 303-459
Segre D, Olson C, Hoerlein AB (1955) Neutralization of bovine papillomavirus with serums from cattle and horses with experimental papillomas. Am J Vet Res 16: 517-520
Severinghaus CW, Cheatum EL (1961) Life and times of the white-tailed deer. In: Taylor WP (ed) The deer of North America. Telegraph, Harrisburg, PA, pp 57-186
Scherneck S, Vogel F, Nguyen HL, Feunteun J (1984) Sequence homology between polyoma virus. Simian Virus 40 and a papilloma producing virus from a syrian hamster: evidences for highly conserved sequences. Virology 137: 41-48
Shope RE (1934) Serial transmission of virus of infectious papillomatosis in domestic rabbits. Proc Soc Exp Biol Med 32: 830-832
Shope RE (1955) An infectious fibroma of deer. Proc Soc Exp Biol Med 88: 533-535
Shope RE, Hurst EW (1933) Infectious papillomatosis of rabbits. J Exp Med 58: 607-624
Shope RE, Mangold R, MacNamara LG, Dumbell KR (1958) An infectious cutaneous fibroma of the Virginia white-tailed deer *(Odocoileus virginianus)*. J Exp Med 108: 797-802
Smith GM, Coates CW, Nigrelli RF (1941) A papillomatous disease of the gall bladder associated with infection by flukes, occurring in the marine turtle, *Chelonia mydas* (Linnaeus). Zoologica 20: 13-16
Smith KT, Campo MS (1985) The biology of papillomaviruses and their role in oncogenesis. Anticancer Res 5: 31-48
Stannard AA, Pulley LT (1978) Tumors of the skin and soft tissue. In: Moulton JE (ed) Tumors in domestic animals. University of California Press, Berkeley, pp 16-74
Stenlund A, Moreno-Lopez J, Ahola H, Pettersson U (1983) European elk papillomavirus: characterization of the genome, induction of tumors in animals, and transformation *in vitro*. J Virol 48: 370-376
Stevens JG, Wettstein FO (1979) Multiple copies of Shope virus DNA are present in cells of benign and malignant non-virus-producing neoplasms. J Virol 30: 891-898
Stone RS, Shope RE, Moore DH (1959) Electron microscopic study of the development of the papillomavirus in the skin of the rabbit. J Exp Med 110: 543-546
Strafuss AC, Cook JE, Smith JE (1976) Squamous cell carcinoma in dogs. J Am Vet Med Assoc 168: 425-427
Strafuss AC, Smith JE, Dennis SM, Anthony HD (1973) Sarcoid in horses. Vet Med 68: 1246-1247
Sugawara K, Fujinaga K, Yamashita T, Ito Y (1983) Integration and methylation of Shope papillomavirus DNA in the transplantable Vx2 and Vx7 rabbit carcinomas. Virology 131: 88-99
Sundberg JP (1987) Animal models for papillomavirus research. In: Hofschneider PH, Mund K (eds) Human tumor viruses. Karger, Basel (submitted for publication)
Sundberg JP, Dunstan RW (1986) Canine papillomas and squamous cell carcinomas. J Am Vet Med Assoc (in preparation)
Sundberg JP, McDonald SE (1984) Cervical squamous cell carcinoma in a blackbuck. J Am Vet Med Assoc 185: 1445-1446
Sundberg JP, Nielsen SW (1981) Deer fibroma: a review. Can Vet J 22: 385-388

Sundberg JP, Nielsen SW (1982) Prevalence of cutaneous fibromas in white-tailed deer *(Odocoileus virginianus)* in New York and Vermont. J Wildl Dis 18: 359-360
Sundberg JP, Olson C (1986) Squamous cell carcinomas induced in dogs by canine oral papillomavirus. J Natl Cancer Inst (submitted for publication)
Sundberg JP, Burnstein T, Page EH, Kirkham WW, Robinson FR (1977) Neoplasms of equidae. J Am Vet Med Assoc 170: 150-152
Sundberg JP, Russell W, Lancaster WD (1981) Papillomatosis in Indian elephants, *Elephas maximus.* J Am Vet Med Assoc 179: 1247-1248
Sundberg JP, Williams E, Thorne ET, Lancaster WD (1983) Cutaneous fibropapilloma in a pronghorn antelope. J Am Vet Med Assoc 183: 1333-1334
Sundberg JP, Junge RE, Lancaster WD (1984) Immunoperoxidase localization of papillomaviruses in hyperplastic and neoplastic epithelial lesions of animals. Am J Vet Res 45: 1441-1446
Sundberg JP, Chiodini RJ, Nielsen SW (1985a) Transmission of the white-tailed deer cutaneous fibroma. Am J Vet Res 46: 1150-1154
Sundberg JP, Hill DL, Williams ES, Nielsen SW (1985b) Light and electron microscopic comparisons of cutaneous fibromas in white-tailed and mule deer. Am J Vet Res 46: 2200-2206
Sundberg JP, Junge RE, El Shazley MO (1985c) Oral papillomas in New Zealand white rabbits. Am J Vet Res 46: 664-668
Sundberg JP, Morris K, Lancaster WD (1985d) Cutaneous fibromas in moose *(Alces alces).* J Wildl Dis 21: 181-183
Sundberg JP, Schmidt E, O'Banion K, Olson C (1985e) Squamous cell carcinoma associated with canine oral papillomavirus. J Cell Biochem 9c: 75
Sundberg JP, Todd KS, DiPietro JA (1985f) Equine papillomatosis: is partial resection of lesions an effective treatment? Vet Med 80: 71-74
Sundberg JP, Williams ES, Hill D, Lancaster WD, Nielson SW (1985g) Detection of papillomaviruses in cutaneous fibromas of white-tailed and mule deer. Am J Vet Res 46: 1145-1149
Sundberg JP, Junge RE, O'Banion MK, Basgall EJ, Harrison G, Herron AJ, Shivaprasad HL (1986a) Cloacal Papillomas in psittacine birds. Am J Vet Res 47: 928-932
Sundberg JP, O'Banion MK, Schmidt-Didier E, Reichmann ME (1986b) Cloning and characterization of a canine oral papillomavirus. Am J Vet Res 47: 1142-1144
Sundberg JP, O'Banion MK, Shima A, Wolf P, Basgall E, Reichmann ME (1986c) Papillomas and papillomaviruses in nonhuman primates. J Natl Cancer Inst (in preparation)
Sundberg JP, O'Banion MK, Reichmann ME (1987) Mouse papillomavirus: pathology and characterization of the virus. Cancer Cells 5 (to be published)
Swan RA, Wilcox GE, Chapman HM, Hawkins CD (1983) Attempted transmission and immunotherapy of squamons cell carcinomas of the vulva of ewes. Aust Vet J 60: 314-315
Swank WG (1958) The mule deer in Arizona chaparral. State of Arizona Game Fish Dept Wildl Bull 3: 109
Syverton JT (1952) The pathogenesis of the rabbit papilloma-to-carcinoma sequence. Ann NY Acad Sci 54: 1126-1140
Tajima M, Gordon DE, Olson C (1968) Electron microscopy of bovine papilloma and deer fibroma viruses. Am J Vet Res 29: 1185-1194
Taylor RL, Hanks MA (1969) Viral isolations from bovine eye tumors. Am J Vet Res 30: 1885-1886
Taylor RL, Hanks MA (1972) Developmental changes in precursor lesions of bovine ocular squamous cell carcinoma. Veterinary Medicine/Small Animal Clinicias 67: 669-671
Theilen G, Wheeldon EB, East N, Madewell B, Lancaster WD, Munn R (1985) Goat papillomatosis. Am J Vet Res 46: 2519-2526
Thomas M, Boiron M, Tanzer J, Levy JP, Bernard J (1964) *In vitro* transformation of mice cells by bovine papillomavirus. Nature 4933: 709-710
Thorsen J, Cooper JE, Warwick GP (1974) Esophageal papillomata in cattle in Kenya. Trop Anim Health Prod 6: 95-98
Tokarnia CH, Döbereiner J, Canella LFC (1969) Incidence of enzootic haematuria and epidermoid carcinoma of the upper digestive tract in cattle. Pesquisa Agropec Brasil 4: 209-224
Tokita H, Konishi S (1975) Studies on canine oral papillomatosis. II. Oncogenicity of canine oral papillomavirus to various tissues of dog with special reference to eye tumor. Jpn J Vet Sci 37: 109-120

Tokita H, Konishi S, Ogata M (1977) Studies on canine oral papillomatosis. III. Cultivation of papilloma cells *in vitro*. Jpn J Vet Sci 39: 619-626
Trainer DO, Knowlton FF, Karstad L (1968) Oral papillomatosis in the coyote. Bull Wild Dis Assoc 4: 52-54
Tustin RC, Thornton DJ, McNaughton H (1982) High incidence of squamous cell carcinoma of the vulva in Merino ewes on a South African farm. J S Afr Vet Assoc 53: 141-143
Tweddle NE, White WE (1977) An outbreak of fibropapillomatosis in cows following rectal examinations. Aust Vet J 53: 492-495
Vandegraaff R (1976) Squamous cell carcinoma of the vulva in Merino sheep. Aust Vet J 52: 21-23
Vanselow BA, Spradbrow PB (1982) Papillomaviruses, papillomas and squamous cell carcinomas in sheep. Vet Rec 110: 561-562
Vanselow BA, Spradbrow PB (1983) Squamous cell carcinoma of the vulva, hyperkeratosis and papillomaviruses in a ewe. Aust Vet J 60: 194-195
Vogel F, Zimmermann W, Krause H, Scherneck S (1984) Characterization of the DNA of the hamster papovavirus: I. Genom length and molecular cloning. Arch Geschwulstforsch 54: 433-441
Voss J (1964) Transmission of equine sarcoid. Proc Conv Am Assoc Equine Pract 10: 199-216
Voss JL (1969) Transmission of equine sarcoid. Am J Vet Res 30: 183-191
Wadsworth JR (1954) Fibrosarcomas in a deer. J Am Vet Med Assoc 124: 194
Watrach AM (1969) The ultrastructure of canine cutaneous papilloma. Cancer Res 29: 2079-2084
Watrach AM, Small E, Case MT (1970) Canine papilloma: progression of oral papilloma to carcinoma. J Natl Cancer Inst 45: 915-920
Watson RE, England JJ, Larson KA (1972) Cultural characterization of a cell line derived from an equine sarcoid. Appl Microbiol 24: 727-731
Watts SL, Ostrow RS, Phelps WC, Prince JT, Faras AJ (1983) Free cottontail papillomavirus DNA persists in warts and carcinomas of infected rabbits and in cells in culture transformed with virus or viral DNA. Virology 125: 127-138
Watts SL, Phelps WC, Ostrow RS, Zachow KR, Faras AJ (1984) Cellular transformation by human papillomavirus DNA *in vitro*. Science 225: 634-636
Weisbroth SH, Scher S (1970) Spontaneous oral papillomatosis in rabbits. J Am Vet Med Assoc 157: 1940-1944
Wellings SR (1970) Biology of some virus diseases of marine fish. Symp Dis Fishes Shellfishes Am Fisheries Soc Special Publ (Washington, DC) 5: 296-306
Wellings SR, Chuinard RG (1964) Epidermal papillomas with virus-like particles in flathead sole, *Hippoglossoides elassodon*. Science 146: 932-934
Wellings SR, Chuinard RG, Gourley RT, Cooper RA (1964) Epidermal papillomas in the flathead sole, *Hippoglossoides elassodon*, with notes on the occurrence of similar neoplasms in other pleuronectids. J Natl Cancer Inst 33: 991-1004
Wettimuny SG de S, Jayasekera MU (1970) The occurrence of cutaneous papillomata (warts) in a dairy farm in Ceylon. Ceylon Vet J 18: 10-14
Wettstein FO, Stevens JG (1980) Distribution and state of viral nucleic acid in tumors induced by Shope papilloma virus. Cold Spring Harbor Conf Cell Proliferation 7: 301-307
Wettstein FO, Stevens JG (1982) Variable-sized free episomes of Shope papillomavirus DNA are present in all non-virus-producing neoplasms and intergrated episomes are detected in some. Proc Natl Acad Sci USA 79: 790-794
Williams RC, Kass SJ, Knight CA (1960) Structure of Shope papilloma virus particles. Virology 12: 48-58
Wolke RE, Murchelano RA (1976) A case report of an epidermal papilloma in *Mustelus canis*. J Wildl Dis Assoc 12: 167-171
Wood AL, Spradbrow PB (1985) Transformation of cultured equine fibroblasts with a bovine papillomavirus. Res Vet Sci 38: 241-242
Yoshikawa T, Oyamada T (1975) Histopathology of papillary tumors in the bovine urinary bladder. Jpn J Vet Sci 37: 277-287

Human Papillomavirus Infections in the Oral Cavity

S. M. Syrjänen

1 Introduction

Benign epithelial tumors and hyperplasias of the oral mucosa have been the subject of extensive discussions. Clinical and histological similarities between verruca vulgaris, squamous cell papilloma, and condyloma acuminatum in the oral cavity are well recognized (Hertz 1972; Shafer et al. 1983; Pindborg 1985). Consequently, it is doubtful whether a consistent morphological distinction among these lesions can be made (Waldron 1970; Hertz 1972; Wysocki and Hardie 1979; Syrjänen et al. 1984b). In textbooks of oral pathology, squamous cell papillomas are always discussed under the general heading of neoplasias or benign oral epithelial tumors (Waldron 1970; Batsakis 1979; Bhaskar 1979; Shafer et al. 1983; Lucas 1984; Pindborg 1985). The viral etiology of oral squamous cell papilloma has been regarded as highly questionable (Shafer et al. 1983). Oral condyloma acuminatum or venereal wart is the only one of these three lesions which is constantly considered to be of infectious (viral) origin, and its association with genital condylomas has been discussed (Fiumara 1984; Syrjänen et al. 1984b; Pindborg 1985).

On the other hand, oral wart or verruca vulgaris in the oral cavity is poorly defined. This topic has been included under a variety of headings, such as neoplasias, oral epithelial tumors (Lucas 1984), or oral manifestations of mucosal and dermal lesions (Waldron 1970; Shafer et al. 1983). It is either found under its own subheading or as one of the variants of squamous cell papilloma (Shafer et al. 1983; Lucas 1984). Some authors consider oral verruca vulgaris as a distinct growth of viral origin (Bhaskar 1979; Pindborg 1985). However, the issue is still the subject of some dispute (Hertz 1972).

Focal epithelial hyperplasia, a term introduced in 1965 (Archard et al. 1965), has been previously described as a papilloma, verruca, or multiple polypoid hyperplasia. Only recently data were presented which implicated human papillomavirus (HPV) type 13 in the development of focal epithelial hyperplasia (Pfister et al. 1983; Syrjänen et al. 1984a), a lesion that was already suspected to be caused by papova group viruses (Praetorius-Clausen and Willis 1971; Buchner et al. 1975; Petzoldt and Dennin 1980).

Much of the recent interest in HPV as a disease-inducing agent in humans is due to its suggested associations with epithelial atypias and even with squamous cell carcinomas (zur Hausen 1977; Meisels et al. 1982; Syrjänen 1986). This concept also seems to be pertinent to lesions of the oral cavity, where some dysplastic changes and squamous cell carcinomas have been ascribed to HPV infections, especially those of the high-risk type HPV16. In addition, recent observations have suggested that more than one HPV type could be responsible for the squamous cell lesions in the oral cavity. Indeed, HPV1, 2, 6, 7, 11, 13, 16, and 32 have been found in different types of oral lesions (Petzoldt and Dennin 1980; Lutzner et al. 1982; Pfister et al. 1983; Syrjänen et al. 1984a, 1986; Löning et al. 1985; Naghashfar et al. 1985; Milde and Löning 1986; de Villiers et al. 1986). In many cases, the HPV types are the same as those seen in the genital and laryngeal areas or in the skin.

This chapter summarizes the current evidence available on HPV infections associated with oral squamous cell lesions in humans. Special emphasis will be placed on epidemiological data, morphology, behavior, and detection of HPV antigens as well as HPV DNA by immunohistochemical and recombinant DNA techniques. Whenever feasible, the therapeutic approaches will be briefly reviewed.

2 Solitary Squamous Cell Papilloma

Squamous cell papilloma is a common, benign tumor of the oral epithelium. Most oral pathology texts list it in the category of benign epithelial neoplasia (Abbey et al. 1980; Shafer et al. 1983; Lucas 1984), and some observers raise the question as to whether the lesion is a reaction to injury rather than a true neoplasia (Batsakis 1979). There is no substantial agreement on the frequency of this disease (Knapp 1971; Spouge 1973; Axell 1976). Bhaskar (1963) reported that papillomas represent nearly 8% of all oral tumors in children. Furthermore, Kohn and associates (1963) suggested that these form the most frequent benign neoplasms of the soft palate and uvula. Jones (1965) found a frequency of 7.5% for papillomas among all oral tumors appearing in Irish children.

Papillomas occur at any age, but they are most frequently seen in patients in the 3rd and 5th decades (Greer and Goldman 1974; Abbey et al. 1980). Papillomas are located most frequently on the palatal complex, dorsum, and lateral borders of the tongue and lower lip (Abbey et al. 1980). In an extensive review of 464 oral papillomas, Abbey and colleagues found a slight male predominance among the reported cases (53.8%). Whites comprised the majority of patients, 87.5% (Abbey et al. 1980).

2.1 Clinical Features

Squamous cell papillomas vary in diameter from 2 mm upwards. Most lesions reported by Abbey and associates (1980) were less than 1 cm in size. Papilloma is an exophytic growth that is either broad-based or pedunculated. The surface may show small fingerlike projections, resulting in a lesion with a rough or cauliflower-like verrucous surface (Fig. 1). The color of the lesions varies from white to pink

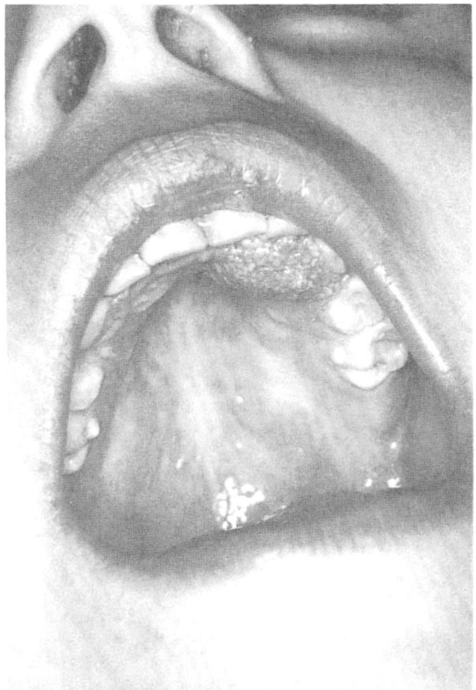

A

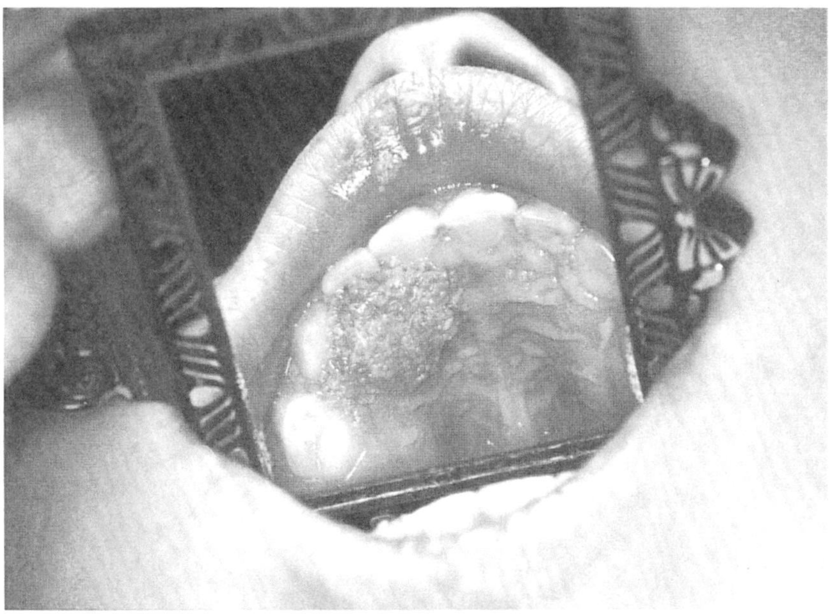

B

Fig. 1 A–D. Oral warts in a 4.5-year-old, healthy, Caucasian boy. The patient was referred for surgical treatment with growths of approximately 9 months duration. The largest lesion, measuring 2×2 cm, extends from the palatal to the gingival mucosa at the region of teeth 21–22 (**A** and **B** viewed by mirror). Two additional small lesions (2×2 mm) are located in the lower *(arrowhead)* and upper *(arrow)* lips (**C, D**). By courtesy of Drs MA Lamberg and A Tasanen

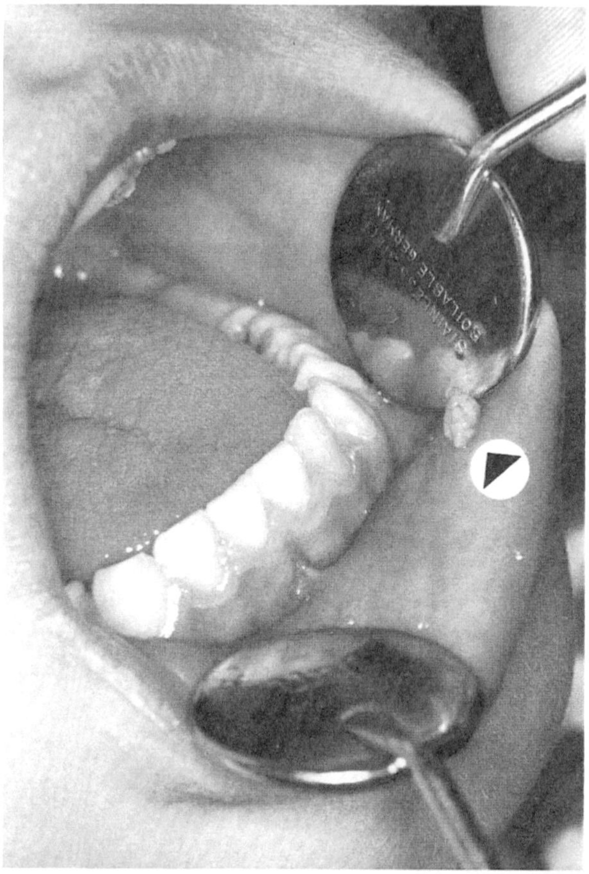

Fig. 1 C

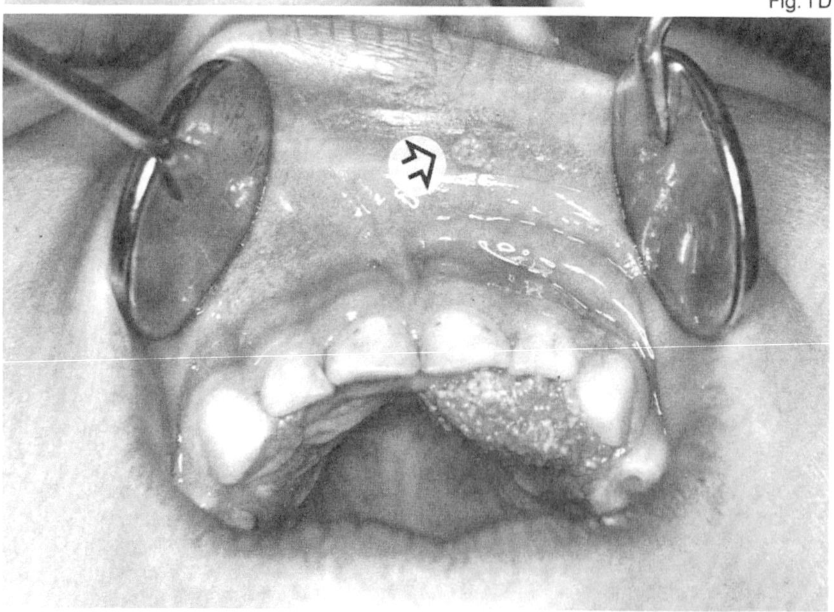

Fig. 1 D

depending on the degree of keratinization and vascularization (Shafer et al. 1983; Lucas 1984). Multiple papillomas are rare (Abbey et al. 1980). Recurrence of oral papillomas is exceptional.

2.2 Histological Features

Under light microscopy, oral papillomas are very similar to squamous cell papillomas at other locations. They consist of thin, long, fingerlike projections of squamous epithelium supported by a connective tissue core. The epithelium is usually hyperkeratotic and shows acanthosis (Shafer et al. 1983; Lucas 1984). Most observers emphasize the benign morphology of the papilloma, although some, notably Shklar (1965), regard this lesion as potentially malignant. Greer and Goldman (1974) examined 110 lesions and could not detect morphological abnormalities of note in any of them. Syrjänen and colleagues (1984b) found, however, changes consistent with mild dysplasia in 20% (14/70) of squamous cell papillomas. This is in agreement with the data reported by others (MacDonald and Rennie 1975; Abbey et al. 1980).

Slight inflammation in the underlying stroma is not uncommon, but a dense inflammatory infiltrate is exceptional. In cutaneous HPV lesions (e.g., common warts) the inflammatory infiltrate has been regarded as a host immune reaction. Analyses of the cellular make-up of the infiltrates in oral papillomas revealed a lymphocyte predominance (Abbey et al. 1980; Syrjänen et al. 1984b), B-lymphocytes far outnumbering the T-lymphocytes (Syrjänen et al. 1984b). The proportion of B cells appeared to be directly related to the density of the infiltrate. Becker and colleagues (1985) found more helper T-lymphocytes and Langerhans' cells in six oral papillomas than in seven samples of normal mucosa.

2.3 Evidence for Involvement of Human Papillomaviruses

The viral etiology of oral papillomatous lesions in animals has been well-established (Shope 1962; Cheville and Olson 1964), but until recently this issue received little attention in human oral papillomas. Although Frithiof and Wersäll demonstrated viral particles closely resembling HPV in oral papillomas in 1967, doubt about the viral etiology of these lesions has been expressed (Greer and Goldman 1974; Shafer et al. 1983). Wysocki and Hardie (1979) demonstrated typical intranuclear viral inclusions in six of ten lesions diagnosed as verruca vulgaris but in none of ten oral papillomas. Recently, the peroxidase-antiperoxidase immunocytochemical (IP-PAP) technique has been used to detect HPV capsid antigens in oral papillomas. Jenson et al. (1982a) reported that three of five oral papillomas were positive for HPV common structural antigens. Syrjänen and coworkers in 1983 confirmed these results by finding five HPV-positive cases among nine lesions analyzed. Eight of the lesions also showed koilocytosis, the specific cytopathic effect of HPV (Syrjänen et al. 1983a).

These preliminary reports (Syrjänen et al. 1983a, b) were followed by an analysis of 70 additional papillomas (Syrjänen et al. 1984b). In this latter study, 41% of the papillomas stained positively for HPV antigens. Recently, a series of 29 oral papillomas were described, all of which failed, however, to show any staining for HPV capsid antigens (Mincer et al. 1985). In yet another study, all five papillomas were shown to express HPV antigens (Löning et al. 1985). This discrepancy might be due to the different antisera used by these investigators. It should be emphasized that the failure to demonstrate viral antigens by the IP-PAP method does not necessarily disprove the role of the virus as an etiological agent. Additional confirmatory evidence on HPV etiology of oral papillomas was provided recently by the in situ DNA hybridization technique (Syrjänen et al. 1986). HPV6 DNA was revealed in four oral papillomas, in formalin-fixed, paraffin-embedded sections. Quite recently de Villiers and coworkers (1986) found HPV11 in two of the three papillomas analyzed by Southern blot method.

3 Multiple Oral Papillomas

Multiple oral papillomas are uncommon except in association with certain rare syndromes, such as dermal hypoplasia syndrome (Gorlin et al. 1963), Cowden's syndrome (Greer et al. 1976; Starink 1984), and nevus unius lateris (Brown and Gorlin 1960; Kelley et al. 1972). Extensive oral papillomatosis has also been reported in Down's syndrome (Eversole and Sorenson 1974). The papilloma-like, warty dyskeratoma affects the skin of the scalp or neck. It has rarely been reported on oral mucosa. Histologically, the lesions are identical with those of Darier's disease. The etiology of these syndromes is unknown, although a genetic origin has been suggested. No data on the possible associations of these papillary lesions with HPV are available as yet.

4 Ductal Papilloma of the Salivary Glands

Papillomas arising from the excretory ducts of the major and minor salivary glands are quite uncommon. Three different forms of ductal papillomas have been described: simple ductal papilloma, inverted ductal papilloma, and sialadenoma papilliferum (White et al. 1982; Shafer et al. 1983).

Several cases of inverted papilloma in the oral cavity have been described. However, the site of origin, i.e., whether in the oral squamous epithelium or in the ductal epithelium of the salivary glands, has not always been clearly indicated. Recently, three cases of ductal papilloma were reported which on histological examination resembled the inverted papillomas of the paranasal sinus/nasal cavity. The latter is a well-known entity which was recently shown to be associated with HPV infections (Syrjänen et al. 1983d; Eavey 1985). The possibility that papillomas of the salivary gland ductal epithelium could also be caused by HPV requires further investigation.

5 Condyloma Acuminatum

Condyloma acuminatum (venereal wart) is generally regarded as a sexually transmitted disease affecting the skin and mucous membranes of the anogenital tract. Recent data suggest that genital condyloma is common in young, sexually active women with a prevalence of as high as 5%-10% (Ludwig et al. 1981; Meisels et al. 1982; Syrjänen 1986). Although orogenital sex is currently known to be a fairly common practice, venereal condylomas transmitted into the oral cavity have been infrequently reported in the literature (Knapp and Uohara 1967; Doyle et al. 1968; Summers and Booth 1974; Shaffer et al. 1980; Choukas and Toto 1982). As early as 1901, Heidingsfield reported a case of a 'puella publica' (a prostitute) believed to have acquired condylomata of the tongue as a result of 'coitus illegitimus' (1901). Additionally, autoinoculation must also be considered as a possible route of infection. The role of oral mucosal trauma has been emphasized as a possible site of entry of HPV infection.

5.1 Clinical Features

The first documented case of an oral condyloma acuminatum was reported by Knapp and Uohara in 1967. Subsequently, only a few additional reports have been published, which are summarized in Table 1. At onset, oral condylomas usually present with multiple, small, white or pink nodules, which proliferate and coalesce to form soft, sessile or pedunculated, papillary growths (Knapp and Uohara 1967; Doyle et al. 1968; Shaffer et al. 1980). Generally, the surface contour is more cauliflowerlike than papillomatous. The lesions are scattered or diffuse, involving the tongue, buccal mucosa, palate, lips, or alveolar ridge. Five of the cases of oral condylomas summarized in Table 1 had anogenital condylomata.

The reported cases are too few to reach conclusions about sex and age predilection. The lesions are usually asymptomatic, and regression from a few months to several years later has been observed (Table 1). Recurrences may occur within a few weeks up to several months. It has been noticed that even in the same patient some lesions may continue to enlarge, while others regress and disappear completely (Swan et al. 1981).

5.2 Histological Features

Histologically, the stratified squamous epithelium is arranged in deep papillary folds and projections, which make up the verrucoid lesion. The surface of the epithelium may be nonkeratinized but generally shows parakeratosis. There is marked acanthosis with thickening and elongation of the rete pegs. In some instances the epithelial organization is disturbed enough to be misinterpreted as a carcinoma. Vacuolization of the cytoplasm of the epithelial cells with deeply hyperchromatic round or ovoid nuclei is a characteristic finding (Praetorius-Clausen 1972). Occasional mitoses can be observed in the basal region (Swan et al. 1981). The supporting connective tissue is usually edematous, with dilated capil-

Table 1. Data concerning oral condylomas

Reference	Origin	Sex	Age	Site of lesion	Duration of lesion	Recurrence time	Additional lesion(s)	Evidence of HPV
Knapp and Uohara 1967	Caucasian	M	18	palate, max gingiva	many years	8 months	-	-
Doyle et al. 1968	NG	F	84	max, alveolar crest	3-4 months	1 month	-	-
Summers and Booth 1974	Australian	F	35	upper lip, cheeks	20 years	NG	anogenital	EM –
Gysland et al. 1976	NG	F	NG	palate	NG	NG	NG	EM +
McClatchery et al. 1979	American	F	22	lower lip	many months	NG	anogenital (1 year)	-
Doran 1980	Caucasian	M	25	labial, gingiva	13 months	10 months	-	-
Shaffer et al. 1980	Caucasian	M	22	palate	2 months	8 weeks	-	EM +
Swan et al. 1981	Caucasian	M	18	tongue	3 months	2-4 weeks	penile	EM –
Choukas and Toto 1982	NG	(6 cases)	NG	NG	NG	NG	NG	EM +
Lutzner et al. 1982	Caucasian	M	49	mand, gingiva	several months	several months	skin, anogenital	IP-PAP +
Jenson et al. 1982a, b	American 4 ×	M	23-31	tongue	NG	NG	NG	IP-PAP 2/5
	American	M	38	palate	NG	NG	anal mucosa	IP-PAP + 1/5
Ashiru et al. 1983	Nigerian	F	6	both lips	several months	-	-	-
Syrjänen et al. 1983a-d	Caucasians	NG		NG	NG	NG	NG	IP-PAP + 2/2
	Caucasian	M	68	NG	NG	NG	NG	IP-PAP –
	Caucasian	F	59	NG	NG	NG	NG	IP-PAP +
Jin and Toto 1984	NG	NG	NG	NG	NG	NG	NG	IP-PAP + 8/11
George and Farman 1984	American	NG	NG	palate	some months	NG	-	EM +
Syrjänen et al. 1984a, b	Caucasians	(16 cases)	NG	NG	NG	NG	NG	IP-PAP + 12/16
Mincer et al. 1985	NG	(3 cases)	NG	NG	NG	NG	NG	IP-PAP –
Naghashfar et al. 1985	NG	(10 cases)	15-30	NG	NG	NG	NG	HPV DNA + 6/10

NG, not given

laries and chronic inflammatory cell infiltration (Knapp and Uohara 1967; Praetorius-Clausen 1972; Shaffer et al. 1980).

The intensity and composition of the inflammatory infiltrate has only recently been analyzed (Syrjänen et al. 1984b). In the 16 condyloma lesions studied, the inflammation was mostly mild; it was moderate in four and marked in only one sample. The cellular composition of the infiltrate was predominately B-lymphocytes (66%). An equal distribution of both T-lymphocytes and mononuclear phagocyte system (MPS) cells was encountered. Five of the specimens showed dysplastic changes classified in three cases as mild and in two cases as moderate in severity (Syrjänen et al. 1984b).

5.3 Electron Microscopic Findings

The first report on the presence of HPV particles in an oral condylomata was published in 1976 by Gysland and coworkers. They found viral particles confined to the nuclei of cells in the layers above the stratum spinosum. The number of infected cells increased toward the surface of the lesion. Nuclei containing virus particles showed peripheral margination of chromatin and an intact nuclear envelope. These nuclei were separated from the peripheral cytoplasm by wide empty zones. The cytoplasm of the affected cells was finely granular and contained regular organelles. These ultrastructural features have been subsequently confirmed in other reports (Shaffer et al. 1980; Choukas and Toto 1982; George and Farman 1984).

5.4 Evidence for Involvement of Human Papillomaviruses

It is generally agreed that condyloma acuminatum in the oral cavity is an infectious disease (Shafer et al. 1983; Pindborg 1985). Evidence for HPV etiology has been provided by detection of viral particles with electron microscopy (Gysland et al. 1976; Shaffer et al. 1980) as well as by demonstration of HPV common structural antigens by immunocytochemical means (Jenson et al. 1982a, b; Syrjänen et al. 1983b, 1984b). The largest series analyzed so far with the IP-PAP technique for HPV has been reported by Syrjänen and coworkers (1983b, 1984b). The data compiled from these two reports indicate that HPV antigen expression is detectable in 15/20 condyloma lesions studied. These results are in agreement with those of other workers (Jenson et al. 1982a, b; Jin and Toto 1984). However, Mincer et al. (1985) could not detect HPV antigens in the three cases studied.

Thus far, HPV DNA has not been found by the DNA hybridization technique. Viral DNA extraction was performed once by Lutzner and coworkers (1982), but this was not successful, most probably due to the small quantity of viral DNA available. In due course it is to be expected that the HPV types found in genital condylomas will be identified in oral condylomas by DNA hybridization procedures.

5.5 Treatment

Topical podophyllin has been widely used in the treatment of oral condyloma without any major benefit. Surgical excision or eradication by electrocautery has been practised, although there are recurrences. As condyloma acuminatum is regarded as a sexually transmitted disease, the sexual partner(s) of the proband should also be examined (Fiumara 1984).

6 Verruca Vulgaris (Oral Warts)

Verruca vulgaris or the common wart is the most prevalent HPV lesion of the skin. The current literature offers conflicting views regarding the existence of verruca vulgaris on oral mucosa (Hertz 1972; Shafer et al. 1983; Lucas 1984). The histological and clinical similarities among verruca vulgaris, squamous cell papilloma, and condyloma acuminatum are well-recognized, to the extent that the validity of differential diagnosis among these lesions has been seriously questioned (Waldron 1970; Hertz 1972). The same authors have also questioned the separation of these lesions on etiological grounds, until definitive evidence concerning their causes is available (Waldron 1970; Hertz 1972; Wysocki and Hardie 1979). The majority of oral warts is seen in children (Fig. 2), who also have warts on their fingers (Shafer et al. 1983; Pindborg 1985). Previously, the oral mucosa has been regarded as an unusual site of warts, but recently, oral involvement has become increasingly com-

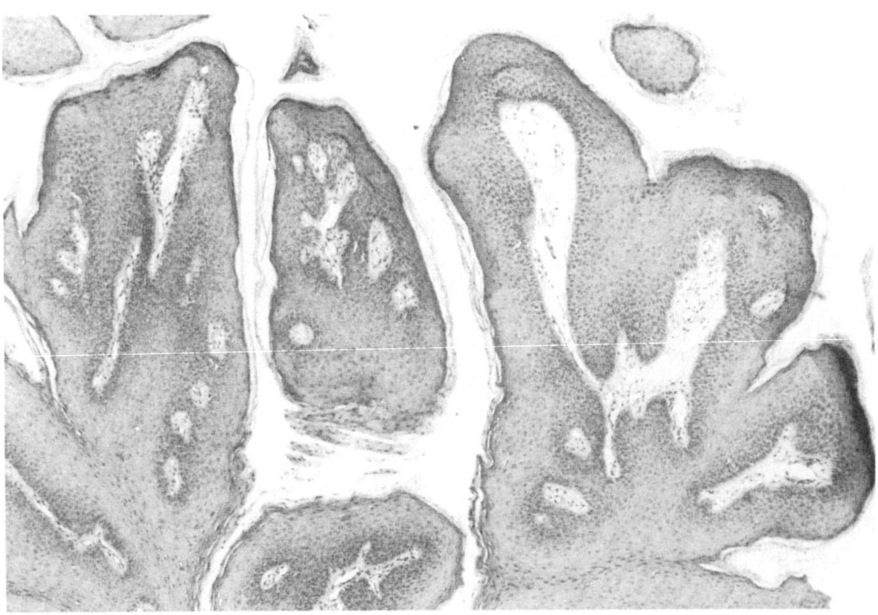

Fig. 2A

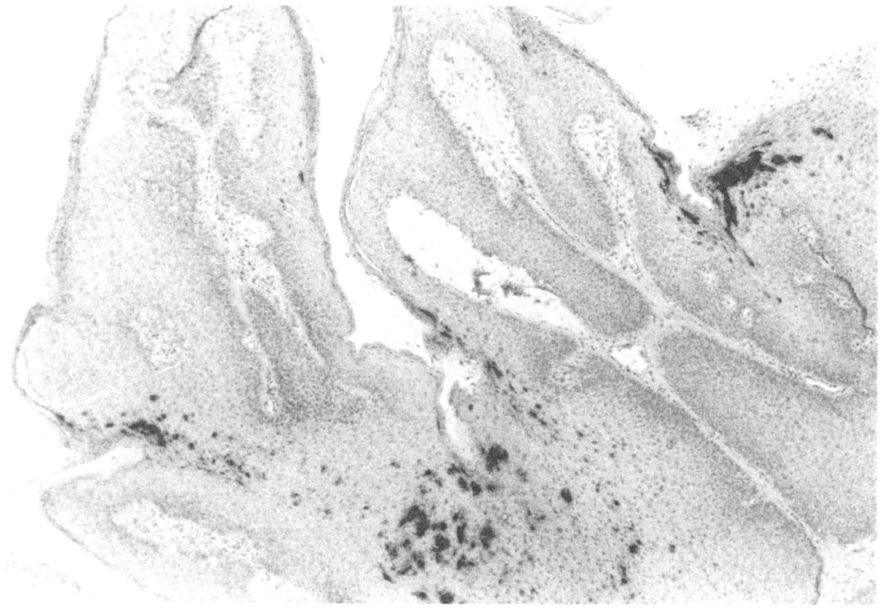

B

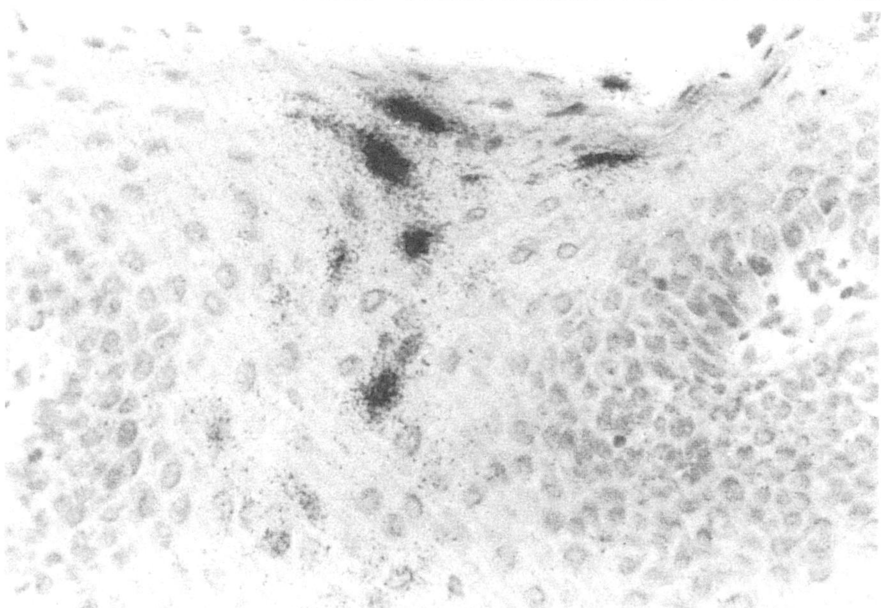

C

Fig. 2 A. A characteristic squamous cell papilloma in the lingual mucosa of a 45-year-old Caucasian man. H & E, original magnification ×100. **B** The same lesion subjected to in situ DNA hybridization with a ^{35}S-labelled DNA probe for HPV11. The *black spots* localized within the nuclei of the epithelial cells demonstrate the presence of viral DNA in these cells. Note the localization of HPV11 DNA close to the surface, while the deeper layers are devoid of HPV DNA-positive cells. Hematoxylin counterstain, original magnification ×100. **C** Medium-power detail of the HPV11 DNA-positive cells. The black silver grains are superimposed on the nuclei of a few superficial cells, indicating the localization of HPV11 DNA. Hematoxylin counterstain, original magnification ×400

mon. Adults have replaced children as the usual patients with intraoral warts (McCarthy and Shklar 1964; Fiumara 1984), although it is possible that condylomata acuminatum may be found among these lesions.

6.1 Clinical Features

Clinically, oral warts usually appear as firm, whitish, sessile, papillomatous, rough-surfaced lesions. According to most authors they are indistinguishable from squamous cell papillomas. Warts appear abruptly in the oral cavity, and their growth is rapid. They may be located at any site, although the lips are preferred (Table 2).

It has been emphasized that the diagnosis of oral verruca should be preserved for lesions showing clinical and histological characteristics of verruca vulgaris of

Table 2. Data concerning oral verruca vulgaris lesions

Reference	Origin	M/F	Age	Site of lesion	Duration of lesion	Recurrence time	Evidence of HPV
Orlean and Ladow 1960	American	M	41	lip, buccal mucosa	6 months	no	Morphology
Hertz 1972	Caucasian	F	56	palate, skin graft	6 months	no	EM+
Wysocki and Hardie 1979	10 cases			NG	NG	NG	6/10 EM+
Lutzner et al. 1982	Caucasian	M	21	hard palate	1 year	NG	EM+, IF+ for HPV2 Southern blot HPV2a
Jenson et al. 1982a, b	29 cases	25/4		upper lip (7 cases)	NG	NG	IP-PAP+
				lower lip (12 cases)	NG	NG	IP-PAP+ (8 cases)
				others (10 cases)	NG	NG	IP-PAP+ (3 cases)
Jin and Toto 1984	11 cases	M	35	buccal mucosa	NG	NG	IP-PAP+
		F	62	upper lip	NG	NG	IP-PAP+
		F	33	lower lip	NG	NG	IP-PAP+
		M	32	upper lip	NG	NG	IP-PAP+
		M	14	lower lip	NG	NG	IP-PAP+
		M	38	anterior palate	NG	NG	IP-PAP+
		F	21	upper lip	NG	NG	IP-PAP+
		M	23	lingual frenulum	NG	NG	IP-PAP+
		F	30	hard palate	NG	NG	IP-PAP−
		M	37	lower lip	NG	NG	IP-PAP−
		F	23	upper lip	NG	NG	IP-PAP−
Fiumara 1984	6 cases, homosexuals						
		M	32	lower lip	5 months		(AIDS)
		M	37	lower lip	9 months	no	NG
		M	27	buccal mucosa	NG	NG	NG
				lower lip	6 weeks	no	NG
		M	31	palate	4 months	NG	NG
		M	44	tip of tongue	3 months	no	NG
		F	22	tip of tongue	6 months	no	partner had penile wart
Mincer et al. 1985	11 cases						5/11 IP-PAP+

NG, not given

the skin. Furthermore, it has been stated that verruca vulgaris could be found in the oral cavity only when skin was present, e.g., after a skin graft (Hertz 1972), and when viral origin could be demonstrated. It is thought that the natural history of oral warts is similar to that of skin warts, that is, self-limiting, benign tumors. Indeed the rate of spontaneous regression is high and, as in cutaneous warts, increases with the age of the lesions.

6.2 Histological Features

Under light microscopy, oral warts are characterized by a papillomatous surface with conspicuous hyperkeratinization. The stratum granulosum is often most pronounced in the grooves between the papillomatous elevations, which are formed around the thin, elongated connective tissue papillae. The rete ridges are elongated and very characteristic in their inward bending at the margin of the verruca, pointing radially toward the center. Inclusion bodies are extremely rare in lesions from oral mucosa (Praetorius-Clausen 1972; Wysocki and Hardie 1979). These inclusion bodies, previously regarded as viral, were recently suggested to be degenerative in nature (Hertz 1972).

6.3 Evidence for Involvement of Human Papillomaviruses

Due to the recent advances in recombinant DNA techniques, different HPV types have also been isolated in oral warty lesions. Both genital tract papillomaviruses, i.e., HPV6 and HPV11, and the skin types, e.g., HPV2, have been identified in oral warts (Lutzer et al. 1982). In the future it will be possible to identify the type of HPV in oral lesions retrospectively, and it may prove possible to trace the origin of the infection (e.g., whether genital or cutaneous) using the in situ DNA hybridization technique.

6.4 Treatment

The management of warts in the oral cavity has been exhaustively reviewed by Fiumara (1984). He concluded that oral warts respond well to 20% podophyllin in ethanol applied at weekly intervals. Light desiccation with curettage is perhaps the most commonly used method for wart removal. Warts of the gums seem to respond most satisfactorily to light application of liquid nitrogen. Although the carbon dioxide laser is widely and successfully used in the treatment of condylomata acuminata in the genital tract, it has not been employed in the oral cavity so far.

7 Oral Florid Papillomatosis (Verrucous Hyperplasia)

Rock and Fisher (1960) introduced the term oral florid papillomatosis (OFP) in their detailed description of three cases. Later Wechsler and Fisher (1962) added another case and suggested the benign nature of the lesion. Samitz and Weinberg (1963) reported one case in which they could not find any malignant changes during the 7-year follow-up period. However, Samitz and associates published in 1967 an additional case, which seemed to be the same patient as reported earlier, who had developed a squamous cell carcinoma at the site of OFP.

OFP is a rare disease of the oral cavity characterized by multiple, exuberant, confluent, cauliflowerlike growths (Rock and Fisher 1960; McClendon 1985). Recently, the name verrucous hyperplasia was suggested instead of OFP (Shear and Pindborg 1980) to differentiate it from verrucous carcinoma, which resembles it clinically and histologically but is a true malignant process. However, there are authors who believe that verrucous carcinoma and OFP may represent the same lesion, advocating the use of the term verrucous carcinoma for both (Shafer et al. 1983).

7.1 Clinical Features

OFP occurs mostly in elderly people. The majority of the reported cases is found in the cheek or gingival/alveolar mucosa, followed by the tongue, floor of the mouth, labial mucosa, and palate (Rock and Fisher 1960; Pindborg 1985). In a series of 68 patients, most of them smokers, with verrucous hyperplasia, two clinical patterns were described by Shear and Pindborg (1980). One, called a 'sharp' variety, comprised long, narrow, heavily keratinized verrucous processes. The other type, referred to as 'blunt' variety, consisted of verrucous processes which were broader, flatter, and not heavily keratinized. According to these authors, verrucous hyperplasia and verrucous carcinoma are indistinguishable clinically, and verrucous carcinoma may develop from verrucous hyperplasia at a later stage. Furthermore, the clinical association with leukoplakia was found to be significant, and evidence was presented that an untreated leukoplakia could develop into a verrucous hyperplasia and/or a verrucous carcinoma (Shear and Pindborg 1980).

7.2 Histological Features

OFP shows a papillomatous configuration with elongated branching papillary processes covered by acanthotic and partially keratinized epithelium. Vacuolization has been seen in the superficial layers, and the rete pegs are thickened and elongated. The basal cells are poorly oriented and their nuclei somewhat hyperchromatic (Rock and Fisher 1960; Wechsler and Fisher 1962; Shear and Pindborg 1980). The submucosal tissue contains a dense inflammatory cell infiltrate composed of lymphocytes. Most of the cases reported so far were associated with epithelial dysplasia (Shear and Pindborg 1980).

7.3 Evidence for Involvement of Human Papillomaviruses

Some evidence implicating the possible viral (HPV) nature of these lesions has been presented (Wechsler and Fisher 1962). Their similarity to canine oral papillomatosis, a known papillomavirus-induced lesion has been emphasized. Ullman (1923) produced comparable lesions in the skin of a child with cell-free filtrate prepared from a laryngeal lesion. So far, HPV particles or antigens have not been demonstrated in OFP lesions. Similarly, no data are available on the existence of HPV DNA in these lesions. It remains to be demonstrated whether one or more of the HPV types found in the other oral lesions or those of the larynx (including the new type HPV30 identified in a laryngeal carcinoma) are associated with OFP.

7.4 Treatment

Attempts have been made to treat OFP in several ways, including surgery and radiotherapy or a combination of the two. However, malignant changes have been observed in lesions of patients receiving X-ray therapy. Local application of various ointments has been used without any significant benefit (Samitz et al. 1967).

8 Verrucous Carcinoma

In 1948, Ackerman first used the term verrucous carcinoma to describe a variant of squamous cell carcinoma in the oral cavity. Since then this lesion, known as Ackerman's tumor, has been recognized as a locally invasive, nonmetastasizing squamous cell carcinoma with a characteristic gross and microscopic appearance, occurring in several locations in the oral cavity, lips and larynx as well as in the genital area (Ackerman 1948; Ackerman and McGavran 1958; Goethals et al. 1963; Jacobson and Shear 1972; Shafer 1972; Väyrynen et al. 1981; Bohmfalk and Zallen 1982). The oral cavity, however, is by far the most common location of this tumor (Kraus and Perez-Mesa 1966; Pindborg 1985). The tumor occurs chiefly in elderly patients, mostly over the age of 60 years, and more often in males than in females. A racial preponderance of this tumor in whites is well-established. A review of the reported cases shows that the site of verrucous carcinoma in the oral cavity is most frequently the cheek, followed by the alveolar or gingival mucosa (Jacobson and Shear 1972; Medina et al. 1984).

A striking relationship between tobacco chewing or snuff taking and verrucous carcinoma has been documented. In a recent study in which 104 patients with verrucous carcinoma were reviewed, 84% of the patients used tobacco products, most of them being cigarette smokers (42%), followed by snuff dippers (32%). However, 16% of patients denied any tobacco use (Medina et al. 1984). A frequent occurrence of verrucous carcinoma in the oral mucosa is reported in Papua-New Guinea with labial commissure involvement. This has been ascribed to repeated application of slaked lime in connection with betel chewing and smoking (Cooke 1969).

Recently, foci of less-differentiated squamous cell carcinomas were found within oral verrucous carcinomas in some 20% of cases (Medina et al. 1984). Such lesions were also observed by Batsakis et al. (1982) in three verrucous lesions of the larynx. Verrucous lesions tend to recur in the form of less differentiated spindle cell carcinomas, especially after irradiation (Väyrynen et al. 1981).

8.1 Clinical Features

Clinically, verrucous carcinoma presents as a very slowly enlarging, warty growth. Infrequently the lesion may invade the adjacent tissues, including the bone. The gross appearance of verrucous carcinoma is characterized by an exophytic, papillary mass with folds and deep clefts (Ackerman 1948; Jacobson and Shear 1972; Pindborg 1985).

8.2 Histological Features

Verrucous carcinoma is defined by the following histological characteristics: a warty, extensively keratinized surface, a sharply circumscribed deep margin, and a pushing rather than an infiltrating type of advancing margin. Bulbous, well-oriented rete ridges with central degeneration composed of proliferating and keratinizing squamous epithelium are seen. The epithelium is well-differentiated and shows little mitotic activity, pleomorphism, or hyperchromatism. Inflammatory cell infiltrate is frequently seen in the underlying stroma (Ackerman 1948; Goethals et al. 1963; Kraus and Perez-Mesa 1966; Medina et al. 1984).

8.3 Evidence for Involvement of Human Papillomaviruses

A viral etiology for laryngeal verrucous carcinomas has been suspected for many years. Quite recently, a new HPV type (HPV30) was identified in a laryngeal carcinoma (Kahn et al. 1985). Many similarities exist between laryngeal and oral verrucous carcinoma. It remains to be seen whether HPV etiology with tobacco as a cofactor can be established in oral lesions. Undoubtedly, this conflicting area with confusing literature on verrucous carcinoma, OFP, and verrucous hyperplasia might be substantially clarified by the evidence of HPV involvement in these lesions.

8.4 Treatment

Local surgery, radiotherapy, or local application with different chemicals have been used in the treatment of these lesions. The management of these lesions was recently reviewed, and surgery was advocated as the treatment of choice (Medina et al. 1984).

9 Focal Epithelial Hyperplasia

The term focal epithelial hyperplasia (FEH) was introduced by Archard, Heck, Stanley, and Gallup in 1965 to describe multiple nodular elevations of the oral mucosa observed among American Indians in the United States of America and Brazil and in an Eskimo boy from Alaska. Since then, many additional cases have been reported (Table 3). Furthermore, it is apparent that some authors had previously described a similar, if not identical, clinical entity (Helms 1894; Stern 1922; Estrada 1956, 1960; Soneira and Fonseca 1964).

Most of the reported FEH cases have come from various parts of South and Central America, Greenland, and Alaska, but individual cases have also been reported from other countries (Table 3). Only a few studies on the prevalence of FEH have been published (Table 3). The highest prevalence (33.8%) has been found in Indian children in Venezuela (Soneira and Fonseca 1964). Other reports, however, concerning Indians in Columbia (Estrada 1956; Gomez et al. 1969), Brazil, El Salvador, Guatemala (Witkop and Niswander 1965), and Paraguay (Fischman 1969) have shown the prevalence of FEH to be less than 3.5%. Studies of the Eskimo populations have shown FEH to be remarkably common, the prevalence varying from 7% to 36% (Praetorius-Clausen et al. 1970; Praetorius-Clausen 1972, 1973). The highest figures are found among the Greenlandic Eskimos on the east coast, where the population is only insignificantly mixed with Caucasians (Praetorius-Clausen 1973). In contrast, the disorder is rare in the Caucasian population. Praetorius-Clausen found only one case among 322 Caucasian Danes living in Greenland, i.e., a prevalence of 0.3%. Notably, he did not find a single case of FEH among 3000 conscript Danish soldiers living in Denmark (Praetorius-Clausen 1973). However, in a study on oral lesions in an adult Swedish population of 20333 subjects, Axell (1976) found a higher prevalence of FEH (0.1%) than expected. During the last few years, this lesion has been reported more frequently in other countries as well (Petzoldt and Dennin 1980; Petzoldt et al. 1982; Pfister et al. 1983; Syrjänen et al. 1984a).

A familial history of FEH has been suggested by several authors. Archard and coworkers (1965) described two families in which more than one child was affected. Perriman and Uthman (1971) reported that in three of the seven cases studied, another sibling was affected as well. Gomez and associates (1969) described seven children in one family, Schock and Wood (1969) reported four affected members in one family, and Buchner and Mass (1973) reported FEH lesions in four children of the same family.

9.1 Clinical Features

FEH appears as multiple, soft, flat or rounded, slightly elevated nodules. The lesions are asymptomatic, their color ranging from pale to normal as compared to the adjacent mucosa. When the involved mucosa is stretched, the lesions tend to disappear. The surface of the papules is smooth, with or without weblike markings. Their size is usually between 1 and 5 mm (Archard et al. 1965; Gomez et al. 1969; Praetorius-Clausen 1972).

Table 3. Data concerning focal epithelial hyperplasia (FEH)

Reference	Origin	Number examined	Age range	Number of FEH found	Duration of lesions	Site of lesions	Detection of HPV	Specific notes
Estrada 1956	Indian	75	all ages	2	–	lower lip	–	–
Reyes 1962	Indian	1	–	1	–	–	–	–
Soneira and Fonseca 1964	Indian	160	4–18	54	–	oral mucosa	–	warts on face (2)
Archard et al. 1965	Indian, Eskimo	19	9.5	19	1–3 years	lower lip	–	1 with warts on face
Bergenholtz 1965	Scandinavian	2	–	2	–	oral mucosa	–	–
Witkop and Niswander 1965	Indian	260	all ages	7	–	–	–	–
	Ladino	1127	5–15	2	–	–	–	–
	Indian	64	6–15	2	–	–	–	–
Hettwer and Rodgers 1966	Polynesian	1	9	1	–	oral mucosa	–	–
Sewerin 1968	Caucasian	1	15	1	2–3 years	oral mucosa	–	–
Waldman and Shelton 1968	Caucasian	1	56	1	–	lower lip, mucosa	–	–
Gomez et al. 1969	Half-breed	11	6–52	7	–	lower labial mucosa	–	–
	Caucasian	80	2–13	0	–	–	–	–
Fischman 1969	Paraquay	8569	all ages	1	–	–	–	–
Decker and de Guzman 1969	Mestizo	4	9–49	4	–	buccal, labial mucosa	–	–
Schock and Wood 1969	Indian	1	11	1	–	lower lip, cheek	–	two sisters, mother
Praetorius-Clausen 1969	Eskimo	10	5–46	10	–	–		–
	Caucasian	1	33	1	–	–		–
Praetorius-Clausen et al. 1970	Eskimo	460	2–79	89	–	oral mucosa	–	–
Anderson 1971	Africa, white	1	53	1	20 years	lips, mucosal surfaces		1 sister
Perriman and Uthman 1971	Iraq	7	6–20	7	1–3 years	lips	–	sisters
El-Khashab and Abd-El-Aziz 1971	Egyptian	1	35	1	2 years	lower, upper lip		–
Jarvis and Gorlin 1972	Eskimo	1571	all ages	169	–	–		–
Hanks et al. 1972	Indian	1	25	1	20 years	labial mucosa	EM+	child
Buchner and Mass 1973	Israel	11	4–23	4	–	lower, upper mucosa	EM–	7/11 skin warts
Praetorius-Clausen 1973	Eskimos	2680	all ages	453	–	–		–
Orfanos et al. 1974	Turkey	3	6–13	3	5 months–3 year	cheeks	–	sisters of 2

Borghelli et al. 1975	Argentina	5	10–27	5	6 months–17 years	lower lip, oral mucosa		–
Wallace 1976	Black, USA	1	16	1	1 year	mucosa of cheek		–
van Wyk 1977	South Africa	1	76	1	–	lingual mucosa	EM+	–
	South Africa	18	6–15	18	–	lower lip	EM+ 6/13	
Thomsson and Hammarström 1978	Turkey	1	12	1	2 years	lower lip	–	–
Knoth and Boepple 1978	Turkey	3	32–52	3	6 months–5 years	lips	–	1 recurrence
Goodfellow and Calvert 1979	West Indian	1	26	1	2 years	lower lip	–	–
Acevedo et al. 1981	Nicaragua	1	12	1	5 years	lip, cheek, tongue	–	1 sibling
Axell et al. 1981	Sweden	20333	16–79	17	–	tongue	EM–	–
Kuhlwein et al. 1981	Turkey	1	6	1	1 year	lips, oral mucosa	EM+	–
Lutzner et al. 1982	Algerian	1	10	1	–	labial mucosa	EM+, IP-PAP+	brothers
	Moroccan	1	6	1	–	buccal mucosa	EM+, IP-PAP+	skin warts
Petzoldt et al. 1982	Turkey, Germany	3	7–30	3	1.5–30 years	oral mucosa	HPV1 in 1 case	mother, father
Pfister et al. 1983	Turkey	1	14	1	3 years	lips, oral mucosa	HPV13	warts on hand
Sawyer et al. 1983	Nigerian	3	8–10	3	4 weeks	lip, oral mucosa	–	cousins
Syrjänen et al. 1984a	Finnish	1	33	1	2 years	buccal mucosa	HPV13A	–
Lang et al. 1984	Turkey	1	11	1	15 months	buccal mucosa	HPV13, 18	–
Bendelac et al. 1984	Turkey	1	67	1	–	–	EM–	
Grung and Bang 1984	Caucasian	6	43–52	6	–	–	–	–
Hallmon et al. 1985	Mexican	1	12	1	18 months	buccal mucosa	–	–
Praetorius et al. 1985	Eskimo	12	5–72	12	–	oral mucosa	IP-PAP+ 8/12	–
	Caucasian	8	5–72	8	–	oral mucosa	IP-PAP+ 8/8	

Although seen most commonly in the lower lip, FEH lesions sometimes extend to the vermilion border, the next most frequent sites being the buccal mucosa, corner of the mouth, and the upper lip (Table 3). The tongue, gingiva, and anterior pillar of fauces are infrequently affected. In Eskimos and adult Swedes, however, more than 50% and 91% of the lesions, respectively, are located on the tongue (Praetorius-Clausen et al. 1970; Axell et al. 1981). The lesions may persist for several years, and they tend to regress spontaneously. Occasionally the lesions recur. FEH appears to occur predominantly in children less than 18 years of age, although an increasing number of adults with this lesion have been described recently (Praetorius-Clausen 1973; Petzoldt and Dennin 1980; Pfister et al. 1983; Syrjänen et al. 1984a).

9.2 Histological Features

The most typical morphological features of FEH consist of localized nodular elevations in the oral epithelium. A varying degree of acanthosis or thickening of the spinous cell layer and mild parakeratosis are frequently seen. The rete pegs show thickening, elongation, and fusion by horizontal outgrowth. Nuclear degeneration with swelling of the cells as well as the presence of intranuclear inclusion bodies have been demonstrated. One of the prominent features of FEH, the so-called FEH cells, have been described by Praetorius-Clausen (1969). These cells can be found at various levels of the epithelium, showing ballooning and nuclear degeneration. Observations concerning the liquefactive degeneration of the basal cell layer and increased mitotic activity have been conflicting (Archard et al. 1965; Praetorius-Clausen 1969). A mild degree of lymphocytic infiltration as well as collections of polymorphonuclear leukocytes in the underlying connective tissue have been described in most cases. The predominance of plasma cells in these inflammatory cell infiltrates has been emphasized by many investigators (Archard et al. 1965; Praetorius-Clausen 1969; El-Khashab and Abd-El-Aziz 1971). In addition, minor mucous salivary glands exhibit duct ectasia and mild chronic inflammation (Archard et al. 1965; Praetorius-Clausen 1969). Dyskeratosis or epithelial atypia have never been described in FEH lesions (Archard et al. 1965; Waldman and Shelton 1968).

9.3 Electron Microscopic Findings

Electron microscopic studies have shown that the nuclei of the affected cells in FEH are typically indented and their nucleoli absent. In more advanced stages there is margination and peripheral clumping of chromatin. Condensations of tonofilaments in the affected cells are usual (Praetorius-Clausen and Willis 1971; Axell et al. 1981). In the ballooning epithelial cells a rim of tonofilaments and other organelles as well as a homogenous granular center interspersed with dense areas are encountered (Praetorius-Clausen and Willis 1971). Numerous lysosomes have been observed in basal and prickle cells, which together with the nuclear characteristics have been regarded as cellular changes induced by viral infection.

In addition, evidence of HPV infection, i.e., the presence of papillomavirus-like particles have been described (Praetorius-Clausen and Willis 1971; Buchner et al. 1975; van Wyk et al. 1977; Petzoldt and Dennin 1980).

9.4 Evidence for Involvement of Human Papillomaviruses

A number of suggestions concerning the etiology of FEH have been given in the literature. Accordingly, local irritating factors such as tobacco (Helms 1894), electrogalvanic currents between amalgam fillings (Bergenholtz 1965), and vitamin A deficiency (Soneira and Fonseca 1964) have been suggested. The familial occurrence of the lesions discussed above has raised some questions regarding the possibility of hereditary factors (Gomez et al. 1969; Buchner and Mass 1973).

A possible viral etiology of FEH has been suggested by many authors (Reyes 1962; Soneira and Fonseca 1964; Archard et al. 1965; Witkop and Niswander 1965; Bergenholtz 1965; Schock and Wood 1969). Some of them have stressed the fact that the arguments in favor of a genetic factor could also apply if the disease was of viral origin (Witkop and Niswander 1965). Papillomavirus-like particles have been repeatedly detected in FEH lesions (Praetorius-Clausen 1969; Praetorius-Clausen and Willis 1971; Hanks et al. 1972; van Wyk 1977; Goodfellow and Calvert 1979; Petzoldt and Pfister 1980). Immunohistochemical techniques have demonstrated papillomavirus structural antigens in FEH (Lutzner et al. 1982; Syrjänen et al. 1983b, 1984b; Praetorius-Clausen et al. 1985). Antigens have been found in the nuclei of superficial cells with morphological changes in a high percentage (80%) of cases.

With modern DNA hybridization techniques, HPV1 DNA was first identified in one FEH lesion (Petzoldt and Pfister 1980) followed by discovery of HPV13 and one of its subtypes (Pfister et al. 1983; Syrjänen et al. 1984a). So far, HPV13 DNA has been found exclusively in oral FEH lesions, suggesting that this is the HPV type most closely related to this disorder. Recently, a new virus HPV32 was found in FEH (Kremsdorf et al. 1985). Thus, these data strongly suggest that FEH is an HPV infection in the oral cavity, with a possible hereditary background.

9.5 Treatment

Most authors agree that no treatment is needed for FEH due to its benign course and frequent, spontaneous regression, mostly seen within 2 months to 3 years (Archard et al. 1965; Orfanos et al. 1974; Knoth and Boepple 1978). Podophyllin (Orfanos et al. 1974) or local application of vitamin A solution (Orfanos et al. 1974; Lang et al. 1984) have been shown to be useless in eradicating these lesions. Similarly, local excisional therapy or cryotherapy have proved ineffective, due to frequent recurrence (Petzoldt et al. 1982; Lang et al. 1984). It remains to be seen whether any of the antiviral therapeutic agents such as interferon or Isoprinosine could be effective.

10 Keratoacanthoma

This benign lesion usually occurs on sun-exposed skin areas, the center of the face being the site of predilection. The lesion was first fully described by McCormac and Scharff (1936) as molluscum sebaceum and later by Rook and Whimster (1950) under its present name. It appears that keratoacanthoma has never been reported in other than the white race. Two principal types of keratoacanthoma exist, multiple and solitary. The solitary keratoacanthoma is most frequently found in the 6th decade of life, whereas the multiple type has its onset usually before the age of 45, and it frequently shows a familial distribution. Keratoacanthoma occurs twice as frequently in men as in women (Ghadially et al. 1963; Scofield et al. 1974).

10.1 Clinical Features

Clinically, keratoacanthoma appears as a small elevated nodule with sloping peripheral borders and a depressed central core. The nodule grows rapidly to its maximum size (usually 10-15 mm) in 4-8 weeks. After a stationary period of about 1-2 months, the nodule may undergo spontaneous regression in the next 6-8 weeks. Usually a slightly depressed scar remains. The lesion is often painful (Scofield et al. 1974; Lucas 1984). While keratoacanthoma is usually benign, some lesions of the cheek may show destructive growth and recurrent behavior. In some cases it is quite impossible to separate them from verrucous carcinoma.

The lesion occurs on the lips in some 8% of the cases, mostly on the vermilion border (Ghadially et al. 1963; Azaz and Lustmann 1974). In an extensive review of the literature, Silberberg and coworkers (1962) found 41 cases involving the lips and noted a 3:1 male predilection. Helsham and Buchanan (1960) reported a keratoacanthoma in the maxillary gingiva. Four intraoral cases were reviewed by Freedman and his associates (1979).

10.2 Histological Features

Keratoacanthoma lesions in the oral mucosa are histologically similar to those of the skin. The characteristic feature is a crater, plugged with keratin and surrounded by pseudocarcinomatous hyperplastic epithelium. The hyperplastic squamous epithelium undermines and forms an acute angle to the surrounding elevated normal epithelium. For this reason, the diagnosis may be impossible if the adjacent border of the specimen is not included in the biopsy (Scofield et al. 1974; Shafer et al. 1983; Lucas 1984).

10.3 Evidence for Involvement of Human Papillomaviruses

Keratoacanthoma is a lesion which both clinically and histologically resembles epidermoid carcinoma but nevertheless is benign. The etiology of keratoacanthoma is still unknown, although a viral agent has been suggested for many years

(Ereaux et al. 1955). Additionally, inflammatory changes in a sebaceous cyst, origin from pilosebaceous or sweat ducts (McCormac and Scharff 1936), immune mechanisms, trauma, chemical carcinogens (Ghadially et al. 1963), genetic predisposition (Whittle and Davis 1957), and sunlight (de Moragas et al. 1958) have all been considered. Zelickson and Lynch (1961) were able to demonstrate intranuclear, virus-like particles in four cases of keratoacanthoma by electron microscopy.

A few keratoacanthomas have been included in the immunhistochemical studies of HPV lesions of the oral cavity and other locations (Jenson et al. 1982a, b; Syrjänen et al. 1983b, 1984b). All these studies were negative. On the other hand, HPV as the potential etiological agent of keratoacanthoma has gained substantial support from the recent discovery of a new type, HPV37, in such lesions (Scheurlen et al. 1985). Recently, two reports have shown an unusually high number of melanocytes in keratoacanthomas, suggesting a close keratinocyte-melanocyte interaction (Mehregan and Plotnick 1984; Sanchez-Yus and Gonzales-Moran 1985). Whether melanoacanthoma, a benign, pigmented skin tumor consisting of proliferating melanocytes and keratinocytes (Mishima and Pinkus 1960), is an entity distinct from keratoacanthoma remains to be seen. Interestingly, a case of intraoral melanoacanthoma was recently described (Schneider et al. 1981). Retrospective analyses of keratoacanthomas using the in situ DNA hybridization technique may shed additional light on the role of HPV in their development.

10.4 Treatment

Surgical excision or electrodesiccation and curettage have been proposed as the treatment of keratoacanthoma. Although spontaneous regression usually occurs, lesions with a duration of several years can be seen. A careful follow-up is justified in all cases because of the difficulties in diagnosis (Scofield et al. 1974). Lesions of the cheek showing destructive growth are indistinguishable from verrucous carcinoma.

11 Oral Leukoplakia, Precancer, and Cancer

For many years, oral leukoplakia has been one of the central topics in dental research. A survey covering the voluminous literature accumulated on oral leukoplakia and cancer is beyond the scope of this chapter. However, the current concepts on oral leukoplakia are reviewed in brief, as far as they are pertinent to the possible role of HPV infections in the etiology of these lesions.

11.1 Clinical Features

Leukoplakias as the most frequent precancerous lesions of the oral cavity have been extensively studied and reviewed by numerous investigators (Mackenzie et al. 1980; Pindborg 1980; Burkhardt and Maerker 1981; Banoczy 1982; and many others). The term 'leukoplakia' was originally proposed by a Hungarian dermatologist Ernst Schwimmer in 1877 and has remained a subject of continuing dispute since. The term has been used (a) as a clinical concept to designate 'a white patch' and (b) as a histopathological diagnosis equivalent to an oral precancerous lesion. According to the WHO (1978), however, the term leukoplakia should be exclusively used with reference to a purely clinical diagnosis. In keeping with this definition, leukoplakia is a whitish patch or plaque that cannot be classified clinically or pathologically as any other lesion and is not associated with any physical or chemical causative agent, except the use of tobacco (Axell et al. 1984). The prevalence of leukoplakia in the world literature varies considerably, from 0.4% to 11.7% (Pindborg 1980; Banoczy 1982). The concept of oral leukoplakia as a premalignant lesion is based on the fact that some proportion of oral squamous cell carcinomas appear to arise within the areas of leukoplakia. The frequency of malignant transformation in oral leukoplakias has been recorded at 0.13%-6% (Pindborg 1980).

The clinical appearance of oral leukoplakia is variable: the lesions may be white and homogenous, patchy or verrucous. Currently, a classification into three forms, as defined by Banoczy and Sugar in 1968, has gained wide acceptance: leukoplakia simplex, leukoplakia verrucosa, and leukoplakia erosiva. While giving an idea (albeit a very rough one) about the behavior of these lesions, this classification has proved to be especially useful for clinicians. Leukoplakia simplex is the most common form (49% of all leukoplakias) and also considered the most benign; the verrucous type comprises 27% and should be regarded as highly suspect of malignant transformation (Burkhardt and Maerker 1981). Erosive leukoplakia represents some 24% of all leukoplakias and ranks as a high-risk lesion with 30% malignant transformation (Burkhardt and Maerker 1981).

In the older literature most leukoplakias were reported as occurring in males. Over the past few decades, however, leukoplakias have been reported with an increased frequency in women (Shafer and Waldron 1961; Waldron and Shafer 1975; Banoczy 1982). Thus, in a study by Waldron and Shafer (1975), which included 3256 leukoplakia patients, only 51% were men. The reason for this shift is not known with certainty, although it has been attributed to changes in the smoking habits among young women (Mackenzie et al. 1980). This is supported by the observations of Shafer and Waldron (1961; Waldron and Shafer 1975), who found a distinct shift towards younger age groups of patients with leukoplakia, which was previously regarded as a lesion of older people. With respect to the most frequent location of leukoplakias, some discrepancy also exists among the various authors. According to Waldron and Shafer (1975), the most common location was the mandibular mucosa, while the tongue proved to be the least common site.

11.2 Histological Features

On biopsy, leukoplakia is known to present a variety of epithelial and stromal changes ranging from harmless epithelial hyperplasia with hyperkeratosis to various degrees of epithelial dysplasia, including in situ and early invasive carcinoma.

Correlations between the clinical grading and histology have been a popular subject for many years. It should be emphasized that the clinical appearance of leukoplakias is not a reliable index for predicting the seriousness of the microscopic changes. Thus, the need for a diagnostic biopsy in all suspicious lesions should be underscored (Waldron and Shafer 1975; Burkhardt and Maerker 1981; Fischman et al. 1982; Lind et al. 1986). This is exemplified by the study of Waldron and Shafer (1975), who reported that of the lesions diagnosed as cancer by clinicians, apparently equal proportions were benign (33%), premalignant (36%), and malignant (31%). Similarly, of the lesions evaluated as leukoplakia or hyperkeratosis by clinicians, 13% proved to be frankly invasive carcinomas on biopsy (Waldron and Shafer 1975). Interestingly, many (if not most) carcinomas in situ of the oral epithelium form red patches, which should be an alarming sign (Koss 1979).

11.3 Etiological Considerations

The etiology of oral leukoplakia is poorly understood. A variety of factors have been considered, including smoking, alcohol, dental restoratives, mechanical irritation, systemic conditions such as scleroderma, sideropenic dysphagia, and *Candida* infection (Mackenzie et al. 1980; Pindborg 1980; Burkhardt and Maerker 1981; Banoczy 1982; Axell et al. 1984; Shafer et al. 1983; Lucas 1984). Smoking, however, is by far the most frequently mentioned etiological agent. Recently, some doubt has been raised about the role of smoking in several reports; the malignant transformation rate of leukoplakia in patients who did not use tobacco was eight times greater than that occurring in tobacco users. This was interpreted as suggesting that leukoplakias developing from causes other than tobacco may have a greater malignant potential (Axell et al. 1984).

So far, the viral etiology of oral leukoplakia or oral cancer has received little or no attention. It is currently thought, however, that HPV infection might well represent a causative agent of cervical dysplasias and carcinomas (zur Hausen 1977; Meisels et al. 1982; Syrjänen 1986). Also, other squamous cell neoplasms including those of the larynx, paranasal sinuses (Syrjänen and Syrjänen 1981; Syrjänen et al. 1983d), and bronchus (Syrjänen 1980) have been speculatively associated with HPV infection. The models of precancer/cancer sequence at these sites are easily applied to the oral cavity as well. The lower female genital tract and oral cavity, although quite distinct anatomically, share a number of features in common with regard to early neoplastic changes. Both the oral cavity and vaginal epithelium are continuously exposed to various microtraumas, irritants, and microorganisms. The gross appearance of leukoplakia associated with HPV in the uterine cervix is identical to that found in the oral cavity, although different nomenclature has been used to describe these lesions.

In 1977 Fejerskov and coworkers described a new type of leukoplakia with features suggestive of viral infection. The similarities of these electron microscopic findings to those of squamous cell papillomas and FEH were emphasized. However, the possibility of HPV infection in the etiology of oral cancer was not appreciated until 1983 by Syrjänen and coworkers (1983c). The well-established morphological manifestations of oral cancer (endophytic, exophytic, and leukoplakia types) were related to the three distinct types of HPV lesions in the uterine cervix; inverted, papillary, and flat condyloma, respectively. Of the 40 biopsies studied, 16 cases fulfilled the criteria of one of these HPV lesions. The papillary type was the most frequent, 22.5% of cases. When subjected to IP-PAP staining, 8 of the 16 specimens with morphological features of HPV infection showed positive reactivity to HPV structural proteins. The positive staining was confined to the epithelium in adjacent regions of carcinomatous changes (Syrjänen et al. 1983c). Subsequently, additional evidence of HPV involvement in oral cancer has been presented (Löning et al. 1985; Milde and Löning 1986; de Villiers et al. 1986; Syrjänen et al. 1986). Löning and coworkers (1985) reported that four of the six oral carcinomas studies reacted with HPV structural antibodies and three with HPV DNA probes. HPV16 DNA was found in one carcinoma and HPV11 DNA in one additional case. Another HPV16-positive oral cancer was reported by Syrjänen et al. (1986) by the in situ DNA hybridization technique.

Quite recently, the first evidence of HPV etiology in oral leukoplakias has been provided as well (Löning et al. 1985; Lind et al. 1986; Syrjänen et al. 1986). In a follow-up study of 20 leukoplakias (Lind et al. 1986), a significant correlation was established between the presence of HPV antigens and the degree of dysplasia and malignant transformation. HPV antigens were commonly found only in the precancerous state of the lesions, and not in invasive cancer. This is in agreement with the earlier studies on cervical dysplastic HPV lesions. Interestingly, there was a relationship between the expression of HPV antigens and the predominance of IgA-secreting plasma cells. However, in connection with malignant transformation, the IgG plasma cells outnumbered the IgA cells in the underlying stroma. Löning and Burkhardt (1979), in their analyses of 202 premalignant and malignant oral lesions, found that the incidence of Ig-positive plasma cells (IgA and IgG) was twice as high in the dysplastic leukoplakias as compared with their nondysplastic counterparts. The number of plasma cells, especially IgA and IgG producing plasma cells, decreased significantly with tumor progression. These authors also found three HPV antigen-positive cases among the five oral leukoplakias studied (Löning et al. 1985). Using the DNA hybridization technique, HPV DNA sequences were detected in four of these lesions. Syrjänen et al. (1986) reported two oral leukoplakias that were positive for HPV11 and HPV16 by in situ hybridization.

A new clinical entity called 'hairy leukoplakia' and connected with AIDS was described by Greenspan and coauthors (1984). In the majority of cases, it was located on the border of the tongue, appearing as a white lesion with papillary projections. In the preliminary series in 17 of the 21 cases the staining with common HPV antigen was interpreted as positive. However, DNA hybridization has not confirmed the presence of HPV DNA in these lesions. It seems most likely that this lesion is a manifestation of Epstein-Barr virus infection (Greenspan et al. 1986a).

The hypothesis that at least some of the oral leukoplakias may be caused by HPV infections remains to be confirmed. Whether or not the clinical appearance and behavior of the lesions are related to different types of HPV can perhaps be established by DNA hybridization techniques as was done with cervical lesions.

11.4 Treatment

The guidelines for treatment of oral leukoplakias are presented in many textbooks (Pindborg 1980; Burkhardt and Maerker 1981; Banoczy 1982; Shafer et al. 1983). If the leukoplakia appears to be caused by local factors, these irritants should be removed. A biopsy should be obtained from any leukoplakia persisting or with a suspicious appearance. Excision is always preferable if the lesion is not too extensive and the location does not pose any technical problems (Burkhardt and Maerker 1981). Cryosurgery has also been suggested as the treatment of choice for oral leukoplakias (Pindborg 1980). High doses of vitamin A proved to be effective in eradicating some forms of oral leukoplakias (Pindborg 1980).

12 Conclusions

With increasing interest in HPV research by oral pathologists, viral lesions in the oral cavity may prove to be more common than previously suspected. Squamous cell lesions in the oral cavity are of especial interest in this regard. The gross appearance of viral lesions is not distinct enough to be readily diagnosed by the clinicians. As pointed out, remarkable morphological similarities exist between different oral epithelial changes, and light microscopy with conventional staining procedures is not an adequate means for scrutinizing these lesions for viral etiology.

Due to the recent advances made in recombinant DNA research substantial progress has been made in virus identification. Thus, in future research it may prove possible to classify oral lesions according to their causative agent, such as HPV, instead of by purely morphological criteria. By identifying the HPV type, the origin of viral transmission can also be traced. The mucosa of the genital tract and oral cavity are similar in many respects, and it is to be expected that HPVs found in the former, e.g., HPV6, 11, 16, 18, and 31, may account for a substantial number of lesions in the oral cavity as well. Preliminary evidence of viral presence has been provided in some recent studies. Dermatotrophic HPV types such as 1, 2, and 7 can also be expected to be found in oral squamous cell lesions, although no significant evidence is available at the moment. Whether additional HPV types exclusively confined to the oral cavity exist, such as HPV13 and its subtypes, remains to be seen.

One of the major challenges for oral pathologists will be to apply modern laboratory methods resulting from research in the field of molecular virology to ensure the proper diagnosis of oral squamous cell lesions.

References

Abbey LM, Page DG, Sawyer DR (1980) The clinical and histopathologic features of a series of 464 oral squamous cell papillomas. Oral Surg 49: 419-428
Acevedo A, Gonzales GM, Nelson JF (1981) Focal epithelial hyperplasia. Oral Surg 51: 524-526

Ackerman LV (1948) Verrucous carcinoma of the oral cavity. Surgery 23: 670-678
Ackerman LV, McGavran MH (1958) Proliferating benign and malignant epithelial lesions of the oral cavity. J Oral Surg 16: 400-413
Anderson DR (1971) Focal epithelial hyperplasia: report of a case in a South African Caucasoid. J Dent Assoc Afr 26: 32-35
Archard HO, Heck JW, Stanley HR, Gallup NM (1965) Focal epithelial hyperplasia: an unusual oral mucosal lesion found in Indian children. Oral Surg 20: 201-212
Ashiru JO, Ogunbanjo BO, Rotowa NA, Adeyemi-Doro FAB, Osoba AO (1983) Intraoral condylomata acuminata. Br J Vener Dis 59: 325-326
Axell T (1976) A prevalence study of oral mucosal lesions in an adult Swedish population. Thesis. Odontol Revy 27 [suppl 36]: 1-77
Axell T, Hammarström L, Larsson A (1981) Focal epithelial hyperplasia in Sweden. Acta Odontol Scand 39: 201-208
Axell T, Holmstrup P, Kramer IRH, Pindborg JJ, Shearn M (1984) Intraoral seminar on oral leukoplakia and associated lesions related to tobacco habits. Community Dent Oral Epidemiol 12: 146-154
Azaz B, Lustmann J (1974) Keratoacanthoma of the lower lip. Review of the literature and report of a case. Oral Surg 38: 918-927
Banoczy J (1982) Oral leukoplakia. Martinus Nijhoff, The Hague
Banoczy J, Sugar L (1968) Control results of patients with leukoplakia. Fogorov Sz 61: 116-123
Batsakis JG (1979) Tumor of the head and neck. Clinical and pathological considerations, 2nd edn. Williams and Wilkins, Baltimore
Batsakis JG, Hybels R, Crissman JD (1982) The pathology of head and neck tumors: verrucous carcinoma. Head Neck Surg 5: 29-38
Becker J, Behem J, Löning TH, Reichart P, Geerlings H (1985) Quantitative analysis of immunocompetent cells in human normal oral and uterine cervical mucosa. Oral papillomas and leukoplakias. Arch Oral Biol 30: 257-264
Bendelac A, Triller R, Pluot M, Orth G, Kalis B (1984) Hyperplasie epitheliale focale, maladie de heck. Ann Dermatol Venereol 111: 671-672
Bergenholtz A (1965) Multiple polypous hyperplasias of the oral mucosa with regression after removal of amalgam fillings. Acta Odontol Scand 23: 111-134
Bhaskar SN (1963) Oral tumors of infancy and childhood. J Pediatr 63: 195-210
Bhaskar SN (1979) Synopsis of oral pathology, 5th edn. Mosby, St. Louis
Bohmfalk C, Zallen RD (1982) Verrucous carcinoma of the oral cavity. Oral Surg 54: 15-20
Borghelli RF, Stirparo MA, Paroni HC, Barros RE, Dominquez FV (1975) Focal epithelial hyperplasia. Oral Surg 40: 107-112
Brown H, Gorlin RJ (1960) Oral mucosal involvement in nevus unius lateris. Arch Dermatol 81: 509
Buchner A, Mass E (1973) Focal epithelial hyperplasia in an Israeli family. Oral Surg 36: 507-511
Buchner A, Bublis JJ, Ramon Y (1975) Ultrastructural study of focal epithelial hyperplasia. Oral Surg 39: 622-629
Burkhardt A, Maerker R (1981) A colour atlas of oral cancers. Wolfe Medical, London
Cheville NF, Olson C (1964) Cytology of the canine oral papilloma. Am J Pathol 45: 849
Choukas NC, Toto PD (1982) Condylomata acuminatum of the oral cavity. Oral Surg 54: 480-485
Colby RA, Kerr NC, Robinson HBG (1983) Color atlas of oral pathology, 4th edn. Lippincott, Philadelphia
Cooke RA (1969) Verrucous carcinoma of the oral mucosa in Papua-New Guinea. Cancer 24: 397-402
Danilov LN (1974) A case of condyloma acuminatum of the oral mucosa in a 5-year-old child. Stomatologiia (Mosk) 53: 84-87
Decker WG, de Guzman MN (1969) Focal epithelial hyperplasia. Oral Surg 27: 15-19
de Moragas JM, Montgomery H, McDonald JR (1958) Keratoacanthoma versus squamous cell carcinoma. Arch Dermatol 77: 390-395
de Villiers EM, Weidauer H, Le J-Y, Neumann C, zur Hausen H (1986) Papillomaviren in benignen und malignen Tumoren des Mundes und des oberen Respirationstraktes. Laryngol Rhinol Otol (Stuttg) 65: 177-179
Doran GA (1980) Oral condyloma acuminatum or venereal wart. Case report. Aust Dent J 25: 212-214

Doyle JL, Grodjesk JE, Manhold JH Jr (1968) Condyloma acuminatum occurring in the oral cavity. Oral Surg 6: 434-440
Eavey RD (1985) Inverted papilloma of the nose and paranasal sinuses in childhood and adolescence. Laryngoscope 95: 17-22
El-Khashab MM, Abd-El-Aziz AEHM (1971) Focal epithelial hyperplasia (Heck's disease). Oral Surg 31: 637-646
Ereaux LP, Schpflocher P, Fournier K (1955) Keratoacanthomata. Arch Dermatol 71: 73-83
Estrada L (1956) Aporte al estudio odontologico de los Indios Katios. Heraldo Dental 2: 5-11
Estrada L (1960) Estudio medico y odontologico de los Indies Katios del Choco. Temas Odontologicas 7: 198-210
Eversole LR, Sorenson HW (1974) Oral florid papillomatosis in Down's syndrome. Oral Surg 37: 202-208
Fejerskov O, Roed-Petersen B, Pindborg JJ (1977) Clinical histological and ultrastructural features of a possibly virus-induced oral leukoplakia. Acta Pathol Microbiol Scand 85: 897-906
Fischman SL (1969) Focal epithelial hyperplasia. Oral Surg 28: 389-393
Fischman SL, Ulmansky M, Sela J, Bab I, Gazit D (1982) Correlative clinico-pathological evaluation of oral premelignancy. J Oral Pathol 11: 283-289
Fiumara NJ (1984) The management of warts of the oral cavity. Sex Transm Dis 11: 267-270
Freedman PD, Kerpel SM, Begel H, Lumerman H (1979) Solitary intraoral keratoacanthoma. Report of a case. Oral Surg 47: 74-77
Frithiof L, Wersäll J (1967) Virus-like particles in human oral papilloma. Acta Otolaryngol 64: 263-266
George DI Jr, Farman AG (1984) Ultrastructural features of oral condyloma acuminatum. J Oral Med 39: 169-172
Ghadially FN, Baron BW, Kerridge DF (1963) The etiology of keratoacanthoma of the lip. Cancer 16: 603-611
Goethals PL, Harrison EG, Devine KD (1963) Verrucous squamous carcinoma of the oral cavity. Am J Surg 106: 845-851
Gomez A, Calle C, Arcilla G, Pindborg JJ (1969) Focal epithelial hyperplasia in a half-breed family of Colombians. J Am Dent Assoc 79: 663-667
Goodfellow A, Calvert H (1979) Focal epithelial hyperplasia of the oral mucosa. A case report from the United Kingdom. Br J Dermatol 101: 341-344
Gorlin RJ, Meskin LH, Peterson WC, Goltz RW (1963) Focal dermal hypoplasia syndrome. Acta Dermatol Venereol 43: 421
Greenspan D, Conant M, Silverman S Jr, Greenspan JS, Petersen V, de Souza Y (1984) Oral hairy leukoplakia in male homosexuals: evidence of association with both papillomavirus and a herpes-group virus. Lancet ii: 831-834
Greenspan J, Greenspan D, Lennette ET (1986a) Replication of Epstein-Barr virus within the epithelial cells of oral 'hairy' leukoplakia, and AIDS-associated lesion. N Engl J Med 313: 1465-1471
Greenspan D, Greenspan J, Pindborg J, Schiødt M (1986b) AIDS and dental team, 1st edn. Munksgaard, Copenhagen
Greer RO (1973) Inverted oral papilloma. Oral Surg 36: 400-403
Greer RO, Goldman HM (1974) Oral papillomas. Clinico-pathologic evaluation and retrospective examination for dyskeratosis in 110 lesions. Oral Surg 38: 435-440
Greer RO, Poppers HA, DeMento FJ (1976) Cowden's disease (multiple hamartoma syndrome). J Periodontol 47: 531-534
Gross GE, Pfister H, Mittermayer C (1980) Papillomavirus particles in a fibroma of the tongue. Acta Derm Venereol (Stockh) 60: 315-318
Grung B, Bang G (1984) Fokal epithelial hyperplasi. Den norske tannlegeforenings tidende 94: 45-47
Gysland WB, Reimann BEF, Shaffer EL Jr (1976) The virus in oral condyloma acuminatum. In: Bailey GM (ed) 34th annual proceedings of the electron microscopy society of America, Miami Beach, pp 246-247
Hallmon WW, Waldropt CT, Houston DG (1985) Focal epithelial hyperplasia (Heck's disease). A case report. J Peridontol 56: 89-96
Haneke E (1984) Klinisches Bild der Virusinfektionen in der Mundhöhle. Dtsch Z Mund Kiefer Gesichtschir 8: 90-92

Hanks CT, Arbor A, Stuart M, Fischman L, de Guzman MN (1972) Focal epithelial hyperplasia. Oral Surg 33: 934-941
Heidingsfield ML (1901) Condylomata acuminata linguata. J Cutan Genitourin Dis 19: 226-234
Helms O (1894) Syfilis i Grönland. Ugeskr Laeg 5 rk. 1bd: 265-276
Helsham RW, Buchanan G (1960) Keratoacanthoma of the oral cavity. Oral Surg 13: 844-849
Hertz RS (1972) The occurrence of a verruca vulgaris on an intraoral skin graft. Oral Surg 34: 934-942
Hettwer KJ, Rodgers MS (1966) Focal epithelial hyperplasia (Heck's disease) in a Polynesian. Oral Surg 22: 466-470
Jacobson S, Shear M (1972) Verrucous carcinoma of the mouth. J Oral Pathol 1: 66-75
Jarvis A, Gorlin RJ (1972) Focal epithelial hyperplasia in an Eskimo population. Oral Surg 32: 227-228
Jenson AB, Lancaster WD, Hartman DP, Shaffer EL Jr (1982a) Frequency and distribution of papillomavirus structural antigens in verrucae, multiple papillomas, and condylomata of the oral cavity. Am J Pathol 107: 212-218
Jenson AB, Link CC, Lancaster WD (1982b) Papillomavirus etiology of oral cavity papillomas. In: Hooks J, Jordan G (eds) Viral infections in oral medicine. Elsevier, Amsterdam, pp 133-146
Jin Y-T, Toto PD (1984) Detection of human papovavirus antigens in oral papillary lesions. Oral Surg 58: 702-705
Jones HJ (1965) Non-odontogenic oral tumours in children. Br Dent J 16: 439-447
Kahn T, Schwarz E, zur Hausen H (1985) Molecular cloning and characterization of the DNA of a new human papillomavirus (HPV30) from a laryngeal carcinoma. Int J Cancer 37: 61-65
Kelley JE, Hibbard ED, Giansanti JS (1972) Epidermal nevus syndrome. Report of a case with unusual oral manifestations. Oral Surg 34: 774-777
Knapp MJ (1971) Oral disease in 181338 consecutive oral examinations. J Am Dent Assoc 83: 1288-1293
Knapp MJ, Uohara GI (1967) Oral condyloma acuminatum. Oral Surg 23: 538-545
Knoth W, Boepple D (1978) Hyperplasia multilocularis oris Heck (sog. focal epithelial hyperplasia. Z Hautkr 53: 675-679
Kohn EM, Dahlin DC, Erich JB (1963) Primary neoplasms of the hard and soft palates and the uvula. Proc Staff Meet Mayo Clin 38: 233-241
Koss LG (1979) Diagnostic cytology and its histopathologic bases, 3th edn. Lippincott, Philadelphia, p 681
Kraus FT, Perez-Mesa C (1966) Verrucous carcinoma: clinical and pathologic study of 105 cases involving oral cavity, larynx and genitalia. Cancer 19: 26-38
Kremsdorf D, Praetorius F, Beaudenon S, Lutzner M, Worsaae N, Pehau-Arnaudet G, Orth G (1985) A new type of human papillomavirus associated with oral focal epithelial hyperplasia. Workshop on papillomaviruses: molecular and pathogenetic mechanisms. Kuopio, Finland, August 25-29
Kuhlwein A, Nasemann T, Jänner M, Schaeg G, Reinel D (1981) Nachweis von Papillomaviren bei fokaler epithelialer Hyperplasie Heck und die Differentialdiagnose zum weißen Schleimhautnaevus. Hautarzt 32: 617-619
Lang E, Zabel M, Ikenberg H (1984) Fokale epitheliale Hyperplasie (Morbus Heck). Dtsch Med Wochenschr 109: 1763-1766
Lind P, Syrjänen S, Syrjänen K, Koppang HS, Aas E (1986) Immunoreactivity and human papillomavirus (HPV) on oral precancer and cancer lesions. Scand J Dent Res 94: 419-426
Löning T, Burkhardt A (1979) Plasma cells and immunoglobulin synthesis in oral precancer and cancer. Virchows Arch (A) 384: 109-119
Löning T, Ikenberg H, Becker J, Gissmann L, Hoepfer I, zur Hausen H (1985) Analysis of oral papillomas, leukoplakias, and invasive carcinomas for human papillomavirus type related DNA. J Invest Dermatol 84: 417-420
Lucas RB (1984) Pathology of tumors of oral tissues, 4th ed. Churchill Livingstone, New York
Ludwig ME, Lowell DM, Livolsi VA (1981) Cervical condylomatous atypia and its relationship to cervical neoplasia. Am J Clin Pathol 76: 255-264
Lund HZ (1957) Tumors of the skin. Atlas of tumor pathology, sect 1, fasc 2. Armed Forces Institute of Pathology, Washington
Lutzner M, Kuffer R, Blanchet-Bardon C, Croissant O (1982) Different papillomaviruses as the causes of oral warts. Arch Dermatol 118: 393-399

MacDonald DG, Rennie JS (1975) Oral epithelial atypia in denture induced hyperplasia, lichen planus and squamous cell papilloma. Int J Oral Surg 4: 40-45
Mackenzie IC, Dabelsteen E, Squier C (1980) Oral premalignancy. Iowa Press, Iowa City
McCarthy PL, Shklar G (1964) Diseases of oral mucosa. McGraw-Hill, New York, pp 93-97
McClatchery KD, Colquitt WN, Robert RC (1979) Condyloma acuminatum of the lip: report of case. J Oral Surg 37: 751-752
McClendon JL (1985) Case of the mouth oral florid papillomatosis. Tex Dent J 102: 14-15
McCormac H, Scharff RW (1936) Molluscum sebaceum. Br J Dermatol 48: 624-626
Medina JE, Dichtel MW, Luna MA (1984) Verrucous-squamous carcinomas of the oral cavity. Arch Otolaryngol 110: 437-440
Mehregan A, Plotnick H (1984) Pigmented keratoacanthoma. Arch Dermatol 120: 1417
Meisels A, Morin C, Casa-Cordero M (1982) Human papillomavirus infection of the uterine cervix. Int J Gynecol Pathol 1: 75-94
Milde K, Löning T (1986) Detection of papillomavirus DNA in oral papillomas and carcinoma: application of in situ hybridization with biotinylated HPV16 probes. J Oral Pathol 15: 292-296
Mincer HH, Coleman SA, Hopkins KP (1972) Observations on the clinical characteristics of oral lesions showing histologic epithelial dysplasia. Oral Surg 33: 389-399
Mincer HH, Jennings BR, Turner JE, Lee WB (1985) Detection of human papovavirus antigens in oral papillary lesions. Letter. Oral Surg 5: 516
Mishima Y, Pinkus H (1960) Benign mixed tumor of melanocytes and Malpighian cells. Arch Dermatol 91: 539-550
Moskow R, Moskow BS (1962) Inverted papilloma. Oral Surg 15: 918-922
Naghashfar Z, Sawada E, Kutcher MJ, Swancar J, Gupta J, Daniel R, Kashima H, Woodruff JD, Shah KV (1985) Identification of genital tract papillomaviruses HPV6 and HPV16 in warts of the oral cavity. J Med Virol 17: 313-324
Orfanos CE, Strunk V, Gartmann H (1974) Focale epitheliale Hyperplasie der Mundschleimhaut: Hecksche Krankheit. Dermatologica 149: 163-175
Orlean SL, Ladow CS (1960) Superficial keratoses verruca vulgaris and pachyderma oris. J Dent Med 15: 108-112
Perriman A, Uthman A (1971) Focal epithelial hyperplasia. Oral Surg 31: 221-225
Petzoldt D, Dennin R (1980) Isolierung virusartiger Partikeln bei fokaler epithelialer Hyperplasie Heck. Hautarzt 31: 35-36
Petzoldt D, Pfister H (1980) HPV1 DNA in lesion of focal epithelial hyperplasia Heck. Short communications. Arch Dermatol Res 268: 313-314
Petzoldt D, Dennin R, Pfister H, Hoffmann C (1982) Fokale epitheliale Hyperplasie Heck. Der Hautarzt 33: 201-205
Pfister H, Hettich I, Runne U, Gissmann L, Chilf G (1983) Characterization of human papillomavirus type 13 from focal epithelial hyperplasia Heck lesions. J Virol 47: 363-366
Pindborg JJ (1980) Oral cancer and precancer. Henry Ling, Dorset Press, Dorchester
Pindborg JJ (1985) Atlas of diseases of the oral mucosa, 4th edn. Munksgaard, Copenhagen
Praetorius F, Praetorius-Clausen P, Mögeltoft M (1985) Immunohistochemical evidence of papillomavirus antigen in focal epithelial hyperplasia. Tandlaegebladet 89: 589-625
Praetorius-Clausen F (1969) Histopathology of focal epithelial hyperplasia. Evidence of viral infection. Tandlaegebladet 73: 1013-1022
Praetorius-Clausen F (1972) Rare oral viral disorders (molluscum contagiosum, localized keratoacanthoma, verrucae, condyloma acuminatum, and focal epithelial hyperplasia). Oral Surg 34: 604-618
Praetorius-Clausen F (1973) Geographical aspects of oral focal epithelial hyperplasia. Path Microbiol 39: 204-213
Praetorius-Clausen F, Willis JM (1971) Papovavirus-like particles in focal epithelial hyperplasia. Scand J Dent Res 79: 362-365
Praetorius-Clausen F, Mogeltoft M, Roed-Petersen B, Pindborg JJ (1970) Focal epithelial hyperplasia of the oral mucosa in a south-west Greenlandic population. Scand J Dent Res 78: 287-294
Reyes DG (1962) Verruca de la cavidad oral. Rev Cal Med Guatemala 13: 223-226
Rock JA, Fisher ER (1960) Florid papillomatosis of the oral cavity and larynx. Arch Otolaryngol 72: 593-598
Rook A, Whimster C (1950) Le keratoacanthome. Arch Belg Dermatol Syphiligr 6: 137-146

Rose HP (1965) Papillomas of the oral cavity. Oral Surg 20: 542-549
Samitz MH, Weinberg RA (1963) Oral florid papillomatosis. Arch Dermatol 87: 478-480
Samitz MH, Ackerman AB, Lantis LR (1967) Squamous cell carcinoma arising at the site of oral florid papillomatosis. Arch Dermatol 96: 286-289
Sanchez-Yus E, Gonzales-Moran A (1985) Proliferation of melanocytes in keratoacanthoma. Letter. Arch Dermatol 121: 968-969
Sawyer DR, Arole G, Mosadomi A (1983) Focal epithelial hyperplasia. Oral Surg 56: 185-189
Scheurlen W, Gissmann L, Ikenberg H, zur Hausen H (1985) Molecular cloning of two new HPV-types from human skin tumors. Workshop on papillomaviruses: molecular and pathogenetic mechanisms. Kuopio, Finland, August 25-29
Schneider LC, Mesa ML, Haber SM (1981) Melanoacanthoma of oral mucosa. Oral Surg 52: 284-287
Schock RK, Wood W (1969) Familial focal epithelial hyperplasia. Oral Surg 28: 598-602
Schwimmer (1877) Die idiopathischen Schleimhautplaques der Mundhöhle (Leukoplakia buccalis). Arch Dermat Syph 9: 511-570
Scofield HH, Werning JT, Shukes RC (1974) Solitary intraoral keratoacanthoma. Review of the literature and report of a case. Oral Surg 37: 889-897
Scully C, Ward-Booth P (1983) Oral carcinoma: evidence for viral oncogenesis. Br J Oral Max Surg 22: 367-371
Seibert JS, Shannon CS Jr, Jacoway JR (1974) Treatment of recurrent condyloma acuminatum. Oral Surg 27: 393-403
Sewerin IB (1968) Fokal epithelial hyperplasi et tilfaelde. Tandlaegebladet 72: 610-618
Shafer M (1972) Verrucous carcinoma. Int Dent J 22: 451-459
Shafer WG (1971) Verruciform xanthoma. Oral Surg 31: 784-789
Shafer WG, Waldron CA (1961) A clinical and histopathologic study of oral leukoplakia. Surg Gynecol Obstet 112: 411-420
Shafer WG, Hine MK, Levy BM (1983) A textbook of oral pathology, 4th edn. Saunders, Philadelphia
Shaffer EL Jr, Reimann BE, Gysland WB (1980) Oral condyloma acuminatum. J Oral Pathol 9: 163-173
Shear M, Pindborg JJ (1980) Verrucous hyperplasia of the oral mucosa. Cancer 46: 1855-1862
Shklar G (1965) The precancerous oral lesions. Oral Surg 20: 58-70
Shope RE (1962) Are animal tumor viruses always virus like? J Gen Physiol [Suppl] 45: 143-151
Silberberg I, Kopf AW, Baer RL (1962) Recurrent keratoacanthoma of the lip. Arch Dermatol 86: 92-101
Soneira A, Fonseca N (1964) Sobre una lesion de la mucosa oral en los ninos Indios de la Mision Los Angeles de Tokuko. Venezuela Odontol 29: 109-119
Spouge JD (1973) Oral pathology. Mosby, St Louis, p 388
Starink M (1984) Cowden's disease: analysis of fourteen new cases. J Am Acad Dermatol 11: 1127-1141
Stern E (1922) Multiple weiche Warzen der Mundschleimhaut. Dermatol Wochen 74: 274-276
Summers L, Booth DR (1974) Intraoral condyloma acuminatum. Oral Surg 38: 273-278
Sundberg JP, Junge RE, El Shazly MO (1985) Oral papillomatosis in New Zealand white rabbits. Am J Vet Res 46: 664-668
Swan RH, McDaniel RK, Dreiman DD, Rome WC (1981) Condyloma acuminatum involving the oral mucosa. Oral Surg 51: 503-508
Syrjänen KJ (1980) Bronchial squamous cell carcinomas associated with epithelial changes identical to condylomatous lesions of the uterine cervix. Lung 158: 131-142
Syrjänen K (1986) Human papillomavirus (HPV) infections of the female genital tract and their associations with intraepithelial neoplasia and squamous cell carcinoma. Pathol Annu 21: 53-89
Syrjänen K, Syrjänen SM (1981) Histological evidence for the presence of condylomatous epithelial lesions in association with laryngeal squamous cell carcinoma. ORL 43: 181-194
Syrjänen K, Syrjänen SM, Lamberg MA, Pyrhönen S (1983a) Human papillomavirus (HPV) involvement in squamous cell lesions of the oral cavity. Proc Finn Dent Soc 79: 1-8
Syrjänen KJ, Pyrhönen S, Syrjänen SM, Lamberg MA (1983b) Immunohistochemical demonstration of human papillomavirus (HPV) antigens in oral squamous cell lesions. Br J Oral Surg 21: 147-153

Syrjänen K, Syrjänen S, Lamberg M, Pyrhönen S, Nuutinen J (1983c) Morphological and immunohistochemical evidence suggesting human papillomavirus (HPV) involvement in oral squamous cell carcinogenesis. Int J Oral Surg 12: 418-424

Syrjänen KJ, Pyrhönen S, Syrjänen S (1983d) Evidence suggesting human papilloma virus (HPV) etiology for the squamous cell papilloma of the paranasal sinus. Arch Geschwulstforsch 53: 77-82

Syrjänen S, Syrjänen K, Ikenberg H, Gissmann L, Lamberg M (1984a) A human papillomavirus closely related to HPV13 found in a focal epithelial hyperplasia lesion (Heck's disease). Arch Dermatol Res 276: 199-200

Syrjänen K, Happonen RP, Syrjänen S, Calonius B (1984b) Human papillomavirus (HPV) antigens and local immunologic reactivity in oral squamous cell tumors and hyperplasias. Scand J Dent Res 92: 358-370

Syrjänen S, Syrjänen K, Lamberg MA (1986) Detection of human papillomavirus DNA in oral mucosal lesions using in situ DNA hybridization applied on paraffin sections. Oral Surg 62: 660-667

Thomsson M, Hammarström L (1978) Focal epithelial hyperplasi. Tandläkartidningen 70: 216-219

Ullman EV (1923) On the aetiology of the laryngeal papilloma. Arch Otolaryngol 5: 317-334

van Wyk CW (1977) Focal epithelial hyperplasia of the mouth: recently discovered in South Africa. Br J Dermatol 96: 381-388

van Wyk CW, Staz J, Farman AG (1977) Focal epithelial hyperplasia in a group of South Africans: its ultrastructural features. J Oral Pathol 6: 14-24

Väyrynen M, Romppanen T, Koskela E, Castren O, Syrjänen K (1981) Verrucous squamous cell carcinoma of the female genital tract: report of 3 cases and survey of the literature. Int J Gynaecol Obstet 19: 351-356

Waldman GH, Shelton DW (1968) Focal epithelial hyperplasia (Heck's disease) in an adult Caucasian. Oral Surg 26: 124-127

Waldron CA (1970) Oral epithelial tumors. In: Gorlin RJ, Goldman HM (eds) Oral pathology, 6th edn. Mosby, St Louis

Waldron CA, Shafer WG (1975) Leukoplakia revisited. A clinicopathologic study of 3256 oral leukoplakias. Cancer 36: 1386-1392

Wallace JR (1976) Focal epithelial hyperplasia (Heck's disease): report of a case. JAMA 93: 118-120

Wechsler HL, Fisher ER (1962) Oral florid papillomatosis. Arch Dermatol 86: 480-492

White DK, Miller AS, McDaniel RK, Rothman BN (1982) Inverted ductal papilloma: a distinctive lesion of minor salivary gland. Cancer 49: 519-524

Whittle CH, Davis RA (1957) Keratoacanthoma of the lower lip red margin. Lancet ii: 1019-1020

WHO collaborating center for oral precancerous lesions (1978) Definition of leukoplakia and related lesions: an aid to studies on oral precancers. Oral Surg 46: 518-539

Witkop CJ, Niswander JD (1965) Focal epithelial hyperplasia in Central and South American Indians and Ladinos. Oral Surg 20: 213-217

Wysocki GP, Hardie J (1979) Ultrastructural studies of intraoral verruca vulgaris. Oral Surg 47: 58-62

Zelickson AS, Lynch FW (1961) Electron microscopy of virus like particles in a keratoacanthoma. J Invest Dermatol 37: 79-83

zur Hausen H (1977) Human papillomaviruses and their possible role in squamous cell carcinomas. Curr Top Microbiol Immunol 78: 1-30

Tumors of the Head and Neck, Larynx, Lung and Esophagus and Their Possible Relation to HPV

H. Kashima and P. Mounts

1 Introduction

This chapter discusses cancers of the upper aerodigestive tract, which are collectively classified as head and neck cancers and include lesions of the nose and paranasal sinuses, pharyngeal spaces, and larynx. Evidence suggestive of a human papillomavirus etiology in lung and esophagus cancer is reviewed. Oral cavity lesions are discussed elsewhere in this text (see chapter by S. Syrjänen, this volume) and thus are omitted here.

Aerodigestive tract lesions occur at sites with unique anatomy and specialized function. Owing to space limitations, the descriptive aspect of anatomy has had to be omitted from this chapter. The interested reader is referred to other sources for a detailed description of these factors (Ackerman 1970; Batsakis 1979).

Although the overwhelming proportion of head and neck cancers are squamous cell in type, other histological varieties occur at virtually every anatomical site, creating an added dimension to problems confronting the physician, pathologist, surgeon, radiotherapist, and medical oncologist.

Head and neck tumors account for 3%-5% of human cancers in the United States of America. Until recently, treatment of these lesions fell variously to the otolaryngologist, the general, plastic, or oncological surgeon, or the radiation therapist. In recent years these experiences have been interchanged so that one body of knowledge with a bearing on epidemiology, staging, and treatment could be developed. The worldwide occurrence of the diseases has stimulated development of uniform staging standards - the first step towards information exchange and problem solving (Barnes 1985; Batsakis 1979; Hyams et al. 1986).

At least two cancers - nasopharynx and esophagus - have well-defined geographical and ethnic tendencies. Most, if not all, upper aerodigestive tract cancers have been associated with exposure to environmental substances, especially chemicals, tobacco products, and alcohol. The prevalence rate has remained constant except in one striking example: esophageal cancer in black males in the United States of America, in which a progressive rise in case numbers is occurring. Treatment results are directly related to the stage of the disease, but only a minority of cases is diagnosed early, when a more favorable outcome is possible. Survival rates do not fully reflect the improvements in treatment techniques; benefits of improved palliative care are not reflected in survival statistics.

The current interest in head and neck cancer coincides with developments of diagnostic and therapeutic value. Although these cancers are relatively uncommon, their high association with recognized carcinogens, distinctive epidemiology, and mutual-risk relationships predisposing to high rates of synchronous and metachronous tumors, combine to identify a subset of cancers of considerable interest for the clinician-scientist.

The role of papillomaviruses in the etiology of human cancers has received increasing attention (for reviews see Howley 1983; Gissmann 1984; Syrjänen 1984). This is largely due to the application of molecular biological techniques both to the study of viruses that induce epithelial cell proliferation during a productive infection and to the diagnosis and investigation of etiological agents in human tumors. The identification of distinguishable types of human papillomaviruses (HPV) that replicate at different anatomical sites, including the respiratory tract, and the availability of techniques to detect viral nucleic acids and proteins has facilitated analysis of tissue specimens for the presence of these viruses. (For the description of the currently used techniques for virus detection and type identification, see the chapter by Schneider, this volume.) Owing to our lack of knowledge on the replication and mechanism of transformation of these viruses, there is no information as yet on how these viruses may induce benign or malignant epithelial tumors.

2 Cancers of the Nasal Cavity and Paranasal Sinuses

2.1 Epidemiology

Cancer of the nose and paranasal sinuses accounts for 0.2%-0.8% of all malignant tumors and 3% of cancers occurring in the head and neck region. According to one analysis, 58% of tumors originate in the antrum, 30% in the nasal cavity, 10% in the ethmoid sinuses, and 1% in the sphenoid and frontal sinuses (Frazell and Lewis 1963). Adenocarcinoma of the nasal cavity occurs with increased frequency among furniture workers in England and Wales (Acheson 1976). A similar high incidence has been observed in French and German wood, asbestos, and textile workers as well as those in the shoe industry (Acheson et al. 1970). The roasting and smelting of nickel is said to be associated with a 40-250 times greater risk of developing nasal cancer (Barton 1977). Typically in the middle turbinate (Virtue 1972) a latency of 25-30 years is described. Formaldehyde (Halperin et al. 1983) and nitrosamines (Stenback 1973) are also implicated as potential carcinogens.

2.2 Pathology

Squamous cell cancers of the nasal cavity occur in the lateral nasal wall and septum. The usual age of occurrence is in the 5th, 6th, and 7th decades, and males predominate. Lymph node metastases are uncommon. Extranasal extension, bone invasion, or lymph note metastases are features with an unfavorable prognosis (Barnes 1985).

Of the paranasal sinus cancers, 80% occur in the maxillary sinus. The disease occurs twice as often in men, and 95% of patients are over 40 years of age. It occurs in association with chronic sinusitis, in which long-standing infection leads to squamous metaplasia of the normal respiratory tract epithelium. Cancers of the maxillary antrum have been described in association with long-standing oroantral fistualae. Other malignant tumors that may occur at these anatomical locations (minor salivary gland tumors, melanomas, sarcomas, etc.) will not be duscussed in this chapter.

2.3 Clinical Features

Common symptoms are those of a nonhealing sore, excessive crusting in the nasal interior, epistaxis, rhinorrhea, or unilateral nasal obstruction. Pain is less frequent, and an ulcer, polyp, or sessile growth may be found on physical examination. Of nasal cavity cancers 10% are said to occur bilaterally. Chaudry (1960) classified symptoms of maxillary sinus cancer into five categories. (1) Oral symptoms are due to lesions causing dental disturbance and swelling of the palate. Trismus may occur when the lesion extends into the pterygoid fossa. (2) Nasal symptoms include stuffiness, discharge, or epistaxis. (3) Ocular signs and symptoms appear with upward extension of the tumor through the floor of the orbit, resulting in displacement of the globe, causing diplopia. (4) Facial symptoms occur in the anterior presentation, with bulging in the cheek, obliteration of the nasolabial fold, or numbness and paresthesia due to disruption of the infraorbital nerve. (5) Neurological symptoms occur with involvement of the cranial nerves III-VIII or evidence of meningeal irritation.

2.4 Treatment

The therapeutic options are radiation therapy or surgical resection, separately or in combination. Due to the usually advanced stage of disease at the time of diagnosis, only about one-third of cases are amenable to surgical resection by conventional operation (Larsson and Martensson 1972). Maxillectomy with or without orbital exenteration is the standard operation for squamous cell cancer. Extension of the tumor into the skull base, destruction of the pterygoid plates, extension into the nasopharynx, or lymph node metastases can be regarded as contraindications to surgical treatment for cure. In recent years some patients with this presentation have been considered suitable for craniofacial resection (Sisson et al. 1976). In resectable cancers, a 45% 5-year cure rate has been reported (Spratt and Mercado 1965). In the most advanced cases, a combination of radiation therapy and surgical resection has been recommended, but experience indicates no greater than 25% 5-year survival. In those patients with nonresectable advanced disease, radiation therapy alone yields cure rates of 12%-19% only (Batsakis 1979).

2.5 Papillomas

The mucous membrane of the nasal cavity and sinuses, also known as the schneiderian membrane, is thought to be formed from the neuroectoderm of the olfactory placode and nasopharyngeal mucosa which is endodermal in origin. The schneiderian mucosa gives rise to at least three distinct types of papilloma. The *fungiform papilloma* or *squamous cell papilloma* develops from the anterior nasal septum and only rarely from the lateral nasal wall or paranasal sinuses. The lesions may be solitary or multiple and recur after excision in approximately 30% of cases. These lesions rarely progress to malignancy. The *inverted papilloma* occurs predominantly in males and constitutes approximately 50% of sinonasal papillomas. These lesions arise from the lateral wall of the nose and may extend into the maxillary or ethmoid sinuses and less often into the sphenoid or frontal sinuses. Less than 10% of inverted papillomas arises from the nasal septum, and they are bilateral in approximately 4% of cases. Painless nasal obstruction, epistaxis, anosmia, or frank purulent rhinitis are the presenting symptoms. The histological picture is typical, and the epithelium frequently exhibits varying degrees of atypia. Unless complete excision is performed, recurrence of the inverted papilloma is not uncommon.

Of greater importance is the risk of carcinoma arising in inverted papilloma; this has been observed in approximately 15% of these tumors according to collected experience. Although most malignant tumors are squamous carcinomas, mucoepidermoid cancer and adenocarcinoma have also been described in this setting.

In the majority of cases, inverted papilloma is a histologically and clinically benign lesion whose total excision is important in preventing recurrence and eliminating the risk of malignant transformation. Total surgical excision may be difficult by a conventional intranasal operation, and for this reason, lateral rhinotomy technique as well as medial subtotal maxillectomy technique and other variations have been proposed. The degloving exposure is regarded by many to be the ideal surgical exposure for assuring complete excision of this lesion (Maniglia 1986).

Finally, *cylindical cell papilloma* is a rare lesion accounting for less than 3%-5% of the sinonasal papillomas, with a male preponderance. This lesion occurs on the lateral nasal wall and in the sinuses; it recurs after inadequate excision, and its association with subsequent carcinoma has been described. The treatment strategy is identical to that for inverted papilloma.

Other tumors of the nasal cavity that may be clinically mistaken for a papilloma are allergic polyps, nasopharyngeal angiofibroma, and the very rare esthesioneuroblastoma. There is no evidence so far of the HPV etiology of these tumors, hence they will not be discussed further.

2.6 Papillomavirus Etiology

Since the lining of the nasal space in the anteriormost portion is squamous epithelium, the nasal vestibule is a potential site for replication of papillomaviruses.

All of the three histological types of papillomas discussed above (i.e., cylindrical cell, inverted, and fungiform or squamous cell) have been suspected of having a viral etiology. For squamous cell papilloma, immunoperoxidase assays to detect papillomavirus capsid antigen have been used to analyze paraffin sections of tissue specimens submitted for routine histological examination (Syrjänen et al. 1983; Wu et al., manuscript in preparation). Syrjänen et al. (1983) examined one squamous cell papilloma of a paranasal sinus and found capsid antigen in the nuclei of a few koilocytotic cells and dyskeratotic cells. Wu et al. (in preparation) found capsid antigen in only 25% of squamous cell papillomas in the nasal vestibule, although most lesions showed koilocytosis and dyskeratosis. Using nucleic acid hybridization techniques to identify the types of HPV present in the lesions, Wu et al. (in preparation) found HPV2 (associated with common warts) in two lesions, HPV6 (associated with genital lesions and laryngeal papillomas) in two lesions, and an uncharacterized HPV genome in one lesion. Brandsma et al. (in press) have found HPV11 DNA in three of four patients. These results and behavioral similarities with squamous cell papillomas that occur at other anatomical sites, particularly frequent recurrence and infrequent progression to malignancy, suggest a papillomavirus etiology for these lesions in the nasal vestibule.

Viral etiology has long been considered for papillomas arising in the pseudostratified columnar epithelium or schneiderian mucosa lining the nasal cavity. Kusiak and Hudson (1970) reported seeing 40-50-nm intranuclear bodies in electron micrographs of two of four inverted papillomas examined. More recently, inverted papillomas of the nasal cavity have been examined for HPV capsid antigen and DNA. Strauss and Jenson (1985) did not find capsid antigen in multiple sections from 13 patients and Wu et al. (in preparation) did not find capsid antigen or HPV DNA in specimens from 3 patients with inverted papilloma. However, Brandsma et al., (Brandsma et al., in press) have found HPV11 DNA and capsid antigen in some cases of inverted papilloma. This virus type is also found in the genital tract (Gissmann et al. 1983) where inverted condylomas of the uterine cervix have been described (Meisels and Fortin 1976; Purola and Savia 1977). In addition, histological changes in the respiratory tract epithelium similar to inverted condylomas have been described in cases of laryngeal squamous cell carcinoma by Syrjänen and Syrjänen (1981).

There is a recognized risk of squamous cell carcinoma arising from inverted papillomas of the sinonasal cavity. Strauss and Jenson (1985) examined a squamous cell carcinoma of the sinonasal tract and did not find papillomavirus capsid antigen in the lesions. Nine cases of cylindrical cell papilloma of the nasal cavity did not contain papillomavirus capsid antigen (Costa et al. 1981). However, more extensive investigations to determine the presence of HPV in the nasal cavity and paranasal sinuses are warranted before conclusions are made concerning their role in the etiology of these benign and malignant lesions.

3 Cancers of the Nasopharynx

3.1 Epidemiology

Nasopharyngeal cancer has a well-recognized prevalence in the Far East, specifically in southern China, where this otherwise uncommon cancer is the most frequently encountered neoplasm. The susceptibility to nasopharyngeal cancer appears to be lessened in Chinese emigrants to other parts of the world, although the nasopharyngeal cancer in Western countries still disproportionately affects patients of Chinese ancestry (Buell 1974). Nasopharyngeal cancer occurs less frequently among the Japanese. In addition, areas of the Middle East, Africa, and India have been identified in which a higher than average frequency of nasopharyngeal cancer is present. Patients with nasopharyngeal cancer have antibodies to Epstein-Barr virus (Andersson-Anvret et al. 1979); elevated titres of IgA viral antigen may be a specific characteristic of patients with nasopharyngeal cancer, especially of the undifferentiated or nonkeratinizing varieties (Henle and Henle 1976). The clinical profile of a male with low nasal index, high frequency of severe nonallergic vasomotor rhinitis, poor nutritional habits associated with excess carbohydrates and deficiencies in vitamins A and B, and scanty masculine body hair has been suggested as representing the high-risk subject for nasopharyngeal cancer (Batsakis 1979).

At birth the epithelium of the nasopharynx is pseudostratified and ciliated. With advancing age the ciliated portion is gradually replaced by a stratified, nonciliated epithelium. In the final pattern squamous epithelium lines the posterior wall, a part of the nasopharyngeal surface of the soft palate, and portions of the lateral walls. Later in life the squamous epithelium becomes keratinized, particularly on the posterior and lateral walls. A nonciliated epithelium of multi-celled thickness is frequently designated as "transitional" epithelium and appears at the junction of the ciliated and nonciliated parts; these squamo-ciliary junctions are sites which may be predisposed to developing papillomas and other lesions.

3.2 Pathology

Nasopharyngeal cancers are grouped into three principal types: undifferentiated, squamous cell, and nonkeratinizing carcinoma. The undifferentiated form occurs in children and younger adults, although the average age at diagnosis clusters around 50 years. Squamous cell carcinoma occurs in the older patient (Fernandez et al. 1976); it has the least tendency to metastasize and has the weakest association with Epstein-Barr virus. The undifferentiated and nonkeratinizing cancers tend to form early metastases, particularly to the cervical lymph nodes (Hoppe et al. 1978). These varieties have a strong relationship with Epstein-Barr virus.

3.3 Papillomavirus Etiology

No evidence either for or against the role of papillomavirus in the etiology of nasopharyngeal cancer could be found in the literature.

4 Tumors of the Larynx

4.1 Cancer of the Larynx

4.1.1 Epidemiology

According to the National Cancer Institute Survey program laryngeal cancer constituted 2.3% of all cancers occurring in men in the United States of America between 1973 and 1977. It is a predominantly male disorder with an age-adjusted incidence of 8.5 per 100000 men and 1.3 per 100000 women. This sex differential has diminished in recent decades, which is attributed to increased tobacco consumption by women; the sex ratio was 15 to 1 in the 1950s, but has changed to 5 to 1 during the 1970s.

Approximately 10000-12000 new cancers of the larynx are diagnosed annually in the United States of America (Silverman 1982).

4.1.2 Pathology

4.1.2.1 Squamous Carcinoma. Most cancers of the larynx are of squamous cell type. The most common location of the tumor is at the true vocal cords where the lesion begins as a carcinoma-in-situ. An inconspicuous tumefaction may develop that can ultimately grow to form a bulky or ulcerated tumor. Invasion into the muscle causes vocal cord fixation; deeper infiltration results in distortion and expansion in the subepithelial compartments. Primary tumors of the laryngeal surface of the epiglottis are more common than those on its lingual surface, and cancers of the subglottis and laryngeal ventricle are the least common. The anterior half of the larynx is more frequently involved with cancer than the posterior half.

Primary squamous cell carcinoma of the true vocal cords and of the laryngeal surface of the epiglottis are usually well-differentiated; those of the lingual surface of the epiglottis are usually poorly differentiated; the remainder are intermediate grade (Jakobsson et al. 1973).

Commonly, laryngeal cancer is an isolated lesion, but in some cases widespread epithelial dysplasia with the potential for multifocal preinvasive and invasive carcinoma is seen. The neoplasms grow by direct invasion and, with the exception of the cancers of the true vocal cord, metastases to the cervical lymph nodes are common, particularly in advanced stages.

4.1.2.2 Verrucous Carcinoma. Verrucous carcinoma is one variety of a well-differentiated squamous cell carcinoma, which exhibits locally aggressive, slow growth and generally does not form metastases. Of 105 cases reported by Krauss and Perez-Mesa (1966), 11% occurred in the larynx, predominantly on the true cords, 73% appeared in the oral cavity, particularly involving the buccal mucosa and the gingiva. Another 4% were present in the nasal fossa, and the remainder were found on the glans penis, with scattered cases in the vulva, vagina, scrotum, and perineum. Lesions have also been described in the esophagus, pyriform sinus, nose, paranasal sinuses, external and middle ear, and skin (Krauss and Perez-Mesa 1966). Some 20%-60% of oral lesions are associated with leukoplakia, and 4%-31% show a risk of developing a second verrucous carcinoma. Also, 10%-15% will develop into anaplastic squamous cell carcinoma, a transformation which has been described as usually occurring after radiotherapy (Goethals et al. 1963; Demian et al. 1973).

4.1.2.3 Other Cancers of the Larynx. Cancers of other histological types, for example, those arising from the minor salivary glands, may also occur in the larynx. The relationship of these lesions to human papillomaviruses has not been reported to date.

4.2 Papillomavirus Etiology

The papillomavirus etiology of carcinoma-in-situ of the larynx has been investigated using an immunocytochemical technique to detect viral capsid common antigen by Kashima et al. (1986b). An examination of 60 biopsy specimens collected from 20 cases demonstrated the presence of HPV in 14. Nuclei staining positive for capsid antigen were found in areas of dysplasia as well as in areas adjacent or superficial to dysplastic cells. In two additional cases not shown to be HPV-positive, histological evidence suggestive of a viral effect (i.e., koilocytosis, parakeratosis, and multinucleation) was identified. Thus, HPV should at least be considered as a cofactor in the etiology of these lesions.

Malignant lesions of the larynx have also been examined for HPV. Histological features consistent with papillomavirus infection (as recognized in the uterine cervix) were found in 49 of 166 cases of laryngeal squamous cell carcinoma by Syrjänen and Syrjänen (1981). Epithelial changes resembling papillomatous condyloma (31%), flat condyloma (19%), and inverted condyloma (51%) were distinguished. The same investigators screened paraffin sections from 36 laryngeal squamous cell carcinomas with histological features consistent with papillomavirus infection and found capsid antigen in 42% of the papillomatous lesions, in 33% of the flat lesions, and in 33% of the inverted lesions (Syrjänen et al. 1982a). Kahn et al. (1986) have recently reported the isolation of a new type of viral genome, HPV30, from a laryngeal carcinoma. However, HPV30 DNA could not be detected in a screen of biopsy specimens from 41 laryngeal carcinomas and 31 other head and neck tumors.

There is also evidence for a viral etiology for verrucous carcinoma of the larynx, a variant of squamous cell carcinoma. Abramson et al. (1985) examined biopsy specimens from five patients with verrucous carcinoma. Histological changes characteristic of HPV infection were present in all cases. No capsid antigen was demonstrable, but HPV16-related sequences were detected in DNA extracted from the lesions.

Thus, there is a growing body of information on the presence of HPV in laryngeal squamous cell carcinomas. The role of this virus in the etiology of these cancers must await additional investigations.

4.3 Laryngeal or Respiratory Papillomatosis

Recurrent respiratory papillomatosis (RRP), also known as juvenile laryngeal papillomatosis, is the persistent regrowth of histologically benign papillomas of the larynx that may affect the lower respiratory tract (for review, see Mounts and Shah 1984). The tumors form mulberrylike nodular masses which on microscopical examination appear as stratified squamous epithelium with a vascular connective tissue core. The most common presenting symptom is hoarseness or voice change which is evident on vocalization, because true vocal cords are the primary site most often affected. Lesions may also occur in the nasal vestibule, nasopharynx, all sites within the larynx, trachea (particularly at the site of prior tracheostomy), and in the lower bronchial tree (Weiss and Kashima 1983). Sites of predilection have been identified at lines of junction of squamous to ciliated epithelium (Kashima et al., unpublished data).

Respiratory papillomatosis may occur at any age, but children under the age of 5 years have the highest risk (for review see Mounts and Shah 1984). Many patients with the juvenile onset form continue to have the disease throughout life, although spontaneous remission from papillomatosis has been described at puberty (Stephens et al. 1979; Strong et al. 1979; Szpunar 1967). There is also an adult onset form in which symptoms begin after the age of 20. The age distribution of adult onset cases shows a peak at 30 years of age.

The severity of symptoms is highly variable, but the need for surgery to maintain an open airway is generally considered to be greater in children and may reflect difficulties in dealing with a narrow airway (Weiss and Kashima 1983). Operations may be necessary as frequently as every 7-10 days, and patients who have had in excess of several hundred endoscopic excisions are not unusual. Tracheostomy is generally regarded as undesirable because of the risk of "seeding" lesions into the distal airway. There is a tendency for papilloma to spread into previously uninvolved areas of the respiratory tract including the bronchi, even in the absence of a tracheostomy. More severe disease, characterized by a need for more frequent surgery and greater spread, has been associated with different subtypes of virus (Mounts and Kashima 1984). Host factors, such as immune deficiency, may be involved; the presence of cutaneous warts in a high proportion of RRP patients has been reported (Bjork and Weber 1956).

Patient management has been by endoscopic excision. The utilization of the carbon dioxide laser has facilitated the surgical management of this condition but has not altered the clinical course. A wide variety of agents have been used in unsuccessful attempts to find an effective therapy for this disease. A recent series of pilot studies suggest that α interferon may be effective as an adjuvant to surgery for controlling respiratory papillomatosis (for review see Weck and Whisnant 1985). However, the persistence of papillomavirus DNA in lesions could be demonstrated during the course of interferon therapy, even when the patient was in clinical remission (Mounts et al. 1985). Viral DNA has also been detected in histologically normal tissue of patients with respiratory papillomatosis and in patients who have been disease free for 2 years (Steinberg et al. 1983).

The papillomavirus etiology of respiratory papillomatosis has recently been firmly established. Although a viral etiology had been suspected for many years, the demonstration of virus particles in laryngeal papilloma by electron microscopy has been difficult. Earlier investigations failed to detect virions in laryngeal papillomas (Cook et al. 1973); others found virions in 2 of 27 cases (Boyle et al. 1973), in 1 of 12 (Spoendlin and Kistler 1978), in 3 of 17 (Incze et al. 1972), and in 5 of 6 (Dmochowski et al. 1964).

These results are consistent with immunocytochemical detection of HPV common capsid antigen, which has demonstrated a paucity of antigen-positive cells in laryngeal papillomas. Lack et al. (1980) found antigen in 26 of 35 juvenile onset cases. Lancaster and Jenson (1981) found antigen in 3 of 4 juvenile onset cases. Costa et al. (1981) found antigen in 11 of 19 juvenile onset but in none of 5 adult onset cases. Braun et al. (1982) found antigen in 7 of 15 cases of juvenile onset but in none of 6 adult onset cases. Mounts et al. (1982) found antigen in 2 of 12 juvenile onset cases and in 2 of 8 adult onset cases. Strauss and Jenson (1985) found antigen in 3 of 8 juvenile onset cases and in 4 of 8 adult onset cases. Thus, the demonstration of papillomavirus common capsid antigen in respiratory papillomata was consistent with the viral etiology of these lesions, but antigen could not be demonstrated in all of them.

Nucleic acid hybridizations have shown that the papillomaviruses in laryngeal papilloma are the same as those which infect the genital tract (Mounts et al. 1982; Gissmann et al. 1983). Mounts et al. found HPV6 DNA in 12 cases of juvenile onset and in 8 cases of adult onset respiratory papillomatosis. Gissmann et al. (1982) cloned the viral DNA from a juvenile onset laryngeal papilloma and classified it as HPV11. They identified HPV11 in 10 laryngeal papillomata as well as in 20% of the genital condyloma and HPV6 in three laryngeal papillomata and in 60% of genital condylomata (Gissmann et al. 1983). Identification of the same HPV in the respiratory genital tract has epidemiological implications.

The manner in which genital papillomaviruses reach the respiratory tract is not established and may be different for cases of juvenile and adult onset disease (for review see Mounts and Shah 1984). Intrapartum infection during fetal passage through the infected genital tract of the mother may be the most common mode of transmission for cases occurring early in childhood, since caesarean delivery is only infrequently correlated with cases of juvenile onset respiratory papillomatosis (Shah et al. 1986). It is possible that cases of adult onset disease acquire the infection at birth, but the virus remains subclinical for several decades. Postpartum infection by direct contact with infected individuals is also possible.

Adult onset laryngeal papillomatosis is regarded by some to be a sexually transmitted disorder.

Malignant conversion of squamous papilloma of the respiratory tract has been reported, although it is an uncommon occurrence (for review see Mounts and Shah 1984). Cofactors such as smoking and radiation therapy have been implicated in the etiology of squamous cell carcinomas in RRP patients. Therefore, as is the case with the human genital tract, there is accumulating evidence for the role of papillomaviruses in the etiology of benign and malignant tumors in the respiratory tract; the mechanism by which HPV transforms these epithelial cells is unknown.

5 Cancers of the Hypopharynx

The carcinomas arising in the hypopharynx include lesions of the posterior and lateral hypopharyngeal mucosa, the pyriform sinus, and the postcricoid pharyngoesophagus. It is estimated that the combined number of cases from these sites totals no more than one-third of all laryngeal cancers (Bryce 1971). Most cancers arising at these sites are squamous cell carcinomas.

Lesions on the posterior wall have a tendency to be exophytic, well-differentiated, and metastasize late whereas those arising from the pyriform sinuses are poorly differentiated, and early metastasis is common. Spindle cell carcinoma and other histological variants also occur at both sites. As with most head and neck cancers, the lesions occur predominantly in men. In postcricoid carcinoma, however, a unique female preponderance is noted. The special realtionship to the Plummer-Vinson syndrome (also known as the Paterson-Kelly) of sideropenic anemia, dysphagia, and glossitis has been recognized among Scandinavian women (Wynder et al. 1957). There is no current evidence to suggest a role for HPV in cancers of the hypopharynx.

6 Lung Cancers

6.1 Epidemiology

Lung cancer is the leading cause of death from cancer in the United States of America. The age-adjusted cancer death rates have doubled every 15 years (Silverberg 1983). In 1983, lung cancer accounted for 22% of cancers in males and 9% in females; it was responsible for 35% of male cancer deaths and 17% of female cancer deaths. A broad range of environmental agents have been identified as etiologically relevant: these include tobacco products, asbestos, polycyclic aromatic hydrocarbons, nickel, chromium, and arsenic, as well as a large number of pollutants encountered in the workplace and in the general environment (Fraumeni 1975). The process of cancer development initiated by some of the above agents is thought to evolve over many years before clinically apparent cancer is detectable. In spite of the concern and awareness of the lung cancer problem, many cases newly diagnosed between 1970 and 1973 showed regional metastasis (22%) and distant metastasis (48%) at the time of diagnosis. Only 17% were diagnosed when the disease was at a localized stage (Cancer patient survival 1977).

6.2 Pathology

Conventionally four principal histological types of lung cancer are recognized: epidermoid or squamous carcinoma, small cell carcinoma, adenocarcinoma, and large cell carcinoma. Other uncommon types of lung cancer may also occur.

Squamous cell carcinoma accounts for 40% of cases. The lesions most frequently arise in the segmental and subsegmental bronchi at the bifurcation sites, where exposure to carcinogens and inflammation is greatest (Auerbach et al. 1961). The initial response to injury is basal cell proliferation. The epithelium undergoes progressive squamous metaplasia that may result in malignant transformation leading to carcinoma-in-situ, whence invasive cancer develops. The spread of squamous cell cancer is centripetal.

Small cell carcinoma accounts for 20%-25% of lung cancers. Several subtypes are recognized including the small cell and intermediate cell types. Small cell carcinoma is characterized by early metastases to a broad variety of organs; it has a striking association with cigarette consumption (Rosenow and Carr 1979).

Large cell carcinoma occurs in 10%-15% of patients. Adenocarcinoma, accounting for 20%-25% of cases, is unique inasmuch as it has a definite female preponderance and has the weakest association with tobacco consumption. An overwhelming percentage of nonsmoking subjects developing lung cancer have adenocarcinoma. Bronchiolalveolar carcinoma, a variant of adenocarcinoma, occurs peripherally and can be solitary or mutlifocal. It is compared to jaagsiekte, a virus-induced lung condition occurring in sheep (Marq and Galy 1973).

For therapeutic reasons small cell carcinoma is considered separately from the other histological types which are grouped as non-small cell-type carcinoma.

6.3 Papillomavirus Etiology

Given the replication of HPV in the respiratory tract, as reviewed in this chapter, and the evidence that supports HPV as an etiological factor in squamous cell carcinoma of the genital tract (for review see Syrjänen 1984), it is reasonable to study squamous cell carcinoma of the lung for the presence of HPV. Syrjänen (1979) observed histological changes in the bronchial epithelium adjacent to an invasive squamous cell carcinoma identical to those described for condyloma in the genital tract. He undertook a larger survey on the histological changes in bronchial squamous cell carcinomas (Syrjänen 1980a, b). Histological lesions fulfilling the criteria of papillary condyloma were found in about 5% of cases, of inverted condyloma in about 6%, and of flat condyloma in about 25%.

Stremlau et al. (1985) used mutliple HPV types under nonstringent conditions of hybridization to screen DNA samples extracted from 24 keratinizing or nonkeratinizing, undifferentiated large cell, bronchoalveolar, and small cell carcinomas. HPV16 DNA was found in a pulmonary lesion of one female patient who had had cervical cancer 9 years previously. It is not known whether the pulmonary lesion was a metastasis or a primary tumor.

A suggestion for HPV involvement in the development of lung tumors comes from the behavior of the virus in RRP patients. The spread of papillomatosis throughout the respiratory tract was suggested by Kirchner (1951) to be the result of virus shedding from the laryngeal lesion. Weiss and Kashima (1983) reported a higher incidence of tracheal involvement in individuals who had detectable papillomavirus capsid antigen in a previous biopsy specimen. The virus appears to replicate in epithelial bronchial cells, because capsid antigen was detected immunocytochemically in paraffin-embedded tissue specimens of lung lesions removed at autopsy of two patients with RRP (P. Mounts, unpublished observation). Although malignant transformation in RRP is extremely rare, usually occurring in patients with therapeutic irradiation or a smoking history (for review see Kashima 1987), there is one case report of malignant transformation in a bronchiolalveolar papilloma in a 6-year-old with RRP of 4 years' duration (Solomon et al. 1985). In the review of the literature by Solomon et al., eight cases of carcinoma arising in RRP patients without a history of smoking or radiotherapy were recorded: three patients developed laryngeal carcinomas, and five had pulmonary primary tumors. Thus, there is a growing body of information that HPV may play a role in the development of bronchogenic carcinoma.

7 Cancers of the Esophagus

7.1 Epidemiology

Esophageal carcinoma has a distinct geographical distribution with a high prevalence in the "esophageal cancer belt" extending from the southern shore of the Caspian Sea in Iran throughout Soviet Central Asia and Mongolia, as far as central People's Republic of China. The highest rates of esophageal cancer are observed in the area surrounding the Caspian Sea and in parts of China (Kmet and Mahboubi 1972). The rate of esophageal cancer in the United States of America is low. Tobacco and alcohol consumption, individually and combined, are felt to be etiologically relevant to esophageal cancer. The disease is more common in blacks regardless of geographical location; in the United States of America esophageal cancer occurs three times more often among blacks. Both the incidence and mortality rates from esophageal cancer in black males has increased at a rapid rate in the past 25 years.

Nutritional factors are also assumed to play a major role. Esophageal carcinoma in females has been associated with the Plummer-Vinson syndrome of sideropenic anemia, glossitis, and esophagitis. Hot foods, heavily seasoned and containing tannin-rich ingredients, have been implicated by some. Dietary deficiencies in riboflavin, nicotinic acid, magnesium, and zinc have also been cited as increasing the risk for esophageal cancer.

7.2 Pathology

Fifteen percent of lesions occur in the upper third of the esophagus, 35% in the lower third, and the remainder in the middle third. Squamous cell carcinoma is the predominant type of esophageal cancer. Adenocarcinoma occurs predominantly in the distal third of the esophagus, often in association with Barrett's esophagus. Of primary adenocarcinomas of the esophagus, 86% are found in conjunction with a columnar-lined lower esophagus (CLLE), and 10% of patients with CLLE have a rsik of developing adenocarcinoma. The CLLE is regarded as a metaplastic event in the esophageal epithelium in response to reflux of gastric juices due to lower esophageal sphincter incompetence (Naef et al. 1975; Haggit et al. 1978). On basis of these associations, close monitoring of patients with CLLE is advised, with frequent biopsies to detect potential malignant degeneration and to correct reflux by fundoplication or another similar procedure. Adenocarcinoma occurs more frequently in whites but, in all other respects, shares the features exhibited by squamous cell cancer. Metastatic spread to regional lymph nodes is common and may occur early in the course of the disease.

7.3 Papillomavirus Etiology

The interest in a papillomavirus etiology in esophageal lesions is twofold. First, the esophagus has a stratified squamous epithelial lining and therefore represents a potential site for replication of these viruses. Secondly, bovine papillomavirus has been identified in papillomas in the alimentary tract of cattle where malignant conversion has been attributed to ingestion of bracken fern (Jarrett et al. 1979). The predominant tumor in the human esophagus is squamous cell carcinoma, which epidemiological studies have associated with dietary factors. Therefore, the presence of papillomavirus in benign and/or malignant lesions of the esophagus may indicate that cancer of the esophagus results from the interaction of a virus with environmental factors, such as tobacco and alcohol, as discussed above.

Syrjänen et al. (1982b) were the first to demonstrate immunocytochemically the presence of papillomavirus capsid antigen in squamous cell papilloma of the esophagus. Reports on squamous papilloma of the esophagus are uncommon, and Weitzner and Hentel (1951) estimated the incidence from autopsy studies at less than 0.04%. A more recent study by Winkler et al. (1985) reviewed 75 cases of esophageal lesions diagnosed as papilloma or hyperplasia of the esophagus. Thirteen of the cases had histological changes characteristic of HPV infection: two were squamous papillomas with koilocytosis, acanthosis, dyskeratosis, and occasional multinucleated cells; eleven were focal epithelial hyperplasia with koilocytosis similar to a flat condyloma in the genital tract. Four of these cases, including both papillary lesions, were positive for papillomavirus capsid antigen.

Syrjänen (1982) also reported histological changes identical to those of condylomatous lesions in esophageal squamous cell carcinoma. In an examination of esophageal specimens from 60 patients, epithelial changes similar to papillary condyloma were found in 1, changes similar to inverted condylomas were found in 3, and changes similar to flat condylomas were found in 20.

The demonstration of HPV in histologically benign lesions and the presence of histological changes in malignant lesions suggest that HPV should be considered as a potential etiological agent in esophageal cancer.

8 Conclusion

Cancers arising in the upper aerodigestive tract are predominantly squamous cell carcinomas and tend to arise at sites where stratified and ciliated epithelium are juxtaposed, or where metaplastic changes occur. HPV has been identified in lesions in the nose, oral cavity, larynx, tracheobronchial trees, lung, and esophagus. Although there is evidence for a differential risk for malignant progression depending on HPV types, the interaction with environmental carcinogens appears to be an important consideration in these lesions. Establishing a viral etiology in cancer is highly relevant to the formulation of appropriate management strategies by the clinician.

References

Abramson A, Brandsma J, Steinberg B, Winkler B (1985) Verrucous carcinoma of the larynx: possible human papillomavirus etiology. Arch Otolaryngol 111: 709-715

Acheson ED (1976) Nasal cancer in the furniture and boot and shoe manufacturing industries. Prev Med 5: 295-315

Acheson ED, Cowdell RH, Jolles R (1970) Nasal cancer in Northamptonshire boot and shoe industry. Br Med J 1: 385-393
Ackerman LV (ed) (1970) Cancer. Diagnosis, Treatment, and Prognosis, 4th ed. Mosby, St Louis
American Joint Committee on Cancer (1983) Beahrs OH, Myers MH (eds) Manual for staging of cancer, 2nd edn. Lippincott, Philadelphia
Anderson I, Lad T (1982) Autopsy findings in squamous cell carcinoma of the esophagus. Cancer 50: 1587
Andersson-Anvret M, Forsby N, Klein G, Hehle W, Bjorklund A (1979) Relationship between the Epstein-Barr virus genome and nasopharyngeal carcinoma in Caucasian patients. Int J Cancer 23: 762-767
Auerbach O, Stout AP, Hammond EG (1961) Changes in bronchial epithelium in relation to cigarette smoking and in relation to lung cancer. N Engl J Med 265: 253-269
Bailey BJ (1966) Partial laryngectomy and laryngoplasty: a technique and review. Trans Am Acad Ophthalmol Otol 70: 559-574
Bailey BJ, Barton S (1975) Olfactory neuroblastoma, management and prognosis. Arch Otolaryngol 101: 1
Baker RR, Ball WC, Carter D, Frost J, Marsh B, Stitik F, Tockman M (1979) Identification and treatment of clinically occult cancer of the lung. Prog Cancer Res Ther 11: 243-249
Barnes L (1985) Surgical pathology of the head and neck. Diseases of the nose, paranasal sinuses and nasopharynx. Dekker, New York, pp 403-451
Barton RT (1977) Nickel carcinogenesis of the respiratory tract. J Otolaryngol 6: 412-422
Batsakis JG (1979) Tumors of the head and neck. Clinical and Pathological Considerations, 2nd edn. Williams and Wilkins, Baltimore, p 177-187
Berman JM (1985) Surgical anatomy of the larynx. In: Bailey BJ, Billar HF (eds) Surgery of the larynx. Saunders, Philadelphia, pp 15-26
Bjork H, Weber C (1956) Papilloma of the larynx. Acta Otolaryngol 46: 499-516
Blakeslee D, Vaughn CW, Simpson GT, Shapshay S, Simpson GT, Strong MS (1984) Excisional biopsy in the selective management of T_1 glottic cancer. A three-year follow-up study. Laryngoscope 94: 488-494
Bloom SM (1969) Cancer of the nasopharynx: a study of 90 cases. J Mt Sinai Hosp 36: 277
Boyle WF, Riggs JL, Oshivo L, Lennette EH (1973) Electron microscopic identification of papovavirus in laryngeal papilloma. Laryngoscope 83: 1102-1108
Brandsma J, Abramson A, Sciubba J, Shah K, Barrezenta N, Galli R (in press). Papillomavirus-infection of the nose. Cancer Cells 5: Papillomaviruses. Cold Spring Harbor Press. Cold Spring Harbor, NY
Braun L, Kashima H, Eggleston J, Shah K (1982) Demonstration of papillomavirus antigen in paraffin sections of laryngeal papillomas. Laryngoscope 92: 640-643
Brown LR, Muhm JR (1983) Computed tomography of the thorax. Current perspectives. Chest 83: 806-813
Bryce D (1971) The conventional surgical management of carcinoma of the hypopharynx. J Laryngol Otol 85: 1221-1226
Buell P (1974) The effect of migration on the risk of nasopharyngeal cancer among Chinese. Cancer Res 34: 1189-1191
Cancer patient survival (1977) Report no 5, DHEW publication no NIH77, p 912
Carbone PP, Frost JK, Feinstein AR (1970) Lung cancer: perspectives and prospects. Ann Intern Med 73: 1003-1024
Chaudry AP, Gorlin RJ, Mosser DJ (1960) Carcinoma of the antrum: a clinical and histopathologic study. Oral Surg 13: 269
Chui MC, Briant TDR, Rotenberg D, Gonsalues CC (1982) Computed tomography and angiofibroma of the nasopharynx. J Otolaryngol 11: 327-330
Cohen MH (1977) Signs and symptoms of bronchogenic carcinoma. In: Strauss MJ (ed) Lung cancer: clinical diagnosis and treatment. Grune and Stratton, New York, pp 85-94
Cook TA, Cohn AM, Brunschwig JP, Goepfert H, Butel JS, Rawls WE (1978) Laryngeal papilloma: etiologic and therapeutic considerations. Ann Otol Rhinol Laryngol 82: 649-655
Cortese DA, Kinsey JH (1982) Hematoporphyrin-derivative phototherapy for local treatment of cancer of the tracheobronchial tree. Ann Otol Rhinol Larnygol 91: 652-655
Costa J, Howley PM, Bowling MC, Howard R, Bauer WC (1981) Presence of human papilloma viral antigens in juvenile multiple laryngeal papilloma. Am J Clin Pathol 75: 194-197

Demian SDE, Buskin FL, Achevarria RA (1973) Perineural invasion and anaplastic transformation of verrucous carcinoma. Cancer 32: 395-401
Dmochowski L, Grey CE, Sykes JA, Dreyer DA, Langford P, Jesse RH, MacComb WS, Ballantyne AJ (1964) A study of the submicroscopic structure of virus particles in cells of human laryngeal papilloma. Tex Rep Biol Med 22: 454-491
Earlam R, Cunha-Melo JR (1980) Oesophageal squamous cell carcinoma. A critical review of surgery. Br J Surg 67: 381
English GM, Hemenway WG, Cundy RL (1972) Surgical treatment of invasive angiofibroma. Arch Otolaryngol 96: 312-318
Epstein MA, Achong B (eds) (1979) The Epstein-Barr virus. Springer, Berlin Heidelberg New York
Fraumeni JF (1975) Respiratory carcinogenesis: an epidemiologic appraisal. JNCI 55: 1039-1046
Fernandez CH, Cangir A, Samaan NA, Rivera R (1976) Nasopharyngeal carcinoma in children. Cancer 37: 2787-2791
Fitzpatrick PJ, Rider WD (1973) The radiotherapy of nasopharyngeal angiofibroma. Radiology 109: 171-178
Fleischer D, Kessler F, Hage O (1982) Endoscopic Nd: YAG laser therapy for carcinoma of the esophagus: palliative approach. Am J Surg 14: 328
Fontana RS (1977) Early diagnosis of lung cancer. Am Rev Respir Dis 116: 399-402
Fowler JF, LaRonde LA (1983) Ronde-Radiation sciences and medical radiology. Radiother Oncol 1: 1
Frazell EL, Lewis JS (1963) Cancer of the nasal cavity and accessory sinuses - a report of the management of 416 patients. Cancer 16: 1293-1301
Gates GA, Ryan NJ, Cooper JC (1982) Current status of laryngectomee rehabilitation: results of therapy. Am J Otolaryngol 3: 1-14
Gissmann L (1984) Papillomaviruses and their association with cancer in animals and in man. Cancer Surv 3: 161-181
Gissmann L, Diehl V, Schultz-Loulon HJ, zur Hausen H (1982) Molecular cloning and characterization of human papillomavirus DNA derived from a laryngeal papilloma. J Virol 44: 393-400
Gissmann L, Wolnick L, Ikenberg H, Koldovsky U, Schnurch HG, zur Hausen (1983) Human papillomavirus type 6 and 11 DNA sequences in genital and laryngeal papillomas and in some cervical cancers. Proc Natl Acad Sci USA 80: 560-563
Goethals PL, Harrison EG, Devine KD (1963) Verrucous squamous carcinoma of the oral cavity. Am J Surg 106: 845-851
Haggit RC, Tryzelaar J, Ellis FH, Colcher H (1978) Adenocarcinoma complicating columnar epithelium-lined (Barrett's) esophagus. Am J Clin Pathol 70: 15
Halperin WE, Goodman M, Stayner L, Elliott LJ, Keenlyside RA, Landrigan PJ (1983) Nasal cancer in a worker exposed to formaldehyde. JAMA 249: 510-512
Heit HA, Johnson LF, Siegel SR, Boyce HW (1978) Palliative dilatation for dysphagia in esophageal carcinoma. Ann Intern Med 89: 629
Henle W, Henle G (1976) Epstein-Barr virus specific IgA serum antibodies as an outstanding feature of nasopharyngeal carcinoma. Int J Cancer 17: 1-17
Hoppe RT, Williams J, Warnke R, Goffinet DR, Bagshaw MA (1978) Carcinoma of the nasopharynx - the significance of histology. Int J Radiat Oncol Biol Phys 4: 199-205
Howley PM (1983) Papovaviruses: search for evidence of possible association with human cancer. In: Phillips LA (ed) Viruses associated with cancer, pp 253-305
Hyams V, Batsakis J, Michaels L (1986) Tumors of the upper respiratory tract and ear. Armed forces Institute of Pathology, Washington DC
Incze JS, Lui PS, Strong MS, Vaughn CW, Clemente MP (1972) The morphology of human papillomas of the upper respiratory tract. Cancer 39: 583-589
Inouye T, Shigematsu Y, Sato T (1973) Treatment of carcinoma. The hypopharynx. Cancer 31: 649-655
Jakobsson PA, Eneroth CM, Killander D, Mobergen G, Martensson B (1973) Histologic classification and grading of malignancy in carcinoma of the larynx. Acta Radiol Ther 12: 1-8
Jarrett WFH, McNeil PE, Laird HM, O'Neil BW, Murphy J, Campo MS, Moar MH (1979) Papillomaviruses in benign and malignant tumors of cattle. In: Essex M, Todaro G, zur Hausen H (eds) Viruses in naturally occurring cancers. Cold Spring Harbor conferences on cell proliferation, vol 7. Cold Spring Harbor, New York, pp 215-222

Jesse RH (1975) The efficacy of combining radiation therapy with a surgical procedure in patients with cervical metastases from squamous cell cancer of the oropharynx and hypopharynx. Cancer 35: 1163-1166

Kahn T, Schwarz E, zur Hausen H (1986) Molecular cloning and characterization of the DNA of a new human papillomavirus (HPV30) from a laryngeal carcinoma. Int J Cancer 37: 61-65

Kashima H, Mounts P, Kuhajda F, Goodstein M, Leventhal B (1987) Sites of predilection in recurrent respiratory papillomatosis (in preparation)

Kashima H, Mounts P, Kuhajda F, Lowry M (1986b) Demonstration of HPV capsid antigen in carcinoma-in-situ of the larynx. Laryngoscope (to be published)

Katlic M, Carter D (1979) Prognostic implications of histology, size and location of primary tumors. Prog Can Res Ther 11: 143-150

Kirchner JA (1951) Papilloma of the larynx with extensive lung involvement. Laryngoscope 61: 1022-1029

Kmet J, Mahboubi M (1972) Esophageal cancer in the Caspian littoral of Iran. Initial studies. Science 75: 846

Koss L (1979) Diagnostic cytology and its histopathologic basis, 3rd edn. Lippincott, Philadelphia

Krauss FT, Perez-Mesa C (1966) Verrucous carcinoma: clinical and pathological study of 105 cases involving oral cavity, larynx and genitalia. Cancer 19: 26-38

Kusiak RJ, Hudson WR (1970) Nasal papillomatosis. South Med J 63: 1277-1280

Lack EE, Jenson AB, Smith HG, Healy GB, Pass F, Vawter GF (1980) Immunoperoxidase localization of human papillomavirus in laryngeal papillomas. Intervirology 14: 148-154

Lancaster WD, Jenson AB (1981) Evidence for papillomavirus genus specific antigens and DNA in larnygeal papilloma. Intervirology 15: 204-212

Larsson S (1976) Mediastinoscopy in bronchogenic carcinoma. A study of 486 cases with special reference to the indications and limitations of the method. Scand J Thorac Cardiovasc Surg 19: 23

Larsson LG, Martensson G (1972) Maxillary antral cancer. JAMA 219: 342

Lederman M (1961) Cancer of the nasopharynx: its natural history and treatment. Thomas, Springfield

Little JB, Schulz MD, Wang CC (1963) Radiation therapy for cancer of the nasopharynx. Arch Otolaryngol 77: 621

Maniglia AJ (1986) Indications and techniques of midfacial degloving. Arch Otolaryngol Head Neck Surg 112: 750-752

Marks J (1978) Carcinoma of pyriform sinus - an analysis of treatment results and patterns of failure. Cancer 41: 1008-1015

Marq M, Galy P (1973) Bronchioloalveolar carcinoma. Clinical pathologic relationships, natural history, and prognosis in 29 cases. Am Rev Respir Dis 107: 621-629

Meisels A, Fortin R (1976) Condylomatous lesions of the cervix and vagina: I. Cytologic patterns. Acta Cytol 20: 505-510

Mountain CF (1977) Assessment of the role of surgery for control of lung cancer. Ann Thorac Surg 24: 365-373

Mounts P, Kashima H (1984) Association of human papillomavirus subtype and clinical course in respiratory papillomatosis. Laryngoscope 94: 28-33

Mounts P, Shah KV (1984) Respiratory papillomatosis: etiological relation to genital tract papillomaviruses. Prog Med Virol 29: 90-114

Mounts P, Shah KV, Kashima H (1982) Viral etiology of juvenile and adult onset squamous papilloma of the larynx. proc Natl Acad Sci USA 79: 5425-5429

Mounts P, Wu T-C, Leventhal BG, Kashima H, Dedo H, Singleton G, Gall S, Weck P, Whisnant J (1985) Analysis of human papillomavirus type 6 in the respiratory and genital tracts during interferon therapy. In: Hawley PH, Broker TR (eds) Papillomaviruses: molecular and clinical aspects. UCLA Symp 32: 137-154

Naef AP, Savary M, Ozzello L (1975) Columnar lined lower esophagus: an acquired lesion with malignant predisposition. J Thorac Cardiovasc Surg 70: 826

Newbill ET, Johns ME, Cantress RW (1985) Ethesioneuroblastoma: diagnosis and management. South Med J 78: 275-282

Novick WH, Shima G, Ryder DR, Pirozyski WJ, Hazel JJ, Bouchard J (1965) Malignant neoplasms of the nasopharynx. Can Med Assoc J 93: 303

Ogura JH (1960) Partial laryngopharyngectomy and neck dissection for pyriform sinus cancer. Larnygoscope 70: 1399-1417

Ogura JH, Biller H (1976) Roles and limitations of conservative surgical therapy for laryngeal cancer. In: Alberti PW, Boyce PD (eds) Centennial conference on laryngeal cancer. Appleton-Century Crofts, New York, pp 392-394
Olofsson J, Renouf JH, Van Nostrand P (1973) Laryngeal carcinoma: correlation of roentgenography and histopathology. AJR 117: 526-539
Ong GB, Lee TC (1960) Pharyngogastric anastomosis after esophageal-pharyngectomy for carcinoma of the hypopharynx and cervical esophagus. Br J Surg 48: 193-200
Orth G, Favre M, Croissant O (1977) Characterization of a new type of human papillomavirus that causes skin warts. J Virol 24: 108-120
Purola E, Savia E (1977) Cytology of gynecologic condyloma acuminatum. Acta Cytol 21: 26-30
Rauch S, Seifert G, Gorlin RJ (1970) Diseases of the salivary glands. In: Thoma KH, Gorlin RJ, Goldman HM (eds) Thoma's oral pathology. Mosby, St Louis, pp 962-1018
Rosenow EC, Carr DT (1979) Bronchogenic carcinoma. CA 29: 233-246
Rush JA (1981) Paralysis of cranial nerves III, IV and VI. Causes and prognosis in 1000 cases. Arch Ophthalmol 99: 76-79
Sanderson DR, Fontana RS (1975) Early lung cancer detection and localization. Ann Otol Rhinol Larnygol 84: 583-589
Schiff M (1959) Juvenile nasopharyngeal angiofibroma: a theory of pathogenesis. Laryngoscope 69: 981-1016
Schuchmann GF, Heydorn WH, Hall RV (1980) Treatment of esophageal carcinoma. A retrospective review. J Thorac Cardiovasc Surg 79: 67
Shah KV, Kashima H, Polk BF, Shah F, Abbey H, Abramson A (1986) Rarity of caesarean delivery in cases of juvenile onset respiratory papillomatosis. J Obstet Gynecol 68: 795-799
Silverberg E (1983) Cancer statistics, 1983. Cancer 33: 9
Silverman DT (1982) Survival of cancers of the respiratory system. NIH publication no. 82: 1540
Sisson GA, Bytell DE, Becker SP (1986) Carcinoma of the paranasal sinuses and craniofacial resection. J Laryngol Otol 90: 59-68
Solomon D, Smith RL, Kahima H, Kramer S, Leventhal BG (1985) Malignant transformation in non-irradiated recurrent respiratory papillomatosis. Laryngoscope 95: 900-904
Som ML (1970) Conservative surgery for carcinoma of supraglottis. JLO 84: 655-678
Spoendlin M, Kistler O (1978) Papovavirus in laryngeal papilloma. Arch Otorhinolaryngol 218: 289-292
Spratt JS, Mercado R (1965) Therapy and staging in advance cancer of the maxillary antrum. Am J Surg 110: 502-513
Steiger Z, Franklin R, Wilson R (1981) Eradication and palliation of squamous cell carcinoma. The esophagus with chemotherapy, radiotherapy and surgical therapy. J Thorac Cardiovasc Surg 82: 713
Steinberg BM, Topp WC, Schneider PS, Abramson A (1983) Laryngeal papillomavirus infection during clinical remission period. N Engl J Med 308: 1256-1264
Stenback F (1973) Glandular tumors of the nasal cavity induced by diethyl nitrosamine in Syrian golden hamsters. JNCI 51: 895-901
Stephens CB, Arnold GE, Botchko GM, Hardy CL (1979) Autogenous vaccine treatment in juvenile laryngeal papillomatosis. Laryngoscope 89: 1689-1696
Strauss M, Jenson AB (1985) Human papillomavirus in various lesions of the head and neck. Otolaryngol Head Neck Surg 93: 342-346
Stremlau A, Gissmann L, Ikenberg H, Stark M, Bannasch A, zur Hausen H (1985) Human papillomavirus type 16 related DNA in anaplastic carcinoma of the lung. Cancer 55: 1737-1740
Strong MS, Vaughan CW, Healy GB, Cooperband SR, Clemente MA (1976) Recurrent respiratory papillomatosis: management with CO_2 laser. Ann Otol Rhinol Laryngol 85: 508-516
Strong MS, Vaughan CW, Healy GB (1979) Recurrent respiratory papillomatosis. In: Healy GB, McGill A (eds) Laryngo-tracheal problems in the pediatric patient. Thomas, Springfield, pp 88-98
Syrjänen K (1979) Condylomatous changes in neoplastic bronchial epithelium. Respiration 38: 299-304
Syrjänen K (1980a) Epithelial lesions suggestive of a condylomatous origin found closely associated with invasive bronchial squamous cell carcinomas. Respiration 40: 150-160
Syrjänen K (1980b) Bronchial squamous cell carcinomas associated with epithelial changes identical to condylomatous lsesions of the uterine cervix. Lung 158: 131-142

Syrjänen K (1982) Histological changes identical to those of condylomatous lesions found in esophageal squamous cell carcinomas. Arch Geschwulstforsch 52: 283-292

Syrjänen K (1984) Current concepts of human papillomavirus infections of the genital tract and their relationship to intraepithelial neoplasia and squamous cell carcinoma. Obstet Gynecol Surv 39: 252-265

Syrjänen KJ, Syrjänen SM (1981) Histological evidence for the presence of condylomatous epitehlial lesions in association with laryngeal squamous cell carcinoma. ORL 43: 181-194

Syrjänen K, Syrjänen SM, Pyrhonen S (1982a) Human papillomavirus antigens in lesions of laryngeal squamous cell carcinomas. ORL 44: 323-334

Syrjänen K, Pyrhonen S, Aukee S, Koskela S (1982b) Squamous cell papilloma of the esophagus: a tumor probably caused by human papillomavirus. Diagn Histopathol 5: 291-296

Syrjänen KJ, Pyrhonen S, Syrjänen SM (1983) Evidence suggesting human papillomavirus etiology for the squamous cell papilloma of the paranasal sinus. Arch Geschwulstforsch 53: 77-82

Szpunar J (1967) Larnygeal papillomatosis. Acta Otolaryngol 63: 74-86

Vermund H (1970) Role of radiotherapy in cancer of the larynx as related to the TNM system of staging. Cancer 25: 485-504

Vincent RG, Chu TM, Lane WW (1979) Carcinoembryonic antigen as a monitor of successful surgical resection in 130 patients with carcinoma of the lung. Prog Cancer Res Ther 11: 191-198

Virtue A (1972) The relationship between the refining of nickle and cancer of the nasal cavity. Can J Otolaryngol 1: 37-42

Wang KP, Terry PB (1983) Transbronchial needle aspiration in the diagnosis and staging of bronchogenic carcinoma. Am Rev Respir Dis 127: 344-347

Weck P, Whisnant J (1985) Clinical approaches to human papillomavirus diseases: the use of interferons. In: Howley PM, Broker TR (eds) Papillomaviruses: molecular and clinical aspects. UCLA Symp 32: 185-200

Weiss M, Kashima H (1983) Tracheal involvement in laryngeal papillomatosis. Laryngoscope 93: 45-48

Weitzner S, Hentel W (1951) Squamous papilloma of the esophagus: case report and review of the literature. Am J Gastroenterol 50: 391-396

Wynder EL, Hultberg S, Jacobson F, Bross IJ (1957) Environmental factors in cancer of the upper alimentary tract: a swedish study with special reference to Plummer-Binson (Patterson-Kelly) syndrome. Cancer 10: 470-487

Winkler B, Capo V, Reumann W, Averill MA, Laporta R, Reilly S, Green PR, Richart RM, Crum C (1985) Human papillomavirus infection of the esophagus. Cancer 55: 149-155

Wu YK, Huang GJ, Shao LF, Zhang YD, Lin XS (1982) Progress in the study of surgical treatment of cancer of the esophagus in China, 1940-1980. J Thorac Cardiovasc Surg 84: 325

Wu T-C, Kashima H, Mounts P (1987) Papillomavirus etiology of nasopapillomata (in preparation)

Papillomavirus-Induced Tumors of the Skin: Cutaneous Warts and Epidermodysplasia Verruciformis

Elke-Ingrid Grußendorf-Conen

1 Introduction

Well known since ancient times, cutaneous warts represent epidermal proliferations of varied clinical appearance. At the beginning of the first century AD, Celsus (25) described three types of warts: acrochordons, usually occurring in children and often disappearing spontaneously; thymion, a vascular papillomatous lesion; and myrmecia, resembling our plantar warts. In the following centuries little attention seems to have been paid to their aetiology. John Payne (1891) was the first to record their infectious nature. Later on, in 1907, their inoculability was demonstrated through Ciuffo's experiments, which suggested the viral nature of a transmissible agent. The virus material itself was isolated in microcrystalline form by Strauss et al. in 1949. In the subsequent 2 decades it was assumed that a single virus - the human wart virus - was responsible for all infectious warty lesions of the skin. It was identified as a double-stranded DNA virus, belonging to the Papovaviridiae family (Pfister 1984). Nowadays, the plurality of human papillomaviruses (HPV) is well-established. More than 40 different types can be isolated and characterised by modern techniques of molecular biology.

1.1 Incidence of Cutaneous Warts

Cutaneous warts are common. Their frequency in the general European and American population is estimated to be 7%-10% (Laurent and Kienzler 1985). Warts are rare before the age of 5 years, their incidence increasing as children become older. It reaches a maximum between the ages of 10 and 14 years, declines rapidly until 20 years old, then gradually increases after 25 years. The age distribution differs somewhat in patients with verrucae vulgares and those with plantar warts, suggesting different modes of acquisition (Barr and Coles 1966, 1969).

1.2 Transmission of Infection

The lack of a convenient cell culture system or of animal models for the replication of HPV makes the study of wart infectivity very difficult. HPV replicate in the benign epithelial tumours they induce. Therefore, the natural main source of the virus is a patient with papilloma. The infected horny skin scales are disseminated in the environment. The transmission of the viruses may be direct or indirect, through the ground or contaminated objects. The length of time a virus can remain active in the environment is unknown. It can be shown that the virus becomes inactivated by heating above 55 °C. The virus can be stored for many months at −25 °C without losing its activity (Bunney 1982). It is assumed that infectious virus particles penetrate through breaks in the epithelium and enter one or more basal cells. Since neither viral DNA nor viral protein has yet been detected in the basal cells of ordinary warts, the replication and expression of the viral genome must be maintained at very low levels in these cells. The migration of epithelial cells from the basal layer to the upper layers of the epidermis and their associated squamous differentiation are accompanied by changes in the cellular environment that apparently promote the replication of the viral DNA to detectable levels and the expression of viral proteins (Lowy and Schiller 1985). The viral proteins and DNA are assembled in the nuclei of granular cells into mature virions that can be seen by electron microscopy (Fig. 1). Release of the infectious virus into the environment occurs with sloughing of the stratum corneum.

The incubation period of cutaneous warts varies from 1 to 20 months.

1.3 Classification

The traditional classification of cutaneous warts is based on clinical appearance and location. It includes the following four clinical types: verruca vulgaris or common wart, deep hyperkeratotic palmoplantar wart or myrmecia, superficial mosaic type palmoplantar wart, and verruca plana or flat wart.

The discovery of the multiplicity of HPV types associated with skin warts raised the problem of the relationship of clinical morphology to distinct types of HPV. A

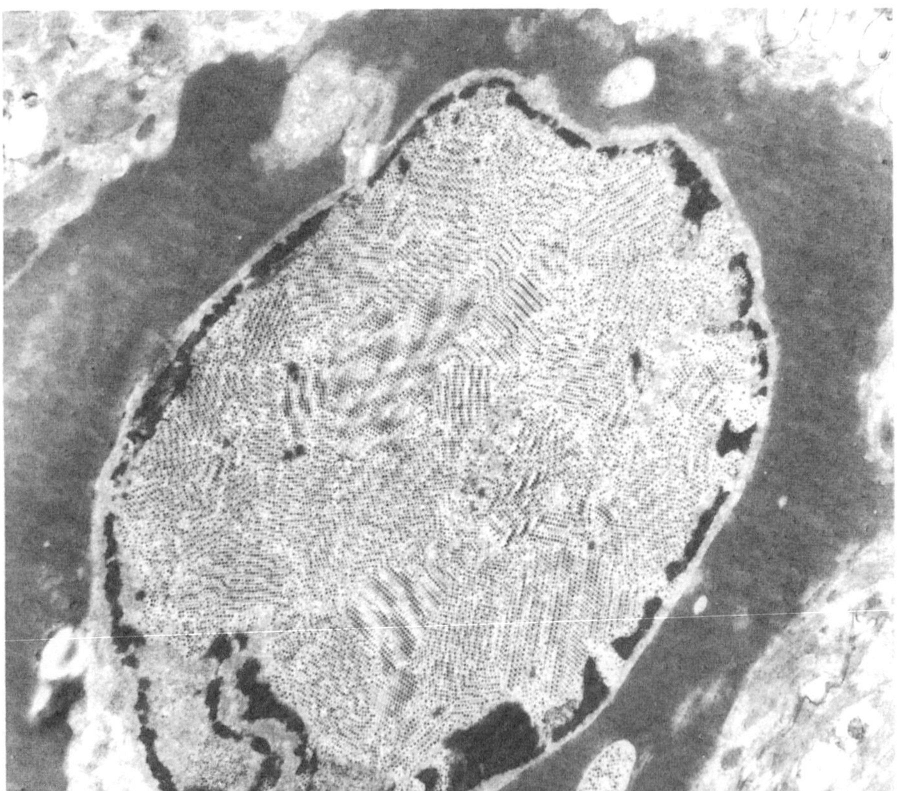

Fig. 1. Virus particles in pseudocrystalline arrangement within a nucleus of a parakeratotic wart cell. Electronmicrograph × approx. 17000

correlation between wart morphology and certain HPV types was first suggested by the characterisation of different HPVs from different types of lesions. Thus HPV1 was found in deep plantar warts, HPV2 could be shown in common hand warts, and HPV3 occurred in plane warts and flat, wartlike lesions of patients with epidermodysplasia verruciformis. At present, more than 40 types of HPV are recognised, with several subtypes for many of them. HPV1-4, 7, 10, 26-28 are specifically associated with cutaneous warts. Owing to the multiplicity of types, the problem of specific or preferential association of distinct HPVs with warts differing in morphology and location is still controversial. A better correlation between virus type and microscopical changes in wart tissue seems to exist if the cytopathic effects are taken into consideration (Jablonska et al. 1985; Croissant et al. 1985). These differences, however, do not modify the clinical appearance of the lesions.

2 Verrucae Vulgares

Common warts begin as little, flat or dome-shaped, firm, skin-coloured papules. With increasing elevation their surface becomes rough, irregular, and hyperkeratotic. Their size varies from a few millimeters to more than 1 cm, and their colour changes to gray or brownish. They may occur singly or in clusters and are sometimes confluent and hypertrophic, especially in a periungual location (Fig. 2). They

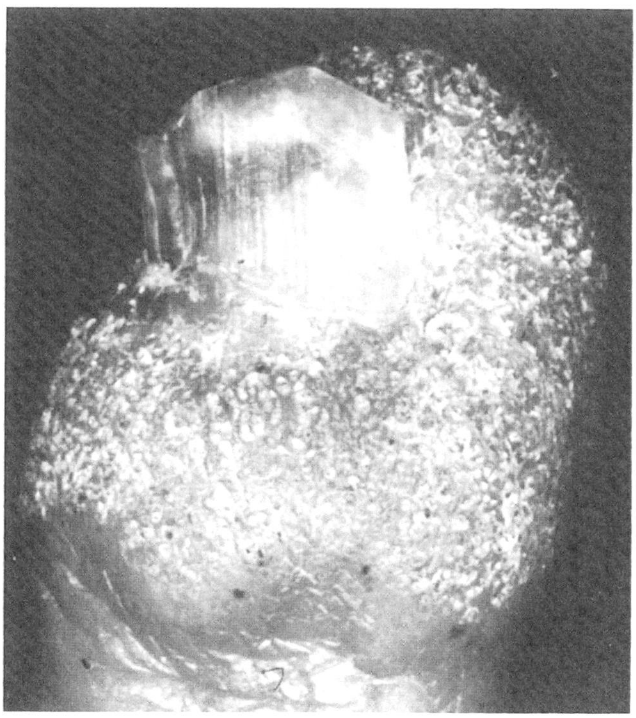

Fig. 2. Confluent wart masses in periungual location

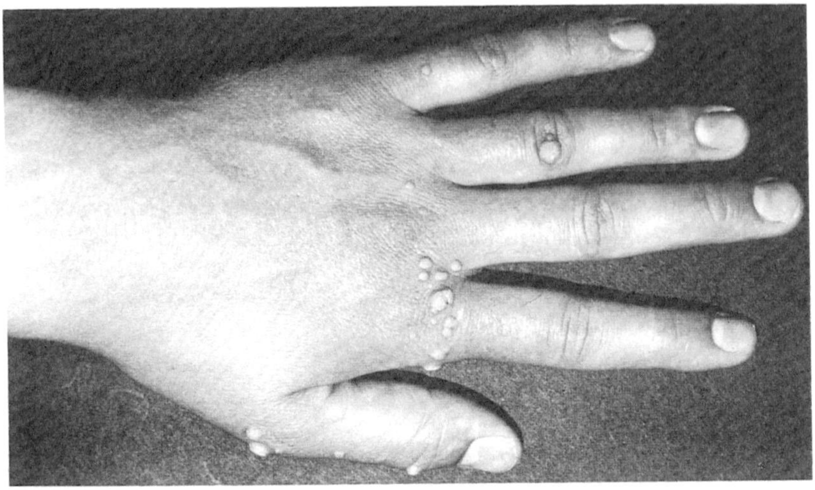

Fig. 3. Multiple common hand warts

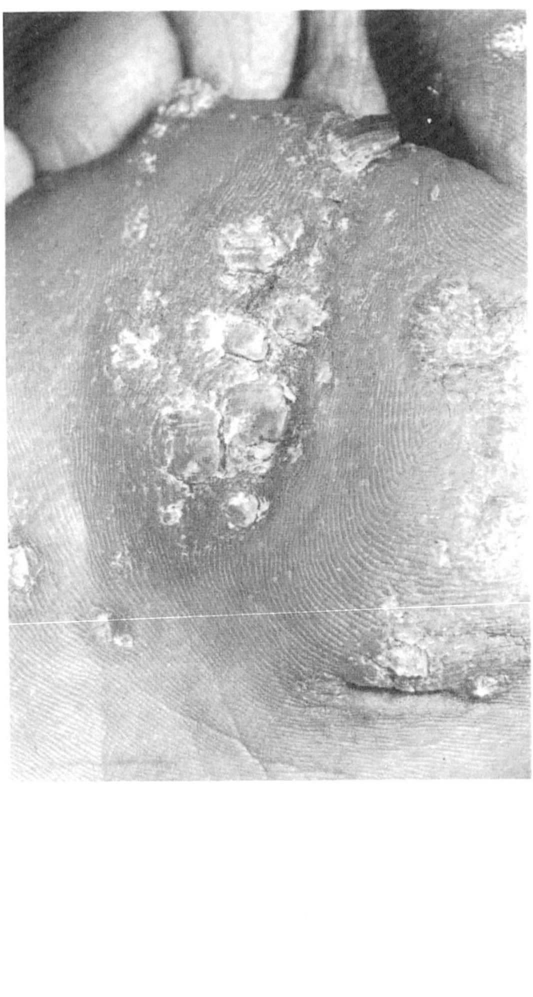

Fig. 4. Mosaic warts of the sole

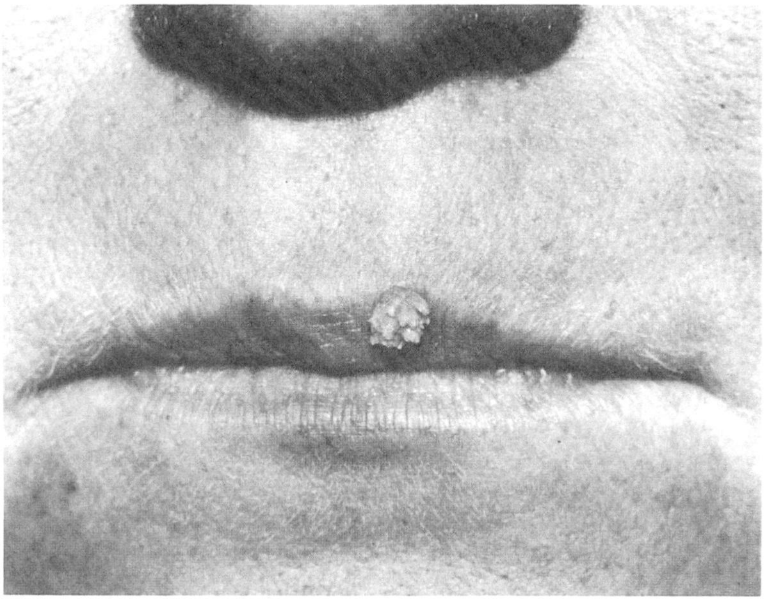

Fig. 5. Multidigitate wart of the lip

may appear anywhere on the skin, including the thin epithelia of the anal and genital regions and the vermillion of the lips and very rarely on the mucous membranes, but they are most commonly localised on the dorsal aspects of the fingers and hands (Fig. 3). On the soles common warts usually appear as *mosaic warts* (Montgomery 1928). They are painless, only slightly raised above the skin level, multiple, and often so closely set that they may impinge one against the other and form a coherent plate of warts (Fig. 4). There may be only one large plaque, but it is common to find several small warts of the same structure in its neighbourhood (Rasmussen 1958). *Endophytic* common warts also appear characteristically in palmar and plantar localisations. They are usually multiple, painless, and have a slightly raised hyperkeratotic surface. Sometimes they are surrounded by a horny wall which defines the size of the lesion. If they are small, they may resemble keratosis punctata (Jablonska et al. 1985).

On the face, head, neck and in flexures, verrucae vulgares can become elongated and papillomatous. Depending on their clinical characteristics and size, they are often referred to as *filiform, digitate, or multidigitate warts* (Fig. 5). Hyperproliferative, papillomatous common warts with a cauliflowerlike, vegetating appearance have been found on the hands of butchers (Orth et al. 1981; Jablonska et al. 1981).

2.1 Virological Findings

Common warts are preferentially associated with HPV2 and HPV4. HPV7 is a common finding in butchers' warts.

2.2 Histological Features

Acanthosis, papillomatosis, and hyperkeratosis are the general architectural hallmarks of cutaneous papillomas such as common warts. The rete ridges are elongated and at the periphery of the verruca; they are sometimes bent inward so that they appear to point radially toward the center. In filiform warts the papillae are conspicuously elongated and contain dilated capillaries; small areas of haemorrhage may be seen in the thickened horny layer.

The characteristic features that distinguish verrucae vulgares from other papillomas are focal parakeratosis, sometimes with columnar arrangement of cells, foci of clumped keratohyalin granules and vacuolated cells, representing the cytopathic virus effect. These virus-induced changes provide the only histological clue for a distinction between warts of different viral origins.

The cytopathic effect characteristic of *HPV2*-induced warts consists in clearing of the cytoplasm of granular cells, which normally contain numerous, condensed, keratohyalin granules of various sizes, shapes, and stainability (Orth et al. 1981; Laurent et al. 1982; Croissant et al. 1985). These clear cells may be widespread within the whole granular layer.

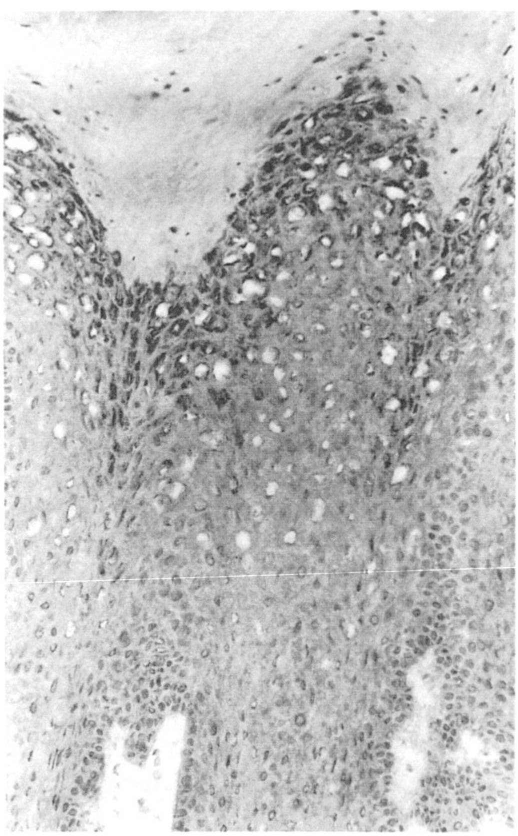

Fig. 6. Histological section of HPV4 wart; H&E × 180

HPV4-induced warts show with routine staining large vacuolated cells with small, crescentic, peripherally located nuclei and almost no detectable keratin granules in the squamous and granular layer (Orth et al. 1981; Croissant et al. 1985). These cells appear isolated in the lower part of the malpighian layers and are often arranged in clusters, especially in the upper parts of the stratum spinosum and stratum granulosum. The clear cells in the granular layer are surrounded by heavily stained cells with very small keratohyalin granules (Fig. 6).

The cytopathic effect of HPV7-induced warts consists in vacuolation of cells with centrally located nuclei and no detectable keratohyalin granules (Orth et al. 1981; Croissant et al. 1985; Jablonska and Orth 1985). The clear cells, sometimes binucleated, are most often isolated, although they may also appear in clusters, and are exclusively limited to the rete ridges. They are surrounded by heavily stained granular cells, containing small- to medium-sized keratohyalin granules.

3 Myrmecia

The prototype of myrmecia is the single, painful palmoplantar wart deeply set in the skin (Fig. 7). It is usually associated with considerable tenderness and may show swelling and redness. When it occurs multiply, the individual lesions do not coalesce. Myrmecia do not only occur on the palms and soles, but also on the lateral aspects of the fingers and toes, under the nails, on the pulp of the digits and, rarely, on the face, scalp and body. Usually the deep palmoplantar wart is covered with a thick callus. When this horny plate is removed, the wart tissue appears soft and crumbly and shows a white opaque colour.

This sort of wart occurs frequently in children between 5 and 15 years of age.

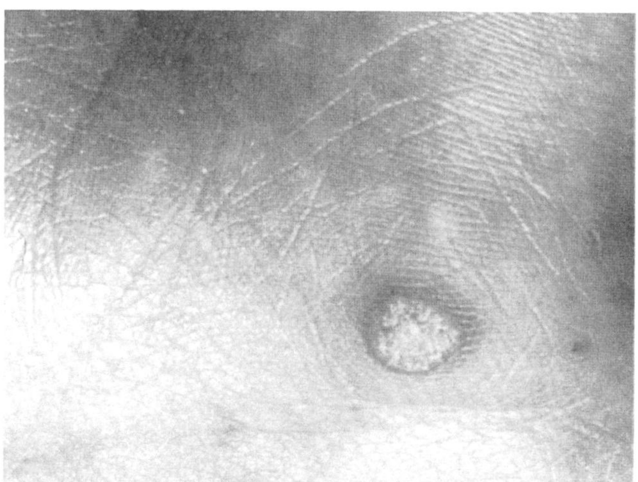

Fig. 7. Myrmecia of the palm

3.1 Virological Findings

Myrmecia are associated with HPV1 (Jablonska et al. 1985).

3.2 Histological Features

First described as an inclusion wart (Strauss et al. 1950; Lyell and Miles 1951), myrmecia have highly characteristic histological features. Even under low power a gross distortion of the architecture of the prickle cell layer is to be seen (Grußendorf 1980). Beginning within the first to third suprabasal cell layer, the cytoplasm of the malpighian cells becomes progressively clearer and the cells larger and irregular. The enlarged nuclei show margination of the chromatin; the nuclear substance changes first from a translucent to a ground glass appearance and then becomes basophilic. An actual granular layer does not exist. Eosinophilic cytoplasmic inclusions, sometimes also within the nuclei are detectable throughout the whole stratum malpighii and become larger and more numerous in the upper layers. Parakeratotic cells contain large, strongly basophilic nuclei bearing the virus. The stratum corneum always shows a pronounced basket weave or honeycomb appearance (Fig. 8).

4 Verrucae Planae and Intermediate Warts

Verrucae planae are slightly elevated, small, skin-coloured papules (Fig. 9). They always appear as multiple, irregularly disseminated or grouped lesions sometimes in linear distribution. The palmar and the dorsal aspects of the hands are most

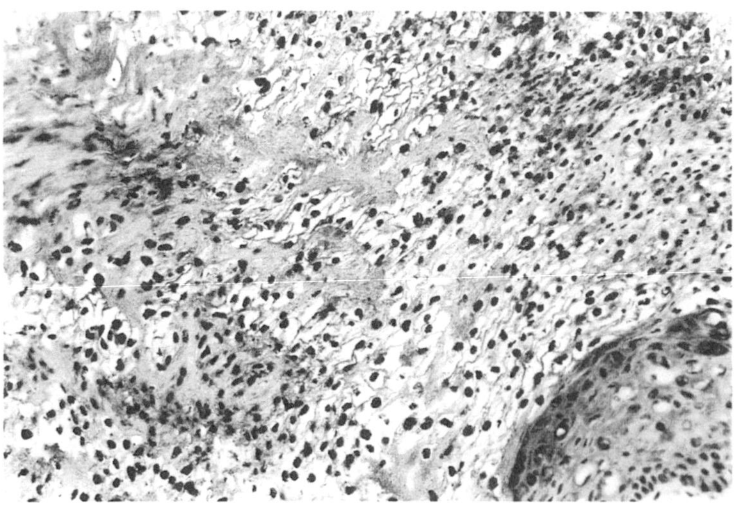

Fig. 8. Histological section of HPV1 wart; H & E × 180

commonly affected; rarely they occur on the forearms and lower extremities. Jablonska et al. (1985) applied the term *intermediate warts* to lesions that cannot be classified as common or plane warts, as they combine clinical features of both. If they are hyperkeratotic, raised, and coalescent, these papillomas differ from common warts by a flatter surface. Typical flat warts are often found between the raised ones. The clinical recognition of intermediate warts and their differentiation from young common warts may be impossible. The histology is of diagnostic significance.

4.1 Virological Findings

Flat warts are induced by HPV3. The recently characterised HPV3-related types HPV10 and HPV28 are often associated with intermediate warts and were demonstrated in some immunosuppressed patients (Pfister et al. 1979; Jablonska and Orth 1983).

4.2 Histological Features

Verrucae planae are characterised by a platelike, thickened epidermis with hypergranulosis and a basket weave-type of horny layer without areas of parakeratosis. They show only slight elongation of some papillae and convergence of the rete ridges. Numerous vacuolated cells lie - often in a bandlike arrangement - in the

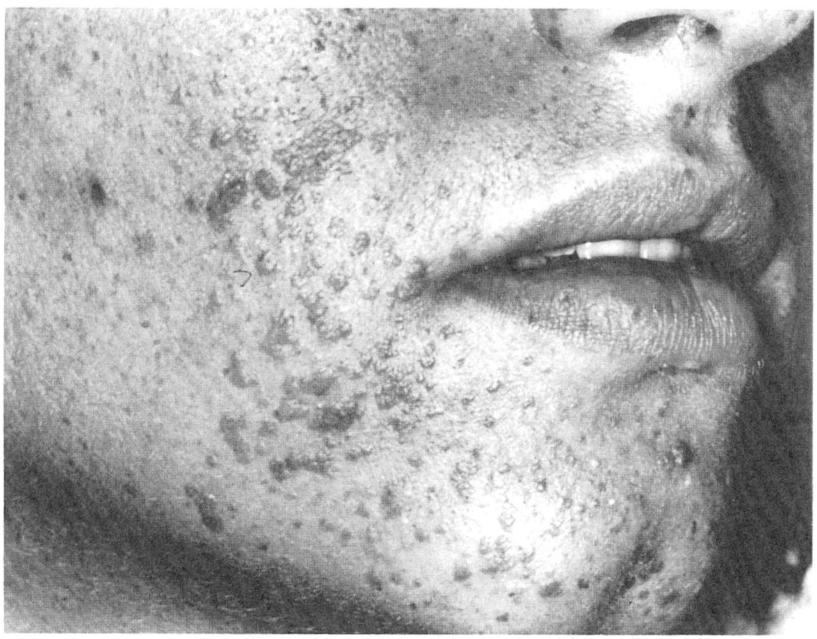

Fig. 9. Flat warts of the face

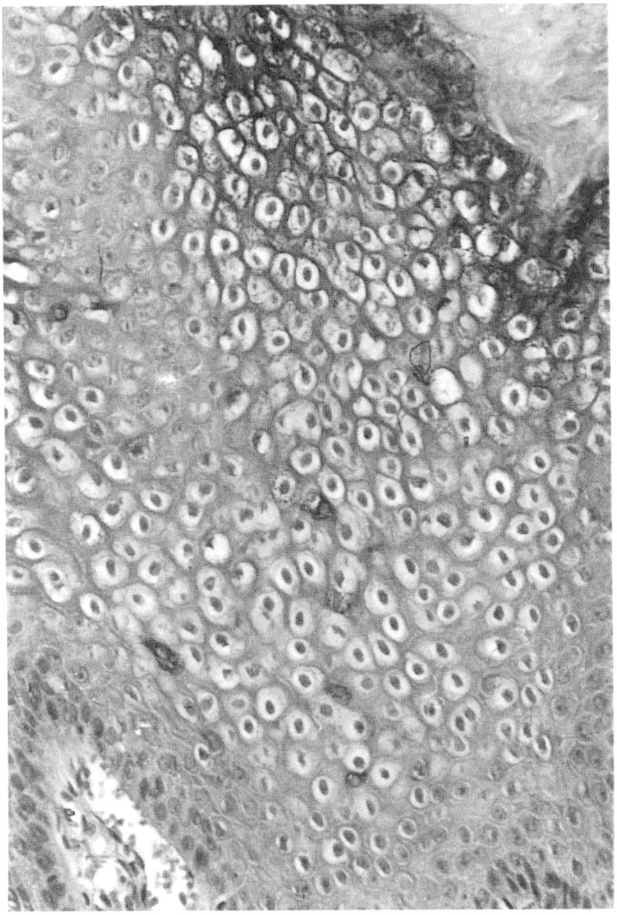

Fig. 10. Histological section of HPV3 wart; H & E × 360

upper stratum malpighii, including the granular layer. The nuclei of the vacuolated cells are sometimes strongly basophilic, often pyknotic, and usually lie in the center of the cells (Fig. 10).

Intermediate warts show more pronounced papillomatosis and more branched rete ridges than plane warts. Hyperkeratosis and parakeratosis are prominent. The cytopathic effect is highly characteristic and similar to plane warts.

5 Special Histological Methods

5.1 In Situ Hybridisation

To identify the virus type and to demonstrate the distribution pattern of viral DNA within the wart tissue, the method of in situ hybridisation with ^{3}H-labelled, type-specific HPV DNA can be applied (Grußendorf and zur Hausen 1979). This technique is very sensitive, but it is limited to frozen tissue sections.

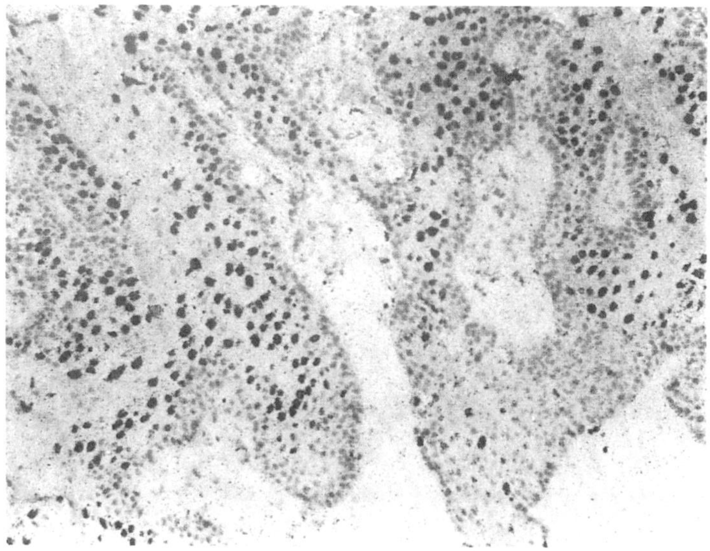

Fig. 11. In situ hybridisation with HPV1. Specific labelling of nuclei in the lower and middle parts of the stratum spinosum; ×90

Specific labelling, strongly restricted to the infected nuclei, can be detected by autoradiography.

Common warts (HPV2- or HPV4-induced) do not show any labelling in the basal cell layer or in the lower layers of the stratum spinosum. Labelling starts with the nonvacuolated cells in the third to fourth suprabasal cell layer, and grain density increases with the cell vacuolation process. Amongst the cells containing labelled nuclei exists a framework of normal-appearing, uninfected cells without any labelling. Amount and distribution pattern of viral DNA in HPV2- or HPV4-induced warts show no differences. In HPV1-induced *myrmecia,* labelling starts in the first to second suprabasal cell layer and appears as dark grains dispersed or in clusters strongly limited to the nuclei (Fig. 11). In the subsequent cell layers the nuclei are entirely covered with grains. In most cases the grain density increases as cells become more vacuolated. Intranuclear vacuolation can be observed sometimes. Here the labelled DNA is pushed to the rim of the nucleus. In the upper horny layer, specific labelling is scattered throughout the whole cell.

In *flat warts* HPV3 DNA can be found first in the second suprabasal cell layer. In the following cell layers, specific labelling appears in a bandlike arrangement, indicating the virus DNA-bearing nuclei.

5.2 Peroxidase-Antiperoxidase Staining

The method of peroxidase-antiperoxidase (PAP)-staining is based on the demonstration of viral protein by an immunological reaction. It may be applied to diagnose the viral origin of a warty lesion. It is a rapid procedure and more efficient

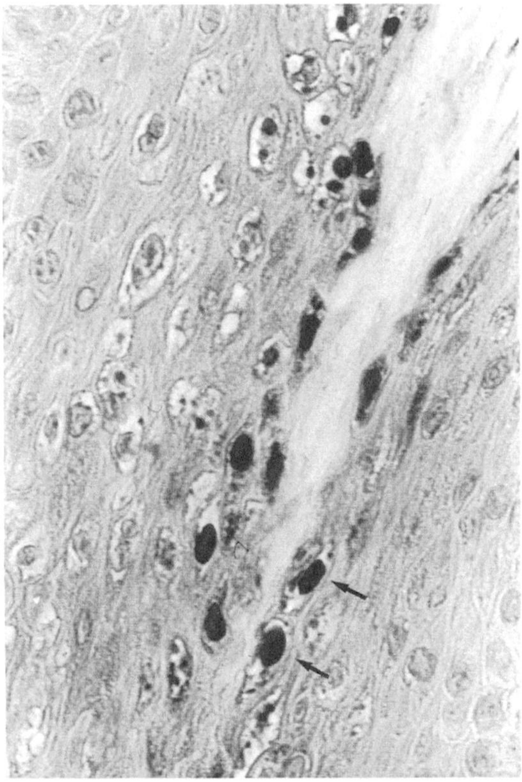

Fig. 12. Peroxidase-antiperoxidase staining. *Arrows* indicate positively stained nuclei in the upper cell layers; ×360

than routine electron microscopy. Papillomavirus-specific common antigens can be made visible in formalin-fixed, paraffin-embedded histological sections by use of antibodies produced by SDS-disrupted virions in rabbits (Jenson et al. 1985). Positive staining for HPV is to be seen as a red oxidation product within the cell nuclei (Fig. 12).

In HPV2- and HPV4-induced common warts, in HPV3 flat warts, as well as in myrmecia due to HPV1, HPV antigen can be found in the nuclei of the stratum granulosum and stratum corneum. The quantity of antigen expression in common warts is less than that in myrmecia, but it is found in the same superficial location. The distribution of common viral antigen correlates with the occurrence and amount of mature viral particles.

5.3 Electron Microscopic Findings

No difference can be noted in the electron microscopic appearance among the virus particles of the different HPV types. Only the quantity varies, being sparse with HPV2 and HPV4 (common warts) and abundant with HPV1 (myrmecia) and HPV3 (plane warts) (Laurent et al. 1975). The wart virus replicates in the nucleus, where the viral particles, when present in large numbers, are located as dense aggregates in a crystalline arrangement (Cornelius et al. 1968). Viral particles are first seen in the upper portion of the stratum malpighii within and around the nucleolus. Their number increases with differentiation of the cells and, in cells just beneath the stratum corneum, nucleoli are no longer detected. Cytoplasmic vacuolation begins in the perinuclear region in the first to third suprabasal cell layers in common warts as well as in myrmecia. The nucleus is surrounded by a circle of membrane-bound vacuoles, which may represent remnants of degenerated mitochondria (Charles 1960; Grußendorf-Conen 1985). Cell vacuolation is accompanied by the early appearance of keratohyalin granules. They are small and sparse in the deeper layers of the stratum spinosum whereas they join into large homogeneous masses in the upper cell layers. Shape and size of the keratohyalin granules vary to a great extent. Generally, they are more numerous in HPV1 warts than in HPV2 or HPV4 warts. Keratohyalin granules of the "single granule" type are to be found within nuclei of HPV4 common warts as well as those of myrmecia. In plane warts, keratohyalin granules do not show any special features (Laurent et al. 1978). Our own electron microscopic investigation of cutaneous warts associated with distinct HPV types did not reveal any differences regarding the ultrastructural morphology of infected cells (Grußendorf-Conen 1985). There are, however, differences with respect to the degree of cellular disturbance: cells rich in viral particles show greater damage than those bearing only small numbers of particles (Grußendorf 1981).

6 Wart Regression and Immunity

Skin warts usually regress spontaneously. It is generally stated that 35% of patients lose their warts within 2-6 months (Massing and Epstein 1963; Bunney 1982), 53% within 1 year and 67% within 2 years. However, there exists a considerable number of people who have had warts for decades.

The mechanism of wart regression has been the subject of much speculation and research. The spontaneous disappearance, frequently observed simultaneously for multiple warts, may indicate that wart rejection is due to humoral and/or cell-mediated immune reactions directed toward the wart cells. Numerous studies have been done by different groups of investigators on the humoral or cellular immunity to HPVs using various techniques (Tagami et al. 1983).

Humoral HPV antibodies seem to have rather little significance in the regression of warts, while there is considerable evidence that cell-mediated immunity plays the major role in the control of HPV infection (Chardonnet et al. 1985). The incidence of warts is increased during immunosuppressive treatment and in persons with cell-mediated immune deficiencies. On the other hand, regression of warts has been reported after inflammation or treatment with dinitrochlorobenzene and with a variety of immune adjuvants. Tagami and his colleagues reported in 1974 the occurrence of mononuclear cell infiltration in the dermis and epidermis of regressing flat warts. This was confirmed by Berman and Winkelman (1977) who have shown similar changes in resolving common warts. In situ investigation of the cellular immune response is a new approach provided by the use of monoclonal antibodies directed against surface differentiation antigens of T cells, B cells and Langerhans' cells. Recent investigations have shown that Langerhans' cells in the epidermis play an important role as antigen carriers, elicitors and targets of immunological reactions. In skin warts Langerhans' cells are markedly reduced in number (Chardonnet et al. 1985; Grußendorf-Conen 1986; Smolle and Kresbach 1986), suggesting a decrease of immunological surveillance which might induce tolerance to HPV. Involution of common warts, myrmecia and plane warts is associated with a mononuclear cell infiltration, exocytosis and degenerative epidermal changes, indicating that regression represents a cell-mediated immune rejection of the warts.

6.1 Clinical Signs of Regression

A deep plantar wart in regression is dark or almost black and becomes painful. The edge sometimes peels slightly. The superficial part of the wart is dry. In advanced regression, the whole wart is dry and friable (Rasmussen 1958). Common warts, however, usually shrink silently. Itching sometimes precedes resolution. They involute individually, in part probably due to an immune response of the host, but mostly due either to simple occlusion of blood vessels in the tumour tissue or to changes in the keratinocytes which are not favourable to the propagation of HPV.

Flat warts in normal people present a dramatic, systemic, regression phenomenon based on the development of a cellular immune response of the host against HPV-transformed tumour cells (Tagami et al. 1985). Before involution the warts suddenly begin to exacerbate. They become reddened and swell up, causing intense pruritus. In severe cases, extensive vesiculation may appear on the top of the warts. After 2-7 weeks, when the inflammation subsides with associated scaling and crusting, all warts will have disappeared as a rule.

7 Treatment of Warts

As warts tend to spontaneous regression without leaving any trace, they should be treated in a gentle, nonscarring way. Any form of therapy should cause no hazard to the patient and only minimal pain, side effects and inconvenience. This limitation excludes certain forms of treatment such as surgical intervention leaving permanent scars, use of general anaesthesia with its small but inherent risk, and systemic application of unspecific immunotherapies with incalculable, severe side effects. Radiotherapy has to be looked upon as malpractice, not only on account of the danger of atrophy and ulceration but also because X-rays have to be considered as a potent cocarcinogen to HPV (Jablonska and Orth 1985).

7.1 Salicylic Acid Preparations

Salicylic acid preparations in varying strengths from 10% to 60% in the form of solutions, ointments and plasters are the best known wart cures and highly effective. Collodion-based solutions are the most convenient preparations as they can be applied quickly and accurately and are cosmetically acceptable when used on the hands. They are the treatment of choice in multiple periungual warts on the fingers of children. Paints should not be applied to the facial or anogenital skin. For the treatment of extensive areas of plantar warts, salicylic acid plasters are suitable. A piece of the plaster is cut to the size and shape of the warts and held in place firmly with strapping. The plaster should be removed after 24 h. The softened keratin is then scraped away and the foot thoroughly washed and dried before a new application is made.

For flat warts, 40% isopropyl alcohol containing 1% tannic acid and 2% salicylic acid may be used twice daily and is tolerated well.

7.2 Topical 5-Fluorouracil

Isolated common warts may be successfully treated with topical application of 0.5% fluorouracil in combination with 10% salicylic acid when applied several times daily over a period of 4 weeks.

7.3 Cryosurgery

Cryotherapy is currently universally popular. Among the cryogenic materials, liquid nitrogen has proved to be especially useful because of its low temperature (−195.6 °C), easy availability, low cost and safety in use: it is not flammable and is chemically inactive. The time taken to freeze a wart varies. It depends on the type of wart, diameter, depth and location and may take from a few seconds to about 1 min. Repeated light freezings at intervals of 2 weeks (up to six treatments) appear to be more effective for multiple warts than more vigorous freezing.

7.4 Curettage

Papillomatous common warts can be scooped off very easily with a sharp spoon currette under local anaesthesia, and haemostasis is obtained by touching the base with a silver nitrate pencil or $FeCl_3$ solution.

8 Epidermodysplasia Verruciformis

The importance of epidermodysplasia verruciformis lies in the fact that this disease represents a natural model of viral oncogenesis in humans.

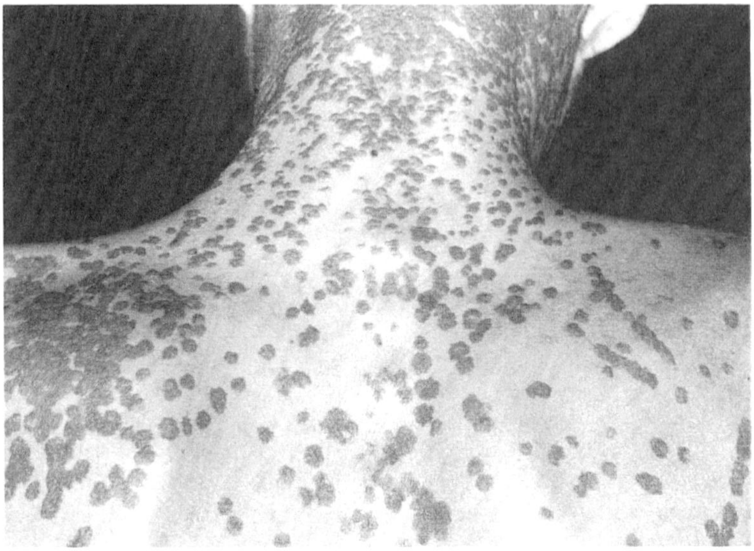

Fig. 13. Epidermodysplasia verruciformis

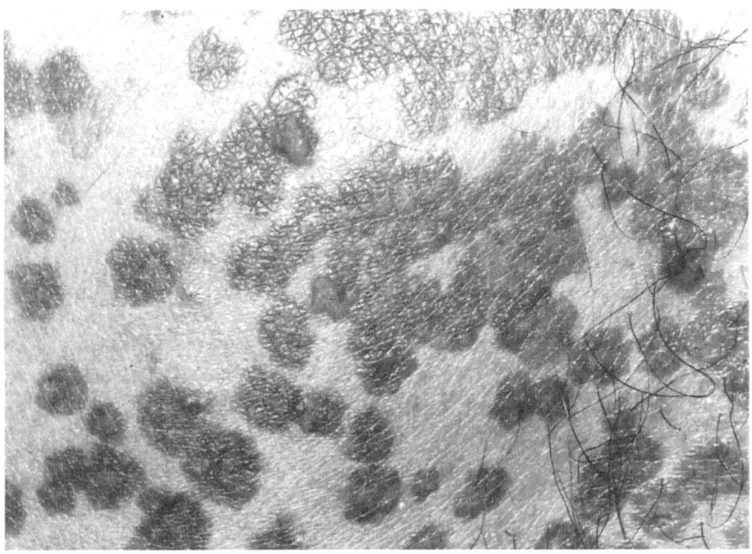

Fig. 14. Confluent macules on the breast (epidermodysplasia verruciformis)

First described in 1922 by Lewandowsky and Lutz, epidermodysplasia verruciformis (EV) is a rare, lifelong disease associated with a diversity of specific human papillomaviruses. It usually starts in early childhood with widespread scaly macules, which most frequently occur on the trunk and upper extremities. These macules may become confluent, developing patches with polycyclic borders (Figs. 13, 14). They are often brown or red in colour and resemble pityriasis versi-

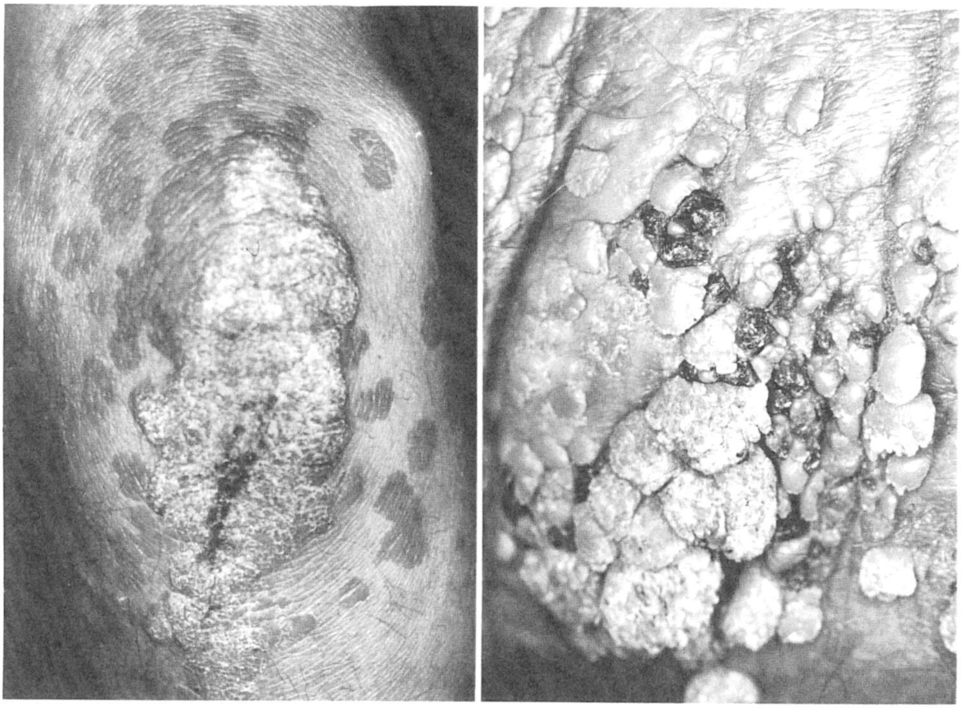

a b

Fig. 15. **a** Hypertrophic verrucous lesions on the knee and **b** papillomatous lesions in the bend of the knees (epidermodysplasia verruciformis)

color. Besides these lesions there exist flat papules, resembling flat warts, preferentially located on the dorsum of the hands and on the back. On elbows and knees the papules may join and become hypertrophic, like confluent common warts (Fig. 15). As a rule the mucous membranes are not involved. A combination of epidermodysplasia verruciformis and condylomata acuminata is exceptional (Jablonska and Orth 1985). The predisposition of patients with epidermodysplasia verruciformis to HPV infections is mainly restricted to specific HPV which seem to be harmless for the general population. In about 30% of patients with EV, cutaneous malignancies develop in the virus-induced lesions early in the life.

8.1 Genetic Factors

Heredity seems to play an important role since a remarkable proportion of patients are descendents from consanguineous marriages and in 10% at least 1 sibling is involved (Lutzner and Blanchet-Bardon 1985). The ratio of male to female incidence is one. Hereditary factors seem to predispose patients for the infection with EV-specific HPV. It is assumed that the mode of heredity is autosomal recessive. Spouses, even of heavily infected patients, never acquire the disease, a fact that underlines the role of genetic factors.

8.2 HPV Types Associated with EV

Recent studies have led to the recognition of a number of HPV types associated with EV-benign lesions in epidermodysplasia verruciformis: HPV3 and HPV10, which are also found in plane warts in the general population, and a multitude of EV-specific HPV such as HPV5, 8, 9, 12, 14, 15, 17 and 19–25. Usually patients are infected by several of these viruses. One patient has been reported to be infected with as many as nine different EV-specific HPV and with HPV3 (Jablonska and Orth 1985). EV has no special racial predisposition or geographic preference. Studying patients from different continents revealed that EV-specific HPV have a worldwide distribution. HPV5, 8, 17 and 20 are encountered most frequently, followed by HPV3 alone or in association with specific EV-specific HPV (Jablonska and Orth 1985).

8.3 Morphology and Pathology of EV Lesions

The morphological character of the lesions differs, depending on the type of the inducing HPV.

8.3.1 EV Induced by HPV3 and/or HPV10

These lesions are of the flat warty type; preferentially they are found on the dorsa of the hands and on the face. Sometimes they are disseminated on the extremities involving the palms and soles (Jablonska et al. 1983). Malignant transformation of lesions of this type occurs extremely rarely.

8.3.2 EV Induced by EV-Specific Virusus

The lesions induced by EV-specific HPV (HPV5, 8, 9, 12, 14, 15, 17, 19–25) differ from those mentioned above in size, appearance, colour and distribution. They may appear as reddish, brownish or non-pigmented plaques resembling pityriasis versicolor. They are widely disseminated, almost generalised, but their most characteristic feature is involvement of the trunk. Not infrequently, pigmented seborrheic keratosis-like lesions are to be found on the forehead or trunk.

Malignancy may develop in lesions induced by EV-specific HPV, indicating the oncogenic potential of these virus types.

8.4 Histological Features

The histology of *HPV3*-induced EV lesions is identical to that of flat warts unrelated to EV. *HPV10*-induced lesions have a pattern of flat warts with a relatively compact horny layer, focal parakeratosis, usually a large, heavily stained granular layer and only slight cytopathogenic effects. The histological hallmarks of EV-lesions induced by *EV-specific HPV* are nests of characteristically pale-staining cells in the epidermis, located in the upper spinous and granular layers, but spar-

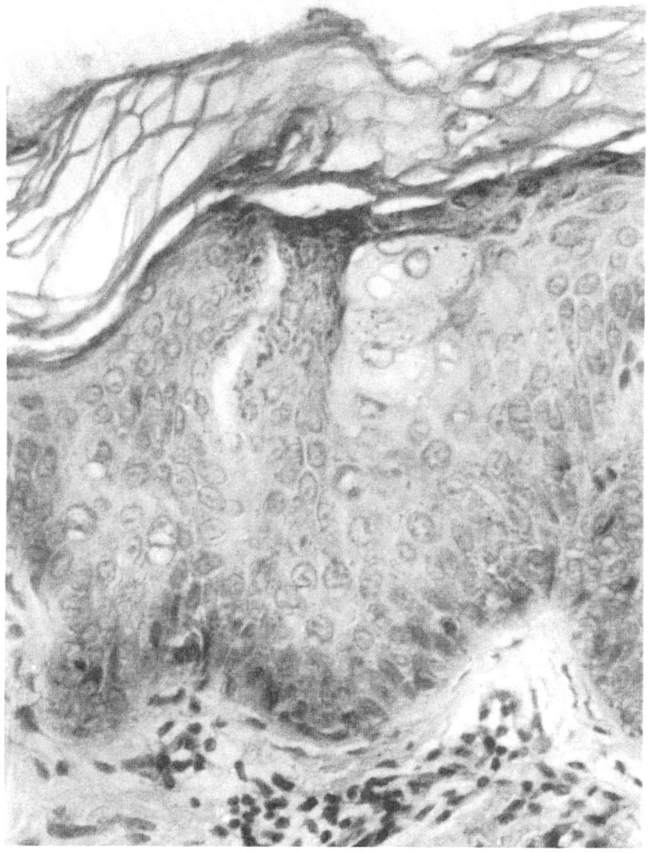

Fig. 16. Histological section of epidermodysplasia verruciformis; H & E × 360

ing the basal layer (Fig. 16). These clear cells are enlarged, with a homogenous, finely granular cytoplasm. The nuclei have empty spaces and show nuclear inclusions in semithin sections. Prominent round or ovoid keratohyalin granules of variable sizes are usually present in the dysplastic clear cells. The vacuolated cells may extend from only a few nests to almost complete replacement of the epidermis in a bandlike arrangement. No differences in the cytopathic effect are to be observed in lesions induced by all the different EV-specific HPV, whereas it is entirely different from the cytopathic effect induced by HPV3 and HPV10.

8.5 Electron Microscopic Features

The electron microscopic findings in EV due to *HPV3* and *HPV10* are identical to those seen in verrucae planae. Marked cytoplasmic oedema is seen. The tonofilaments are dislodged to the periphery of the cells. Numerous virus particles lie in the nuclei of the vacuolated cells (Laurent et al. 1975). In EV induced by *EV-specific HPV,* the clear cells are almost completely devoid of cytoplasmic organelles. The cytoplasm is filled with ribosomes which are responsible for its fine granular appearance. Tonofilaments are reduced and not associated with keratohyalin which appears in isolated granules. The enlarged nuclei have a highly disorganised chromatin and are filled with virions.

8.6 Malignant Transformation

In EV the rate of malignant conversion differs considerably in various series from none in black people from Nigeria (Jacyk and Subbuswanny 1979) to 100% in white people from Colombia (Rueda and Rodriquez 1972). An average of one-third of EV patients develop multiple skin cancers within their warty lesions. Malignancy may first be seen in the 2nd decade of life but usually appears in the late 20s or early 30s. As a rule the time interval between the onset of benign lesions and the onset of first malignant changes is about 20 years, but occasionally malignant transformation may occur within 2 years or take place much later (Lutzner and Blanchet-Bardon 1985). Cancers are usually invasive squamous cell carcinomas with features of Bowen's disease. They occur most frequently on sun-exposed skin areas, preferentially on the forehead; in rare instances they develop in the retroauricular area, in the genital region or elsewhere on the body. Metastases are rare. Viral particles are sometimes observed in carcinoma-in-situ, but never in invasive cancers (Jablonska and Orth 1985). Molecular hybridisation experiments have demonstrated the presence of multiple copies of the genomes of HPV5 and the HPV5-related types HPV8 and HPV14 in primary carcinomas, and HPV5 in metastatic tumours. In contrast to the great heterogeneity of HPV types found in the benign lesions of the same patient, carcinomas harbour only HPV5, HPV8 or HPV14, strongly suggesting that these HPV types have a higher oncogenic potential than other EV-specific HPV.

8.6.1 Extrinsic Cocarcinogens

The most potent cocarcinogen in EV seems to be ultraviolet (UV) irradiation, for malignancies usually develop in sun-exposed and traumatised areas. Although DNA repair in EV patients was found to be normal or only slightly reduced, the interaction of a specific virus, UV damage and trauma may be of pathogenic significance. The low rate of malignant conversion in black patients could be due to the protective role of melanin (Lutzner et al. 1984). X-rays, too, belong to the most effective cocarcinogens. Jablonska et al. (1983) observed deeply invasive carcinomas in patients whose early lesions had previously been irradiated. Otherwise, cancers developing from EV lesions are only superficially destructive.

9 Role of Immunity

Cell-mediated immunity is lowered in most EV patients. The number of T-lymphocytes may be decreased, the lymphocyte responsiveness to mitogens is reduced, and an anergy to dinitrochlorobenzene can be found, irrespective of the inducing HPV. Normal immune responses were exclusively found in abortive cases, mainly due to HPV3. It is supposed that a (inherited?) defect of cell-mediated immunity is responsible for the persistence of HPV infections, but malignant transformation depends on the oncogenic potential of the infecting HPV type.

Humoral immunity in patients with EV appears to be undisturbed. General or specific anomalies in antibody reactivity have been excluded (Lutzner 1978; Pyrhönen et al. 1980). Patients with EV have a normal resistance to other viral and to bacterial infections.

10 Treatment

Patients with EV should be under constant observation. They must be protected against UV irradiation. All precancers and developing cancers are to be removed or treated with 5-fluorouracil ointments.

References

Barr A, Coles RB (1966) Plantar warts: a statistical survey. Trans St John's Hosp Derm Soc 52: 226

Barr A, Coles RB (1969) Warts on the hands: a statistical survey. Trans St John's Hosp Derm Soc 55: 69-73

Berman A, Winkelman RK (1977) Flat warts undergoing involution: histopathological findings. Arch Dermatol 113: 1219-1221

Bunney MH (1982) Viral warts: their biology and treatment. Oxford University Press, Oxford

Celsus A, Cornelius (c. 25 AD) De medicina book V 28, 14 B-E. English translation by Spencer WG (1961) vol II. Heinemann, London, pp 160-163

Chardonnet Y, Viac J, Staquet MJ, Thivolet J (1985) Cell-mediated immunity to human papillomavirus. Clin Dermatol 3: 156-164

Charles A (1960) Electron microscope observations on the human wart. Dermatologica 121: 193-203

Cornelius CE, Witkowski JA, Wood MG (1968) Viral verruca, human papova virus infection. Arch Dermatol 98: 377-384

Croissant O, Breitburd F, Orth G (1985) Specifity of cytopathic effect of cutaneous human papillomaviruses. Clin Dermatol 3: 43-55

Ciuffo G (1907) Innesto positivo con filtrato di verruca volgare. J Ital Mal Vener Pelle 48: 12-17

Grußendorf E-I (1980) Lichtmikroskopische Untersuchungen an typisierten Viruswarzen (HPV-1 und HPV-4). Arch Dermatol Res 268: 141-148

Grußendorf E-I (1981) Morphologie verschiedener typisierter Virusakanthome des Menschen (HPV-1 und HPV-4 induzierte Warzen). Habilitationsschrift Aachen

Grußendorf-Conen E-I (1985) Zur Frage typspezifischer Morphologie von Viruspapillomen (Elektronenmikroskopische Befunde an HPV-1 und HPV-4 Warzen). Z Hautkr 60: 1745-1749

Grußendorf-Conen E-I (1986) Langerhanszellen in HPV-induzierten Warzen: II. Benjamin Lipschütz Symposium/Saulgau (to be published)

Grußendorf E-I, zur Hausen H (1979) Localisation of viral DNA-replication in sections of human warts by nucleic acid hybridization with complementary RNA of human papillomavirus type 1. Arch Dermatol Res 264: 55-63
Jablonska S, Orth G (1983) Human papillomaviruses. In: Rook A, Maibach H (eds) Recent advances in dermatology. Churchill Livingstone, London, pp 1-36
Jablonska S, Orth G (1985) Epidermodysplasia verruciformis. Clin Dermatol 3: 83-96
Jablonska S, Orth G, Glinski W (1981) Morphology and immunology of human warts and familial warts. In: Bachmann PA (ed) Leukemias, lymphomas and papillomas: comparative aspects. Taylor and Francis, London, pp 107-131
Jablonska S, Orth G, Obalek S, Croissant O, Jarzabek-Chorzelska M, Favre M, Kremsdorf D (1983) Oncogenic potential of human papillomaviruses epidermodysplasia verruciformis. A counterpart of Shope papilloma-carcinoma complex. Arch Geschwulstforsch 53: 207-215
Jablonska S, Orth G, Obalek S, Croissant O (1985) Cutaneous warts: clinical, histologic, and virologic correlation. Clin Dermatol 3: 71-82
Jacyk WK, Subbuswanny SG (1979) Epidermodysplasia verruciformis in Nigerians. Dermatologica 159: 256-265
Jenson AB, Kurman J, Wayne D, Lancaster WD (1985) Detection of papillomavirus common antigens in lesions of skin and mucosa. Clin Dermatol 3: 56-70
Laurent R, Kienzler JL (1985) Epidemiology of HPV infections. Clin Dermatol 3: 64-70
Laurent R, Agache P, Coume-Marquet J (1975) Ultrastructure of clear cells in human viral warts. J Cutan Pathol 2: 140-148
Laurent R, Coume-Marquet S, Kienzler JL (1978) Comparative electron microscopic study of clear cells in epidermodysplasia verruciformis and flat warts. Arch Dermatol Res 263: 1-12
Laurent R, Kienzler JL, Croissant O (1982) Two anatomoclinical types of warts with plantar localisation: specific cytopathogenic effects of human papillomavirus type 1 (HPV-1) and type 2 (HPV-2). Arch Dermatol Res 274: 101-111
Lewandowsky F, Lutz W (1922) Ein Fall einer bisher nicht beschriebenen Hauterkrankung (Epidermodysplasia verruciformis). Arch Dermatol Syph 141: 193-203
Lowy DR, Schiller JT (1985) Papillomaviruses, Clin Dermatol 3: 1-7
Lutzner MA (1978) Epidermodysplasia verruciformis. An autosomal recessive disease characterised by viral warts and skin cancer. A model for viral oncogenesis. Bull Cancer (Paris) 65: 169-182
Lutzner MA, Blanchet-Bardon C (1985) Epidermodysplasia verruciformis. Curr Probl Dermatol 13: 164-185
Lutzner MA, Blanchet-Bardon C, Orth G (1984) Clinical observations, virologic studies and treatment trials in patients with epidermodysplasia verruciformis, a disease induced by specific papillomaviruses. J Invest Dermatol 83: 185-255
Lyell A, Miles JAR (1951) The myrmecia - a study of inclusion bodies in warts. Br Med J 1: 912-915
Massing AM, Epstein WL (1963) Natural history of warts. Arch Dermatol 87: 306-310
Montgomery AH (1928) Paper read before the West Side Clinical Society, March 1928. Cited in Montgomery AH, Montgomery RM (1937) NY State J Med 37: 1978
Orth G, Jablonska S, Favre M (1981) Identification of papillomaviruses in butchers warts. J Invest Dermatol 76: 97-102
Payne J (1891) On the contagious rise of common warts. Br J Dermatol 3: 185
Pfister H (1984) Biology and biochemistry of papillomaviruses. Rev Physiol Biochem Pharmacol 99: 112-168
Pfister H, Gross G, Hagedorn M (1979) Characterization of human papillomavirus 3 in warts of renal allograft patient. J Invest Dermatol 73: 349-353
Pyrhönen S, Jablonska S, Obalek S, Kuismanen E (1980) Immune reactions in epidermodysplasia verruciformis. Br J Dermatol 102: 247-254
Rasmussen KA (1958) Verrucae plantares: symptomatology and epidemiology. Acta Derm Venereol (Stockh) 38 (Suppl 39): 1-146
Rueda LA, Rodriquez G (1972) Comparaction de la virogénesis en la epidermodisplasia verruciforme y en las verrugas planas. Med Cutan Iber Lat Am 6: 451-458
Smolle J, Kresbach H (1986) Viruspapillome - quantitative Immunhistochemie des entzündlichen Infiltrates, vol II. Benjamin Lipschutz Symposium, Saulgau (to be published)

Strauss MJ, Shaw EW, Bunting H, Melnick JL (1949) "Crystalline" virus-like particles from skin papillomas characterized by intranuclear inclusion bodies. Proc Soc Exp Biol Med 72: 46
Strauss MJ, Bunting H, Melnick JL (1950) Virus-like particles and inclusion bodies in skin papillomas. J Invest Dermatol 15: 433-444
Tagami H, Ogino A, Takigawa M (1974) Regression of plane warts following spontaneous inflammation. Br J Dermatol 90: 147-154
Tagami H, Oguchi M, Ofuji S (1983) Immunological aspects of wart regression with special reference to regression phenomena of numerous flat warts: an experiment on tumor immunity in man by nature. J Dermatol 10: 1-12
Tagami H, Setsuya A, Masakuzu R (1985) Regression of flat warts and common warts. Clin Dermatol 3: 170-178

Genital and Anal Papillomavirus Infections in Human Males

J. D. Oriel

1 Genital Papillomavirus Infections

1.1 History

Genital warts were familiar to physicians in ancient times, who called them 'condylomas' or 'figs' (Bafverstedt 1967). In the eighteenth and nineteenth centuries the disease was common, and was wrongly attributed first to syphilis, then to gonorrhoea, then to irritation from secretions "disturbed by venery" – hence the synonym 'venereal' warts. Knowledge about the true etiology of genital warts came slowly. Their histological similarity to cutaneous warts was noted in the 1890s, and subsequently some experimental inoculations of extracts of penile warts into nongenital epithelia resulted in the development of skin warts (Oriel 1971a). The viral etiology of cutaneous warts was confirmed by electron microscopy in the late 1940s, and it was assumed that genital warts were caused by the same virus. Not until the advent of molecular virology was this opinion shown to be incorrect.

The concept of genital warts as a sexually transmitted disease (STD) was first mentioned by Barrett et al. (1954); it has been confirmed in many studies and is now generally accepted. In the United Kingdom, genital warts have been included in the returns from STD clinics to the Department of Health and Social Security since 1971.

1.2 Epidemiology

1.2.1 Incidence

In 1972, 45.54 new cases of genital warts in males per 100000 population were treated in STD clinics in England (Chief Medical Officer 1974). In 1982, this figure had increased to 90.59 new cases per 100000 population; in this year 20639 men with genital warts were treated in STD clinics (Chief Medical Officer 1985). The total national incidence of genital warts in England is unknown, but it is probably much higher than these data from STD clinics indicate, as they do not include cases treated by general practitioners, urologists, dermatologists, or in the armed forces. In the USA there has been a large increase in the number of new cases. A study from the Centers for Disease Control (Editorial note 1983) shows that the number of consultations for genital warts with office-based physicians increased fivefold between 1966 and 1981. In a study in Rochester, Minnesota, it was noted that the number of new cases of condyloma acuminatum increased almost every year between 1950 and 1975 (Chuang et al. 1984). In American public health clinics, 3.4 cases of genital warts are diagnosed in men for every 100 visits, compared with 3.4 cases of genital herpes and 24.0 of gonorrhoea, the incidence being higher in white than in black men (Editorial note 1983). All this evidence suggests that genital papillomavirus infections in men are becoming increasingly common.

1.2.2 Age of Onset

Genital warts are a disease of sexual maturity. In a study from the United States of America, the peak incidence rate in men occurred in the age group 20-29 years, and the median age of onset was 26 years (Chuang et al. 1984). It is likely that the age incidence is similar throughout Western industrialised societies (Oriel 1971a); the age incidence in developing countries is not known.

1.2.3 Infectivity

Most genital warts in men are acquired by sexual contact with an infected partner, and conversely they can be transmitted to others. In one investigation it was found that 31 (54%) of 57 current female sex partners of men with penile warts had vulval warts (Oriel 1971a). Subclinical penile lesions are clearly of epidemiological importance. Levine et al. (1984) examined the sex partners of 34 women with either warts of the lower genital tract or cervical dysplasia, and detected external lesions in 18 (53%). These were mostly arranged as small, slightly raised clusters on the shaft of the penis, but one man had a small condyloma in the urethral meatus. These findings emphasise the need for the careful examination of sex contacts of all patients with genital warts.

1.2.4 Associated Infections

The association of genital warts with other STDs is well known. Kinghorn (1978) detected synchronous gonorrhoea in 10%, and nongonococcal urethritis in 17% of a group of men attending an STD clinic. Associated infections may be symptomatic or asymptomatic. Harahap (1979) reported from Indonesia that 23 of 67 men (34%) with penile warts but no urethral symptoms or signs yielded *Neisseria gonorrhoeae* on urethral culture.

Concomitant infections are also present in patients with genital warts seen in dermatological practice (Farris et al. 1984).

1.2.5 Genital Human Papillomavirus Infections in Children

While vulval and perianal condylomas in young children are seen occasionally, penile lesions are rare. Hajek (1956) described papillomatous growths on the glans penis in a child of 18 months who also had laryngeal papillomatosis; the penile lesions had apparently been noticed since he was 6 months old. His mother had vulval warts late in pregnancy which may have been the source of the papillomavirus. Penile condylomata acuminata in prepubertal boys have been reported by de Jong et al. (1982); transmission of the virus in children may occur during parturition, from close non-sexual contact with infected individuals, and by sexual molestation.

1.2.6 Genital Warts and Skin Warts

The alleged relationship between these two types of lesion was formerly much discussed, but there is no epidemiological evidence of any close association (Oriel 1971a). Nevertheless, a few men with common skin warts develop lesions of verruca vulgaris on the shaft of the penis. In these men the same virus has probably caused the two diseases, having been accidentally transferred to the genital area by the hands. The virus from these genital lesions has not yet been typed.

1.3 Lesions of the Glans Penis, Shaft and Scrotum

1.3.1 Clinical Features

Genital warts in men are pleomorphic. The commonest variety are the soft fleshy condylomata acuminata, which are usually seen on the corona and glans penis and on the inner lining of the prepuce (Fig. 1). They may sometimes develop on the shaft of the penis and are common on the scrotum. Condylomata acuminata are the predominant type of lesion in moist genital areas. Plane warts may appear on drier areas, particularly the shaft of the penis, sometimes associated with condylomata acuminata elsewhere. Common warts (Fig. 2) also affect the shaft of the penis, often in association with similar lesions on non-genital skin.

Warts often appear first on areas which are subject to trauma during intercourse, namely the frenum, coronal sulcus, inner surface of the prepuce and the preputial opening. No doubt small fissures and abrasions in these areas facilitate the entry of the infecting agents. Studies of vulval papillomavirus infection have revealed that subclinical lesions are common, and vulval inspection with a colposcope after the application of acetic acid shows that the extent of infected epithelium is often greater than naked eye inspection suggests. There already exists good evidence that subclinical condylomas occur on the penis (Levine et al. 1984). It

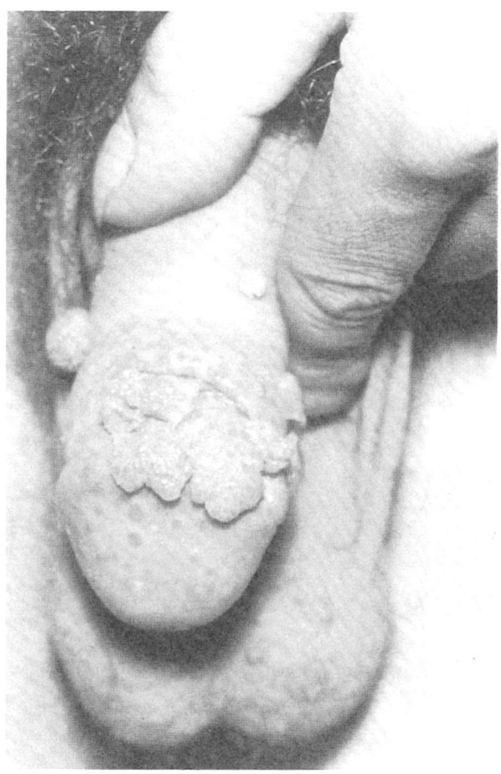

Fig. 1. Condylomata acuminata of the penis (reproduced by the courtesy of Baillière Tindall, London

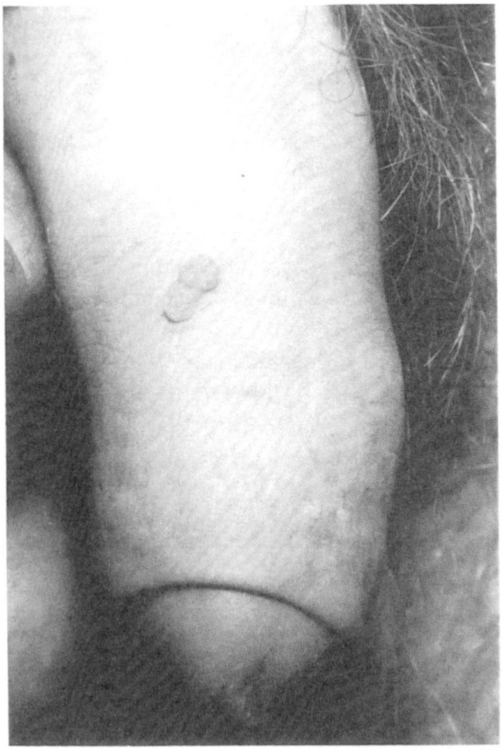

Fig. 2. Common wart on the shaft of the penis. This patient also had a common wart on one of his fingers

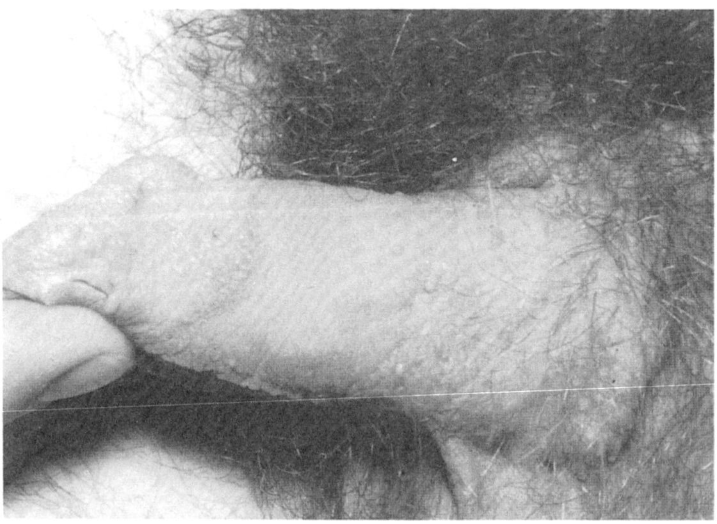

Fig. 3. Multiple small warts are scattered over the shaft of the penis, making it difficult to define the extent of the disease

may then be difficult to decide the extent of the disease (Fig. 3), and examination of the penis with a colposcope may be helpful.

1.3.2 Evidence for Involvement of Human Papillomaviruses

Intranuclear virus particles have been demonstrated in penile warts (Dunn and Ogilvie 1968; Oriel and Almeida 1970). Immunochemical methods have shown the presence of viral antigen in tissue sections treated with common papillomavirus (HPV) antibodies. Approximately one-half of penile condylomas show the presence of virus particles or viral antigen by these methods.

Recently, DNA-DNA hybridisation has been used to study the types of HPV associated with condylomata acuminata, bowenoid papulosis, and malignant disease of the penis. HPV6 is associated with benign condylomas of the shaft, prepuce and urinary meatus. HPV16 has been recovered from penile condylomata acuminata, but also from bowenoid papulosis and invasive carcinoma. HPV18 has been recovered from penile carcinoma tissue (McCance 1986).

1.3.3 Diagnosis

The diagnosis of penile warts is usually clinical. They need to be distinguished from other papular lesions of the area. Fordyce's spots are minute, ectopic sebaceous glands which often occur on the inner lining of the prepuce and in the coronal sulcus and may be mistaken for warts. Another minor papular condition is 'hirsutes papillaris penis,' in which parallel rows of lesions resembling filiform warts appear around the corona; the condition, like Fordyce's spots, is an anatomical variant which does not need treatment. The umbilicated lesions of molluscum contagiosum are usually not difficult to distinguish from warts. Among other infective diseases which enter the differential diagnosis the mucocutaneous lesions of secondary syphilis are the most important; dark-field microscopy for *Treponema pallidum* and serological tests for syphilis will allow the correct diagnosis to be made. In the tropics donovanosis (granuloma inguinale) of the penis may resemble condylomata acuminata; the demonstration of Donovan bodies in tissue smears is diagnostic.

The definitive diagnosis of penile warts depends on histology. Biopsy is sometimes done as a routine procedure, usually if the diagnosis is in doubt because of atypical appearances or poor response to treatment. The histological features of genital warts are well known (Lever 1967). The essential features are elongation of the dermal papillae, acanthosis and the presence of vacuolated cells (which often contain nuclear viral inclusions) in and below the stratum granulosum. Hyperkeratosis is not usually a feature of genital warts, but may develop as the lesions age or are unsuccessfully treated. If carcinoma-in-situ (Bowen's disease and bowenoid papulosis) is superimposed on condylomata acuminata, biopsy will show areas where the epithelium of the wart shows dyskeratosis, abnormal mitosis, giant and multinucleate cells, and often a dermal inflammatory infiltrate. Areas of multifocal invasive carcinoma may also appear in association with condylomata acuminata (Bender and Pass 1981).

1.4 Lesions of the Urethra

1.4.1 Clinical Features

Involvement of the urethra occurs in about 5% of men with genital warts (Morrow et al. 1952). Urethral condylomas may accompany warts elsewhere on the penis, or they may be the only manifestation of genital HPV infection. Nearly 80% of lesions are in the distal 3 cm of the urethra, but the growths can affect any part (de Benedictis et al. 1977). Patients with urethral condylomas may complain of urethral bleeding or discharge, dysuria, and reduction of the urinary stream, but some men are symptomless. Examination reveals the presence of characteristic bright red condyloma acuminatum tissue in the urinary meatus (Fig. 4); it is often difficult, particularly in men with a narrow meatus, to ascertain the extent of the disease by simple inspection. Condylomas may develop anywhere in the urethra, including the prostatic portion, but it is believed that the posterior urethra is not affected without anterior involvement (Bissada et al. 1974; de Benedictis et al. 1977). Owing to their often inconspicuous nature, it has been suggested that urethral condylomas may play an important role in the sexual transmission of HPV.

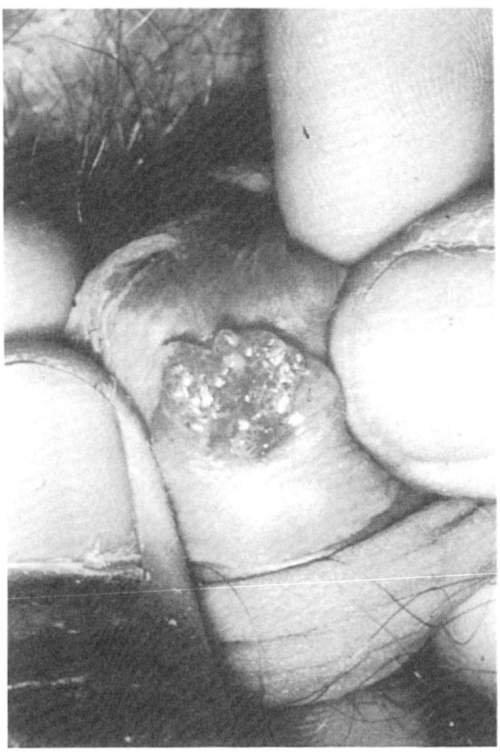

Fig. 4. Condylomata acuminata of the terminal urethra

1.4.2 Diagnosis

The diagnosis of proximal infection of the urethra may be made my endoscopy, but it is conceivable that infected material might be spread up the urethra by this procedure (Dretler and Klein 1975). This might also apply to retrograde urethrography. An alternative procedure, voiding cystourethrography after the intravenous injection of contrast medium, has been advocated by Pollack et al. (1978).

1.4.3 Evidence for Involvement of Human Papillomaviruses

Dean et al. (1983) examined 25 squamous papillomas of the penile meatus histologically and for the presence of HPV antigen with an immunochemical technique. All the tumours had the histological characteristics of HPV-associated changes in squamous epithelium, particularly koilocytosis, and viral antigen was demonstrated in 11 of the 25 (44%).

1.5 Lesions of the Bladder

Condylomata acuminata of the bladder are rare. They are usually associated with urethral condylomas. Bissada et al. (1974) described several cases of vesical involvement with condylomas, including a male patient with massive involvement of the urethra and bladder who eventually needed total cystourethrectomy to control the disease. A similar patient, who had condylomas affecting the entire urethra, bladder neck, trigone, and posterior bladder wall, was described by de Benedictis et al. (1977); although prolonged treatment was necessary, radical surgery was avoided in this case. Vesical condylomas have also been reported in a post-transplant immunosuppressed patient (Nielsen 1975).

Masse et al. (1981) described a patient with condyloma acuminatum of the bladder without urethral involvement. He was an elderly man who presented with recurrent urinary tract infection following prostatectomy. A sessile tumour of the anterior wall of the bladder, consisting of condyloma acuminatum tissue, was excised but recurred 11 months later. Cytological examination of urine samples showed a moderate number of atypical cells which were chiefly of transitional type.

1.6 Complications

1.6.1 Giant Condyloma

Large penile warts may cause great discomfort to the patient because of their size or because of secondary infection or haemorrhage, but the most serious clinical problems come from the association of these lesions with various types of genital neoplasm. Giant condyloma (Buschke-Löwenstein's tumour) is an exuberant warty growth which, while it causes extensive tissue destruction, does not metastasise and is apparently histologically benign, although later malignant change has been described (Dawson et al. 1965). Over 100 cases have now been reported, most of them in males. The nature of the disease remains uncertain. The histological distinction of giant condyloma from squamous-cell carcinoma of the penis has been discussed by Davies (1965) and Ananthakrishnan et al. (1981); on the other hand, Tessler and Applebaum (1982) believe that giant condyloma is probably a well-differentiated carcinoma from the onset, and that areas of anaplasia which develop subsequently represent a loss of differentiation in a previously well-differentiated carcinoma rather than malignant degeneration of a benign lesion.

The close histological resemblance between giant condyloma and condyloma acuminatum suggests that they may have a common viral aetiology. Papillomavirus particles have not been seen in giant condylomas, but recently Gissmann et al. (1982) identified HPV6 DNA in each of three tumours. Whether other factors are involved is unknown.

1.6.2 Premalignant and Malignant Disease

Penile intraepithelial neoplasia (IN) presents with lesions which may be single or multiple, erosive or papular, and pigmented or non-pigmented. These lesions have traditionally received various names, including Bowen's disease, bowenoid papulosis, carcinoma-in-situ, and erythroplasia of Queyrat. Histologically, vulval intraepithelial lesions are now graded 1–3 according to the amount of pleomorphism within the epithelium (Ridley 1983), and it is likely that these penile lesions will follow suit.

Clinically, it is often difficult to distinguish penile IN from condyloma acuminatum (Lupulescu et al. 1977; Chapel and Rahbari 1980), and indeed penile IN is often associated with HPV. Wade et al. (1979) reported 34 cases of bowenoid papulosis, 28 of them in males, and obtained a history of antecedent genital warts from 12 of them. Gross et al. (1985) examined 15 males with bowenoid papulosis. In 11 the lesions were solitary and in 10, multiple. Of the 15 males 6 gave a history of previous condylomata acuminata, and in 1 condylomata were present at the time of examination.

Invasive carcinoma of the penis has been described in association with condylomata acuminata (Rhatigan et al. 1972). In the laboratory, HPV DNA has been found in penile bowenoid papulosis and verrucous carcinoma by the Southern blot technique; zur Hausen et al. (1984) have reported the presence of HPV16 in bowenoid papulosis of the penis, and HPV16 and HPV18 sequences in carcinoma of the penis. Ikenberg et al. (1983) have found HPV16 in 80% of cases of bowenoid papulosis of the penis. Carcinoma of the penis is rare in Europe and North America, whereas it is more frequent in South America and parts of Africa (Singer and McCance 1985). In a recent study it was found that 49% of penile carcinomas contained HPV16, mostly integrated into the host chromosomes (McCance et al., quoted by Singer and McCance 1985).

Studies of vulval IN and invasive cancer have revealed that HPV16 and related sequences can be identified in the majority of lesions, and although the data for the corresponding penile diseases is more scanty, they appear to be aetiologically similar. The analogy between vulval and penile IN is shown in a case report by Stein (1980). A woman had been treated for vulval condylomata acuminata for 6 years until a biopsy revealed that she had developed carcinoma-in-situ. Two years before, her husband had developed extensive carcinoma-in-situ of the prepuce. Campion et al. (1985) reported that one-third of women who were the sole sexual partners of males who had had penile condylomas for a year or more developed CIN. HPV16 was isolated from 66% of the males and from 77% of the CIN lesions of their sex partners.

2 Anal Papillomavirus Infections

2.1 History

Anal warts were well-known in the ancient world and were linked with male homosexuality and anal intercourse by popular writers of the period. Like genital warts, they were confused with syphilis in the eighteenth century, and in the nineteenth century were described as a disease almost exclusively of women (Oriel 1971b). In the twentieth century the association with anal intercourse has again been acknowledged, but although they are common, anal warts have received remarkably little attention.

2.2 Epidemiology

Approximately 85% of males with genital warts have associated anal warts (Oriel 1971). Lesions which are confined to the anus are commoner. Males with anal warts are more likely to give a history of anoreceptive intercourse than those without anal warts (Carr and William 1977). The average age of onset of anal warts is similar to that of genital warts, and associated infections are common. It is generally assumed that anal warts are transmitted by sexual contact from an infected partner, but in homosexual males, for reasons which are obscure, anal warts are much more frequent than penile warts (Carr and Williams 1977).

Perianal condylomata acuminata have been described in male infants and prepubertal children (de Jong et al. 1982; Baruah et al. 1984). Because of the long incubation period of some mucosal HPV infections, it may be difficult in an individual case to determine the source of the infection. Since sexual molestation is a possibility, adequate medical and social evaluation of these children is essential.

2.3 Clinical Features

Perianal warts are usually of condyloma acuminatum type (Fig. 5), and histologically are identical to genital condylomas. In the moist conditions which prevail around the anus they may grow rapidly and reach a large size. Internal condylomas develop in over 50% of males with perianal warts (Schlappner and Shaffer 1978); most of these are in the anal canal (Fig. 6), but they may also appear in the rectum above the pectinate line.

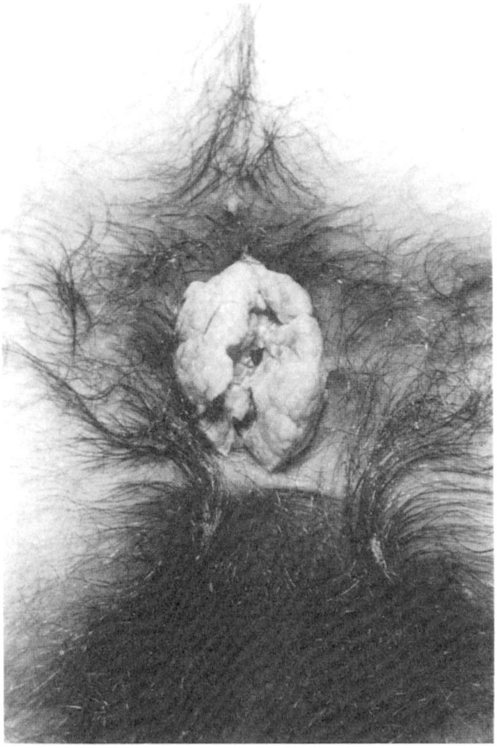

Fig. 5. Perianal condylomata acuminata

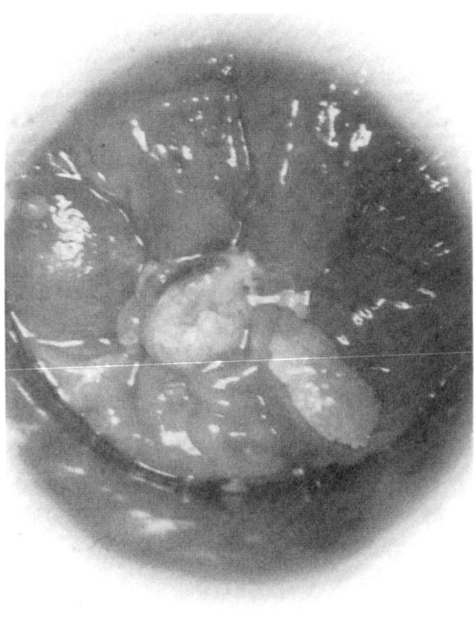

Fig. 6. Condylomata acuminata of the anal canal

Medley (1984) performed a cytological study to determine the incidence of non-condylomatous wart virus infection in the anal canal. Anal smears were processed by the Papanicolaou technique. It was found that of 102 males, 45 (44%) showed cytological evidence of HPV infection; many of them had received treatment in the past for anal warts, but only 5 still had them. This study suggests that, as on the penis, subclinical HPV infection of the anal canal is not uncommon.

The differential diagnosis of anal warts is with other raised perianal lesions. Fibroepithelial polyps are common in this area. Acquired syphilis is often seen in homosexual males, and anal condylomata lata may resemble condylomata acuminata; indeed, the two diseases may coexist (Dexter and Rockwell 1951). Squamous cell carcinoma of the anus must also be differentiated from anal warts.

2.4 Evidence for Involvement of Human Papillomavirus

Anal condylomata acuminata have not been extensively studied. Their histological identity with genital condylomas has been noted above. Limited virological studies have shown that HPV6 and HPV11 are present in these lesions (McCance 1986).

2.5 Complications

Giant condyloma can occur in the anorectal region (Shah and Hertz 1972; Lock et al. 1977; South et al. 1977); radical surgery is often required, and the disease may prove fatal. As on the penis, giant condyloma of the anus may be complicated by squamous cell carcinoma (Sturm et al. 1975).

Both intraepithelial neoplasia and invasive carcinoma arising in anal warts have been reported (Oriel and Whimster 1971; Fitzgerald and Hamit 1974; Ejeckam et al. 1983). These neoplastic changes may be insidious, and biopsy of lesions which are persistent or in any way atypical is essential.

3 Treatment of Anogenital Papillomavirus Infections

The therapy of HPV infections is discussed in another chapter in this volume. In this section only some special problems relating to male infections will be discussed. Anogenital warts are a sexually transmitted disease, thus the first essential step is to exclude, by appropriate microbiological tests, the presence of associated infections. It is also important to examine sex partners of males with anogenital HPV infections, a careful clinical examination being supplemented by cervical cytology.

Before commencing treatment, it is important to define the extent of the disease. The penis should be carefully inspected, under magnification if possible, and the terminal urethra examined for meatal condylomas. The anus should always be examined and, if anal warts are present, anoscopy performed. Condylomas of the terminal urethra are often inaccessible, and for this reason the optimal treatment is often ablation through an operating urethroscope under general anaesthesia; even then, a meatotomy may be needed to give adequate exposure. Instillation of 5-fluorouracil cream has been recommended for the treatment of urethral condylomas (Dretler and Klein 1975), but this may cause severe urethritis (Cetti 1984). The sheer size of some condylomatous masses, particularly around the anus, may give therapeutic problems. Podophyllin should not be applied to these large lesions, because absorption may lead to severe toxic effects (Chamberlain et al. 1972). The carbon dioxide laser has proved very useful for the removal of extensive condylomas. An alternative surgical treatment for extensive anal warts is the scissor excision procedure introduced by Thomson and Grace (1978); condylomas of the anal canal may be dealt with similarly.

Recurrence of anogenital warts after treatment may take place regardless of the modality used, and many patients require multiple courses of treatment. Systemic therapy for HPV infections in men would have many advantages, but no safe and effective regimen has yet been developed.

References

Ananthakrishnan N, Ravindran R, Veliath AJ, Parkash S (1981) Loewenstein-Buschke tumour of penis - a carcinomimic. Report of 24 cases and review of literature. Br J Urol 53: 460-465

Bafverstedt B (1967) Condylomata acuminata - past and present. Acta Derm Venereol (Stockh) 47: 376-381

Barrett TJ, Silbar JD, McGinley JP (1954) Genital warts - a venereal disease. JAMA 154: 333-334

Baruah MC, Sardari L, Selvaraju M, Veliath AJ (1984) Perianal condylomata acuminata in a male child. Br J Vener Dis 60: 60-61

Bender ME, Pass F (1981) Papillomavirus and cutaneous malignancy. Int J Dermatol 20: 468-474

Bissada NK, Cole AT, Fried FA (1974) Extensive condylomata acuminata of the entire male urethra and the bladder. J Urol 112: 201-203

Campion MJ, Singer A, Clarkson PK, McCance DJ (1985) Increased risk of cervical neoplasia in the consorts of men with penile condylomata acuminata. Lancet i: 943-945

Carr G, William DC (1977) Anal warts in a population of gay men in New York City. Sex Transm Dis 4: 56-57

Cetti NE (1984) Condyloma acuminatum of the urethra: problems in eradication. Br J Surg 71: 57

Chamberlain MJ, Reynolds AL, Yeoman WB (1972) Toxic effect of podophyllum in pregnancy. Br Med J 3: 391-392

Chapel TA, Rahbari H (1980) Genital bowenoid papulosis - squamous-cell carcinoma in situ. Sex Transm Dis 7: 139-141

Chief Medical Officer (1974) Sexually transmitted diseases. Extract from the annual report of the Chief Medical Officer of the Department of Health and Social Security for the year 1972. Br J Vener Dis 50: 73-79

Chief Medical Officer (1985) Sexually transmitted diseases. Extract from the annual report of the Chief Medical Officer to the Department of Health and Social Security for the year 1982. Br J Vener Dis 60: 199-203

Chuang T-Y, Perry HO, Kurland LT, Ilstrup DM (1984) Condylomata acuminata in Rochester Minnesota 1950-1978: 1. Epidemiology and clinical features. Arch Dermatol 120: 469-475

Davies SW (1965) Giant condyloma acuminatum: incidence among cases diagnosed as carcinoma of the penis. J Clin Pathol 18: 142-149

Dawson DF, Duckworth JK, Bernhardt H, Young JM (1965) Giant condyloma and verrucous carcinoma of the genital area. Arch Pathol 79: 225-231

de Benedictis TJ, Marmar JL, Praiss DE (1977) Intraurethral condylomata acuminata: management and a review of the literature. J Urol 118: 767-769

de Jong AR, Weiss JC, Brent RL (1982) Condylomata acuminata in children. Am J Dis Child 136: 704-706
Dean P, Lancaster WD, Chun B, Jenson AB (1983) Human papillomavirus structural antigens in squamous papilloma of the male urethra. J Urol 129: 873-875
Dexter HLT, Rockwell EM (1951) Simultaneous condylomata acuminata and condylomata lata. Report of a case. Arch Dermatol Syph 64: 205-207
Dretler SP, Klein LA (1975) The eradication of intraurethral condylomata acuminata with 5 per cent 5-fluorouracil cream. J Urol 113: 195-198
Dunn AEG, Ogilvie MM (1968) Intranuclear virus particles in human genital wart tissue: observations on the ultrastructure of the epidermal layer. J Ultrastruct Res 22: 282-295
Editorial note (1983) Condyloma acuminatum - United States 1966-1981. JAMA 250: 336
Ejeckam GC, Idikio HA, Nayak V, Gardiner JP (1983) Malignant transformation in an anal condyloma acuminatum. Can J Surg 26: 170-173
Farris GM, Statham BN, Waugh MA (1984) The investigation of patients with genital warts. Br J Dermatol 111: 736-738
Fitzgerald DM, Hamit HF (1974) The variable significance of condylomata acuminata. Ann Surg 179: 328-331
Gissmann L, de Villiers EM, zur Hausen H (1982) Analysis of human genital warts (condylomata acuminata) and other genital tumours for human papillomavirus type 6 DNA. Int J Cancer 29: 143-146
Gross G, Hagedorn M, Ikenberg H, Rufli T, Dahlex C, Grosshans E, Gissmann L (1985) Bowenoid papulosis. Presence of human papillomavirus (HPV) structural antigens and of HPV 16-related sequences. Arch Dermatol 121: 858-863
Hajek EF (1956) Contribution to the etiology of laryngeal papilloma in children. J Laryngol Otol 70: 166-168
Harahap M (1979) Asymptomatic gonorrhoea among patients with condylomata acuminata. Br J Vener Dis 55: 450
Ikenberg H, Gissmann L, Gross G, Grußendorf-Conen E, zur Hausen H (1983) Human papillomavirus type 16-related DNA in genital Bowen's disease and in bowenoid papulosis. Int J Cancer 32: 563-565
Kinghorn GR (1978) Genital warts: incidence of associated genital infections. Br J Dermatol 99: 405-409
Lever WF (1967) Histopathology of the skin, 4th edn. Pitman Medical, London, pp 375-382
Levine RU, Crum CP, Herman E, Silvers D, Ferenczy A, Richart RM (1984) Cervical papillomavirus infection and intraepithelial neoplasia: a study of male sexual partners. Obstet Gynecol 64: 16-20
Lock MR, Katz DR, Sammoorian S, Parks AC (1977) Giant condyloma of the rectum: report of a case. Dis Colon Rectum 20: 154-157
Lupulescu A, Mehregan DH, Rahbavi H (1977) Bowen's disease of genital areas. J Cutan Pathol 4: 266-274
Masse S, Tosi-Kruse A, Carmel M, Elhilali M (1981) Condyloma acuminatum of bladder. Urology 17: 381-382
McCance D (1986) Human papillomaviruses infecting the genital tract. In: Oriel JD, Harris JRW (eds) Recent advances in sexually transmitted diseases. Churchill Livingstone, Edinburgh
Medley G (1984) Anal smear test to diagnose occult anorectal infection with human papillomavirus in men. Br J Vener Dis 60: 205
Morrow RP, McDonald JR, Emmett JL (1952) Condylomata acuminata of the urethra. J Urol 68: 909-917
Nielsen HV (1975) Condylomata acuminata of the bladder (case report). Scand J Urol Nephrol 9: 169-170
Oriel JD (1971a) Natural history of genital warts. Br J Vener Dis 47: 1-13
Oriel JD (1971b) Anal warts and anal coitus. Br J Vener Dis 47: 373-376
Oriel JD, Almeida JD (1970) Demonstration of virus particles in human genital warts. Br J Vener Dis 46: 37-42
Oriel JD, Whimster I (1971) Carcinoma in situ associated with virus-containing anal warts. Br J Dermatol 84: 71-73
Pollack HM, de Benedictis TJ, Marmar JL, Praiss DE (1978) Urethrographic manifestations of venereal warts (condylomata acuminata). Radiology 126: 643-646

Rhatigan RM, Jimenez S, Chopskie EJ (1972) Condyloma acuminatum and carcinoma of the penis. South Med J 65: 423-428
Ridley CM (1983) Vulval dysplasia. J Hosp Med 30: 223
Schlappner OLA, Shaffer EA (1978) Anorectal condylomata acuminata: a missed part of the condyloma spectrum. Can Med Assoc J 118: 172-173
Shah IC, Hertz RE (1972) Giant condyloma acuminatum of the anorectum. Dis Colon Rectum 15: 207-210
Singer A, McCance D (1985) The wart virus and genital neoplasia; a casual or causal association. Br J Obstet Gynaecol 92: 1083-1085
South LM, O'Sullivan JP, Gazet JC (1977) Giant condyloma of Buschke and Loewenstein. Clin Oncol 3: 107-115
Stein DS (1980) Transmissible venereal neoplasia: a case report. Am J Obstet Gynecol 137: 864-865
Sturm JT, Christenson CE, Uecker JH, Perry JF (1975) Squamous-cell carcinoma of the anus arising in a giant condyloma acuminatum: report of a case. Dis Colon Rectum 18: 147-151
Tessler AN, Applebaum SM (1982) The Buschke-Loewenstein tumour. Urology 20: 36-39
Thomson JFS, Grace RH (1978) Treatment of perianal and anal condylomata acuminata. A new operative technique. J R Soc Med 71: 181-185
Wade TR, Koppf AW, Ackerman AB (1979) Bowenoid papulosis of the genitalia. Arch Dermatol 115: 306-308
zur Hausen H, Gissmann L, Schlehofer JR (1984) Viruses in the aetiology of human genital cancer. Prog Med Virol 30: 170-186

Lesions of the Male and Female External Genitalia Associated with Human Papillomaviruses

G. Gross

1 Introduction

Genital warts (condylomata acuminata), the most common benign genital tumor in both sexes, have been considered for many centuries to be related to venereal diseases such as gonorrhea and syphilis (Oriel 1981). Until some 15 years ago condylomata acuminata were thought to be trivial lesions of little importance to the patient. There is no doubt today that these lesions are sexually transmissible and, in view of the striking increase in prevalence, they are recognized as one of the most important sexually transmitted diseases (STD). According to the Center for Disease Control (United States of America) genital warts are currently three times more common than genital herpes. In 1984 there were approximately 1 million

Table 1. HPV types in squamous cell tumors of the lower genital tract[a]

Clinical manifestation	HPV type					
	6a-f	11a,b	16	18	31	33,34,35
Condyloma acuminatum (cervix, vagina, vulva, penis)	+++	++	+			
Giant condyloma/Buschke-Löwenstein's tumor (vulva, penis)	++	+				
Flat condyloma						
cervix (CIN I-III)	+	++	+		+	+
vagina		+				
Flat condylomatous lesion						
vulva (VIN I-II)	+	++	+			
penis (PIN I-II)	+	++	+			
Pigmented papular lesion						
vulva (VIN I-II)	+	++	+			
penis (PIN I-II)	+	+	+			
Pigmented bowenoid papulosis						
vulva (VIN III)			++			
penis (PIN III)			++			+
Bowen's disease						
vulva (VIN III)			+			
penis			+			
Carcinoma						
cervix uteri			++	+	+	+
vagina			+			
vulva			+			
penis			+	+		
Adenocarcinoma (cervix)			+			

[a] Additional virus types found in a few biopsies of malignant lesions: HPV2 (Bowen's disease), HPV10 (cervical carcinoma), and HPV1,2,3,5,8,10,12,19-22,25,30,31 in condylomata acuminata/condylomata plana.

new cases in the United States of America (Chuang et al. 1984). Human papillomaviruses (HPV) have been identified as the agents responsible for genital warts (Gissmann and zur Hausen 1980). The importance of HPV infection increased still further because of the recognized linkage between HPV and the development of malignant and premalignant conditions affecting the cervix, vagina, vulva, penis, and anus (zur Hausen 1982; Gissmann 1984) (Table 1). Another disease wherein HPV plays a significant role is epidermodysplasia verruciformis (EV). There are well-documented case reports on the malignant conversion of skin warts in this rare disorder.

By means of molecular hybridization techniques distinct HPV types and subtypes have been discovered. Based on the principle that distinct HPV types show less than 50% nucleic acid concordance with each other (Coggin and zur Hausen 1979), more than 40 HPV genotypes have been identified (Table 2; Gissmann 1984; McCance 1986). An important contribution has come from the discovery that certain genotypes of HPV are preferentially associated with squamous cell

Table 2. Human papillomavirus types and associated clinical manifestations

HPV type	Associated clinical manifestations
HPV 1 a,b,c	Plantar warts
HPV 2 a–e	Common warts
HPV 3 a,b	Flat warts/juvenile warts
HPV 4	Plantar/palmar warts
HPV 5 a,b	Macules in EV patients
HPV 6 a–f	Condylomata acuminata/CIN I–II/VIN I–II/giant condylomata/ Buschke-Löwenstein's tumor/laryngeal papillomas
HPV 7	"Butchers" warts
HPV 8	Macules in EV patients
HPV 9	Warts and macules in EV patients
HPV 10 a,b	Flat warts
HPV 11 a,b	Condylomata acuminata/CIN I–III/VIN I–II/giant condylomata/ Buschke-Löwenstein's tumor/laryngeal papillomas
HPV 12	Warts and macules in EV patients
HPV 13	Oral focal hyperplasia (Heck lesions)
HPV 14 a,b	Skin lesions of EV patients
HPV 15	Skin lesions of EV patients
HPV 16	Condylomata plana/CIN I–III/VIN I–III/Bowenoid papulosis and precursors[b]; Bowen's disease; carcinoma of cervix, vulva, and penis; verrucous carcinoma of the larynx
HPV 17 a,b	Skin lesions from EV patients
HPV 18	Carcinoma of cervix and penis
HPV 19–29	Various warty and hyperplastic lesions on the skin of EV patients
HPV 30	Laryngeal carcinoma
HPV 31	CIN/malignant carcinoma of the cervix
HPV 32	Oral focal hyperplasia (Heck lesions)
HPV 33	Bowenoid papulosis/CIN/carcinoma of the cervix
HPV 34	Bowenoid papulosis
HPV 35	CIN/carcinoma of the cervix
HPV 36	Actinic keratosis
HPV 37	Keratoacanthoma
HPV 38	Malignant melanoma (1 case)
HPV 39	Cutaneous lesions
HPV 40	Laryngeal carcinoma
HPV 41	Multiple flat warts of skin from a young girl

[a] *CIN*, Cervical intraepithelial neoplasia; *EV*, epidermodysplasia verruciformis; *VIN*, vulvar intraepithelial neoplasia.
[b] flat condylomalike lesions and pigmented papular lesions.

cancer and precursor lesions (Dürst et al. 1983; Gissmann 1984). Data on the natural histology of vulvar and penile warts as well as on the transmissibility and clustering of both cancers and precursor lesions of the vulva and of the penis also speak in favor of HPV as an important agent in genital cancer (Martinez 1969; Singer 1973; Campion et al. 1985; Gross et al. 1985a, b, 1986).

It is generally assumed today that the knowledge of the clinical findings and the determination of the virus type are important in early diagnosis and treatment and may contribute to the prevention of HPV infection and squamous cell malignancy in patients and their sexual partner(s). Further improvement in the diagnosis of genital squamous cell disease may be achieved by the adoption of a standard terminology by gynecologists and by dermatologists. Terms such as Bowen's disease

Table 3. Proposal for a nomenclature of squamous cell precancer[a] of the vulva and penis

Old nomenclature	Carcinoma in situ simplex - Bowen's disease - erythroplasia of Queyrat	
Proposal of the ISSVD 1976	Carcinoma-in-situ: undifferentiated, moderate differentation, marked differentation	
Proposal of the ISSVD 1983[b] (Friedrich 1983)	Vulvar intraepithelial neoplasia (VIN)	Penile intraepithelial neoplasia (PIN)
	grade I grade II grade III	

[a] Condylomatous dysplasias and bowenoid papulosis included.
[b] PIN modified (see text).

or erythroplasia of Queyrat are still in use in dermatology whereas in gynecology they have been deleted from the vocabulary of vulvar diseases. In 1983 a new nomenclature for precursor lesions of vulvar squamous carcinoma was proposed by the International Society for the Study of Vulvar Disease (ISSVD; Friedrich 1983) (Table 3).

In view of the similarities in morphology, epidemiology, and perhaps also etiology of these conditions, the same terminology should be used for penile lesions (penile intraepithelial neoplasia, PIN I-III). This would contribute to the uniformity and comparability of studies on such diseases in sexual partners. This chapter focuses on epidemiological, morphological, and diagnostic aspects of papillomavirus infections and associated diseases of the external genitalia of the female and of the male. Special reference will be made to the veneral transmission of papillomavirus-associated premalignant conditions.

2 Epidemiology

Although genital warts were suspected for centuries to be a venerally transmitted disease, this theory could not be confirmed until 1954, when Barrett et al. reported on the occurrence of such lesions in the wives of infected soldiers returning from the Korean War about 4-6 weeks after exposure.

As a rule HPV infections appear to be transmitted by skin to skin contact. Thus, genital warts occur primarily as a result of sexual activity. In general, HPV infections are associated with long incubation periods (zur Hausen 1977). In a study by Oriel (1971a), approximately two-thirds of the sexual partners of individuals having genital warts developed lesions after a mean incubation period of 3 months. In contrast with skin warts that occur mainly in children (zur Hausen 1977; Pfister and zur Hausen 1978), genital HPV infections are prevalent in groups with high degrees of sexual activity, having a peak incidence between the ages of 20 and 30. This age distribution curve is also seen with other STDs such as gonorrhea (Oriel 1971a) and is one more fact in support of the notion that genital warts are indeed a venereal disease. Since sexual contact does not produce genital warts in all cases, some degree of individual resistance has been postulated. It has not been fully clarified whether or not the rate of infectivity depends on the age of the lesions and thus on the quantity of infecting virus. There is no longer any doubt that endogenous factors such as cellular immunity or local factors may influence genital HPV infection. Another nonvenereal mode of transmission merits discussion. Theoretically, HPV-positive epithelial cells may be transmitted by means of repeated use of gloves or endoscopic instruments utilized in colposcopy, anoscopy, or urethroscopy.

Warts confined exclusively to the anus are found predominantly in homosexual men and have a peak incidence at 30 years of age (Marino 1964; Oriel 1971b). There is limited evidence for an association between anal warts and penile infections in sexual partners (Oriel 1971b; Judson et al. 1980). On the other hand, a link is suspected between juvenile laryngeal papillomatosis and a perinatal infection from a mother with condylomata acuminata (zur Hausen 1977). This theory has been supported by the documented presence of HPV6 and HPV11 in condylomata acuminata and in laryngeal papillomas (Gissmann et al. 1983).

The outcome of the papillomavirus-associated lesions may be a function of cell-mediated immunity. This is suggested by a series of observations demonstrating the high prevalence of skin warts in either congenitally or iatrogenically immunosuppressed individuals (zur Hausen 1977; Spradbrow, this volume). Some evidence has been provided by recent studies suggesting that the same also holds true for HPV lesions in the genital tract (Schneider et al. 1983; Shokri-Tabibzadeh et al. 1981; Koss, this volume). There is an increased incidence of cervical condylomas and intraepithelial neoplasia (CIN) in female renal transplant recipients and in patients with Hodgkin's disease. In the same group of patients, an overall decrease in the latency period has been noticed (Sillmann et al. 1984). Under these circumstances the HPV-associated neoplasia often persists, recurs, and extends to adjacent areas of the cervix, vagina, vulva, and anus in spite of conventional therapy (Sillmann et al. 1984). In pregnant women, condylomata acuminata exhibit enhanced growth, probably due to the lowered immune functions (Petrucco et al. 1976). The same has been observed in cases with chronic hepatic failure and in patients suffering from diabetes mellitus (zur Hausen 1977). Spontaneous regression of genital warts is seen only in rare instances; after delivery, however, such events are quite common, probably as a result of restored immunity. The rising frequency of anogenital warts in children in the past few years is far from being understood (Stumpf 1980; Dejong et al. 1982). Taking into account that girls are affected twice as often as boys (Dejong et al. 1982) and that the same HPV types are predominant in such warts as are observed in the adult population (del Mistro et al. 1986), transmission must have taken place at least in some of these children as a consequence of sexual abuse. However, other mechanisms of transmission cannot be ruled out, since infections occur also in neonates.

It is generally agreed that during the last 2 decades the incidence of genital warts (condylomata acuminata) has been increasing like that of other STDs (Oriel

1971a, 1981; Powell 1978). It is noteworthy that over the same period the incidence of cervical neoplasia has also risen (Chief Medical Officer 1983; Roberts 1982), particularly among young women (Draper and Cook 1983); likewise sexual activity in the same age group has been on the rise, coupled with a decrease in the age of first sexual experience (Campion et al. 1985). At this time condylomata acuminata constitute the most important STD after gonorrhea and chlamydia in the United States of America (Chuang et al. 1984). Bearing in mind that flat condylomata and subclinical HPV lesions have not been included in these studies, there is no doubt that the frequency of HPV infections in the population is even higher than reported up to now. The prevalence of HPV DNA found in cervical samples by means of filter in situ hybridization techniques (Wagner et al. 1984) is about 10%-13% of randomly selected women from an outpatient clinic (Wagner and de Villiers, personal communication).

Special interest has been focused on epidemiological factors that may predispose to genital HPV infections. There is growing evidence that sexual behavior has a striking influence on the transmission and fate of the infection. Sexual promiscuity, poor hygiene, associated recurrent infections with yeast, herpes simplex virus, gonorrhea, Trichomonas vaginalis, and chlamydia are significantly more frequent among HPV-positive individuals than among healthy persons (Syrjänen et al. 1984). As reported by Kinghorn (1978), genital infections with Candida albicans (25%), gonorrhea (21%), and Trichomonas (12%) were seen in women affected by condylomata acuminata. In male patients with genital warts, nonspecific urethritis (17%) and gonorrhea (12%) could be detected. Up until very recently viruses related to condylomata acuminata have been suspected to be the causative agents of cancer. Now it is generally accepted that the risk of condylomata acuminata progressing to malignancy is relatively low. Virus types present in such benign tumors are rarely found within genital squamous cell cancers (Dürst et al. 1983). On the other hand high-risk viruses have been observed in flat condyloma-like lesions, especially in the so-called bowenoid papulosis of the external genitalia (Gross et al. 1986). There is ample evidence at present that squamous intraepithelial neoplasias of the vulva have many features in common with premalignant lesions of the penis and the perianal area, such as epidemiology, pathomorphology, and a putative etiological relationship with specific HPV types. Squamous intraepithelial neoplasias (Table 4) are a challenge both to the gynecologist and the dermatologist because of their increasing incidence, their uncertain relationship to the development of invasive cancer, and especially the current uncertainty about optimal management. Together with the increasing incidence of vulvar carcinoma in situ (VIN III) during the past 3 decades, a striking decrease has been observed in the average age of affected women (di Paola et al. 1982). In 1958 the average patient with VIN was reported to be 54 years of age and no patient was younger than 33 years (Woodruff and Hildebrandt 1958). Today younger women (peak incidence between 48 and 51 years of age) are affected with VIN and in about one-third of cases the disease is diagnosed before the age of 40 (Böcker and Stegner 1981). This implies a bimodal prevalence curve of VIN. Furthermore, the hypothetical malignant conversion of the premalignant lesions of the vulva and of the penis seems to follow a biphasic course with a long latency ranging from 20 to 50 years (zur Hausen 1986). Clinically, VIN in young women differs from Bowen's disease, a condition occurring later in life (Abell and Gosling 1961; Woodruff et al. 1983). It is assumed that VIN in the young may be related to bowenoid papulosis, a generally benign, in situ lesion which was first described 15 years ago at vulvar and penile sites (Lloyd 1970; Kimura et al. 1978; Wade et al. 1978, 1979).

Squamous cell carcinomas of the cervix, vulva, vagina, anus, and penis may be histologically similar and appear to share common risk factors (Peters et al. 1984).

3 Diagnostic Methods

In most cases the diagnosis of genital papillomavirus infections can be made clinically. In vaginal and/or cervical condylomas discharge, pruritus, and postcoital bleeding are commonly observed, especially in florid lesions. Nevertheless, there

Table 4. Differential diagnosis of HPV-associated squamous cell lesions of the lower genital tract

flat condylomata	Carcinoma-in-situ of the vagina/cervix Vaginal/cervical carcinoma Herpes genitalis (cervix, vagina)
Condylomata acuminata	Syphilitic condylomata lata (lues II) Common warts Papillae coronae glandis Mollusca contagiosa Pemphigus vegetans
	In patients from tropical regions: Lymphogranuloma venereum Granuloma venereum Chancroid Amebiasis Schistosomiasis
Giant condylomata	Buschke-Löwenstein's tumor Squamous cell carcinoma Lymphogranuloma venereum
Flat condyloma-like lesions	Syphilitic condylomata lata (lues II) Bowenoid papulosis (VIN III, PIN III of the young) Bowen's disease (VIN III, PIN III of the elderly) Queyrat's erythroplasia Lichen simplex chronicus Lichen ruber Psoriasis Balanitis Vulvitis Paget's disease
Pigmented papular lesions	Multicentric pigmented Bowen's disease (Bowenoid papulosis) Pigmented (melanotic) nevus Seborrheic keratosis Basal cell carcinoma Malignant melanoma

are cases of long-standing genital warts without any discomfort. In contrast to condylomata acuminata and giant condylomata, flat lesions do not cause clinical symptoms. At times the clinical picture does not suggest HPV infection or may resemble different conditions, thus calling for supplementary laboratory investigations. The differential diagnosis of anogenital condylomata acuminata and of flat condylomas includes syphilitic condylomata lata, mollusca contagiosa, papillomatosis of the penis, and common warts (Table 4).

HPV lesions present on vaginal, urethral, or rectal orifices are often associated with internal involvement. In homosexual men with perianal warts, 70% have an additional anorectal disorder (Oriel 1971b). Cervical HPV infection can be dis-

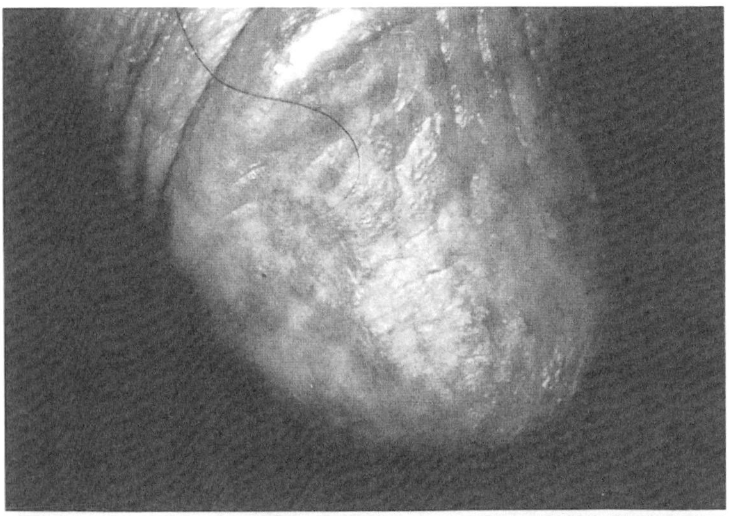

A

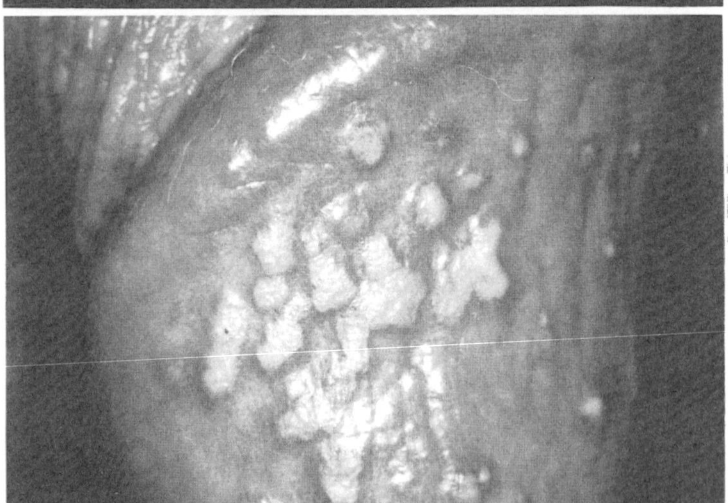

B

Fig. 1 A, B. Flat condylomalike lesions of the glans penis before **(A)** and 5 min after **(B)** application of 5% acetic acid

covered in 50% of all women with vulval or vestibular lesions (Walker et al. 1983). In view of the infectious nature of genital warts, digital palpation or endoscopy should be performed only after successful treatment or spontaneous healing of external lesions. Otherwise the transmission of HPV-positive cellular material may occur giving rise to multifocal disease.

Generally, colposcopy is necessary for the detection of condylomas in the vagina. The same holds true for subclinical lesions of the vulva and of the penis, which are invisible to the naked eye. The application of 3%–5% acetic acid leads to a white epithelium and to the visualization of pathological vessels, both characteristic of HPV infection (Fig. 1). If one is faced with the difficult problem of

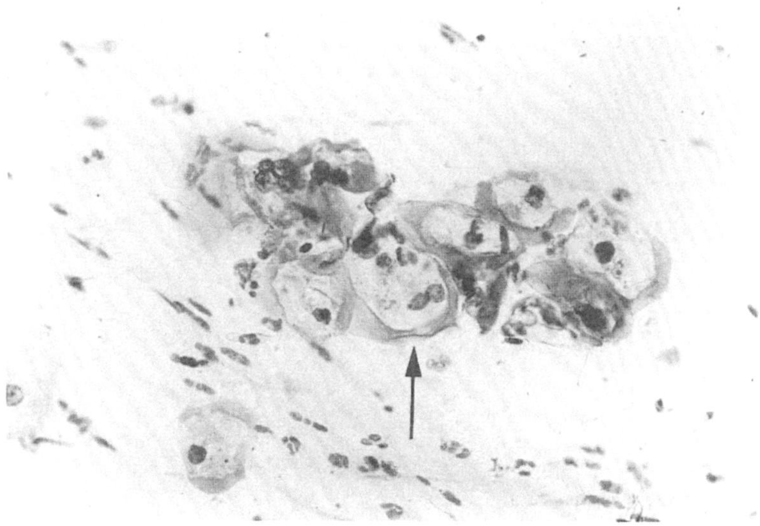

A

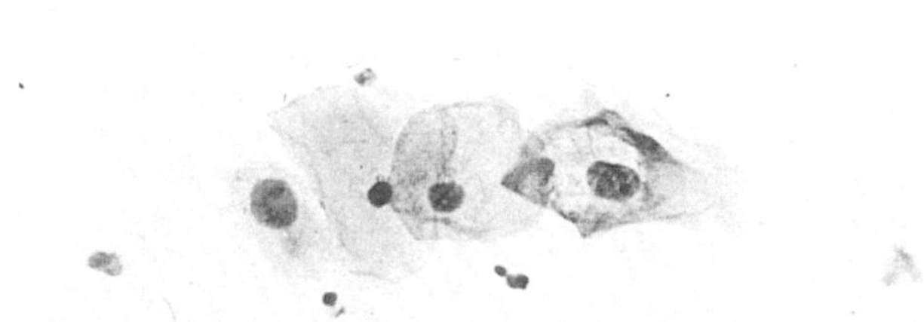

B

Fig. 2. A Papanicolaou smear from the vagina: koilocytes with perinuclear vacuolization and binucleated koilocyte *(arrow)* present (H & E, ×560); **B** Papanicolaou smear class III D of the cervix (HPV16/18 DNA positive on filter in situ hybridization), dyskeratinocytes, and koilocytes (H & E, ×560)

where to biopsy a diffuse penile or vulvar lesion, it may be helpful to stain the area with 2% toluidine blue, which is rinsed with acetic acid (Collins assay). Toluidine blue - positive areas should be biopsied. Toluidine is a nuclear stain, and thus any process, cancerous or inflammatory, in which nucleated cells are on the surface will give positive results. Colposcopy can also be used to monitor lesions: their size, location, and morphological variability can be defined by this procedure. Cytological techniques can be used for the detection of vaginal lesions which shed cells pathognomonic of HPV infection (Fig. 2) (Meisels and Fortin 1976; Meisels et al. 1977). However, Papanicolaou smears are inadequate for the diagnosis of vulvar or penile HPV infections. For a further description of cytological findings see chapter by Koss, this volume, while the methods of HPV identification are described by Schneider, this volume. Figures 3, 4, and 15 are examples of these methods.

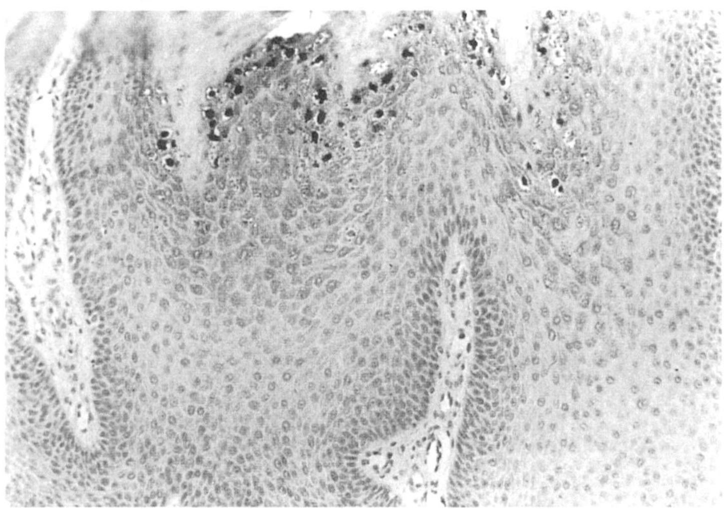
Fig. 3A

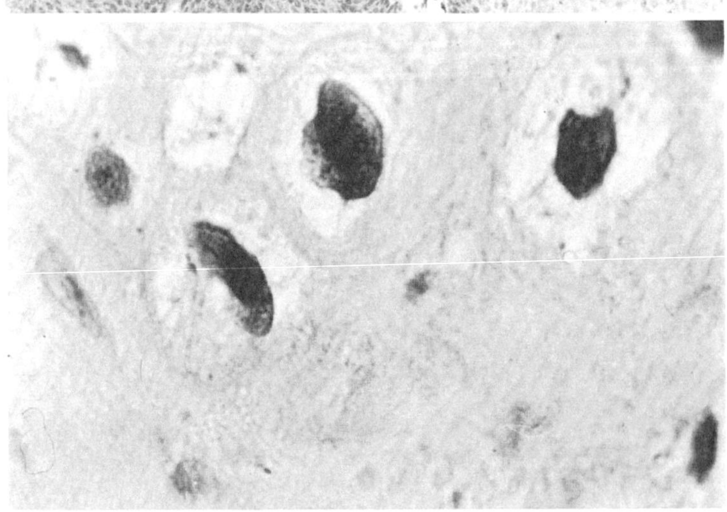
Fig. 3B

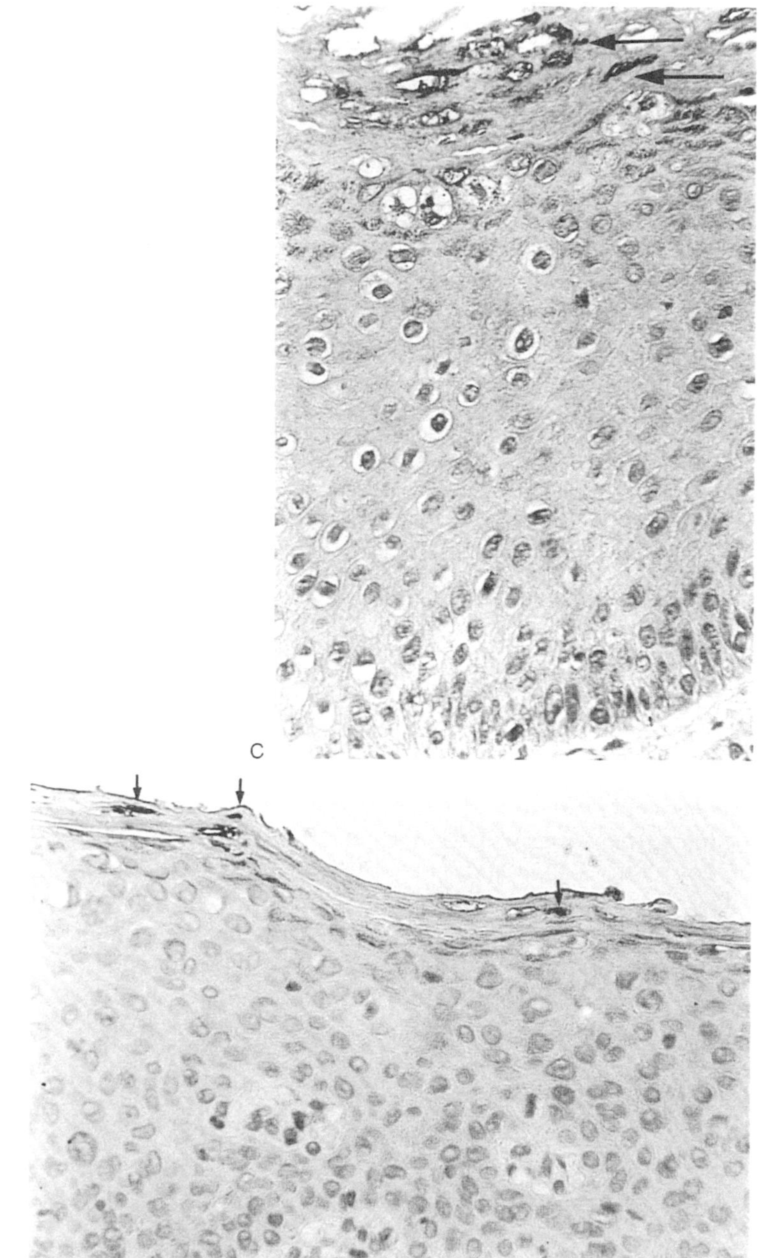

Fig. 3 A–D. PAP-stained section of HPV-associated genital lesions. **A** Condyloma acuminatum showing positive papillomavirus common antigen (PCA) within clusters of nuclei of keratinocytes in the upper epithelial layers (H & E, × 30); **B** magnification of dark-brown precipitates confined to nuclei of koilocytes (H & E, × 400); **C** flat condyloma-like lesion, PCA present in some granular and corneal nuclei *(arrows)* (H & E, × 100); **D** bowenoid papulosis, PCA-positive nuclei in the most superficial corneocytes *(arrows)* (H & E, × 100)

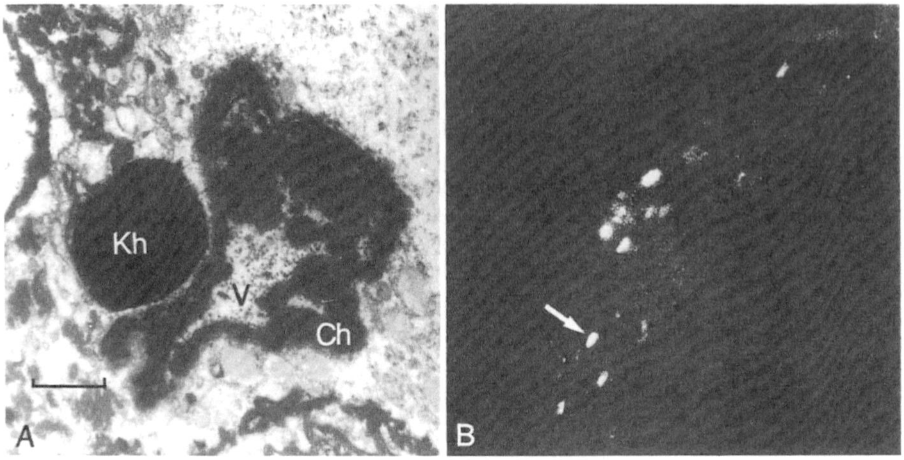

Fig. 4. **A** Electron micrograph of a koilocytotic granular cell with condensed keratohyalin material *(Kh)* and marginated chromatin *(Ch)* from a HPV6-related condyloma. Virions *(V)* are present in the nucleus (*bar* represents 1 μm); **B** indirect immunofluorescence on frozen section of the same condyloma acuminatum using group-specific common rabbit papillomavirus antiserum followed by rabbit-IgG antiserum. Papillomavirus-positive nuclei are present in the corneal and granular layer *(arrow)* (× 100)

4 Clinical Manifestation

Human papillomaviruses are epitheliotropic and induce in the host epithelial proliferations (papillomas) harboring mature virus in different concentrations. In addition to the skin, a wide range of different mucosal sites such as the cervix, vulva, penis, anus, larynx, esophagus, tongue, and buccal mucosae have been found to be affected (reviewed by McCance 1986). HPV-induced tumors are primarily benign and show only limited growth. Malignant conversion occurs only in rare instances after a long latency period of years or even decades (zur Hausen 1986). It is well-known that skin warts develop in microlesions or along scratches. In the case of genital warts it seems likely that the various forms of tissue irritation prevalent in the anogenital area, such as eczema, discharge, vaginitis, balanitis, and phimosis, lead to a continuous local trauma that promotes viral infection.

4.1 Benign Lesions

The most common HPV-related disorder of the lower genital tract in both sexes is condyloma acuminatum. The vulva has been recognized as the prevalent site of such lesions in the female. With the identification of flat condylomas in the cervix uteri (Meisels and Fortin 1976; Purola and Savia 1977), it became clear that the cervix uteri is the most frequently affected site.

In addition there are, in both sexes, subclinical genitourinary and perianal lesions which are different from the "classical" papillary form and which have

Table 5. Clinical comparison between bowenoid papulosis and Bowen's disease

	bowenoid papulosis	Bowen's disease
Mean age (years)	30	45
Number of lesions	multiple	single
Distribution	skin and mucous membranes	skin
Spontaneous regression/association with pregnancy	yes	no
Clinical appearance of lesions	lichenoid/pigmented papules erythematous maculae leukoplakialike lesions	irregular plaque slightly raised erythematous plaque
Color of lesions	pink reddish-brown brown to black	white red
Symptoms	none (slight pruritus)	50% pruritus

been linked to HPV only in recent years (Gross et al. 1985a). As summarized in Table 1, five distinct, benign HPV-related, squamous cell lesions in the lower genital tract have been classified to date: (1) the papillary "classical" genital wart (condyloma acuminatum), (2) the giant condyloma, (3) the Buschke-Löwenstein's tumor, (4) the flat condyloma-like lesion, and (5) the pigmented papular lesion (Gross et al. 1985a). With the exception of the flat condylomas of the type exclusively present in the vagina and in the uterine cervix, all these varieties are found both on female and male external genital sites and in the adjacent transitional mucosal sites.

The five types of lesions have in common a more or less prominent acanthosis which can result in a tumorous thickening of the epithelium, as in the case of condyloma acuminatum and giant condyloma. Koilocytes, i.e., cells showing the cytopathic effects of the permissive HPV infection (Koss and Durfee 1956), are regularly present in condylomata acuminata and in the giant forms; the flat condyloma-like and the pigmented papular lesions, however, are often devoid of such cells (Fig. 5).

Four common HPV types are associated with the cutaneous and (muco)cutaneous genital lesions cited: HPV6, HPV11, HPV16, and HPV18 (Gissmann et al. 1982; Dürst et al. 1983; Gissmann 1984; Boshart et al. 1984; de Villiers et al. 1986a). It is known that apart from an obvious target cell specificity, the individual virus types found in the skin and in the genital tract vary in their oncogenic potential (zur Hausen 1982; Gissmann 1984; K. Syrjänen, this volume) (Table 1). Virological analysis of the two recently described papillomavirus-related clinical conditions, the flat condyloma-like lesions and the pigmented papules, has revealed an association between HPV6 and HPV11 and lesions with low epithelial atypia and an association between HPV16 and lesions with severe atypia or carcinoma-in-situ (CIS) (bowenoid papulosis). In the case of vulvar lesions this disorder has been shown to be identical with multicentric severe VIN III (Gross et al. 1985a, b; Ikenberg et al. 1983). Without any doubt additional papillomavirus types exist in

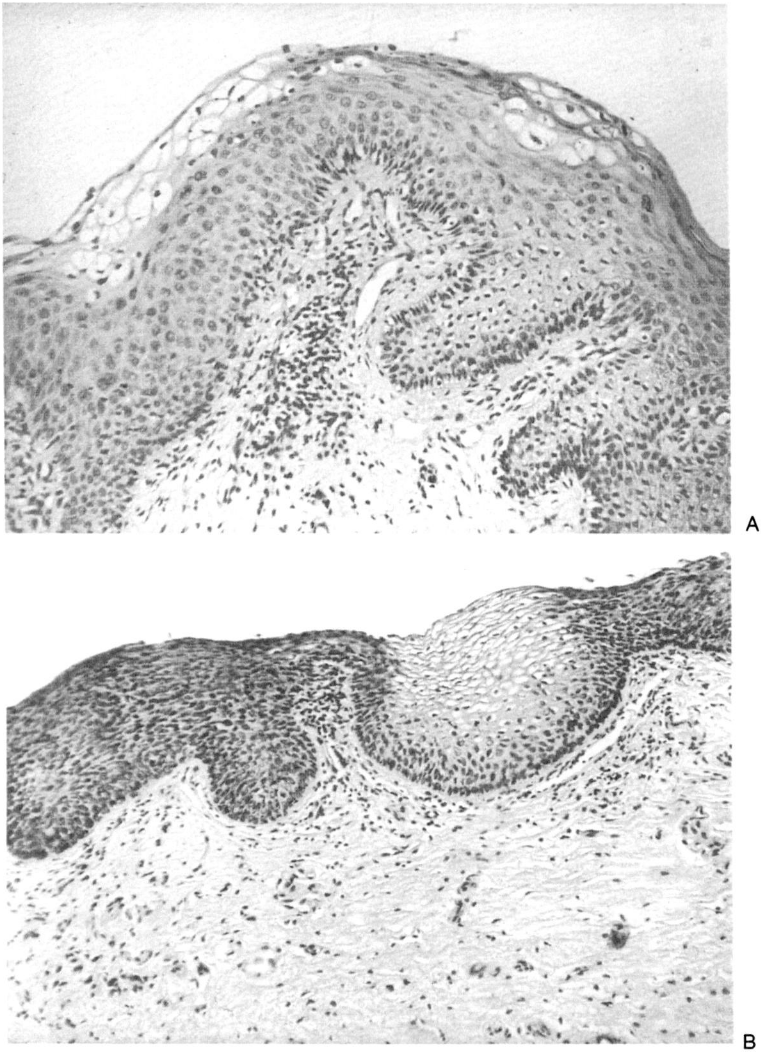

Fig. 5. A Histology of a flat condyloma-like lesion from the penile shaft: acanthosis and clusters of koilocytes in superficial layers (H & E, ×120). **B** Histology of another case from the vulva: focal akanthosis and koilocytosis (H & E, ×120)

the anogenital mucosae. The detection of HPV1, HPV2, and HPV10 sequences within anogenital papillomas has been occasionally reported (Green et al. 1982; Krzyzek et al. 1980; Zachow et al. 1982; Gissmann, personal communication). HPV10, an "EV-virus", deserves special attention since 2 of 31 cervical carcinomas and 2 condylomata acuminata harbored sequences related to this HPV DNA, which had been cloned from cutaneous EV lesions (Green et al. 1982). Recently HPV3 DNA and those DNAs which are also related to a group of skin-associated HPVs present in EV lesions (HPV5, 8, 12, 19, 21, 22, and 25; Table 1) were dis-

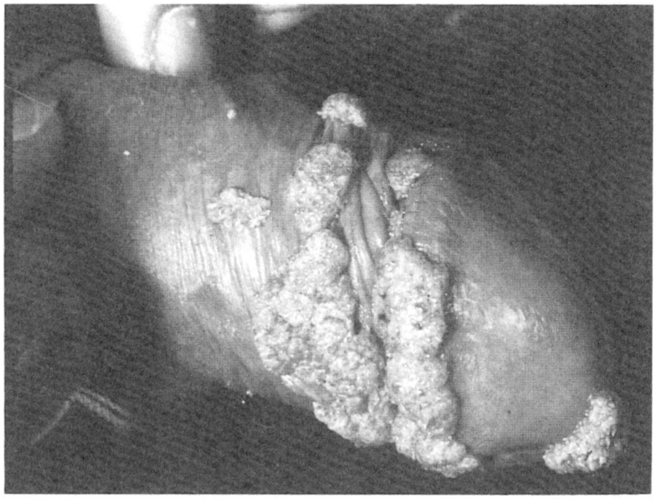

Fig. 6. Condylomata acuminata of the penis

covered in samples of condylomata acuminata and condylomata plana, respectively (de Villiers et al. 1986a).

By means of molecular hybridization under low-stringency conditions, at least four further viral genotypes have been identified in a few cervical cancer tissue specimens: HPV31, HPV33, HPV34, and HPV35 (Lorincz et al. 1985; Beaudenon et al. 1986). HPV33 and HPV34 have been detected in some lesions of bowenoid papulosis of the external genitalia (Beaudenon et al. 1986).

4.1.1 Condylomata Acuminata

Condylomata acuminata are papillary genital warts familiar to anyone involved in the clinical diagnosis and management of genital disease. In general they are clinically diagnosable with the naked eye, appearing grouped around mucocutaneous junctions. They are found primarily on the shaft of the penis, on the prepuce, and on the glans (Fig. 6); in the female they are located mainly in the periurethral region, on the fourchette, and on the labial folds. Rather frequently such lesions are seen in the perianal skin and on the adjacent sites of the inner thighs and groins. Condylomata acuminata occurring in the anal mucosa are seen predominantly among homosexual men and among females who engage in anal intercourse (Oriel 1971b).

Grossly, they appear either soft and bulky, or in the form of pedunculate papules or tumors. They are composed of pink-reddish or yellow-brownish, soft, sessile papules. Fingerlike projections can be seen on the tip of the warts, from the pointed appearance of which the term "condyloma acuminatum" has been derived. In the groins and perianal and labial folds condylomas may imitate a cock's comb arrangement. In such cases the surface appears hyperkeratotic and verrucous. Condylomata acuminata are usually multiple and grow in clusters. A small percentage develops into cauliflowerlike tumors. In the vagina condylomata acuminata are localizable by means of acetic acid and colposcopy, appearing as a

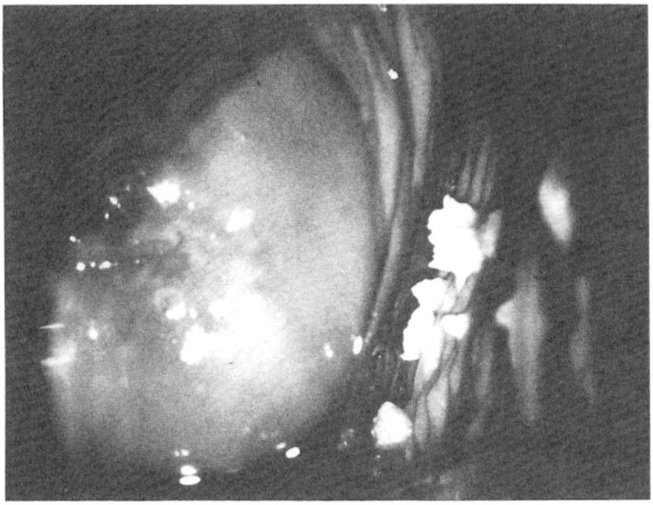

Fig. 7. Condylomata acuminata of the vagina

raised, white lesion (Fig. 7) with characteristic fingerlike projections bearing prominent capillaries. In these cases multifocal lesions affecting the portio, vulva, vagina, and perianal skin are common.

The exophytic or papillomatous condyloma acuminatum is characterized histologically by acanthosis, papillomatosis, focal parakeratosis, and various degrees of koilocytosis (Fig. 8) (Gross et al. 1982; 1985a). The basal cell layers regularly exhibit some proliferation. Elongated dermal papillae, intercellular edema, dilated vessels, and round cell infiltrates of varying extent are a constant finding. In longstanding condylomata the connective tissue appears to be increased, thus giving the impression of a fibroepithelial tumor (Fig. 8).

Both flat condylomas and condylomata acuminata exhibit nuclear atypia to some degree. The nuclei appear enlarged and hyperchromatic. Additionally anisokaryosis, pyknosis, and atypical mitotic figures are present. These conditions are often difficult to differentiate from severe dysplasia or CIS (Fig. 9). The regularity of the basal cell layers and the low nucleocytoplasmic ratio indicate a benign condition (Meisels et al. 1981).

4.1.2 Giant Condylomata

Condylomata acuminata of the vulva may become very large, particularly during pregnancy or in cases of immunosuppression or Hodgkin's disease (Fig. 10). These unusual venereal warts are characterized by large size and cauliflowerlike appearance. The largest ones of these (more than 3 cm in diameter) have a high incidence of histological atypia (Lee et al. 1981), in which case they resemble Buschke-Löwenstein's tumors of the penis.

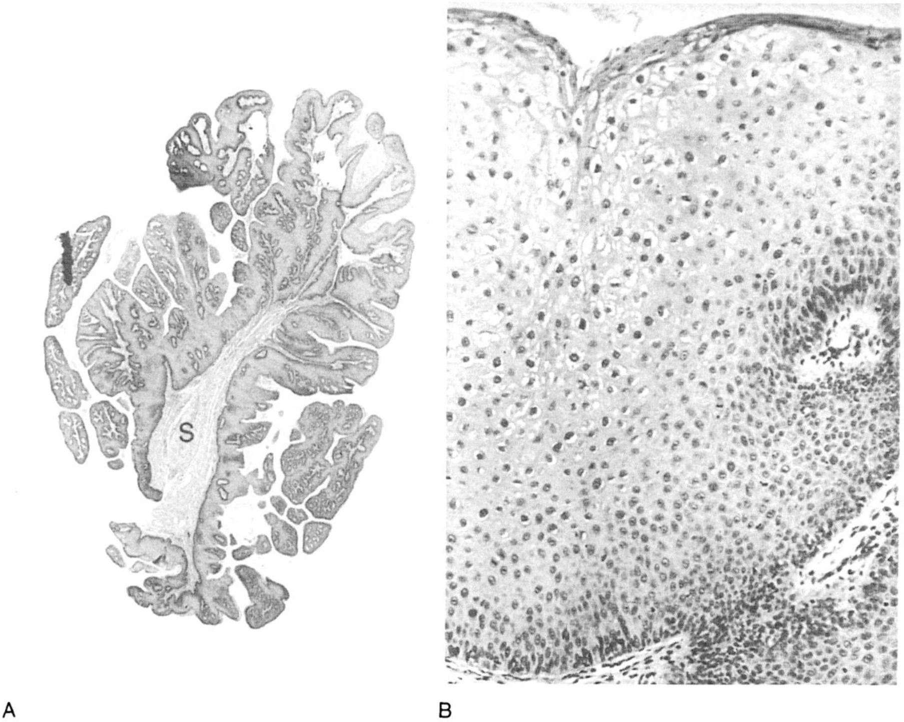

Fig. 8 A, B. Histological section of a vulvar condyloma acuminatum. **A** Prominent papillomatosis and acanthosis, stromal fibrous connective tissue appears increased (H & E, × 1); **B** higher magnification showing koilocytosis (H & E, × 120)

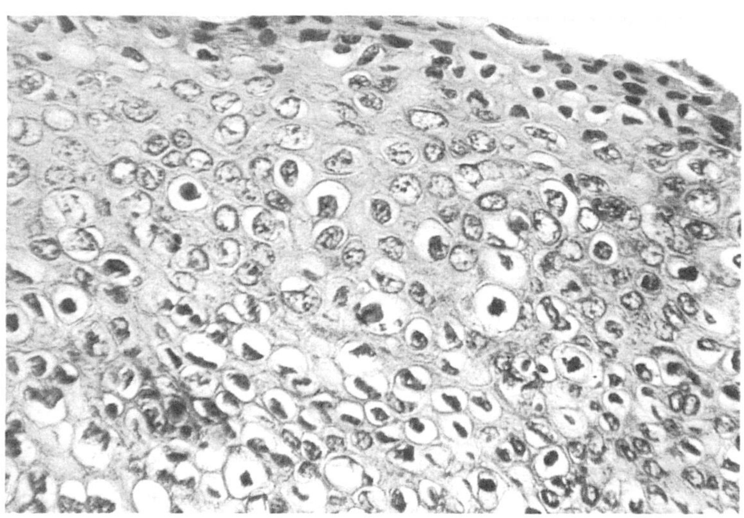

Fig. 9. Histological section of a vulvar "atypical condyloma" with large hyperchromatic nuclei (H & E, × 300)

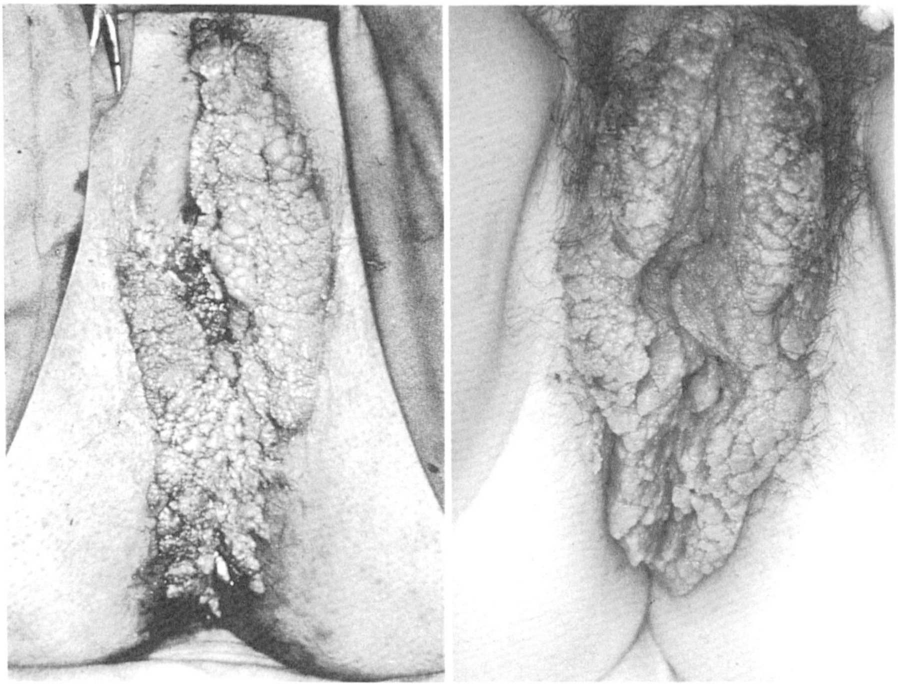

A B

Fig. 10 A, B. Disseminated anogenital giant condylomata **(A)** in a 35-year-old female suffering from severe diabetes mellitus, and **(B)** in a 19-year-old female with Hodgkin's disease

4.1.3 Buschke-Löwenstein's Tumor

There is no longer any doubt concerning the viral etiology of Buschke-Löwenstein's tumors, which may be regarded as a lesion intermediate between a condyloma and carcinoma. Hybridization experiments have demonstrated the presence of HPV6 DNA and to a lesser degree of HPV11 DNA sequences in a number of such tumors (Gissmann et al. 1982; Boshart et al. 1984; de Villiers et al. 1986a). DNAs from two verrucous carcinomas of the vulva also hybridized to HPV6 DNA under stringent conditions (Zachow et al. 1982).

The Buschke-Löwenstein's tumor, first described on the penis in 1925, is an uncommon, cauliflowerlike neoplasm (Fig. 11). Similar tumors may occur on the vulva (Judge 1969; Weed et al. 1983) and less frequently in the perineoanal area, groins, urethra, and in rare instances, the anal canal. The tumor usually arises in males after years of poor hygiene. It has been argued that it represents an intermediate lesion between benign condyloma acuminatum and squamous cell carcinoma, although the relationship between these tumors is poorly understood. The Buschke-Löwenstein's tumor can be differentiated from benign giant condylomata by characteristic exophytic and endophytic growth properties leading to local expansion with invasion of adjacent structures such as the urethra or rectum. Furthermore, these tumors are not responsive to podophyllin, a standard agent in

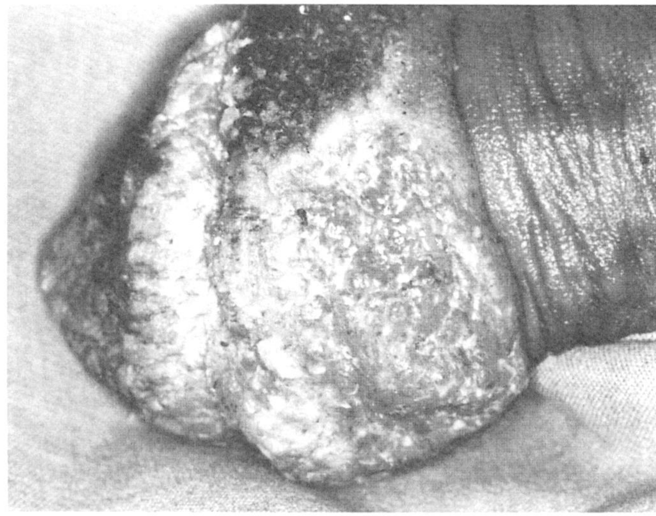

Fig. 11. Buschke-Löwenstein's tumor of the glans penis

local therapy of genital warts. The usual absence of local lymph node involvement or of distant metastases as well as the close histological resemblance to genital warts distinguishes this condition from invasive squamous cell carcinomas. Most of the innumerable case reports describe the semimalignant growth patterns of this disorder. Nevertheless, there is a very small series of documented cases in which a Buschke-Löwenstein's tumor transformed into a metastasizing carcinoma (Boxer and Skinner 1977; zur Hausen 1977) (Fig. 12).

Thus, the clinical course is all-important and close surveillance of suspicious cases, vital. A thorough histological investigation of the tumors in multiple sections is necessary in order to rule out carcinoma. Again, both the clinical picture and the histology must be taken into consideration when this tumor is diagnosed.

Morphologically, there is some similarity to verrucous carcinoma (Ackerman tumor) (Ackerman 1948), i.e., oral florid papillomatosis, and to carcinoma cuniculatum (Aird et al. 1954), aggressive proliferations of squamous epithelia of the upper airways, oral cavity, and sole of the foot. Although virus particles were not discernible in these conditions by means of electron microscopy or immunohistochemistry, HPV16 and HPV2 DNAs were detected within a verrucous carcinoma of the larynx (Brandsma et al. 1986) and of the oral mucosa (Adler-Storthz et al. 1986).

Histologically the Buschke-Löwenstein's tumor is differentiated from the simple giant condyloma acuminatum by the downgrowth of the acanthotic epithelium into the underlying subcutaneous fat. Malignant inversion, however, is usually absent. There is sometimes a complex histological pattern with large areas of benign giant condyloma (koilocytosis present) intermixed with areas of atypical epithelial cells or even differentiated squamous cell carcinoma (koilocytosis absent) (Boxer and Skinner 1977) (Fig. 12).

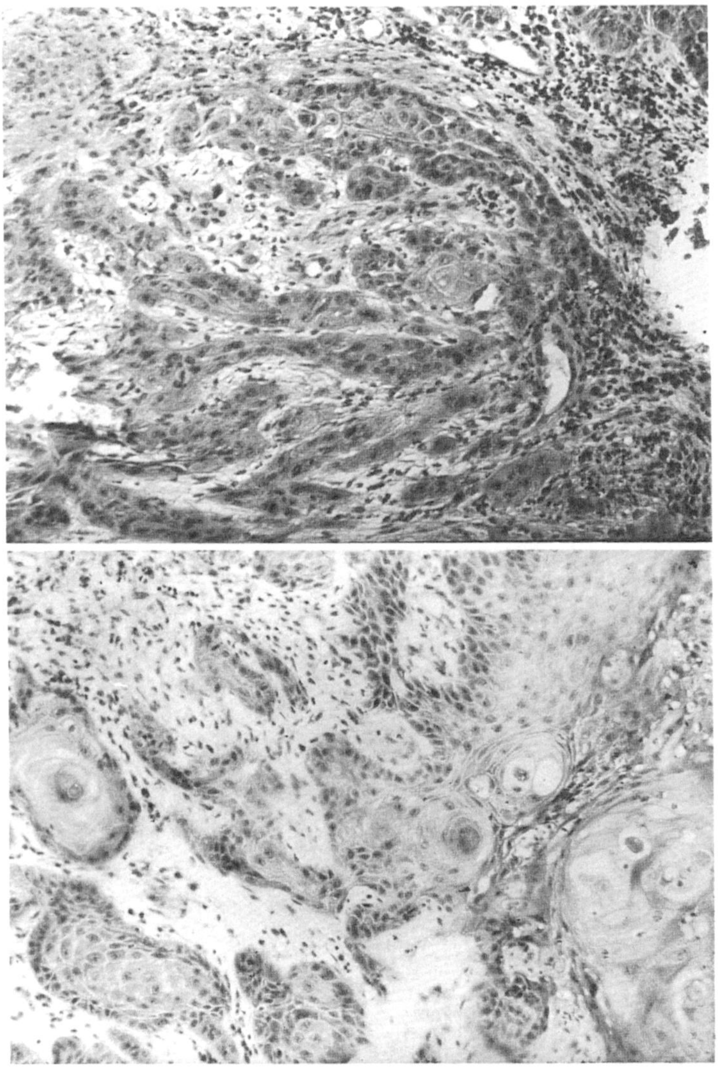

A

B

Fig. 12 A, B. Histological section of a penile Buschke-Löwenstein's tumor with an early malignant change. **A** Area of atypical epithelial cells with nest formation and spread into dermis (H & E, ×120); **B** under higher magnification dyskeratinocytes, perl formation can be seen (H & E, ×300)

4.1.4 Flat Condyloma-like Lesions

As in the case of flat condylomata and pityriasis versicolor-like maculae in patients with EV at external genital sites, flat erythematous lesions and leukoplakia-like plaques were found to harbor HPV and were designated as "flat condyloma-like lesions" (Fig. 1) (Gross et al. 1985a, b). Often these conditions remain subclinical. They are hardly visible with the naked eye. In the vaginal vestibulum, prepuce, and glans penis, they appear as whitish, maculopapular, slightly raised lesions which, in contrast to the so-called spiked conylomata, possess no asperities or spikes but have a soft, velvety surface (Gross et al. 1985a). Often these lesions may become visible only by means of colposcopy after acetic acid treatment (Fig. 1). In these cases carefully directed punch biopsies must be taken in order to rule out intraepithelial neoplasia. In most cases there are no symptoms. These lesions change widely in clinical appearance from one day to the next. Formerly such lesions were not thought to be associated with papillomavirus infection since they resemble other dermatological conditions such as lichen ruber or psoriasis (Table 4). In view of the synchronous detection of such lesions in male partners of females with cervical flat condyloma, and vice versa, it has been suggested that this disorder is the counterpart of the flat condyloma of the cervix uteri (Gross et al. 1986).

4.1.5 Pigmented Papular Lesions

This fifth benign clinical variety is a small, flat-topped papule of black-brownish pigmentation which mostly appears in the form of disseminated multiple lesions on the skin of the vulva, penile shaft, pubic area, groins, and at perianal sites. There is no similarity to condyloma acuminatum, and keratosis is always absent (Fig. 13).

Fig. 13. Disseminated pigmented papular lesions in the inguinocrural folds (histologically benign acanthosis)

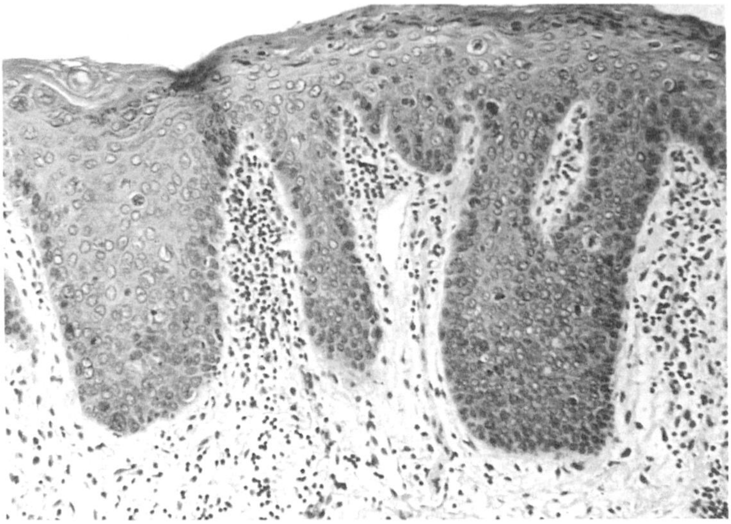

Fig. 14A

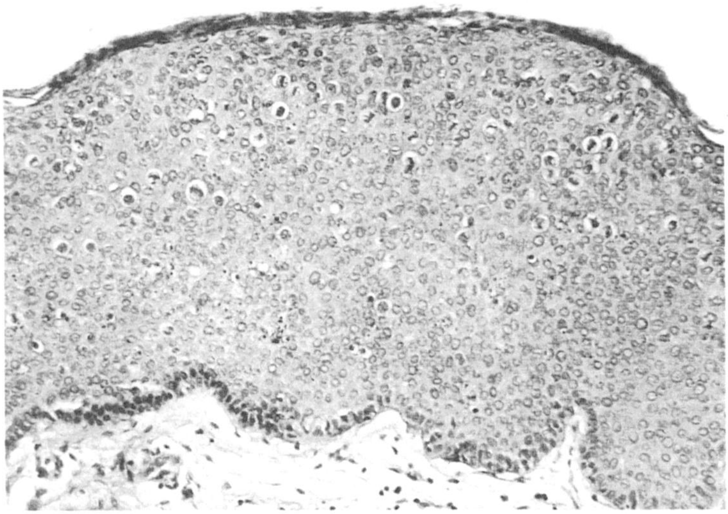

Fig. 14B

These pigmented papules have been recognized only recently as a HPV-related disease, no doubt because their clinical appearance resembles that of seborrheic keratosis, fibrosis, or pigmented nevi (Table 4), or even because the lesions have simply been overlooked. Furthermore, their papillomavirus particle content is rather minimal and at times entirely lacking. Flat condyloma-like lesions and pigmented papular lesions are characterized pathologically either by (a) the orderly arrangement of the epithelial layers and preserved cell polarity or (b) with features of an intraepithelial neoplasia:

(a) There is usually just a slight papillomatosis. In rare instances clusters of koilocytes are present in the uppermost layer (Fig. 5). Otherwise, there are no features characteristic of a HPV infection. Epithelial cells show no or only a very slight degree of atypia. The number of mitotic figures is not increased.

Fig. 14 A-C. Histology of bowenoid papulosis. **A** Histology from the case shown in Fig. 16 A. Epithelial hyperplasia, features of a carcinoma-in-situ, vesicular chromatin present. Moderate stromal infiltrate of mononuclear cells (H & E, × 120). **B** Histology from case shown in Fig. 16 B. Slight epithelial thickening, typical bowenoid atypia (H & E, × 120). **C** Section of a multifocal leukoplakia-like plaque. Epithelial thickening, dysplastic cellular changes of carcinoma-in-situ (dyskeratotic, pleomorphic, and atypical cells) (arrow) in close connection with benign acanthosis (two arrows) (H & E, × 120)

C

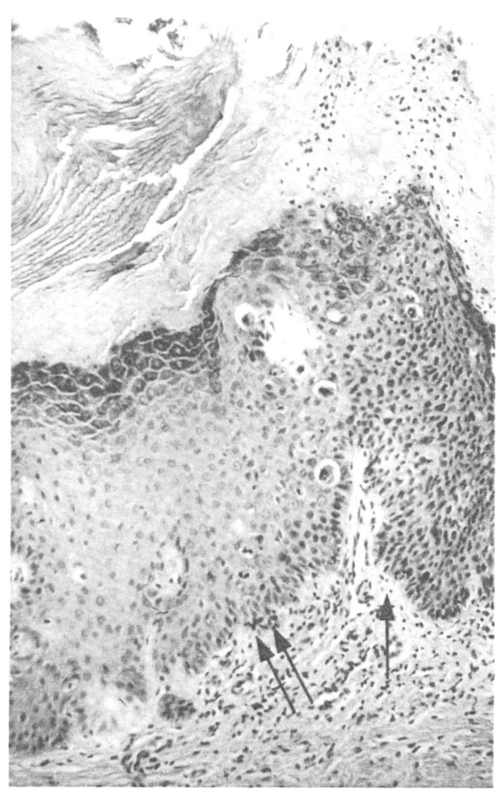

(b) Although grossly the lesion appears benign, histologically such lesions may show areas of epithelium with normal appearance intermixed with foci of increased cellular atypia or even with microscopic similarities to Bowen's disease. Such lesions have been designated "bowenoid papulosis" (see below; Wade et al. 1978, 1979) and in pigmented cases "multicentric pigmented Bowen's disease" (see below Lloyd 1970; Kimura et al. 1978) (Figs. 13, 14, and 16).

4.1.6 Further Reservoirs for HPV and Subclinical Infections

The existence of subclinical infections has to be taken into account. In recent years increasing evidence has accumulated that HPV DNA and even viral particles continue to persist in clinically and histologically normal laryngeal and genital mucosae and in the skin (Steinberg et al. 1983; Ferenczy et al. 1985). This would imply that papillomaviruses are produced independently of epithelial disease. Thus, HPV DNA was detected in approximately 10%-13% of cellular swabs taken from the cervix uteri of 2000 women in an outpatient clinic (Wagner and de Villiers, personal communication). A further reservoir for HPV seems to be the male urethra, which is very likely to be responsible for symptomless transmission of certain virus types such as HPV11 or HPV16, leading to genital warts or precancer (de Villiers et al. 1986a; Gross et al. 1985c, 1986).

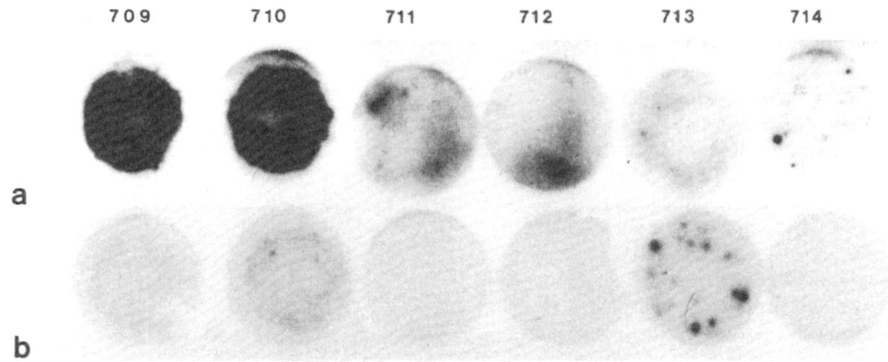

Fig. 15. Autoradiography after filter in situ hybridization. The cells were obtained from smears of a perianal condyloma acuminatum *(709)* and a vulvar condyloma acuminatum *(710)* of the same patient, a perianal condyloma acuminatum *(711)*, a penile flat condyloma-like lesion *(712)*, a bowenoid papulosis lesion from the vestibulum vaginae *(713)*, and from the urine sediment of a patient after successful α-interferon treatment of penile condylomata acuminata *(714)*. The nitrocellulose filters in *lane a* were hybridized with HPV11 under stringent conditions and exposed for 10 days. Filters 709, 710, and 714 are HPV11-positive. Filters 711, 712, and 713 show background intensity due to high cell number. Rehybridization with HPV16 was followed by exposure for 10 days *(lane b)*. Filter 713 is HPV16-positive

Filter in situ hybridization can also be used to demonstrate the presence of HPV DNA in the urine of patients with genital warts and their partners (Gross et al. 1985c; Fig. 15). This is an additional tool to identify subclinical HPV lesions in the urethra and may help to disclose further reservoirs of asymptomatically infected individuals.

4.2 Premalignant Lesions

Although viruslike inclusions have been observed in some lesions of Bowen's disease (Wade et al. 1978, 1979), the involvement of viruses in the etiology of this disorder and of bowenoid papulosis remained unresolved until recently. Using molecular hybridization techniques a consistent association of bowenoid papulosis, including the pigmented forms (multicentric pigmented Bowen's disease) with HPV16 has now been established (Ikenberg et al. 1983; Gross et al. 1986; Obalek et al. 1985). In rare instances the presence of HPV6 and HPV11 DNAs (Zachow et al. 1982) as well as coinfections of HPV16 and HPV6 (Gross et al. 1986) have been reported. Molecular hybridization experiments under nonstringent conditions indicate the existence of additional unclassified HPV types in rare instances (Ikenberg et al. 1983; Obalek et al. 1985).

HPV33 and HPV34, members of a group of lately characterized viral genotypes, have been detected in some bowenoid papulosis lesions (Beaudenon et al. 1986). Sequences of the same viruses have been shown in a small percentage of CIN. It is of interest that HPV16 DNA was found in six of ten biopsies (four of five from genital sites and two of five from extragenital sites) of Bowen's disease (Ikenberg et al. 1983). This points to common etiological elements in both disorders.

4.2.1 Bowenoid Papulosis

In view of the currently emerging relationship between bowenoid papulosis and Bowen's disease, which is substantiated by the regular presence of HPV16 DNA (Table 1) (Ikenberg et al. 1983; Gross et al. 1985b, 1986), it is important to draw clinical, histological, and prognostic distinctions between these conditions (Table 5). Most importantly, lesions referred to as bowenoid papulosis show a striking discrepancy between their invariably benign clinical appearance and their pathology, which is indistinguishable from a squamous cell CIS, as seen in Bowen's disease. (1) The disorder affects predominantly young adults, thus following the age distribution of STDs and condylomata acuminata, with which it may occasionally be associated (Oriel 1971a; Wade et al. 1978). The average age in our series of 22 patients (15 men and 7 women) was 28 and 32 years, respectively (Gross et al. 1985b), confirming earlier reports by Wade et al. (1978, 1979). (2) The lesions are usually multiple and have a multicentric origin. (3) Bilateral distribution in the genital and inguinocrural folds is frequently observed. (4) Hyperpigmentation may be present. (5) There is a tendency to spontaneous regression, especially in mothers after delivery (Friedrich 1972; Berger and Hori 1978).

The clinical appearance of bowenoid papulosis includes a spectrum of clinically inconspicuous, noncondylomatous manifestations (Fig. 16): (1) lichenoid pink-colored or reddish-brown papules sometimes coalescing to form small plaques; (2) macular erythematous lesions, especially located at the border of the skin and mucous membranes; (3) leukoplakia-like lesions. The surfaces of all these lesions are smooth or slightly papillomatous.

Table 6. Prevalence of papillomavirus common antigen in genital lesions (immunohistology)

Lesion	Percent with antigen
Flat condyloma	70
Condyloma acuminatum	60–70
Giant condyloma	40
Flat condyloma-like lesion	40
Pigmented papular lesion	30
Bowenoid papulosis	5
Genital Bowen's disease	0
Extragenital Bowen's disease	0
Paget's disease	0
Buschke-Löwenstein's tumor of the penis	0
Penile carcinoma	0
Vulvar carcinoma	0

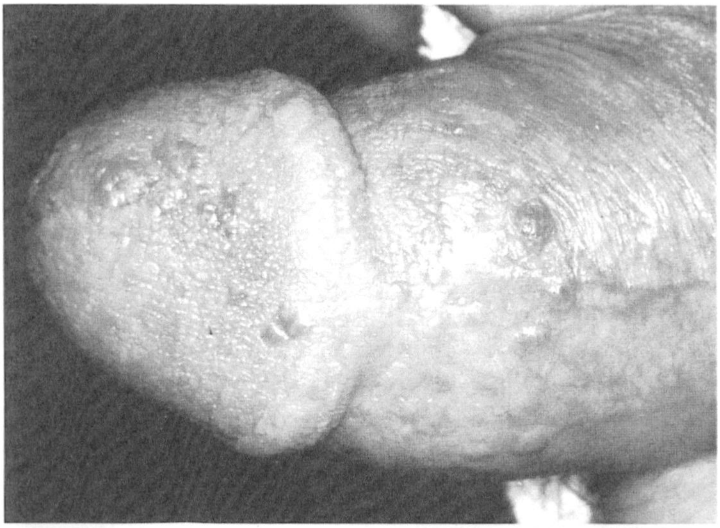

Fig. 16A

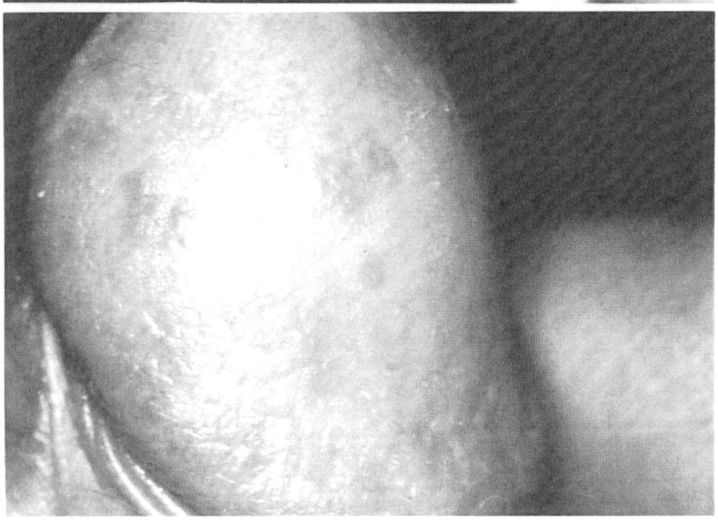

Fig. 16B

In general bowenoid papulosis affects the anogenital skin and adjacent mucosal sites in males and females. In our series the glans and the vaginal vestibulum as well as the labia minora and majora have been the most commonly involved sites (Gross et al. 1985b). The incidence of bowenoid papulosis is certainly much higher than so far estimated. Due to the symptomless course and the inconspicuous clinical appearance, biopsies are rarely done. The clinical diagnoses most often submitted with the biopsy specimens are condyloma acuminatum, psoriasis, lichen ruber, melanotic nevus, and seborrheic keratosis. Occasionally fungal infections, balanitis, vulvitis, lichen simplex chronicus, and even basal cell carcinoma and malignant melanoma are suspected (Wade et al. 1978, 1979). The heterogeneity of the clinical appearance is paralleled by a large variety of different eponyms which have been applied to lesions of bowenoid papulosis (Obalek et al. 1985).

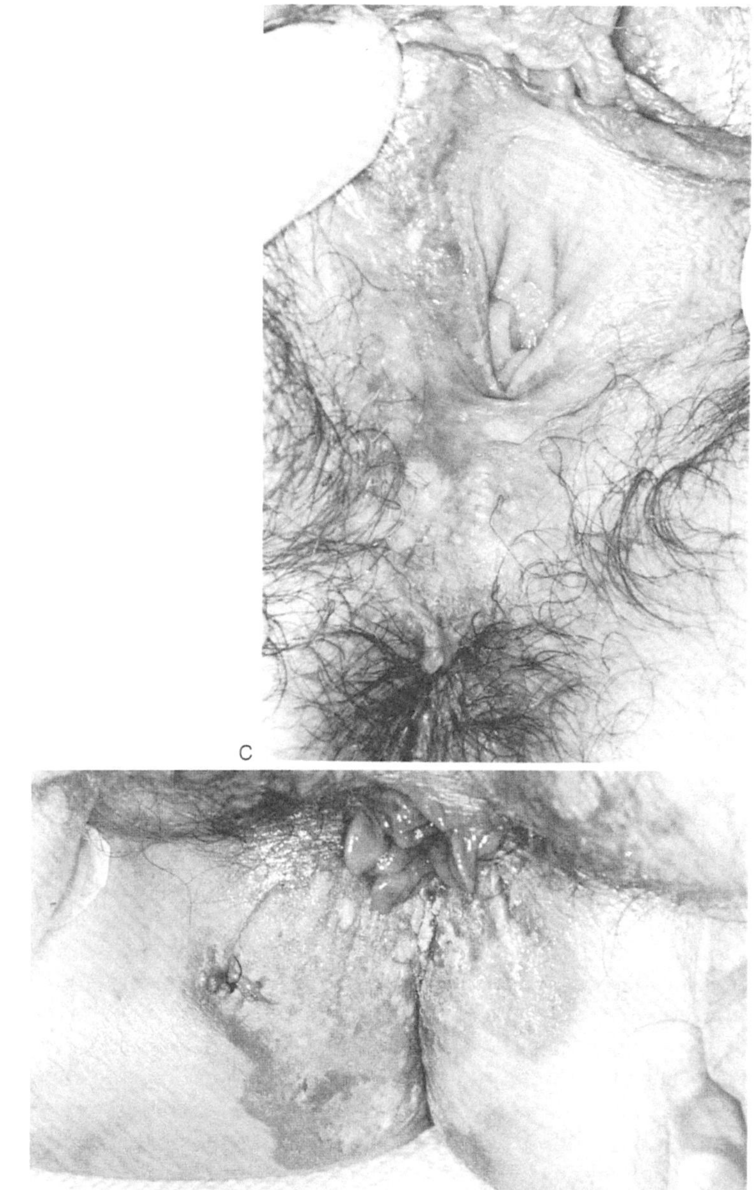

Fig. 16A-D. Bowenoid papulosis and Bowen's disease. **A** Lichenoid, pink-colored papules; **B** macular lesions on the glans penis; **C** multifocal leukoplakia-like plaques in the perineoperianal area; **D** Bowen's disease with flat moist plaque of the perianal and gluteal skin

4.2.2 Genital Bowen's Disease

This condition usually affects patients older than 45 years. The clinical appearance of Bowen's disease is mostly that of a single nonpigmented, fairly well-demarcated, flat, moist, or scaly plaque. It is itchy and appears red, granular, or velvety. The lesions range from a few millimeters to several centimeters in diameter, tend to spread, and do not regress. Such lesions may become very large, especially in moist areas (Fig. 16). Bowen's disease may affect the female external genitalia and involve the adjacent mucosa. Bowen's disease of the penis is rare: the disorder is localized preferentially on the foreskin or glans penis. Erythematous shiny lesions involving the glans of uncircumcised men are referred to as Queyrat's erythroplasia, a pathological process which is fundamentally the same as that seen in Bowen's disease (Pinkus and Mehregan 1976).

The histological features of Bowen's disease are consistent with those of a squamous cell CIS. Thus a clearcut differentiation between bowenoid papulosis and Bowen's disease is hardly possible by histological means. The microscopic features consist of varying amounts of hyperkeratosis, parakeratosis, and psoriasiform epidermal hyperplasia. Further findings include nuclear enlargement, hyperchromasia, pleomorphism with the presence of multinucleated giant cells, increased number of mitoses, and bizarre-shaped dyskeratotic cells spread throughout the epithelium. Rarely, koilocytosis is present in the outermost superficial cell layer (Fig. 14).

4.2.3 Comparison Between Bowen's Disease and Bowenoid Papulosis

In contrast to Bowen's disease pigmented bowenoid papulosis contains abundant melanocytes in the basal cell layer. Again in contrast with Bowen's disease, bowenoid papulosis usually shows a lesser degree of cytological atypia. A further feature of possible value in differentiating between the two conditions is the vesicular chromatin pattern and the focal distribution of dysplastic cells in the case of bowenoid papulosis (Fig. 14). In contrast to Bowen's disease and invasive carcinoma of the vulva and penis, bowenoid papulosis lesions exhibit human papillomavirus common antigens in 5% of cases (Fig. 3) (Braun et al. 1983; Gross et al. 1985a, b; Guillet et al. 1984; Jenson et al. 1980). The prevalence of these viral antigens is much higher in ordinary condylomas and in flat condyloma-like as well as pigmented papular lesions (Table 6). Bowenoid papulosis represents a "clinicopathological entity." Thus, essentially, the diagnosis has to be made by clinical examination, biopsy, and histological analysis in order to avoid unnecessary surgery in such cases.

The current virological, epidemiological, and clinical data reveal bowenoid papulosis to be the primary manifestation of HPV16 infection in both male and female genitalia. Bowenoid papulosis may be considered as an equivalent lesion of CIS of the uterine cervix and an important source of infection with HPV16.

An epidemiological study of the sexual partners of 50 individuals with genital HPV infections showed that in 24% HPV16 DNA was present in tissue from both partners. In these instances the male exhibited genitourinary bowenoid papulosis lesions (Gross et al. 1986). These findings along with case reports on the concomitant presence of bowenoid papulosis in the male and cervical neoplasia or precursor lesions in the female partner (Hauser et al. 1985; Gross et al. 1986; McCance and Singer 1986; Obalek et al. 1985) support the idea that this condition probably is not only a reservoir for HPV16 but also represents both a venereally transmissible disease and a high-risk lesion in the etiology of cervical cancer. Moreover, condylomata acuminata must be regarded as indicator lesions for the presence of papillomaviruses other than the low-risk types HPV6 or HPV11.

Such knowledge is being used to screen for type-specific HPV DNA sequences by molecular hybridization on filters to identify patients with a risk of developing genital squamous cell neoplasia and to attempt early diagnosis with careful clinical, cytological, and virological follow-up and thorough treatment of VIN, CIN, and corresponding bowenoid papulosis lesions (PIN) in the partner.

4.2.4 Vaginal Intraepithelial Neoplasia

Simultaneous involvement of the external genitalia as well as of the vagina and cervix with genital warts has been documented in a number of case studies (Syrjänen et al. 1984). In a study using filter in situ hybridization for the detection of HPV DNA, 42% of women with positive cervical cytology yielded HPV-positive smears from the vaginal and vestibular areas as well (Schneider et al. 1985). As in the case of the vulva, subclinical lesions have also been described in the vaginal epithelium (McCance and Singer 1986). In keeping with the flat condyloma-like lesions of the vulva, these acetic acid-white vaginal lesions are also associated with intraepithelial neoplasia (VAIN) and frequently harbor HPV16.

CIS of the vagina (VAIN III) has been reported rather frequently in the literature recently. In comparison to CIN III, however, VAIN III is a rare disease. The most common site of manifestation is the upper third of the vagina. In 1977 the annual age-adjusted incidence rate per 100000 women in the United States of America was calculated to be 0.2–0.31. The majority of these cases occurred in women with a prior history of CIS or invasive cervical cancer. A review of literature in 1981 noted fewer than 300 recorded cases (Woodruff 1981). Based on a review of 136 cases of VAIN III, an increasing incidence between 1953 and 1982 has been suspected (Benedet and Sanders 1984).

VAIN occurs mostly after menopause. The mean age of the affected patients is 55 years. The incidence rate in women under 40 years of age, however, has considerably increased in cases diagnosed since 1972 (Benedet and Sanders 1984). The multifocality of genital HPV infections fits well with the high association of VAIN III with other neoplasms of the lower genital tract, especially with those of the cervix. It sustains the theory of a multicentric origin of genital carcinoma.

4.3 Cancer

There are three major parallels in the epidemiology of squamous cell carcinoma of the penis, vulva, vagina, cervix, and anus. The incidence of carcinomas at each site shows an increase with decreasing social class, is low among Jews, and is elevated among separated and divorced invididuals (Peters et al. 1984). There is a large body of evidence that malignant squamous neoplasia of the five distinct sites is associated with a common sexual etiological factor such as HPV16 and/or HPV18.

Sequences of these two virus types, consistently associated with the majority of cervical carcinomas (Dürst et al. 1983; Boshart et al. 1984), have also been discovered in penile, vulvar, and vaginal squamous cell neoplasms. The risk of these malignancies, however, is by far lower than that of cervical cancer, although HPV16-related premalignant lesions exist in the entire female lower genital tract and on the penis. This is perhaps due to co-carcinogenic factors which may be more effective in the cervix uteri (zur Hausen 1982) or to differing levels of susceptibility for HPV of the penile, vulvar, or vaginal epithelium compared to the immature metaplastic cells at the transformation zone of the cervix.

In this context it is important to recognize, however, that HPVs such as types 6 and 11, known to be consistently related to condylomata acuminata, are also involved in a small proportion of invasive cancers of the genitalia, estimated at less than 5%.

4.3.1 Penile Carcinoma

Cancer of the penis accounts for 2% of all malignant tumors of the male genitourinary tract, but represents less than 0.1% of all male cancers in North America. Most penile cancers are squamous carcinomas, and grossly they may be nodular, exophytic, or fungating.

The tumor's geographic distribution is truly fascinating. Although it is a rare disease in the United States of America and in Europe (0.8-2 new cases per 100000 annually; Bracken 1981), it is fairly common in many developing countries in Africa, Central, and South America wherever circumcision is not practised and poor hygiene abounds (Waterhouse et al. 1982).

Thus far smegma, phimosis (75%-90% of all cases), inflammation, and a history of venereal disease (27%) have been regarded as contributing factors. The epidemiological data of Segi (1977) suggest that some of the world's high-risk areas for cancer of the penis are in Brazil. Here, generally younger patients have been shown to be affected (33-35 years) compared with cases in the United States of America and Europe (mean age 65 years; Berg and Lampe 1981). Extremely high prevalence rates have been found in rural regions of the northern part of Brazil (6.8 per 100000) in comparison to industrialized urban regions in the south (2.9 per 100000).

Currently available data clearly show the presence of HPV16 (Dürst et al. 1983; Gissmann and Schneider 1986; de Villiers et al. 1986a; McCance and Singer 1986) and HPV18 DNAs (Boshart et al. 1984) in a high percentage of penile carcinoma. Lately, the prevalence of HPV18 DNA was recorded in 7 of 11 biopsies originating from Brazilian men (Villa and Lopes 1986). This implicates possible geographical and/or racial associations. Analysis of precursor lesions from the penis as well as of CIN and invasive cervical cancer from Brazilian patients is mandatory in order to evaluate the association of HPV18 with genital neoplasias in this geographical area. The existence of additional cofactors is suspected, since carcinoma of the penis is uncommon in other countries where female genital cancer is prevalent.

4.3.2 Vulvar Carcinoma

Malignant tumors of the vulva account for less than 4% of all cancers of the female genital tract (Shingleton et al. 1970). Most of these tumors are squamous cell carcinomas, comprising 82% of the total. Vulvar carcinomas occur predominantly in postmenopausal women. Apart from venereally transmissible infections, further risk factors include nulliparity, obesity, and diabetes as well as vulvar dystrophies (Franklin and Rutledge 1972).

As pointed out before, HPV DNA sequences could be found in more than 90% of vulvar squamous cell carcinoma biopsies ($n = 57$) tested so far under stringent and nonstringent hybridization conditions (Gissmann and Schneider 1986). About half of the tumors exhibited unclassified DNAs, whereas HPV16 and HPV18 were present in 36% and 12.5%, respectively; HPV6 and HPV11 DNAs were not detected in the invasive cancers investigated. In two cases from a series of de Villiers (1986a), multiple infections were observed: HPV16 together with HPV18 DNA was present in one, while the other harbored HPV6, HPV16, and HPV18. The simultaneous presence of vulvar bowenoid papulosis and invasive squamous cell carcinoma of the vulva in two recently reported cases (Bergeron et al. 1986) contributes evidence for the above-mentioned hypothetical role of bowenoid papulosis in the pathogenesis of genital squamous cell carcinoma. In the two cases cited nuclear aneuploidy was present, and HPV16 DNA appeared in an integrated form in tissues both from bowenoid papulosis and from invasive carcinoma.

Multifocality (multiple lesions at one site) and/or multicentricity (involvement of more than one site) are common findings in women with vulvar condylomas and different grades of VIN (Jimerson and Merrill 1970). Invasive carcinoma, however, is rarely multifocal (Hammond and Monagham 1983). In this case multiple areas of neoplasia may arise simultaneously or over a prolonged period of time. In addition there are a number of reports on separate primary squamous cell cancer of the cervix which appear prior to, concurrent with, or subsequent to carcinoma of the vulva, vagina, or anus (reviewed by Peters et al. 1984). At present it seems likely that anatomical proximity and exposure to an agent such as HPV, HSV II, and/or radiation and other environmental factors may explain the relationship of multifocal and multicentric genital neoplasms (Woodruff et al. 1983).

4.3.3 Vaginal Carcinoma

Malignant tumors of the vagina, squamous cell carcinoma being the most common, tend to occur in the elderly (Perez et al. 1973) and comprise only about 1%-2% of all female genital cancers (Kanbour et al. 1974). The current concept that the HPV infection of the female lower genital tract is a multicentric process is paralleled by the increased frequency of vaginal carcinoma in women who have been treated successfully for cervical cancer (Kanbour et al. 1974; Franklin and Rutledge 1972).

The incidence of cancer of the vagina and of the cervix is, however, very different, in spite of the fact that the vaginal squamous epithelium is continuous with that of the cervix. Again this hypothetically can be explained by assuming different susceptibilities of the cervical and vaginal epithelium (zur Hausen 1986).

Data on the distribution of HPV DNA in invasive carcinomas of the vagina are scanty because the total number of cases investigated by Southern blot analysis is rather low (de Villiers et al. 1986a; Smotkin et al. 1986). Retrospective examination of formalin-fixed tissues for viral DNA by in situ hybridization has led to the identification of HPV16 DNA in a number of invasive vaginal cancers (Ostrow et al. 1986). This has provided evidence that HPV16 is equally involved in vaginal squamous cell carcinoma and in corresponding malignancies of the cervix and vulva.

The data obtained from patients suffering from EV and from experimental work in animals lead one to expect that special HPV strains may act to produce malignant growth only in conjunction with other extrinsic physical or chemical cofactors. In addition, intrinsic components such as the individual genetic and immunological background certainly may contribute to the process of malignant conversion (zur Hausen 1977, 1982). The relationship between cancer and hormonal imbalance, as in the case of prenatal exposure to diethylstilbestrol (DES), has been considered since 1971, following reports on the rare development of clear cell adenocarcinoma of the vagina and the cervix in daughters of DES-exposed women (Herbst et al. 1971). DES, a synthetic nonsteroidal estrogen, was administered from the 1940s through the 1960s to women with high-risk pregnancies. Benign conditions such as vaginal adenosis (glandular epithelium in the vagina) and cervical ectropion have been described subsequently in daughters of women who received DES during pregnancy.

Studies of DES-exposed individuals are of special interest since the glandular epithelium in adenosis and ectropion is susceptible to the same range of pathological changes encountered in the vagina and cervix of women not exposed to DES. Preliminary data suggest that the incidence rate of dysplasia and infection with herpes simplex virus type II and HPV is increased in women with a DES-history (Adam et al. 1986) although the views on this association vary. This may provide an example of synergistic effects between viral factors and a DES-modified terrain (endocervix and transformation zone in the vagina) in the genesis of benign and premalignant lesions of the vagina.

5 Bowenoid Papulosis - High-Risk Lesion in Cervical Carcinogenesis?

Recent investigations have provided evidence that certain types of HPV are involved equally in premalignant squamous cell lesions of the cervix uteri, throughout the whole female lower genital tract, and on the penis and that they are present in invasive carcinomas of these sites. HPV16 is the genotype most frequently identified within tissues from malignant genital tumors and precursor lesions in both sexes. This HPV type is very rarely found in benign condylomata acuminata, more than 90% of which are associated with HPV6 and HPV11 (Gissmann 1984).

HPV16-associated, disseminated, inconspicuous, and subclinical lesions occur at external genital sites and in the introitus vaginae, exhibiting histological features of a CIS akin to CIS of the uterine cervix. These lesions are termed bowenoid papulosis of the penis and vulva. They are fundamentally identical with VIN III of young women. The hypothesis that bowenoid papulosis is a high-risk lesion in cervical carcinogenesis is supported by the concurrent presence of HPV16 DNA in genital CIS of the sexual partners (Gross et al. 1986; Obalek et al. 1985). Additional data were provided by a study on venereal transmission of HPV between sexual partners, in 24% of which HPV16 DNA was discovered in both the female and the male (Gross et al. 1985c, 1986).

Although bowenoid papulosis may regress spontaneously, the possibility exists that this disease is also a risk factor with respect to the etiology of carcinoma of the penis, vulva, and vagina (Table 7). The risk of cancer development, however, is by far higher at the cervix uteri than at other sites (zur Hausen 1986). Additional factors seem to be important since only a small number of premalignant lesions convert to invasive disease and since there is an overall striking difference between the incidence of virus-associated genital cancer in females and in males (zur Hausen 1986).

From the clinical point of view, early diagnosis and treatment of suspicious lesions are mandatory (Singer et al. 1976). Attention must focus on the detection of "high-risk males" bearing lesions such as bowenoid papulosis and subclinical infections which regularly harbor HPV16. Since patients at risk cannot yet be recognized serologically, screening for such individuals is done by in situ hybridization on filters of epithelial smears using a probe which distinguishes among virus types (Wagner et al. 1984; Gissmann 1984; Schneider et al. 1985). The identification of HPV16 DNA and the establishment of the relationship between its presence and malignancy as being either causal or casual will possibly enable us to identify patients at risk and thus will certainly contribute to early detection and prevention of genital cancer.

Acknowledgements. The author is grateful to M. Hilgarth, H. Ikenberg, and D. Wagner for their collaboration, and B. Luplow (Freiburg, FRG) for preparation

Table 7. Hypothetical course of HPV16 infection of the vulva and the penis

HPV16 ↓				
Normal → epithelium	VIN I–II → PIN I–II	VIN III → PIN III of the young	VIN III → PIN III of the elderly	Invasive cancer
	Precursor lesions	Bowenoid papulosis	Bowen's disease	
	Flat condyloma-like lesions	→ Lichenoid papule		
	Pigmented papular lesion	→ Pigmented papule	Red or white plaque	
	Subclinical HPV infection?	↗ Erythematous macule ↘ Leukoplakialike lesion		
Mean age	30 years	30 years	>45 years	>65 years

of the manuscript. Original work cited in this paper was supported by the Deutsche Forschungsgemeinschaft (SFB 31: Tumorentstehung und -entwicklung).

References

Abell MR, Gosling JR (1961) Intraepithelial and infiltrative carcinoma of the vulva: Bowen's type. Cancer 14: 318-329

Ackerman LV (1948) Verrucous carcinoma of the oral cavity. Surgery 23: 670-678

Adam E, Kaufman RH, Adler-Storthz K, Melnick JL, Dreesman GR (1986) A prospective study of association of herpes simplex virus and human papillomavirus infection with cervical neoplasia in women exposed to diethylstilbestrol in utero. Int J Cancer 35: 19-26

Adler-Storthz K, Newland JR, Tessin BA, Yeudall WA, Shillitoe EJ (1986) HPV2 DNA in oral verrucous carcinoma. International workshop on papillomaviruses. Cold Spring Harbor, September

Aird I, Johnson HD, Lennox B, Stansfeld AG (1954) Epithelioma cuniculatum. A variety of squamous carcinoma, peculiar to the foot. Br J Surg 42: 245-250

Barrett TJ, Silbar BH, McGinley JP (1954) Genital warts, a venereal disease. JAMA 154: 333-334

Beaudenon S, Kremsdorf D, Croissant O, Jablonska S, Wain-Hobson S, Orth G (1986) A novel type of human papillomavirus associated with genital neoplasias. Nature 321: 246-249

Benedet JL, Sanders BH (1984) Carcinoma in situ of the vagina. Am J Obstet Gynecol 148: 695-700

Berg JW, Lampe JG (1981) High-risk factors in gynecologic cancer. Cancer 48: 429-441

Berger BW, Hori Y (1978) Multicentric Bowen's disease of the genitalia. Spontaneous regression of lesions. Arch Dermatol 114: 1698-1699

Bergeron C, Naghasfar Z, Shah K, Fu Y, Ferenczy A (1986) Human papillomavirus type 16 in intraepithelial neoplasia (bowenoid papulosis) and coexistent invasive carcinoma of the vulva. International workshop on papillomaviruses, Cold Spring Harbor, September

Böcker W, Stegner HE (1981) Präneoplasien der Vulva. Arch Gynecol 232: 1-12

Boshart M, Gissmann L, Ikenberg H, Kleinheinz A, Scheurlen W, zur Hausen H (1984) A new type of papillomavirus DNA, its presence in genital cancer biopsies and in cell lines from cervical cancer. Int J Cancer 3: 1151-1157

Boxer RJ, Skinner DG (1977) Condylomata acuminata and squamous cell carcinoma. Urology 9: 72-78

Bracken RB (1981) Genitourinary cancer. Cancer Chemother 2: 199-242

Brandsma JL, Steinberg BM, Abramson AL, Winkler B (1986) Presence of human papillomavirus type 16 related sequences in verrucous carcinoma of the larynx. Cancer Res 46: 2185-2188

Braun L, Farmer ER, Shah KV (1983) Immunoperoxidase localization of papillomavirus antigen in cutaneous warts and bowenoid papulosis. J Med Virol 12: 187-193

Buschke A, Löwenstein L (1925) Über carcinomähnliche Condylomata acuminata des Penis. Klin Wochenschr 4: 1726-1728

Campion MJ, Singer A, Clarkson PK, McCance DJ (1985) Increased risk of cervical neoplasia in consorts of men with penile condylomata acuminata. Lancet i: 943-946

Chief Medical Officer of the Department of Health and Social Security (1983) Sexually transmitted diseases. Br J Vener Dis 59: 134-137

Chuang TY, Perry HO, Kurland LT, Ilstrup DM (1984) Condyloma acuminatum in Rochester, Minn, 1950-1978: I. Epidemiology and clinical features. Arch Dermatol 120: 469-475

Coggin JR, zur Hausen H (1979) Workshop on papillomaviruses and cancer. Cancer Res 39: 545-546

Crum CP, Braun LA, Shah K, Fu YS, Levine RS, Fenoglio CM, Richart RM, Townsend DE (1982) Vulvar intraepithelial neoplasia: correlation of nuclear DNA content and the presence of a human papillomavirus (HPV) structural antigen. Cancer 49: 468-471

Dejong AR, Weiss JC, Brent RL (1982) Condylomata acuminata in children. Am J Dis Child 136: 704-706

Del Mistro A, Vallejos H, Kleinhaus S, Braunstein JD, Halwer M, Koss LG (1986) HPV types in condylomata acuminata in prepubertal children. International workshop on papillomaviruses, Cold Spring Harbor, September

De Villiers EM, Schneider A, Gross G, zur Hausen H (1986a) Analysis of benign and malignant urogenital tumors for human papillomavirus infection by labelling cellular DNA. Med Microbiol Immunol 174: 281-284
De Villiers EM, Weidauer H, Otto H, zur Hausen H (1986b) Papillomavirus DNA in human tongue carcinomas. Int J Cancer 36: 575-577
Di Paola GR, Gomez Rueda-Leverone H, Belardi MG, Vighi S (1982) Vulvar carcinoma in situ: a report of 28 cases. Gynecol Oncol 14: 236-241
Draper GJ, Cook GA (1983) Changing pattern of cervical cancer rates. Br Med J 287: 510-514
Dürst M, Gissmann L, Ikenberg H, zur Hausen H (1983) A papillomavirus DNA from a cervical carcinoma and its prevalence in cancer biopsies from different geographical regions. Proc Natl Acad Sci USA 80: 3812-3815
Ferenczy A, Mitao M, Nagai N, Silverstein SJ, Crum CP (1985) Latent papillomavirus and recurring genital warts. N Engl J Med 313: 784-788
Franklin EW, Rutledge FD (1972) Epidemiology of epidermoid carcinoma of the vulva. J Obstet Gynecol 39: 165-172
Friedrich EG (1972) Reversible vulvar atypia. A case report. Obstet Gynecol 39: 173-181
Friedrich EG (1983) Vulvar disease, 2nd edn. Saunders, Philadelphia
Gissmann L, zur Hausen H (1980) Partial characterization of viral DNA from human genital warts (condylomata acuminata). Int J Cancer 25: 605-609
Gissmann L, de Villiers EM, zur Hausen H (1982) Analysis of human genital warts (condylomata acuminata) and other genital tumours for human papillomavirus type 6 DNA. Int J Cancer 29: 143-146
Gissmann L, Wolnik L, Ikenberg H, Koldovsky U, Schnurch HG, zur Hausen H (1983) Human papillomavirus types 6 and 11 DNA sequences in genital and laryngeal papillomas and in some cervical cancers. Proc Natl Acad Sci USA 80: 560-563
Gissmann L, Boshart M, Dürst M, Ikenberg H, Wagner D, zur Hausen H (1984) Presence of human papillomavirus (HPV) DNA in genital tumors. J Invest Dermatol [Suppl] 83: 26-28
Gissmann L (1984) Papillomaviruses and their association with cancer in animals and in man. Cancer Surv 3: 161-181
Gissmann L, Gross G (1985) Association of HPV with human genital tumors. Clin Dermatol 3: 124-129
Gissmann L, Schneider A (1986) Human papillomavirus in preneoplastic and neoplastic genital lesions. In: zur Hausen H, Peto R (eds) Origins of female genital cancer: virological and epidemiological aspects. Banbury Report, Cold Spring Harbor, New York, pp 217-224
Green M, Brackmann KH, Sanders PR, Löwenstein PM, Freel JH, Eisinger M, Switlyk SA (1982) Isolation of a human papillomavirus from a patient with epidermodysplasia verruciformis: presence of related viral DNA genomes in human urogenital tumors. Proc Natl Acad Sci USA 79: 4437-4441
Gross G, Pfister H, Hagedorn M, Gissmann L (1982) Correlation between human papillomavirus (HPV) type and histology of warts. J Invest Dermatol 78: 160-164
Gross G, Ikenberg H, Gissmann L, Hagedorn M (1985a) Papillomavirus infection of the anogenital region: correlation between histology, clinical picture and virus type. Proposal of a new nomenclature. J Invest Dermatol 85: 147-152
Gross G, Hagedorn M, Ikenberg H, Rufli T, Dahlet C, Grosshans E, Gissmann L (1985b) Bowenoid papulosis. Presence of human papillomavirus (HPV) structural antigens and of HPV 16-related DNA sequences. Arch Dermatol 121: 858-863
Gross G, Schneider A, Hauser-Brauner B, Wagner D, Ikenberg H, Gissmann L (1985c) Transmission of genital papillomavirus infections: a study of sexual partners. J Cell Biochem [Suppl] 96: 71
Gross G, Ikenberg H, de Villiers EM, Schneider A, Wagner D, Gissmann L (1986) Bowenoid papulosis: a venereally transmissible disease as reservoir for HPV 16. In: zur Hausen H, Peto R (eds) Origins of female genital cancer: Virological and epidemiological aspects. Banbury Report, Cold Spring Harbor, New York, pp 149-165
Guillet GY, Braun L, Masse R, Altimos J, Geniaux M, Texier L (1984) Bowenoid papulosis. Demonstration of human papillomavirus (HPV) with anti-HPV immune serum. Arch Dermatol 120: 514-516

Hammond JG, Monagham JM (1983) Multicentric carcinoma of the female lower genital tract. Br J Obstet Gynecol 90: 557-561
Hauser B, Gross G, Schneider A, de Villiers EM, Gissmann L, Wagner D (1985) HPV 16 related bowenoid papulosis (letter). Lancet ii: 106
Herbst AC, Ulfelder H, Poskauzer DC (1971) Adenocarcinoma of the vagina: association of maternal stilbestrol therapy with tumor appearance in young women. N Engl J Med 284: 878-881
Ikenberg H, Gissmann L, Gross G, Grußendorf-Conen EI, zur Hausen H (1983) Human papillomavirus type-16-related DNA in genital Bowen's disease and in bowenoid papulosis. Int J Cancer 32: 563-565
Jenson AB, Rosenthal JD, Olson C, Pass F, Lancaster WD, Shah K (1980) Immunologic relatedness of papillomaviruses from different species. JNCI 64: 495-500
Jimerson GK, Merrill JA (1970) Multicentric squamous malignancy involving both cervix and vulva. Cancer 26: 150-153
Judge JR (1969) Giant condylomatum involving vulva and rectum. Arch Pathol 88: 46-49
Judson FN, Penley KA, Robinson ME, Smith JK (1980) Comparative prevalence rates of sexually transmitted diseases in heterosexual and homosexual men. Am J Epidemiol 112: 836-843
Kanbour AI, Klionsky B, Murphy AI (1974) Carcinoma of the vagina following cervical cancer. Cancer 34: 1838-1841
Kimura S, Hirai R, Harada R, Nagashima M (1978) So-called multicentric Bowen's disease. Report of a case and a possible etiologic role of human papillomavirus. Dermatologica 157: 229-237
Kinghorn GR (1978) Genital warts; incidence of associated genital infections. Br J Dermatol 99: 405-409
Koss LG, Durfee GR (1956) Unusual patterns of squamous epithelium of the uterine cervix. Cytologic and pathologic study of koilocytotic atypia. Ann NY Acad Sci 63: 1245-1261
Krzyzek RA, Watts SL, Anderson DL, Faras AJ, Pass F (1980) Anogenital warts contain several distinct species of human papillomavirus. J Virol 36: 236-244
Lee SH, McGregor DH, Kuzier MN (1981) Malignant transformation of perianal condyloma acuminatum. Dis Colon Rectum 24: 462-467
Lloyd KM (1970) Multicentric pigmented Bowen's disease of the groin. Arch Dermatol 101: 48-51
Lorincz AI, Lancaster WD, Temple GF (1985) Detection and characterization of a new type of human papillomavirus. J Cell Biochem [Suppl] 9c: 75
Marino AW Jr (1964) Proctologic lesions in male homosexuals. Dis Colon Rectum 7: 121-125
Martinez I (1969) Relationship of squamous cell carcinoma of the cervix uteri to squamous cell carcinoma of the penis. Cancer 24: 777-780
McCance DJ (1986) Human papillomaviruses and cancer. Biochim Biophys Acta 823: 195-205
McCance DJ, Singer A (1986) The importance of HPV infections in the male and in the male genital tract and their relationship to cervical neoplasia. In: zur Hausen H, Peto R (eds) Origins of female genital cancer: histological and epidemiological aspects. Banbury Report, Cold Spring Harbor, New York, pp 311-319
Meisels A, Fortin R (1976) Condylomatous lesions of cervix and vagina: I. Cytologic patterns. Acta Cytol (Baltimore) 20: 505-509
Meisels A, Fortin R, Roy M (1977) Condylomatous lesions of cervix: II. Cytologic, colposcopic and histopathologic study. Acta Cytol (Baltimore) 21: 379-390
Meisels A, Roy M, Fortier M, Morin C, Casas-Cordero M, Shah KV, Turgeon H (1981) Human papillomavirus infection of the cervix. The atypical condyloma. Acta Cytol (Baltimore) 25: 7-16
Obalek S, Jablonska S, Orth G (1985) HPV associated intraepithelial neoplasia of external genitalia. Clin Dermatol 3: 104-113
Oriel JD (1971a) Natural history of genital warts. Br J Vener Dis 47: 1-13
Oriel JD (1971b) Anal warts and anal coitus. Br J Vener Dis 47: 373-376
Oriel JD (1981) Genital warts. Sex Transm Dis 8: 326-329
Ostrow R, Manias D, Clark B, Okagaki T, Twiggs L, Faras A (1986) The analysis of malignant tumors of the vulva, vagina and cervix for HPV DNA. International workshop on papillomaviruses, Cold Spring Harbor, September

Perez CA, Arneson AN, Galakatos A, Samanth HR (1973) Malignant tumors of the vagina. Cancer 31: 36-41
Peters RK, Mack TM, Bernstein L (1984) Parallels in the epidemiology of selected anogenital carcinomas. JNCI 72(3): 609-615
Petrucco DM, Seamark RF, Holmes K, Forbes IJ, Symons RG (1976) Changes in lymphocyte function during pregnancy. Br J Obstet Gynecol 83: 245-250
Pfister H, zur Hausen H (1978) Seroepidemiological studies of human papillomavirus (HPV I) infections. Int J Cancer 21: 161-165
Pfister H (1984) Biology and biochemistry of papillomaviruses. Rev Physiol Biochem Pharmacol 99: 111-181
Pinkus H, Mehregan AH (1976) A guide to dermatohistopathology, 2nd edn. Appleton-Century-Crofts, New York
Powell L (1978) Condyloma acuminatum. Recent advances in development, carcinogenesis and treatment. Clin Obstet Gynecol 21: 1061-1079
Purola E, Savia E (1977) Cytology of gynaecologic condyloma acuminata. Acta Cytol (Baltimore) 21: 26-31
Roberts A (1982) Cervical cytology in England and Wales. 1965-1980. Health Trends 14: 41-43
Schneider A, Kraus H, Schuhmann R, Gissmann L (1985) Papillomavirus infection of the lower genital tract: detection of viral DNA in gynecological swabs. Int J Cancer 35: 443-448
Schneider V, Kay S, Lee HM (1983) Immunosuppression as a high-risk factor in the development of condyloma acuminatum and squamous neoplasia of the cervix. Acta Cytol (Baltimore) 27: 220-224
Segi M (1977) Graphic presentation of cancer incidence by site and by area and population. Segi Institute for Cancer Epidemiology, Nagoya
Shingleton HM, Fowler WC, Palumbo L, Koch GG (1970) Carcinoma of the vulva: influence of radical operation on cure rate. Obstet Gynecol 35: 1-8
Shokri-Tabibzadeh S, Koss LG, Molinar J, Romney MD (1981) Association of human papillomavirus with neoplastic processes in the genital tract of four women with impaired immunity. Gynecol Oncol 12: S129-S140
Sillmann F, Stanek A, Sedlis A, Rosenthal J, Lanks KW, Buchhagen D, Nicastri A, Boyce J (1984) The relationship between human papillomavirus and lower genital intraepithelial neoplasia in immunosuppressed women. Am J Obstet Gynecol 150: 300-308
Singer A (1973) A male factor in aetiology of cervical cancer? Oxford Med School Gaz 25: 18-19
Singer A, Reid BL, Coppleson M (1976) A hypothesis: the role of the high-risk male in the etiology of cervical carcinoma. Am J Obstet Gynecol 126: 110
Smotkin D, Fu YS, Wettstein FO (1986) Distribution of human papillomavirus types 16 and 18 in lesions of the female lower genital tract. International workshop on papillomaviruses. Cold Spring Harbor, September
Steinberg BM, Topp WC, Schneider PS, Abramson AL (1983) Laryngeal papillomavirus infection during clinical remission. N Engl J Med 308: 1261-1264
Stumpf PG (1980) Increasing occurrence of condylomata acuminata in premenarchal children. Obstet Gynecol 56: 262-264
Syrjänen KJ, Vayrynen M, Castren O, Yiskoski M, Mantyjarri R, Pyrhonen S, Saarikoski S (1984) Sexual behaviour of females with human papillomavirus lesions in the uterine cervix. Br J Vener Dis 60: 243-248
Villa LL, Lopes A (1986) Human papillomavirus. DNA sequences in penile carcinomas in Brazil. Int J Cancer 37: 853-855
Wade TR, Kopf AW, Ackerman AB (1978) Bowenoid papulosis of the penis. Cancer 42: 1890-1903
Wade TR, Kopf AW, Ackerman AB (1979) Bowenoid papulosis in the genitalia. Arch Dermatol 115: 306-308
Wagner D, Ikenberg H, Boehm N, Gissmann L (1984) Identification of human papillomavirus in cervical swabs by deoxyribunucleic acid in situ hybridization. Obstet Gynecol 64: 767-772
Walker PG, Singer A, Dyson J, Oriel JD (1983) Natural history of cervical epithelial abnormalities in patients with vulval warts. Br J Vener Dis 159: 327-329
Waterhouse J, Muir C, Correa P, Powell J (1982) Cancer in five continents, vol 4. International Agency for Research on Cancer, Lyon

Weed JC, Lozier C, Daniel SJ (1983) Human papillomavirus in multifocal invasive female genital tract malignancy. Obstet Gynecol 62: 583-587

Woodruff JD (1981) Carcinoma in situ of the vagina. Clin Obstet Gynecol 2: 485-488

Woodruff JD, Hildebrandt EE (1958) Carcinoma in situ of the vulva. Obstet Gynecol 12: 414-424

Woodruff JD, Julian C, Puray T, Mermut S, Katayama P (1983) The contemporary challenge of carcinoma in situ of the vulva. Am J Obstet Gynecol 115: 667-684

Zachow KR, Ostrow RS, Bender M, Watts S, Okagaki T, Pass F, Faras AJ (1982) Detection of human papillomavirus DNA in anogenital neoplasias. Nature 300: 771-773

zur Hausen H (1977) Human papillomaviruses and their possible role in squamous cell carcinomas. Curr Top Microbiol Immunol 78: 1-30

zur Hausen H (1982) Human genital cancer: synergism between two virus infections or synergism between a virus infection and initiating events. Lancet ii: 1370-1372

zur Hausen H (1986) Intracellular surveillance of persisting viral infections. Human genital cancer results from deficient cellular control of papillomavirus gene expression. Lancet i: 489-491

Carcinogenesis in the Uterine Cervix and Human Papillomavirus Infection

L. G. Koss

1 Epidemiological Considerations

In 1842 an Italian physician, Rigoni-Stern, published a paper on the statistical evaluation of death records from cancer in the city of Verona for the years 1760-1839. In this remarkable document it was pointed out that deaths due to "cancer of the uterus" were rare among virgins and nuns and quite common among married women and widows. Since it is known from other sources that cancer of the cervix was by far the most common form of uterine cancer in the 19th century, it may safely be assumed that Rigoni-Stern was the first epidemiologist to point out that "marital" or "sexual" events played an important role in the development of this disease.

These observations were repeatedly confirmed in the extensive epidemiological studies of carcinoma of the uterine cervix and its precursor lesions conducted during the past 50 years (summary in Kessler 1976; Koss 1979, 1981; Rotkin 1981). It is quite evident that this disease behaves not unlike a venereally transmitted infection in which an agent or agents are transmitted to the woman by a male partner(s). The two most important high risk factors for cervical carcinoma include: young age at the onset of sexual activity and multiplicity of sexual partners, or sexual promiscuity. Pridan and Lillienfeld (1971) also suggested that association with promiscuous males may constitute a risk factor. The disease is virtually never observed among virgins (although I have seen an exception to this rule) and is relatively common among professional prostitutes.

A number of possible transmissible agents that could trigger the disease have been examined, usually on the basis of the statistical association of cervical carcinoma or its precursors with another sexually transmitted disease. Until the recent evidence linking human papillomavirus infection with neoplastic events in the epithelia of the female genital tract, syphilis, gonorrhea, *Trichomonas vaginalis*, and, during the past 10 years, herpesvirus 2 were considered as potentially important transmissible agents (summary in Koss 1979; Roitkin 1981; Kessler 1984). One of the more intriguing theories, first proposed by Coppleson and Reid (1968) suggested that spermatozoa may trigger neoplastic events in the cervical epithelium. This theory is not without merit (Koss 1979). It is, of course, quite possible that these various agents may coexist and somehow work in synchrony with each other or with yet other agents such as human papillomavirus.

2 Sequence of Events in the Genesis of Epidermoid Carcinoma of the Uterine Cervix

The concept of origin of invasive epidermoid or squamous carcinoma of the uterine cervix from abnormal cervical epithelium may be traced to a paper published by Schauenstein in 1908. Based on similarity of histological patterns, Schauenstein proposed that cancerous changes confined to the epithelium of the cervix (Oberflächenkrebs, or surface carcinoma) represented a precursor stage of invasive cancer. Subsequently, the term "carcinoma-in-situ" was proposed for such precancerous epithelial lesions (summary in Koss 1979).

Mass screening of asymptomatic women conducted for the past 40 years by means of vaginal and cervical smears, as first proposed by Papanicolaou and Traut (1943), has led to a large number of studies of the events preceding or leading to cervical carcinoma; the cervix is now the best known model of human epithelial carcinogenesis. A number of unexpected observations has come to light:

a. Extensive cytological and colposcopic studies documented that most initial neoplastic events occur within a rather small segment of metaplastic cervical squamous epithelium adjacent to the endocervical epithelium (the squamocolumnar junction). Why this area (also known today as the "transformation zone") is most likely to be affected is a matter for speculation. It appears possible that during the process of transformation of columnar epithelium into squamous epithelium (squamous metaplasia) a temporary state of genetic imbalance may occur that renders the young squamous epithelium susceptible to neoplastic events. It has also been shown that the transformation of an epithelium composed of a single layer of cells into a multilayered squamous epithelium is associated with production of a new and different species of intermediate keratin filaments (summary in Sun et al. 1984). The possibility that these events somehow contribute to the susceptibility of the transformation zone to neoplastic changes, particularly in reference to infection with human papillomavirus, must be considered. It may be noted here that the proliferation and maturation of papillomavirus is related in an unknown fashion to maturation of squamous epithelium (summary in Taichman et al. 1984). Regardless of these considerations, subsequent epithelial abnormalities spread from this small zone to adjacent squamous epithelium of the portio vaginalis of the cervix and the endocervical epithelium.

b. Morphological differences that were observed among the precancerous lesions were thought to represent stages in the development of cervical cancer. It was proposed that the disease progressed from relatively minor epithelial changes in the squamous epithelium to moderate and severe changes, usually located in the endocervical canal, to invasive cancer. The term "dysplasia" was used to characterize the lesser changes and "carcinoma-in-situ", the more severe changes. A number of studies (notably by Koss et al. 1963; Richart 1967, 1973; Richart and Barron 1969; and Burghardt 1973) documented that the behavior of the precancerous lesions is not related to the type of morphological abnormality. The term "cervical intraepithelial neoplasia" (CIN) proposed by Richart (1967) and encompassing all intraepithelial neoplastic lesions, regardless of degree of histological or cytological abnormality, has been nearly universally accepted as a replacement for the "dysplasia/carcinoma-in-situ" system of nomenclature. Further, it has been proposed (Koss et al. 1963; Koss 1979, 1981) that the morphological expression of the precancerous lesions is closely related to their anatomic location: more mature, keratin-forming lesions occur mainly on the squamous epithelium of the portio vaginalis of the cervix whereas less mature lesions composed of small cells occur mainly within the endocervical canal. Hence, the alleged "progression" of the precancerous lesions reflects in reality the involvement of sequential segments of cervical epithelium by the cancerous process. These views received strong support from careful mapping studies by Burghardt (1973).

c. The frequency of precancerous lesions in asymptomatic women has been shown to be much higher than the expected rate of invasive cancer - hence, not all precancerous lesions could be expected to advance to invasion. In fact, many precancerous lesions may disappear either spontaneously or after minor diagnostic procedures (Koss et al. 1963). Nasiell et al. (1976) estimated after 12 years of follow-up that not more than 30% of "moderate dysplasias" will persist or "progress."

d. Attempts to establish the prognostic value of DNA measurements in precursor lesions (Fu et al. 1981) suggested that diploid or diploid-tetraploid intraepithe-

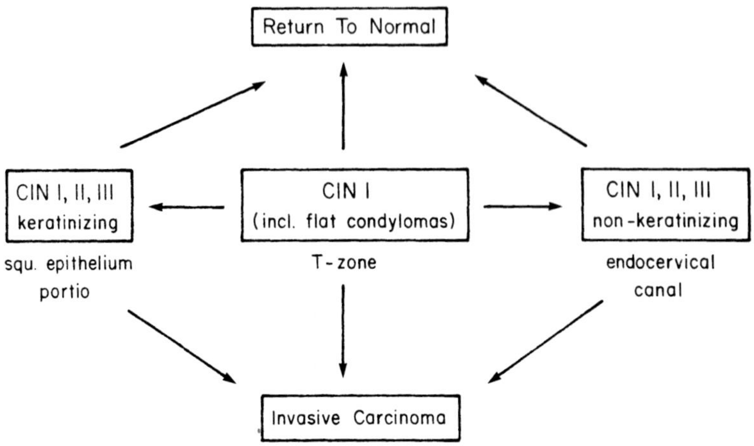

Fig. 1. Sequence of events in carcinogenesis of the uterine cervix

lial lesions were more likely to disappear than aneuploid lesions. The study confirmed that aneuploid lesions may occur over the entire morphological spectrum of CIN, ranging from CIN 1 (mild dysplasia) to CIN 3 (carcinoma-in-situ).

Figure 1 proposes a sequence of events in the genesis of carcinoma of the cervix. It is quite evident that the understanding of the biological mechanisms of these events still eludes us to a large extent. The possible role of human papillomavirus infection and other factors that may either trigger, inhibit, or promote cervical carcinogenesis will be discussed in the concluding paragraphs of this chapter.

3 Morphological Manifestations of Papillomavirus Infection of Cervix and Vagina

The recognition that human papillomavirus (HPV) infection of the genital tract may have manifold clinical manifestations including flat lesions of the epithelium of the vulva, vagina, and uterine cervix is relatively recent. The common lesion associated with HPV infection is condylomata acuminata, appearing as grossly visible, usually multiple, wartlike lesions of the external genitalia, perineum, and anus, that have been recognized as a venereally transmitted disease for many years (summary in Oriel 1971a, b). The presence of intranuclear viral particles in these lesions has been documented by electron microscopy (Dunn and Ogilvie 1968; Oriel and Almeida 1970). The occurrence of similar, visible, wartlike papillary lesions on the moist vaginal and cervical squamous epithelium was also noted but was considered a rarity (Marsh 1952; Raferty and Payne 1954; Kazal and Long 1958). Still, long-term persistence and, in some cases, progression of these "squamous papillomas" to squamous cancer was repeatedly observed (Koss and Durfee 1956; Kazal and Long 1958; Koss 1979). The cytological manifestations of warty, grossly visible condylomata acuminata have been repeatedly described. Such lesions, even if transient in nature, were known to shed abnormal squamous cells akin to those observed in some precancerous lesions and in squamous cancer of the uterine cervix (summary in Koss 1979).

In 1976 and 1977 Meisels and Fortin, and Purola and Savia simultaneously suggested that certain epithelial abnormalities of the uterine cervix, hitherto considered as a form of intraepithelial neoplastic lesions (CIN grade 1 or 2 or dysplasia) were, in fact, flat mucosal equivalents of the wartlike condylomas. This suggestion was based on similarities in the cytological presentation between the clinically obvious condylomas and the flat lesions, mainly the presence of cells known as koilocytes. The support for this concept was provided by Laverty et al. (1978), who were the first to document by electron microscopy the presence of nuclear viral particles in tissue biopsies of such lesions. These observations were subsequently repeatedly confirmed by other observers using various techniques (see below).

3.1 Historical Note: The Origin of the Concept of Koilocytosis and "Warty" Lesions of Cervical Epithelium

In the early 1950s my coworker, Grace Durfee, and I were attempting to catalogue cell and tissue abnormalities that came to light in the course of cytological screening of well women for precancerous lesions of the uterine cervix. A pattern of tissue and cell abnormalities came to light that was described by us as follows (Koss and Durfee 1956, cited with permission of the New York Academy of Sciences):

The (pattern in cervical biopsies) is characterized by the presence in histologic sections of large cells with relatively small but irregular and hyperchromatic nuclei surrounded by clear and transparent cytoplasm. Thus, the nucleus seems to be suspended in an empty space. For descriptive purposes, we have coined the term koilocytotic atypia, from the Greek word 'koilos' meaning hollow or cavity, to designate this lesion. Many of these lesions were recognized primarily in smears before their recognition in tissues. The term 'warty dyskeratosis' was used . . . when referring to this type of epithelial change.

The cell changes observed in cervical smears were described as follows:

All cytologic observations reported below were made on material stained according to the method of Papanicolaou.

Cytologically, all the cases of koilocytotic atypia were characterized by the presence of atypical cells approaching the size and general appearance of superficial and parabasal cells of the squamous epithelium. The striking features of such cells were:

(1) Abnormal, moderately enlarged, and hyperchromatic single or multiple nuclei. The degree of hyperchromasia varied from slight to marked. The majority of the nuclei displayed a granular structure of chromatin, but some took the strain in a diffuse, dense manner obscuring structural details. Although occasionally irregular in shape, the nuclei were usually sharply outlined.

(2) Abundant cytoplasm that was predominantly basophilic, although eosinophilic cells were not uncommon. It appears to the authors that, in most instances, the

increase in the amount of cytoplasm paralleled nuclear enlargement. Moreover, the cytoplasm was not uniform and presented one feature that is most characteristic of the cells under discussion, namely:

(3) Perinuclear zones that were distinctly different from the peripheral cytoplasmic areas. In some cells, the perinuclear zones were clearly outlined and transparent, obviously consisting of a very thin layer of cytoplasm. In other cells, the perinuclear zones were less transparent taking the stain just a shade lighter than the peripheral cytoplasm. Variations in thickness of the perinuclear zone, at times, were also noticeable within one cell, the thinnest portion being always perinuclear in position.

The area occupied by the perinuclear zone varied in size and in position. In most cases, the clear zone occupied one-fourth to two-thirds of the total cytoplasmic surface, but sometimes it was so wide that only a rim of thicker peripheral cytoplasm remained. For the most part, the perinuclear zone was centrally located, but occasionally, it was eccentric in position. In these latter cases, the nucleus was also eccentrically located. The significance of the perinuclear zones is not clear at this point. It is possible that the zones result from altered cellular metabolism.

In three cases of moderate, and in two cases of marked koilocytotic atypia, the above described cells were supplemented by a few small cells with atypical nuclei and scanty cytoplasm. By our standards, these cells were morphologically indistinguishable from cancer cells.

Observations in 16 cases of koilocytotic atypia were reported in the same paper. In four cases the degree of nuclear abnormality was minimal, in eight moderate, and in four, marked. In addition, smears and tissues from 40 cases of classical carcinoma-in-situ and tissue sections from 110 cases of invasive carcinoma were reviewed for the presence of coexisting koilocytosis. In smears from carcinomas-in-situ, koilocytes were present in 18 cases. In tissues the association of koilocytosis and carcinoma-in-situ was noted in four cases; in an additional seven cases koilocytosis was present within carcinoma-in-situ. In 57 cases of invasive carcinoma, no surface epithelium remained; in 53 cases of invasive carcinomas with surface epithelium present in adjacent areas, koilocytotic atypia was definitely present in 3, and probably present in an additional 6 cases. It was also noted that koilocytes were present in four cases of atypical squamous metaplasia and two papillomas of the cervix.

Although the viral origin of the koilocytic atypia was not suspected in 1956, the description of the wartlike epithelial change pointed in this direction. The conclusions of the original paper warrant repeating because they are still valid today.

The koilocytotic atypia is an alteration of the surface squamous epithelium of the cervix and, when uncomplicated, it cannot be called cancer by our present criteria. The recognition of this epithelial pattern by pathologists will undoubtedly clarify the origin of many an abnormal cell seen in smears. Koilocytotic atypia can be associated with in situ or invasive cancer of the uterine cervix. Therefore, recognition of this pattern, either in smears or in sections, should not preclude search for cancer. The role of koilocytotic atypia as a stepping stone in the genesis of cervix cancer is not clear, and at this point, we cannot exclude the possibility that some of these lesions may progress to cancer. The evidence accumulated so far, however, indicates that some of these lesions may regress.

As noted in the original paper, the koilocytotic atypia was previously observed by others, notably by Ayre and Ayre (1949) who initially named it "precancer cell complex". They observed subsequent development of carcinoma-in-situ in one such patient. The entity was later renamed "nearocarcinoma" or early carcinoma (Ayre 1951) but unfortunately the observations in support of this concept were anecdotal. Papanicolaou (1954) also observed abnormal squamous cells "with extreme cavitation" and classified them among "dyskaryotic cells," i.e., cells showing early malignant change. The specific example in Papanicolaou's atlas was apparently derived from a patient with carcinoma-in-situ of the cervix, and the matter was not further pursued.

In a subsequent major paper on the behavior of precancerous lesions of the uterine cervix followed without treatment, several examples of koilocytotic atypia were included (Koss et al. 1963). Some of the lesions disappeared, some remained stationary, and some progressed either to carcinoma-in-situ or to invasive carcinoma. As the unpredictable behavior of koilocytotic atypia did not differ significantly from other forms of intraepithelial neoplasia (including carcinoma-in-situ), the lesions were henceforth considered part of the morphological spectrum of cervical intraepithelial neoplasia (dysplasia and carcinoma-in-situ).

3.2 Cytology and Histology

There is little doubt today that the "warty" intraepithelial lesions and the corresponding koilocytosis in smears of the uterine cervix described in detail in 1956 by Koss and Durfee (and similar lesions occurring in the epithelium of the vagina) are associated with HPV infection. There is also little to add to the original description of the cells and tissue patterns cited above, and only a brief summary is provided below.

In *cervicovaginal smears* the key cell is the koilocyte, a mature intermediate or superficial squamous cell with a large, clear perinuclear zone, often sharply demarcated against the peripheral cytoplasm. Such cells may occur singly but quite often form clusters. In these cells the nuclear abnormalities in the form of enlargement and hyperchromasia may be moderate (Fig. 2 A) or marked (Fig. 3 A). Double or multiple nuclei are commonly observed. Cells showing transition between slight perinuclear clearing and a fully developed clear zone may be frequently observed (Fig. 2 A). Condylomatous lesions may also shed squamous cells with enlarged dark pyknotic nuclei, derived from the superficial keratinized layers of the squamous epithelium (Fig. 4 A). Meisels et al. (1984) named such cells "dyskeratinocytes" and documented the presence of the virus in some of them. In light microscopy the dyskeratinocytes cannot be differentiated from squamous cancer cells. Thus, the possibility must be considered that these are, in fact, cancer cells with secondary viral infection. Some condylomatous lesions also shed highly atypical squamous cells, sometimes forming squamous "pearls" with a keratinized center, identical to those shed from a squamous carcinoma (Fig. 5).

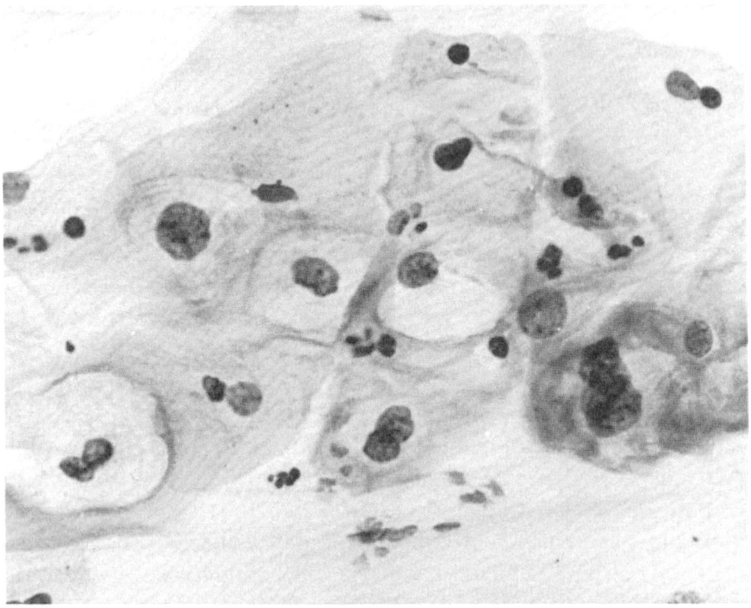

A

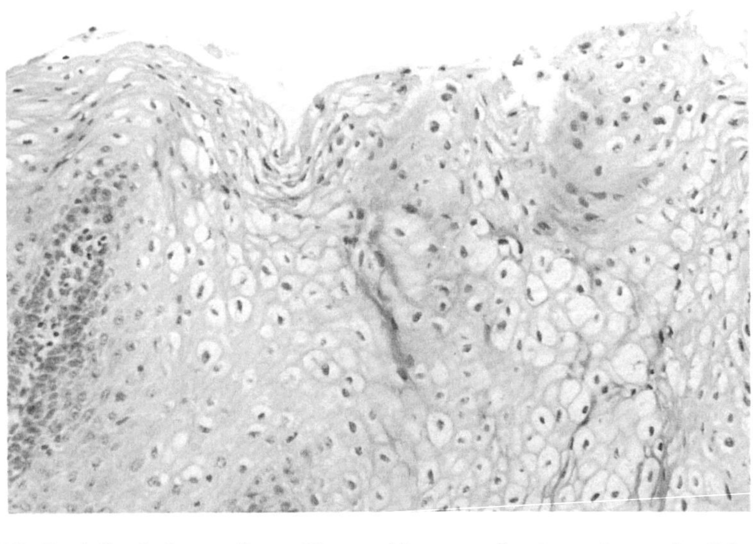

B

Fig. 2. **A** Cervical smear from a 20-year-old woman, showing a cluster of well-formed koilocytes. Note the sharply demarcated, wide, perinuclear clear zones in some of the cells. In other cells the clear zone is not fully developed and these cells may represent earlier stages of HPV infection. Nuclear abnormalities in the form of moderate to marked, enlargement, hyperchromasia, and double or multiple nuclei may also be noted (× 500). **B** Cervical biopsy corresponding to **A**. Papillomavirus infection of squamous epithelium of the uterine cervix with koilocytosis (CIN 1 with features of flat condyloma). Enlargement and doubling of the nuclei and the perinuclear halos are shown. The epithelium is thickened but its growth pattern is not significantly disturbed (× 140)

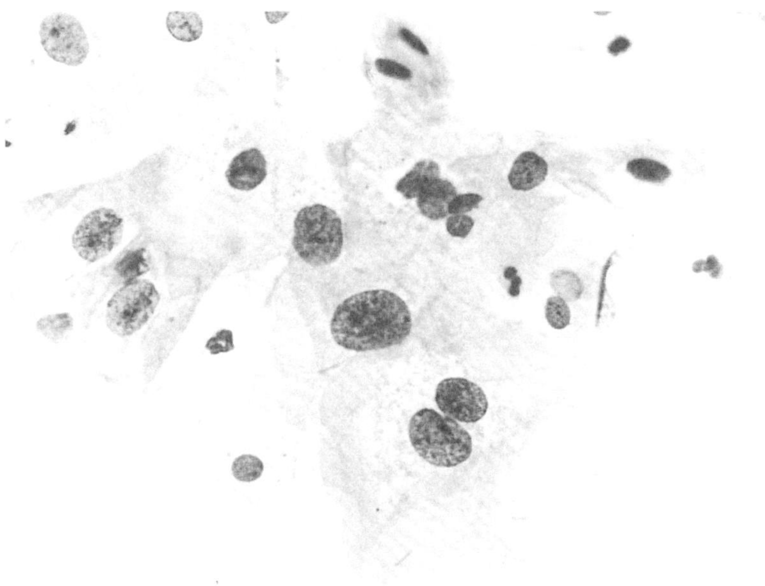

A

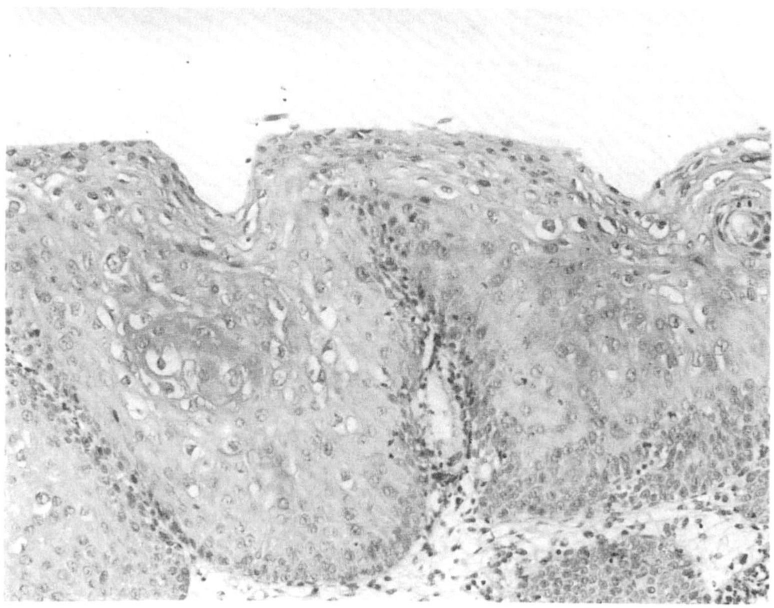

B

Fig. 3. **A** Cervical smear from a 22-year-old woman. Squamous cells and koilocytes with marked nuclear atypia. The perinuclear halo is well-developed in some of the cells. In other cells there is a perinuclear loss of cytoplasmic density (×500). **B** Cervical biopsy corresponding to **A**. The epithelium shows neoplastic changes of considerable magnitude (CIN 2, moderate dysplasia with features of HPV infection) yet shows clear evidence of koilocytosis. Such lesions were named "atypical condylomas" by Meisels et al. (1981) (×140)

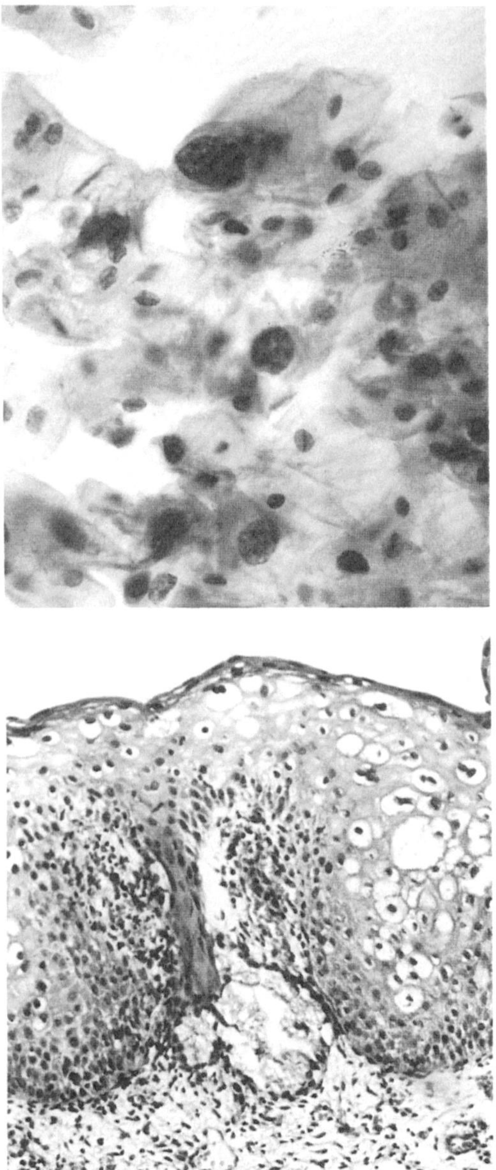

A

B

Fig. 4. A Cervical smear in a 17-year-old girl with a cluster of squamous cells showing several enlarged pyknotic nuclei. The perinuclear halos are difficult to see. Such cells were named "dyskeratinocytes" by Meisels et al. (1984) and are very difficult to differentiate from squamous cancer cells (×500). **B** The corresponding cervical biopsy shows CIN 1 with features of flat condyloma. Note the layer of keratinized cells on the surface. The lesion disappeared without further treatment (×140)

The corresponding tissue lesions do not always reflect the pattern of the smears. The extensive and marked koilocytosis shown in Fig. 2A is poorly reflected in the biopsy (Fig. 2B), which shows only slight changes within the nearly normal squamous epithelium with koilocytosis and minor nuclear abnormalities such as slight nuclear enlargement and binucleation. In more severe changes the koilocytosis is accompanied by nuclear enlargement and marked hyperchromasia (Fig. 4B). Still

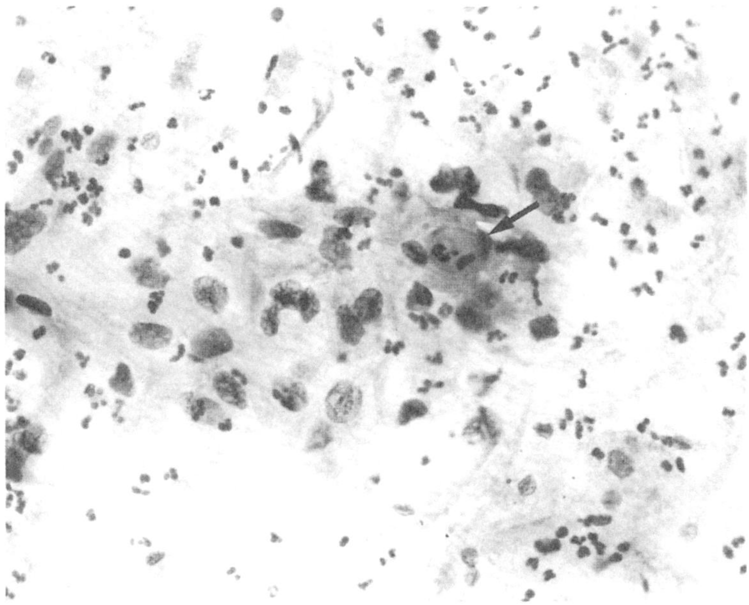

Fig. 5. Smear from a vulvar condyloma acuminatum in a 16-year-old girl. The sheet of abnormal cells and the presence of a squamous "pearl" *(arrow)* mimic to perfection the cytological changes in squamous carcinoma (× 300)

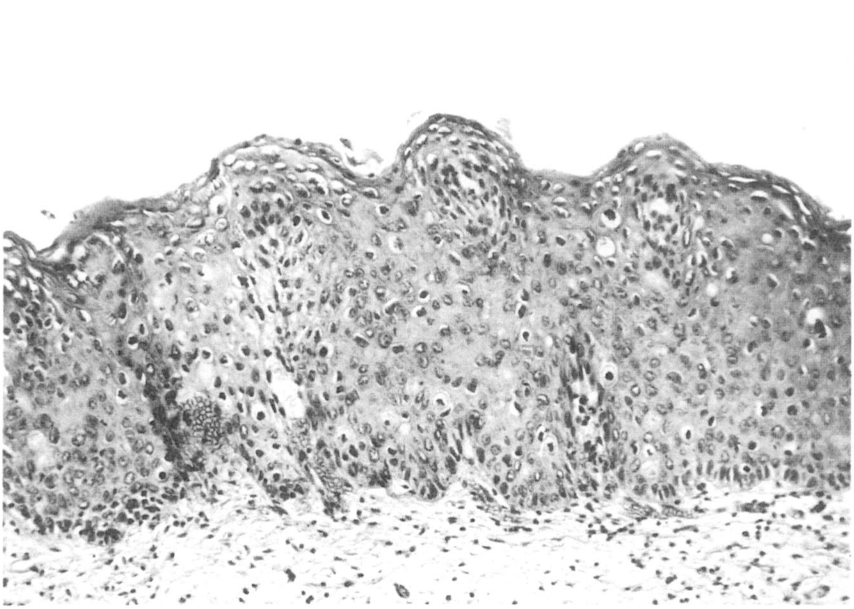

Fig. 6. Cervical biopsy from a 32-year-old woman classified as CIN 2-3 with features of HPV infection. The surface of the lesion shows protrusions or "spikes." Koilocytosis within the superficial layers of the epithelium is accompanied by marked nuclear abnormalities. × 160

more severe tissue changes are shown in Fig. 3 B, wherein there is a marked generalized nuclear enlargement with relatively minor evidence of koilocytosis. Another example of CIN 2-3 with features of HPV infection is shown in Fig. 6. The surface shows protrusions or "spikes." The bulk of the epithelium shows marked nuclear abnormalities. Meisels et al. (1981) named such lesions "atypical condylomas" and documented the presence of the virus in some of them. The average observer may not, however, be capable of differentiating these lesions from other forms of intraepithelial neoplasia, grade 3 (severe dysplasia or carcinoma-in-situ), particularly when the lesion extends into the endocervical glands. Such lesions may prove to be a particularly important stepping stone to invasive carcinoma of the cervix, as was suggested by Meisels et al. (1984).

3.3 The Nature of Koilocytosis

In light microscopy the formation of the perinuclear clear zone in squamous cells appears to proceed in stages. The earliest event is the nuclear enlargement and

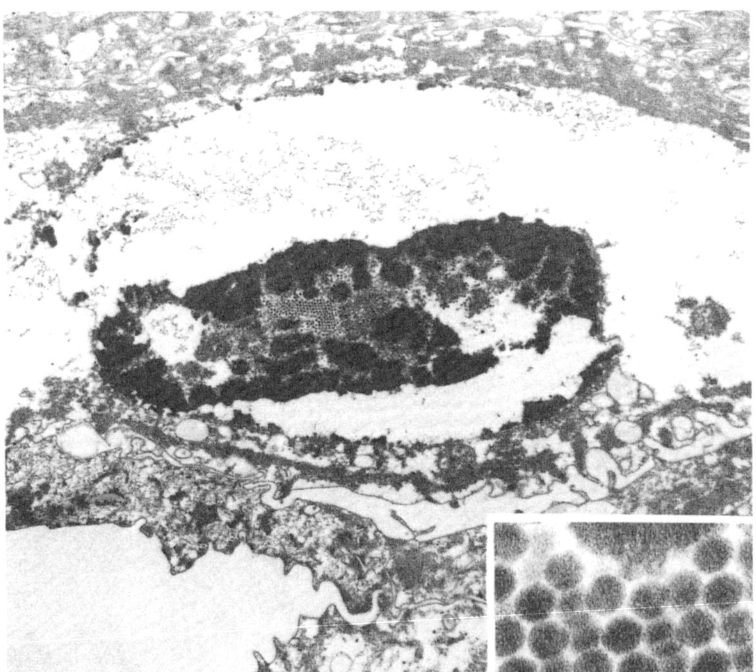

Fig. 7. Electron micrograph of a koilocyte. The nucleus shows a crystalline array of viral particles each about 50 nm in diameter. The perinuclear cytoplasm has been destroyed, accounting for the halo in light microscopy. At the periphery of the cell, there is a condensation of cytoplasmic fibrils, accounting for the dense peripheral cytoplasm in light microscopy (×3900). **Inset** Viral particles under light magnification. A dense core and lighter periphery may be observed (×78400). Photo taken by Dr. Shokri-Tabibzadeh, Montefiore Medical Center.

hyperchromasia which is accompanied by an ill-defined perinuclear zone of cytoplasmic clearing. The clearing progresses in size until it reaches the periphery of the cell (Fig. 2 A). Ultrastructural studies of fully developed koilocytes disclosed the presence of viral particles in the nuclei of some of the cells and a zone of perinuclear cytoplasmic necrosis accounting for the perinuclear clear zone in light microscopy (Fig. 7). The cytoplasmic fibrils are condensed at the periphery of the cells, accounting for the sharp demarcation of the perinuclear zone. Transitional forms of koilocytosis show that the cytoplasmic necrosis progresses from the perinuclear area to the periphery of the cell. Thus, koilocytosis appears to represent a specific cytopathic effect of HPV. The biomechanism of this effect is unknown at this time because as yet no tissue culture system could support the proliferation of the virus in vitro (Taichman et al. 1984). One recently described experimental system (Kreider et al. 1985), based on implantation of fragments of human squamous epithelium into nude mice, appears to be capable of supporting the growth of HPV and offers excellent hope for unraveling the sequence of these events.

It must be stressed that the specificity of koilocytosis as an expression of HPV infection is confined to morphological images associated with conspicuous nuclear abnormalities, whether present within the epithelium or in cytological preparations derived therefrom. Nonspecific, narrow perinuclear clear zones may be observed in squamous cells in cytological preparations from the cervix or vagina in a variety of inflammatory processes, notably in *Trichomonas vaginalis* infection. In such cells, however, the nuclear enlargement and hyperchromasia are absent (Koss 1979).

It is also worthy of note that in tissue sections from invasive epidermoid carcinoma, cancer cells with large perinuclear clear zones are fairly common. There is no evidence at this time that such cells are infected with HPV. More likely, large deposits of glycogen may mimic koilocytosis. This issue clearly warrants further investigation.

3.4 Documentation of Human Papillomavirus Infection in Cervical Smears and Tissue Biopsies (see also Schneider, this volume)

3.4.1 Electron Microscopy

As described above, the first supporting evidence that koilocytosis was associated with the presence of HPV particles was provided by Laverty et al. (1978) by electron microscopy of cervical biopsy material in two patients, one of whom was immunosuppressed. Shortly thereafter, using a modification of a technique developed by Coleman et al. (1977), Hills and Laverty (1979) documented the presence of viral particles in koilocytes in a cervical smear. Within the past 8 years there have been several additional reports pertaining to electron microscopic observations in cervical smears and biopsies (summary in Meisels et al. 1984). It has been documented that in about 50% of the patients afflicted with intraepithelial cervical lesions showing koilocytosis, viral particles may be observed in *mature* squamous cells. In tissue sections the HPV-positive cells are confined to the superficial layers of the epithelium. As discussed above, Meisels et al. (1984) also emphasized the presence of virus in atypical squamous cells with homogeneous and hyperchromatic nuclei but without the perinuclear clear zone (dyskeratocytes). In intact nuclei of koilocytes, the viral particles usually form crystalline lattices located in between coarsely clumped nuclear chromatin (Fig. 7). On close inspection the particles measure about 50 nm in diameter and show a denser core and somewhat lighter periphery (Fig. 7, inset). The crystalline particles have been shown to be icosahedral, i.e., have 20 faces. In damaged nuclei the viral particles may be dispersed; if the nuclear membrane is broken, dispersed viral particles may also be observed in the cytoplasm. This ultrastructural configuration is characteristic of the entire family of papovaviruses to which HPV belongs. A similar appearance has been observed in human polyomavirus infection monitored in cells of urinary sediment, although the particles were slightly smaller (Coleman et al. 1977; Coleman 1981). Thus, by ultrastructural studies alone, the precise type of virus cannot be determined, except that it belongs to the family of papovaviruses.

3.4.2 Immunocytology and Immunohistology

The observation that bovine papilloma virus (which can be sustained in tissue culture) has common antigenic properties with human papillomavirus (Jenson et al. 1980) has led to the development of commercially available broad spectrum antibodies that can be used to document the presence of viral capsular proteins in smears and biopsies (Morin et al. 1981; Gupta et al. 1983). Human viral particles derived from a variety of condylomas and warts may also be used to raise a broad spectrum antiserum (Syrjänen and Pyrhonen 1982). The presence of mature viral particles may be demonstrated by the peroxidase-antiperoxidase (PAP) reaction (Sternberger et al. 1970) or similar systems. A positive PAP reaction results in golden-brown staining of nuclei (Fig. 8 A, B). The results in cervical biopsy material, summarized by Meisels et al. (1984), document that in about 50% of suspect tissues with CIN, viral particles can be demonstrated. It is of interest that positive immune reactions occur mainly in the superficial epithelial layers and virtually never in deep layers. Another observation of note is that viral particles can be demonstrated mainly in lesions displaying marked koilocytosis (Ferenczy et al. 1981). With diminishing koilocytosis and decreasing degree of epithelial maturation, the presence of virus is increasingly difficult to document. In precancerous lesions composed of immature abnormal cells (CIN 3, severe dysplasia or carcinoma-in-situ) or in invasive carcinoma, the immune reactions are generally negative, and viral proteins cannot be demonstrated.

Thus, there is remarkable similarity in the demonstration of viral particles by electron microscopy and by immune reaction. The results suggest that the same targets are elucidated by these two methods. In fact, Morin et al. (1981) were able to demonstrate viral particles by electron microscopy in cells with positive immune reaction.

These observations suggest that maturation of the virus leading to the development of cross-reactive proteins occurs only in the superficial layers of the squamous epithelium where it is associated with formation of koilocytes. Hence, the formation of capsular proteins is bound in an unknown fashion to the biological events occurring in maturation of squamous epithelium. In some epithelia the

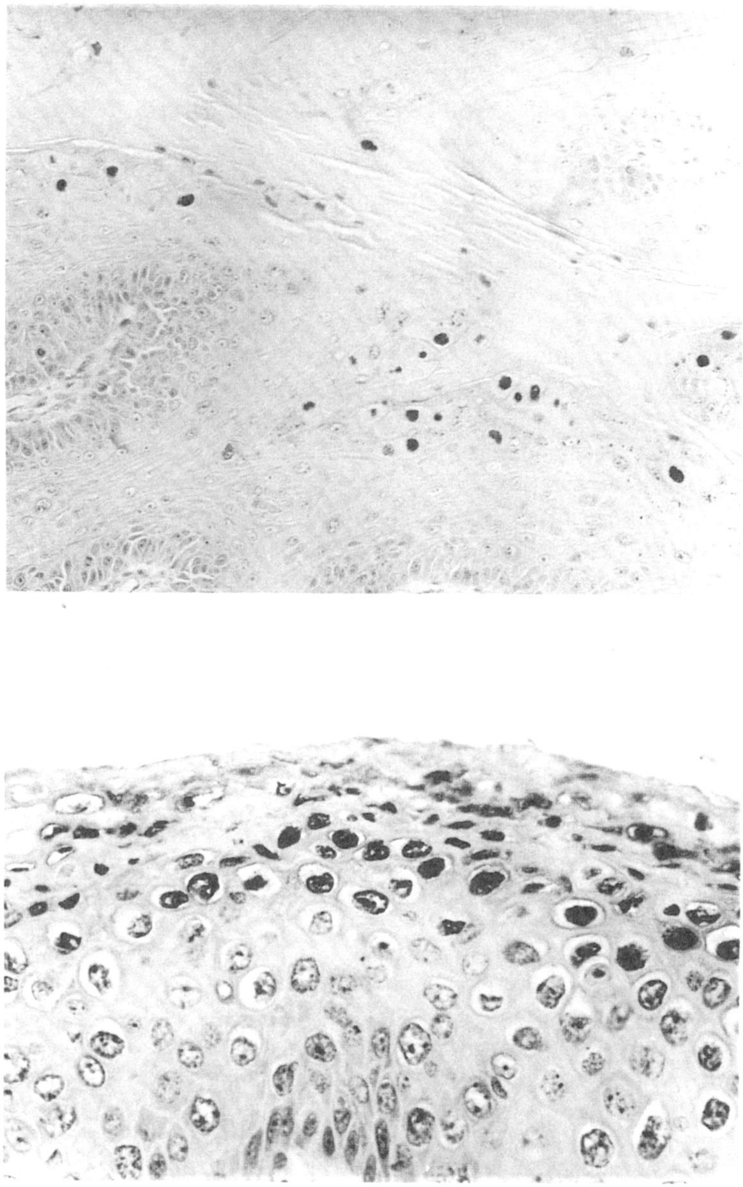

A

B

Fig. 8. **A** Vulvar condyloma acuminatum. Immunoperoxidase reaction to a broad spectrum antiserum to human papillomavirus. The positive-staining dark nuclei are located in the surface layers of the epithelium (×140). **B** Immunoperoxidase reaction to the same serum as in **A** in markedly atypical flat condyloma of the anus in a 28-year-old patient with acquired immunodeficiency syndrome (AIDS). Note that the positive nuclei are located predominantly in the superficial one-third of the epithelium (×300). Case courtesy of Dr. Heidrun Rotterdam, Lenox Hill Hospital, New York.

viral capsular proteins never develop, although the presence of viral DNA can be documented by other means (see below).

3.4.3 Demonstration of Viral DNA by Molecular Biological Techniques

Developments in molecular virology, described in detail elsewhere in this book (see Pfister and Fuchs, this volume) have led to analysis of viral DNA and to classification of HPV into more than 40 types. All HPV DNAs share the basic configuration of open reading frames but differ from each other in nucleotide sequences, as can be demonstrated by molecular biological techniques (Gissmann and zur Hausen 1976). It has been shown that specific types of HPV have a remarkable affinity for certain target tissues, with most types observed in a broad variety of skin lesions (see Pfister and Fuchs, this volume).

HPV6, 11, 16, and 18 have been shown to be commonly associated with anogenital lesions (recent summary in Syrjänen et al. 1985a). Possibly other as yet inadequately characterized types of HPV (31, 33, and 35) may also play a role in this regard. Laryngeal papillomatosis, often first observed in childhood, is also associated with HPV6 and 11, suggesting that this chronic disease may be transmitted from the infected mother to the child (Gissmann et al. 1983, 1984). Isolation of viral DNA of various types, and its propagation in plasmids, has led to new methods of identification of viral DNA, first by the Southern (1975) blot technique in tissue fragments and subsequently by in situ blot hybridization of cells from the uterine cervix (Dürst et al. 1983; Crum et al. 1984; Gissmann et al. 1984; Wagner et al. 1984). The data suggest that DNA from HPV6, 11, 16, and 18 is present in nonintegrated episomal form in many of the CIN lesions of all grades. Integrated HPV DNA sequences of types 16 and 18 were also shown to be present in at least some invasive carcinomas of the uterine cervix (Dürst et al. 1983), although the frequency of occurrence in invasive cancer is a subject of conflicting reports (Fukushima et al. 1985). It is also of note that in established cell lines derived from human cervical cancer such as Hela, C4-1, and Caski, integrated HPV DNA sequences of types 16 and 18 were also identified by DNA hybridization (Boshart et al. 1984; Yee et al. 1985).

Unfortunately, the DNA extraction required for Southern blot technology precludes the localization of viral DNA in tissue sections or intact cells. Thus, the need for HPV DNA identification in morphologically well-preserved new or archival material suitable for microscopic examination becomes apparent. We have had considerable experience with a modification of a technique of in situ hybridization developed by Drs. Kerti Shah and Jean Gupta. The principle is based on DNA hydrolysis of material spread on glass and hybridization with nick-translated viral DNA labelled with radioactive sulfur. Autoradiography of the tissues and cells will reveal a positive reaction in the form of a concentration of granules in the area of the nucleus (Fig. 9). Although the technique is tedious and time-consuming, it appears promising: viral DNA could be demonstrated in superficial and deeper layers of the epithelium in a broad spectrum of CIN lesions and in condylomata acuminata. The use of biotinated DNA probes instead of radioactive markers has been recently described from McDougall's laboratory with similar results (McDougall et al. 1986). These techniques allow matching of

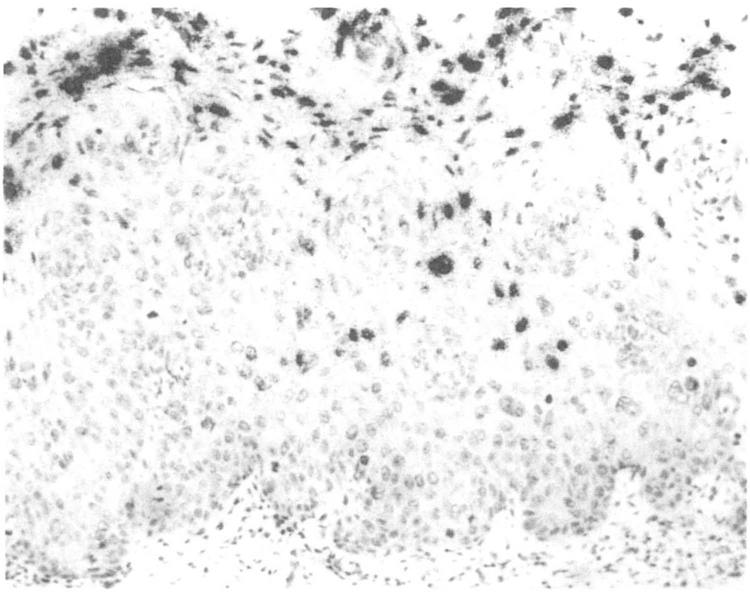

Fig. 9. In situ hybridization with HPV DNA. Autoradiography of a cervical biopsy with CIN 2, using an S-labelled probe derived from HPV DNA of type 16. The presence of viral DNA may be documented in epithelial cells in superficial and deeper layers. ×300

the type of HPV with the morphological type of lesion and may prove extremely helpful in correlating morphological manifestations of HPV infection with their behavior. Occasionally a positive reaction is also observed in stromal cells, probably macrophages. It remains to be determined whether this reaction is an artifact or a repository of viral DNA.

3.5 The Relationship of Human Papillomavirus Type to Morphology and Behavior of CIN

The identification of viral types in preinvasive epithelial lesions of the cervix (CIN), in similar lesions of the vulva and vagina (Ikenberg et al. 1983), and in invasive cervical carcinoma has led to the assessment of the correlation between morphological manifestations, behavior of CIN, and type of virus. At the time of writing this assessment must be considered preliminary because the long-term follow-up that is essential to the determination of the true naturel history of these lesions has not yet been completed.

Ever since the observation that HPV16 and HPV18 could be identified in some invasive cancers of the uterine cervix (and in a number of cell lines of similar origin such as HeLa, C4-1, Caski, etc. Dürst et al. 1983; Yee et al. 1985), it was assumed that CIN lesions associated with HPV6 and HPV11 are less menacing to the patient than lesions with HPV16 and HPV18. At this time there is no evidence that the morphology of the lesions classified as CIN 1 or 2 differs according to viral type. Still, in lesions classified as CIN 3 (severe dysplasia and carcinoma-in-situ) HPV16, and less commonly HPV18, appear to be more frequent (Crum et al. 1984; Wagner et al. 1984).

Prospective studies of CIN lesions with known HPV infection are still very few and the follow-up has been of short duration. The most notable observations were recently provided by Syrjänen et al. (1985a, b). A cohort of 343 patients with cytological evidence of HPV infection were followed by colposcopy with or without biopsies for a mean period of 18.7 months (S.D. ±15.2 months). During this brief follow-up period 25% of the lesions regressed, 61% persisted, and 14% progressed to carcinoma-in-situ. Morphologically more advanced lesions (CIN 2 or 3) were somewhat more likely to progress than CIN 1 lesions, although the follow-up period was too short to draw definitive conclusions. In a subsequent paper Syrjänen et al. (1985b) correlated the behavior of 56 lesions according to HPV and type. The regression was somewhat less likely (45%) and progression more likely (39%) in lesions associated with HPV16 and HPV18 than in lesions with HPV6 or HPV11. Still, 55% of the lesions in the latter category persisted, and some progressed within the average follow-up period of 20 months (S.D. ±15 months).

Syrjänen's preliminary conclusions suggested that the behavior of the HPV-associated CIN lesions did not differ from data derived from other studies on the behavior of CIN and were in keeping with data generated from a large retrospective study of 2466 patients by de Brux et al. (1983).

Recently Ferenczy et al. (1985) documented the presence of HPV DNA of types 6 and 11 in morphologically normal peripheral epithelium in 9 of 20 patients with treated anogenital condylomatous lesions. In 6 of these 9 patients, the lesions recurred within 6 months after therapy. Recurrent disease was observed in only 1 of 11 patients without evidence of residual viral DNA. Thus HPV may infect morphologically normal tissues and be the source of recurrent lesions. Similar observations were reported by Wickerden et al. (1985) in cervical scrapes. Steinberg et al. (1983) also reported the presence of HPV11 in morphologically normal biopsies of the larynx in patients with laryngeal papillomatosis.

3.6 Prevalence of Human Papillomavirus Infection of the Uterine Cervix

In 1977 Hein et al. described the findings from cervical smears in 403 sexually active adolexcent girls between the ages of 12 and 16. In 14 girls there was evidence of epithelial abnormalities classified as CIN of low grade. On review 12 of the 14 smears contained koilocytes, and hence features of human papillomavirus infection (Koss et al., unpublished data). It was of particular interest that the duration of sexual activity that could be elicited in 11 of the 14 girls was less than 2 years in 5 of them, and 2-4 years in 6. This suggests that given appropriate circumstances only a short period of sexual exposure is required for the lesions to occur. Unfortunately, because of the nature of the survey that took place in a detention center in New York City, no follow-up of these girls could be conducted.

We also observed condylomata acuminata of the anogenital area in nine children, 1.5-12 years of age. The source of transmission could be traced to sexual abuse in three or possibly four children; past history of treated condylomas in the mother was noted in two children; in one case the mother had "warts" on her hand. In two cases no clear source of infection could be identified. It is of particular note that in the two children whose mothers had treated vaginal condylomas, the possibility of transmission at birth could not be ruled out (Vallejos et al., 1981). By in situ hybridization of tissue sections it was shown that the dominant HPV types were the same as in anogenital condylomata in adults, namely types 6, 11, and, in one case, also 16 and 18.

Because morphological changes associated with human papillomavirus infection in the uterine cervix epithelium have been generally classified as dysplasia or the equivalent CIN designation, there is limited information on the prevalence of the lesions. Perhaps the best data are available from Meisels' laboratory in Quebec City (Meisels and Morin 1983). In a population of 234715 women screened from 1975 to 1979, there were 3977 instances (1.69%) of papillomavirus infection recognized in cervical smears. It must be stressed that the true rate of papillomavirus infection is probably higher. Thus, in Syrjänen's experience (unpublished data), evidence of papillomavirus infection was observed in 2.6% of a group of 22-year-old women. There is an unknown failure rate in cytological screening as it is not uncommon to find in a tissue biopsy changes suggestive of papillomavirus infection not reflected in smears.

The full scope of prevalence of HPV infection in the general population became apparent only recently. A study of 10000 cervical swab samples by filter in situ hybridization technique revealed the presence of HPV in 11% of disease-free women. HPV16/18 were present in 8.7% and HPV6/11, in 6.9%. Thus simultaneous infection with several viral types was common. Similar observations were recorded in 544 penile swabs in disease-free blood donors: the presence of HPV was recorded in 5.5% of them. The lower figure for normal males was attributed to technical difficulties (Grußendorf-Conen et al., 1987).

3.7 Human Papillomavirus Infection and Age of Patients

The survey of teenage girls by Hein et al. (1977), cited above, strongly suggested that young adolescents may be particularly susceptible to HPV infection. In fact, surveys by Meisels and Morin in Canada (1983) and Syrjänen in Finland (1979) implied that sexually active young women, below the age of 25, are more likely to develop HPV-related changes than older women. Syrjänen suggested that all neoplastic events in women age 20 or less are triggered by HPV infection. In Meisels' survey the rate of the morphological manifestations of papillomavirus infection reached its peak between the ages of 21 and 25 years. The rate diminished with age in favor of other forms of cervical intraepithelial neoplasia.

In our own laboratories nearly all young women (below the age of 25) with abnormal cervical smears show some features of koilocytosis often accompanying other abnormal fingings. With advancing age the evidence of koilocytosis is less frequent.

3.8 The Male as Carrier of Human Papillomavirus

In view of the association of human papillomavirus infection with sexual activity, it is natural to suspect the male as the carrier of the virus. Visible warty condylomata of the penis are a known manifestation of HPV infection and the high rate of the association of these lesions with warty condylomata of the external genitalia in women was recognized many years ago (summary in Oriel 1971a). A more recent survey documented that penile or urethral condylomata were present in over 50% of male sexual partners of 34 women with CIN all grades (Levine et al. 1984). It is of note that many of the penile lesions were quite small and thus difficult to detect. Intraurethral condylomas are another possible source of infection with HPV. While they are more commonly seen in distal urethra, proximal urethra may also be affected (Oriel 1977). HPV virus antigen was documented by the immunoperoxidase method in about 50% of the urethral lesions (Murphy et al. 1983). By in situ hybridization technique we were able to show the presence of HPV DNA in 16 of 17 penile and urethral condylomata. Simultaneous infection with HPV6 and HPV11 was observed in 12 patients, and with HPV6 and HPV18 in 1 (Del Mistro et al., to be published). Unfortunately, the deeply seated urethral condylomata may be difficult to diagnose. Urinary sediment rarely shows koilocytes (Levine et al. 1984), although I have observed several cases of bladder condylomata with koilocytes in the urine. It must also be noted that Ostrow (personal communication) observed HPV DNA in ejaculated and seminal fluid of several patients, a point of some interest because of the suggestion by Coppleson and Reid (1968) that spermatozoa may contribute to carcinoma of the cervix.

Another possible source of infection is condylomata acuminata of the bladder. Several examples were personally seen, mainly in immunocompromised patients treated for malignant lymphoma (see below).

Male-to-male transmission of HPV infection is quite common in homosexuals who develop anal warts (Carr and William 1977). Anal condylomata may also occur in heterosexuals through anal coitus (Oriel 1971b).

Until recently, bowenoid papulosis was thought to be a very rare skin disorder of the external genitalia of both sexes. Close attention to minor skin changes suggests that the disease may be more common than previously thought. HPV16 was shown to be present in the majority of these lesions, which may thus constitute yet another important source of HPV infection (Gross et al. 1985). Surprisingly, HPV16 was also documented in six of ten examples of Bowen's disease (carcinoma-in-situ) of the skin of external genitalia and several other sites (Ikenberg et al. 1983). For further discussion see Gross (this volume).

Perhaps the most important conclusion from these data is that HPV is a highly infectious agent that may be transmitted among sexual partners with great ease and after a short exposure. This confirms the observation cited above that only a short period of exposure is necessary for young women to develop evidence of HPV infection of the uterine cervix.

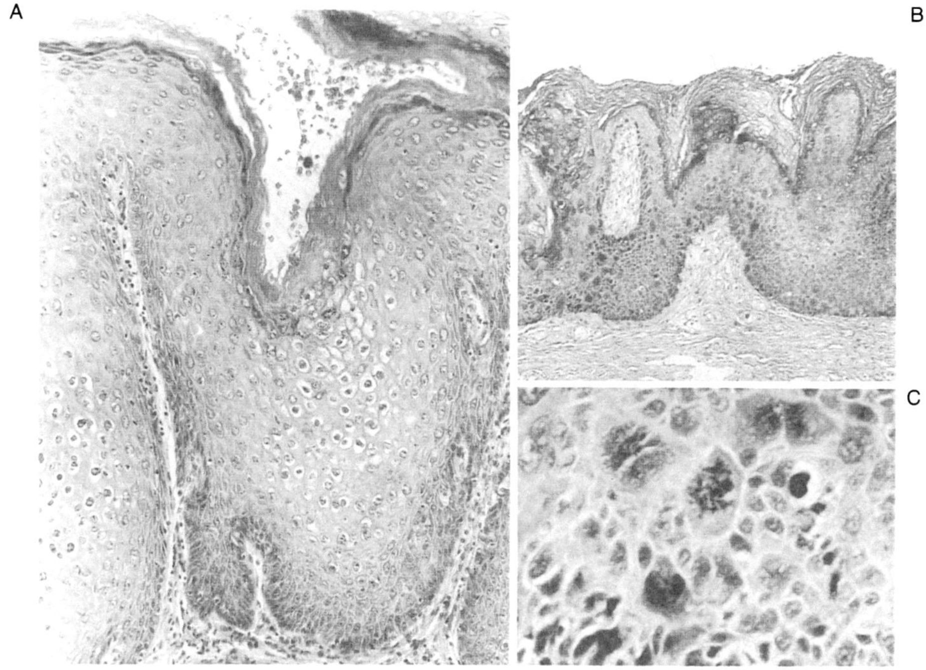

Fig. 10. **A** A vulvar condyloma with nuclear atypia observed in 1976 in a 27-year-old woman with treated Hodgkin's disease (×50). **B** Bowen's disease of the vulva observed in 1980 (×50). **C** Higher magnification of **B** to show abnormal mitoses (×250)

3.9 Human Papillomavirus Infection in Immunocompromised Patients

In 1981 we described human papillomavirus infection, documented by electron microscopy, in four young women with impairment of immunity, due to treated Hodgkin's disease in three and to an unknown cause in one (Shokri-Tabibzadeh et al. 1981). The sequence of events in one of the patients with Hodgkin's disease is illustrated in Figs. 10 and 11. A vulvar condyloma observed in 1976 and not responsive to conservative treatment (Fig. 10A) evolved into a carcinoma-in-situ (Bowen's disease) in 1980 (Fig. 10B, C). In 1976 the patient also displayed koilocytes in her cervical smear (Fig. 11A); 1 year later a carcinoma-in-situ of the cervix was demonstrated (Fig. 11B). In one of the four patients, an invasive carcinoma of the vulva, perineum, and the vestibulum of the vagina developed 10 years after the initial condyloma acuminatum of the vulva was observed (Fig. 12).

Recently Schneider et al. (1983) reported on a group of renal transplant patients, eight of whom developed papillomavirus infection of the cervix. One of these patients developed an invasive and metastatic squamous carcinoma and died of it.

As mentioned above, evidence linking the immune status with papillomavirus infection has been observed by us in the urinary bladder. In several immunocompromised patients condylomata acuminata of the bladder were observed. In yet

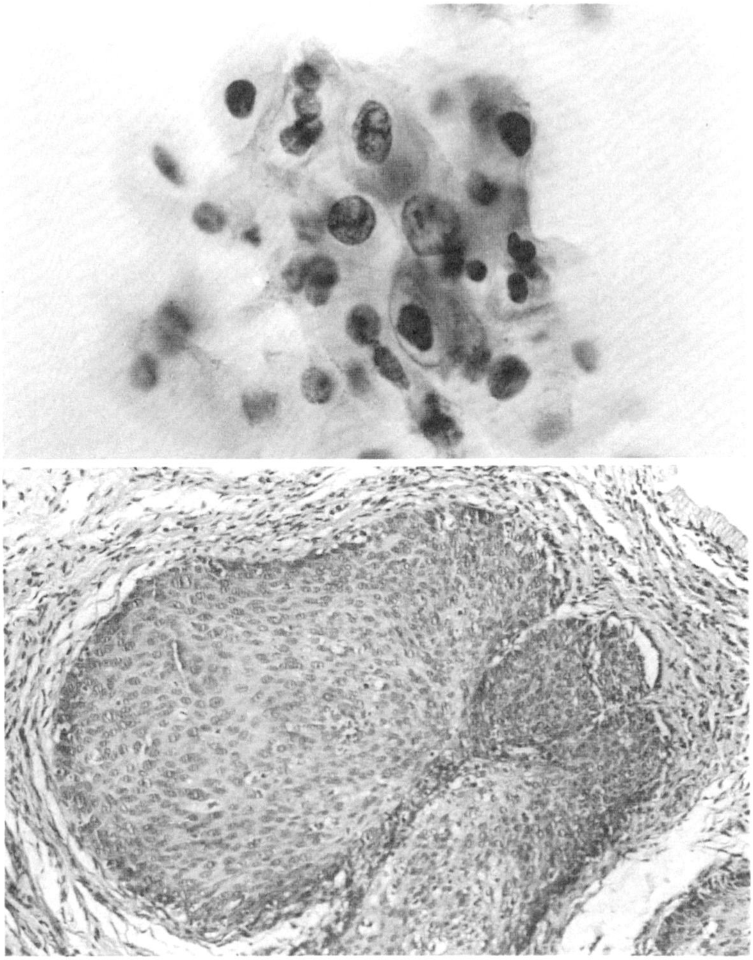

A

B

Fig. 11 A, B. Same patient in Fig. 9. **A** Cervical smear with atypical koilocytes in 1976 (× 500). **B** Area of carcinoma-in-situ in endocervical glands seen in 1977 (× 140)

another patient typical koilocytes were observed in the urinary sediment without clinical evidence of condyloma. Petterson et al. (1976) also reported two patients on immunosuppressive therapy after renal transplant who developed condylomata acuminata of the bladder.

In patients with the acquired immunodeficiency syndrome (AIDS), markedly atypical anal condylomata may be observed (Fig. 8 B). In one specific example the condyloma was accompanied by Kaposi's sarcoma.

There is also anecdotal evidence that patients with a past history of condylomata acuminata and who become pregnant, and hence temporarily immunosuppressed, may develop recurrent condylomata which disappear again at the conclusion of pregnancy. These casual observations were confirmed by Schneider et al. (1986) using Southern blot hybridization technique on cervical smears. HPV DNA

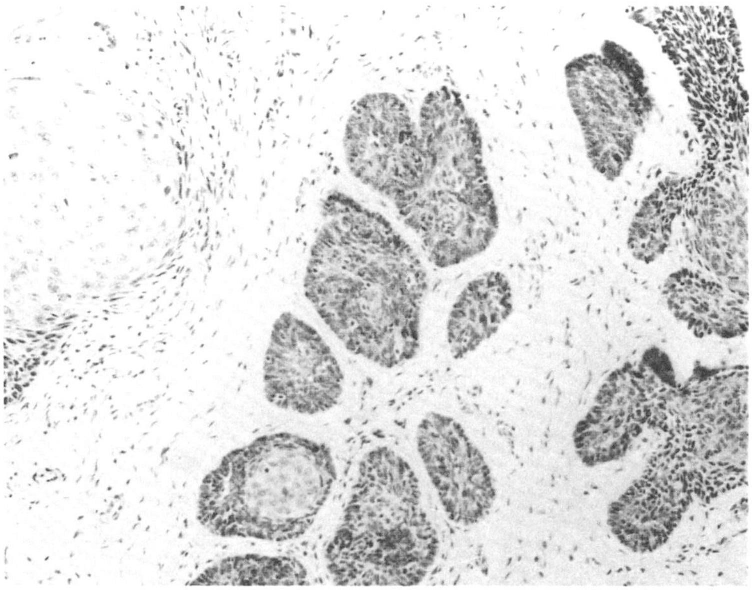

Fig. 12. Invasive carcinoma of the vaginal antrum developing in an immunocompromised young woman 10 years after the development of an initial condyloma. Viral particles were demonstrated by electron microscopy in the squamous epithelium of the vulva at the time of diagnosis of cancer. × 140

was observed in 26 (28%) of 92 pregnant women and in 12 (12.5%) of 96 nonpregnant women. HPV16 was more often observed and showed more active replication during pregnancy. The data were cited in support of the thesis that HPV replication may be related to natural immunosuppression occurring during pregnancy.

4 The Association of Human Papillomavirus Infection with Precancerous and Cancerous Lesions in the Uterine Cervix

As discussed in Sect. 3.1, Koss and Durfee (1956) observed that koilocytosis in smears or tissues was associated with carcinoma-in-situ in nearly half of the 40 cases studied. In invasive cancer similar changes in adjacent peripheral epithelium were observed in 9 of 53 cases (16%).

With the renewed interest in koilocytosis as a pathognomonic effect of HPV infection, its relationship with other forms of CIN and invasive cancer became the subject of increasing attention (zur Hausen 1977). Various authors reported such associations in variable proportions of women with CIN, ranging from 25.6% (Meisels et al. 1984) to nearly 50% (Syrjänen 1979, 1981). Perhaps the most dedicated advocates of this association have been Reid et al., who in a series of papers (Reid 1983; Reid et al. 1982, 1984a, b) suggested that in virtually every instance of CIN or invasive epidermoid carcinoma there is some morphological evidence of infection with HPV.

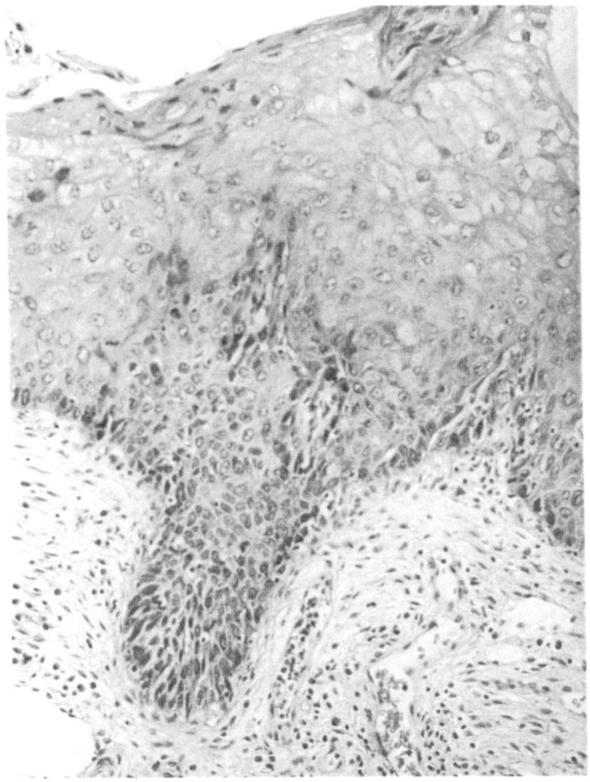

Fig. 13. Biopsy of the uterine cervix showing a well-differentiated carcinoma-in-situ with local extension into the stroma. Koilocytosis may be noted within the superficial layers of the adjacent epithelium. ×140

Morphological manifestations of the association of koilocytosis with CIN are twofold: koilocytosis may be present in lesions that otherwise show sufficient cell abnormalities to be classified as CIN 1-CIN 3 (Figs. 3B and 6). Figure 13 documents that such lesions appear capable of early invasion. Alternatively, koilocytosis may also be present in an epithelial segment adjacent to CIN lesions of high grade (Fig. 14). In long-term follow-up studies of intraepithelial lesions, several, retrospectively recognized as flat condylomas of the cervix, progressed to carcinoma-in-situ or to invasive cancer (Koss et al. 1963). Meisels et al. (1984) also recorded the progression of 4.7% of condylomas and 10.2% of atypical condylomas to carcinoma-in-situ or even to microinvasive carcinoma. Boon and Fox (1981) compared the behavior of CIN in 50 women with and 50 without koilocytosis and failed to document any differences of behavior during 3-4 years of follow-up; both types of lesions were judged to put the women at risk for further neoplastic lesions. Similar conclusions were more recently reached by Syrjänen et al. (1985a, b) in prospective studies.

On the other hand, there is also excellent evidence that a significant proportion of lesions suggestive of HPV infection may disappear (see Fig. 4). Meisels and Morin (1983) estimated that nearly 70% of these lesions disappear after a short follow-up time of 15 months. Syrjänen et al. (1985a, b) reported a much lower figure of only 30% regression. Still the behavior of HPV-induced lesions is unpredictable and in every way similar to other forms of CIN.

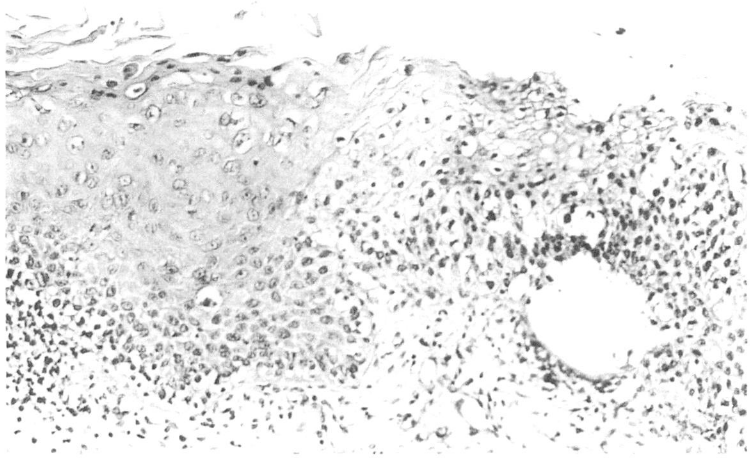

Fig. 14. Biopsy of the uterine cervix with a koilocytotic change *(right)* adjacent to an intraepithelial neoplastic lesion (CIN 2, moderate dysplasia); *left*. ×140

The search for objective prognostic factors that would allow a priori determination of the behavior of CIN, whether or not associated with HPV infection, has led to studies of the DNA content in these lesions. Fu et al. (1981) proposed that aneuploid lesions were more likely to persist or progress than diploid or polyploid lesions. It was also pointed out that aneuploid lesions are more likely to display abnormal mitoses than diploid or polyploid lesions. In a subsequent study (Winkler et al. 1984) it was shown that the majority of the HPV-induced lesions displayed abnormal mitoses and an aneuploid DNA pattern. DNA measurements in classical papillary condylomata acuminata yielded conflicting results: Shevchuck and Richart (1982) found polyploid (tetraploid and octaploid) DNA patterns in five such lesions studied. Jagella and Stegner (1974) observed aneuploid DNA patterns in several of 50 condylomata. It is possible that tetraploidy or polyploidy is the initial DNA change induced by HPV whereas, in selected cases, and for reasons unknown, aneuploid clones of cells may develop. The latter lesions may persist and lead to further carcinogenic events.

It appears important at this time to summarize the current evidence on the relationship of HPV infection as a factor in the genesis of neoplastic events in the uterine cervix:

a. Carcinoma of the uterine cervix behaves in a manner similar to a venereally transmitted disease: the risk factors are young age at the onset of sexual activity and sexual promiscuity. The evidence gathered to date supports the role of HPV in this regard. Studies of teenage girls with sexual exposure, the short time elapsed between the exposure and the development of koilocytosis, and the high rate of koilocytosis among young women with CIN speak in favor of this possibility.

b. Association of koilocytosis with CIN of various grades and occasionally with invasive carcinoma has been repeatedly demonstrated.

c. A significant proportion (at least 30%, probably more) of CIN lesions disappear spontaneously or after small biopsies. However, such women remain at risk for future neoplastic events. This observation also fits into the concept of infection with HPV. Persistence of the virus without morphological manifestation of disease may account for recurrent infections under favorable circumstances. This was reported by Wickenden et al. (1985) and by Ferenczy et al. (1985) and is known to occur in pregnant women with a past history of condylomata acuminata: recurrent condylomata acuminata may be observed during the second half of pregnancy, but the lesions usually disappear after delivery.

d. Dependence of outcome on viral type. The possibility that the infection with HPV6 and HPV11 has a more favorable outcome than infection with HPV16 and HPV18 is clearly deserving of further prospective studies. This thesis is supported by finding HPV DNA sequences of types 16 and 18 in some CIN lesions in episomal form, and in integrated form in invasive cervical carcinomas and in established cell lines derived therefrom (Dürst et al. 1983; Yee et al. 1985). Yet Yee raises the question whether the viral DNA is in fact an active participant in tumor proliferation or merely represents evidence of archival infection. It is of great significance in this regard that Lehn et al. (1985) failed to observe transcriptional activity of HPV DNA in three of four invasive cervical carcinomas. The question whether the virus is an initiating agent that leads to early cell changes (as recently suggested by the experimental data provided by Kreider et al. 1985) but does not directly participate in the cascade of events leading to invasive cancer cannot be answered at this time.

e. Condylomata acuminata and flat condylomas occur in immunocompromised patients (latter stages of pregnancy, AIDS, patients treated for malignant lymphoma). In such patients the presence of the virus is readily demonstrated. This observation is in keeping with data on immunocompromised patients, such as renal transplant recipients, known to be susceptible to the development of carcinoma of the uterine cervix (Porreco et al. 1975).

f. Precancerous states of the uterine cervix involve adjacent segments of the epithelium. The morphological spectrum of abnormalities depends on the epithelium of origin: highly differentiated lesions occur in the squamous epithelium of the portio vaginalis; poorly differentiated lesions are derived from the endocervical epithelium. It may be conceived that the viral infection spreads along the epithelium and triggers various morphological events depending on the type of epithelium that becomes infected (Fig. 15).

g. Viral particles are demonstrable by immune techniques and electron microscopy only in mature lesions of squamous epithelium and not in poorly differentiated lesions. Current evidence suggests that various species of keratin are present within epithelia of various degrees of maturity (summary in Sun et al. 1984). It is possible that the availability of certain species of keratin (or mechanisms leading to their production) are essential to virus maturation.

h. Why does HPV rather than Herpesvirus II trigger the neoplastic events? Disre-

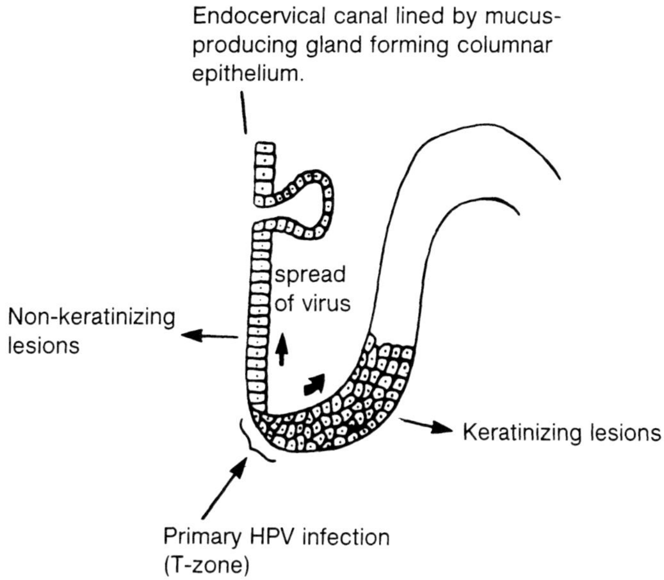

Fig. 15. Possible sequence of events in HPV infection of the uterine cervix

garding molecular biological evidence (zur Hausen 1982), pure pathological evidence suggests that HPV rather than Herpesvirus II is a more likely neoplastic agent. Herpesvirus produces ulcerative lesions whereas HPV is known to produce proliferative lesions of squamous epithelium in many different body sites.

There are also a great many unresolved questions that must be addressed in order to strengthen the hypothesis of HPV infection as a factor in the genesis of carcinoma of the uterine cervix:

a. The mechanism of viral transformation of the epithelium
b. The relationship of the viral types to behavior of the lesions
c. The relationship of disappearing CIN lesion to those persisting and progressing
d. The mechanism of selection of cell clones capable of invasion
e. Clarification of the relationship of episomal viral DNA observed in CIN to the integrated DNA observed in invasive carcinoma
f. Clarification of major differences in the frequency of HPV infections in various geographic locations (Dürst et al. 1983; Fukushima et al. 1985). Papillomavirus infection implies at least some degree of sexual license. However, during a recent visit to the People's Republic of China, where sexual license is not officially tolerated and presumably not exercised, it was noted that in some areas of this vast country, carcinoma of the uterine cervix is a common and frequent disease *(Atlas of Cancer Mortality in the People's Republic of China).* It would be of great interest to study the frequency of papillomavirus infection in China for comparison with the Western world.
g. The role of other viruses in HPV infection: zur Hausen (1982) proposed that in cancer of the human genital tract, herpesvirus may play an initiating role whereas HPV may play the role of a promoter.

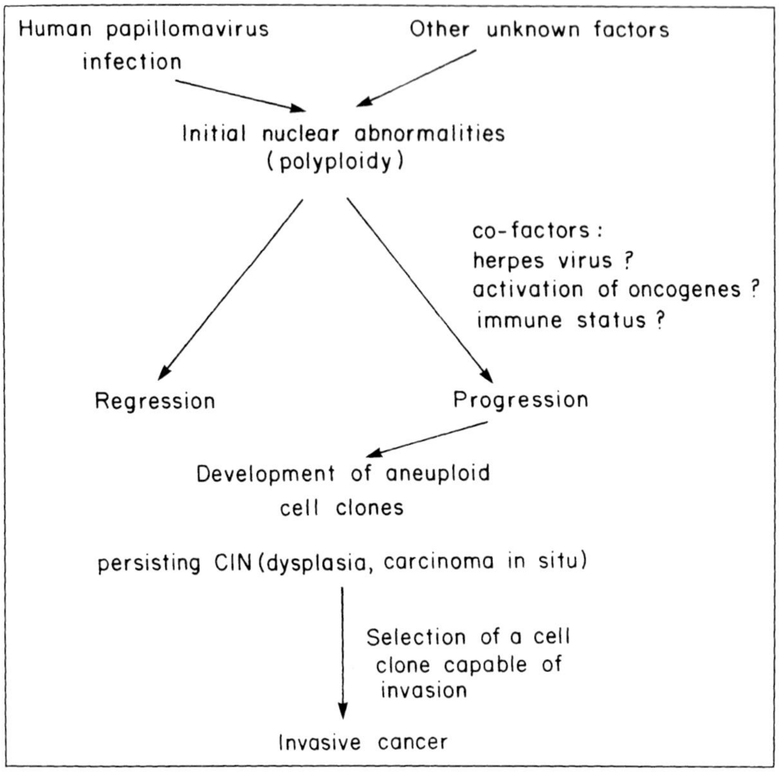

Fig. 16. Possible sequence of events in carcinogenesis of the uterine cervix, vagina, and vulva

On the assumption that HPV infection is, in fact, a major factor in the genesis of carcinoma of the uterine cervix and presumably also of the vagina and vulva (Buscema et al. 1980; Crum et al. 1982), the sequence of events outlined in Fig. 16 may be considered. This is a personal view of these events based on the current state of our knowledge, notwithstanding numerous gaps which require further investigation.

The model suggests that HPV infection is, in fact, the primary but not the only event in the genesis of carcinoma of the cervix. The data cited above on prevalence of HPV in asymptomatic men and women (de Villiers et al. 1986) suggest that the presence of the virus is not a sufficient condition per se to induce neoplastic lesions. It is suggested that carcinogenesis is a complex, multistep process with several options at each stage of development. It is quite evident that a great deal of additional work is needed to clarify many of the questions raised.

5 Clinical Implications

Regardless of the many unresolved issues concerning the relationship of HPV infection to the precancerous states and invasive carcinoma of the uterine cervix, the presence of koilocytes in cervical smears identifies women at high risk. In fact, the cytological sampling is often not representative of the lesion, and often times colposcopic biopsies may reveal the presence of lesions not identified by cytology in adjacent epithelial segments. Therefore, we recommend that all women with koilocytosis in cervical smears be referred for further evaluation by colposcopy and biopsies. As HPV hybridization techniques are not yet readily available in clinical laboratories and since the evidence as to the behavior of lesions associated with various types of HPV is not fully documented, there are no secure laboratory methods to separate women at lesser risk for the development of carcinoma of cervix from those at high risk. It is to be hoped that at some future time the identification of the two groups will become possible with resulting significant savings in the costs of health care.

Acknowledgements. We thank Drs. Shah and Gupta of the Johns Hopkins School of Public Health, Baltimore, MD, for their courtesy in sharing with us the details of their procedure of in situ hybridization.

References

Atlas of Cancer Mortality in the People's Republic of China. Not dated. China Map Press, Maps 47-50

Ayre JE (1951) Cancer cytology of the uterus. Grune and Stratton, New York

Ayre JE, Ayre WB (1949) Progression from precancer stage to early carcinoma of cervix within 1 year; combined cytologic and histologic study with report of case. Am J Clin Pathol 19: 770-778

Boon ME, Fox CH (1981) Simultaneous condyloma acuminatum and dysplasia of uterine cervix. Acta Cytol 25: 393-399

Boshart M, Gissmann L, Ikenberg H, Kleinheinz A, Schuerlen W, zur Hausen H (1984) A new type of papillomavirus DNA, its presence in genital cancer biopsies and in cell lives derived from cervical cancer. EMBO 3: 1151-1157

Burghardt E (1973) Early histological diagnosis of cervical cancer. (Translated by EA Friedman). Thieme, Stuttgart

Buscema J, Woodruf JD, Parmley TH, Genadry R (1980) Carcinoma in situ of the vulva. Obstet Gynecol 55: 225-230

Carr G, William DC (1977) Anal warts in a population of gay men in New York City. Sex Transm Dis 4: 56-57

Coleman DV (1981) Diagnosis of human polyomavirus infection. In: Koss LG, Coleman DV (eds) Advances in clinical cytology, vol 1. Butterworths, London, pp 136-159

Coleman DV, Russel WJI, Hodgson J, Tun Pe, Mowbray JF (1977) Human papovavirus in Papanicolaou smears of urinary sediment detected by transmission electron microscopy. J Clin Pathol 30: 1015-1020

Coppleson M, Reid B (1968) The etiology of squamous carcinoma of the cervix. Obstet Gynecol 32: 432-436

Crum CP, Braun LA, Shah KV, Fu YS, Levine RU, Fenoglio CM, Richart RM, Townsend D (1982) Vulvar intraepithelial neoplasia. Correlation of nuclear DNA content and the presence of human papillomavirus (HPV) structural antigen. Cancer 49: 468-471

Crum CP, Ikenberg H, Richart RM, Gissmann L (1984) Human papillomavirus type 16 and early cervical neoplasia. N Engl J Med 310: 880-883

de Brux J, Orth G, Croissant O, Cohard E, Ionesco M (1983) Lesions condylomateuses du col utérin, évolution chez 2466 patients. Bull Cancer 70: 410-422

de Villiers EM, Wagner D, Schneider A, Miklaw H, Grussendorf-Conen E, zur Hausen H (1986) A survey on the infection rate in a normal population with genital papillomaviruses. Meeting abstract, Cold Spring Harbor symposium on papillomaviruses, September
Del Mistro A, Braunstein JD, Halwer M, Koss LG (1987) Identification of human papillomavirus types in male urethral condylomata acuminata by in situ hybridization. Hum Pathol (to be published)
Dunn AEG, Ogilvie NM (1968) Intranuclear virus particles in human genital wart tissue. Observations on the ultrastructure of the epidermal layer. J Ultrastruct Res 22: 282-295
Dürst M, Gissman L, Ikenberg H, zur Hausen H (1983) A papillomavirus DNA from a cervical carcinoma and its prevalence in cancer biopsy samples from different geographic regions. Proc Nat Acad Sci USA 80: 3812-3815
Ferenczy A, Braun L, Shah KV (1981) Human papillomavirus (HPV) in condylomatous lesions of cervix. Am J Surg Pathol 5: 661-670
Ferenczy A, Mitao M, Nagai N, Silverstein S, Crum C (1985) Latent papillomavirus and recurring genital warts. N Engl J Med 313: 784-788
Fu YS, Reagan JW, Richart RM (1981) Definition of precursors. Gynecol Oncol 12: S220-S231
Fukushima M, Okagaki T, Twiggs LB, Clark BA, Zachow KR, Ostrow RS (1985) Histological types of carcinoma of the uterine cervix and the detectability of human papillomavirus DNA. Cancer Res 45: 3252-3255
Gissman L, zur Hausen H (1976) Human papillomaviruses DNA: physical mapping and genetic heterogeneity. Proc Natl Acad Sci USA 73: 1310-1313
Gissmann L, Wolnik L, Ikenberg H, Koldovsky U, Schnurch HG, zur Hausen H (1983) Human papillomavirus types 6 and 11 DNA sequences in genital and laryngeal papillomas and in some cervical cancers. Proc Natl Acad Sci 80: 560-563
Gissman L, Boshart M, Dürst M, Ikenberg H, Wagner D, zur Hausen H (1984) Presence of human papillomavirus in genital tumors. J Invest Dermatol 83: 26-28
Gross G, Hagedorn M, Ikenberg H, Rufli T, Dahlet C, Grosshans E, Gissmann L (1985) Bowenoid papulosis. Presence of human papillomavirus (HPV) structural antigens and of HPV 16-related DNA sequences. Arch Dermatol 121: 858-863
Grußendorf-Conen E-I, Meinhof W, de Villiers EM, Gissmann L (1987) Occurrence of HPV genomes in penile smears of healthy men. Arch Dermatol Res 279: 573-574
Gupta JW, Gupta PK, Shah KV, Kelly DP (1983) Distribution of human papillomavirus antigen in cervico-vaginal smears and cervical tissues. Int J Gynecol Pathol 2: 160-170
Hein K, Schreiber K, Cohen MI, Koss LG (1977) Cervical cytology. The need for routine screening in sexually active adolescent. J Pediatr 91: 123-126
Hills E, Laverty CR (1979) Electron microscopic detection of papillomavirus particles in selected koilocytic cells in a routine cervical smear. Acta Cytol 23: 53-56
Ikenberg H, Gissmann L, Gross G, Grußendorf-Conen E-I, zur Hausen H (1983) Human papillomavirus type-16-related DNA in genital Bowen's disease and in Bowenoid Papulosis. Int J Cancer 32: 563-565
Jenson AB, Rosenthal JR, Olson C, Pass F, Lancaster WD, Shah K (1980) Immunological relatedness of papillomaviruses from different species. J Natl Cancer Inst 64: 495-500
Jagella HP, Stegner HE (1974) Zur Dignität der Condyloma Acuminata. Klinische, histopathologische and cytophotometrische Befunde. Arch Gynäkol 216: 119-132
Kazal HL, Long JP (1958) Squamous cell papilloma of the uterine cervix. Cancer 11: 1049-1059
Kessler II (1976) Human cervical cancer as a venereal disease. Cancer Res 36: 783-791
Kessler II (1984) Natural history and epidemiology of cervical cancer with special reference to the role of herpes genitalis. In: McBrien DCH, Slater TF (eds) Cancer of the uterine cervix. Biochemical and clinical aspects. Academic, London
Koss LG (1979) Diagnostic cytology and its histopathologic bases, 3rd edn. Lippincott, Philadelphia
Koss LG (1981) Pathogenesis of carcinoma of the uterine cervix. In: Dallenbach-Hellweg G (ed) Cervical cancer. Springer, Berlin Heidelberg New York
Koss LG, Durfee GR (1956) Unusual patterns of squamous epithelium of the uterine cervix; cytologic and pathologic study of koilocytotic atypia. Ann NY Acad Sci 63: 1245-1261
Koss LG, Stewart FW, Foote FW, Jordan MJ, Bader GM, Day E (1963) Some histological aspects

of behavior of epidermoid carcinoma in situ and related lesions of uterine cervix. A long term prospective study. Cancer 16: 1160-1211

Kreider J, Howett M, Wolfe SA, Bartlett G, Zaino R, Sedlacek T, Mortel R (1985) Morphological transformation in vivo of human uterine cervix with papillomavirus from condylomata acuminata. Nature 317: 639-641

Laverty CR, Russell P, Hills E, Booth N (1978) The significance of non-condyloma wart virus infection of the cervical transformation zone. A review with discussion of two illustrative cases. Acta Cytol 22: 195-201

Lehn H, Kreig P, Sauer G (1985) Papillomavirus genomes in human cervical tumors: Analysis of their transcriptional activity. Proc Natl Acad Sci USA 82: 5540-5544

Levine RU, Crum CP, Herman E, Silvers D, Ferenczy A, Richart RM (1984) Cervical papillomavirus infection and intraepithelial neoplasia: a study of male sexual partners. Obstet Gynecol 64: 16-20

Marsh MR (1952) Papilloma of cervix. Am J Obstet Gynecol 64: 281-291

McDougall JK, Myerson D, Beckmann AM (1986) Detection of viral DNA and RNA by in situ hybridization. J Histochem Cytochem 34: 33-37

Meisels A, Fortin R (1976) Condylomatous lesions of the cervix and vagina. Cytologic patterns. Acta Cytol 20: 505-509

Meisels A, Morin C (1983) Human papillomavirus and cancer of the uterine cervix. Gynecol Oncol 12: S111-S123

Meisels A, Roy M, Fortier M et al. (1981) Human papillomavirus infection of the cervix: the atypical condyloma. Acta Cytol 25: 7-16

Meisels A, Morin C, Casas-Cordero M (1984) Lesions of the uterine cervix associated with papillomavirus and their clinical consequences. In: Koss LG, Coleman DV (eds) Advances in clinical cytology, vol 2. Masson, New York, pp 1-32

Morin C, Braun I, Casas-Cordero M, Shah KV, Roy M, Fortier M, Meisels A (1981) Confirmation of papillomavirus etiology of condylomatous cervix lesions by the peroxidase-antiperoxidase technique. J Natl Cancer Inst 66: 831-835

Murphy WM, Fu YS, Lancaster WD, Jenson AB (1983) Papillomavirus structural antigen in condyloma acuminatum of the male urethra. J Urol 130: 84-85

Nasiell K, Nasiell M, Vadavinkova V, Roger V, Hjerpe A (1976) Follow-up studies of cytologically detected precancerous lesions (dysplasia) of the uterine cervix. In: Health control in detection of cancer. Skandia International Symposia. Almqvist and Wiksell, Stockholm, pp 244-252

Oriel JD (1971a) A natural history of genital warts. Br J Vener Dis 47: 1-13

Oriel JD (1971b) Anal warts and anal coitus. Br J Vener Dis 47: 373-376

Oriel JD (1977) Genital warts. Sex Transm Dis 4: 153-159

Oriel JD, Almeida JD (1970). Demonstration of virus particles in human genital warts. Br J Vener Dis 46: 37-42

Papanicolaou GN (1954) Atlas of exfoliative cytology. Commonwealth Fund by Harvard University Press, Cambridge, Mass

Papanicolaou GN, Trant HF (1943) Diagnosis of uterine cancer by the vaginal smear. Commonwealth Fund, New York

Petterson S, Hansson G, Blohme I (1976) Condyloma acuminatum of the bladder. Am J Urol 115: 535-536

Porreco R, Penn I, Droegemueller W, Greer B, Makowski E (1975) Gynecologic malignancies in immunosuppressed organ homograft recipients. Obstet Gynecol 45: 359-364

Pridan H, Lillienfeld AM (1971) Carcinoma of the cervix in Jewish women in Israel (1960-67). An epidemiological study. Is J Med Sci 7: 1465-1470

Purola E, Savia E (1977) Cytology of gynecologic condyloma acuminatum. Acta Cytol 21: 26-31

Raferty A, Payne WE (1954) Condyloma (sic) acuminata of the cervix. Obstet Gynecol 4: 581-584

Reid R (1983) Genital warts and cervical cancer. II. Is human papillomaviral infection the trigger to cervical carcinogenesis? Gynecol Oncol 15: 238-252

Reid R, Stanhope CR, Herschman BR, Booth E, Phibbs GD, Smith JP (1982) Genital warts and cervical cancer. I. Evidence of an association between subclinical papillomaviral infection and cervical malignancy. Cancer 50: 377-387

Reid R, Crum C, Herschman BR et al. (1984a) Genital warts and cervical cancer. III. Subclinical papillomavirus infection and cervical neoplasia are linked by a spectrum of continuous morphologic and biologic change. Cancer 53: 943-953

Reid R, Fu YS, Herschman BR, Crum CP, Braun L, Shah KF, Agronow SJ, Stanhope CR (1984b) Genital warts and cervical cancer. VI. The relationship between aneuploid and polyploid cervical lesions. Am J Obstet Gynecol 150: 189-199

Richart RM (1967) The natural history of cervical intraepithelial neoplasia. Clin Obstet Gynecol 10: 748-784

Richart RM (1973) Cervical intraepithelial neoplasia. In: Somers SC (ed) Pathology Annual. Appleton-Century-Crofts, New York, pp 301-328

Richart RM, Barron BA (1969) A follow-up study of patients with cervical dysplasia. Am J Obstet Gynecol 105: 386-393

Rigoni-Stern (1842) Fatti statistici relativi alle malattie cancerose che servirono di base alle poche cose dette dal dott ... Gior Servire Progr Path Therap 2: 507-517

Rotkin ID (1981) Etiology and Epidemiology of cervical cancer. In: Dallenbach-Hellweg G (ed) Cervical cancer. Springer, Berlin Heidelberg New York

Schauenstein W (1908) Histologische Untersuchungen über atypisches Plattenepithel an der Portio und an der Innenfläche der Cervix uteri. Arch Gynäkol 85: 576-616

Schneider A, Hotz M, Gissmann L (1986) Prevalence of genital HPV infections in pregnant women. Meeting abstract, Cold Spring Harbor Symposium on papillomaviruses, September

Schneider V, Kay S, Lee HM (1983) Immunosuppression: high risk factor for the development of condyloma acuminata and squamous neoplasia of the cervix. Acta Cytol 27: 220-224

Shevchuck MM, Richart RM (1982) DNA content of condyloma acuminatum. Cancer 49: 489-492

Shokri-Tabibzadeh S, Koss LG, Molnar J, Romney S (1981) Association of human papillomavirus with neoplastic processes in genital tract of four women with impaired immunity. Gynecol Oncol 12: S129-S140

Southern EM (1975) Detection of specific sequences among DNA fragments separated by gel electrophoresis. J Mol Biol 98: 503-517

Steinberg BM, Topp WC, Schneider PS, Abramson AL (1983) Laryngeal papillomavirus infection during clinical remission. N Engl J Med 308: 1261-1264

Sternberger LA, Hardy PH, Cuculis JJ, Meyer HC (1970) The unlabeled antibody enzyme method of immunohistochemistry: preparation and properties of soluble antigen-antibody complex (horseradish peroxidase-antihorseradishperoxidase) and its use in identification of spirochetes. J Histochem Cytochem 18: 315-333

Sun TT, Eichner R, Schermer A, Cooper D, Nelson WG, Weiss RA (1984) Classification, expression, and possible mechanisms of evolution of mammalian epithelial keratins: a unifying model. In: Levine AJ, Vande Wonde GF, Topp WC, Watson JD (eds) Cancer cells. 1. The transformed phenotype. Cold Spring Harbor Laboratory, Cold Spring Harbor, New York, pp 169-176

Syrjänen K, Väyrynen M, Saarikoski S, Mantyjarvi R, Parkkinen S, Hippelainen M, Castren O (1985a) Natural history of cervical human papillomavirus (HPV) infections based on prospective follow-up. Br J Obstet Gynaecol 92: 1086-1092

Syrjänen K, Parkkinen S, Mantyjarvi R, Väyrynen M, Syrjänen S, Holopainen H, Saarikoski S, Castren O (1985b) Human papillomavirus (HPV) type as as important determinant of the natural history of HPV infections in uterine cervix. Eur J Epidemiol 1: 180-187

Syrjänen KJ (1979) Morphologic survey of the condylomatous lesions in dysplastic and neoplastic epithelium of the uterine cervix. Arch Gynecol 227: 153-161

Syrjänen KJ, Pyrhönen S (1982) Immunoperoxidase demonstration of human papillomavirus (HPV) in dysplastic lesions of the uterine cervix. Arch Gynecol 233: 53-61

Syrjänen KJ, Heinonen UM, Kauraniemi T (1981) Cytologic evidence of the association of condylomatous lesions with dysplastic and neoplastic changes in the uterine cervix. Acta Cytol 25: 17-22

Taichman LB, Breitburd F, Croissant O, Orth G (1984) The search for a culture system for papillomavirus. J Invest Dermatol 83: 2s-6s

Vallejos H, Del Mistro A, Kleinhaus S, Braunstein JD, Halwer M, Koss LG (1987) Characterization of human papillomavirus types in condylomata acuminata in children by in situ hybridization Lab Invest 56: 611-615

Wagner D, Ikenberg H, Boehm N, Gissmann L (1984) Identification of human papillomavirus in cervical swabs by deoxyribonucleic acid in situ hybridization. Obstet Gynecol 64: 767-772

Wickenden C, Malcolm ADB, Steele A, Coleman DV (1985) Screening for wart virus infection in normal and abnormal cervices by DNA hybridization of cervical scrapes. Lancet 1: 65-67

Winkler B, Crum CP, Fujii T, Ferenczy A, Boon M, Braun L, Lancaster WD, Richart RM (1984) Koilocytotic lesions of the cervix. The relationship of mitotic abnormalities to the presence of papillomavirus antigens and nuclear DNA content. Cancer 53: 1081-1087

Yee C, Krishnan-Hewlett I, Baker CC, Schlegel R, Howley P (1985) Presence and expression of human papillomavirus sequences in human cervical carcinoma cell lines. Am J Pathol 119: 361-366

zur Hausen H (1977) Human papillomaviruses and their possible role in squamous cell carcinomas. Curr Top Microbiol Immunol 78: 1-30

zur Hausen H (1982) Human genital cancer: synergism between two virus infections or synergism between a virus infection and initiating events? Lancet 2: 1370-1372

Colposcopic Appearances of Human Papillomavirus of the Uterine Cervix

E. C. Pixley

1 Introduction

Since it was first recognised that human papillonaviruses (HPV) caused lesions of the cervix other than the classical condyloma, the colposcope has proved an essential aid in their recognition. The instrument enables clinicians to identify sites infected by the virus and to provide accurately selected tissue for laboratory studies (Fig. 1). Were the method unavailable, it is doubtful whether our current understanding of the HPV disorders of the genital tract could have been attained.

When introduced as a clinical method, colposcopy was used to examine systematically the epithelia of the uterine cervix with a view to identifying cervical cancer in its earlier stages. Cervical intraepithelial neoplasia (CIN), the precursor of cervical cancer, was shown to have a distinctive appearance, and the method became essential in the evaluation of individuals whose vaginal smears contained abnormal cells.

In addition to its use in practice, the instrument provided new perspectives on morphology of cervical lesions and to new concepts of order and disorder in the small epithelial field of the cervical surface. A terminology evolved which proved reliable in morphological descriptions of static and changing states (see Figs. 2-5). When the female genital tract was recognised as a common target of HPV, this terminology easily accommodated descriptions of the resultant lesions.

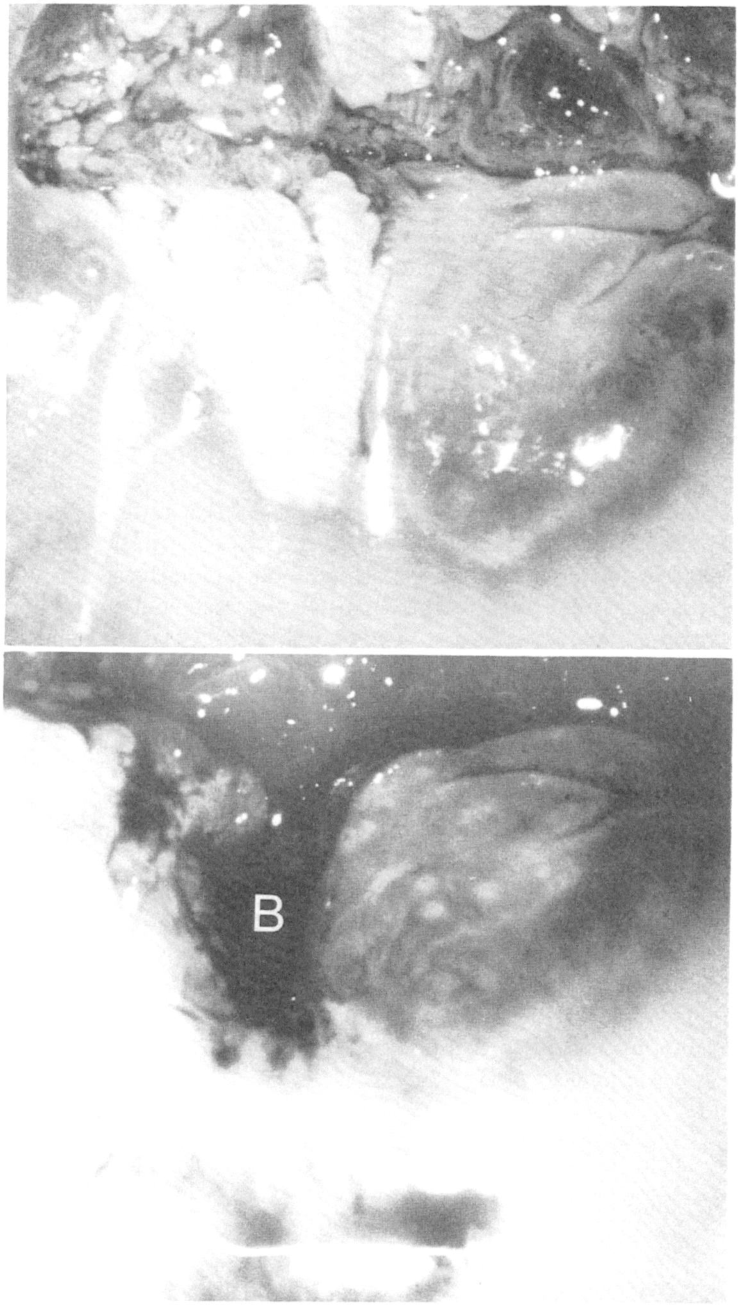

Fig. 1 A, B. The colposcope allows delineation of areas of epithelium for accurate targetting of biopsy **(A)**; a portion of the atypical transformation zone (ATZ) is chosen. The site of the biopsy **(B)** is shown, the procedure having produced little or no discomfort

This chapter uses the nomenclature and concepts fully developed in Coppleson, Pixley, and Reid (1971, 1978,1986).

2 Basic Colposcopic Morphology

The colposcope provides magnification and high intensity illumination for direct examination of the cervix through a vaginal speculum. The vulva and vagina can also be viewed.

In the genital tract the epithelia are somewhat transparent, so that tissue structures subjacent to the epithelia are available for examination. The character of normal epithelial forms and states can be instantly appreciated, and the role of subsequent structural modifications can be uniquely interpreted in the living subject. Longitudinal studies in individuals are possible, and results documented in high resolution photographs (see figures).

The lower genital tract epithelia and related structures are best described using three fundamental morphological features - colour, surface configuration and vascular structure - which from the recognisable architectural components of the visible tissues. The areas occupied by specific epithelial forms and disorders are defined by the fourth feature - topography. Three specific epithelia are proper to the lower genital tract:

1. Original squamous epithelium (OSE) extends cephalad from the vulva, convering a portion of the vaginal introitus, the whole vagina and a variable portion of the ectocervix (Fig. 2).
2. Columnar epithelium (CE) extends cephalad from its site of junction with the original squamous variety (the original squamocolumnar junction, OSCJ) (Fig. 2). It often covers portion of the ectocervix and forms an essential functional component of the cervix in providing mucus secretion (Fig. 2).
3. Metaplastic epithelium, the central focus of the colposcopist, evolves in the caudal extent of the columnar epithelium. It is usually present in individuals of all ages and extends cephalad from the OSCJ to a second, new squamocolumnar junction (NSCJ). The plate of metaplastic epithelium and the surviving columnar epithelial scaffold of its origin generate the colposcopic appearances of the *transformation zone* (Fig. 2).

The common, normal appearances, exhibiting a great variety of features, constitute the *typical transformation zone* (TTZ). (Fig. 2).

Transformation Zone. The cellular activity of squamous metaplasia is of signal significance. In the cervix in most individuals, the new cell population emerges from the original columnar cells and perhaps other multipotential cells to constitute a distinctive new surface, which is destined only for senescence. In some individuals the new population emerges with a potential for neoplastic behaviour and exhibits the distinctive morphology embodied in the term *atypical transformation zone* (ATZ) (Fig. 3). The transformation zone is the preferred target of HPV, bringing into coincident focus the interests of clinicians and laboratory workers who are interested in oncogenic mechanisms.

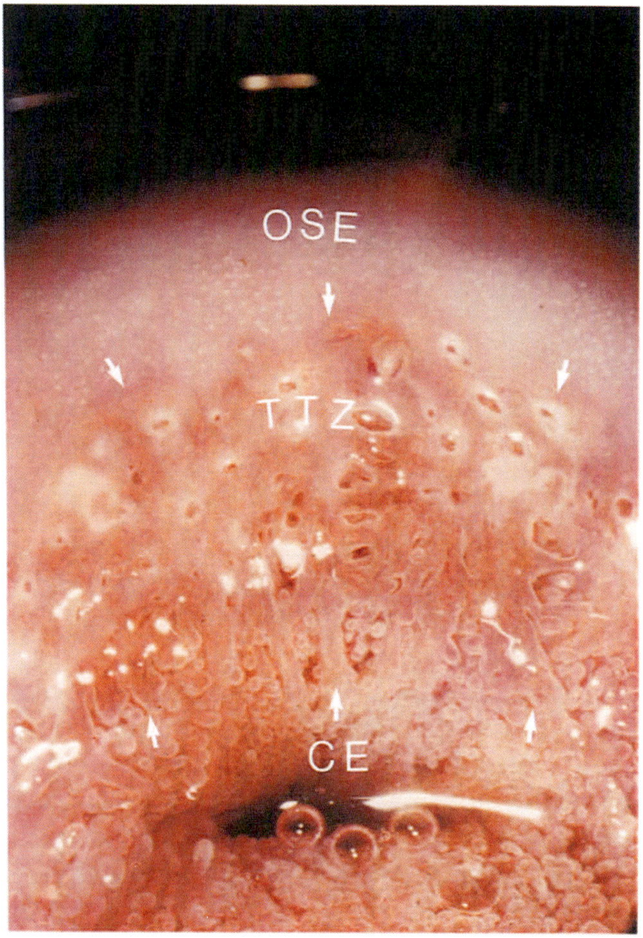

Fig. 2. Basic colposcopic morphology. This photograph and the others which follow are enlargements of 1:1 macrophotographs taken with a Leisegang stereo camera-colposcope. This study illustrates the basic morphology of the three essential epithelia of the cervix and vagina. The appearances are revealed after careful cleansing with a dilute (3%) solution of acetic acid, which removes mucus to reveal the prime morphological features of colour, surface sonfiguration, vascular structure and topography of the epithelia and subjacent tissues. Original squamous epithelium *(OSE)* is a smooth, pink, featureless epithelium, within which intraepithelial capillary loops and networks may be seen. Histological examination during reproductive life reveals a highly differentiated, stratified, squamous epithelium. Columnar epithelium *(CE)* is characteristically seen as villiform and papillary projections, with clefts extending into the cervical stroma. A single layer of mucus-secreting epithelium covers the stromal tissue. Transformation zone *(TZ)* between the original squamous columnar junction (OSCJ) *(downward arrows)* and the new squamous columnar junction (NSCJ) (upward arrows) has an appearance generated by the metaplastic epithelium. Characteristics of colour and surface configuration are distinctive, and remnants of columnar epithelium persist as gland openings and inclusion cysts. Histological findings are as variable as are the colposcopic findings in this zone. Usually multilayered immature and mature squamous metaplasia is observed in association with the original glandular remnants

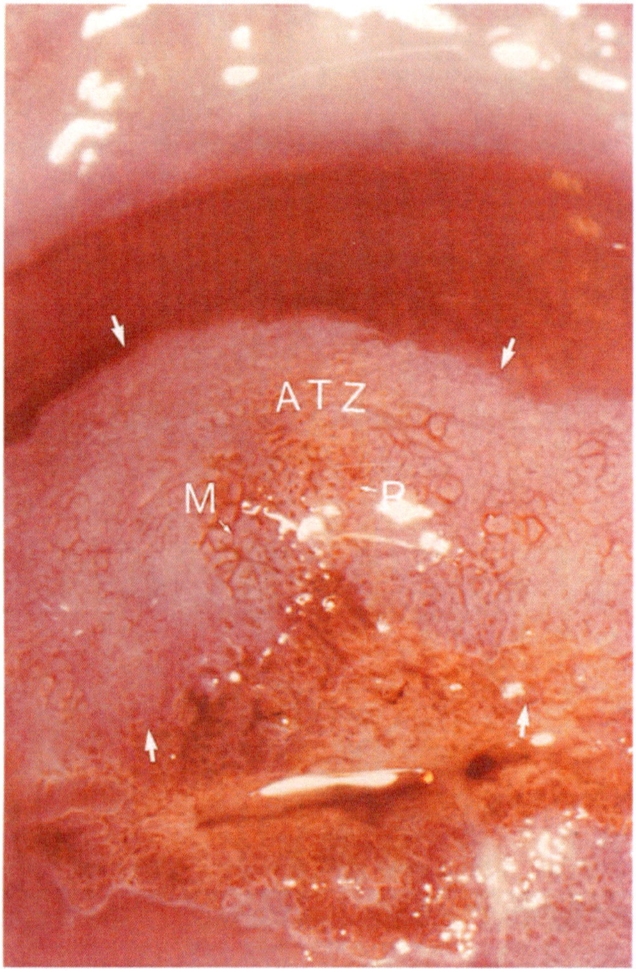

Fig. 3. Application of acetic acid has resulted in whiteness developing in the transformation zone. This change is transient, appears within a minute or so, and fades within minutes. The characteristics of the atypical transformation zone *(ATZ)* are shown, *Acetowhite epithelium:* The response to acetic acid application is apparent in CIN, several normal forms of epithelium, and in many forms of HPV infection. *Mosaic (M):* Horizontal terminal vascular structures are frequently arranged in regular mosaic forms, giving an easily recognised pattern to the white background. *Punctation (P):* Vertically disposed capillary loops are seen as points within the background. Acetowhiteness in the common characteristics of ATZ. The vascular features may be absent, present singly or in combination. The brilliant white points and patterns are reflections of the strobe light used in exposures

In addition to CIN, a number of normal cellular states and epithelial types generate features which can be classified as ATZ. The colposcopic appearances of the cervical epithelia, of the transformation zone (typical and atypical) and basic morphological terms are provided in Figs. 2-5).

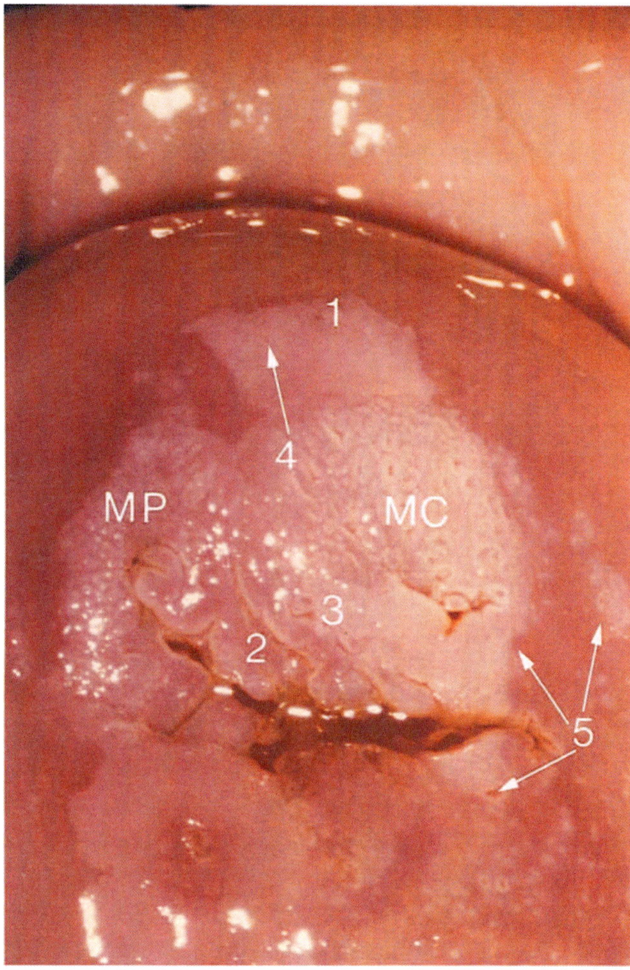

Fig. 4. Appreciation of surface characteristics allows useful predictions of the nature of disorders to be made. Smooth, featureless atypical transformation zone (ATZ) *(1)* is generally characteristic of HPV infections of the original squamous epithelium (OSE). Physiological contours of the original surface structures usually persist when transformation zone (TZ) or ATZ develops *(2)*. Specific surface features are evoked by certain disorders, including neoplasia (Figs. 17 and 18) and in this instance HPV infection, which exhibits micropapillary *(MP)* and microconvoluted *(MC)* structures. The site, extent and confluence of epithelial states constitute the topographic characteristics. Lesions may be present in the TZ *(3)*, extend from the TZ into the OSE *(4)*, and be multifocal *(5)*

3 Human Papillomavirus Disorders

During the development of current descriptions of HPV disorders, histological methods were initially the standard means of diagnosis. Transmission electron microscopy and later recombinant DNA technology have been used to mark the presence of virus in cell populations. Of the two methods, the latter may well be

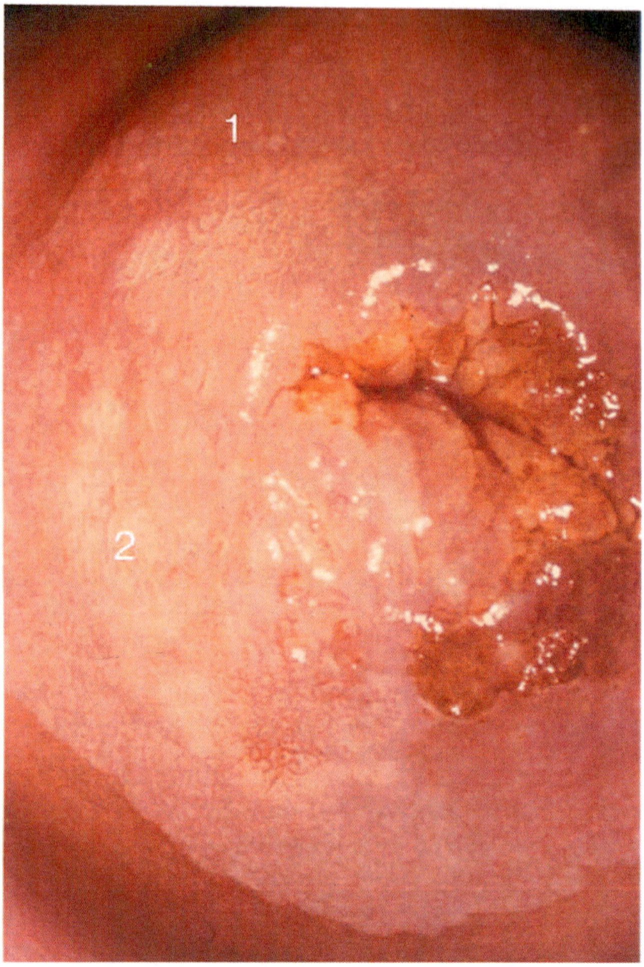

Fig. 5. Grading of the atypical transformation zone (ATZ). Variations in the prime morphological features of the ATZ create differences in appearance which allow delineation of normal and abnormal forms of epithelia. Within the abnormal group the variations reflect differences in the significance of the lesions and in expected histological findings, allowing a system of grading to be developed. In this example two distinct forms of ATZ are present. The original squamous columnar junction circumscribes an extensive area of acetowhite epithelium, the outermost portion *(1)* being less strikingly white than the inner *(2)*. The lesser grade, exhibiting a faint mosaic, is the original transformation zone, which is metaplastic epithelium that has persisted since the intrauterine period. Whiter and with a more pronounced mosaic and distinctive micropapillary surface formation is an area of HPV infection. The grading systems allow useful but not specific or sensitive correlations with histological findings. The existence of several forms of ATZ of physiological significance bearing some resemblance to CIN and HPV infections makes for difficulties of interpretation; these have to be resolved by the more sensitive methods of molecular biology

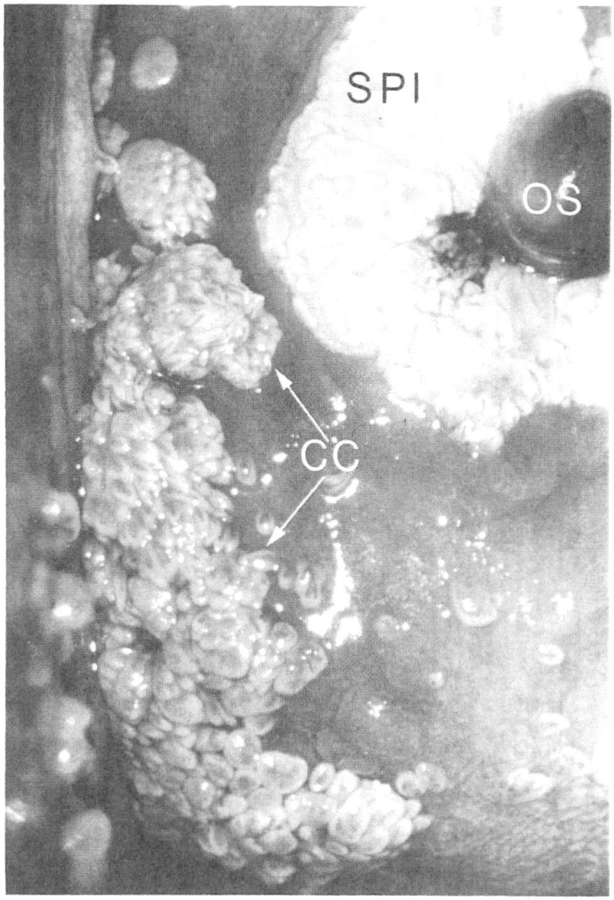

Fig. 6. A portion of the cervix and adjacent vagina with condylomata evident on clinical inspection *(CC)* shows the coexistence of clinical and subclinical infections. The most minute changes are not apparent. The abnormalities of the area surrounding the external *os* of the cervix became evident only after acetic acid cleansing and colposcopy. It is designated as a subclinical papillomavirus infection *(SPI)*

the most sensitive. It has been shown that virus particles, virus proteins or specific nucleotide sequences can be demonstrated in tissues which may appear normal, in obvious condylomas, in CIN, in invasive cancer and in lesions which are composites of all the above. One view suggests invasive squamous cancer of the cervix as an end point of a lengthy continuum, possibly initiated by HPV. Translation of this biological process into a morphological system is not possible. Diagnostic examinations with colposcopy and microscopy of removed tissue provide material which characterise a single instant in the continuum, and this, the element of timing, introduces the first of the difficulties in providing a comprehensive dynamic of the disorders. The second relates to the target epithelium, which is the original terrain upon which the host-virous response fashions recognisable change. As the surface morphology in each individual cervix is as unique as a fingerprint, virus-

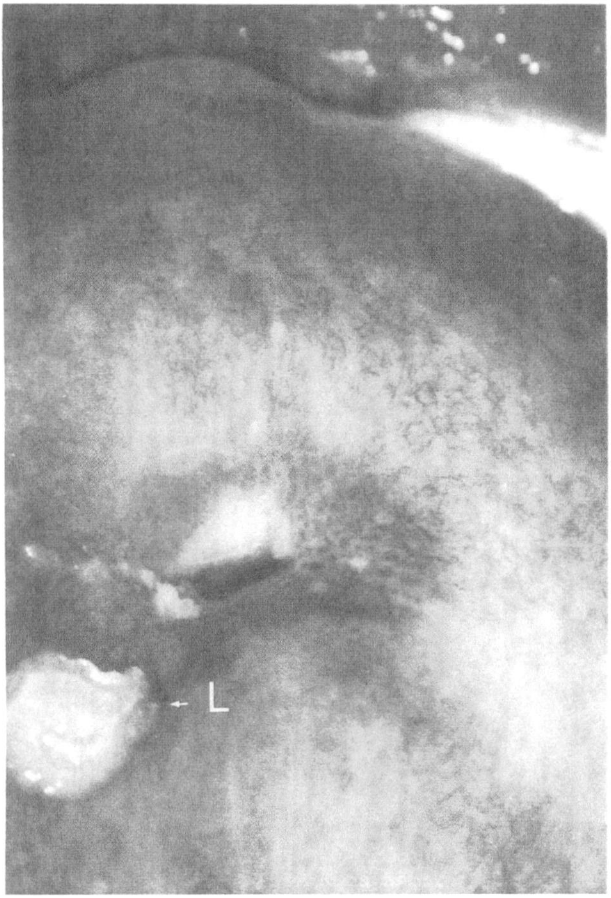

Fig. 7. Leukoplakia *(L)*. When SPI is suspected from cytological screening reports, a white patch is often evident before acetic acid application. Here, other morphological features are not obvious, apart from the capillary network structures in the original squamous epithelium

induced changes are extremely variable. A third difficulty is introduced by the fact that several types of HPV may act as infecting agents. It is possible that certain colposcopic appearances will be shown to result from involvement with specific virus types. Perhaps the most significant difficulty lies in the existence of several normal and abnormal states which appear similar. This, acting in association with the other variables mentioned above provides for an array of colposcopic images that are difficult to describe systematically.

As the host-viral reaction develops to the point of recognition, morphological expression is fashioned by fibroepithelial proliferation and angiogenesis. When these changes are recognisable by unaided vision, they constitute the *clinical condyloma*. Lesions which stop short of such development but which are demonstrable with colposcopy form the group described as *subclinical papillomavirus infection* (SPI).

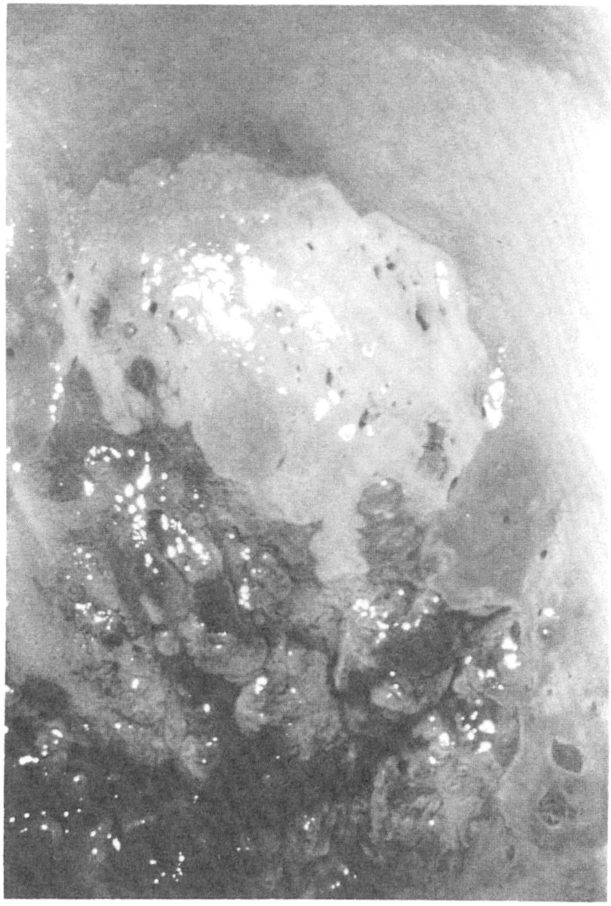

Fig. 8. Acetowhite epithelium. A considerable portion of the anterior transformation zone is atypical, becoming white after acetic acid application. The usual features, gland openings and islets of columar epithelium are present. In most areas no vascular structures are discernible

Both clinical and subclinical lesions may coexist in the individual (Fig. 6), and changes found on any one occasion may recede and return to normal or present new features after an interval (Figs. 35 and 36). The relative prevalence of clinical and subclinical forms depends to some extent upon the clinical setting. Clinical condylomata are more common in clinics for sexually transmitted diseases, whilst subclinical varieties are seen more often during examination of individuals with cytological abnormalities.

4 Prime Colposcopic Features of Subclinical Papillomavirus Infection

Colour. Proliferative changes in epithelial cell populations are usually expressed as a colour variation of the translucent red hues of normal tissues.

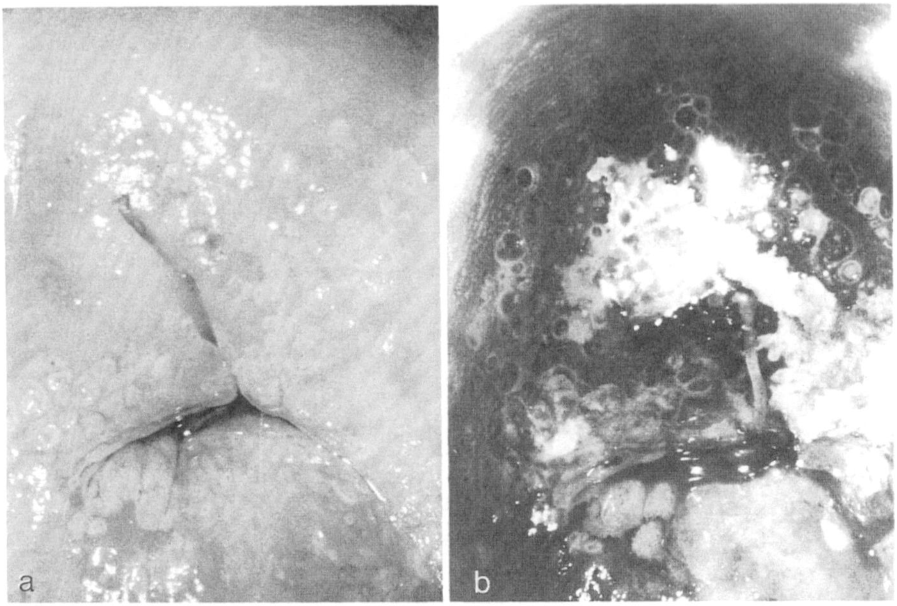

Fig. 9 a, b. Iodine response. Acetowhite epithelium displaying circular patterns is evident in **a, b.** Subsequent application of aqueous iodine solution results in deep staining of much of the cervical surface. Most areas previously acetowhite fail to stain, and circular non-staining patterns are also present in the original squamous epithelium

Leukoplakia. Careful colposcopic examination performed after minimal cleansing of the cervical surface often reveals a white patch (Fig. 7). Such a feature is suggestive but not pathognomic of subclinical papillomavirus infection (SPI).

Acetowhite Epithelium. The characteristic whiteness of the atypical transformation zone (ATZ), developing after careful cleansing of the cervix with diluted acetic acid may be slow to appear in SPI. There is some correlation between the degree of whiteness and histological findings, an appreciation of which allows some form of grading of the level of abnormality (see p. 270 and Fig. 5). Acetowhite epithelium is the commonest presentation of SPI and forms the essential basis of its recognition (Fig. 8).

Iodine Response. It has been known for decades that fully differentiated squamous epithelium of the cervix stains deeply when iodine solutions are applied. This response is due to the presence of glycogen, which is absent in glandular epithelium, immature squamous epithelium, and in most precancerous states and cancers. SPI usually fails to stain and sometimes exhibits stippling and circular patterns which are markers of the viral presence (Fig. 9). Again, variations in the degree of the reaction are said to be useful in grading the abnormalities (see p. 270).

Surface Configuration. A surprising development in this past decade has been the identification of non-condylomatous forms of HPV infections, and the term flat

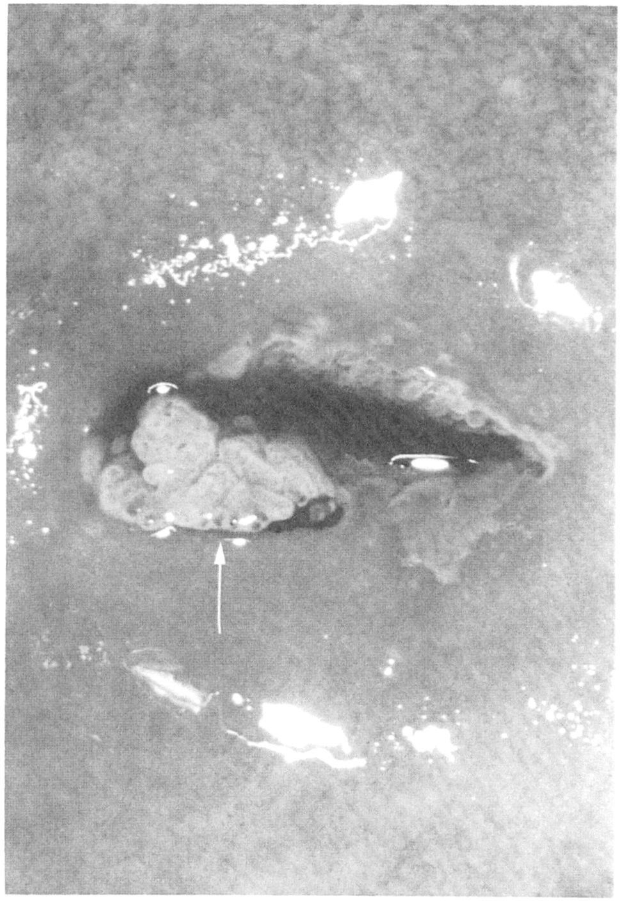

Fig. 10. Covert condyloma. In some instances during a search for the site of origin of atypical cells, a small condyloma, invisible to the unaided eye, is instantly seen through the colposcope *(arrow)*

condyloma has been used to describe them. However, a considerable variety of surface features is encountered, which are not necessarily evident when biopsy material is studied histologically. Under colposcopic examination the most obvious feature is the covert condyloma disclosed by the superior illumination and magnification gained (Fig. 10).

Micropapillary Structures. In many instances, acetowhite epithelium exhibits minute papillary projections, each containing a stromal core. These reflect fibro-epithelial proliferation and vary in degree (Figs. 11 and 12). One particular response to viral infection is seen as a diffuse change in the original squamous epithelium (OSE) of the cervix and adjacent vagina. Multiple micropapillary structures of this type form the condition sometimes termed condylomatous vaginitis (Fig. 13).

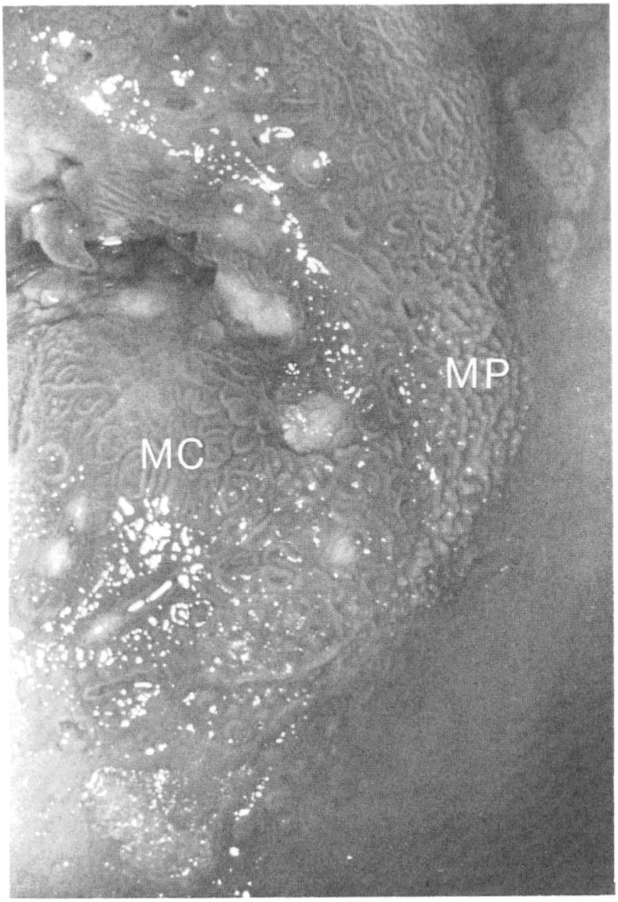

Fig. 11. Micropapillary and microconvoluted structures. In a woman suspected of harbouring CIN, the HPV-infected areas display striking characteristics. Minute surface projections *(MP)* and patterned ridges *(MC)* form the diagnostic features of subclinical papillomavirus infection

Microconvoluted Structures. Perhaps the most striking and characteristic appearance of SPI is that of corrugation and convolution resembling the surface of the brain (Fig. 14). Equally striking are the features seen when the surface ridges develop a polygonal pattern (Figs. 11 and 15). Faint circular outlines are sometimes seen in the OSE, indicating the presence of SPI (Fig. 16). When iodine solutions are applied, the circular structures become even more apparent (Fig. 9). It is

→

Fig. 12. Acetowhite epithelium, present in both the transformation zone and original squamous epithelium, exhibits a small micropapillary focus. Punctation and mosaic are also present (see also Figs. 2 and 21)

Fig. 13a, b. „Condylomatous vaginitis": a specific form of subclinical papillomavirus infection (SPI) affects the original squamous epithelium of the cervix and vagina. A diffuse change over a considerable area causes micropapillary elevations *(a)*. A similar appearance, perhaps an earlier stage, is seen with obvious SPI in *b*

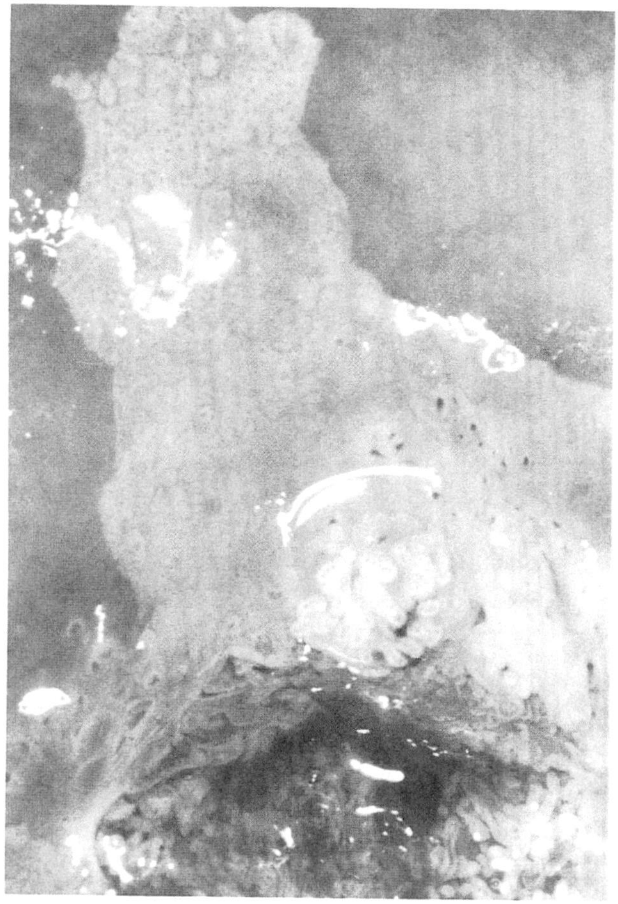

Fig. 12

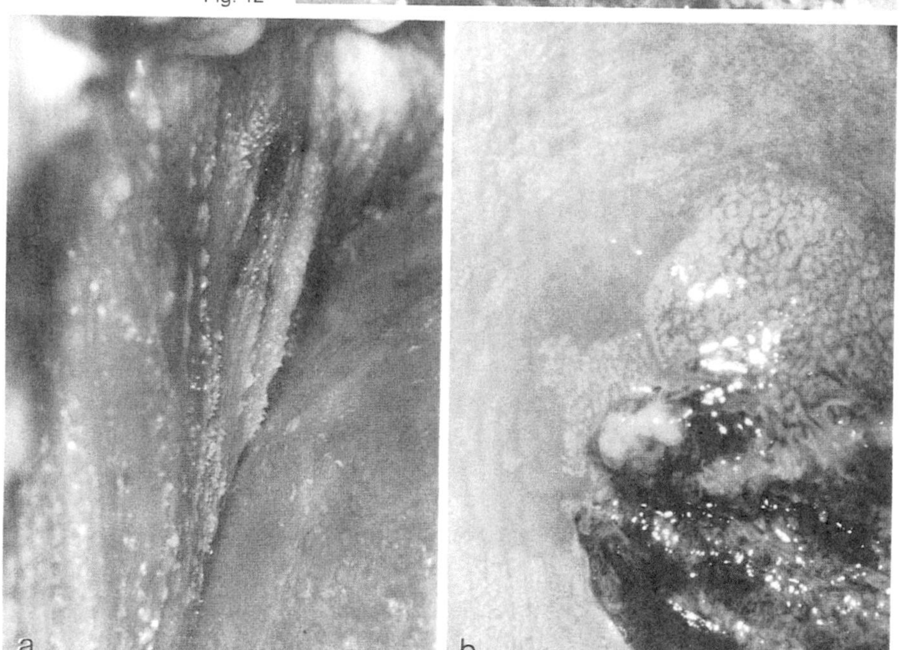

Fig. 13a, b

of the greatest importance to realise that invasive cancer can exhibit similar micropapillary surface structures (Figs. 17 and 18).

The contours and other features of the epithelium in which the response occurs constitute a base for the overall structure of the lesions. The common features of the original columnar and metaplastic elements usually persist in SPI (Fig. 19).

4.1 Angioarchitecture

Fibroepithelial proliferation usually evokes changes in the pre-existent vascular structures of the surface and may cause new vessel formation. In SPI vessels resembling those described in classical ATZ are seen, usually within the background of the acetowhite epithelium (Fig. 20). However, there are some vascular structures said to be characteristic of SPI. Vertical vessels present as small, single, intraepithelial loops, described as punctation. (Figs. 2 and 21). Horizontally disposed vessels usually form a mosaic (Figs. 3 and 21) and, in particular, circular arragements (Figs. 21 and 22). Particularly characteristic vascular arrangement of SPI is that of a modified mosaic with each component demonstrating a central vessel (Fig. 23).

4.2 Topography

The virus may affect all lower genital tract epithelia, singly or together, the transformation zone being the most common site. Portions of, or the entire zone, may be involved, and there is considerable variation in the extent of SPI lesions. In some instances only minute areas can be identified (Fig. 24), whereas in others the lesion is extensive, replacing the original epithelia (Figs. 25 and 26).

Multicentric lesions are most characteristic of HPV infections including SPI, especially when seen in OSE (Fig. 27) extending across the original squamocolumnar junction from the transformation zone to the OSE (Fig. 28). The margin of the lesions is considered to be a feature subject to variations that can be used in grading systems.

4.3 Grading

The concept of assigning significance to individual variations in the prime morphological features of ATZ has been advanced (Fig. 5). Reid, Stanhope, Herschman, et al. (1984) have advocated extension of the concept into evaluation of SPI. Assessment is made of five features including thickness, colour, surface contour, vascular structure and iodine staining. Care must be taken in the application of grading because the subjectivity of the morphological evaluation and the several independent variables that contribute to the final picture of SPI are immensely difficult for the observer to judge.

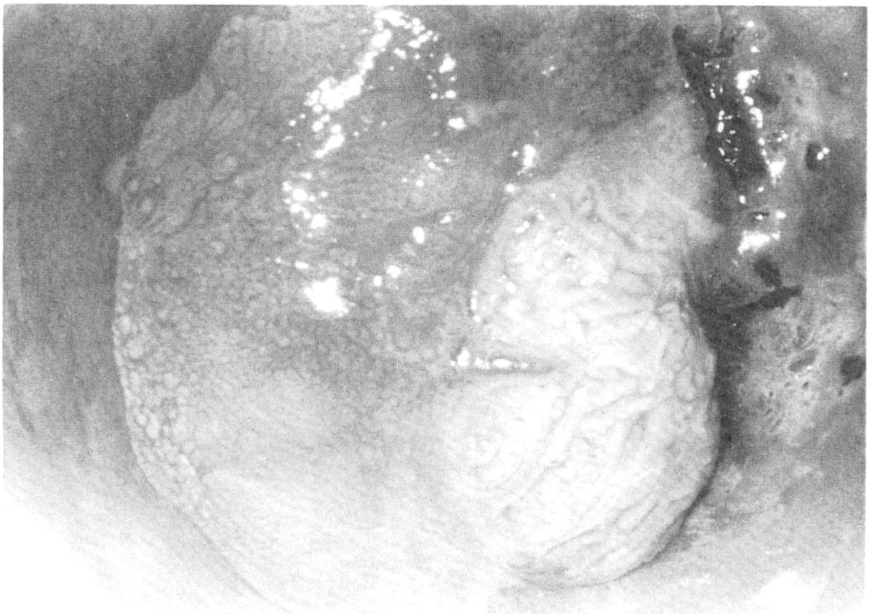

Fig. 14. Microconvoluted structure. Ridges develop in a form resembling the sulci and gyri of the brain

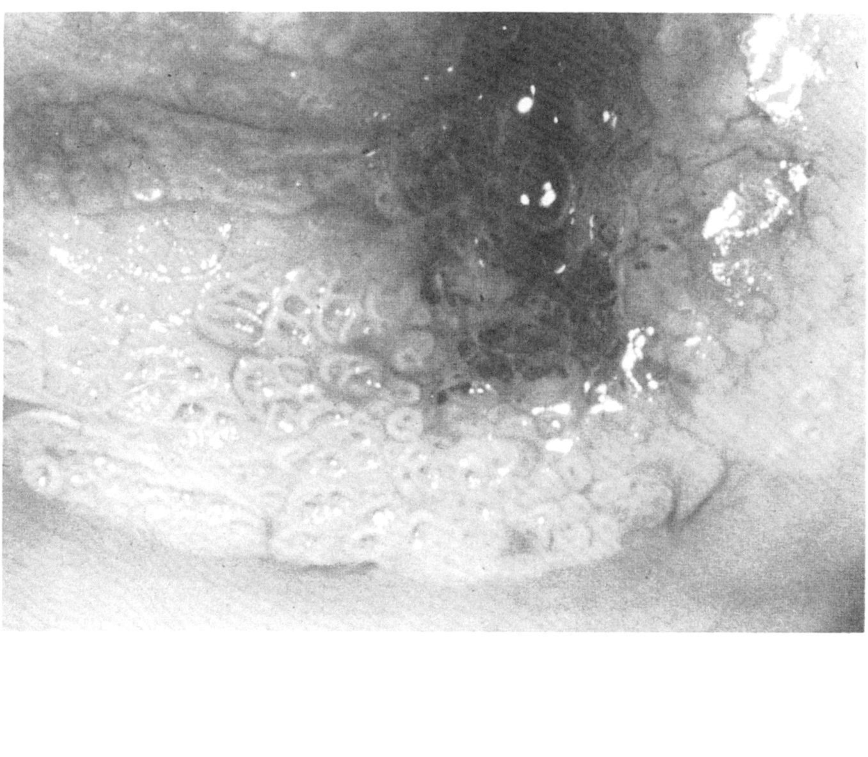

Fig. 15. Microconvolution. In a girl aged 14, the surface arragements are again strikingly characteristic, forming a mosaic. Prior to the application of acetic acid, circular vessels were apparent

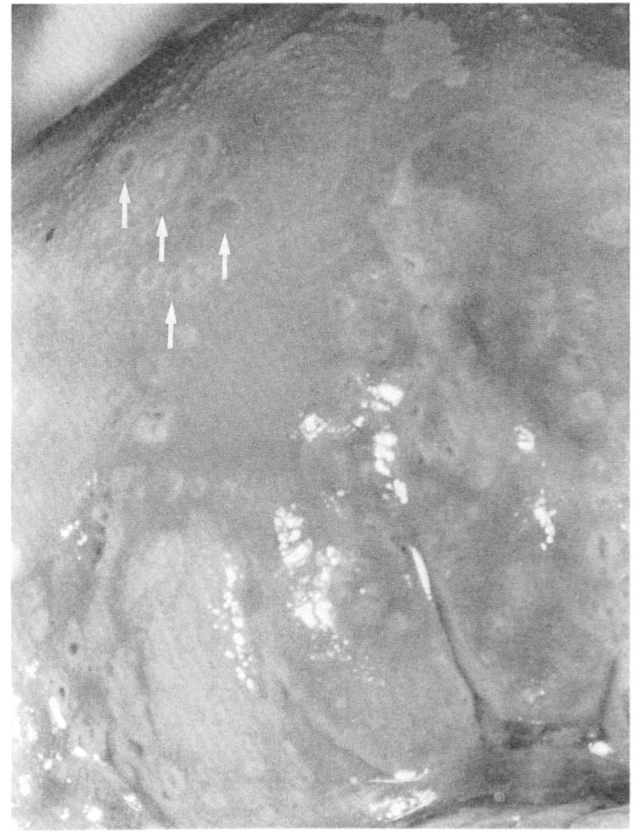

Fig. 16. Circular fields are an occasional appearance of SPI seen in the OSE. Faint circular patterns resembling the lunar surface can be identified *(arrows)*. After iodine application, the pattern is more striking (see also Fig. 9b)

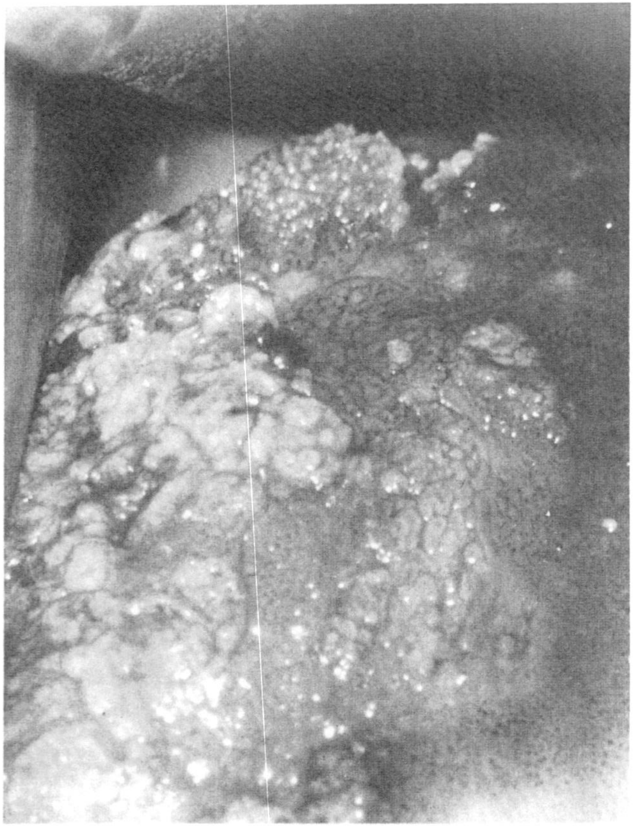

Fig. 17. Invasive cancer: micropapillary surface. In the presence of surface elevations the possibility of invasive cancer should always be suspected. This example was seen some years after conization showed a lesion with subclinical papillomavirus infection and CIN. The lesion, still stage 1 A cancer, probably evolved in a persistent area of the original lesion

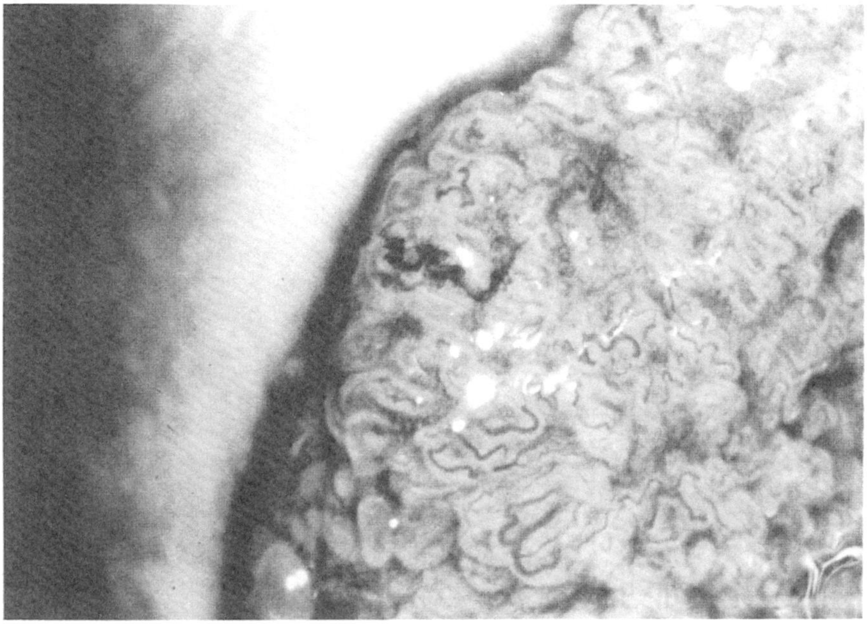

Fig. 18. Invasive cancer: micropapillary structure. Stage 1 B cancer was present, demonstrating surface features that can be confused with subclinical papillomavirus infection and some forms of clinical condyloma

Fig. 19. In some lesions the principal surface feature is that conferred by the original epithelial terrain. Glandular ridges, clefts, and gland openings persist in the area affected by subclinical papillomavirus infection

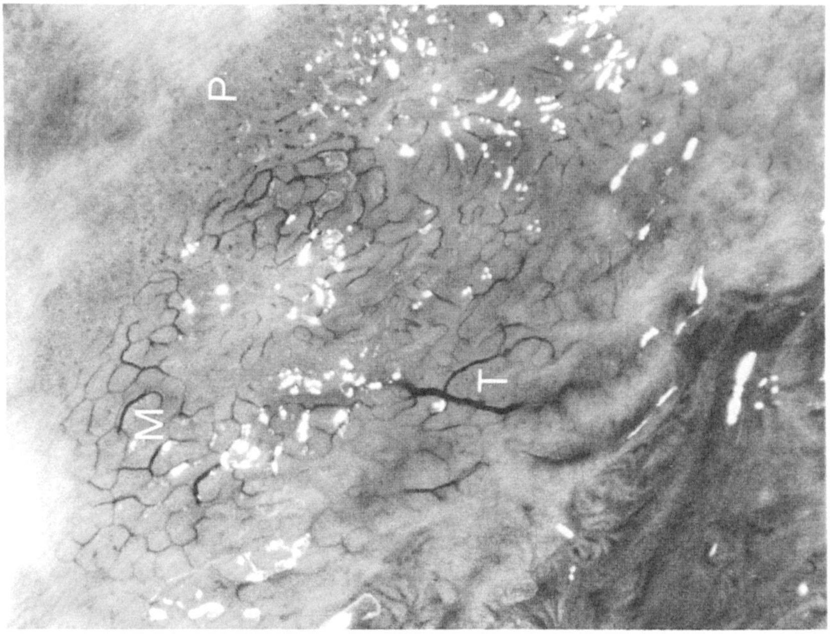

Fig. 20. Characteristic of the atypical transformation zone is the presence of specific arrangements of the terminal vascular structures. This example, seen before the application of acetic acid, shows a normal branching vessel *(T)*, mosaic *(M)*, and punctation *(P)*

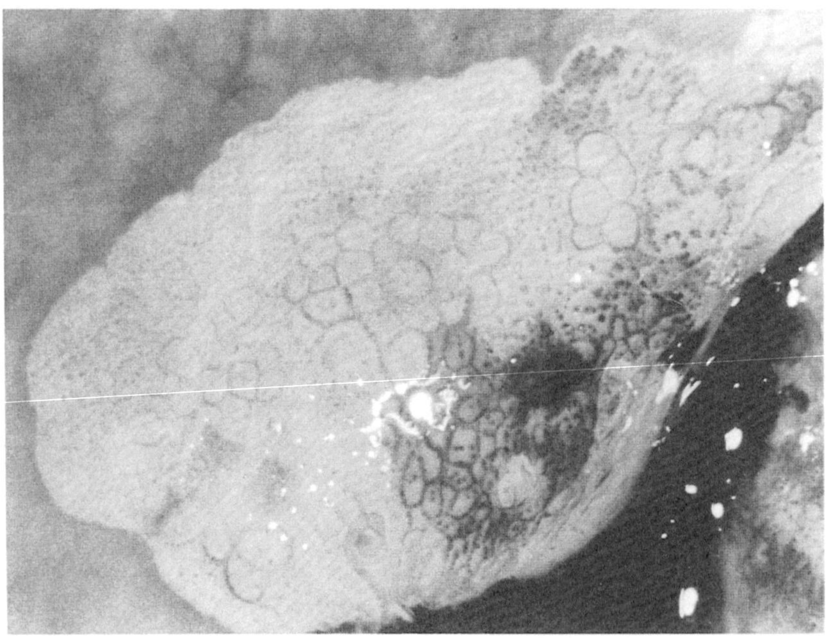

Fig. 21. A variety of vascular formations is seen within the background of aceto-white epithelium (mosaic, punctation, and circular vessels)

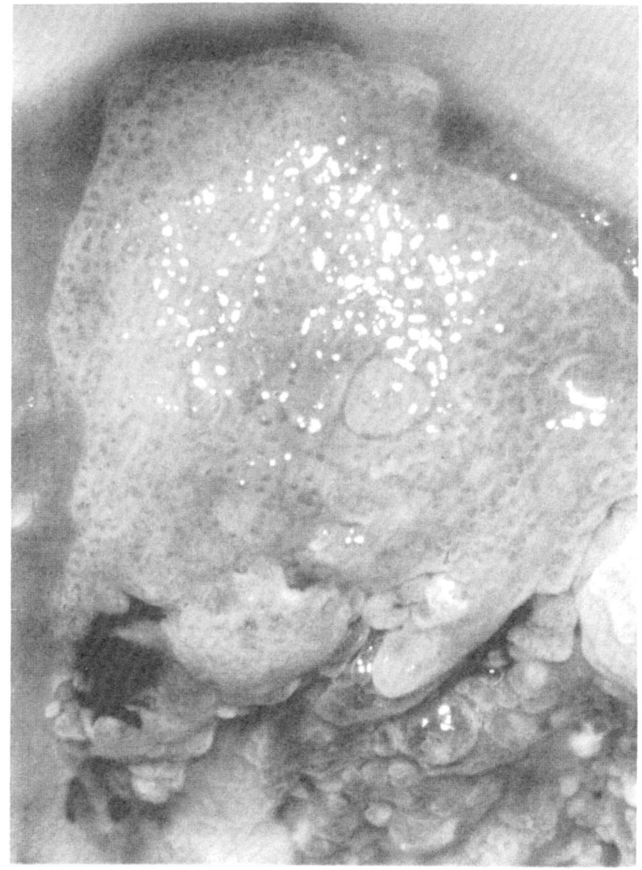

Fig. 22. Characteristic of a subclinical papillomavirus infection lesion is the large circular vessel seen here within acetowhite epithelium otherwise exhibiting punctation

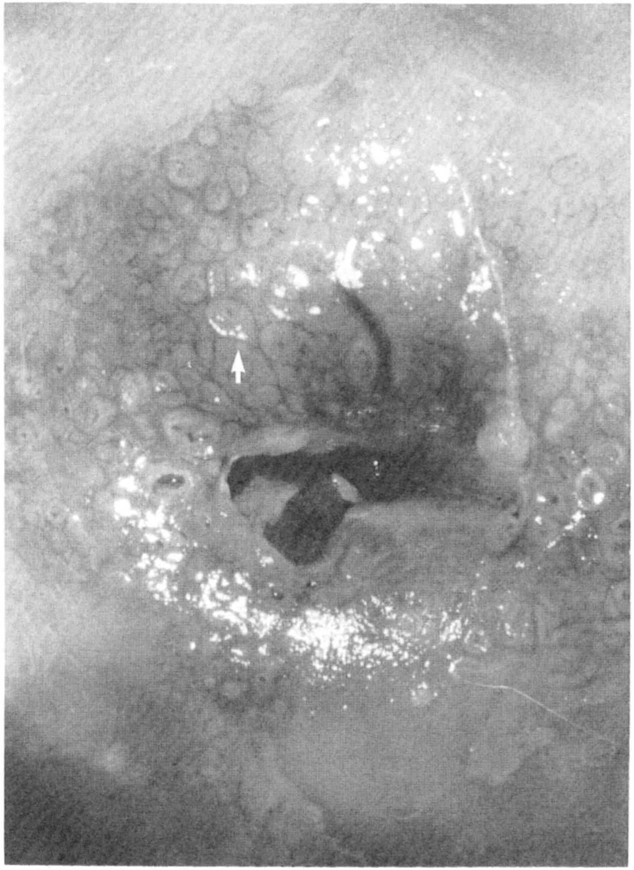

Fig. 23. Mosaic and circular structure with central vessel. This appearance is considered to be characteristic of subclinical papillomavirus infection. The familiar circular and mosaic horizontal vessels frequently enclose a single vertical capillary loop (arrow)

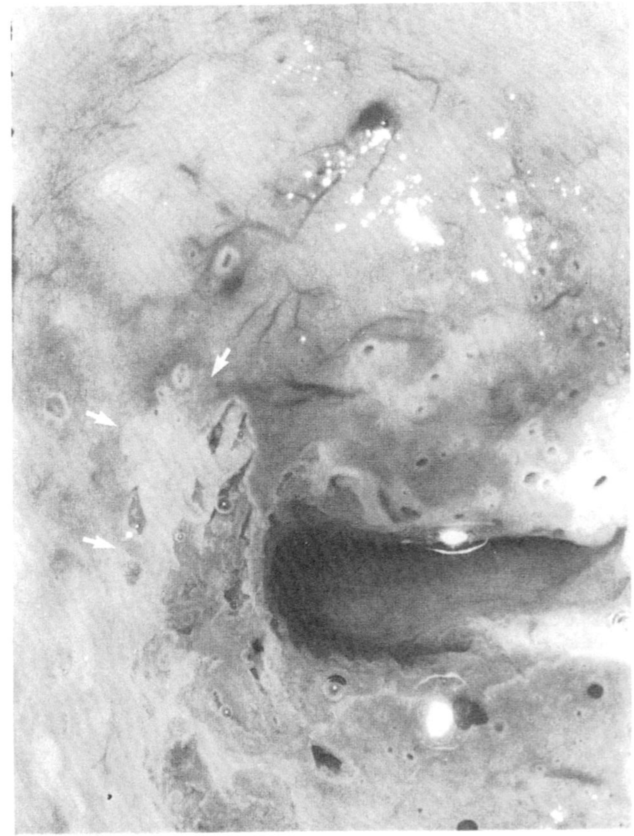

Fig. 24. Transformation zone. A variation is shown which emphasises the care needed in the search for the source of atypical cells. A small portion of the otherwise normal transformation zone is acetowhite *(arrows)*. In this area the subclinical papillomavirus infection was confirmed on histological study

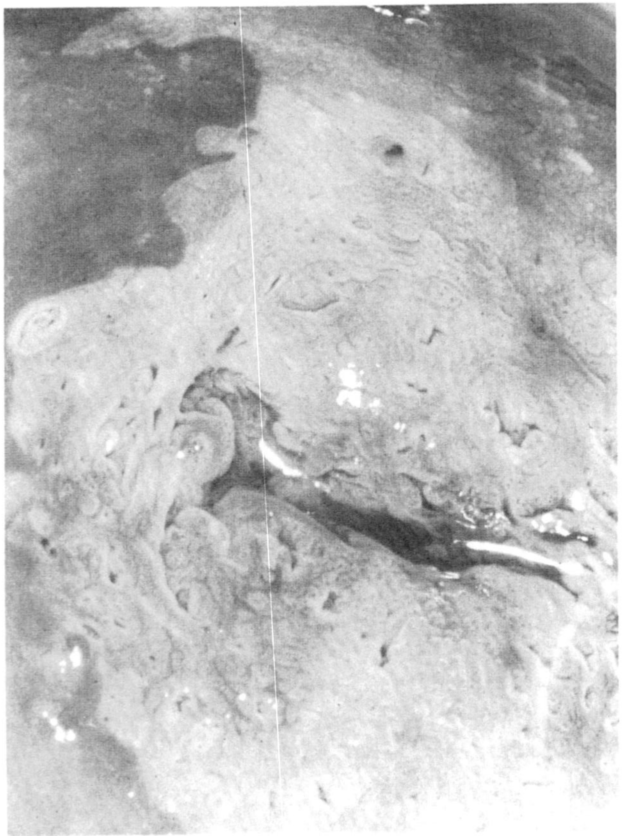

Fig. 25. Transformation zone. All of an extensive transformation zone may be involved. Here, characteristic surface features and vascular structures are evident in acetowhite epithelium. Glandular elements persist in the form of gland openings

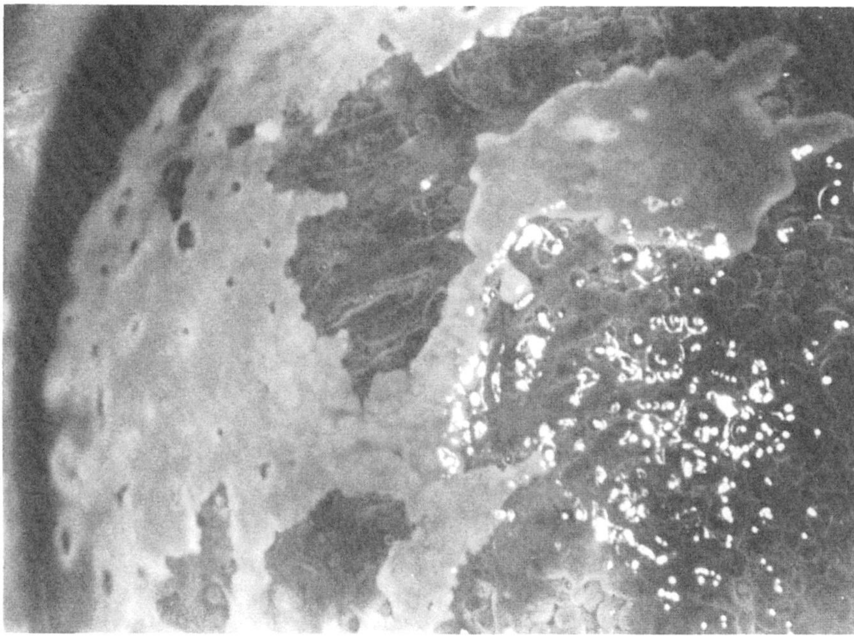

Fig. 26. Transformation zone. Many variations in the distribution of subclinical papillomavirus infection exist, such as a devious course of the new squamous columnar junction enclosing a large area of infected epithelium

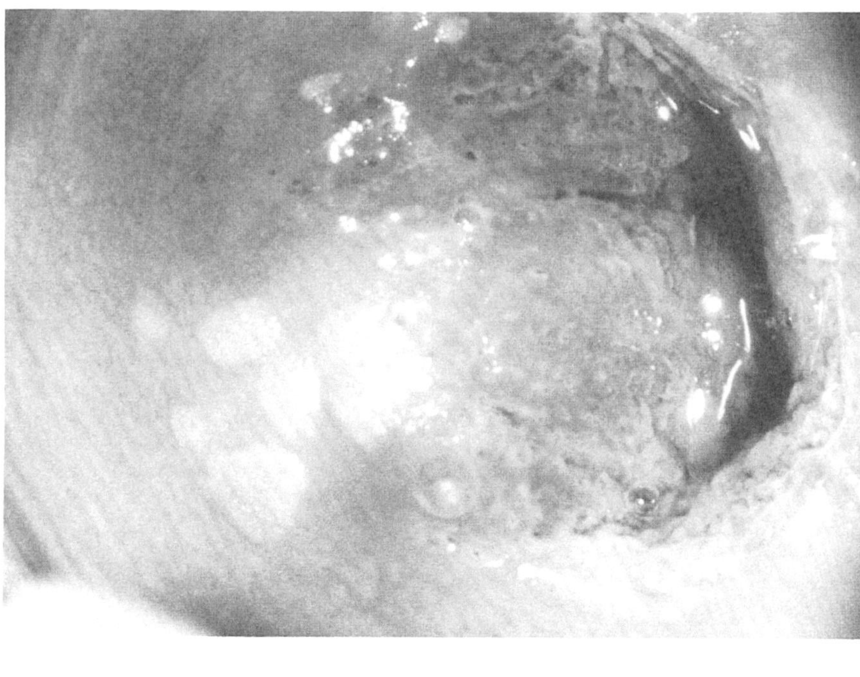

Fig. 27. Multicentric foci. Characteristic of subclinical papillomavirus infection is the presence of several areas of acetowhite epithelium in the original squamous epithelium adjacent to the transformation zone

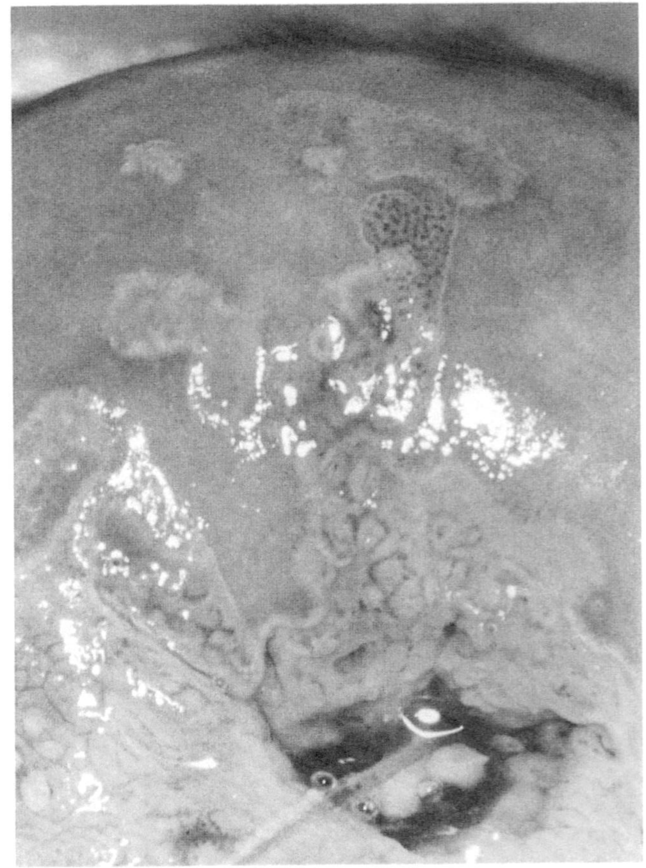

Fig. 28. Acetowhite epithelium of subclinical papillomavirus infection is frequently continuous across and beyond the original squamous columnar junction. Multicentric foci are also present

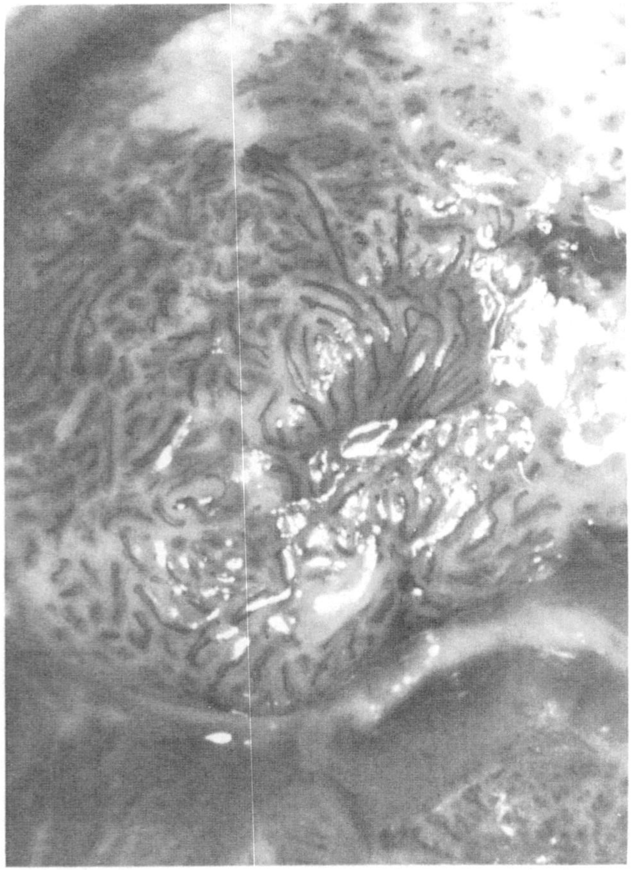

Fig. 29. The angiogenesis evoked by HPV may mimic true neoplasia. Bizarre vascular structures are evident in this exophytic lesion (clinical condyloma), which cannot be satisfactorily distinguished from those of cancer

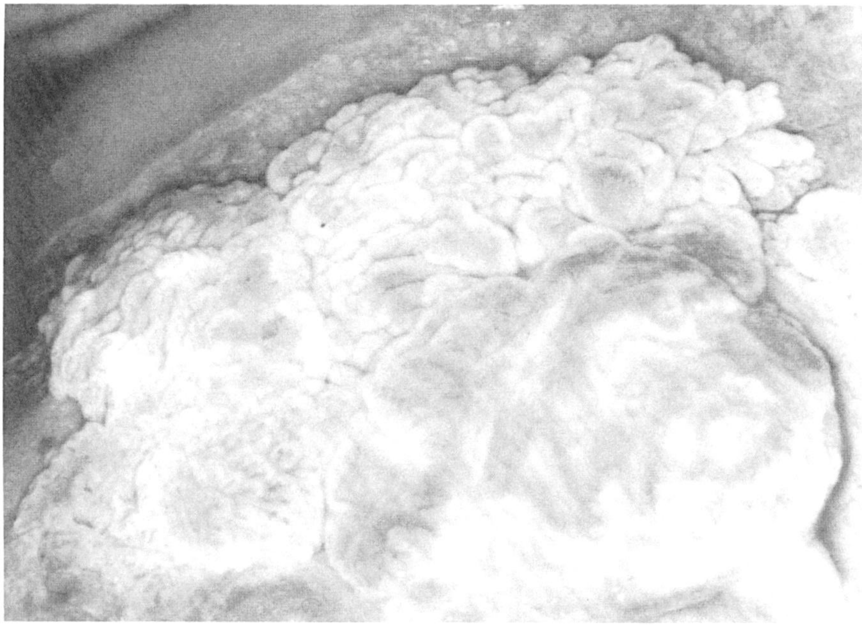

Fig. 30. Surface structure of clinical condyloma: Villiform and papillary projections of varying sizes form the outstanding characteristic. This study followed the application of acetic acid which renders the surface pale and opaque, obscuring the vascular structures

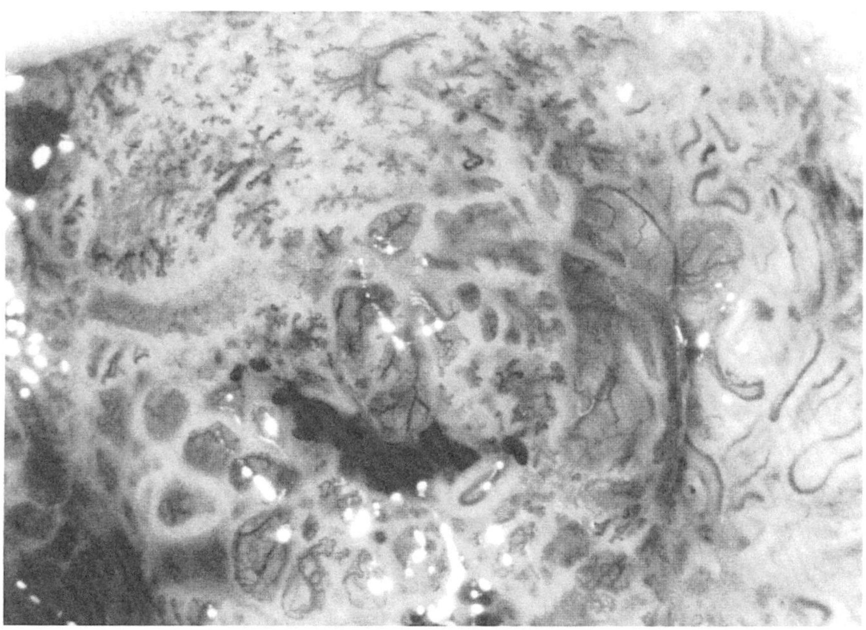

Fig. 31. Vascular structures in clinical condyloma: Within each component of the exophytic mass, single and multiple coiled capillaries are seen. Some bizarre vessels are seen on the lower portion

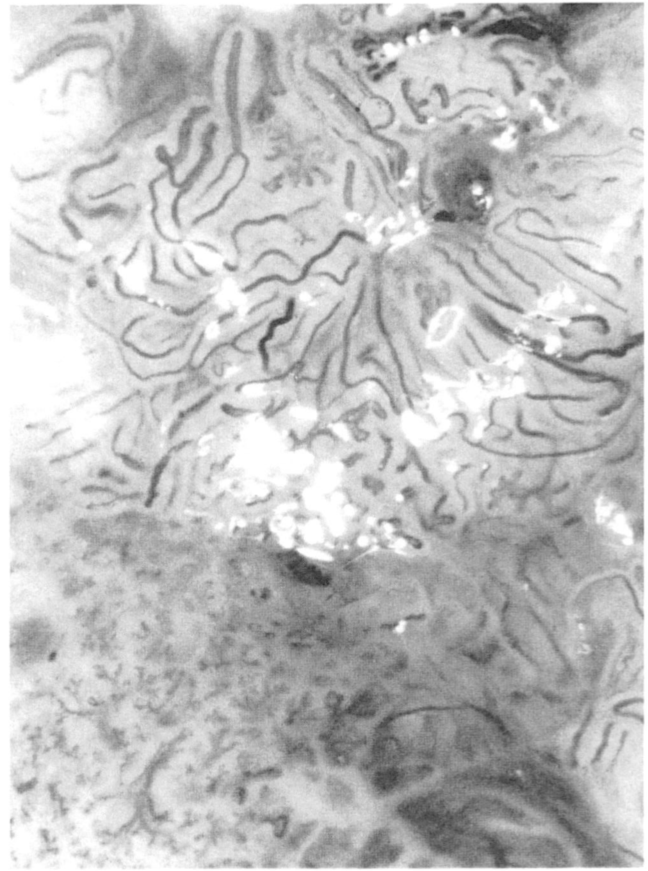

Fig. 32. Vascular structure in clinical condyloma: Large, bizarre-shaped, looped vessels are present in this condyloma. They are indistinguishable from those seen in some cancers (see also Fig. 29)

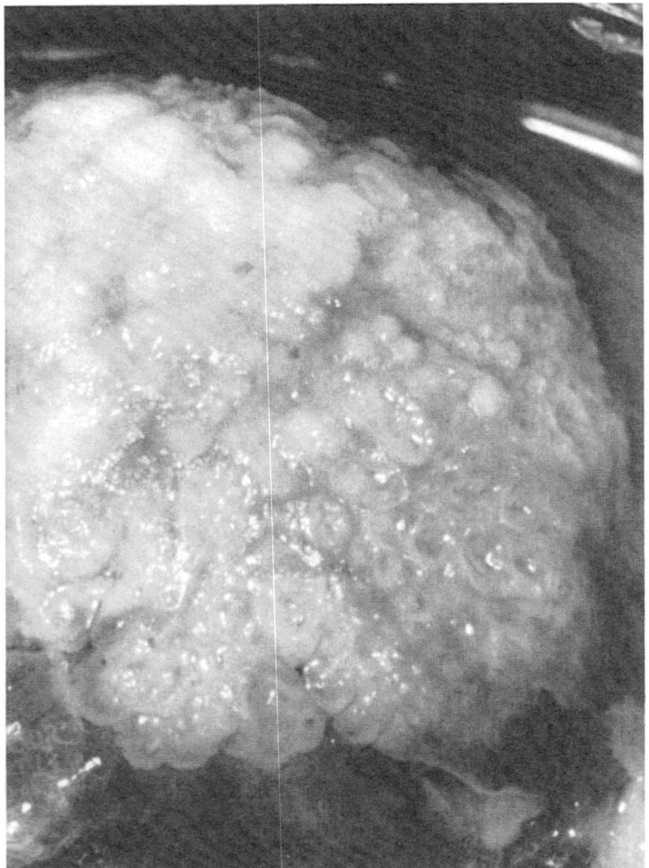

Fig. 33. Some forms of HPV infection are clinically apparent because of gross leukoplakia. Colposcopic examination reveals appearances often seen in SPI

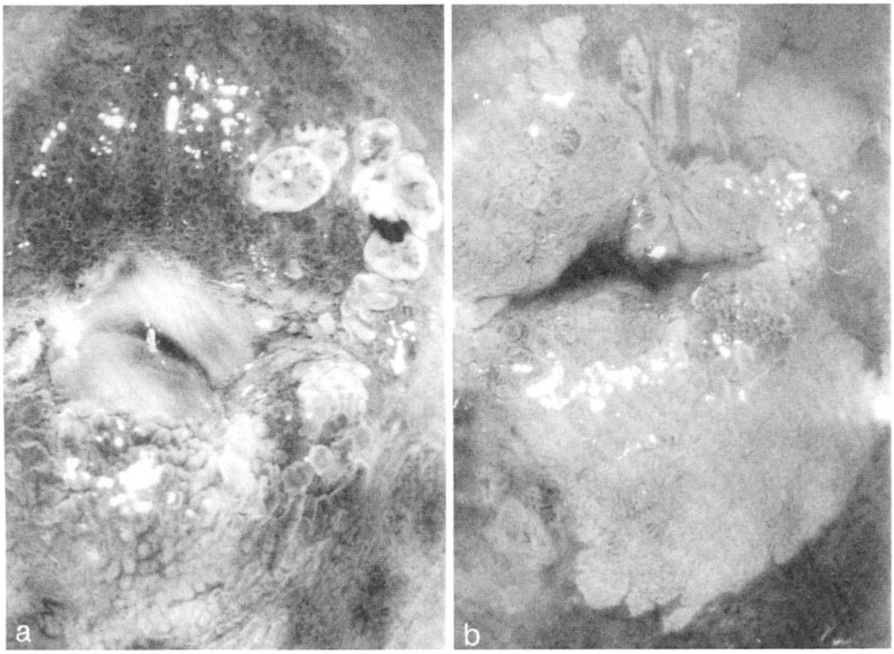

Fig. 34 a, b. Two studies have been made, the second *(b)* 3 weeks after the first. *(a)* Covert condylomata are present in the anterior portion of the columnar epithelium and early changes are evident in the posterior area. *(b)* Virtually all areas display acetowhite epithelium with punctation

4.4 Human Papillomavirus Type and Colposcopic Morphology

It might be expected that the type or mixture of types of HPV involved in the host-viral response should create a characteristic morphology, facilitating the recognition of the more significant SPI. However, it has not yet been possible to identify any colposcopic appearances which are indicative of HPV type.

5 Clinical Condyloma

Clinically evident condyloma acuminatum is frequently associated with cytological abnormality and may be the source of atypical cells. Condylomata restricted to the cervix are relatively uncommon except in sexually transmitted disease (STD) clinic settings. They may coexist with SPI (Fig. 6), and the clinical appearances sometimes give rise to the impression of malignant disease (Fig. 29), from which they must be distinguished (Fig. 18).

The morphological features vary considerably, but, characteristically, the complicated papillary or villiform surface features form an exophytic mass of varying size (Fig. 30). Within the structures many vascular patterns are present, including

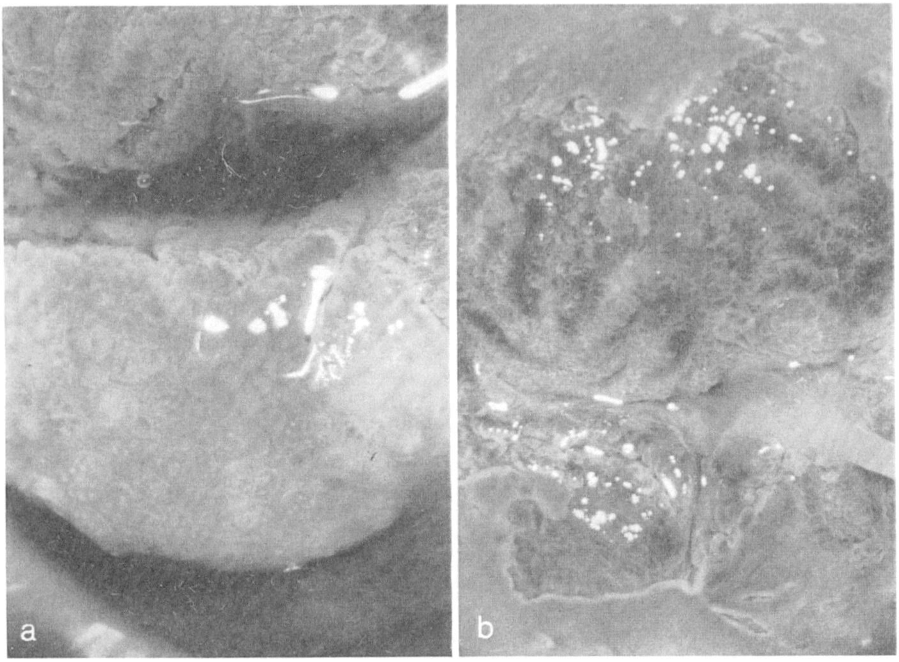

Fig. 35. **a** The posterior area displays subclinical papillomavirus infection with polygonal fields. **b** One month later, only columnar epithelium and a small transformation zone were present in the same site. The midline cleft is the site of the biopsy after healing

multiple looped capillaries, elongated loops and occasional bizarre irregular vessels which are indistinguishable from malignancy (Figs. 31 and 32).

A striking and characteristic feature of overt HPV infection is seen when the whole transformation zone exhibits gross hyperkeratosis (Fig. 33).

6 Progression and Regression of Human Papillomavirus Lesions

The colposcope materially aids the study of the clinical course and significance of HPV infections. Serial observations are easily made and recorded during examinations when cytology, biopsy, and other investigations are performed. The spontaneous regression and disappearance of cutaneous warts is common knowledge. SPI may behave in a similar fashion (Figs. 34 and 35).

When, however, the viral infection persists, continuing progression may be monitored by regular cytological and colposcopic examinations, if patient cooperation can be assured. The transition from SPI to CIN is then reflected in the emergence of characteristic colposcopic features in virus-affected epithelia (Fig. 36).

Application of colposcopy to the study of HPV-associated disorders has proved fruitful. The method has consolidated its place in the evaluation of suspect disorders uncovered by cytological screening. It has made it possible for the investiga-

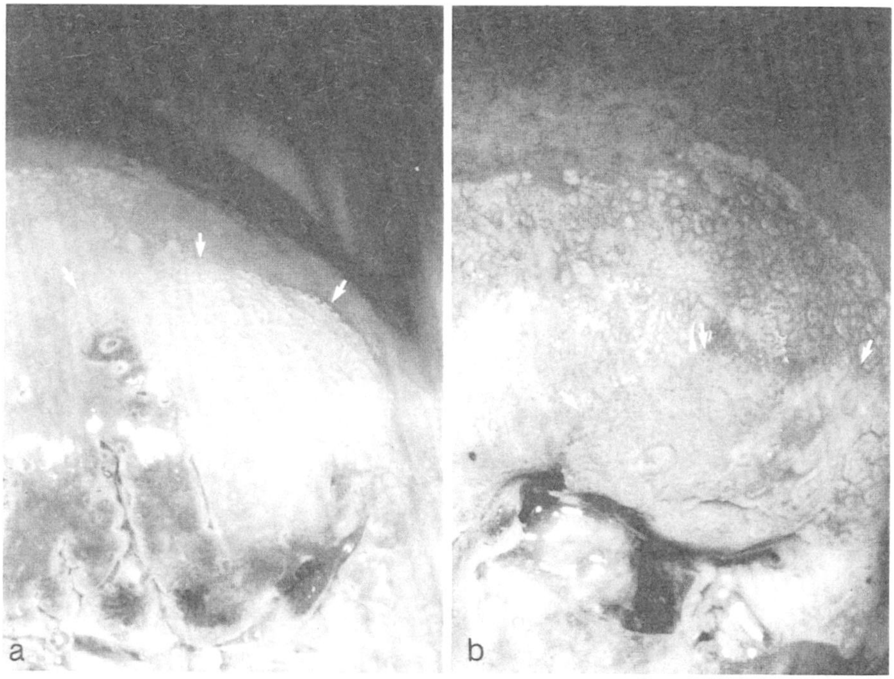

Fig. 36. **a** An area of classical multipapillary subclinical papillomavirus infection is present; histological studies showed the characteristic appearances without cellular atypia. **b** Three years later punctation and mosaic vessels have evolved in the site, and CIN III was present

tion of the role of HPV in oncogenesis to be advanced, acting as a bridge between the affected individual and the molecular biologist.

In our times when human sexual behaviour results in an increasing incidence of HPV infections, those clinicians participating in the evaluation of patients caught in the screening net should be thoroughly familiar with the colposcopic method and morphology. The laboratory worker should be aware of the opportunities for research that the susceptible cervical epithelium allows. All involved should be aware of the limitation of morphology as a tool in the description and definition of dynamic, biological processes.

References

Coppleson M, Pixley EC, Reid BL (1971) Colposcopy. A scientific and practical approach to the cervix and vagina in health and disease, 1st edn. Thomas, Springfield

Coppleson M, Pixley EC, Reid BL (1978) Colposcopy. A scientific and practical approach to the cervix and vagina in health and disease, 2nd edn. Thomas, Springfield

Coppleson M, Pixley EC, Reid BL (1986) Colposcopy. A scientific and practical approach to the cervix and vagina in health and disease, 3rd edn. Thomas, Springfield

Reid R, Stanhope CR, Herschman BR, Crum CP, Agronow SJ (1984) Genital warts and cervical cancer: IV. A colposcopic index for differentiating subclinical papillomaviral infection from cervical intraepithelial neoplasia. Am J Obstet Gynecol 149: 815-826

Treatment of Human Papillomavirus-Induced Lesions of the Skin and Anogenital Region

G. von Krogh

1 Introduction

Since ancient times warts have represented a therapeutic problem. A high prevalence of warts in the general population and a universal distaste for these lesions have instigated various remedies and manipulations associated more with faith than with science. The emergence of warts from the mists of folklore into the light of modern scientific technology is indeed a recent event. Although a number of methods used in the past may very well be dismissed as pure fraud, they to illustrate the importance of a potential placebo effect for wart cure that may be mediated by sophisticated and as yet poorly defined immunomodulatory mechanisms. Although most available methods for wart eradication are based on traditional clinical empiricism and have not been submitted to trials with controlled placebo studies, most treatment regimes favored today have been subjected to critical evaluation in carefully performed comparative studies.

This chapter is concerned basically with therapeutic modalities that are currently favored by most clinicans in the field but some newer, still experimental approaches in particularly recalcitrant cases of human papillomavirus (HPV) infections will also be discussed. It should be appreciated that there is no certain, once-only treatment for warts; the statement, "that all work in some cases but none work in all" certainly holds some truth (Bunney 1982). Skin warts are mostly a cosmetic annoyance; nevertheless, their management is important because of their high prevalence in the population. The importance of an aggressive approach to anogenital warts will be emphasized because of their potential association with malignant transformation.

2 Clinical Features and Management of Cutaneous Warts

2.1 Anatomical Distribution and Gross Appearance

The various HPV types that induce skin warts, or verrucae, in immunocompetent individuals are listed in Table 1. These HPV types show a clearcut tropism for fully keratinized epithelia.

Table 1. Cutaneous warts in immune-competent individuals

HPV type	Wart type
1	Verruca plantaris (myrmecia)
2	Verruca vulgaris (common wart) Verruca plantaris (mosaic wart)
3	Verruca plana (plane/flat wart)
4	Verruca plantaris/palmaris (punctate wart)
7	Butchers' wart

Although each HPV type may induce papillomata that present in a variety of sizes and shapes, the individual type of virus determines, generally, the clinical features of the corresponding warts (see Grußendorf-Conen, this volume). Three different growth patterns may be distinguished: endophytic, exophytic, and flat.

HPV1 characteristically afflicts the soles of the feet (Fig. 1), inducing highly keratinized papillomata with an endophytic growth pattern *(myrmecia)*. Such flask-shaped warts, which press into the epidermis, may be exquisitely painful if they are on pressure points, particularly when extensive. HPV1 may also induce warts in the palms of the hands. Endophytic lesions in the soles or in the palms may also be induced by HPV4; they tend to remain as punctate warts (Fig. 2) with a diameter of a few millimeters. Nevertheless, they can be painful if located, for example, on the pulpar part of the fingertips.

HPV2 and HPV7, on the other hand, predominantly cause exophytic warts. HPV2 causes common warts or verrucae vulgaris usually on the back of hands (Fig. 3), although they may occur on any area of the skin, including the face and genitals. The hyperkeratotic surface is characteristically rough and irregular (warty).

However, when HPV2-induced lesions are localized to the back of the heels, palms of the hands, and around the nails, they tend to be relatively flat *(mosaic*

Fig. 1

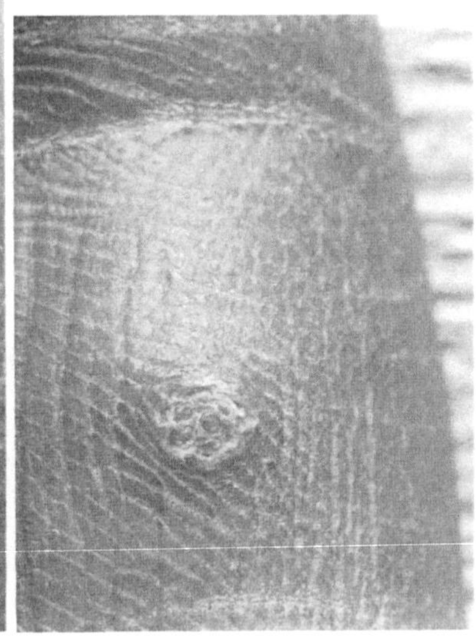

Fig. 2

Fig. 1. Deep verruca plantaris (»myrmecia«) associated with a painful fissure (courtesy of Department of Photography, Södersjukhuset, Stockholm, Sweden)

Fig. 2. Punctate wart of a finger (courtesy of Department of Photography, Södersjukhuset, Stockholm, Sweden)

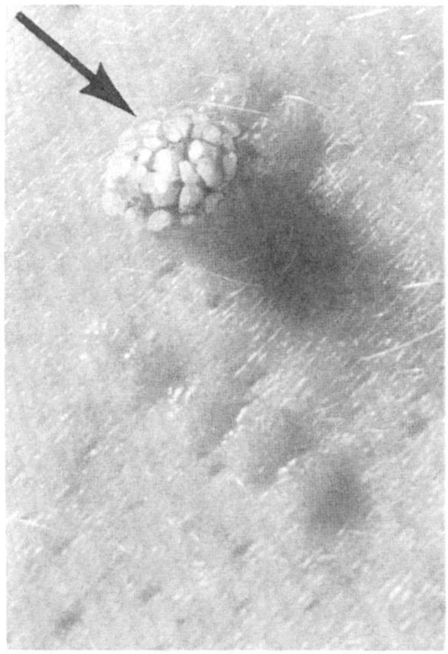

Fig. 3

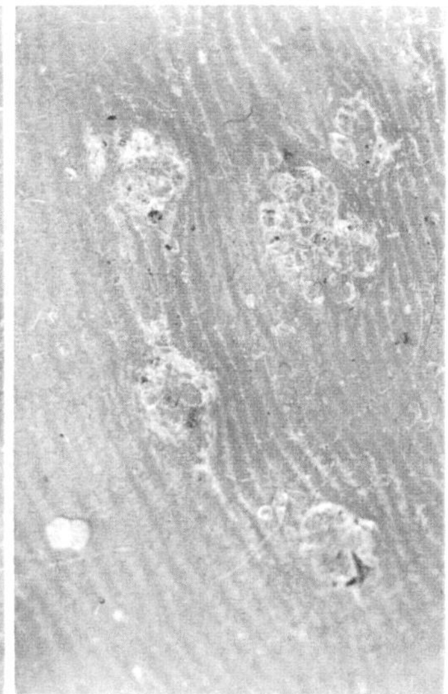

Fig. 4

Fig. 3. Concurrent existence of exophytic verruca vulgaris *(arrow)* and verruca plana (courtesy of Department of Photography, Södersjukhuset, Stockholm, Sweden)

Fig. 4. Superficial verruca plantaris (»mosaic wart«) (courtesy of Department of Photography, Södersjukhuset, Stockholm, Sweden)

warts; Fig. 4) and are often extensive. They are usually painless unless associated with fissures and extensive callus formations.

Butchers' warts, associated with HPV7, are, as the name indicates, exclusively seen in butchers and meat handlers. Hitherto undefined environmental factors may predispose to their occurrence.

HPV3 induces exclusively flat-topped papular lesions lacking a prominent hyperkeratotic surface. These flesh-colored to yellowish-brown *verrucae planae* (flat warts; Fig. 3) are always multiple and usually seen in profusion on the forehead, face, and extensor surfaces of the limbs.

2.2 Natural Course

Although no satisfactory studies exist on the natural course of various wart types, it is generally stated that most warts will regress spontaneously sooner or later. In one study the skin warts in one-half of the afflicted children disappeared spontaneously within 1 year after onset (Massing and Epstein 1963). However, in light of the recently revealed diversity of HPV types causing skin warts, more prospective studies will be required to elucidate the true spontaneous regression rates for different types of warts at various stages of life. Although the data are mostly based on anecdotal reporting, it is believed that 20% of patients lose their warts within 6 months (Bunney 1982). Spontaneous disappearance is particularly common for verrucae planae (Tagami 1984). On the other hand, there is a considerable number of people whose warts persist for many years in spite of aggressive therapy.

2.3 When to Treat?

The most comprehensive survey and critical evaluation of available treatment regimens has been conducted by Bunney (1982); for details beyond the present synopsis, the reader is referred to her outstanding monograph.

As skin warts are so common and often self-limiting, they should not be treated except when they cause a serious cosmetic annoyance or a major discomfort such as pain or disability. The painful hyperkeratotic deep plantar warts may be a true hindrance to personal or professional live or athletic performance. Painful plantar warts in children may cause permanent postural defects. Fissured warts are painful and liable to secondary infections. Punctate warts on the tips of the fingers, although objectively unimpressive, may cause a serious disability for example in a musician. The attitude towards active therapy must be individualized with respect to the needs of each patient including the appreciation of cosmetic disturbances.

2.4 Where to Treat?

The majority of warts may be treated at home by the patient or his family. The need for careful instruction in the procedures to be used is a prerequisite for a sucessful result. Somewhat more aggressive treatment modes need not be reserved for specialists such as the dermatologist or the surgeon but should rather be carried out in a health center or school clinic. The facilities of specialized hospital clinics ought to be reserved for those cases with "problem warts" requiring particular attention.

2.5 Treatment at Home

Many remedies for self-treatment of skin warts have been proposed over the years. None is entirely satisfactory. All involve a chemical cautery approach in order to produce a gradual destruction of the wart tissue through repeated daily applications of the remedy combined with a mechanical removal of excessive keratin and dead tissue. According to the Edinburgh model monitored by Bunney et al. (1976), which represents a standardized comparative assessment of available modalities, three different substances are available: salicylic acids 10%–40%), formalin (5%–20%), and glutaraldehyde (10%).

Patients must be aware that 12 weeks is a realistic period of time for successful therapy. A properly used treatment should not be deemed a failure before the end of this time period. The most common cause of recurrence is incomplete cure. Furthermore, success depends on correct and adequate application of the remedy.

Wart paints are best put on with a pointed applicator such as a match stick. Creams should be rubbed into the area. The treatment is best carried out in the evening when patients have enough time. Strapping or occluding plasters facilitate penetration and should be routinely used when warts are on the sole of the foot. A regular use of soaking the area in warm water prior to treatment is also recommended. Keratin, dead tissue, and old layers of the remedy should be removed with the aid of a pumice stone or a callus file. Paring down of the warts by the patient with a sharp instrument such as scissors may be useful for plantar warts but only after careful medical instruction. The use of a scalpel or a razor blade should not be encouraged for home use but may be added to the hospital regimen when patients are evaluated in the course of therapy, for example, once every month. Such clinical controls may be performed by a skilled nurse.

Numerous basic formulations exist for topical wart treatment, and proprietary preparations are available under different trade names in various countries. Therefore, only a few principles for the composition of different remedies will be mentioned in the present context. *Topical therapy of this type should never be used on facial or anogenital warts.*

2.5.1 Salicylic Acid

The main ingredient of most remedies is salicylic acid in strengths varying from 10% to 40% as tinctures and plasters. A *collodion*-based tincture containing salicylic acid and lactic acid will cure more than two-thirds of common and deep plantar warts within 3 months when applied nightly. Such tinctures are convenient as they can be applied quickly and accurately and are relatively free of side effects. The collodion hardens rapidly, and occlusive strapping is required only for warts on the sole of the foot. The salicylic acid concentration varies in different proprietary preparations but is usually within the range of 10%-15%. Some stronger ointments and solutions may be used for removal of calluses. Routine use of ointments is not recommended as it is difficult to confine them to the target area or to get a protective dressing to adhere to the resultant greasy skin.

Salicylic acid may also be incorporated into adhesive plasters in concentrations of 20%-40%. Such plasters are suitable for the treatment of plantar warts; an appropriate piece of the plasters is cut to the size of the wart and held in place with adhesive. The plaster is removed after 24 h, the area treated mechanically as indicated previously, and a new piece of plaster subsequently applied.

2.5.2 Formalin

Formalin-containing preparations are particularly suitable for warts occurring on weight-bearing areas of the feet where the keratin layer is thick. Various remedies contain 3%-20% formalin in water or may be prepared in a cream base such as Merck ointment. The efficacy against plantar warts is comparable to salicylic acid; this treatment may possibly be the most effective home treatment for mosaic warts. Nightly applications produce hardening and drying of the skin surface, following which the warts may be shelled out. However, formalin can also produce excessive irritation of the skin leading to painful fissures.

2.5.3 Glutaraldehyde

A water-based solution of 10% glutaraldehyde has the same indication, way of action, and efficacy as formalin.

Salicylic acid, formalin, and glutaraldehyde may all have sensitizing properties in some patients. This may also occur with collodion-based preparations due to the presence of colophony.

2.6 Treatment by Physician

The primary goals in more aggressive wart therapy should be minimal discomfort or inconvenience and avoidance of permanent scarring. These goals cannot always be achieved.

Radiotherapy has definitely been abandoned; efficacy is low despite dosages that produce postradiation damage in a high proportion of cases.

2.6.1 Topical Applications

Strong acids such as *nitric acids, monochlor- and trichloracetic acids* as well as caustic compounds such as *silver nitrate* and *copper sulphate* have all been tried, but their effect is poorly documented.

Podophyllin (20%–25%) may be effective against deep plantar warts. The remedy is rubbed into the wart base after removal of excessive keratin and the foot is covered for 1 week with stretch strapping. After repeated applications at 1–3 week intervals, 80%–90% of deep plantar warts may disappear. Somewhat inferior results are to be expected for mosaic warts.

2.6.2 Cryosurgery

Cryotherapy is at present the most popular treatment for skin warts and is highly favored by us. Various agents are available: *carbon dioxide, nitrous oxide,* and *liquid nitrogen.* The effect is due to destruction of the tumor cells following the formation of intracellular water crystals. The ensuing cellular necrotic process is enhanced if freezing is performed twice at the same session and a spontaneous thawing is allowed in between. Cellular degeneration is initiated within a few hours and terminated within a few days. Destruction of epidermal cells is associated with vesiculation, and damage of capillaries in the dermal papillae may cause a hemorrhagic exudate (Zocarian 1977).

Care must be taken to avoid freezing of major digital vessels and nerves, as otherwise long-lasting or permanent paresthesias may ensue. Likewise, caution should be exercised when freezing near the nail matrix to avoid nail dystrophy. When used on large areas of the thin facial skin, a pronounced edema may follow, which may, however, be reduced by a single subsequent application of a potent topical steroid ointment (Hindson et al. 1985).

A burning sensation, usually well-tolerated, is experienced whilst the warts are being frozen. Soon after thawing, the area begins to throb and continues to do so for a day or two. When large warts are treated, and the freezing time is relatively long, it is advisable to give the patient a mild analgesic before leaving the office. A peripheral flare and an edematous wheal develop almost immediately, and within a few hours a serous or hemorrhagic blister may form. Describing this to the patient as part of the therapeutic response is important in order to avoid subsequent "emergency calls". Also, it must be mentioned that walking may be uncomfortable for a day or two after freezing of plantar warts. A mild analgesic should be taken as required against any pain. Any blisters developing should not be ruptured unless they become large enough to be a nuisance: then a small puncture with a sterile needle may be performed to empty out the fluid.

One cosmetic complication is the depigmentation of the treated areas, which may take months to repigment. A permanent hyperpigmentation may occur but is very rare.

On average three to four treatments are usually required for the complete cure of all warts. The interval between freezings should be no longer than 3 weeks; if extended to 4, the cure rate is cut by half (Bunney et al. 1976).

Cryoglobulinemia or cold urticaria may prove to be a contraindication to cryosurgery and patients with a history of these disorders should preferably be treated by other methods.

Carbon dioxide in the form of snow or, slush has been used successfully by dermatologists for 70 years. Snow is obtained by collecting carbon dioxide gas from a tank into a chamois leather bag. The snow is compressed firmly into funnel-shaped molds of various sizes. The resultant rod, which should be slightly smaller than the wart, is pressed firmly against the area to be treated. Carbon dioxide slush is produced by adding a few drops of acetone to the snow immediately before use. An applicator with a cotton wool ball of appropriate size is rolled in the slush until the formation of an ice ball that must be used at once as it melts rapidly. A slight pressure against the wart is sufficient.

For several reasons the alternative use of *liquid nitrogen* has become more and more common. First, the rapidity and efficacy of freezing is superior due to its much lower temperature. The liquid can be transported in metal-cased vacuum flasks which is convenient for smaller dermatology units. For the *"open system"* use, cotton wool balls adjusted to the size of the wart are dipped into the nitrogen. Evaporation occurs rapidly, so the ball needs to be applied immediately to the lesion and to be re-dipped every 5–10 s. Pressure is not necessary. Use of several applicators allows for rapid replacement and permits more extensive application. However, the depth of freeze is limited to 1.5–2 mm because of the rapid evaporation.

During the past 2 decades major improvements of cryotherapy technique with **nitrous oxide** or with **liquid nitrogen** have taken place. Convenient, portable, hand-held spray units have been developed, and for some products *closed cryosystem* tips are available. A continuous flow of nitrous oxide or nitrogen is delivered through flexible cables to and from hollow cryoprobe tips of various sizes. As the depth of freezing is dependent upon the continuous application of any cryogen, this approach is most rational for eradication of warts with a pronounced endo- or exophytic growth pattern. More sophisticated and expensive closed-system units have also been constructed and gradually improved in recent years. I have found this type of treatment suitable to most types of warts and have been especially pleased with their efficacy in eradicating plantar and facial warts.

In general, the wart is frozen until a pale halo of 1-2 mm appears around the base, indicating that the full depth has been frozen. This will take a variable amount of time depending on the anatomical site, the size and thickness of the wart, and also according to the method of freezing. In the case of facial warts no more than about 5-10 s should be spent on the first freezing with liquid nitrogen. Flat warts in other areas may be frozen for 10-20 s. Verrucae vulgares seldom require more than 30-40 s, while large plantar warts may need 1 min. When mosaic warts cover large areas, it is best to freeze them in segments to obtain an even effect. For extensive plantar warts a stepwise treatment of different areas is recommended, leaving untreated areas for body support between sessions. Underfreezing is better than overfreezing in order to minimize unnecessary blister formation and potential subsequent scarring; one can always refreeze at a later date (Crumay 1985).

With any form of wart treatment, any excessive surface keratin must first be removed. A careful paring down, preferably with curved scissors, is a prerequisite for the successful treatment of plantar warts. The sole is rather sensitive to this type of manipulation, and whenever more than minor lesions are treated, local anesthesia is advantageous. In most other areas anesthesia is not required.

2.6.3 Surgical Intervention

Surgical excision is often demanded by the patient and may sometimes represent the only available treatment for certain skin warts when other modalities have failed. However, distinct disadvantages do exist. The scarring produced on any pressure points may be very painful; this is particularly true for plantar warts where irreversible scarring may be more painful than the original wart. The alternative use of cryotherapy only causes minor scarring, if any at all. Therefore, most dermatologists have abandoned surgical intervention completely for plantar warts. The alternative use of curettage with a sharp spoon-curette, followed by electrocautery of the base is advocated (Bunney 1982). Any bleeding may be treated with a silver nitrate stick followed by application of a pad of gauze kept in place with elastic strapping. It takes 2-3 weeks for the cavity to fill in; in the meantime the area should be cleaned regularly with an antiseptic solution, dressed with an antimicrobial powder or cream, and kept covered. Periungual warts extending under the nail plate can be scooped out rather easily without damaging the nail.

Small filiform or papillomatous warts can be removed very easily by burning them off at their base with electrocautery. Such warts may alternatively be treated with electrodesiccation by high-frequency sparking.

All surgical intervention requires local anesthesia.

2.6.4 Psychotherapy

Early reports on the efficacy of skin wart treatment by suggestion and charms are anecdotal. Subsequent large-scale studies have been difficult to evaluate due to lack of comparative groups receiving placebo or other treatment types. Clarke

(1965) reviewed a considerable number of these reports and was dissatisfied by the methods used. He considered the cure to be attributable to suggestion in less than 1% of cases and that spontaneous regression occurred in 20% of patients not receiving any therapy. Stankler (1967) conducted similar studies and proposed that the success of suggestion depended upon a fortuitous temporal relationship between treatment and spontaneous resolution.

Likewise, results from treatment by hypnosis have been conflicting and subject to controversy. Ullman and Dudek (1960) claimed that patients who were good hypnotic subjects responded significantly better than those who were not easily put into a deep hypnotic state. This study was criticized by Tenzel and Taylor (1969), who found no evidence that suggestion was effective in curing warts, whether associated with hypnosis or not.

2.6.5 Bleomycin

The antibiotic bleomycin inhibits DNA synthesis by preventing thymidine incorporation and causing single strand scission in DNA. As it has a high affinity for squamous epithelium, intralesional injections of the drug in saline have been tried in otherwise recalcitrant cases of skin warts. Cure rates on the order of 63%-99% have been reported after 1-3 repeated injections of up to 2 mg of bleomycin per session. The lowest cure rates have been obtained for plantar warts, while periungual warts and lesions located elsewhere react favorably (Bremner 1976; Hudson 1976; Olson 1977; Shumack and Haddock 1979; Cordero et al. 1980). However, recent reports have given conflicting results: while Shumer and O'Keefe (1983) found no effect from saline-treated patients but an 81% cure rate from 1-2 bleomycin injections, Munkvad et al. (1983) found no difference between bleomycin and placebo-treated groups. After having tried the method on a few patients, our view concords with that of Czarnecki (1984), who stated that the method should be reserved for use when other treatments have failed because it is expensive, may not work, and is not without significant side effects in some patients. Thus, bleomycin may induce exquisite pain either upon injection or during the 1st week after treatment when a necrotic eschar develops. When used against periungual warts, a temporary nail loss may follow. However, in the hands of a well-trained dermatologist this treatment should definitely be considered for motivated patients with large and recrudescent warts.

3 Clinical Features and Management of Anogenital Lesions Associated with HPV Infections[1]

3.1 Clinical Features

3.1.1 Gross Appearance

Anogenital papillomavirus infections may show considerable morphological variations. Four types can be distinguished: the verruca vulgaris-like variants, exophytic (condyloma acuminatum), papular, and flat warts. To some extent, the appearance of a particular type depends on the site affected.

Verruca vulgaris-like lesions, representing the least common type, are most likely caused by HPV1 or HPV2. They occur only on keratinized parts of the anogenital skin and are usually hyperkeratotic, skin-colored, and/or somewhat pigmented.

Exophytic acuminate warts predominate in moist areas, i.e., the preputial cavity, urinary meatus, inner aspects of the vulva, vestibulum, vagina, and anal area; they seem to be induced predominantly by HPV6 and/or HPV11. They are usually pedunculated and papillomatous with fingerlike projections in which the richly vascularized dermal core is distinctly apparent (Fig. 5). The color is pinkish-red to grayish-white. In females individual elongated papillae in the vestibular and periurethral area may have a "wormlike" appearance. Acuminate warts often coalesce into confluent plaques which may occasionally proliferate into cauliflowerlike tumors, a phenomenon common during pregnancy. In very rare instances, condylomata of the outer anogenital areas form a semimalignant giant variant, the Buschke-Löwenstein's tumor (Fig. 6), that grows simultaneously in an exo- and endophytic manner, penetrating the underlying tissues (Harvey et al. 1983).

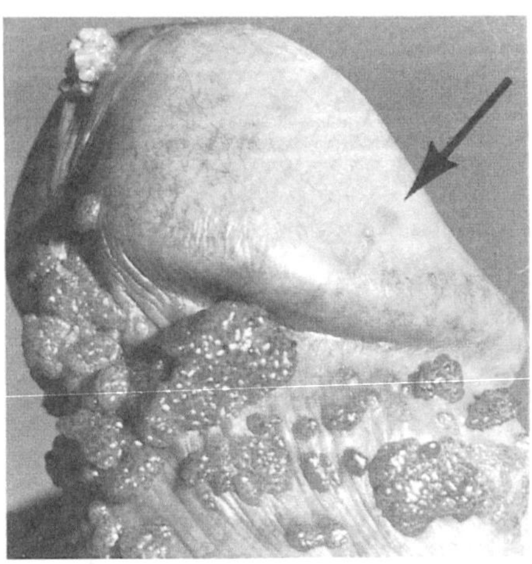

Fig. 5. Typical acuminate warts in the preputial cavity. Punctate capillaries are prominent. Note that the patient also exhibits the concurrent presence of papular lesions *(arrow)* (courtesy of Department of Photography, Södersjukhuset, Stockholm, Sweden)

[1] This chapter was written with the collaboration of Dr. E. Rylander, Department of Gynecology, University Hospital of Umeå, S-901 85 Umeå, Sweden

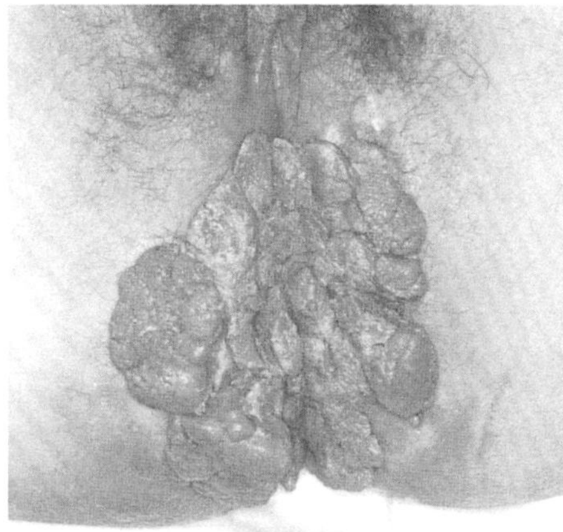

Fig. 6. Buschke-Löwenstein's tumor of the perineal and anal area (courtesy of Department of Photography, Södersjukhuset, Stockholm, Sweden)

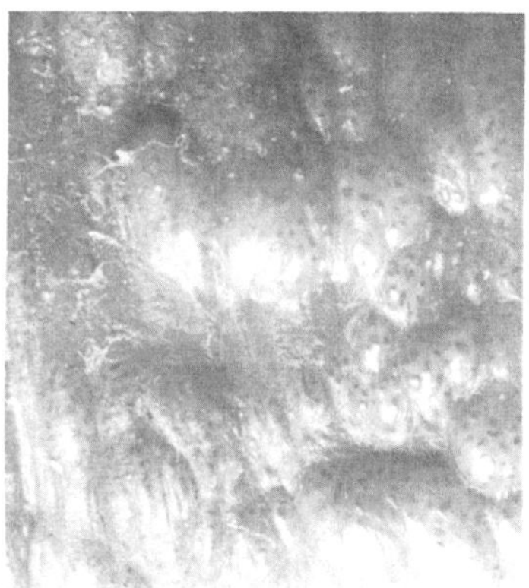

Fig. 7

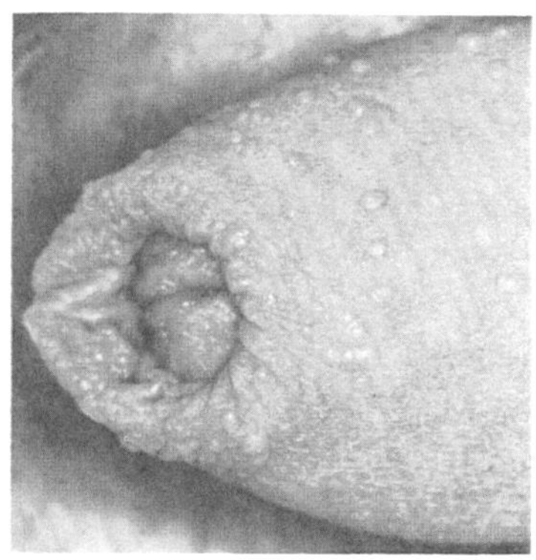

Fig. 8

Fig. 7. Confluent papular warts in the preputial cavity. Punctate capillaries are visible (courtesy of Department of Photography, Södersjukhuset, Stockholm, Sweden)

Fig. 8. Papular, slightly pigmented warts of the outer aspect of the prepuce (courtesy of Department of Photography, Södersjukhuset, Stockholm, Sweden)

Exophytic lesions often occur together with other types of lesions (Fig. 5). Papular warts (Fig. 7) are most common on the outer parts of the genitals, i.e., the penile shaft, the perineum, and the outer aspects of the vulva. In some instances, exophytic warts may change into lesions of papular type. Papular warts may be grayish-white or reddish-brown. On the outer parts of the genitals they may be somewhat pigmented (Fig. 8).

Vulvovaginal warts may become extremely hypertrophic when such lesions coalesce into clusters, revealing a spiky or dropletlike surface structure. This type of lesion has been designated "papillomavirus vulvovaginitis" (Rylander et al. 1986) and "pruritic vulvar squamous papillomatosis" (Growdon et al. 1985).

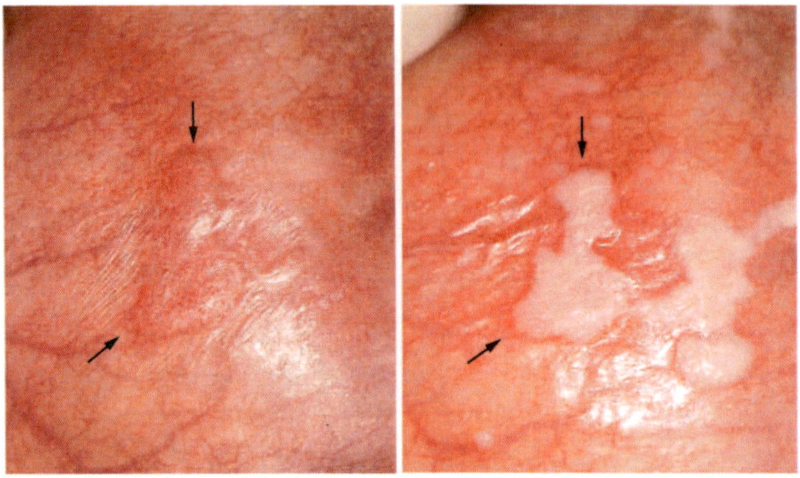

a b

Fig. 9 a, b. The use of acetic acid application in the male. **a** Flat wart in the preputial cavity. Upper and lower demarcation is indicated by *arrows*. Capillaries are hardly visible. **b** After a few minutes' soaking in acetic acid, previously unidentified flat warts become apparent as tongues of whitish epithelium adjacent to the warts that were visible in **a** *(arrows)*. (Courtesy of Department of Photography, Södersjukhuset, Stockholm, Sweden)

Flat lesions can be present in the same area as exophytic warts. Such lesions are usually subclinical, and their prevalence seems to be considerably more common than was appreciated until quite recently. They are usually multiple and disseminated, but occasionally they may merge into larger plaques. Flat lesions require application of acetic acid solution, followed by colposcopy or other magnifying equipment to be fully appreciated and clearly outlined (Figs. 9 and 10; von Krogh and Rylander 1986). Flat lesions of the uterine cervix often contain varying degrees of histologically atypical epithelium (Syrjänen et al. 1983).

Bowenoid papulosis (see Gross, this volume) is an entity afflicting the epithelium of the outer anogenital area of males and females 15–30 years of age. The lesions (Fig. 11) appear as solitary or multifocal, lichenoid, erythematous, and/or pigmented macules and papules with histological changes suggestive of carcinoma in situ. However, the clinical course appears to be benign in most cases, and a spontaneous regression within 1 year has been reported in some. HPV16 DNA has been identified in such lesions (Ikenberg et al. 1983). Bowenoid papulosis may be indistinguishable clinically from histologically benign, flat and papular warts.

3.1.2 Anatomical Distribution

Those parts of the anogenital skin and mucosa subjected to trauma during coitus are most frequently affected by HPV-associated lesions (Oriel 1971 a). In *uncircumcised males,* the preputial cavity is affected most often (i.e., the glans penis, coronal sulcus, frenum, and inner aspects of the prepuce), followed by the urinary meatus. Dissemination to proximal parts of the urethra or to the bladder is extremely rare. Keratinized genital skin (i.e., the outer aspects of the prepuce, penile shaft, pubic area, groin, and perineum) may also be involved. In *circumcised*

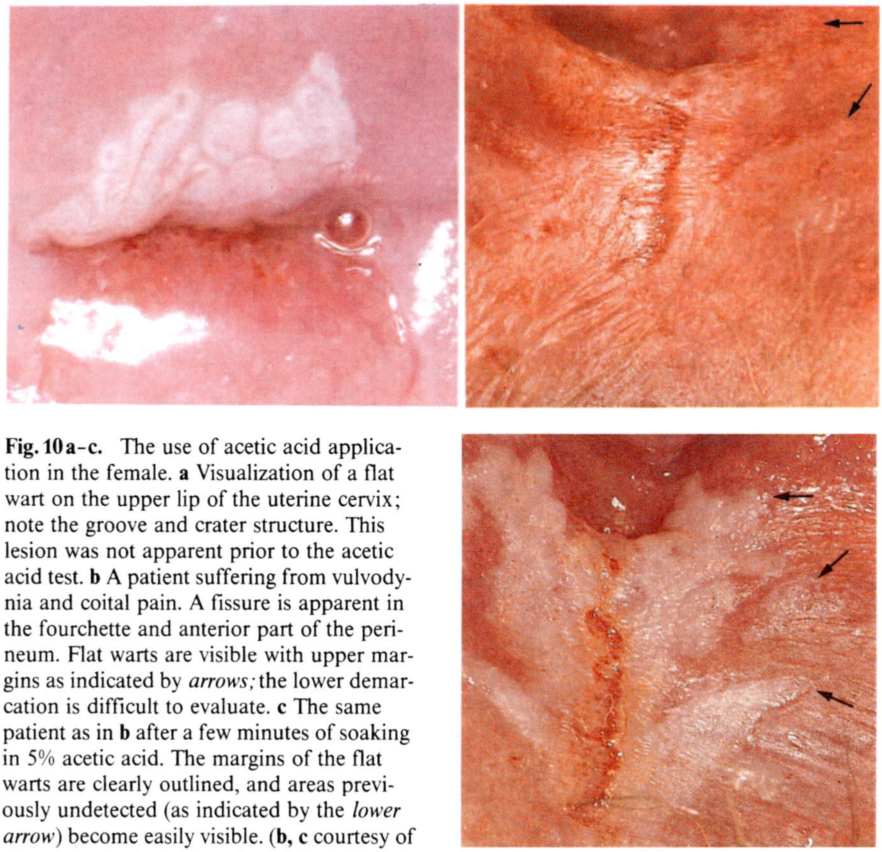

a b c

Fig. 10 a–c. The use of acetic acid application in the female. **a** Visualization of a flat wart on the upper lip of the uterine cervix; note the groove and crater structure. This lesion was not apparent prior to the acetic acid test. **b** A patient suffering from vulvodynia and coital pain. A fissure is apparent in the fourchette and anterior part of the perineum. Flat warts are visible with upper margins as indicated by *arrows;* the lower demarcation is difficult to evaluate. **c** The same patient as in **b** after a few minutes of soaking in 5% acetic acid. The margins of the flat warts are clearly outlined, and areas previously undetected (as indicated by the *lower arrow*) become easily visible. (**b, c** courtesy of Department of Photography, Södersjukhuset, Stockholm, Sweden)

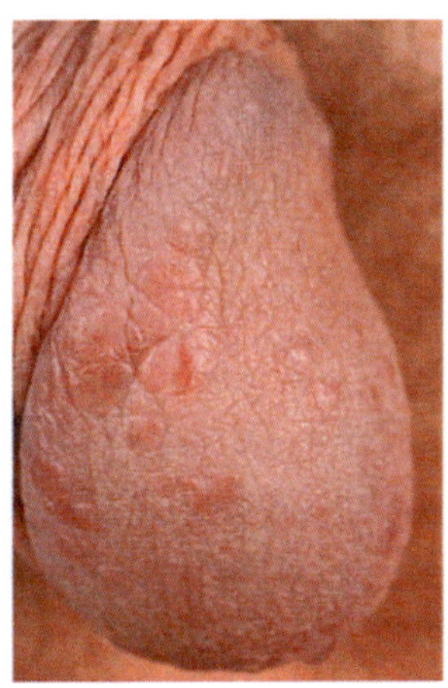

Fig. 11. Bowenoid papulosis; multiple disseminated brownish-red papules of the glans penis (courtesy of Department of Photography, Södersjukhuset, Stockholm, Sweden)

males, keratinized penile skin is more commonly afflicted than the glans penis. Peri- and intraanal warts are most common in *homosexual males* (Oriel 1971 b).

In *women,* genital warts most commonly manifest clinically at the posterior fourchette, in the perineum, on the vulva, in the vestibulum, in the vagina, and on the cervix uteri. Vaginal and/or cervical lesions occur much more frequently than was previously appreciated (Meisels et al. 1981) and may appear with or without simultaneous HPV infection of the external anogenital areas. On the cervix uteri they mostly have a micropapillary or flat surface structure and are easily overlooked by naked eye examination. The urinary meatus is less frequently involved than in males. Anal/perianal lesions occur in up to one-third of the cases.

3.1.3 Symptomatology

Anogenital HPV infection is often asymptomatic. When it first appears, patients may notice some itching. Exophytic warts can cause slight bleeding when submitted to trauma, such as intercourse, and such lesions represent an important point in the differential diagnosis of postcoital bleedings in women. Flat warts may contribute to the spectrum of causes of vulvodynia and other types of coital dysfunctions. The recently identified conditions designated "papillomavirus vulvovaginitis" or "pruritic vulvar squamous papillomatosis" may cause itching, burning, and/or pain that is probably due to repetitive fissure formation (Fig. 10b, c).

3.2 Clinical Evaluation

3.2.1 Differential Diagnosis

In the *male,* exophytic warts are easily distinguished from physiological anatomical variants such as filiform Tyson's sebaceous glands appearing on the corona glandis. These glands are typically pearly white and appear in parallel rows. Warts are also easily distinguished from yellowish sebaceous glands on the inner and/or outer aspect of the prepuce (Fig. 12).

In the *female,* the appreciation of physiological squamous papillae and glands of the vulva is very much dependent on the experience of the investigator. Text-

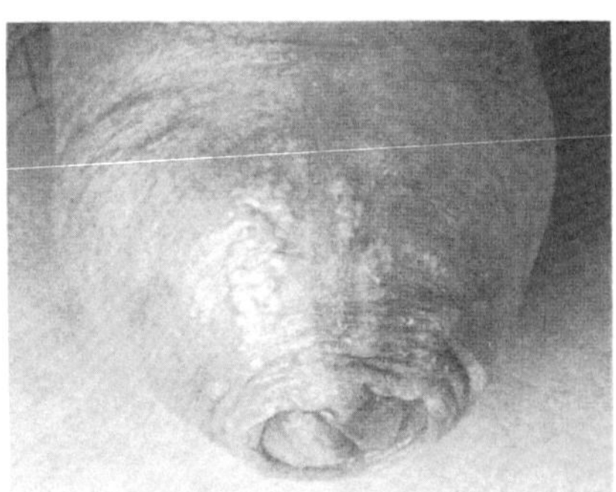

Fig. 12. Sebaceous glands on the outer aspect of the prepuce (courtesy of Department of Photography, Södersjukhuset, Stockholm, Sweden)

books and specialized texts on vulvar disease fail to cover this topic. However, it is apparent that quite a number of females normally have variably prominent, symmetrical, epithelial papillae in various areas of the vestibulum. Also, sebaceous glands are present on the lateral parts of the vulva and in the perineal area.

3.2.2 Diagnostic Tools

Since anogenital lesions associated with HPV may be clinically insignificant, the use of certain diagnostic tools improves the accuracy of identification.

Acetic Acid Staining. Application of 5% acetic acid will induce a whitish color and swelling due to a coagulation of epidermal proteins, which is more marked in HPV-induced lesions. The acetic acid test often requires at least 2 min to be effective. It is very helpful in determining the full extent of warts in both sexes (Fig. 9 and 10) and is particularly valuable for visualizing benign lesions and/or cervical intraepithelial neoplasia (CIN) of the uterine cervix prior to biopsy. On the cervix, application with a cotton-wool swab is usually sufficient, while on the outer genitals the afflicted parts should be soaked with an acetic acid-moistened gauze. It should be emphasized, however, that acetic-white areas may also represent inflammatory conditions unrelated to HPV infection (von Krogh, personal observations).

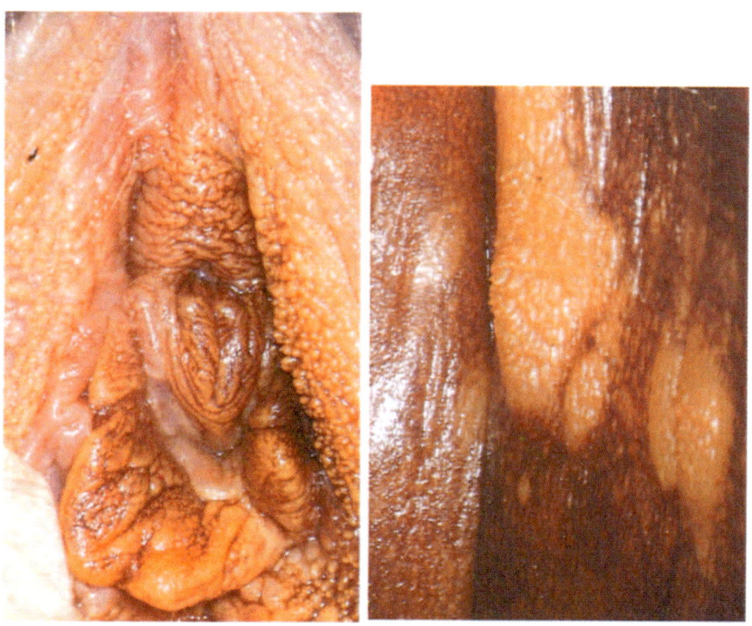

a b

Fig. 13 a, b. The use of iodine application in the female. **a** Painting with iodine outlines the papillomatous structure of HPV vulvitis, appearing as striped yellowish-brown areas that contrast to the brownish color of physiological epithelium of the labia minora. **b** Painting with iodine reveals the appearence of clearly outlined yellow areas indicating the presence of flat HPV lesions that were invisible at first inspection. Note the existence of micropapillary spikes

Therefore, a biopsy is required for a diagnostic confirmation whenever the acetic acid test is positive.

Iodine Staining. Normal squamous epithelium of the vagina and the portio are stained dark-brown by iodine due to the presence of glycogen. In papillomavirus-infected as well as in neoplastically transformed epithelium the glycogen content is often reduced. Also, such an epithelium may show surface keratinization. Therefore, when applying a 5% iodine solution (Lugol's solution) to the region, HPV-afflicted areas will appear patchy or striped brownish-yellow in color (Fig. 13). Intraepithelial neoplasia particularly becomes yellowish and sharply demarcated. As for the acetic acid test, a positive outcome represents an indication for subsequent histological evaluation of a biopsy specimen.

Colposcopy. Adequate magnification and a good light source are required for a satisfactory clinical evaluation of any patient with condylomata. For the investigation of the uterine cervix, colposcopy is a prerequisite for an optimal appraisal and for obtaining a correctly directed biopsy specimen. However, the colposcope is also valuable for proper identification of HPV lesions in any part of the anogenital area (von Krogh, personal observations). Therefore, colposcopic equipment should be available at any modern gynecology and venereology unit managing anogenital HPV infections. The colposcopic features suggestive of the presence of HPV infections of the uterine cervix are described by Pixley (this volume). In cases

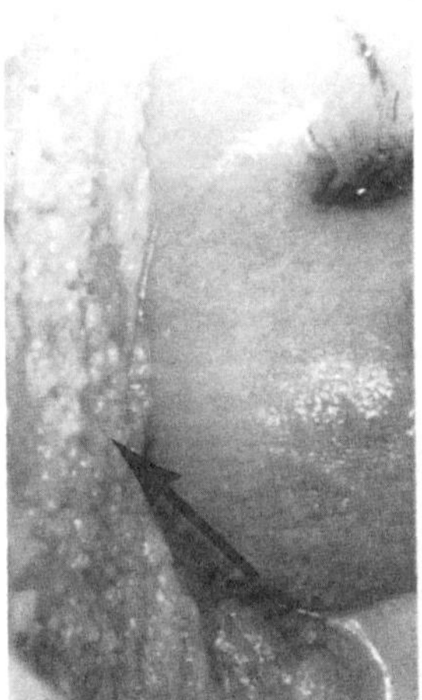

Fig. 14

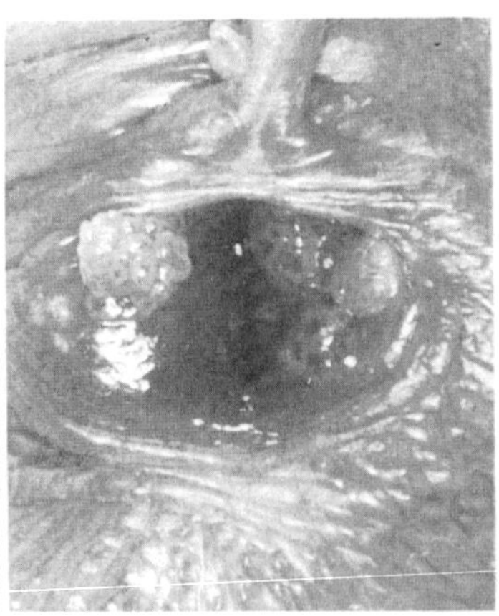

Fig. 15

Fig. 14. HPV vaginitis: adjacent to the smooth epithelium of the portio, the vaginal wall is covered with a coarsely granulated HPV-induced hyperplasia *(arrow)* appearing as spikes and dropletlike papules

Fig. 15. Acuminate warts in the urinary meatus of a male. The use of a small nose speculum usually allows inspection of the proximal demarcation (courtesy of Department of Photography, Södersjukhuset, Stockholm, Sweden)

with "papillomavirus vulvovaginitis" (Fig. 14), a brain-cortexlike, spiky, or droplet-like hypertrophy of the mucosa may be observed. Such patients also often have a palpable roughness of the vaginal wall (Rylander et al. 1985, 1986).

Investigation of Urethral Warts. The use of a small nose speculum facilitates an optimal inspection of the urinary meatus (Fig. 15) up to the distal part of the fossa navicularis. When meatus warts are present, the anterior part of the urethra should be routinely palpated. If any mass is present or if the patient has voiding problems such as reduced urinary stream or hematuria, endoscopy and/or cysturethrography should be performed for the identification of potential intraurethral and/or bladder papillomata. Such procedures are very rarely required.

Proctoscopy. In homosexual men proctoscopic examination must be performed; in those practising receptive anal intercourse, the prevalence of intraanal warts is high with or without the coexistence of perianal lesions. Condylomata are seldom found proximal to the linea dentata. However, proliferation up to the internal sphincter and to distal parts of the rectum occurs in some cases. Therefore, a digital palpation should be included in the workup.

Histopathology and Cytology. The histopathologic criteria for HPV infection are discussed by Koss (this volume). HPV-transformed epithelium often reveals a typical and highly diagnostic pattern of hyperplasia associated with the pathognomonic cellular degenerative changes designated as "koilocytosis".

However, two major problems exist. First, the absence of koilocytes in a tissue specimen or a cytological smear does not preclude the existence of HPV infection; a technically adequate biopsy specimen may in some instances merely indicate the existence of a histologically nonspecific, benign, papillomatous, epithelial hyperplasia. Second, in a number of cases, and in particular for cervical lesions, various degrees of dysplasia may coexist either directly adjacent to or merging with areas of benign hyperplasia. Thus, cervical biopsies and smears often show simultaneous evidence of HPV infection and CIN. For these reasons routine biopsies of the uterine cervix are necessary whenever HPV infection is suspected; biopsies should be obtained with colposcopic guidance after application of acetic acid. However, the absence of pathognomonic, diagnostic, histological criteria of HPV infection in a number of instances requires the use of additional diagnostic tools for clinical use. The contribution of type-specific probes for the detection of viral DNA through hybridization assays (see Schneider, this volume) represents a major breakthrough, enabling not only an accurate diagnosis of anogenital HPV infection but also providing guidance to optimal therapy and follow-up programs. Furthermore, the in situ hybridization technique, having both a high sensitivity and specificity, ought to be as valuable for epidemiological work on anogenital lesions associated with HPV as bacteriological studies are for patients with gonorrhea. Thus, tracing of "high-risk males," i.e., those infected with HPV16 and HPV18, may prove to be as important as treating the female partner.

It seems reasonable to believe that current technology will shortly result in the availability of HPV DNA probes for routine clinical use.

3.3 Therapy

Anogenital HPV infection is not only a psychologically distressing ailment but may in some cases also contribute to the development of anogenital carcinoma. Until the tools are available that allow a clear distinction at routine examination between patients harboring HPVs associated with high risk (types 16 and 18) and low risk (types 6 and 11) for malignant transformation, all lesions should be treated aggressively and the patients carefully followed up until a complete cure has been obtained. Whenever possible, sexual partners should also be examined, and, if necessary, treated until cured, as evaluated after at least a 3-month follow-up investigation. The use of condoms is recommended for the male partner/patient. Any concurrent sexually transmitted diseases (STDs) should be diagnosed and treated, preferably prior to wart therapy.

Through the years, a great number of therapeutic modalities has been tried. Such methods fall into three categories: cytotoxic destruction, surgical removal, and immunotherapeutic approaches. As is the case for skin warts, no single modality alone is fully satisfactory. The present recommendations for routine treatment are mostly based on personal preference.

3.3.1 Cytotoxic Destruction

3.3.1.1 Podophyllin treatment suffers from the disadvantages of limited clinical efficacy, a marked risk of severe side effects, heavy demands on the medical and nursing staff, and considerable inconvenience to the patient. The remedy represents a resinic extract from the rhizome of the *Podophyllum* plant family. *Podophyllum peltatum* grows in North and South America and *Podophyllum emodi* in the Himalayan area. The quantitative and qualitative contents differ between the two plants. Unfortunately, only a semiquantitative method is used for determining the content in various plant batches of the active ingredients, designated lignans, of which *podophyllotoxin* predominates in both species.

Podophyllin is usually applied to the warts as a 20% solution in ethanol or as a tincture of benzoin, and the treated areas are allowed to dry for a minute or two. The therapeutic effect is due to a necrotic involution of the condylomata which is maximal 3-5 days after application. Owing to the risk of local side effects, the warts are painted maximally at 1-week intervals by a doctor or an experienced nurse, and the preparation must be washed away within a few hours. Residual warts are usually papular or flat and are in general refractory to podophyllin. More than four repeated applications are of little value, and surgical removal is indicated instead (von Krogh 1981a; Jensen 1985; Dolezal 1985).

The long-term efficacy from 20% *P. peltatum* and *P. emodi* preparations was studied prospectively in a comparative investigation on previously untreated penile warts of 105 noncircumcised men, as evaluated at 3-month follow-up (von Krogh 1978). In accordance with other reports (Simmons 1981; Gabriel and Thin 1983), it was confirmed that efficacy is considerably lower than was claimed originally (Culp and Kaplan 1944). After a single application only 22% of cases were completely cured, and the cumulative effect from a second treatment on residual warts given about a week after the first was 38%.

A major problem with podophyllin is its obvious systemic toxic potential when a podophyllin solution is freshly made (von Krogh 1982). A few patients treated for very large warts have developed systemic symptoms within a few hours. High vascularization of the lesions and occlusive factors in the anogenital areas may enhance absorption. As lignans act as antimitotic agents, sensitive tissues with a high degree of cellular proliferation such as bone marrow and intestinal mucosa are at risk. Nervous cell functions are also susceptible, due to inhibition of axoplasmic flow. Symptoms include dizziness, weakness, lethargy, emesis, and diarrhea. More alarming signs may develop with stupor, coma, respiratory failure, vascular crisis, peripheral neuropathy, urine retention, and paralytic ileus. Hemoperfusion may be life-saving in such instances (Heath et al. 1982).

Taking account of toxicity margins for 20% podophyllin preparations, it is recommended that the maximums, volume for topical use should be in the range of 0.9-1.2 ml for freshly made *P. peltatum* - based preparations and 0.4-0.5 ml for freshly made *P. emodi*-based remedies (Fig. 16). The use of podophyllin during pregnancy is contraindicated, due to the potential risk for fetal death and teratogenic influence (Chamberlain et al. 1972; Karol et al. 1980).

3.3.1.2 Purified Podophyllotoxin: Self-Treatment. A method for self-treatment with 0.5% podophyllotoxin of penile warts was evaluated by von Krogh (1981b). Patients were given a 5-ml bottle of 0.5% podophyllotoxin in 70% ethanol, to which 0.05% methylrosaniline was added as a color indicator for visualization of the areas being treated. The solution was painted on the wart-afflicted areas twice daily for 3 days using a cotton wool swab and keeping the prepuce retracted for a minute or two until the preparation had dried.

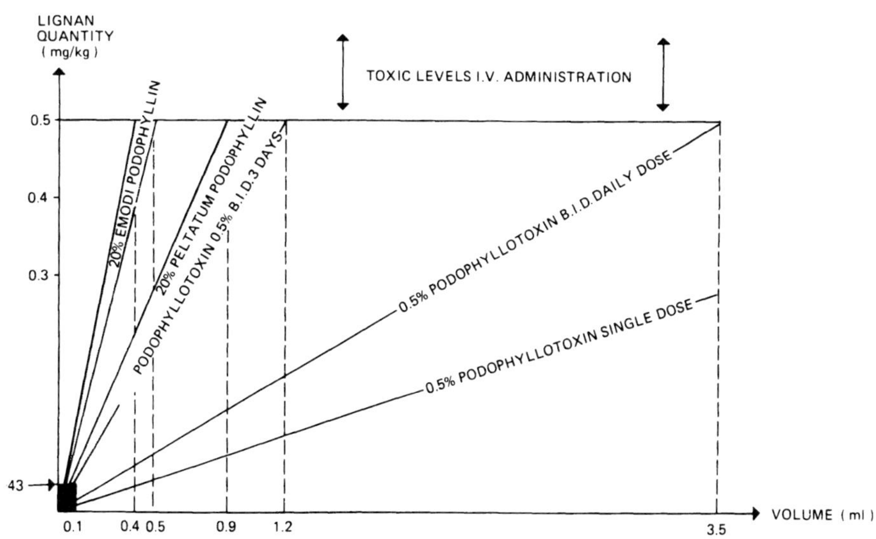

Fig. 16. Reciprocity between various volumes (ml) of 20% podophyllin or 0.5% podophyllotoxin preparations and corresponding lignan content (mg/kg with reference to 70 kg body weight). *Arrow* indicates total cumulative dose for 0.5% podophyllin preparation (0.1 ml) applied twice daily for 3 days (von Krogh 1982).

A complete wart eradication, as evaluated at 3-month follow-up, was accomplished in 49% (35 of 71) of the men. Preputial cavity lesions reacted best and disappeared in 70%. Just as for podophyllin, warts in the urinary meatus and on the penile shaft were quite resistant to treatment. The cumulative effect from two courses of self-treatment was 82%; this was a highly significant improvement when compared with two applications with 20% podophyllin, which induced a complete cure in only 38% of the cases. In subsequent studies by Lassus et al. (1984), it was confirmed that this modality for self-treatment of penile warts is significantly more effective than the use of 20% podophyllin, and that the potential for a cumulative effect from repeated cycles of treatment by the patient himself at home is a great advantage. It should be noted that the wart necrosis occurring on days 3-5 after start of treatment is associated with some mild local irritation in 17%-50% of patients; these side effects disappear spontaneously within a few days. As with podophyllin, the potential for a mild balanoposthitis must be kept in mind when very large or numerous warts are encountered (von Krogh 1981a; Lassus et al. 1984). The use of 0.5% podophyllotoxin satisfies any reasonable demands on systemic toxicity safety margins, which is not the case with 20% podophyllin (Fig. 16; von Krogh 1982). In Sweden the modality is registered under the trade name Wartec. The effect from self-administration of podophyllotoxin preparations to vulvar warts is under investigation. Just like podophyllin, the use of podophyllotoxin is contraindicated for pregnant women.

3.3.1.3 Other Chemotherapeutic Modalities. As the shortcomings of traditional podophyllin have been recognized, other chemotherapeutic modalities for topical treatment of anogenital warts have been sought. Success with topical *colchicine* has been reported in some small series. We have evaluated topical colchicine treatment in a large group of patients with penile warts (von Krogh and Rudén 1980). Although colchicine has in vitro pharmacokinetic properties that are similar to those of podophyllotoxin, our investigations demonstrated that the therapeutic activity of colchicine is lower than that of podophyllotoxin and that unacceptable local adverse effects occur considerably more often. Accordingly, colchicine is not recommended in the routine treatment of anogenital warts.

Newer agents, such as thiotepa and bleomycin, have only limited usefulness. *Thiotepa* therapy has been used experimentally for the management of extensive intraurethral lesions, which are otherwise particularly hard to manage, even with repeated surgery (Halverstadt and Parry 1969). Topical application of *bleomycin* has been tried with poor results. Up to 70% cure rates have been achieved for anal warts that were injected directly; however, repeated administrations over a period of several months may be required, and local soreness is common when the warts necrotize (Figueroa and Gennaro 1980).

A number of workers have reported on the use of *5-fluorouracil* (5-FU). The results have been somewhat controversial; cure rates in the range of 33%-70% have been reported after daily applications for up to 8 weeks. However, cure is usually obtained at the cost of painful ulcerations. For this reason, primary use of 5-FU is not justified. However, a 5% 5-FU cream may be tried against lesions of the urinary meatus; these disappear in up to 90% of cases when the cream is applied after each voiding for 2-3 weeks. On this site the medication can be confined to a limited area, and subsequent ulcerations only cause dysuria (von Krogh 1976). Furthermore, intraurethral installations of the cream may be performed using a syringe when extensive intraurethral growths are encountered (Dretler and Klein 1975; Wein and Benson 1977; Cetti 1984). In such cases the patient must be monitored carefully for the development of urinary retention.

Over the years various caustic agents have also been used in the therapy of venereal warts. Their efficacy has been poorly documented. The most commonly used compounds are *bi- and trichloracetic acid,* both of which rapidly penetrate and cauterize the skin. Maximum benefit from such treatment occurs when application is performed at weekly intervals. However, numerous treatments are required, and the procedure is in no way superior to podophyllin with the exception of its lack of systemic toxicity (Swerdlow and Salvati 1971; Willcox 1977). Therefore, it may be used during pregnancy.

3.3.2 Surgical Intervention

3.3.2.1 Scissor Excision. Simple surgical excision using sharp-pointed scissors, with or without concomitant **electrocautery/electrodesiccation,** is not practical for routine treatment of relatively large or numerous genital lesions. Therefore, surgical removal should be used primarily in cases exhibiting solitary or only a few warts. Some reports indicate that excision produces faster clearance and lower recurrence rates of condylomata acuminata than podophyllin does, particularly when anal warts occur (Jensen 1985; Khawaja 1986). Indeed, some authors advocate scissor excision as the treatment of choice for anal warts, especially for intra-anal warts (Samenius 1983). The technique can be successfully applied to anogenital warts resisting treatment and does not tend to traumatize or scar the skin or mucous membranes to any significant degree (Simmons and Thomson 1986; von Krogh, personal observations). Surgical methods are often required during pregnancy, when podophyllin derivatives must be avoided. Usually, surgery may be performed with local anesthesia; Xylocaine 0.5%-1% is injected subcutaneously or into the submucosa, allowing an optimal separation of individual warts so that as much healthy epithelium as possible can be preserved for rapid re-epithelialization of the excision sites. It may be argued that the administration of local anesthesia in the anogenital area may cause unnecessary discomfort. However, most patients are satisfied with the procedure (Jensen 1985; von Krogh, personal observations). Therefore, general anesthesia is usually only required in patients with extraordinarily excessive wart formation.

Bowenoid papulosis is treated with excision of individual lesions and the adjacent 1-2 mm of grossly normal epithelium; the margin of the excision site should be cauterized subsequently.

When intra-anal warts are encountered, the use of Xylocaine with adrenalin 1:200000 provides an excellent hemostasis. Infiltration with approximately 30-40 ml will usually provide an adequate relaxation of the anal sphincters, and allow the use of a bivalve proctoscope. Individual warts are removed by lifting the lesions with a pair of toothed forceps and by using fine-pointed, slightly curved scissors. Electrocoagulation is only required to stop any persistent bleeding. Subsequent application of absorbable hemostatic material (oxycel, sorbacel, spongostan) that may be held firmly in position by a T-bandage is advantageous. Stool softeners promote a minimum of postoperative discomfort. Using these procedures, even large confluent anal warts may be removed in a single session (Samenius 1983).

3.3.2.2 Cryosurgery. In carefully performed comparative studies on large patient groups with previously untreated warts of the outer genitals, the long-term efficacy of cryotherapy has been found to be superior to podophyllin treatments (Bashi 1985). The method is also useful for the treatment of anorectal warts (Dodi et al. 1982) and may be considered during pregnancy (Bergman et al. 1984) and for the treatment of bowenoid papulosis (Mortimer et al. 1983). The use of "open-system" freezing may certainly be tried (Bergman et al. 1984). However, more convenient, "closed-system", cryosurgical applicances for the use of nitrous oxide or liquid nitrogen are also available. Some cryoprobes have been constructed that are particularly fit for the portiocervical area. Cryosurgery may be performed without local anesthesia when the lesions are few and relatively small; however, most authors advocate the use of a local anesthetic when large lesions are treated. Freezing should be performed until an iceball 1-2 mm larger than the diameter of the wart is produced (Simmons et al. 1981). The number of cryotherapy sessions required for a complete cure varies for each patient and is correlated to the number and size of the lesions; the average number of sessions is 2-3 (Simmons et al. 1981; Dodi et al. 1982).

Given the current state of knowledge and understanding of the natural history of cervical intraepithelial neoplasia (CIN) and improved techniques (colposcopy, punch biopsy, endocervical curettage) of evaluating its distribution and size in the cervix, a trend toward a more conservative treatment of CIN has developed in recent years. Therefore, the use of cryotherapy not only for benign cervical HPV infections but for CIN as well has been evaluated (Ferenczy 1985). The unequivocal advantages of cryotherapy for such lesions are its low cost and complication rates. Furthermore, as for laser therapy, it does not interfere with fertility. In the most large and recent series the freeze failure rate after one cryoprocedure was 12%, and refreezing persistent CIN lesions provides up to 97% overall therapeutic success. Freeze failure rates are related to the size rather than the histological grade of CIN. When compared to CO_2 laser therapy, a relatively unfavorable response has been found with the cryotherapy approach only when large and/or endocervical lesions are encountered. Thus, CIN lesions mesuring less than 3 cm in diameter and without extension into the endocervix have similar failure rates (4%-5%) using either method. The CO_2 laser produces comparatively better results only for lesions larger than 3 cm in diameter and those with up to 5 mm extension into the endocervical canal. Therefore, as complications are considerably higher after laser than cryotheraphy, the latter may be prefered for less than 3-cm large CIN lesions. Such lesions comprise about two-thirds of all CIN. When cervical lesions are treated, local anesthesia using a cervical block with 1% Xylocaine is recommended, and the cryo-iceball should extend 5 mm beyond the lesional margins and 5 mm deep into the cervical stroma (Ferenczy 1985). However, because cryotherapy of the portio of the cervix may be accompanied by vasomotor symtoms and uterine cramps, prolonged cryotherapy is poorly tolerated by most patients. Therefore, large lesions must be treated during multiple separate visits.

3.3.2.3 Laser Surgery. In recent years the use of laser equipment under colposcopic guidance has found great applicability in the treatment of genital warts of the female, in whom benign warts frequently coexist with neoplastic transformation of the cervix uteri. One advantage of this type of surgical therapy is that the instrument may be adjusted in a way that allows a complete removal during the same session not only of benign warts but of precancerous lesions as well. Furthermore, as laser evaporation/excision is performed entirely under colposcopic guidance, an excellent overview of the diseased areas is accomplished simultaneously while moving the laser beam by a micromanipulator (Dorsey and Diggs 1979). Furthermore, optimal visualization is facilitated by the fact that the laser beam minimizes bleeding as the heat tends to seal off small severed blood vessels. Therefore, subsequent suturing is not required. Local infiltration anesthesia is sufficient, and most cases can be treated as outpatients. A laser smoke evacuation system is required for suction of smoke and scattered particles arising from the laser wound.

The infrared CO_2 laser beam is strongly absorbed by living tissues. In superficial epithelial layers 90% of the light energy is converted into heat that induces boiling of intracellular water, followed by an explosive disruption of cells. As the heat conductivity is minimal, a layer of necrosis of only a few tenths of a millimeter may be produced; this represents a major advantage compared to electrocautery which also tends to damage the surrounding tissues. Therefore, the development of scar tissue is usually minimal, and healing is rapid. Accordingly, repeated laser surgery may be easily performed.

Furthermore, treatment of extensive areas is possible due to remarkable healing conditions after laser surgery; this is frequently of great advantage in women as HPV may simultaneously afflict the vulvoanal area, the vagina, and the uterine cervix. The depth of tissue destruction can be carefully controlled by adjusting the power setting and regulating the diameter of the light spot as well as the speed of beam motion. A small beam will penetrate into the tissue and is, accordingly, used as a cutting instrument. A larger light beam will only cause vaporization of superficial cell layers. Thus, during the same sitting the instrument can be used alternatively for cutting and vaporization. Through proper manipulations and repetitive use of the laser beam in the same area an accurate depth of tissue destruction can be accomplished. When crater formation is intended, the depth can be carefully monitored. Due to the intense heat formation, the crater surface is sterilized (Wright et al. 1983; Baggish and Dorsey 1985).

Laser treatment represents the treatment of choice for eradication of relatively large cervical and vaginal lesions as well as of warts and/or CIN during pregnancy. The method is also of great value for the treatment of large and otherwise difficult to treat lesions on any part of the genitoanal in both sexes (Anderson et al. 1984; Ferenczy 1984a, b).

Exophytic warts are most easily cut close to the base, which is then treated again in order to remove the lesion radically and to control bleeding. Papular and flat lesions are entirely destroyed, unless there are any indications for obtaining a histopathological specimen, for example in the presence of bowenoid papulosis or if other dysplasias are suspected.

It has been suggested that recurrences after removal of clinically manifest HPV-associated lesions may be related to HPV DNA persisting in normal-looking, adjacent epithelium (Ferenczy et al. 1985). Therefore, a treatment that extends 1-2 cm beyond grossly apparent tumors might improve the cure rates. Lasers may prove to be the best instrument for this purpose. Vaporization of adjacent epithelium can be performed superficially to a depth of 0.2-0.5 mm, which destroys epidermal cells but still allows quick re-epithelialization. In using a large focal spot size and moving the beam rapidly, a superficial depth of laser ablation can be ensured, which is of particularly great value when treating lesions on the thin epithelia of vulvar, penile, urethral, and anal areas.

Once CIN has been diagnosed on biopsy, the remaining tissue should be removed using the cutting technique, and the specimen should be evaluated histopathologically with respect to need for further treatment. This is accomplished through excision of a cone or a cylinder using high effect output and a small focal spot size. The aim is to achieve a cutting depth of 6-30 mm, depending on the size of the lesion. Destruction/excision in the uterine cervix should always include the cervical crypts that usually attain a depth of 5 mm (Anderson and Hartley 1980). Both benign warts and CIN may reach to this depth. The adjacent epithelium should be vaporized thoroughly in order to destroy any HPV-containing epithelium and/or residual neoplastic lesions.

However, neither destructive nor excisional methods will eradicate completely papillomavirus-induced or neoplastic changes in all cases. A failure rate of at least 2%-10% may be predicted even under optimal conditions. Therefore, if CIN is present, a follow-up over a long period of time is mandatory. As for cryotherapy, the use of laser surgery preserves the anatomy of the uterine cervix: this is propitious for subsequent follow-up investigations including colposcopy and cytology. Furthermore, the use of a technique that minimizes any scarring of the cervix is favorable for the treatment of women in the reproductive age group, as it preserves the reproductive function. This is highly pertinent due to the fact that HPV lesions and the associated CIN are becoming increasingly more common in younger women.

Laser surgery seems superior to cryotherapy in treating cervical lesions larger than 3 cm in diameter (Ferenczy 1985). Cure rates for CIN have been as high as 90%-97% after 1-3 sittings; for condylomata of the remaining part of the female tract the corresponding figures are in the range of 63%-95% (Baggish 1982; Dorsay and Diggs 1979; Baggish 1980; Burke 1982; Bellina 1983; Grundsell at al. 1984; Rylander et al. 1984). Studies on the applicability of laser treatment for penile warts have been initiated, and cure rates of 98% after 1-3 sessions have been reported for otherwise resistant warts in the urethra (Lundquist and Lindstedt 1985).

Laser surgery may also be used for intra-anal warts. However, if methane gas from the bowel is allowed to accumulate during treatment, a flaming up may occur. Protective retention of gas and feces may be accomplished by using wet gauze packs, and additional use of adequate suction is required (Billingham and Lewis 1982; Grundsell et al. 1984).

Laser instruments are still expensive. However, less expensive models will probably become available shortly, and their use outside of larger medical centers may become increasingly popular. It is important to stress that the use of these instruments requires special clinical experience and skill and training in colposcopy.

4 Immunotherapeutic Approaches to HPV Infection

Some major immunological events that may potentially be elicited by HPV infection and associated wart formation are illustrated in Fig. 17. It is generally agreed that spontaneous wart regression is conditioned by immunological mechanisms.

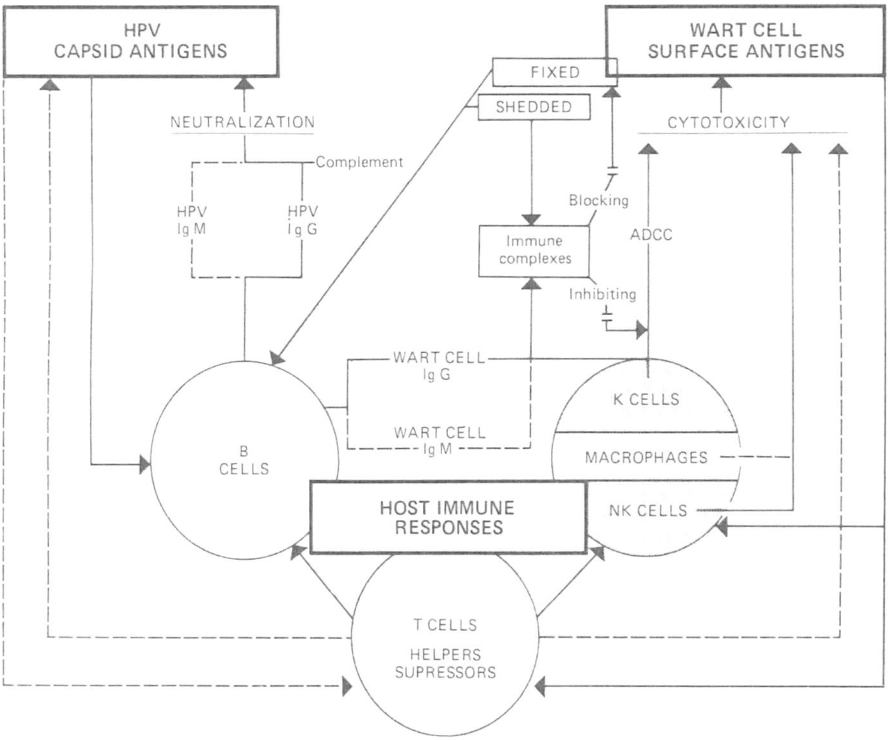

Fig. 17. Major immunological events potentially elicited during the progression of warts. HPV antibodies, predominantly of the IgG class, may contribute to limit the tumor bulk during wart establishment in various epithelia through complement-dependent neutralization of the virus. Humoral and cell-mediated immune responses directed against tumor surface antigens are believed to represent the major effector mechanisms inducing a cytotoxic response. In this respect, NK cells and macrophages are the predominant components of cell-mediated resistance; macrophages may contribute nonspecifically, while clones of NK cells are coded for specific cognizance of tumor neoantigens and other surface markers. K cells effectuate antibody-dependent, cell-mediated cytotoxicity *(ADCC)* through cognizance of tumor-cell-specific IgG on the wart cells, to which K cells attach due to their possession of Fc receptors; the reaction may be inhibited by the occurrence of circulating immune complexes due to shedding of wart cell antigens, or blocked by immune complexes coating the tumor. This figure does not account for the immunity-modulating influence of, e.g., interferons and lymphokines and does not include the concept of skin- and mucous membrane - associated lymphoid tissue.

Thus, the incidence of spontaneous regression decreases as a function of any systemic suppression of cell-mediated immunity (Jablonska et al. 1982; Kreider and Bartlett 1985). Although the mechanisms of immunological rejection are not understood as yet, nevertheless studies of man as well as of some animal models clearly indicate that the predominant immunological events are those of cell-mediated immune responses directed against the wart cell. As shown in the Shope rabbit papilloma model, antiviral humoral immunity does not appear to be of primary importance for papilloma rejection. Thus, a vaccine prepared from papilloma, but not from the virus alone, may augment significantly the regression rate of Shope papillomata (Kreider and Bartlett 1981). The favorable prognostic association with the presence of HPV-IgG that has been demonstrated in wart patients (Pyrhönen and Johansson 1975; Pyrhönen and Penttinen 1972) may, accordingly, very well reflect secondary events during wart cure. Specific antiviral antibodies are perhaps of importance in preventing the spread of virus during the process of regression (von Krogh 1979).

According to the current concept, antitumor surveillance and rejection phenomena are associated predominantly with activation of cell-mediated immune mechanisms through a specific recognition and subsequent destruction of transformed cells. The theory implies that such cells, including those induced by viruses, possess cell surface antigens that distinguish them from normal cells. Such "neoantigens" may in part be related to the transplantation (HLA) antigens and in part to virus-specific changes of the cell membrane in the transformed cell. However, tumor-specific antigens have not been conclusively recognized for any human tumor to date (Safai and Leffell 1985).

Generation of tumor-specific immunity is a complex phenomenon in which several cell types cooperate (Safai and Leffell 1985; Cottier et al. 1986). Natural killer (NK) cells represent one of the main effectors of cell-mediated immunity against virus-infected and neoplastic cells. Specific clones probably develop through the course of tumor formation and seek their target by recognition of neoantigens on tumor-cell surfaces. Assisted by macrophages, NK-cells seem to contribute significantly to cell-mediated tumor surveillance. Humoral responses may also play some role in this process. Thus, NK-cells, representing the predominant effector cell for antibody-dependent, cell-mediated cytotoxicity (ADCC), may exert a cytotoxic effect on IgG-coated target cells due to their possession of membrane Fc receptors with avidity for the activated Fc region of IgG. Target-cell selectivity of ADCC resides in the specificity of the coating antibody and not in the effector lymphocyte (Herberman 1980). The histopathological features of regressing warts are similar to those observed in delayed hypersensitivity reactions. The primary change entails a dermal accumulation of mononuclear cells which are predominantly of the T-lymphocyte type. As regression continues, mononuclear cells tend to infiltrate the epidermis and a degeneration of epidermal cells occurs. Both suppressor and helper T lymphocytes as well as killer cells seem to take part in this process (Iwatsuki et al. 1986; Bender 1986). The crucial question is why this process seems to be inhibited or delayed in some instances; it is a well-recognized phenomenon that various degrees of similar inflammatory responses in the dermis may occur in nonregressing, recalcitrant warts as well (Chardonnet et al. 1986; Väyrynen et al. 1985).

Stimulation of tumor-specific immunity depends on a proper processing and presentation of tumor-specific antigens. It is beyond the scope of the current presentation to elucidate this complex issue. However, it deserves note that the process entails an expression of HLA-DR antigens when T-cells, macrophages, and Langerhans' cells (LC) are being activated. Furthermore, stimulation of these immune functions are mediated in part by a chemical communication involving interleukin I and II as well as various interferons.

The concept has evolved in recent years that an integrated immunological entity may exist in the skin as well as in mucous membranes. Designated "skin-associated lymphoid tissue" (SALT) and "mucosal-associated lymphoid tissue" (MALT) respectively, these systems should be regarded as part of the local environment. They are probably capable on their own of accepting, processing, and presenting antigens that have been conditioned by an interplay between LC, keratinocytes, and immunocompetent lymphocytes (Poulter 1983; Streilein 1983; Breathnach 1986). Recently, it has been demonstrated that a dysbalance might exist with respect to the distribution as well as the morphology of LC within the SALT and MALT systems. Thus, in nonregressing warts either on the skin or anogenital mucosa, the quantity of LC is frequently reduced, and in some warts they are completely absent. Also, some of the LC present have lost their normal dendritic shape (Chardonnet et al. 1986). HLA-DR antigens are only expressed in a few LC and also have a reduced expression in the subepidermal cellular infiltrate. It is yet not possible to interpret the dynamic implications of these findings. However, the issue of persistent warts in some individuals may certainly turn out to be one of a local immune defect in the skin and mucous membranes.

Various immunotherapeutic approaches to HPV infection are believed to act predominantly through a stimulation of cell-mediated mechanisms by activating NK-cells and macrophages. However, other immune functions may be activated as well, for example an enhancement of HLA-DR expression on LC (Sontheimer et al. 1986).

4.1 Interferons

The cytokines designated as interferons (IFNs) comprise a group of secretory proteins produced in vivo by a large number of cells. Leukocytes, lymphocytes, macrophages, and fibroblasts are particularly productive in response to various stimuli such as exposure to virally infected or neoplastic cells. Through binding to specific receptors on the cell surface, IFNs may augment HLA gene expression two- to fivefold and thereby enhance recognition and destruction of tumor cells. Intracellular replication of viruses may also be suppressed. Furthermore, IFNs have a direct, nonspecific, immunoregulatory effect through enhancement of phagocytosis by macrophages and of NK-cell functions (Herberman 1980; Revel et al. 1984; Lau et al. 1985).

DNA recombinant techniques have advanced the study of the molecular structure of IFNs and the understanding of the genes encoding for these proteins. Thereby, the chemical heterogeneity of IFNs has become increasingly evident. The biological implications of this diversity remain to be established. An ongoing expansion of biological and therapeutic work has been made possible by the availability of pure IFNs produced through recombinant DNA technology in bacterial vectors. Classification of IFNs is, in part, based on cell of origin and type of inducer. Type 1 interferons are induced by viruses and include leukocyte *(α)* and fibroblast *(β)* IFNs, while type 2 includes immune *(γ)* IFN which is secreted by lymphocytes following a specific antigenic or mitogenic stimulus (Friedman 1977). The *α*- and *β*-IFNs are closely related chemically, while no overall amino acid homology exists with *γ*-IFN. *γ*-IFN seems most effective for the induction of HLA expression and, accordingly, is most interesting with regard to tumor immunology. Furthermore, it also potentiates the effect of the other interferon types (Berman and Jaliman 1985). Nevertheless, most progress has been made in dissecting the biochemistry of *α*-IFN; 20 different but structurally closely related subtypes have been identified. Therefore, purified *α*-IFNs have been most extensively used in clinical trials so far.

The amount of IFN available in a particular preparation is determined by an IFN assay based on measurements of antiviral activity in cell culture. The units so obtained are compared to an International Reference Standard IFN Preparation, and the titre is expressed in International Units (IU) (Strander and Einhorn 1982). IFNs are usually administered at dose levels of $3-10 \times 10^6$ IU. When administered intraveneously, a rapid clearing from the blood occurs. Therefore, intramuscular inoculation is preferred, ensuring that significant levels are present in most body fluids for the following 24 h. The most important toxic side effects include transient flulike symptoms such as fever, muscle pain, and fatigue for the first few hours after injection, as well as reversible leukopenia, thrombocytopenia, and hepatic toxicity. These phenomena tend to decrease in frequency and severity with continuing administration of the drug.

The bovine papillomavirus (BPV)-transformed mouse fibroblast system has provided a useful model for examining the biological and biochemical effect of IFN on cells persistently infected with a papillomavirus (Turek et al. 1982). Thus, interferon treatment may reduce the efficiency of cell transformation and is also capable of inducing morphological reversion of cells already transformed. Furthermore, a reduction or complete elimination of BPV DNA was accomplished.

Recently an increasing interest has been generated concerning IFNs as attractive possible agents for treating HPV infections, in particular treatment-resistant lesions of the anogenital areas and the respiratory tract. The results from some of the preliminary trials reported so far are encouraging, indicating that interferon therapy has a niche in the therapeutic arsenal for warts. However, definite limitations exist with respect to efficacy as well as systemic adverse effects from the use of high doses. The use of exogeneous IFN must be considered highly experimental until more information is available in response to some crucial questions: which IFN type gives optimal results, and what is the proper dosage and way of administration.

4.1.1 Treatment of Anogenital HPV Infection

Lymphoblastoid *(α)* and fibroblast *(β)* IFN have been administered intramuscularly in some pilot studies on patients with treatment-resistant genital warts. Doses in the range of 1-6 $\times 10^6$ IU have been given either daily or three times a week for a total of 1.5-6 weeks, reflecting initial lack of knowledge on optimal dosage and treatment duration (Alawattegama and Kinghorn 1984; Schonfeld et al. 1984; Gall et al. 1985; Olsen et al. 1985). The results have been variable with respect to cure rates, being in the range of 25%-100%. However, these preliminary trials have clearly demonstrated a therapeutic effect from IFN on genital warts and seem to indicate that individual doses exceeding 2-3 $\times 10^6$ IU are not required for such treatment. Furthermore, administration 2-3 times a week seems as effective as daily injections. There are no indications that α-IFN is superior to β-IFN.

With the goal of reaching a high IFN concentration on the sites of condylomata without evoking significant toxic systemic reaction, a number of recent studies have focused on the effect of relatively low doses of α- or β-IFN injected into the tumor base. Injections were performed 2-3 times a week in most studies, and the duration of therapy varied from 3 to 14 weeks (McEwen et al. 1983; Friedman-Kien et al. 1985; Gall et al. 1985; Hatch et al. 1985; Peets and Eron 1985). These studies have produced encouraging results and added some new information of clinical importance. Using recombinant α-2-IFN (injections with 1-5 $\times 10^6$ IU three times a week for 3 weeks), resistant lesions have disappeared in about half the patients, as documented after up to 2 years of follow-up. Friedman-Kien et al. (1985), who injected warts twice a week for 8 weeks, reported a cure rate of 83% with a dosage of about 600000 IU per attendance. In some of the responders the wart size started diminishing after the first few treatments and continued to show a gradual regression subsequent to cessation of therapy; a complete disappearance was observed within the first 3 months after therapy was initiated. Mild flulike symptoms or fever occurred in about half the cases, indicating that some systemic effects from injections into the lesions cannot be avoided. However, such adverse effects tended to disappear after the first few injections. A few patients complained of some pain at the injection site.

A major advantage from successful IFN treatment is the absence of ulcerations or subsequent scarring. Relatively small lesions react more favorably than large tumors. Nevertheless, even in cases not responding with a complete cure, some reduction in the tumor bulk has been observed. Therefore, IFN therapy may very well prove to be valuable as an adjuvant to surgical treatment whenever the lesions are too large for primary excision in a single session. However, because of the requirement for numerous visits, IFN treatment should only be considered in particularly well-motivated patients.

Topical treatment with IFN incorporated in creams and ointments has also been tried for genital warts. Ikić et al. (1981) reported that a complete remission of exophytic vulvar warts could be obtained in most cases when applications were performed 4-5 times daily for 4-12 weeks. However, the method is tedious and lengthy; therefore, it cannot be considered practical for routine use. Vesterinen et al. (1984) treated extensive flat vaginal warts with a cream containing leukocyte IFN (2 $\times 10^6$ IU per g); the cream was applied nightly during four 2-week treatment courses. Five of eight patients showed complete clearance. No overt side effects have been observed from these studies. The use of self-administered topical IFN preparations may become a practical possibility in the future; however, much more work needs to be done in order to establish stable preparations of sufficient potency.

A few preliminary reports of a favorable effect of IFN on various degrees of CIN and bowenoid papulosis must as yet be interpreted with caution; further data are required for any conclusions to be reached (de Palo et al. 1984; Hsu et al. 1984; Choo et al. 1985).

4.1.2 Treatment of Skin Warts

Very few studies have been performed on the effect of IFNs on skin warts in immuno-competent individuals, and the results are not conclusive, although some favorable data have been presented (McEwen et al. 1983; Yamazaki 1983).

4.2 Autogenous Vaccine Therapy

Potential use of vaccine therapy with wart tissue components represents an old concept which was first tested by Biberstein in 1925. High cure rates were achieved for patients with warts after injection with crude wart extracts. A dispute in the literature ensued when these results could not be reproduced by others, and the technique was forgotten. During the past decade, however, the method has been reevaluated in patients with condylomata who received subcutaneous injections of autogenous tumor homogenates repeatedly for about 6 weeks. Although initial reports indicated cure rates of more than 80% (Abcarian et al. 1976; Abcarian and Sharon 1977; Eftaiha et al. 1982), a dispute has again arisen with respect to the scientific background for current use of such techniques. It has recently been documented that autogenous wart vaccine does not seem to induce a higher frequency of wart rejection than use of 'placebo' vaccine prepared from the patients' own normal skin (Malison et al. 1982). This is not to say that the immune system does not play a role in the resolution of wart disease; it rather focuses on specific cellular factors capable of enhancing the efficacy of the immune system. Injections of homogenates from normal epidermis may stimulate a focal attraction of nonspecific cytotoxic cells through induction of interferon production (Herbeman 1980). Recently, it has been advocated that patients should no longer be treated with autogenous vaccines because of the potential oncogenic effect of the viral DNA (Bunney 1986). However, the concept of vaccine therapy is fascinating; the development of an HPV subunit vaccine through isolation of the surface antigen gene from the genome is being explored at present (Faras and Pass 1985). Research in this area seems to be reaching a crescendo phase, providing hope that more scientifically monitored vaccine programs may eventually be developed for various HPV-associated disorders.

4.3 Treatment of Skin Warts in Immune-Deficient Individuals

Immunodeficiency can be associated with the development of extensive persistent warts (Reid et al. 1976; Jablonska et al. 1982). The occurrence of generalized treatment-resistant warts is also the rule in the severe form of epidermodysplasia verru-

Table 2. Cutaneous warts in immune-compromised individuals

	HPV type
Epidermodysplasia verruciformis	3, 5, 8, 9, 10, 12, 14, 15
Renal transplants	5, 27
Others	5, 26

ciformis (Lutzner 1978) where a number of specific HPV types have been identified (Table 2). In such patients the lack of therapeutic alternatives represents a major problem due to severe cosmetic implications and because HPV-induced warts in such patients may be potentially oncogenic.

Systemic administration of *α-IFN* may have a favorable effect when given for 4 weeks. However, recurrences seem to be inevitable after cessation of treatment (Androphy et al. 1984). The risks and benefits of long-term, prophylactic IFN in these patients merit investigation.

4.3.1 Oral Retinoids

Tigason may give a temporary substantial improvement when administered in doses starting at about 100 mg a day. A reduction in the number and size of warts may be seen within a few weeks of therapy, at which point the dose may be reduced. Unfortunately, as with interferon, most warts tend to reappear when treatment is stopped. Therefore, continuous medication is required; in such cases, the dosage should be lowered gradually and maintained at the lowest possible level in order to minimize any toxic side effects such as cheilitis, hair loss, and shift in laboratory values. Also, due to their potential teratogenic properties, they cannot be given during pregnancy (Boyle et al. 1983; Gross et al. 1983).

4.3.2 Levamisole

Levamisole may increase cellular immunity; a thymosin-like effect is possible, inducing lymphocyte proliferation as well as augmented lymphocytotoxicity and lymphokine production. The optimal dosage is unknown. Oral treatment with 5 mg/kg body weight on 3 consecutive days every 2 weeks has been given in up to 12 cycles. However, the results in various publications have been conflicting. Thus, Moncada and Rodriguez (1979) claimed that the cure rate was as high as 77% when patients with multiple resistant warts were treated for 1-4 months, while Schou and Helin (1977) found no difference when compared with placebo. Subsequent experiences revealed that levamisole did not live up to its early promise (Bunney 1986).

4.3.3 2,4-Dinitrochlorobenzene

2,4-Dinitrochlorobenzene (DNCB) application in sensitized individuals has been used as an immunomodulatory therapeutic approach to attract immunologically active lymphocytes focally to the site of warts. The method has no place in routine treatment but may be of some value in patients with chronic warts and depressed, cell-mediated, immune responses, in which case half of the patients may be cured. However, the method is tedious and should be reserved for particularly distressing cases (Johansson and Forström 1984). The fact that the magnitude of challenge doses required for cure will induce a dermatitis in the sensitive anogenital areas indicates that the regimen is quite unsuitable for therapy of condylomata (von Krogh, unpublished data).

Somewhat more encouraging results have been achieved with methods that attempt to attract potentially cytotoxic effector cells to the site of tumor growth. Injection of *Bacille Calmette Guérin (BCG)* into condylomatous lesions is technically difficult and associated with intense local pain; alternative injection into adjacent anogenital areas is better tolerated but does not assure cure (Malison and Salkin 1981). The possibility exists of using tuberculin jelly for topical application in future trials. This procedure has led to the cure of treatment-resistant common warts in 57% of cases (Lahti and Hannuksela 1982). The method is not attractive for routine use because applications several times a week for many months seem necessary.

References

Abcarian H, Sharon N (1977) The effectiveness of immunotherapy in the treatment of anal condylomata acuminatum. J Surg Res 22: 231-236

Abcarian H, Sharon N (1982) Long-term effectiveness of the immunotherapy of anal condyloma acuminatum. Dis Colon Rectum 25: 648-651

Abcarian H, Smith D, Sharon N (1976) The immunotherapy of anal condylomata acuminatum. Dis Colon Rectum 19: 237-244

Alawattegama AB, Kinghorn GR (1984) Bowenoid dysplasia in human papillomavirus-16-DNA positive flat condylomas during interferon treatment. Lancet i: 1467-1468

Anderson MC, Hartley RB (1980) Cervical crypt involvement by intraepithelial neoplasia. Obstet Gynecol 55: 546-550

Anderson MC, Horwell D, Broby Z (1984) Outcome of pregnancy after laser vaporization conization. Colposcopy Gynecol Laser Surg 1: 35-40

Androphy EJ, Dvoretzky I, Maluish AE, Wallace HJ, Lowy DR (1984) Response of warts in epidermodysplasia verruciformis to treatment with systemic and intralesional alpha interferon. J Am Acad Dermatol 11: 197-202

Baggish MS (1980) Carbon dioxide laser treatment for condylomata acuminata venereal infection. Obstet Gynecol 55: 711-714

Baggish M (1982) Management of cervical intraepithelial neoplasia by carbon dioxide laser. Obstet Gynecol 60: 378-384

Baggish MS, Dorsex JH (1985) Carbon dioxide laser for combination excisional-vaporzation conization. Am J Obstet Gynecol 151: 23-27

Bashi SA (1985) Cryotherapy versus podophyllin in the treatment of genital warts. Int J Dermatol 24: 535-536

Bellina J (1983) The use of the carbon dioxide laser in the management of condyloma acuminatum with 8-year follow up. Am J Obstet Gynecol 147: 375-378

Bender ME (1986) Concepts of wart regression. Arch Dermatol 122: 644-647

Bergman A, Bhatia NN, Broen EM (1984) Cryotherapy for treatment of genital condylomata during pregnancy. J Reprod Med 29: 432-435

Berman B, Jaliman D (1985) The human interferon system. In: Stone J (ed) Dermatologic immunology and allergy. Mosby, St Louis, pp 899-909

Biberstein H (1925) Versuche über Immunotherapie der Warzen und Kondylome. Klin Wochenschr 4: 638-641

Billingham RP, Lewis FG (1982) Laser versus electrical cautery in the treatment of condylomata acuminata. Surg Gynecol Obstet 155: 865-867
Bodén E, Rylander E, Evander U, Wadell G, von Schultz B (1986) Papilloma virus infection of the vulva. Int J Reprod Med (to be published)
Boyle J, Dick DC, Mackie RM (1983) Treatment of extensive virus warts with etretinate (Tigason) in a patient with sarcoidosis. Clin Exp Dermatol 8: 33-36
Breathnach SM (1986) Do epidermotropic T cells exist in normal human skin? A re-evaluation of the SALT hypothesis. Br J Dermatol 115: 389-392
Bremner RM (1976) Warts: treatment with intralesional bleomycin. Cutis 18: 264-266
Bunney MH (ed) (1982) Viral warts: their biology and treatment. Oxford University Press, Oxford
Bunney MH, Nolan M, Williams D (1976) An assessment of methods of treating viral warts by comparative treatment trials based on a standard design. Br J Dermatol 94: 667-669
Bunney MH (1986) Viral warts: a new look at an old problem. Br Med J 293: 1045-1047
Burke L (1982) The use of the carbon dioxide laser in the therapy of cervical intraepithelial neoplasia. Am J Obstet Gynecol 144: 337-340
Cetti NE (1984) Condyloma acuminatum of the urethra: problems in eradication. Br J Surg 71: 57
Chamberlain MJ, Reynolds AL, Yeoman WB (1972) Toxic effect of podophyllum application in pregnancy. Br Med J 3: 391-393
Chardonnet Y, Viac J, Thivolet J (1986) Langerhans' cells in human warts. Br J Dermatol 115: 669-675
Choo YC, Hsu C, Seto WH, et al. (1985) Intravaginal application of leucocyte interferon gel in the treatment of cervical intraepithelial neoplasia (CIN). Arch Gynecol 237: 51-54
Clarke GHV (1965) The charming of warts. J Invest Dermatol 45: 15-20
Cordero AA, Guglielmi HA, Woscoff A (1980) The common wart: intralesional treatment with bleomycin sulfate. Cutis 26: 319-324
Cottier H, Hess MW, Walti ER (1986) Immunodeficiency and cancer: mechanisms involved. Schweiz Med Wschr 116: 1119-1126
Crumay HM (1985) Physical modalities of therapy, Sect 1. In: Moschella SL, Hurley HJ (eds) Dermatology, vol 2. WB Saunders, Philadelphia, pp 1996-2001
Culp OS, Kaplan IW (1944) Condylomata acuminata. Two hundred cases treated with podophyllin. Ann Surg 120: 251-256
Czarnecki MB (1984) Bleomycin and periungual warts. Med J Aust 141: 40
Dodi G, Infantino A, Moretti R, Scalco G, Lise M (1982) Cryotherapy of anorectal warts and condylomata. Cryobiology 19: 287-288
Dolezal JF (1985) Podophyllum resin and podophyllin. J Am Acad Dermatol 12: 728-730
Dorsey J, Diggs E (1979) Microsurgical conization of the cervix by carbon dioxide laser. Obstet Gynecol 54: 565-570
Dretler SP, Klein I (1975) The eradication of intra-urethral condylomata acuminata with 5% fluorouracil cream. J Urol 113: 195-197
Eftaiha MS, Amshel AL, Schonberg IL, et al. (1982) Giant and recurrent condyloma acuminatum. Dis Colon Rectum 25: 136-140
Faras AJ, Pass F (1985) Applications of biotechnology to human papillomavirus-induced diseases. Clin Dermatol 3/4: 200-203
Ferenczy A (1984a) Comparison of 5-fluorouracil and CO_2 laser for treatment of vaginal condylomata. Obstet Gynecol 64: 773-778
Ferenczy A (1984b) Treating genital condyloma during pregnancy with the carbon dioxide laser. Am J Obstet Gynecol 148: 9-12
Ferenczy A (1985) Comparison of cryo- and carbon dioxide laser therapy for cervical intraepithelial neoplasia. Obstet Gynecol 66: 793-797
Ferenczy A, Mitao M, Nagai N, Silverstein SJ, Crum CP (1985) Latent papillomavirus and recurring genital warts. N Engl J Med 313: 784-788
Figueroa S, Gennaro A (1980) Intralesional bleomycin injection in treatment of condyloma acuminatum. Dis Colon Rectum 23: 550-552
Friedman R (1977) Antiviral activity of interferons. Bacteriol Rev 41: 543-567
Friedman-Kien AE, Plasse TF, Cremin P, Castro B, Badiak H, Geffen JR, Fedorczyk D, Trout R (1985) Natural leukocyte interferon for treatment of condylomata acuminata: a randomized,

double-blind, placebo controlled study. Workshop on papillomaviruses, Kuopio, Finland, August 25–29
Gabriel G, Thin RNT (1983) Treatment of anogenital warts. Comparison of trichloracetic acid and podophyllin versus podophyllin alone. Br J Vener Dis 59: 124–126
Gall SA, Hughes CE, Weck P, Whisnant J (1984) Interferon for the treatment of resistant condyloma acuminata. Gynecol Oncol 17: 264–266
Gall SA, Hughes CE, Trofatter K (1985) Interferon for the therapy of condyloma acuminatum. Am J Obstet Gynecol 153: 157–163
Gibson JR, Harvey SG, Kemmett D, Salisbury J, Marks P (1986) Treatment of common and plantar viral warts with human lymphoblastoid interferon-α – pilot studies with intralesional, intramuscular and dermojet injections. Br J Dermatol 115 [Suppl 31]: 76–79
Gross G, Pfister H, Hagedorn M, Stahn R (1983) Effect of oral aromatic retinoid (RO 10-9359) on human papilloma virus-2-induced common warts. Dermatologica 166: 48–53
Growdon W, Fu Y, Lebherz T, Rapkin A, Mason GD, Parks G (1985) Pruritic vulvar squamous papillomatosis: evidence for human papillomavirus etiology. Obstet Gynecol 66: 564–568
Grundsell H, Larsson G, Bekassy Z (1984) Treatment of condylomata acuminata with the carbon dioxide laser. Br J Obstet Gynaecol 91: 193–196
Halverstadt RB, Parry WL (1969) Thiotepa in the management of intraurethral condylomata acuminata. J Urol 101: 729–730
Harvey JM, Glen E, Watson GS (1983) Buschke-Loewenstein tumour of the penis. Br J Vener Dis 59: 273–276
Hatch KD, Bart JB, Hansen RC, Millikan LE, Reichman RC, Berman B (1985) Evaluation of interferon alpha 2 in the treatment of condyloma acuminatum. 14th international congress on chemotherapy, Kyoto, Japan, June 23–28
Heath A, Mellstrand T, Ahlmén J (1982) Treatment of podophyllin poisoning with resin hemoperfusion. Hum Toxicol 1: 373–378
Herberman RB (ed) (1980) Natural cell-mediated immunity against tumors. Academic Press, New York
Hindson TC, Spiro J, Scott LV (1985) Clobetasol proprionate ointment reduces inflammation after cryotherapy. Br J Dermatol 112: 599–602
Hsu C, Choo Y-C, Seto W-H, Pang S-W, Tan CY, Merigan TC, Ng MH (1984) Exfoliative cytology in the evaluation of interferon treatment of cervical intraepithelical neoplasia. Acta Cytol 28: 111–117
Hudson AL (1976) Treatment of plantar warts with bleomycin. Arch Dermatol 112: 1179
Ikenberg H, Gissman L, Gross G, Grußendorf-Conen E, zur Hausen H (1983) Human papillomavirus type 16-related DNA in genital Bowen's disease and in bowenoid papulosis. Int J Cancer 32: 563–565
Ikić D, Trajer D, Čupak K, Petricevic I, Prazic M, Soldo I, Jusic D, Smerdel S, Soos E (1981) The clinical use of human leukocyte interferon in viral infections. Int J Clin Pharmacol Ther Toxicol 19: 498–505
Iwatsuki K, Tagami H, Takigawa M, Jamada M (1986) Plane warts undergo spontaneous regression. Immunopathologic study on cellular constituents leading to the inflammatory reaction. Arch Dermatol 122: 655–659
Jablonska S, Orth G, Lutzner MA (1982) Immunopathology of papillomavirus-induced tumors in different tissues. Springer Semin Immunopathol 5: 33–62
Jensen SL (1985) Comparison of podophyllin application with simple surgical excision in clearance and recurrence of perianal condylomata acuminata. Lancet ii: 1146–1148
Johansson E, Forström L (1984) Dinitrochlorobenzene (DNCB) treatment of viral warts. A 5-year follow-up study. Acta Derm Venereol (Stockh) 64: 529–533
Karol MD, Conner CS, Murphrey KJ (1980) Podophyllum: suspected teratogenicity from topical application. Clin Toxicol 16: 283–285
Khawaja HT (1986) Treatment of condyloma acuminatum. Lancet i: 208–209
Kreider JW, Bartlett GL (1981) The Shope papilloma-carcinoma complex of rabbits: a model system of neoplastic progression and spontaneous regression. Adv Cancer Res 25: 81–110
Kreider JW, Bartlett GL (1985) Shope rabbit papilloma carcinoma complex: a model system of human papillomavirus infections. Clin Dermatol 3: 20–26
Lahti A, Hannuksela M (1982) Topical immunotherapy with tuberculin jelly for common warts. Arch Dermatol Res 273: 153–155

Lassus A, Haukka K, Forsström S (1984) Podophyllotoxin for treatment of genital warts in males. A comparison with conventional podophyllin therapy. Eur J Sex Transm Dis 2: 31-33

Lau AS, Hannigan GE, Freedman MH, Williams BRG (1985) Regulation of interferon receptor expression on human lymphocytes. Pediatr Res 19: 299A

Lundquist SB, Lindstedt EM (1983) Laser treatment of condylomata acuminata in the male (Abstr). Laser Surg Med 2 (3): 177

Lutzner MA (1978) Epidermodysplasia verruciformis. An autosomal recessive disease characterized by viral warts and skin cancer. A model for viral oncogenesis. Bull Cancer (Paris) 65: 169-182

Malison MD, Salkin D (1981) Attempted BCG immunotherapy for condylomata acuminata. Br J Vener Dis 57: 148-150

Malison MD, Morris R, Jones LW (1982) Autogenous vaccine therapy for condylomata acuminatum. A double-blind controlled study. Br J Vener Dis 58: 62-67

Massing AM, Epstein WL (1963) Natural history of warts. Arch Dermatol 87: 306-309

McEwen CJ, Millikan LE, Goswitz MS (1983) Parallel group study comparing alpha 2 interferon to placebo in the treatment of warts. Clin Res 31: 921A

Meisels A, Roy M, Fortier M, Morin C, Casas-Cordero M, Shah KV, Turgeon H (1981) Human papillomavirus infection of the cervix. Acta Cytol (Baltimore) 25: 7-16

Moncada B, Rodriguez ML (1979) Levamisole therapy for multiple warts. Br J Dermatol 101: 327-330

Mortimer PS, Sonnex TS, Dawber RPR (1983) Cryotherapy for multicentric pigmented Bowen's disease. Clin Exp Dermatol 8: 319-322

Munkvad M, Genner J, Staberg B, Kongsholm H (1983) Locally injected bleomycin in the treatment of warts. Dermatologica 167: 86-89

Olsen EA, Trofatter KF, Gall SA, Medoff JR, Hughes CE, Weiner MS, Kelly FF (1985) Human lymphoblastoid interferon-alpha in the treatment of refractory condyloma acuminata. Clin Res 33: 673A

Olson RL (1977) Plantar warts yield to DNA inhibitor. JAMA 237: 940-941

Oriel D (1971a) Natural history of genital warts. Br J Vener Dis 47: 1-13

Oriel D (1971b) Anal warts and anal coitus. Br J Vener Dis 47: 373-376

Palo G de, Stefanon B, Rilke F, Pilotti S, Ghione M (1984) Human fibroblast interferon in cervical and vulvar intraepithelial neoplasia associated with papilloma virus infection. Int J Tissue React 6: 523-527

Peets E, Eron L (1985) Treatment of venereal warts with recombinant alpha 2 B interferon. Nordic round-table conference, Copenhagen, Denmark, October 24

Poulter LW (1983) Antigen presenting cells in situ: their identification and involvement in immunopathology. Clin Exp Immunol 53: 513-520

Pyrhönen S, Johansson E (1975) Regression of warts. An immunological study. Lancet i: 592-595

Pyrhönen S, Penttinen K (1972) Wart-virus antibodies and the prognosis of wart disease. Lancet ii: 1330-1332

Reid TMS, Fraser NG, Kernohan IR (1976) Generalized warts and immune deficiency. Br J Dermatol 95: 559-564

Revel M, Mory Y, Chernajovsky Y, Vaks B, Chebath J, Kimchi A, Schonfeld A (1984) Human interferons: production, action and therapeutic use. Isr J Med Sci 20: 467-477

Rylander E, Isberg A, Joelsson I (1984) Laser evaporization of cervical intra-epithelial neoplasia. A five-year follow-up. Acta Obstet Gynecol Scand [Suppl] 125: 33-36

Rylander E, Eriksson A, Ingelman-Sundberg H, von Schultz B (1985) Classification of colposcopic findings associated with human papilloma virus infection of cervix uteri and vagina. Cervix Lower Female Genital Tract 3: 123-132

Safai B, Leffell D (1985) Tumor immunology. In: Stone J (ed) Dermatologic immunology and allergy. Mosby, St Louis, pp 825-833

Samenius B (1983) Perianal and ano-rectal condyloma acuminatum. Schweiz Rundsch Med 72: 1009-1014

Schonfeld A, Schattner A, Crespi M, et al. (1984) Intramuscular human interferon-beta injections in treatment of condylomata acuminata. Lancet i: 1038-1041

Schou M, Helin P (1977) Levamisole in a doubleblind study. No effect on warts. Acta Derm Venereol 57: 449-454

Shumack DH, Haddock MJ (1979) Bleomycin: an effective treatment for warts. Austr J Dermatol 20: 41-42
Shumer SM, O'Keefe EJ (1983) Bleomycin in the treatment of recalcitrant warts. J Am Acad Dermatol 9: 91-96
Simmons PD (1981) A comparative double-blind study of 10% and 25% podophyllin in the treatment of anogenital warts. Br J Vener Dis 57: 208-209
Simmons PD, Thomson JPS (1986) Scissor excision of penile warts: case report. Genitourin Med 62: 277-278
Sontheimer RD, Stastny P, Nũnez G (1986) HLA-D region antigen expression by human epidermal Langerhans' cells. J Invest Dermatol 87: 707-710
Stankler L (1967) A critical assessment of the cure of warts by suggestion. Practitioner 198: 690-694
Strander H, Einhorn S (1982) Interferon in cancer - faith, hope and reality. Am J Clin Oncol 5: 297-301
Streilein JW (1983) Skin-associated lymphoid tissue (SALT): origins and functions. J Invest Dermatol 80: 12s-16s
Swerdlow DB, Salvati EP (1971) Condyloma acuminatum. Dis Colon Rectum 14: 226-230
Syrjänen K, Väyrynen M, Hippeläinen M, Castrén O, Saarikoski S, Mäntyjärvi R (1983). The in situ immunological reactivity and its significance in the clinical behavior of cervical human papillomavirus lesions. Neoplasma 32: 181-190
Tagami H (1984) Regression phenomenon of numerous flat warts - an experiment on the nature of tumor immunity in man. Int J Dermatol 23: 570-571
Tenzel JH, Taylor RL (1969) An evaluation of hypnosis and suggestion as treatment for warts. Psychosomatics 10: 253-257
Townsend DE, Ostergard DR, Lickrish GM (1971) Cryosurgery for benign disease of the cervix. J Obstet Gynecol 78: 667-669
Turek LP, Byrne JC, Lowy DR, Dvoretzky I, Friedman RM, Howley PM (1982) Interferon induces morphological reversion with elimination of extrachromosomal viral genomes in bovine papillomavirus-transformed mouse cells. Proc Natl Acad Sci USA 79: 7914-7918
Ullman M, Dudek S (1960) On the psyche and warts: II. Hypnotic suggestion and warts. Psychosom Med 22: 68-76
Väyrynen M, Syrjänen K, Mäntyjärvi R, Castrén O, Saarikoski S (1985) Immunophenotypes of lymphocytes in prospectively followed up human papillomavirus lesions of the cervix. Genitourin Med 61: 190-196
Vesterinen E, Meyer B, Cantell K, Purola E (1984) Topical treatment of flat vaginal condyloma with human leucocyte interferon. Obstet Gynecol 64: 535-538
von Krogh G (1976) 5-fluorouracil cream in the successful treatment of therapeutically refractory condylomata acuminata of the urinary meatus. Acta Derm Venereol (Stockh) 56: 297-300
von Krogh G (1978) Topical treatment of penile condylomata acuminata with podophyllin, podophyllotoxin and colchicine. A comparative study. Acta Derm Venereol (Stockh) 58: 163-168
von Krogh G (1979) Warts: immunologic factors of prognostic significance. Int J Dermatol 18: 195-204
von Krogh G (1981a) Podophyllotoxin for condylomata acuminata eradication. Clinical and experimental comparative studies on *Podophyllum* lignans, colchicine and 5-fluorouracil. Acta Derm Venereol (Stockh) Suppl 98
von Krogh G (1981b) Penile condylomata acuminata: an experimental model for evaluation of topical self-treatment with 0.5%-1.0% ethanolic preparations of podophyllotoxin for three days. Sex Transm Dis 8: 179-186
von Krogh G (1982) Podophyllotoxin in serum: absorption subsequent to three-day repeated applications of a 0.5% ethanolic preparation on condylomata acuminata. Sex Transm Dis 9: 26-33
von Krogh G (1983) Condylomata acuminata 1983: an up-dated review. Semin Dermatol 2: 109-129
von Krogh G (1987) Topical self-treatment of penile condylomata with 0.5% podophyllotoxin twice versus once daily for four or five days. Sex Transm Dis. In Press
von Krogh G, Rudén A-K (1980) Topical treatment of penile condylomata acuminata with colchicine at 48-72 hours intervals. Acta Derm Venereol (Stockh) 60: 87-89

von Krogh G, Rylander E (1986) Genital papilloma virus infections (GPVI). Epidemiology, clinical presentation and treatment. In: Thorén L, Beerman B, Ljunggren H, Lönnerholm G (eds) Treatment of sexually transmitted diseases. National board of health and welfare drug information committee, Sweden, vol 4, pp 85-106

Wein AJ, Benson GS (1977) Treatment of urethral condylomata acuminatum with 5 FU cream. Urology 9: 413-415

Willcox RR (1977) How suitable are available pharmaceuticals for the treatment of sexually transmitted diseases? II. Conditions presenting as sores or tumors. Br J Vener Dis 53: 340

Wright VC, Davies E, Riopelle MA (1983) Laser surgery for cervical intraepithelial neoplasia: principles and results. Am J Obstet Gynecol 145: 181-184

Yamazaki S (1983) Current status of clinical interferon research in Japan (Abstr). Antiviral Res 2

Zocarian SA (1977) The observation of freeze-thaw cycles upon cancer-cell suspension. J Dermatol Surg Oncol 3: 173-174

zur Hausen H (1985) Currently established facts and still speculative aspects in the role of papillomaviruses in human carcinogenesis. Workshop on papillomaviruses, Kuopio, Finland, August 25-29

Immune Response to Papillomavirus Infection

P. B. Spradbrow

1 Introduction

1.1 Immune Responses to Viral Infections and to Neoplasms

It is conventional, and convenient, to recognise both antibody-mediated and cell-mediated immune responses to viral infections. Both processes are complex, and there are complex interactions between them, but detailed descriptions of the immune response to some viral infections are available. The antigenic targets are both virions and virus-infected cells that are marked with viral products. The immune response acts to combat initial lodgement of the virus and to prevent reinfection, to modulate the clinical response during an episode of acute infection, and to influence the development of persistent infection. Events that impair immune responsiveness modify the course of infections; sometimes the infecting virus itself contributes to immunosuppression. We understand no infectious disease unless we understand the immune events that determine the clinical outcome. Immune responsiveness is exploited in the diagnosis of viral disease and in the induction of immunity with viral vaccines.

The immune system also responds to many neoplasms but the process is not as well understood for neoplasms as for viral infections. A similar diversity of immune cells is involved and the target is the neoplastic cell. The process is further complicated when the neoplasm is induced by a virus and the neoplastic cell expresses viral antigens. Again the behaviour of neoplasms is altered in immunosuppressed individuals, and certain neoplasms have immunosuppressive activity. There is a potential for using immune reactions in diagnosing and treating neoplastic disease.

Papillomaviruses have been amongst the most neglected of the viral groups. Recent developments have allowed detailed study of the molecular biology of papillomaviruses, but the biology of the diseases they cause is still poorly understood. Papillomaviruses are associated with benign and sometimes with malignant proliferations. Neither the viruses nor the lesions are readily amenable to immunological study. The large but fragmentary literature on this subject is considered in the present review.

1.2 Special Features of Papillomavirus Infections

Papillomaviruses, because they fail to replicate in cultured cells, have posed special problems for virologists. Infected hosts are still the only source of virions. There are no type cultures and no sources of standardised antigens for use in diagnostic tests, epidemiological studies, or vaccines. There is no convenient neutralisation test, and no method for establishing serotypes. The resources of molecular biology have recently provided means for a classification of papillomaviruses based on comparisons of sequence homologies over the entire viral genome. The viral types distinguished in this way may not correspond to serotypes defined by tests that measure antigenic differences in the surface components of the virion. Many of these difficulties will be overcome as monoclonal antibodies to papillomaviruses become available and as portions of viral genome cloned in expression vectors yield viral polypeptides that will serve as reference reagents. For the present, the serology of papillomaviruses in not a very exact science.

Infections caused by papillomaviruses have also raised problems for the immunologist. The lack of standard reagents, the lack of a serological classification, and the lack, until recently, of any means of classification lead to imprecision. Papillomaviruses used in earlier studies cannot be equated with viral types as they are presently classified. The frequent derivation of antigen from pooled lesions, in the mistaken belief that any particular host species harboured but a single type of papillomavirus, must have introduced confusion. Not only are the papillomaviruses difficult to study, but the lesions they produce are unusual. The most common lesion is a benign proliferation of epithelial cells with the production of virions restricted to the terminally differentiating keratinizing cells. There is no good opportunity for interaction between the immune system and the most abundant source of viral antigen. Cells closer to the basal layer contain viral genomes and probably produce some viral polypeptides or express unusual cellular antigens as a consequence of the viral infection. These again are not readily accessible to cells of the immune system. Papillomaviruses are believed to reach the target epithelial cells through abrasions; there is no viraemia, and no alerting of immune cells, as the virus reaches susceptible cells or as the virus is shed. Nor is there spread of virion through body fluids from infected foci.

Certain papillomaviruses that infect ruminant animals also induce a fibroblastic response, and the resulting lesion is a fibropapilloma. The fibroblasts contain viral genomes, although they produce no virions, and contribute to the immunological stimulus to the host.

Biological analogy is often a powerful tool when studying a poorly understood infection (Beveridge 1972), but there are few other viral diseases that produce superficial epithelial proliferations without systemic infection. The most useful comparisons might be with certain diseases caused by poxviruses that do not generalise, e.g. avian poxes or the infections of ruminant animals with parapoxviruses (ovine contagious ecthyma and bovine papular stomatitis). Even these analogies are not very fruitful as very little is known about the immune responses to these poxviruses, and they do not produce chronic infections. However the ready control of some of these diseases by attenuated vaccines might be instructive to those working with papillomaviruses.

The conditions chosen for consideration have a demonstrated association with papillomavirus and have been the subject of immunological investigations. Not all are typical papillomas. The benign epithelial lesions that produce papillomavirus virions have been studied most effectively in human, bovine and canine hosts (Pfister 1984). The epithelial portions of bovine fibropapillomas contain virions, but only viral genome is found in the cells of the fibrous component. Viral genome also persists in fibromas induced in laboratory animals by infection with bovine fibropapilloma virus and in the naturally occurring equine fibrotic tumours termed sarcoids that are probably caused by infection with bovine fibropapilloma viruses (Pfister 1984). Of special interest are the carcinomas that develop from some benign lesions caused by papillomaviruses. These carcinomas do not produce virions but often contain the genomes of papillomaviruses.

The Shope rabbit papillomas and the carcinomas that develop from them were long regarded as biological oddities. However, detailed immunological studies were made and extended as further papilloma-carcinoma complexes were recognised in human patients and in domestic animals. Carcinomas develop from benign, papillomaviral lesions on the sun-exposed skin of human patients with epidermodysplasia verruciformis (Orth et al. 1980) and in the human genital tract (Syrjänen 1984). In Scotland cattle grazing on pastures that are infested with bracken develop viral papillomas and then carcinomas of the digestive tract (Jarrett et al. 1978). Cattle of the Hereford breed develop carcinomas when nonpigmented tissues of the conjunctival sac and eyelid are chronically exposed to strong sunlight (Spradbrow and Hoffmann 1980). These ocular carcinomas develop from benign precursor lesions that contain papillomaviruses (Ford et al. 1982), and papillomavirus genomes have been demonstrated in cells of the carcinomas (unpublished data). Figures 1, 2 and 3 illustrate a plaque, a papilloma and a squamous cell carcinoma in bovine eyes. A similar disease is seen in Holstein-Friesian cattle but is usually restricted to the nonpigmented tissues of the third eyelid. Papillomaviruses are also present in benign proliferative lesions on the sun-exposed skin of sheep (Vanselow and Spradbrow 1983; Vanselow et al. 1982) and these lesions frequently transform into carcinomas.

Are there immune reactions at all to papillomaviruses or the lesions they induce? The demonstration of specific antibody and of specific cell-mediated immune responses indicates that there is. Has the immune response any clinical significance? The frequent regression of papillomas is assumed to have an immunological basis but this is difficult to prove, and there may be variations between papillomaviruses. Are recovered hosts resistant to reinfection? The recognition of multiple viral types in human patients and in cattle makes difficult the interpretation of numerous older observations in these species. Bagdonas and Olson (1953), for example, made detailed observations on a herd of cattle over 2.5 years, during

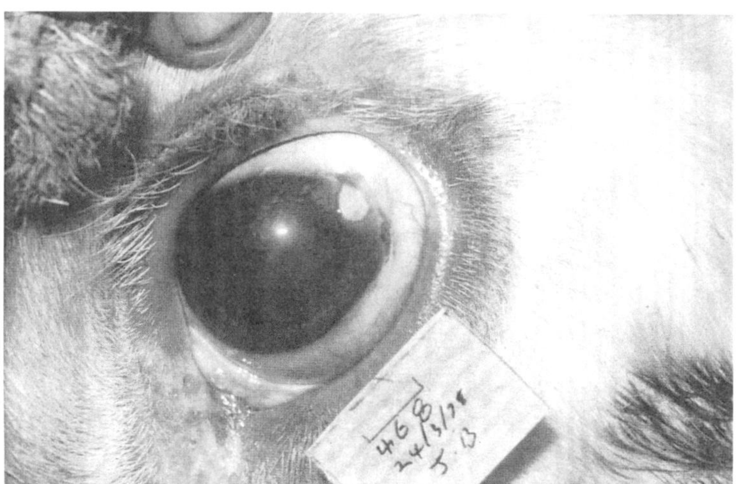

Fig. 1. A plaque on the eye of a cow. The virions of an uncharacterised papillomavirus have been demonstrated in such lesions, which may regress or which may progress to a papilloma

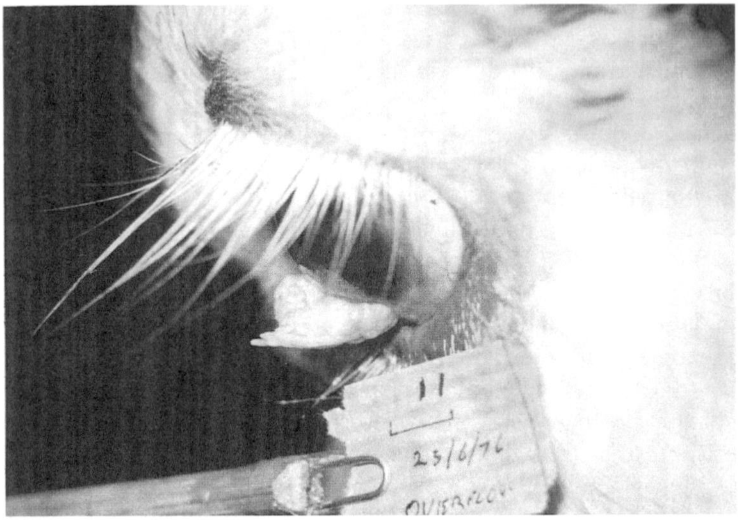

Fig. 2. A papilloma on the eye of a cow. The virions of an uncharacterised papillomavirus have been demonstrated in such lesions, which may regress or which may transform to squamous cell carcinoma

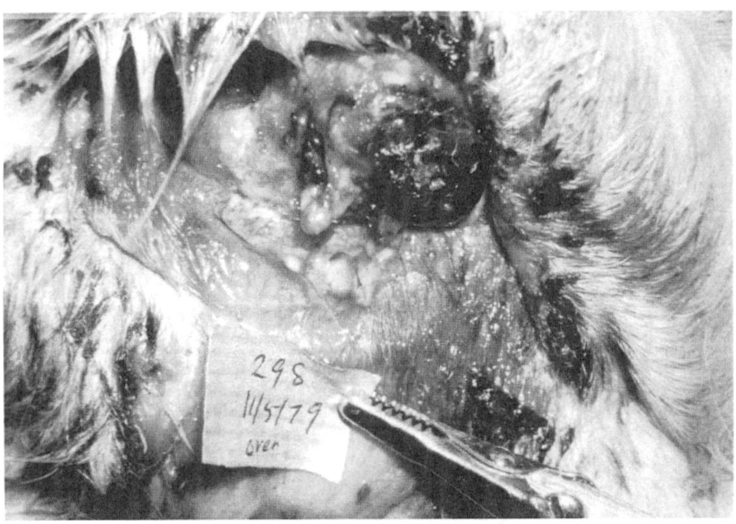

Fig. 3. A squamous cell carcinoma in the eye of a cow

which time two outbreaks of papillomatosis occurred. The second outbreak involved 32 of 60 animals that had been infected 2 years earlier. Although it is probable that the same virus was involved, this could not be demonstrated. Recovery does seem to lead to immunity in certain diseases where single papillomavirus types are currently recognised - in canine oral papillomatosis (DeMonbreun and Goodpasture 1932), in rabbit oral papillomatosis (Parsons and Kidd 1943), and in infection of rabbits with the Shope virus (Shope 1933). Equine cutaneous papillomatosis also leaves the recovered host immune (Cook and Olson 1951), but virological studies on this condition are few. Certainly the immunology of papillomavirus infection warrants some attention.

1.3 Tests for Antigen, Antibody and Cell-Mediated Immunity

Tests for detecting antibody against papillomaviruses are difficult and not well standardised. Many have been developed as experimental techniques, but none is in routine laboratory use. Serological tests that rely on replication of viruses in cultured cells are not possible, and there is no conventional neutralisation test. Neutralising antibodies directed against bovine fibropapilloma viruses have been assayed in vivo using cattle or horses (Segre et al. 1955) as the indicator systems. Similar tests have been performed with Shope cottontail rabbit papillomavirus (Kidd et al. 1936) and with canine oral papillomavirus (Chambers et al. 1960) and would be possible in other species. The ability of the viruses causing bovine fibropapillomas to transform cultured cells has been used in serological tests, using mouse cells (Dvoretzky et al. 1980) or bovine cells (Meischke 1979) as substrate. The mechanism of inhibition of transformation is not known; possibly the antibodies prevent the attachment of virions to target cells, and the test is basically a neutralisation test. The transformation inhibition test reported by Meischke (1979), however, was complement dependent, and the activity of the serum was adsorbed by both virions and transformed cells.

The virions causing bovine fibropapillomas are the only papillomaviruses for which a haemagglutinin has been demonstrated (Favre et al. 1974). Full or empty virions agglutinate mouse erythrocytes, and the reaction is inhibited by specific antiserum. This test and the test for inhibition of transformation could be expected to detect antibodies directed against surface determinants on the virion, similar to the determinants that distinguish serotypes in other viruses.

These tests, and others adapted from general virological techniques for the detection of papillomavirus-specific antibodies, are listed in Table 1. A major problem with all these tests is the lack of standard antigens. At present, antigens are either virions purified from lesions or crude homogenates of lesions. Papilloma cells or cells from other papillomavirus-associated lesions have also been used as antigens in some immunological studies (Table 1). Some of these tests are also available for detecting viral antigen in suspension or for comparing viral antigens. The antiserum used in these tests may come from animals naturally or experimentally infected with papillomaviruses, or it may be hyperimmune serum produced by repeated immunisation of laboratory animals.

Other tests are available for the detection of viral antigens in cells. The common tests are indirect immunofluorescence on frozen sections or the peroxidase-antiperoxidase (PAP) test on sections from formalin-fixed tissues. The latter test is coming into wide usage, possibly because of the availability of commercial reagents.

Some techniques have been borrowed from tumour immunology to explore specific cellular recognition of antigens in papillomas or carcinomas. These tests include antigen-specific blastogenesis, leucocyte adherence inhibition, and erythrocyte-rosette augmentation.

Table 1. Examples of tests used to detect antibodies to papillomavirus antigens

Test	Antigen	Reference
Neutralisation in vivo	Infectious virions	Segre et al. (1955)
Inhibition of transformation in cell culture	Infectious virions	Meischke (1979)
Complement fixation	Virions	Ogilvie (1970)
Haemagglutination inhibition	Virions	Favre et al. (1974)
Immunodiffusion	Virions	Koller et al. (1974)
Passive haemagglutination	Virions	Ogilvie (1970)
Radioimmunoassay	Virions	Pfister et al. (1979)
ELISA	Virions	Baird (1983)
Immune electron microscopy	Virions	Almeida and Goffe (1965)
Counter current immunoelectrophoresis	Virions	Cubie (1972)
Cytotoxicity	Cultured cells	Chung et al. (1977)
Indirect immunofluorescence	Cells	Matthews and Shirodaria (1973)

1.4 Antigens of Papillomaviruses

The proteins of papillomaviruses are not yet well-defined. The virion contains a major capsid protein (Favre et al. 1975; Meinke and Meinke 1981), and in some viruses four host-derived histones that are associated with viral DNA in chromatin-like structures have been observed (Favre et al. 1977). Additional minor proteins of uncertain significance have been found in some studies.

Antigenic analysis of papillomaviruses is little further advanced. Studies on intact virions by precipitin test yielded no evidence of cross-reaction between papillomaviruses from different species and no diversity within species (Le Bouvier et al. 1966). Other studies with different serological techniques - haemagglutination inhibition (Favre et al. 1974), PAP staining (Jenson et al. 1980) and radioimmunoassay (Pfister et al. 1979) - have confirmed this lack of cross-reactivity when whole virions are used as antigen.

Serum directed against sodium dodecyl sulphate (SDS)-disrupted virions is broadly cross-reactive. This has been found with a number of serological tests including indirect immunofluorescence (Jenson et al. 1980), PAP (Jenson et al. 1980; Sundberg et al. 1984), and immune electron microscopy (Sundberg et al. 1984). The broadly reactive antisera are held to recognise shared antigenic determinants of structural viral proteins that are not exposed at the surface of the virion. This is in keeping with the recognition of conserved nucleotide sequences in the genomes of many papillomaviruses. The conserved sequences and the supposed group antigen have not yet been detected in all papillomaviruses. Orth et al. (1978) presented evidence for a second antigenic determinant shared between human papillomavirus (HPV) type 1 and Shope papillomavirus.

If the concept of a protected, shared antigen is correct, intact virions injected into unnatural hosts to produce antiserum must not be degraded to expose the common antigen. Nor can papilloma-bearing animals have contact with the shared antigen that is produced in their lesions, although they produce antibodies against the surface epitopes of the same antigen. It may be that the antibody response in natural infection is directed solely against the virions involved in the infection event and that the components of progeny virions have no impact on the immune system.

1.5 Antigens on Cells Transformed by Papillomaviruses

Viral structural antigens, present in the differentiating epithelial cells of papillomas, interfere with the demonstration of new cellular antigens. These are more readily sought in cells free of virions - in cultured cells, in fibromas and carcinomas, and in the basal layers of papillomas. However, caution is necessary because of the presence in some lesions of other viruses including another papovavirus (Hartley and Rowe 1964) and rabbit fibroma virus (Shope 1933) in Shope papillomas.

Antigens associated with Shope papillomavirus have been detected in biopsies (Noyes and Mellors 1957) and in cells cultured from papillomas (Shiratori et al. 1967) but nonvirion antigens have also been detected. Ishimoto and Ito (1969), using serum from papilloma-bearing cottontail rabbits, detected membrane antigen in a few ($<5\%$) unfixed, cultured papilloma cells by indirect immunofluorescence. Antiviral antiserum produced no reaction. In further studies involving three cell lines established from Shope papillomas (Ishimoto et al. 1970) new, nonviral surface antigens were detected. Viral antigen was detected in the cytoplasm of the cells although no virions could be detected.

New antigens, both nuclear and cell surface, have also been demonstrated in human papilloma cells by immunofluorescence (Pass and Marcus 1973). The antigens, neither of which was a viral structural protein, were found by the same test in squamous cell carcinomas, fetal skin and psoriatic epidermis but not in normal skin or basal cell carcinomas. However, concentrated extracts of normal skin did contain a similar antigen detectable not by immunofluorescence but by gel diffusion. The antigens were apparently a feature of rapidly growing epithelial cells. Viac et al. (1978a) used antibodies from patients with warts to detect aggregates of tissue antigens within some human wart cells. These aggregates, which did not stain with rabbit antiviral antiserum, were evidently not viral proteins but had elicited a host immune response. In this study no surface antigens were detected. Barthold and Olson (1974a) used serum from fibroma-bearing calves to detect new membrane antigens on cultured, unfixed bovine fibroma cells by indirect immunofluorescence. The test sera failed to stain cultures of normal fibroblasts, and antiserum directed against virions failed to stain the cultures of fibroma cells. Similar studies detected common membrane antigens in the cells of cultured bovine fibromas and of cultured equine fibromas induced by infection with a bovine papillomavirus (BPV) (Barthold and Olson 1978). Hamster cells transformed by BPV have not yielded new membrane antigens although Geraldes (1970) detected cytoplasmic antigen.

Naturally occurring equine sarcoids also contain unusual surface antigens. Watson and Larson (1974) detected these in a cultured cell line by indirect immunofluorescence with autochthonous serum and with serum from other horses with sarcoids. The serum was not cytotoxic but the autochthonous lymphocytes were, again indicating altered surface antigenicity.

Bovine ocular squamous cell carcinomas have been used in several examinations for new antigens. Serum from cattle with precursor lesions or with carcinomas was used to stain cultured plaque and carcinoma cells, either allogeneic or autochthonous, and antibody was detected by indirect immunofluorescence (Atluru et al. 1982a) or with protein A labelled with radio-iodine (Atluru et al. 1982b). Sera from cattle with lesions reacted with both cultured plaque and carcinoma cells but control serum did not. Cultures of normal epithelial cells were not stained. Absorption studies indicated shared antigens in carcinomas and in precursor lesions. In another study, sera from cattle with ocular squamous cell carcinomas were allowed to react with various cultured cells and attached immunoglobulin was detected with a fluorescein-labelled antiglobulin. Sera from most cattle with carcinomas reacted with autologous and allogeneic ocular carcinoma cells and also with cultures of bovine cutaneous papillomas and equine sarcoids. There was no reaction with cultures of normal bovine skin cells, xenogeneic tumours, and allogeneic tumours other than ocular carcinomas (Kuchroo and Spradbrow 1985). The cattle serum was evidently detecting an antigen common to several types of cell altered by infection with papillomavirus.

Tests for cell-mediated immunity, discussed in Sect 3, also give evidence for new antigens in papilloma and carcinoma cells.

2 Systemic Humoral Immunity

2.1 Animal Papillomas

In considering the humoral immune response to papillomaviruses, one deficiency becomes obvious - the inability readily to detect neutralising antibody. Correlations have been made between the presence of antibody, measured by various techniques, and resistance to reinfection or the persistence or regression of lesions. The response to experimental infection may not accurately mimic that to natural infection, the former involving large quantities of virus and artificial trauma. The antibody response to natural infection with papillomavirus appears to be unspectacular.

Neutralising antibody has been demonstrated to some viruses. Shope (1933) found that serum from rabbits with papillomas would fully or partially destroy the infectivity of Shope papillomavirus although serum from two rabbits whose warts had regressed had no neutralising activity. The rabbits with warts were resistant to reinfection, possibly a function of neutralising antibody, or developed very few warts after a prolonged incubation period, suggesting an immune reaction directed against papilloma cells. As papillomas in domestic rabbits produce few virions, any antiviral antibody is probably a response to the original inoculum. Seto et al. (1977) found that even newborn rabbits developed neutralising antibody when challenged with Shope papillomavirus, and they also developed antibody against papilloma cells, detectable by indirect immunofluorescence. Antiviral antibody is probably not concerned with the regression of Shope papillomas, as virus persists in regressing papillomas (Evans and Rashad 1967), and transfer of serum from rabbits with regressed papillomas did not influence the rate of regression in papilloma-bearing recipients (Evans et al. 1963). Antiviral antibody did not protect against challenge with papilloma cells nor was it present at higher levels in rabbits with regressing warts than in rabbits with progressing papillomas (Kreider 1963).

Antibody responses to bovine papillomaviruses are meagre. Olson and Segre (1955) and Segre et al. (1955) used equine and bovine skin as indicators in an in vivo neutralisation test. They failed to detect antibodies in the sera of 21 calves with papillomas, but sera from two horses with lesions experimentally induced with BPVs did neutralise. A heat-labile, non-antibody, neutralising substance was present in preinoculation serum from one horse. Rosenburger and Gründer (1959) did detect neutralising antibody in the sera of cattle with warts, but it was present at very low levels.

Precipitating antibody is easier to measure. Lee and Olson (1969b) infected 21 experimental calves which developed 19*S* precipitating antibody after 1 week and 7*S* precipitating antibody after 2 weeks. Only the 7*S* form was detected after 16 weeks. Precipitins were present before the calves became resistant to reinfection and there was no correlation between the presence of antibodies and the persistence or regression of lesions. Using a sensitive radioimmunoassay, Pfister et al. (1979) detected antibodies to BPV1 or BPV2 in 19% of bovine sera but not in human sera.

Neutralising antibodies to papillomavirus were demonstrated in dogs whose oral papillomas had regressed (Chambers et al. 1960). Such dogs were immune to reinfection.

If circulating antiviral antibody is protective, young animals from immune dams would be passively protected. It is usual to keep calves until they are some months old to avoid the supposed risk of antibody interference in transmission experiments. Lee and Olson (1968), however, had no difficulty in infecting newborn calves and serum transferred from cattle with papillomas did not completely protect calves against experimental infection with papillomavirus (Rosenburger and Gründer 1959).

2.2 Animal Carcinomas and Fibromas

Calves infected with BPV developed antibodies against membrane antigens on the developing fibroma cells. The antibodies did not influence progression or regression of the lesions (Barthold and Olson 1974a).

Chung et al. (1977) demonstrated antibodies in the serum of some cattle with ocular squamous cell carcinoma that were cytotoxic for cultured allogeneic ocular carcinoma cells but not for other allogeneic cells. However the reaction required large amounts of xenogeneic complement and was not regarded as physiological. Jennings (1979) also demonstrated complement-dependent cytotoxic antibodies. Sera from 6 of 12 cattle with ocular carcinomas were toxic for one or more cultures of allogeneic carcinoma cells but not for cultures of allogeneic plaque or papilloma cells. Sera from two of three cattle with carcinomas at nonocular sites were also toxic for cultured ocular carcinoma cells. Kuchroo and Spradbrow

(1985) found that sera from cattle with ocular carcinomas contained antibodies that attached to autologous and allogeneic cultured cells, to cultured bovine papilloma cells and to cultured equine sarcoid cells, but not to normal skin cells or to other types of tumour cells.

2.3 Human Papillomas

Not all human patients with warts have detectable antibody against papillomavirus antigen. It may be that tests are still relatively insensitive or, more probably, there is minimal exposure of the immune system to viral antigen during the establishment of infection and the initial development of the lesion. Cubie (1972) noted that antibody could take several months to develop but that it was still detectable 9 years after the appearance of lesions. Reinfection was occasionally noted in the presence of antibodies. The prevalence of antibody was highest in subjects with a history of past warts. Pyrhönen (1978) measured precipitating antibodies in consecutive serum samples from large numbers of patients with skin warts and condylomas. Fourfold or greater increases in titres over 2–35 months were noted in 13% of patients with skin warts and in 3% with condylomas. The author interpreted this as a weak antibody response to a chronic virus infection. More antibodies were detected in age-matched but lesion-free controls, possibly indicating that antibodies are protective.

Complement-fixing (CF) antibodies have been detected in some patients with common warts. Genner (1971) detected antibody in 20% of patients with warts, but not in condyloma patients, and in 3% of controls. The levels of antibodies rose as warts were removed. Pyrhönen and Penttinen (1972) found CF antibodies, always associated with IgG, in 12% of 182 patients, and these were associated with a good chance of regression. In a subsequent study (Pyrhönen and Johansson 1975) CF antibodies in 20% of patients were indications of rapid healing. Precipitating antibodies detected by Pass and Maizel (1973) in fractionated human sera were almost all IgG.

Matthews and Shirodaria (1973) believed that regressing warts released virus into the circulation. They found that most patients with regressing warts had antibody (IgM, IgG, and IgA) against viral antigen and IgM against cells infected with papillomavirus. In patients whose warts were not regressing, there were very few with virus-specific IgM, none with virus-specific IgG or IgA, and about half with IgM against virus-infected cells. Viac et al. (1977b) demonstrated a serological response to injected wart particles. Control patients with no history of warts were tested for delayed-type hypersensitivity by intradermal injection of formalin-inactivated purified virions obtained from plantar warts. Nearly half the patients had developed IgM antibodies, detectable by indirect immunofluorescence, 8 days after injection.

2.4 Human Carcinomas

Baird (1983) used an ELISA test to detect IgG reacting with disrupted BPV2 virions in women with cervical carcinoma. Sera from 48 children and from 108 nonselected symptomless adults were used as controls. Test sera falling more than 3 standard deviations above the mean of the control sera were regarded as positive. It was found that 95% of patients with anogenital warts, 60% of patients with cervical intraepithelial neoplasia and 93% of patients with cervical carcinoma were positive. The author did not claim that the control groups had no antibody but that many patients in the test groups had significantly more. The levels of antibody against group antigen were highest in the patients with invasive carcinoma and the source of antigen was presumed to be the carcinoma. Even more striking results were obtained when the ELISA test was modified to detect IgM (Baird 1985).

2.5 Lesions and Humoral Immunity in the Immunosuppressed Individual

Although suppression of cell-mediated immunity (Sect. 3.5) influences the response to infection with papillomaviruses, there appear to be no specific studies on papillomaviruses in individuals with depressed humoral immunity. Orth et al. (1980) did note that individuals with epidermodysplasia verruciformis, in addition to their cell-mediated immune defects, had produced little or no precipitating antibody to the types of papillomavirus that were present in their lesions. The levels of antibody to HPV1 were similar to those in the general population, and the authors considered the possibility of a specific inability to react to some types of papillomavirus.

3 Systemic Cell-Mediated Immunity

3.1 Animal Papillomas

Cell-mediated immunity is usually more difficult to demonstrate than antibody-mediated immunity. Tests have usually been designed to demonstrate reactions against cellular antigens. Hosts often develop no lesions when reinfected with a papillomavirus, and it is often difficult to determine whether this results from humoral antiviral immunity or a cell-mediated rejection of transformed cells. This problem has been overcome in studies on Shope papillomavirus in rabbits. Kreider (1963) washed skin fragments free of antibody, exposed them to virus in vitro and then used the fragments as autografts. Evans and Ito (1966) infected with viral DNA instead of virions. In both studies, rabbits with growing papillomas were receptive to the growth of new papillomas, but rabbits whose papillomas had regressed were resistant. In the latter study rabbits whose papillomas had transformed to carcinomas were still susceptible to the growth of papillomas. It seemed that the regression of papillomas was accompanied by an efficient immune response directed against papilloma cells.

Kreider and Breedis (1969) demonstrated that fetal rat skin was susceptible to Shope papillomavirus. Infected fetal skin, when grafted onto syngeneic rats, developed papillomas that always regressed. Regression appeared to be produced by an immune response and was preceded by and accompanied by infiltration of the lesion with lymphocytes. More papillomas developed when the fetal skin was protected from immune responses by nurturing it in the cheek pouch of hamsters.

Hellström et al. (1969) detected cells in the lymph nodes of rabbits infected with Shope papillomavirus that reduced the plating efficiency of cultured papilloma cells but not of normal cells from the same rabbits. The active lymph node cells were present in rabbits still bearing papillomas and in rabbits whose papillomas had regressed. Although the studies were performed before the recognition of natural killer cells, the active cells were probably some form of lymphocyte specifically recognising an antigen on papilloma cells.

Kuchroo et al. (1983b) demonstrated cell-mediated immunity in cattle with experimentally induced papillomas caused by BPV3. In the erythrocyte-rosette augmentation test, lymphocytes from calves with papillomas reacted to extracts of papillomas but not to extracts of bovine ocular squamous cell carcinomas. The activity was associated with a soluble factor having the properties of a lymphokine and capable of increasing the rosetting activity of normal bovine lymphocytes.

Hypersensitivity reactions present a means for detecting cell-mediated immunity against papillomavirus virions, but such tests have found little application. Dogs with actively growing oral papillomas showed no hypersensitivity reactions to extracts of oral papillomas injected into the conjunctiva (Chambers et al. 1960). On the other hand, newborn rabbits infected with Shope papillomavirus developed delayed-type hypersensitivity to injected virions, although their papillomas contained neither infectivity nor detectable viral particles (Seto et al. 1977).

3.2 Animal Carcinomas and Fibromas

Transplant rejection is regarded as an indication of a cell-mediated immune response, but unless inbred animals are available, it is difficult to know whether the response is directed against tumour-specific antigens or histocompatibility antigens. The cultured carcinoma cell line VX7, derived from a lesion induced by Shope papillomavirus, is readily transplantable in normal rabbits. Vaccination with fresh allogeneic papilloma tissue induced a considerable degree of resistance to transplantation in 2–4 weeks (Evans et al. 1962a). Suppression of the growth of VX7 cells was not correlated with the progression or persistence of papillomas.

Equine sarcoids have been transplanted autologously (Olson 1948; Voss 1969) and to other sarcoid-bearing and sarcoid-free horses (Voss 1969). In the study by Voss occasional regression of original and transferred sarcoids was found, but there was little indication of an effective immune response directed against a sarcoid-specific antigen or a histocompatibility antigen. Bovine ocular squamous cell carcinomas readily formed autografts, but allografting was successful on only one animal and the allografts later regressed although autografts on the same animal persisted (Hoffmann et al. 1981b).

Barthold and Olson (1974b) studied the natural regression of fibromas induced by a bovine fibropapilloma virus in calves. Resistance to challenge developed after the onset of regression of fibromas and was not correlated with levels of antiviral antibody. A mononuclear cell infiltrate in regressing fibromas was indicative of a delayed-type hypersensitivity reaction.

Indirect evidence for an influence of cell-mediated immunity on the transformation of papilloma to carcinoma comes from the work of Kreider et al. (1979). Rabbit skin, infected with Shope papillomavirus and grafted onto nude mice, underwent malignant transformation earlier than would be expected on rabbits.

In vitro studies give clearer evidence of committed lymphocytes of the types that participate in cell-mediated immune responses. Autologous lymphocytes were cytotoxic in vitro for cultured equine sarcoid cells (Watson and Larson 1974), and lymphocytes of horses with sarcoids gave slight reactions with cultured sarcoid cells in mixed lymphocyte-target cell assays and in chromium release assays (Broström et al. 1979).

Extensive immunological studies of bovine ocular squamous cell carcinoma have revealed specific cell-mediated immune responses with certain tests. Assays based on leucocyte migration inhibition, lymphocyte transformation and lymphocyte-mediated cytotoxicity have usually yielded negative results (Lindsay et al. 1978). However the leucocyte-adherence inhibition test gave a clear indication of cell-mediated immunity. Of 18 cattle with squamous cell carcinoma, 14 reacted with saline extracts of squamous cell carcinoma but not with extracts of skin, while leucocytes from normal cattle and cattle with ocular papillomas gave no reactions (Jennings et al. 1979a). The erythrocyte-rosette augmentation test gave similar results (Kuchroo et al. 1983b), leucocytes of 19 of 21 carcinoma-bearing cattle reacting with carcinoma extracts. There were very few nonspecific reactions with extracts of skin or other tumours, including cutaneous papillomas, and only 1 of 21 control animals reacted to carcinoma extract.

Different approaches have demonstrated cell-mediated immunity to ovine carcinoma cells. Jun et al. (1979) demonstrated blastogenic responses of peripheral ovine lymphocytes to extracts of ovine carcinomas. Al-Yaman and Willenborg (1984) challenged sheep whose carcinomas has been resected with autochthonous carcinoma cells and found cells in the lymph that were specifically cytotoxic for ovine carcinoma cells.

3.3 Human Papillomas

Lymphocyte migration inhibition (Morison 1974, 1975c) and lymphocyte transformation (Ivanyi and Morison 1976) have been used to demonstrate the responsiveness of human lymphocytes to papillomavirus antigens. Reactive lymphocytes were detected both in patients with actively growing warts and in patients whose warts had regressed, but the reactions were strongest in the latter group and persisted for at least some months after the warts had regressed.

Lee and Eisinger (1976) carried out lymphocyte-transformation and leucocyte-migration inhibition studies on human patients, using both wart tissue extract and purified virions as antigen. Most patients with warts of less than 1 year's duration responded to both antigens as did those whose warts had regressed. Patients with warts of more than 1 year's standing responded to neither antigen (although phytohaemagglutinin (PHA) responses were normal), and patients with no history of warts responded only to the viral antigen. The authors suggested that there are distinct wart tissue-associated antigens and viral antigens, and that cell-mediated immunity is important in causing warts to regress. Patients responsive to viral antigen but with no history of warts had probably undergone unsuccessful infection with virus.

The immune elimination of infected cells is probably complete. Massing and Epstein (1963) studied 1000 institutionalised children over 2 years and noted that reappearance of warts at sites of natural involution was very rare while reappearance after treatment was common.

Viac et al. (1977b) purified papillomaviruses from plantar warts for use as a skin test antigen after inactivation with formalin. Of patients with regressing warts or with a history of past warts 76% developed delayed-type hypersensitivity reactions at the site of challenge. Many of the patients with positive skin reactions had no detectable papillomavirus-specific antibody.

Thivolet et al. (1977) reported a similar trial with similar results, 76% of patients with a history of past warts, 56% of patients with current warts, and only 7% of patients with no history of warts responding to the intradermal test. Levels of antibody detectable by indirect immunofluorescence were initially low but increased after the intradermal test. The amount of antigen (5 μg of viral protein) was probably much greater than that encountered during natural infection. Of patients with genital warts 60% also responded to this skin test antigen (Viac et al. 1978b). In this group of patients, circulating antibodies were more prevalent after skin testing than before skin testing, and there was better seroconversion in the patients with positive skin tests.

3.4 Human Carcinomas

Reviews such as this can indicate areas where the research effort has been scant. Why do we know so little of the cell-mediated immune response to human carcinomas that arise from benign lesions caused by papillomaviruses? That there is such a response is suggested by studies on the cellular infiltration of lesions (Sect. 4). Any detailed analysis of the response appears to be lacking.

3.5 Lesions and Cell-Mediated Immunity in the Immunosuppressed Individual

In the human patient cell-mediated immunity is reduced in certain diseases and is purposely suppressed in some therapies. Observations have been made on the prevalence of papillomas and on their conversion to carcinomas in such patients. There is a paucity of similar studies in animals.

Human patients with defects in cell-mediated immunity associated with diseases such as Hodgkin's disease, malignant lymphoma, and chronic lymphatic leukaemia had an increased prevalence of warts, and their lesions were often not responsive to treatment (Morison 1975a; Perry and Harman 1974). Similar observations have been made on patients with systemic lupus erythematosis (Johansson et al. 1977).

Epidermodysplasia verruciformis is a rare disease with features that indicate a role for genetic factors, reduced cell-mediated immune reactivity, and the combined action of certain human papillomaviruses and sunlight in producing carcinomas (Orth et al. 1980). The nature of the immune defect is not understood, but affected patients have been described as having low numbers of T-lymphocytes, the T-lymphocytes being poorly responsive to mitogens, and a reduced ability or an inability to produce cutaneous hypersensitivity reactions (Obalek et al. 1980). Two sibling patients examined by Prawer et al. (1977) had normal serum IgG levels, normal numbers of T- and B-lymphocytes, very poor blastogenic responses to mitogens and antigens and cutaneous anergy to many antigens. It is possible that progressing papillomavirus infection in such patients exacerbates an initial immunosuppression.

The immunosuppression that is practised on allograft recipients is associated with large numbers of persistent papillomas and with carcinomas on sun-exposed skin. Spencer and Andersen (1970) reported that 42% of renal allograft recipients who had been immunosuppressed for more than a year developed warts and that no patient who developed warts subsequently became free of them. However the warts were not more serious nor more widely disseminated than in normal patients. Ingelfinger et al. (1977) noted warts on 18 of 49 transplant patients, and most of the warts were refractory to treatment. In a series of renal transplant patients studied by Hoxtell et al. (1977), "skin cancer" occurred at 7 times the expected rate and squamous cell carcinoma at 36 times the expected rate. HPV5 DNA has been detected in benign lesions (Lutzner et al. 1980) and skin cancers (Lutzner et al. 1983) on allograft patients.

The immunosuppressed patient is also more susceptible to the genital consequences of infection with papillomaviruses. Shokri-Tabibzadeh et al. (1981) studied over several years genital lesions on four women with impaired immunity, three who had Hodgkin's disease and one with a skin condition resembling epidermodysplasia verruciformis. All four patients developed vulval condyloma acuminatum. In one patient there was no progression of the lesion, but in three patients there was progression to more serious lesions, namely flat condyloma, carcinoma in situ and squamous cell carcinoma. Koilocytic changes and electron microscopic detection of typical virions served to link these lesions with infection by papillomaviruses. Schneider et al. (1983) studied the response to immunosuppression in a large group of female recipients of renal transplants. Of 132 patients considered to be at risk, 11 developed condylomata of the cervix and 6 of these went on to become cervical neoplasia. These prevalences were calculated to be higher than those expected for women who are not immunosuppressed, by a factor of 5.5 for condyloma and by a factor of 7 for neoplasms.

Even "normal" patients with warts often have defective cell-mediated immunity. In a series reported by Obalek et al. (1980) the extent of defectiveness varied with different types of warts, and patients with plantar or genital warts had almost normal responses. Patients with common warts and particularly patients with flat warts had a lower percentage of T cells than controls (as judged by erythrocyte rosetting) and a poorer blastogenic response to mitogen. These patients also developed skin sensitivity to dinitrochlorobenzene (DNCB) less readily and reactions were less intense than in controls. The defects probably preceded infection and might determine susceptibility to viruses of low intrinsic infectivity. Patients with condylomas refractory to treatment were examined by Seski et al. (1978). Their lymphocytes had lower mitogenic responses than controls to several mitogens and this led the authors to postulate some underlying immune defect.

Jarrett et al. (1978) noted a relationship between the ingestion of bracken by cattle and the development of large numbers of papillomas caused by BPV4 in the digestive tract. Many of these transformed to squamous cell carcinomas. Bracken fern contains carcinogens and is also supposed to be immunosuppressive for cattle, but the nature of the immunosuppression is apparently yet to be documented. In acute bracken poisoning the haematological changes feature a loss of granulocytes and of thrombocytes (Evans et al. 1982).

Embryonic rat skin, infected in vitro with Shope papillomavirus and grafted onto adult rats, sometimes develops papillomas. These regress rapidly, and regression is accompanied by an intense lymphocyte infiltration. If the graft recipient is immunosuppressed chemically or with anti-lymphocyte serum and thymectomy, the number of papillomas is not increased, indicating that the initial resistance is not immunological. However, the papillomas that do form develop no lymphocytic infiltration, and they persist. Although this degree of immunosuppression prevented regression of the papillomas, it did not permit transformation to carcinoma (Kreider et al. 1971).

McMichael (1967) made similar observations in rabbits immunosuppressed with large doses of methylprednisolone and infected with Shope papillomavirus. Papillomas on the steroid-treated rabbits developed minimal mononuclear cell infiltrates and regressed much less frequently than they did in control animals. There was, however, no increased tendency to transform to malignancy and no increase in virus content of the lesions. It was also noted that the immunosuppressed rabbits developed papillomas at sites other than the inoculation sites and had a decreased resistance to reinfection. Rabbits whose papillomas had regressed (usually resistant to reinfection) developed papillomas after steroid treatment and reinfection, but the papillomas did not persist.

There has been a case report of a bull with defective cell-mediated immunity that developed severe and protracted papillomatosis (Duncan et al. 1975). The animal had normal humoral immune responses.

4 Local Immune Reactivity

Wart virus infections do not generalise. The only opportunity for immune reactivity is within, or close to, the localised lesions. Some indication of effector mechanisms has been gained by studying cellular responses within lesions.

4.1 Observations on Lesions in Regression

Papillomas usually regress, as do some papillomavirus-associated fibromas, but carcinomas associated with papillomaviruses rarely regress. Several workers have taken the opportunity to study involution in human flat warts and human common warts. There is sometimes a history of chemical treatment of the warts but no indication that the involution is attributable to treatment.

Tagami et al. (1974) observed rapid spontaneous involution of flat warts in 10 patients after the abrupt development of inflammatory changes. Histologically there was an early appearance of degeneration in the upper epidermis, so that the lesion no longer resembled a wart, and infiltration of mononuclear cells into the dermis, exocytosis, epidermal spongiosis, and necrosis of cells. These results were confirmed in a study on a further 25 patients (Tagami et al. 1977) and the scarcity of neutrophils and eosinophils in the cellular infiltrate was taken as evidence for the absence of a humoral immune response. A detailed study of the recovery process (Takigawa et al. 1977) revealed two phases: initial changes in the infected cells leading to degeneration and later rejection of the wart from the level of the basal layer, and regeneration of normal epithelium that contained lymphocytes and macrophages. The presence of melanocytes in degenerating warts was an indication of rejection of the total lesion including the basal portions.

Berman and Winkelmann (1977) noted that erythema and oedema at the base of flat warts, itching and the efflorescence of small new warts were clinical indications of impending regression. The histological appearance of the involuting flat warts was again suggestive of delayed cutaneous hypersensitivity with infiltration of mononuclear cells, exocytosis of mononuclear cells into the epidermis and degenerative changes leading to focal areas of epidermal necrosis. Amongst the early events in the involution of flat warts Oguchi et al. (1981a, b) described the reaction of macrophages to degenerating epidermal cells and the activation of Langerhans cells.

Common warts, by contrast, often give no clinical indication of imminent regression. Berman and Winkelmann (1980) obtained regressing common warts by observing patients with multiple lesions and removing those that persisted as others regressed. The histological changes - infiltration with mononuclear cells, exocytosis and degeneration of the epidermis - were similar to those observed in regressing flat warts.

There may be a second path to regression in common warts. Common warts sometimes turn black or develop dark spots and such changes are taken to indicate regression. Common warts with these changes had no cellular infiltration (Brodersen and Genner 1973). Matthews and Shirodaria (1973) also found thrombosis but no cellular infiltration in regressing non-genital warts. Kossard et al. (1980) studied randomly selected biopsies of common warts (not necessarily in regression) and noted inflammatory changes in very few (41 of 500).

Kidd (1938) noted lymphocytes and macrophages in the connective tissue underlying regressing rabbit papillomas, with reduced cell proliferation in the lesion so that loss of cells by differentiation was more rapid than replacement. Kreider (1980) reported a detailed study of rabbit papillomas in regression. He noted marked leucocytic infiltration of the dermis and the lower parts of the germinal layer. There was reduced replication of keratinocytes in germinal and spiny layers, as indicated by reduced labelling with tritiated thymidine, and the proximity of leucocytes was apparently not important in this process. Progressively growing papillomas had few infiltrating leucocytes. It was suggested that a locally produced lymphokine was suppressing replication.

Cattle fibropapillomas have also been observed during natural regression (Lee and Olson 1968). The features of note were keratinization and atrophy of the epidermis, infiltration of the dermis with lymphocytes, and separation of the basal layer from overlying layers.

4.2 Analysis of the In Situ Cellular Infiltrates

With modern reagents it is possible to identify specialised cells in histological sections. An examination of the infiltrates present in progressing or regressing lesions may indicate the type of immune response that is taking place. This could have predictive potential, although several meticulous trials have left this hope unfulfilled.

Syrjänen (1983) stained 97 papillomavirus-associated genital lesions and 53 not associated with papillomavirus for T-lymphocytes, B-lymphocytes, and macrophages. In the papillomavirus-associated lesions there was, with increasing atypia, a decrease in the proportion of B cells and an increase in the proportion of T cells (possibly suppressor T cells). The suggestion was made that an immune response was directed against HPV-associated cell surface antigens. The studies were extended by the use of monoclonal antibodies to identify subsets of T cells (Syrjänen et al. 1984). Sequential observations failed to distinguish between the cellular infiltrates in lesions that regressed, progressed or did not alter. A similar study was made on oral lesions of possible papillomavirus origin (Syrjänen et al. 1983). The highest proportion of T cells and macrophages occurred in condylomas and squamous papillomas that also contained viral antigen. The cell-mediated immune response in oral lesions appeared to have similar features to that seen in genital lesions.

Chardonnet et al. (1983) identified Langerhans' cells in human warts and in normal epithelium. Most of the 28 warts had a cellular infiltrate, although this was usually mild, and most contained fewer Langerhans' cells in the epidermis than were found in normal epidermis. In some lesions there were T cells (OKT4 and OKT8) and high numbers of Langerhans' cells both in the epidermis and the dermis, suggestive of a local cellular immune reaction. The authors postulated that the paucity of Langerhans' cells in most lesions might lead to tolerance of antigen. Väyrynen et al. (1984) used a monoclonal antibody to identify Langerhans' cells in cervical lesions. These cells were present, but in low numbers, in biopsies of lesions associated with human papillomaviruses. No variations were detected between regressing, persisting or progressing lesions, and the Langerhans' cells appeared not to be influencing the natural history of the disease.

Bovine ocular squamous cell carcinomas and their precursor lesions have been examined for antibody-forming cells. Hamir et al. (1980) found no immunoglobulin-containing cells in normal ocular tissues, but these cells were found in both cancerous and precancerous ocular lesions. Cells producing IgG were more prevalent than cells staining for IgM and IgA, and the IgG-secreting cells were most numerous in papillomas.

5 Immunosuppression Associated with Papillomavirus Infection

5.1 Papillomas

There are several indications that human patients with papillomas, and especially patients with persisting papillomas, have defective immune responses. It has not been determined whether the immune deficiency is a primary lesion that precedes papillomatosis or whether the growth of papillomas results in immunosuppression. The deficiencies are severe in patients with epidermodysplasia verruciformis but are not confined to them. In epidermodysplasia verruciformis cases studied by Glinski et al. (1976, 1981), there were reduced numbers of erythrocyte-rosetting cells (T-lymphocytes), a lowered blastogenic response to PHA, and no skin sensitivity to DNCB. The impairment was greater with longer duration of the disease and more extensive lesions.

Brodersen et al. (1974) compared the results of tuberculin tests in 100 bacille Calmette-Guérin (BCG)-vaccinated children who had common warts with 400 control children. Reactions were significantly reduced in the group with warts. About 30% of 78 patients with warts had abnormally low responses to PHA stimulation of leucocyte-migration inhibition in a survey conducted by Morison (1975b). All 16 controls had normal values. The response to purified protein derivative (PPD) also appeared to be different in the test group, but tests and controls had not been matched.

Patients with warts had significantly fewer T cells and a lower percentage of T cells than control subjects in a study by Chretien et al. (1978) in which erythrocyte rosetting was used to enumerate T cells. The morphology of rosettes was different in wart patients, with smaller numbers of erythrocytes binding, and the binding was not as strong as in controls.

Jablonska et al. (1980) studied the clinical and histological responses of patients to the human papillomaviruses that were recognised at that time and evaluated nonspecific cellular immunological reactivity. They measured sensitivity to DNCB, the percentage of erythrocyte rosette-forming cells, and the mitogenic response of lymphocytes to PHA. A high degree of immunosuppression was associated with HPV3 and HPV5 infections in epidermodysplasia verruciformis; immunosuppression was also a feature of patients with HPV2-associated skin papillomatosis. Patients with plantar warts and with genital warts had nearly normal cell-mediated immunity. Disease-free family contacts of patients with familial epidermodysplasia verruciformis had no antibodies to HPV3 or HPV5 and normal cell-mediated immunity. The authors argued for the importance of cell-mediated immunity in protection against infection with papillomavirus.

Jablonska et al. (1982) observed a patient whose HPV3 lesions, associated with epidermodysplasia verruciformis, regressed over 2 pregnancies. Cell-mediated immunity, which had been impaired, became normal and antibodies to HPV3 developed.

Blocking factors were indicated in the vicinity of persisting warts on a single patient by Freed and Eyres (1979). Intralesional injection of either DNCB or PPD did not invoke a sensitivity reaction although distant skin developed a normal delayed-type hypersensitivity to such challenge. Leucocyte-migration inhibition reactions were detected with PPD but not with wart antigen nor with mixtures of wart antigen and PPD. This was taken to indicate blocking factors in the wart extract, but the negative tests may have been an indication of the toxicity of the wart extract.

Glinski et al. (1976) noted that serum from a patient with epidermodysplasia verruciformis blocked the mitogenic reaction of lymphocytes to PHA.

Hellström et al. (1969) recognised serum factors that protected cultured Shope papilloma and carcinoma cells from the destructive action of cells derived from immune lymph nodes. The serum factors, which were believed to be antibody, were present in rabbits whose papillomas or carcinomas persisted but not in rabbits whose papillomas had regressed. These factors, apparently dependent on the presence of a lesion, might be specifically immunosuppressive in vivo.

Cattle with cutaneous papillomas caused by BPV3 had lymphocytes that were reactive with papilloma antigen in an erythrocyte-rosette augmentation test (Kuchroo et al. 1983b) and serum-blocking factors that interfered with this reaction (Kuchroo et al. 1983a).

5.2 Carcinomas

Immunosuppression is associated with the development of carcinomas from papillomas. This has been demonstrated in rabbits (Hellström et al. 1969; Sect. 5.1) and in cattle and sheep.

Leucocytes present in diluted whole-blood cultures of cattle with ocular squamous cell carcinoma had a reduced mitogenic response to stimulation with PHA when compared with those from lesion-free cattle. The degree of impairment increased as carcinoma size increased. The specific response of leucocytes to a saline extract of carcinoma, measured by leucocyte adherence inhibition, also fell in parallel with the fall in responsiveness to PHA (Jennings et al. 1979b). Lyndsay et al. (1978) had previously argued that cattle with ocular carcinomas were immunocompetent, on the basis of lymphoproliferative responses induced in isolated lymphocytes of four lesion-bearing cattle after challenge with autochthonous tumour tissue. Lymphocytes were responsive to phytomitogens and to carcinoma extracts, but no comparisons were made with control cattle. The immunosuppression here was thought to be due to suppressive factors in the serum of carcinoma-bearing animals, and a search was made for these. Such factors were demonstrated by Kuchroo et al. (1983a) using the erythrocyte-rosette augmentation test. Autologous serum from 16 cattle with ocular carcinomas blocked the augmentation produced by exposure of leucocytes to extracts of carcinomas. Blocking also occurred in some allogeneic combinations, but not when serum from lesion-free cattle or serum from papilloma-bearing cattle was used. Serum from carcinomatous cattle also reduced the augmenting activity of PHA and was broadly alloreactive. One cow, whose carcinoma had regressed following immunotherapy, had lymphocytes that were reactive with carcinoma antigen but lacked blocking factor, indicating a dependence of blocking factor on persistence of carcinoma.

The blocking factors have not been fully identified. They are complex and appear in two different molecular weight forms, about 140K and 20K-50K. Their relationship to immunoglobulins was shown by their removal by protein A or by antibovine IgM or antibovine IgG antisera (Kuchroo 1984).

There is also good evidence that sheep with skin carcinomas are immunosuppressed and that the carcinomas contribute to the immunosuppression. Sheep with carcinomas had reduced mitogenic responses to PHA compared with control sheep and the reduction became greater as the carcinomas increased in size (Jun et al. 1979). Treatment of cancerous sheep with cyclophosphamide restored blastogenic responses of lymphocytes, restored skin sensitivity to carcinoma antigen and resulted in complete or partial regression of the carcinomas (Jun and Johnson 1979a). Suppressive factors were demonstrated by Jun and Johnson (1979b) in the serum from sheep with squamous cell carcinomas. Sheep with a primary carcinoma reacted to challenge with autochthonous carcinoma cells by producing specific, but non-cytotoxic antibodies; those whose carcinomas had been resected responded to a similar challenge by producing cytotoxic antibody and cytotoxic cells (Al-Yaman and Willenborg 1984). The presence of the primary carcinoma was evidently specifically immunosuppressive. Direct evidence for the production of immunosuppressive substances by cultured ovine carcinoma cells comes from studies on mixed lymphocyte reactions reported by Al-Yaman and Willenborg (1985). Allogeneic lymphocytes could not be stimulated in the presence of live tumour cells.

6 Immunological Aspects of Therapy

6.1 Papillomas

There is probably an immunological basis for the involution of papillomas. Kreider (1980) argued that this was so in rabbits, and there is a similar pattern of spontaneous involution of multiple papillomas in other species. Therapies exploiting this putative immunological response have been used in man and other animals. Indeed most methods of therapy, other than complete surgical excision, could expose the host to viral antigens or antigens present on papilloma cells and could constitute an immunological intervention. Therapies that seek to initiate or augment an immune response can be termed immunotherapy.

Not all therapies use physical, chemical or immunological agents. Human warts are sometimes believed to respond to suggestion, charming, magic, hypnosis or what Thomas (1981) described as "thinking, or something like thinking". Although there is now a great interest in interactions between the nervous system and the immune system, many still doubt that warts are susceptible to such influences, which might conceivably act through immune mechanisms or through local interruption to the blood supply. Clarke (1965) in a critical review of the charming of warts found no evidence for any effect, and suggested that 30% of warts would regress without treatment in 3–6 months.

Nonspecific therapy with immune stimulants is designed to augment host responses. Levamisole did not prove useful in a controlled study of human warts (Schou and Helin 1977), while local injection of BCG into human warts is regarded as hazardous (Bunney 1982). Human warts are sometimes susceptible to local treatment with DNCB after skin sensitisation with this substance (Greenberg et al. 1973; Russo et al. 1975). The local delayed hypersensitivity reaction apparently destroys the warts. Some wart patients, however, demonstrate local or systemic anergy.

Nonspecific immune therapy has not been used extensively against papillomas in animals. Shope papillomas regressed significantly earlier in rabbits that received repeated intralesional doses of *Corynebacterium parvum* than in control rabbits (Kreider and Bartlett 1981).

Immune serum has been used unsuccessfully in the treatment of bovine papillomas (Rosenburger and Gründer 1959), canine oral papillomas (Chambers et al. 1960) or rabbit papillomas (Evans et al. 1963).

Specific immunotherapy has been attempted with various preparations of papillomas. The early literature, with many claims for success, has been reviewed by Biberstein (1944). Concern for other infectious agents might now restrict such treatments to autogenous preparations or to inactivated preparations. Viac et al. (1977a), in their studies on the production of cutaneous hypersensitivity with formalin-inactivated purified virions, noted that previously resistant warts regressed in some patients who developed hypersensitivity. The responsive patients also developed IgG antibodies, but these were judged to be irrelevant to the regression.

There are several recent reports on the response of human condyloma acuminatum to immunotherapy, which were reviewed by Abcarian and Sharon (1982) when they also reported a trial involving 200 patients with anal condyloma acuminatum. Autogenous vaccines were prepared from homogenised tissues that were frozen and thawed four times "to kill the live virus" and heat inactivated (1 h, 56 °C) before the administration of six subcutaneous doses at weekly intervals. Excellent results were recorded: 84% of the patients remained free of disease for an observation period that averaged 46 months. The nature of the immunological stimulus is not known. It is apparently specific as urogenital warts regressed in 40 of 45 patients, but coexisting warts on other parts of the body persisted. There were no local reactions; however, the suspicion must remain that if the original tissue contained active papillomavirus, then so did the vaccine.

Bovine fibropapillomas are frequently treated by immunotherapy, usually with a crude formalin-inactivated vaccine prepared from lesions. Evidence for efficacy is difficult to obtain in clinical trials because of the high rate of spontaneous regression, and in some trials the numbers of controls and the methods for selecting these have not been stated. These criticisms do not apply to a study by Olson and Skidmore (1959) in which formalised vaccine had no therapeutic action on experimentally induced fibropapillomas. Homogenates of fibropapillomas, suspended in glycerol saline and not treated with inactivating agents, have also been used. Such preparations would probably contain active papillomaviruses and could be contaminated with other viruses. In an apparently successful trial reported by Pearson et al. (1958), a few cattle did develop connective tissue tumours at the site of subcutaneous vaccination. Du Casse (1961) used similar preparations, and only 6 of 106 calves given two subcutaneous doses of vaccine failed to respond.

Chambers et al. (1960) have used various immunological treatments for canine oral papillomas. Transfer of whole blood from dogs whose papillomas had regressed naturally did not hasten regression; neither did injections of papilloma extracts or papilloma cells with or without adjuvant. Transfer of autologous lymphoid cells had an equivocal effect - there was a possible hastening of regression in 2 of 11 recipients.

The Shope papilloma is susceptible to immunological intervention. Evans et al. (1962b) vaccinated domestic rabbits with freshly prepared allogeneic papilloma tissue at the time of viral infection. Papillomas appeared simultaneously in vaccinated and control rabbits, but regressed with increased frequency in the vaccinated rabbits. Shope papillomas in domestic rabbits produce very few virions, and the regression is presumably mediated by an immune response directed against cellular antigens. Autologous vaccines administered after papillomas were apparent also increased the frequency of regression.

Evans et al. (1963) were unable to influence the rate of regression by transfer of splenocytes and lymph node cells from immune rabbits. However, in outbred rabbits allogeneic immune cells would be unlikely to survive.

The results of immunotherapy have been variable and often disappointing. Sufficiently encouraging results are available to suggest that it may be possible to apply some immunological leverage in the treatment of papillomas. The preparation of effective antigens will be important and there is probably a role for improved adjuvants.

The treatment of papillomas with interferon might also be considered a form of immunotherapy. The topic is considered in detail elsewhere in this volume (see Kashima and Mounts; von Krogh).

6.2 Carcinomas and Fibromas

Nonspecific and specific immunotherapy has been used in the treatment of malignancies that are associated with infection with papillomaviruses. Equine sarcoids (Murphy et al. 1979) have regressed following intralesional injection of modified BCG. Bovine ocular squamous cell carcinomas are extremely responsive to both nonspecific and specific immunotherapy. Nonspecific therapy has usually involved the application of various preparations of BCG cell walls and has induced regressions in the majority of treated animals (Kleinschuster et al. 1977). In the long-term study reported by Kleinschuster et al. (1981), 6 of 23 treated animals survived 2.5 years but only 1 of 18 control animals did. Klein et al. (1982) reported similar encouraging results with either BCG cell walls or with live BCG introduced into the lesions. Regressions occurred in 60%-70% of the treated animals. Van Kampen et al. (1973) observed temporary regressions in carcinomas in two animals treated with endotoxin of *Escherichia coli* but Jennings (1979) obtained no

Table 2. Dose response of bovine ocular squamous cell carcinomas to immunotherapy with phenol saline extract of allogeneic carcinomas[a]

Dose (mg of protein)	Number responding[b] (number treated)	Complete regressions (number treated)
4	3/10	0/10
8	2/3	0/3
12	2/2	0/2
16	8/10	3/10
20	16/17	11/17
38	1/1	0/1
60	1/1	1/1
Controls[c]	1/47	0/47

[a] Compiled from Hoffmann et al. (1981a).
[b] Responding means cessation of growth of carcinomas, reduction in bulk of carcinomas or complete regression of carcinomas.
[c] No treatment or phenol saline extracts of control tissues.

response in four cattle whose lesions were treated with DNCB. Specific immunotherapy uses substances obtained from ocular carcinomas. These have usually been derived by some form of extraction with saline and phenol. Manilla et al. (1972) and van Kampen et al. (1973) used single intramuscular injections of phenol saline extracts prepared from carcinomas that were presumably allogeneic. In the latter study regressions of carcinomas were observed in 37 of 39 treated cattle. Spradbrow et al. (1977) reported regressions in 39 of 46 carcinomas in cattle treated with a single intramuscular injection of phenol saline extracts of allogeneic carcinomas, and Fivaz (1978) reported similar successful experiments. Hoffmann et al. (1981a) refined the technique and use lyophilised preparations of phenol saline extracts of allogeneic carcinomas. The results of a dose response study by these workers are summarised in Table 2. When the dose contained a protein equivalent of 8 mg or greater, 30 of 34 cattle gave some response to therapy; with a protein equivalent of 16 mg or greater, 15 of 29 cattle lost their carcinomas completely. The specificity of the reaction is suggested by the failure of phenol saline extracts of bovine cutaneous papilloma, fetal skin or normal cornea and conjunctiva to induce regressions. Regressing carcinomas have the histological appearance of a cell-mediated rejection with infiltration by lymphocytes, plasma cells and macrophages (Spradbrow et al. 1977). Figures 3 and 4 demonstrate the response of a bovine ocular squamous cell carcinoma to specific immunotherapy.

The active components of the phenol saline extract have not been defined. There seems to be a requirement for alloantigens as in a study reported by Jennings and Spradbrow (1980) in which extracts of autochthonous carcinomas rarely produced regression although immunosuppressive activity was shown when some of the same extracts were used allogeneically. Assays in cattle are cumbersome and expensive, but they have indicated that the active ingredient in producing tumour regression is resistant to heat (56 °C for 2 h), is unaffected by deoxyribonuclease and ribonuclease, but is inactivated on exposure to proteolytic enzymes (Jennings 1979). Gel filtration indicates a protein with a molecular weight less than 23000.

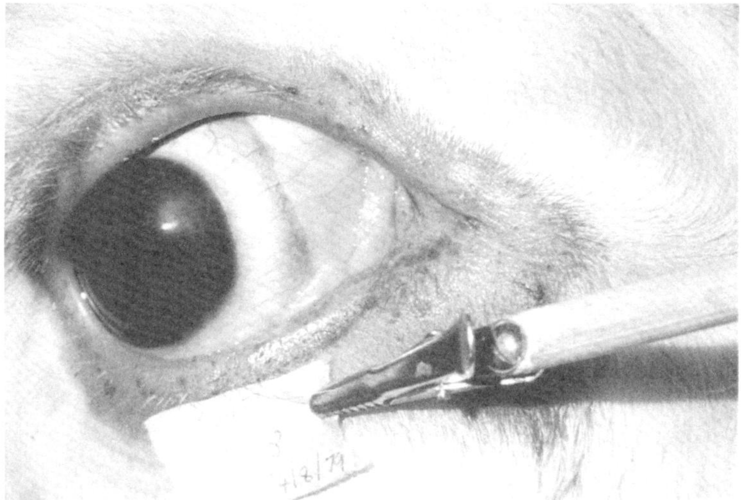

Fig. 4. The eye shown in Fig. 3, about 3 months later. The carcinoma regressed following a single intramuscular treatment with a phenol saline extract of an allogeneic carcinoma

Kuchroo (1984) studied the in vitro reactivity of phenol saline extracts of ocular squamous cell carcinomas in parallel with their immunotherapeutic behaviour. The phenol saline extracts strongly suppressed the erythrocyte-rosette augmentation reaction usually induced in leucocytes of cancerous cattle by 3 *M* KCl extracts of carcinomas. The suppressive factor was apparently a heat-resistant protein of low molecular weight. There is an apparent contradiction in extracting from carcinomas a factor that is immunosuppressive in vitro but that causes tumour regression in vivo. A possible explanation comes from studies by Nelson et al. (1985). Mice sensitised to sheep red blood cells produce a delayed-type hypersensitivity reaction on challenge with sheep red blood cells. Factors in cancer cells, including factors in saline extracts of bovine ocular squamous cell carcinomas, depress this hypersensitivity reactivity. The depression was overcome in two ways, by vaccinating mice with phenol saline extracts of bovine carcinomas or by treating the mice with monoclonal antibody directed against the retroviral envelope protein p15E. Profound immunosuppression is associated with retroviral p15E and with similar proteins that are expressed on many neoplastic cells not infected with retroviruses (Snyderman and Cianciolo 1984). The Nelson study suggested that bovine ocular squamous cell carcinomas contain such a protein, and that successful immunotherapy results from an immune reaction directed against the immunosuppressive products, thereby allowing operation of host anti-tumour defences.

Sheep carcinomas responded differently to immunotherapy – with enhancement (Jun et al. 1978). After sheep were treated with various extracts of carcinomas, their carcinomas grew more rapidly and metastasised more frequently. Extracts of fetal sheepskin produced similar effects. In another study (Swan et al. 1983) sheep with carcinomas did not respond at all to attempted immunotherapy.

7 Papillomavirus Vaccines

Perhaps genetic engineering will lead to the production of acceptable papillomavirus vaccines. The current vaccines are crude and, except in cattle, they find little use. The only current sources of vaccine are lesions harvested from infected hosts. They contain both viral and cellular antigens, they cannot be standardised, and they cannot be produced by a seedlot system in specific pathogen-free substrates. Most are inactivated vaccines, although dose response curves to the inactivating agents have not been produced. Some "vaccines" used in immunotherapy (Sect. 6) apparently contain unmodified virions. Attenuated vaccines are usually the most effective of viral vaccines but attenuated papillomavirus vaccines have not been developed. Suspicions of oncogenicity will probably limit the use of papillomavirus vaccines in human patients.

7.1 Vaccination of Cattle

Vaccines have been prepared against the viruses causing bovine fibropapillomas but not against those causing bovine papillomas. The usual vaccine is a crude homogenate of bovine fibropapillomas, treated with formalin to destroy the infectivity of the virus. The evidence considered below, allowing for the almost certain multiplicity of viral types, indicates that the vaccines give adequate protection. Chicken embryo-adapted vaccines have also been used but they were judged ineffective (Olson et al. 1959, 1960), and there must now be doubt about the ability of bovine papillomaviruses to replicate in chicken embryos. The method of production of these vaccines ensures that they are not available in large quantities, and this restricts their use to herds with special problems.

Olson et al. (1959) tested the efficacy of vaccination by challenge with homologous virus. Two doses of tissue vaccine gave reliable protection against challenge by intradermal inoculation or by scarification. Although a single dose of vaccine produced significant immunity, immunity was enhanced by multiple doses (Olson et al. 1960). This is the usual response that is produced by inactivated viral vaccines. The effector mechanism is presumably antibodies directed against virions - inactivated vaccines do not usually induce cell-mediated immunity. Indeed, Barthold et al. (1976) showed that cattle vaccinated with formalin-inactivated vaccine produced precipitating antibody, and the levels of antibody increased with repeated administration of vaccine. Similar precipitin responses were noted in calves exposed to uninactivated papillomavirus (Lee and Olson 1969 a, b). However in the latter studies the production of precipitating antibodies preceded the development of resistance to reinfection and the presence of antibodies was not correlated with regression of lesions.

Further observations have been made on the efficacy of vaccines against bovine fibropapillomas. Olson et al. (1962) investigated four preparations of formalised vaccines made from viruses that were apparently immunologically related. They tested immunity at various sites by challenge and found that resistance was most obvious on the skin and least well-developed in the vaginal mucosa. Bagdonas and Olson (1954) noted that the resistance of the epidermis was more complete after vaccination than was the resistance of the dermis. Calves that developed no papillomas when challenged by scarification could still develop fibroblastic lesions after intradermal challenge. An epizootic of penile and cutaneous fibropapillomas in a large bull-rearing establishment was brought under control by the use of formalin-inactivated vaccine (Olson et al. 1968). Three intradermal doses of vaccine were used.

Not even in cattle have attenuated papillomavirus vaccines been evaluated. However it has been shown that immunity ensues from a single actively growing wart in most cattle (Olson et al. 1960), and attenuated or even unaltered virus could find use as a vaccine. Olson and Segre (1955) had noted variations in virulence among field strains of bovine fibropapilloma virus, and selection for attenuation could be possible.

7.2 Vaccination of Other Animals

There appear to be no records of the commercial production of papillomavirus vaccines for use in species other than cattle. A few experimental vaccine procedures have been described.

Cook and Olson (1951) found that the immunity produced by artificial induction of a few warts on horses was less than that resulting from the natural acquisition of numerous warts. The experimental initiation of a single oral papilloma in dogs led to resistance to further infection within 3 weeks although the initial papilloma was still actively growing (Chambers et al. 1960). The resistance was attributed to neutralising antibody, which was demonstrated in dogs whose papillomas had regressed but which was unfortunately not sought in dogs with active lesions. Subcutaneous injection of dogs with the oral papillomavirus preparations with adjuvant also led to resistance to experimental infection within 2-3 weeks.

Kidd (1938) was unable to protect rabbits against infection with Shope papillomavirus by vaccination with embryonic rabbit skin cells.

There has been very little investigation of vaccination procedures to protect against papillomavirus-associated carcinomas. Any procedure that prevents the formation of the benign precursor lesion will eliminate the structure in which transformation occurs. Only if carcinomas possess antigens that are absent on papilloma cells could specific immunological processes be expected to interfere with malignant transformation. None of the papillomavirus-associated carcinomas is known to regress spontaneously. Possibly the only immune action to be taken against them is the one that has proved ineffective against the precursor papilloma. If this is so, when regression of carcinomas is induced by immunotherapy, both carcinomas and precursor lesions should regress. This does not happen. Alternatively, there may be specific antigens on carcinomas and specific immune responses that are impaired by blocking factors.

Some observations have been made with the VX7 transplantable rabbit carcinoma that contains papillomaviral antigen (Evans et al. 1962a). If rabbits were vaccinated with fresh allogeneic papilloma tissue, only about half would accept a transplant of VX7 cells 2-4 weeks later. This was taken as evidence of shared rejection antigens in VX7 cells and papilloma cells. However, Shope virus applied at the time of vaccination led to the formation of papillomas, some of which regressed but with no correlation with suppression of carcinoma growth.

8 Conclusions

Papillomaviruses seem to be well-adapted parasites and to provoke minimal immune responses. It is as if the watchdog in the immune system growls at their presence, but does not bite. Lesions caused by papillomaviruses are tolerated, often for long periods, in the presence of some degree of immune reactivity. This muted reactivity has proved difficult to study in the laboratory, but results have accumulated that indicate an immunological basis for resistance to reinfection and for involution of benign lesions. No great use has been made of immune responses of the host either diagnostically or therapeutically. Certain biological aspects of papillomavirus infection, specificity and latency that might conceivably have an immunological basis have been almost totally neglected.

Papillomaviruses contain specific surface antigens but these have not been sufficiently defined to allow a taxonomy based on serotypes. Neutralising antibody is probably directed against these specific surface antigens and may be the basis for acquired resistance to reinfection. However neutralising antibody is difficult to detect in the absence of in vitro neutralisation tests, and where it is detectable, the neutralising antibody response to natural infection is poor. Artificial infection, with trauma and large inoculums, seems to induce a better response. The biological significance of the antibodies to surface antigens that are detected by numerous other tests is not known. Antibody to the "common" internal antigen of papillomaviruses is produced by immunisation with disrupted virions and not by natural infection. The one exception of profound interest is the demonstration by Baird (1983) of antibody, apparently directed against common antigen, in human patients with genital carcinomas. There is a need for experiments with passively transferred antisera to clarify the role of antiviral antibodies in diseases caused by papillomaviruses.

The cells of papillomas contain antigens, possibly virus-specific, that provoke an immune response. Cell-mediated immunity directed against these antigens is probably important in causing the regression of papillomas and in preventing the growth of papillomas from reinfected basal cells. Individuals with papillomas sometimes have defective cell-mediated immunity and specific blocking factors are found in the serum in some papillomavirus infections. Papillomas should be utilised more often in oncological research. They grow rapidly, they invoke a cell-mediated immune response, they give rise to blocking factors, and, unlike their malignant counterparts the squamous carcinomas, they can be induced at will.

Carcinomas arising from papillomas contain papilloma-specific and possibly carcinoma-specific antigens. Those that have been studied in animals induce a cell-mediated immune response, but the carcinomas are immunosuppressive, apparently through the production of specific blocking factors, and they grow progressively. Ocular carcinomas in cattle are remarkably susceptible to specific and nonspecific immunotherapy. Both forms of immunotherapy probably function by restoring the immunocompetence of the host, and this model should find wider use in comparative medicine. Many of the lesions associated with papillomaviruses seem to exploit immunosuppressed hosts and to be themselves immunosuppressive.

Papillomatosis should be susceptible to control by vaccination. Only the crudeness of the vaccines that are at present available has confined their use to cattle. These vaccines, prepared from homogenates of lesions, should invoke responses to both viral antigens and to altered cellular antigens. They appear to be effective. Prevention of carcinomas should also be possible where they arise from papillomavirus-induced benign lesions that could themselves be prevented by vaccines. Some of the fruits of molecular biology could be effective papillomavirus vaccines, be these viral polypeptides from expression vectors, infectious viral nucleic acid derived by molecular cloning or even cultured cells bearing specific papilloma antigens. In veterinary medicine papillomavirus genomes could be used as vehicles for genes from other viruses to produce genetically engineered mixed vaccines.

Papillomaviruses are species- and often site-specific. The basis of this specificity is possibly not immunological. Much virus specificity is determined by cellular receptors, and these have been little studied with papillomaviruses. Kreider et al. (1967) showed that Shope papillomavirus attached well, not only to fetal and adult skin cells from rabbits, but also to adult rabbit kidney cells. The virus adsorbed poorly to rabbit erythrocytes or mouse embryo cells, and the ability to attach to rabbit cells was destroyed when the virus was treated with antibody or heat-inactivated. Apparently cells other than the normal target cells had receptors. Susceptibility may depend not only on receptors in basal epithelial cells, but on the ability of these cells to express some viral genetic functions and on the ability of the terminally differentiating progeny cells to produce virions. If there is an immunological component to site specificity, it is likely to be cell-mediated and not antibody-mediated.

Latency is a feature of infection with some types of papillomavirus, the viral genome being present in apparently normal epithelium and causing lesions after successful removal of an adjacent lesion or under conditions of immunosuppression. It is not known why infected cells, capable under some circumstances of pathological proliferation, remain quiescent, but the constraint may be immunological. The immunology of the latency of papillomaviruses would be a fascinating study.

References

Abcarian H, Sharon N (1982) Long-term effectiveness of the immunotherapy of anal condyloma acuminatum. Dis Colon Rectum 25: 648-651

Almeida JD, Goffe AP (1965) Antibody to wart virus in human sera demonstrated by electron microscopy and precipitin tests. Lancet II: 1205-1207

Al-Yaman F, Willenborg DO (1984) Immune reactivity to autochthonous ovine squamous cell carcinomata. Vet Immunol Immunopathol 7: 153-168

Al-Yaman FM, Willenborg DO (1985) Mixed lymphocyte reaction suppression by tumour cell lines from naturally occurring squamous cell carcinomata. Aust J Exp Biol 63: 183-193

Atluru D, Johnson DW, Muscoplat CC (1982a) Tumor associated antigens of bovine cancer eye. Vet Immunol Immunopathol 3: 279-286

Atluru D, Kleinschuster SJ, Zupancici ML, Muscoplat CC (1982b) Detection of cell-surface antigens of bovine ocular squamous cell carcinoma. Am J Vet Res 43: 1156-1159

Bagdonas V, Olson C (1953) Observations on the epizootiology of cutaneous papillomatosis (warts) of cattle. J Am Vet Med Assoc 122: 393-397

Bagdonas V, Olson C (1954) Observations on immunity in cutaneous bovine papillomatosis. Am J Vet Res 15: 240-245

Baird PJ (1983) Serological evidence of the association of papillomavirus and cervical neoplasia. Lancet II: 17-18

Baird PJ (1985) The role of human papilloma and other viruses. Clin Obstet Gynaecol 12: 19-32

Barthold SW, Olson C (1974a) Membrane antigen of bovine papilloma virus-induced fibroma cells. JNCI 52: 737-742

Barthold SW, Olson C (1974b) Fibroma regression in relation to antibody and challenge immunity to bovine papilloma virus. Cancer Res 34: 2436-2439

Barthold SW, Olson C (1978) Common membrane neoantigen on bovine papilloma virus-induced fibroma cells from cattle and horses. Am J Vet Res 39: 1643-1645

Barthold SW, Olson C, Larson LL (1976) Precipitin response of cattle to commercial wart vaccine. Am J Vet Res 37: 449-451

Berman A, Winkelmann RK (1977) Flat warts undergoing involution: histopathological findings. Arch Dermatol 113: 1219-1221

Berman A, Winkelmann RK (1980) Involuting common warts. Clinical and histopathologic findings. J Am Acad Dermatol 3: 356-362

Beveridge WIB (1972) The logic of the comparative method. In: Frontiers in comparative medicine. University of Minnesota Press, Minneapolis, pp 21-40

Biberstein H (1944) Immunization therapy of warts. Arch Dermatol Syph 50: 12-22

Brodersen I, Genner J (1973) Histological and immunological observations on common warts in regression. Acta Dermatovener (Stockholm) 53: 461-464

Brodersen I, Genner J, Brodthagen H (1974) Tuberculin sensitivity in BCG vaccinated children with common warts. Acta Derm Venereol (Stockh) 54: 291-292

Broström H, Bredberg-Råden U, England J, Obel N, Perlmann P (1979) Cell-mediated immunity in horses with sarcoid tumors against sarcoid cells in vitro. Am J Vet Res 40: 1701-1706

Bunney MH (1982) Viral warts: their biology and treatment. Oxford University Press, Oxford, p. 70

Chambers VC, Evans CA, Weiser RS (1960) Canine oral papillomatosis II. Immunologic aspects of the disease. Cancer Res 20: 1083-1093

Chardonnet Y, Beauve P, Viac J, Schmitt D (1983) T-cell subsets and Langerhans cells in wart lesions. Immunol Lett 6: 191-196

Chretien JH, Esswein JG, Garagusi VF (1978) Decreased T cell levels in patients with warts. Arch Dermatol 114: 213-215

Chung YS, Spradbrow PB, Wilson BE (1977) Cytotoxic antibody in the serum of cattle with squamous cell carcinoma. Res Vet Sci 22: 263-264

Clarke GHV (1965) The charming of warts. J Invest Dermatol 45: 15-21

Cook RH, Olson C (1951) Experimental transmission of cutaneous papilloma of the horse. Am J Pathol 27: 1087-1097

Cubie HA (1972) Serological studies in a student population prone to infection with human papilloma virus. J Hyg (Lond) 70: 677-690

DeMonbreun WA, Goodpasture EW (1932) Infectious oral papillomatosis of dogs. Am J Pathol 8: 43-55

Du Casse FBW (1961) Bovine papillomatosis, with special reference to treatment with tissue vaccines. J S Afr Vet Med Assoc 32: 59-63

Duncan JR, Corbeil LB, Davies DH, Schultz RD, Witlock RH (1975) Persistent papillomatosis associated with immuno deficiency. Cornell Vet 65: 205-211

Dvoretzky I, Shober R, Chattopadhyay SK, Lowy DR (1980) A quantitative in vitro focus assay for bovine papilloma virus. Virology 103: 369-375

Evans CA, Ito Y (1966) Antitumor immunity in the Shope papilloma-carcinoma complex of rabbits. III Response to reinfection with viral nucleic acid. JNCI 36: 1161-1166

Evans CA, Rashad AL (1967) Virus content of Shope papillomas of cottontail rabbits. Cancer Res 27: 1011-1015

Evans CA, Gorman LR, Ito Y, Weiser RS (1962a) Antitumor immunity in the Shope papilloma-carcinoma complex of rabbits II. Suppression of a transplanted carcinoma, VX7, by homologous papilloma vaccine. JNCI 29: 287-292

Evans CA, Gorman LR, Ito Y, Weiser RS (1962b) Antitumor immunity in the Shope papilloma-carcinoma complex of rabbits I. Papilloma regression induced by homologous and autologous tissue vaccines. JNCI 29: 277-285

Evans CA, Weiser RS, Ito Y (1963) Antiviral and antitumor immunologic mechanisms operative in the Shope papilloma-carcinoma system. Cold Spring Harbor Symp W4, Quant Biol 27: 453-462

Evans WC, Patel MC, Kooky Y (1982) Acute bracken poisoning in homogastric and ruminant animals. Proc R Soc Edinb 81 B: 29-64

Favre M, Breitburd F, Croissant O, Orth G (1974) Hemagglutinating activity of bovine papilloma virus. Virology 60: 572-578

Favre M, Breitburd F, Croissant O, Orth G (1975) Structural polypeptides of rabbit, bovine, and human papilloma viruses. J Virol 15: 1239-1247

Favre M, Breitburd F, Croissant O, Orth G (1977) Chromatin-like structures obtained after alkaline disruption of bovine and human papillomaviruses. J Virol 21: 1205-1209

Fivaz BH (1978) The immunotherapy of bovine ocular squamous cell carcinoma. Rhod Vet J 9: 24-27

Ford JN, Jennings PA, Spradbrow PB, Francis J (1982) Evidence for papillomaviruses in ocular lesions in cattle. Res Vet Sci 32: 257-259

Freed DLJ, Eyres KE (1979) Persistent warts protected from immune attack by a blocking factor. Br J Dermatol 100: 731-733

Genner J (1971) Verrucae vulgares II. Demonstration of a complement fixation reaction. Acta Derm Venereol (Stockh) 51: 365-373

Geraldes A (1970) New antigens in hamster embryo cells transformed in vitro by bovine papilloma extracts. Nature 226: 81-82

Glinski W, Jablonska S, Langner A, Obalek S, Haftek M, Proniewska M (1976) Cell-mediated immunity in epidermodysplasia verruciformis. Dermatologica 153: 218-227

Glinski W, Obalek S, Jablonska S, Orth G (1981) T cell defect in patients with epidermodysplasia verruciformis due to human papillomavirus type 3 and 5. Dermatologica 162: 141-147

Greenberg JH, Smith TL, Katz RM (1973) Verrucae vulgaris rejection. A preliminary study of contact dermatitis and cellular immunity response. Arch Dermatol 107: 580-582

Hamir ANJ, Ladds PW, Boland PH (1980) An immunopathological study of bovine ocular squamous cell carcinoma. J Comp Pathol 90: 535-549

Hartley JW, Rowe WP (1964) New papovavirus contaminating Shope papillomata. Science 143: 258-260

Hellström I, Evans CA, Hellström KE (1969) Cellular immunity and its serum-mediated inhibition in Shope-virus-induced rabbit papillomas. Int J Cancer 4: 601-607

Hoffmann D, Jennings PA, Spradbrow PB (1981a) Immunotherapy of bovine ocular squamous cell carcinomas with phenol-saline extracts of allogeneic carcinomas. Aust Vet J 57: 159-162 and 250

Hoffmann D, Jennings PA, Spradbrow PB, Wilson BE (1981b) Autografting and allografting of bovine ocular squamous cell carcinoma. Res Vet Sci 31: 48-53

Hoxtell EO, Mandel JS, Murray SS, Schuman LM, Goltz RW (1977) Incidence of skin carcinoma after renal transplantation. Arch Dermatol 113: 436-438

Ingelfinger JR, Grupe WE, Topor M, Levey RH (1977) Warts in a pediatric renal transplant population. Dermatologica 155: 7-12

Ishimoto A, Ito Y (1969) Specific surface antigens in Shope papilloma cells. Virology 39: 595-597

Ishimoto A, Oota S, Kimura I, Miyake T, Ito Y (1970) In vitro cultivation and antigenicity of cottontail rabbit papilloma cells induced by the Shope papilloma virus. Cancer Res 30: 2598-2605

Ivanyi L, Morison WL (1976) In vitro lymphocyte stimulation by wart antigen in man. Br J Derm 94: 523-527

Jablonska S, Orth G, Glinski G, Obalek S, Jarzabek-Chorzelska M, Croissant O, Favre M, Rzesa G (1980) Morphology and immunology of human warts and familial warts. In: Bachmann PA (ed) Leukaemias, lymphomas and papillomas: comparative aspects. Taylor and Francis, London, pp 107-131

Jablonska S, Obalek S, Orth G, Haftek M, Jarzabek-Chorzelska M (1982) Regression of the lesions of epidermodysplasia verruciformis. Br J Dermatol 107: 109-116

Jarrett WFJ, McNeil PE, Grimshaw WTR, Selman IE, McIntyre WIM (1978) High incidence area of cattle cancer with a possible interaction between an environmental carcinogen and a papilloma virus. Nature 274: 215-217

Jennings PA (1979) Immunological and aetiological studies on bovine ocular squamous cell carcinoma. PhD thesis, University of Queensland

Jennings PA, Spradbrow PB (1980) Immunotherapy of bovine ocular squamous cell carcinoma with extracts of either autochthonous or allogeneic tumour. Proc 4th Int Congr Immunol Paris Abs 10.5.39

Jennings PA, Halliday WJ, Hoffmann D (1979a) Tumor associated immunity in bovine ocular squamous cell carcinoma detected by leukocyte adherence inhibition microassay. JNCI 63: 775-779

Jennings PA, Lavin MF, Hughes DJ, Spradbrow PB (1979b) Bovine ocular squamous cell carcinoma. Lymphocyte response to phytohaemagglutinin and tumour antigen. Br J Cancer 410: 608-614

Jenson AB, Rosenthal JD, Olson C, Pass F, Lancaster WD, Shah K (1980) Immunological relatedness of papillomaviruses from different species. JNCI 64: 495-500

Johansson E, Pyrhonen S, Rostila T (1977) Warts and wart virus antibodies in patients with "systemic lupus erythematosis". Br Med J 1: 74-76

Jun MH, Johnson RH (1979a) Effect of cyclophosphamide on tumour growth and cell-mediated immunity in sheep with ovine squamous cell carcinoma. Res Vet Sci 27: 155-160

Jun MH, Johnson RH (1979b) Suppression of blastogenic response of peripheral lymphocytes by serum from ovine squamous cell carcinoma bearing sheep. Res Vet Sci 27: 161-168

Jun MH, Johnson RH, Maguire DJ, Hopkins PS (1978) Enhancement and metastasis after immunotherapy of ovine squamous-cell carcinoma. Br J Cancer 38: 382-391

Jun MH, Johnson RH, Mills JM (1979) In vitro response of lymphocytes of normal and ovine squamous cell carcinoma-bearing sheep to phytomitogens and tumour extract. Res Vet Sci 27: 144-148

Kidd JG (1938) The course of virus-induced rabbit papillomas as determined by virus, cells, and host. J Exp Med 6: 551-574

Kidd JG, Beard JW, Rous P (1936) Serological reactions with a virus causing a rabbit papilloma which becomes cancerous. II. Tests of the blood of animal carrying various tumors. J Exp Med 64: 63-78

Klein WR, Buitenberg EJ, Steerenberg PA, de Jong WH, Kruizinga I-W, Misdorp W, Bier J, Tiesjema RH, Kreeftenberg JG, Teppema JS, Rapp JH (1982) Immunotherapy by intralesional injection of BCG cell walls or live BCG in bovine ocular squamous cell carcinoma. A preliminary report. JNCI 69: 1095-1103

Kleinschuster SJ, Rapp HJ, Lueker DC, Kainer RA (1977) Regression of bovine ocular squamous cell carcinoma by treatment with mycobacterial vaccine. JNCI 58: 1807-1814

Kleinschuster SJ, Rapp HJ, Green SB, Bier J, van Kampen K (1981) Efficacy of intra-tumorally administered mycobacterial cell wall in the treatment of cattle with ocular carcinoma. JNCI 67: 1165-1171

Koller LD, Barthold SW, Olson C (1974) Quantitation of bovine papilloma virus and serum antibody by immunodiffusion. Am J Vet Res 35: 121-124

Kossard S, Xenias SJ, Palestine RF, Scheen SR III, Winkelmann RK (1980) Inflammatory changes in verruca vulgaris. J Cutan Pathol 7: 217-221

Kreider JW (1963) Studies on the mechanism responsible for the spontaneous regression of the Shope rabbit papilloma. Cancer Res 23: 1593-1599

Kreider JW (1980) Neoplastic progression of the Shope rabbit papilloma. Cold Spring Harbor Conf Cell Prolif 7: 283-300

Kreider JW, Bartlett GL (1981) The Shope papilloma-carcinoma complex of rabbits: a model system of neoplastic progression and spontaneous regression. Adv Cancer Res 35: 81-110

Kreider JW, Breedis C (1969) The susceptibility of fetal rat skin in different immunologic environments to neoplastic induction with Shope papilloma virus. Cancer Res 29: 989-993

Kreider JW, Breedis C, Curran JS (1967) Interactions of Shope papilloma virus and rabbit skin cells in vitro 1. Immunofluorescent localization of virus inocula. JNCI 38: 921-931

Kreider JW, Benjamin SA, Pruchnic WF, Strimlan CV (1971) Immunologic mechanisms in the induction and regression of Shope papilloma virus-induced epidermal papillomas of rats. J Invest Dermatol 56: 102-112

Kreider JW, Bartlett GL, Sharkey FE (1979) Primary neoplastic transformation in vivo of xenogeneic skin grafts on nude mice. Cancer Res 39: 273-276

Kuchroo VK (1984) Immunological aspects of bovine ocular squamous cell carcinoma. PhD thesis, University of Queensland

Kuchroo VK, Spradbrow PB (1985) Tumour-associated antigens in bovine ocular squamous cell carcinoma: studies with sera from tumour bearing animals. Vet Immunol Immunopathol 9: 23-36

Kuchroo VK, Halliday WJ, Jennings PA, Spradbrow PB (1983a) Serum blocking factors in bovine ocular squamous cell carcinoma demonstrated by inhibition of erythrocyte rosette augmentation. Cancer Res 43: 1325-1329

Kuchroo VK, Jennings PA, Spradbrow PB (1983b) Bovine ocular squamous cell carcinoma: tumor associated immunity detected by E-rosette augmentation test. JNCI 70: 305-309

Le Bouvier GL, Sussman M, Crawford LV (1966) Antigenic diversity of mammalian papillomaviruses. J Gen Microbiol 45: 497-501

Lee AKY, Eisinger M (1976) Cell-mediated immunity (CMI) to human wart virus and wart-associated tissue antigens. Clin Exp Immunol 26: 419-424

Lee KP, Olson C (1968) Response of calves to intravenous and repeated intradermal inoculation with bovine papilloma virus. Am J Vet Res 29: 2103-2112

Lee KP, Olson C (1969a) A gel precipitin test for bovine papilloma virus. Am J Vet Res 30: 725-731

Lee KP, Olson C (1969b) Precipitin response of cattle to bovine papilloma virus. Cancer Res 29: 1393-1397

Lindsay GC III, Heck FC, England RB (1978) Ocular squamous cell carcinoma: immunological responses to tumor tissue and phytomitogens. Res Vet Sci 24: 113-117

Lutzner M, Croissant O, Ducasse M-F, Kreis H, Crosnier J, Orth G (1980) A potentially oncogenic human papillomavirus (HPV-5) found in two renal allograft recipients. J Invest Dermatol 75: 353-356

Lutzner MA, Orth G, Dutronoquay V, Ducasse MF, Kreis H, Crosnier J (1983) Detection of human papillomavirus type 5 DNA in skin cancers of an immunosuppressed renal allograft recipient. Lancet II: 422-424

Manilla GT, van Kampen KR, Marcus S (1972) Nucleoprotein immunotherapy of ocular squamous cell carcinoma in cattle. Fed Proc 31: 768

Massing AM, Epstein WL (1963) Natural history of warts. A two-year study. Arch Dermatol 87: 306-310

Matthews RS, Shirodaria PV (1973) Study of regressing warts by immunofluorescence. Lancet I: 689-691

McMichael H (1967) Inhibition by methylprednisolone of regression of the Shope rabbit papilloma. JNCI 39: 55-65

Meinke W, Meinke GC (1981) Isolation and characterization of the major capsid protein of bovine papilloma virus type 1. J Gen Virol 52: 15-24

Meischke HRC (1979) In vitro transformation by bovine papilloma virus. J Gen Virol 43: 473-487

Morison WL (1974) In vitro assay of cell-mediated immunity to human wart antigen. Br J Dermatol 90: 531-534

Morison WL (1975a) Viral warts, herpes simplex and herpes zoster in patients with secondary immune deficiencies and neoplasms. Br J Dermatol 92: 625-630

Morison WL (1975b) Cell-mediated immune response in patients with warts. Br J Dermatol 93: 553-556

Morison WL (1975c) In vitro assay of immunity to human wart antigen. Br J Dermatol 93: 545-551

Murphy JM, Severin GA, Lavack JD, Hepler DI, Lueker DC (1979) Immunotherapy in ocular equine sarcoid. J Am Vet Med Assoc 174: 269-272

Nelson M, Nelson DS, Spradbrow PB, Kuchroo VK, Jennings PA, Cianciolo GJ, Snyderman R

(1985) Successful tumour immunotherapy: possible role of antibodies to antiinflammatory factors produced by neoplasms. Clin Exp Immunol 61: 109-117
Noyes WF, Mellors RC (1957) Fluorescent antibody detection of the antigens of the Shope papilloma virus in papillomas of the wild and domestic rabbit. J Exp Med 106: 555-562
Obalek S, Glinski W, Haftek M, Orth G, Jablonska S (1980) Comparative studies on cell-mediated immunity in patients with different warts. Dermatologica 161: 73-83
Ogilvie MM (1970) Serological studies with human papova (wart) virus. J Hyg 68: 479-490
Oguchi M, Komura J, Tagami H, Ofuji S (1981 a) Ultrastructural studies of spontaneously regressing plane warts. Langerhans' cells show marked activation. Arch Dermatol Res 271: 55-61 (cited by Pfister 1984)
Oguchi M, Komura J, Tagami H, Ofuji S (1981 b) Ultrastructural studies of spontaneously regressing plane warts. Macrophages attack verruca-epidermal cells. Arch Dermatol Res 270: 403-411 (cited by Pfister 1984)
Olson C (1948) Equine sarcoid, a cutaneous neoplasm. Am J Vet Res 9: 333-341
Olson C, Segre D (1955) Neutralization of bovine papillomavirus with serums from papilloma bearing horses and cattle. Am J Vet Res 16: 517-520
Olson C, Skidmore LV (1959) Therapy of experimentally produced bovine cutaneous papillomatosis with vaccines and excision. J Am Vet Med Assoc 135: 339-343
Olson C, Segre D, Skidmore LV (1959) Immunity to bovine cutaneous papillomatosis produced by vaccine homologous to the challenge agent. J Am Vet Med Assoc 135: 499-502
Olson C, Segre D, Skidmore LV (1960) Further observations on immunity in bovine cutaneous papillomatosis. Am J Vet Res 21: 233-242
Olson C, Leudke AJ, Brobst DF (1962) Induced immunity of skin, vagina, and urinary bladder to bovine papillomatosis. Cancer Res 22: 463-468
Olson C, Robl MG, Larson LL (1968) Cutaneous and penile bovine fibropapillomatosis and its control. J Am Vet Med Assoc 153: 1189-1194
Orth G, Breitburd F, Favre M (1978) Evidence for antigenic determination shared by the structural polypeptides of (Shope) rabbit papillomavirus and human papillomavirus type 1. Virology 91: 243-255
Orth G, Favre M, Breitburd F, Croissant O, Jablonska S, Obalek S, Jarzabek-Chorzelska M, Rzesa G (1980) Epidermodysplasia verruciformis: a model for the role of papillomaviruses in human cancers. In: Essex M, Todaro G, zur Hausen H (eds) Viruses in naturally occurring cancers. Cold Spring Harbor Laboratory, New York, pp 259-282
Parsons RJ, Kidd JG (1943) Oral papillomatosis of rabbits: a virus disease. J Exp Med 77: 2233-2250
Pass F, Maizel JV (1973) Wart-associated antigens. II Human immunity to viral structural proteins. J Invest Dermatol 60: 307-311
Pass F, Marcus DM (1973) Wart-associated antigens. I. Isolation of tissue antigens using antibody immunoadsorbents. J Invest Dermatol 60: 301-306
Pearson JKL, Kerr WR, McCartney WDJ, Steele THJ (1958) Tissue vaccines in the treatment of bovine papillomas. Vet Rec 70: 971-973
Perry TL, Harman L (1974) Warts in diseases with immune defects. Cutis 13: 359-362
Pfister H (1984) Biology and biochemistry of papillomaviruses. Rev Physiol Biochem Pharmacol 99: 111-181
Pfister H, Huchthausen B, Gross G, zur Hausen H (1979) Seroepidemiological studies of bovine papillomavirus infections. JNCI 62: 1423-1425
Prawer SE, Pass F, Vance JC, Greenberg LJ, Yunis EJ, Zelickson AS (1977) Depressed immune function in epidermodysplasia verruciformis. Arch Dermatol 113: 495-499
Pyrhönen S (1978) Human wart-virus antibodies in patients with genital and skin warts. Acta Dermat Venereol (Stockh) 58: 427-432
Pyrhönen S, Johansson E (1975) Regression of warts. An immunological study. Lancet I: 592-596
Pyrhönen S, Penttinen K (1972) Wart virus antibodies and the prognosis of wart disease. Lancet II: 1330-1332
Rosenberger VG, Gründer HD (1959) Untersuchungen über die Immunitabildung und Immunotherapie bei der Papillomatose des Rindes. Dtsch Tierärztl Wochenschr 66: 661-666
Russo L, Russo A, Russo V (1975) Immunotherapy of warts. Lancet I: 921
Schneider V, Kay S, Lee HM (1983) Immunosuppression as a high-risk factor in the development

of condyloma acuminatum and squamous neoplasia of the cervix. Acta Cytol (Baltimore) 27: 220-224
Schou M, Helin P (1977) Levamisole in a double blind study. No effect on warts. Acta Derm 57: 449-454
Segre D, Olson C, Hoerlein AB (1955) Neutralization of bovine papilloma virus with serums from cattle and horses with experimental papillomas. Am J Vet Res 16: 517-520
Seski JC, Reinhalter ER, Silva J Jr (1978) Abnormalities of lymphocyte transformations in women with condylomata acuminata. Obstet Gynecol 51: 188-192
Seto A, Notake K, Kawanishi M, Ito Y (1977) Development and regression of Shope papillomas induced in newborn domestic rabbits. Proc Soc Exp Biol Med 156: 64-67
Shiratori O, Osato T, Ito Y (1967) Immunofluorescent studies of virus-induced rabbit papilloma (Shope) in vitro. Proc Soc Exp Biol Med 125: 435-438
Shokri-Tabibzadeh S, Koss LG, Nolnar J, Romney S (1981) Association of human papillomavirus with neoplastic process in the genital tract of four women with impaired immunity. Gynecol Oncol 12: 5129-5140
Shope RE (1933) Infectious papillomatosis of rabbits, with a note on the histopathology by E Weston Hurst. J Exp Med 58: 607-624
Snyderman R, Cianciolo GJ (1984) Immunosuppressive activity of the retroviral envelope protein P15E and its possible relationship to neoplasia. Immunology Today 5: 240-244
Spencer ES, Andersen HK (1970) Clinically evident non-terminal infections with herpesviruses and the wart virus in immunosuppressed renal allograft patients. Br Med J III: 251-254
Spradbrow PB, Hoffmann D (1980) Bovine ocular squamous cell carcinoma. Vet Bulletin 50: 449-459
Spradbrow PB, Wilson BE, Hoffmann D, Kelly WR, Francis J (1977) Immunotherapy of bovine ocular squamous cell carcinoma. Vet Rec 100: 376-378
Sundberg JP, Junge RE, Lancaster WD (1984) Immunoperoxidase localization of papillomaviruses in hyperplastic and neoplastic epithelial lesions of animals. Am J Vet Res 45: 1441-1446
Swan RA, Wilcox GE, Chapman HM, Hawkins CD (1983) Attempted transmission and immunotherapy of squamous cell carcinoma of the vulva of ewes. Aust Vet J 60: 314-315
Syrjänen KJ (1983) Immunocompetent cells in uterine cervical lesions of human papillomavirus origin. Gynecol Obstet Invest 16: 327-340
Syrjänen KJ (1984) Current concepts of human papillomavirus infections in the genital tract and their relationship to intraepithelial neoplasia and squamous cell carcinoma. Obst Gyn Survey 39: 253-265
Syrjänen KJ, Syrjänen SM, Lamberg MA, Happonen RP (1983) Local immunological reactivity in oral squamous cell lesions of possible HPV (human papillomavirus) origin. Arch Geschwulstforsch 53: 537-546
Syrjänen K, Väyrynen M, Castrén O, Mäntyjärvi R, Yliskoski M (1984) The relation between the type of immunoreactive cells found in human papillomavirus (HPV) lesions of the uterine cervix and the subsequent behaviour of these lesions. Arch Gynecol 234: 189-196
Tagami H, Ogino A, Takigawa M, Imamura S, Ofuji S (1974) Regression of plane warts following spontaneous inflammation. A histopathological study. Br J Derm 90: 147-154
Tagami H, Takigawa M, Ogino A, Imamura S, Ofugi S (1977) Spontaneous regression of plane warts after inflammation. Arch Dermatol 113: 1209-1213
Takigawa M, Tagami H, Watanabe S, Ogino A, Imamura S, Ofugi S (1977) Recovery processes during regression of plane warts. Arch Dermatol 113: 1214-1218
Thivolet J, Hegazy MR, Viac J, Chardonnet Y (1977) An in vivo study of cell-mediated immunity in human warts. Acta Derm Venereol (Stockh) 57: 317-319
Thomas L (1981) On warts. In: The medusa and the snail. More notes of a biology watcher. Penguin, Harmondsworth, p 66
Van Kampen KR, Crisp WE, De Martini JC, Ellsworth HS (1973) The immunologic therapy of squamous cell carcinoma - a preliminary report. Am J Obstet Gynaecol 116: 569-574
Vanselow BA, Spradbrow PB (1983) Squamous cell carcinoma of the vulva, hyperkeratosis and papillomavirus in a ewe. Aust Vet J 60: 194-195
Vanselow BA, Spradbrow PB, Jackson ARB (1982) Papillomaviruses, papillomas and squamous cell carcinomas in sheep. Vet Rec 110: 561-562
Väyrynen M, Syrjänen K, Mäntyjärvi R, Castrén O, Saarikoski S (1984) Langerhans' cells in

human papillomavirus (HPV) lesions of the uterine cervix identified by the monoclonal antibody OKT-6. Int J Gynaecol Obstet 22: 375-383
Viac J, Thivolet J, Chardonnet Y (1977a) Specific immunity in patients suffering from recurring warts before and after repetitive intradermal tests with human papilloma virus. Br J Dermatol 97: 365-370
Viac J, Thivolet J, Hegazy MR, Chardonnet Y, Dambuyant C (1977b) Comparative study of delayed hypersensitivity skin reactions and antibodies to human papilloma virus (HPV). Clin Exp Immunol 29: 240-246
Viac J, Schmitt D, Thivolet J (1978a) An immunoelectron microscopic localization of wart associated antigens present in human papilloma virus (HPV) infected cells. J Invest Dermatol 70: 263-266
Viac J, Staquet MJ, Miguet M, Chabanon M, Thivolet J (1978b) Specific immunity to human papilloma virus (HPV) in patients with genital warts. Br J Vener Dis 54: 172-175
Voss JL (1969) Transmission of equine sarcoid. Am J Vet Res 30: 183-191
Watson RE Jr, Larson KA (1974) Detection of tumor-specific antigens in an equine sarcoid cell line. Infect Immun 9: 714-718

Cell Culture Systems for the Study of Papillomaviruses

F. V. Breitburd

The viral origin of warts (Ciuffo 1907) and the oncogenic potential of a papillomavirus (PV), the cottontail rabbit (Shope) papillomavirus (CRPV) (Rous and Beard 1935), were recognized early, pointing to the potential importance of the PV group in human pathology. The capacity of bovine papillomavirus (BPV) to transform fibroblastic cells in vitro (Thomas et al. 1964) and the malignant conversion of some PV-associated lesions in patients with epidermodysplasia verruciformis or genital lesions and in cattle with alimentary tract papillomatosis argued for the possible role of PV in the origin of some cancers (see Orth et al. 1977). The lack of a system allowing the propagation of PVs in vitro precluded, however, investigations by standard virological studies and genetic manipulations (Butel 1972). It is thus mainly by the use of recombinant DNA technology that most of the human papillomavirus (HVP) types have been identified, their great plurality established, their pathogenic properties defined, and the association of specific HPV types with some skin and genital cancers demonstrated (Orth and Favre 1985; Orth 1986, 1987; Gissmann and Schwarz 1986). In spite of their great genetic heterogeneity, PVs have a common genomic organization. Two domains have been recognized: the E region with up to eight open reading frames (ORFs E1 to E8), which is considered to be expressed early in the virus life cycle and to contain the transformation functions, and the L region (ORFs L1, L2) encoding the viral structural polypeptides (for review see Giri and Danos 1986). Little is known, however, about the biology of PVs, which is important because of the correlation between PVs and human cancers. In that respect, the lack of simple cell culture system has proved to be a major impediment.

In the first part of this chapter, some of the known aspects of the biology of PVs in vivo and the characteristics of in vitro growth and differentiation of their host cell, the keratinocyte, will be summarized. These characteristics may relate to the difficulties encountered for the development of in vitro systems for the study of PVs. In the second part, the main tissue culture models and experimental animal systems available so far for the study of the mechanisms involved in keratinocyte infection and transformation by PVs will be reviewed.

1 Factors Complicating the Development of a Culture System

The capacity of BPV1 to transform fibroblasts in vitro has provided a good model for the genetic analysis of the transforming region of the BPV1 genome (Howley et al. 1986; Turek and Haugen, this volume). In view of the common genomic organization of PVs, the characterization of the BPV1 genes and their functions expressed in transformed fibroblasts should provide some clues to the mechanisms involved in keratinocyte infection and transformation by PVs under natural conditions.

In contrast, after many years, the search for a culture system for the propagation of PVs is still a challenge. Several reports on such culture systems were published but remained unconfirmed, and many unsuccessful attempts have been reported (for review see Taichman et al 1984). The strict conditions that PVs require for their multiplication as defined by in vivo observations and the difficulty of obtaining in vitro indefinite growth and terminal differentiation of their host cell may relate to this failure.

1.1 Conditions of PV Multiplication In Vivo

Studies on the infectivity of PVs (see Orth et al. 1977; Jarrett 1985; Kreider and Bartlett 1985) and in situ methods allowing the detection of vegetative viral DNA replication, viral capsid antigens, and virus particles in tissue sections (see Croissant et al. 1985) have defined some of the main features of PV productive infection.

1.1.1 Host Specificity

PVs generally show a strict host specificity which for example has precluded the study of human PVs in animals. Some animal PVs have a wider host range. CRPV may induce skin warts evolving into carcinomas in domestic rabbits (Kreider and Bartlett 1985). BPV, just like sheep, deer, and European elk PVs, induce connective tissue tumors in a number of laboratory rodents (Jarrett 1985; Howley et al. 1986). No progeny virus, however, is produced in BPV1-associated connective tissue tumors and only in a few cells, if any, in domestic rabbit CRPV-associated skin papillomas. Infection results only in the abnormal proliferation of keratinocytes in rabbit papillomas or of fibroblasts in connective tissue tumors, providing particularly good models for the study of the viral mechanisms involved in the formation of a benign tumor. In that respect, viral genomes persist in tumors as free episomes at high copy number (from 10 to more than 200 genome copies per cell), and the viral early transforming regions, in particular the CRPV *E6* and *E7* genes in domestic rabbit papillomas (Georges et al. 1984; Nasseri and Wettstein 1984a), is transcribed (Howley et al. 1986).

1.1.2 Tissue Affinity

PVs are epitheliotropic viruses which infect the skin and some mucous membranes. Several distinct types of PV have been distinguished in infected humans (Orth and Favre 1985; Broker and Botchan 1986) and in cattle (Jarrett 1985). The genetic heterogeneity of HPVs and BPVs is reflected in particular biological properties, their tissue affinity, among them. Some HPV types infect the skin specifically while others infect preferentially the mucocutaneous regions or the mucous membranes of the genital area and the orolaryngeal tract (Orth and Favre 1985; Broker and Botchan 1986). BPV3, BPV4 and BPV6 induce epithelial papillomas of the skin, alimentary tract, and udder, respectively. In contrast, BPV1, BPV2, and BPV5 induce fibropapillomas with a lesser degree of site restriction (Jarrett 1985). When inoculated across rabbit lip, CRPV induces papillomas exclusively on the hair-bearing skin of the outer lip and not on the vermilion border or the buccal mucosa (Kreider and Bartlett 1985). Such affinities may have to be considered in the choice of PV type and host cell for in vitro studies.

1.1.3 Cell Type Specificity and Interaction with Terminal Differentiation

PVs replicate exclusively in keratinocytes, i.e., the epithelial cells of the tissues they infect. PVs inducing fibropapillomas and fibromas are capable of also infecting the fibroblasts of the subepithelial connective tissue, but virus production is exclusively detected in the epithelium covering the connective tissue portion of the tumors (Jarrett 1985). The presence of viral transcripts, containing sequences corresponding to the transforming region of the viral genome in the connective tissue component of BPV1-induced papillomas, supports the role of some viral gene product(s) in the abnormal proliferation of fibroblasts (Howley et al. 1986).

PV vegetative growth occurs exclusively in the terminally differentiating, infected keratinocytes of the papillomas (Croissant et al. 1985; Jarrett 1985; Kreider and Bartlett 1985), Although there is no direct evidence for the infection of cells forming the basal epithelial layer, their modified response to the mechanisms regulating their growth, which results in the formation of a papilloma, suggests such an event. It has been hypothesized that infected basal keratinocytes are nonpermissive to virus production and behave like transformed cells and that the nature of the transformation does not result in the loss of their capacity to express their differentiation program. With the onset of the differentiation process, vegetative viral DNA replication takes place, followed by viral capsid protein synthesis and virion formation as the cells migrate towards more differentiated papilloma layers (Orth et al. 1977). In turn, virus multiplication leads to some alterations in the expression of the terminal differentiation program. In human skin warts, these modifications give rise to recognizable cytological and ultrastructural features which were found to be specific for the virus type (Croissant et al. 1985; Schneider, this volume). These alterations were observed as early as the onset of vegetative viral DNA replication (Croissant et al. 1985) and, in HPV1-induced warts, appeared together with the expression of the viral *E4* gene product (Croissant et al. 1985; Breitburd et al. 1987).

Finally, PVs no longer replicate when benign lesions become malignant, and this may be a consequence of the changes in the terminal differentiation program of carcinoma cells (Orth et al. 1977).

1.1.4 Cell Competence to PV Infection and Transformation

An additional difficulty in finding an appropriate culture system is that PV infection and transformation are usually limited. It is commonly accepted that tumors arise from the clonal growth of one or a few infected cells (Orth and Favre 1985). Specific genetic make-up, immunological defects, or particular diet conditions are involved in extensive PV infection (Jarrett 1985; Orth 1986, 1987). Moreover, PV-induced lesions have a self-limited growth. Lesions of patients suffering from epidermodysplasia verruciformis, a generalized HPV infection of the skin, can remain stable for many years (Orth 1987). There is no spread of infection to neighboring cells in the absence of additional traumas, suggesting that the physiological state of the cell plays a role in the susceptibility to PV infection. Another striking feature of PV-induced lesions is that they do not contain solely productively infected cells but also a hyperplastic cell population showing no evidence of productive infection and undergoing an apparently normal process of terminal differentiation. It is not known whether these cells are infected and nonpermissive to virus production (Croissant et al. 1985) and whether they might overgrow productively infected cells in vitro. Finally, malignant conversion of lesions associated with certain types of PVs occurs only at low frequency, after a long delay, and through different tumor stages of increasing severity, suggesting a multistep process of carcinogenesis. It has been shown for CRPV-induced tumors that chemical carcinogens increased markedly the progression rate of warts (Kreider and Bartlett

1985), and no difference has been found so far in the expression of the CRPV genome in non-virus-producing benign or malignant tumors (Georges et al. 1984; Nasseri and Wettstein 1984a). This indicates that some cellular changes, which would not be sufficient for malignant conversion in the absence of infection with certain PV types, are required for the expression of the oncogenic potential of these PVs.

1.2 Growth and Differentiation of Keratinocytes In Vitro

Several methods have been developed in the recent past which, by providing suitable substrates and convenient growth factors, permit the growth of keratinocytes in vitro and some expression of their terminal differentiation.

1.2.1 Methods

Cultures may be initiated by seeding keratinocyte suspensions or by fixing tissue fragments either on collagen substrates in the form of hydrated gels (Orth et al. 1974; Karasek 1983), films (Hawley-Nelson et al. 1980), lattices incorporated with fibroblasts (Bell et al. 1983), or in the presence of irradiated 3T3 mouse fibroblasts as feeder layers (Green et al. 1981). Chemically defined media have also been designed which allow the selective growth of keratinocytes and the regulation of their terminal differentiation, while preventing dermal fibroblast proliferation (see Boyce and Ham 1983).

1.2.2 Limits of Keratinocyte Cultures as Related to the Propagation of PVs

Keratinocytes from normal epithelia have a limited lifespan in vitro corresponding to 50-150 doublings. The cells can be serially grown for 2-5 subcultures (Green et al. 1981) or maintained as long-term cultures for over a month (Orth et al. 1974). Both conditions would provide a sufficient period of time for the development of a PV infection, as indicated by experimental infection of rabbits (Kreider and Bartlett 1985) or cattle (Jarrett 1985). Depending on seeding densities, keratinocytes form colonies or confluent cultures, first growing as monolayers with a typical epithelial morphology. The cells subsequently form multiple layers and become a mixed population of proliferating and terminally differentiating cells. As in the natural epithelium, proliferating cells occupy the basal position and terminally differentiated cells slough into the medium. Growth curves established for long-term cultures account well for these different phases of proliferation, pluristratification, and sloughing of differentiated cells (Fig. 1) (Orth et al. 1974; Taichman et al. 1984). Cell differentiation is marked by the synthesis of keratin filaments and of involucrin followed, in a proportion of the cells, by the formation of cross-linked cornified envelopes and the loss of nuclei. Keratohyalin is scarce, however, and the expression of keratin polypeptides is modified when compared to normal tissue (Green et al. 1981; Sun et al. 1984). In particular, keratins occur in smaller amounts, and high-molecular weight keratin polypeptides specific for terminally differentiating keratinocytes are undetectable (Fig. 2). Thus well-defined

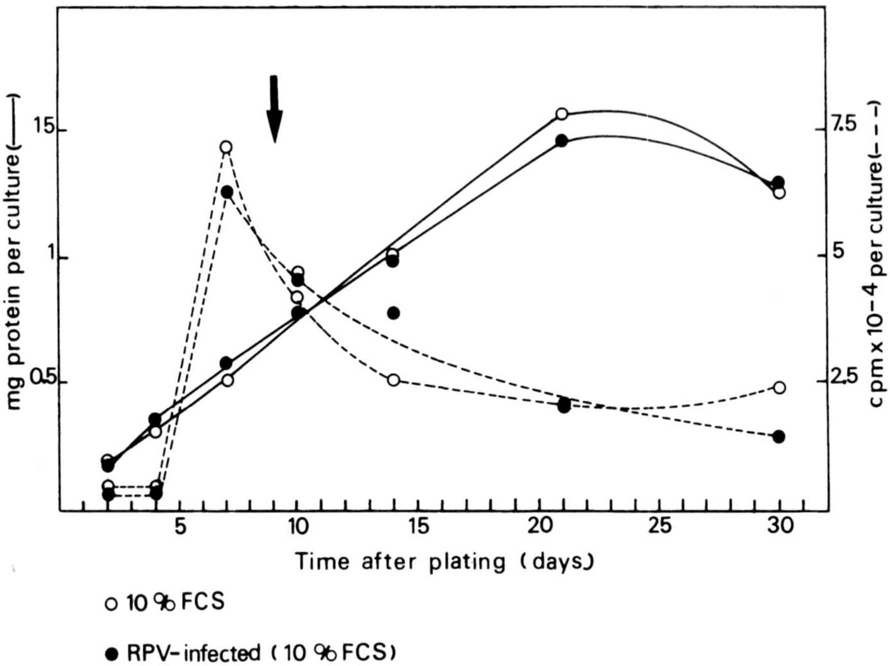

Fig. 1. Effect of CRPV infection on the growth of domestic rabbit keratinocytes in culture. Keratinocytes from domestic rabbit epidermis were seeded in the presence (●) or the absence (○) of purified CRPV particles, at high multiplicity of infection, on coverslips bearing collagen gels set in Leighton tubes (2.5×10^5 cells/tube) (Orth et al. 1974). The rate of cellular DNA synthesis (----) was measured by the incorporation of tritiated thymidine, and growth was evaluated by the DNA (not shown) and the protein (——) contents of the culture. Each value corresponds to duplicate experiments. Note that after the outburst of DNA synthesis following plating, the rate of DNA synthesis remains significant after confluency *(arrow)*, and DNA (not shown) and protein contents increase in pluristratifying cultures until a plateau value is reached, when sloughing of differentiated cells compensates for cell multiplication (from Taichman et al. 1984)

granular and cornified layers, in which papillomavirus vegetative growth is usually detected, are not formed.

The possibility of regulating the terminal differentiation of cultured keratinocytes may provide an environment closer to the conditions necessary for virus production. Suspension of cells in semisolid media, or in the presence of an ionophore, leads to the rapid formation of cornified envelopes and the loss of nuclei, which is the last step of keratinocyte terminal differentiation (Green et al. 1981). In chemically defined media, the expression of terminal differentiation of human keratinocytes may be triggered by higher calcium concentrations (Boyce and Ham 1983), as first shown for mouse keratinocytes (Yuspa et al. 1985). Finally, cultivation in a medium devoid of vitamin A leads to the formation of a more highly keratinized epithelial structure containing high-molecular weigth keratin polypeptides (Fuchs and Green 1981). There is, however, no information as yet on the event(s) involved in the triggering of vegetative viral DNA replication in vivo, and it is particularly unclear whether expression of the keratins specific for terminal differentiation is required. In situ studies have shown the absence of some keratin

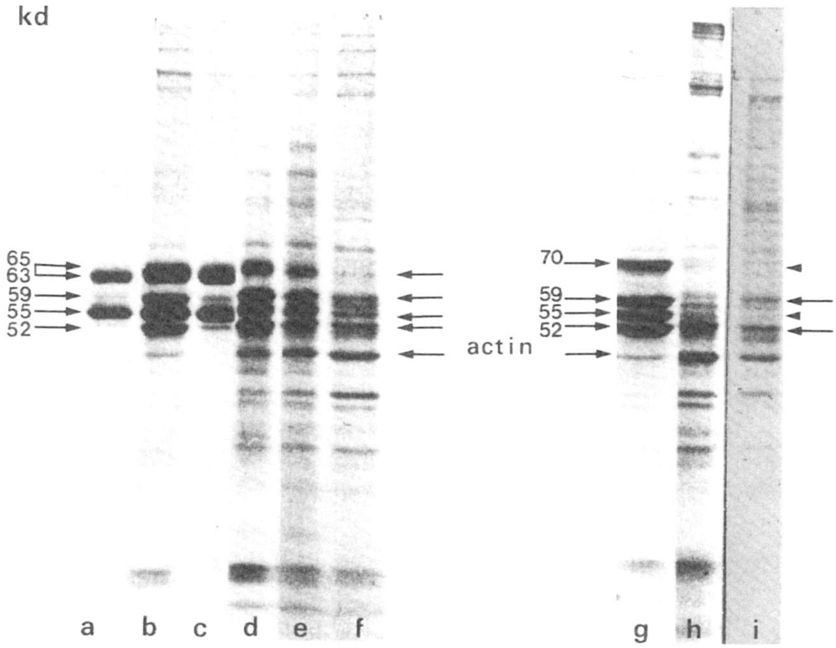

Fig. 2. Polypeptide patterns of rabbit epidermal keratinocytes in vivo and in vitro. Trypsinization of epidermis from dermis allows sheets of fully keratinized cells and of suspensions of keratinizing and proliferating cells to be obtained. After plating, only basal keratinocytes remain attached to the collagen substrate. Samples were heat-dissociated in the presence of sodium dodecyl sulfate (SDS), urea, and 2-mercaptoethanol and analyzed by electrophoresis in the presence of SDS in polyacrylamide gradient slab gels. (lanes *a–f*) Domestic rabbit keratinocytes: *(a)* keratinized cells, *(b)* keratinocyte suspension, *(c)* unattached cells, and attached cells of *(d)* 1-day culture, *(e)* 4-day culture, and *(f)* 7-day culture. (lanes *g–i*) Cottontail rabbit keratinocytes: *(g)* keratinocyte suspension, *(h)* 28-day culture, *(i)* [^{14}C]polypeptides in a 28-day culture. Note that the keratin polypeptide subset specific for terminally differentiating keratinocytes (*55–63/65 k* for domestic rabbit, *55–70 k* for cottontail rabbit) are little, if at all, expressed after some time in vitro (*arrows* on the right side of the gels). Molecular weights are indicated in kilodaltons *(kd)*

subsets or of keratin filaments as early as the onset of vegetative viral DNA replication in cells productively infected by some skin HPV types (Croissant et al. 1985; Breitburd et al. 1987). These data confirm a most specific interference of virus replication with the expression of cell terminal differentiation but do not answer the question of whether this reflects a mechanism deviating from terminal differentiation in favor of viral replication (Breitburd et al. 1987).

Another limit of keratinocyte cultures is that keratinocytes from different anatomical sites (epidermis, conjunctiva, esophagus, oral cavity, genital tract) all form in vitro incompletely keratinized epithelia that resemble one another (see Taichman et al. 1984). In view of the site specificity of some PVs, it might be anticipated that PV types specific for tissue showing little or no granular and cornified layers would be more adapted than others to in vitro conditions. Keratinocytes are however capable of reversible adaptation to their environment. Cultures reinjected subcutaneously into nude mice reexpress their original tissue-specific differentiation to varying degrees (Doran et al. 1980). This may be utilized to reproduce site-specific keratinocyte terminal differentiation. A third difficulty arises from the heterogeneity of the cell population of the basal layer of epithelia. Studies of clone-forming ability in vitro of keratinocytes isolated from fresh tissue have confirmed that some basal cells are not capable of proliferation and even fewer are clonogenic (Barrandon and Green 1985). The existence of nonclonogenic cells giving rise to only a restricted number of progeny, the so-called transit amplifying population (Lavker and Sun 1983), could not be demonstrated in vitro (Barrandon and Green 1985). Prior trauma is required for PV infection, and it has been shown that the target cells for CRPV infection were those stimulated to proliferate in the migrating "wound" epithelium (Kreider and Bartlett 1985). It is not known whether these cells are clonogenic in vitro.

1.2.3 In Vitro Selection of PV-Associated Papilloma and Carcinoma Cells

In the two-stage mouse skin carcinogenesis model, it has been shown that benign skin papillomas result from a single initiation event (virus infection or carcinogen application) followed by repeated treatments with a promoting agent (the most commonly used being 12-O-tetradecanoyl phorbol 13-acetate or TPA). Further cellular changes are required for progression to carcinoma (see Brown et al. 1986). The first visible step in the interaction of a PV with a keratinocyte is the hyperproliferative response resulting in the growth of a papilloma (Orth et al. 1977). It is not known whether this benign transformation of keratinocytes is necessary for vegetative viral growth (Taichman et al. 1984). Selection of PV-transformed keratinocytes from normal cells may not be readily achieved on the basis of growth rate in vitro. Cultured keratinocytes are in a hyperproliferative state compared to normal tissue as suggested by the high rate of DNA synthesis (Orth et al. 1974; Fig. 1), the synthesis of keratin subsets typical of hyperproliferating keratinocytes (Sun et al. 1984; Fig. 2), and the persistence of the colony-forming ability over a range of cell sizes normally reserved for terminally differentiating cells (Barrandon and Green 1985).

It has been shown that chemical or viral initiation of murine basal keratinocytes blocked the cells at an early stage of terminal differentiation and that initiated cells showed continued growth in vitro under conditions in which growth arrest and expression of terminal differentiation were observed for normal keratinocytes (see Yuspa et al. 1985). Benign transformation of keratinocytes by PV could thus be assessed by continued growth under conditions of commitment to terminal differentiation such as a vitamin A-deprived medium or semisolid medium. It is not known what properties papilloma cells share with malignant cells, and thus it is not clear how to select for cells undergoing PV-induced benign or malignant transformation. It must be stressed in addition that keratinocytes obtained from naturally occurring carcinomas or transformed in vitro by SV40 or murine sarcoma viruses (for review see Taichman et al. 1984) do not exhibit a single typical phenotype. Transformed keratinocytes are immortalized and have reduced requirements for serum and growth factors, but some cell lines retain the dependence on 3T3 feeder layers for growth. They form either rapidly growing, well-differentiated, squamous cell carcinomas or progressively growing squamous cysts in nude mice.

They have a reduced requirement for anchorage but, for most of them, show only abortive growth in semisolid medium, precluding the use of this test for selection. Finally, they show a marked decrease in their commitment to terminal differentiation which is manifested to varying extents by little to no stratification, subnormal rates of formation of cross-linked cornified envelopes upon anchorage deprivation, and different alterations in the expression of keratin polypeptides (Rheinwald and Beckett 1981; Rheinwald et al. 1983; Taichman et al. 1984).

2 Tissue Culture Systems for PVs

Keratinocyte culture systems allowing PV replication or susceptible to PV transformation have not yet been found, but cell cultures derived from PV-associated tumors and introduction of PV genomes into cultured cells have provided useful tools for progress in the field.

2.1 Tumor-Derived Cell Cultures

The recent availability of cell lines from CRPV-associated skin tumors and from human genital tumors associated with HPV16 and HPV18 have offered new possibilities for the analysis of the structural organization, transcription, and expression of PV genomes in PV-infected and transformed keratinocytes, and also for the study of the phenotype of keratinocytes undergoing benign or malignant transformation by PVs.

2.1.1 Wart and Papilloma Cells

Early studies had shown that rabbit papilloma cells grew in vitro like epidermal cells while maintaining the main ultrastructural features of keratinocytes, but no antigenic or ultrastructural evidence had been obtained for vegetative viral replication (Orth and Croissant 1968). Keratinocytes growing out from fragments of human skin warts (Niimura et al. 1975) and laryngeal papillomas (Steinberg et al. 1982) maintain a normal morphology and senesce like normal keratinocytes, with no ultrastructural or antigenic evidence for vegetative viral replication. This does not result from an absence of the viral genomes since, for instance, HPV6 genomes are present at the same copy number as in the original tumor in keratinocytes cultured from a laryngeal papilloma and passaged once (Steinberg et al. 1983). Similarly, free CRPV genomes (10-100 copies per cell) are retained in domestic rabbit papilloma cells maintained in vitro for a limited time (G. Orth, N. Jibard, and E. Georges, unpublished results). No evidence for keratinocyte immortalization or transformation was found in any of these studies. Immortalized cell lines have been derived, however, from papillomas. A cell line with no evidence for transformation was established from a human laryngeal papilloma but showed no antigenic evidence of virus production (Leventon-Kriss et al. 1983). Three fully transformed cell lines obtained from BPV4-induced alimentary tract papillomas of cattle no longer harbored viral genomes (Campo et al. 1985), but in vivo studies indicate that the presence of the BPV4 genome is not required for tumoral progression of BPV4-induced papillomas and for the maintenance of their transformed state (Campo et al. 1985). Finally, an epithelial cell line (sf 1Ep) derived from the ear epidermis of a cottontail rabbit with papillomatosis has been recently examined for the presence and expression of CRPV genome (Nasseri and Wettstein 1985). Ten to twenty copies of CRPV DNA are maintained extrachromosomally in sf1Ep cells, and viral transcripts are detectable only after treatment with TPA. RNA levels are much lower than those detected in domestic rabbit papillomas, and they have not allowed characterization of the viral genes transcribed. No antigenic evidence has been found for the expression of viral capsid antigens. Since sf1Ep cells are nontumorigenic for nude mice, it has been proposed that they could represent a line of infected basal keratinocytes (Nasseri and Wettstein 1985). Further studies should determine whether sf1Ep cells have retained the differentiated properties of an epidermal keratinocyte and can be triggered to allow virus replication or can progress towards a more malignant phenotype after treatment with cocarcinogens.

2.1.2 Carcinoma Cells

2.1.2.1 Human Cell Lines. Cultures or cell lines from primary skin carcinomas associated with HPV5 and related types (Orth 1986, 1987) have not been reported. In contrast a number of cell lines have been derived in the past from human carcinomas ot the uterine cervix, and in view of the association of a majority of these cancers with HPV16 and HPV18 (Gissmann and Schwarz 1986), a search for the presence and expression of genital HPVs in the cell lines was an important step in assessing the etiological role of HPVs in cervical cancers and in providing tissue culture models. HPV DNA sequences were detected in 11 of the 14 lines studied, including morphologically distinct variants of two cell lines (C4I and II, SKGIIIa and b) (Boshart et al. 1984; Schwarz et al. 1985; Yee et al. 1985; Pater and Pater 1985; Tsunokawa et al. 1986b). Integration has been demonstrated in most cells, if not all (Schwarz et al. 1985; Yee et al. 1985; Tsunokawa et al. 1986b). HPV18 DNA sequences correspond to about 10-50 copies of the viral genome per cell in Hela and SW756 cells but to 1 or less than 1 copy of the viral genome per cell in C4I and II, ME180, MS751, and SKGI and II cells. HPV16 DNA sequences are present in amounts greater than 500 copies per cell in CaSki cells, 1 copy per cell in SiHa cells, and less than 1 copy per cell in SKGIIIa, b (Schwarz et al. 1985; Yee et al. 1985; Tsunokawa et al. 1986b; Baker et al. 1987). Analysis of viral DNA integration patterns indicates that HPV18 sequences are amplified together with cellular flanking sequences in Hela and SW756 cells (Schwarz et al. 1985, 1986), while HPV16 sequences are in head-to-tail tandem arrays in CaSki cells (Yee et al. 1985; Smotkin and Wettstein 1986; Baker et al. 1987). Transcription of viral sequences has been demonstrated in all cells, except ME180 (Schwarz et al. 1985; Yee et al. 1985); it was found unaffected by cycloheximide treatment in Hela and SiHa cells in contrast to BPV1-transformed mouse fibroblasts (Kleiner et al. 1986).

Characterization of the structural organization and transcription patterns of the HPV18 sequences in Hela, SW756, and C4I cells (Schwarz et al. 1985, 1986;

Schneider-Gädicke and Schwarz 1986) have revealed strikingly common features (see also Schwarz, this volume). (1) Integration of viral sequences interrupts the early region of HPV18 within a segment of ORFs E1 and E2. (2) HPV18-positive messenger RNAs are virus-cell fusion transcripts which contain 5′-terminal sequences all transcribed from ORFs E6, E7, and E1, and 3′-terminal host cell sequences different for each cell line. (3) Three types of mRNAs, recognized by nucleotide sequence analysis of cDNA clones, are present in different relative amounts in the cell lines. The most common species contain ORFs E6* (generated by splicing of E6) and E7. The others contain ORFs E6 and E7 and ORFs E6*, E7, and E1. This would give rise to an E6 or E6* protein and possibly to an E7 protein. A 12K HPV18 E7 protein has been identified in all three cell lines (Seedorf et al. 1987), and an HPV18 E6 protein has been detected recently in Hela cells (Banks et al. 1987), using antibodies to E6 or E7 fusion proteins expressed in bacteria. The structural organization of HPV16 DNA sequences in CaSki and SiHa cells has been characterized recently (Baker et al. 1987). Integration of a single HPV16 genome in SiHa cells disrupts ORFs E2 and E4, with a small deletion fusing them. As for CaSki cells, the three most abundant HPV16 genomic structures are present in large tandem arrays and consist of full-length genomes, genomes with a 1.4kb deletion in the control region and genomes with a 2.6kb tandem repeat of the 3′-early region (Baker et al. 1987). Viral transcripts in both cell lines were found composed principally of sequences derived from ORFs E6 and E7 (Smotkin and Wettstein 1986; Baker et al. 1987). The analysis of viral transcripts in CaSki cells indicates that they contain sequences from the E6-E7 region, spliced to the E2-E4 region (Smotkin and Wettstein 1986). The major and one minor RNA species have a coding capacity for E7 and introns within E6, while a minor species would code for a slightly truncated E6 protein. The HPV16 E7 protein has been identified in CaSki cells using antisera to an E7 fusion protein expressed in bacteria (Smotkin and Wettstein 1986; Seedorf et al. 1987) and in SiHa cells (Seedorf et al. 1987). In vitro translation of hybrid-selected HPV16- and HPV18-specific poly(A^+) RNA from SiHa, CaSki, and HeLa cells allowed, in addition, the identification of an 11K E6 and a 10K E4 protein in CaSki cells and of a 70K E1 protein in HeLa cells (Seedorf et al. 1987).

Altogether, the data on the integration and transcription patterns of HPV DNA sequences in different cervical carcinoma cell lines (Schwarz et al. 1985, 1986; Schneider-Gädicke and Schwarz 1986; Smotkin and Wettstein 1986; Baker et al. 1987), the integration of HPV DNA sequences in cervical cancer biopsies (Dürst et al. 1986; Dürst, this volume; see also Orth 1986, 1987), and the characterization of HPV16 transcripts coding for an E7 and an E6 protein in a cervical cancer (Smotkin and Wettstein 1986) argue for an essential role of the *E6* and *E7* genes in the maintenance of the malignant state of cervical carcinoma cells. Transcription of ORFs E6 and E7 has been demonstrated also in skin carcinomas associated with HPV5 (Orth 1986; 1987) and in CRPV-associated skin carcinomas and transplantable VX2 and VX7 carcinomas (Georges et al. 1984; Nasseri and Wettstein 1984a, b), although in primary skin carcinomas integration of PV sequences in to the cell genome is rather infrequent (see Orth 1986, 1987). Genetic studies with BPV1-transformed cells have shown that the BPV1 *E6* gene encodes a transforming protein and that the *E7* gene product is required to maintain plasmids at a high number (Howley et al. 1986; Turek and Haugen, this volume). It is not clear at present how *E7* expression is related to progression towards malignancy and long-term maintenance of the malignant state, but the availability of cell lines should help to characterize the E6 and E7 proteins and their functions. Also, the disruption of ORF E2 by integration may interfere with its transcriptional regulatory function involved in the control of viral gene expression and may play a role in malignant progression. Further support for the validity of these models is the recent isolation of a keratinocyte line (SK-v) from bowenoid papules of the vulva containing predominantly free HPV16 genomes. After a number of passages, SK-v cells exclusively harbor integrated HPV16 DNA sequences which are amplified five- to ten-fold together with cellular flanking sequences and transcribed (Orth et al., manuscript in preparation). Interestingly, it has been shown that integration which interrupts ORF E2 had already taken place in the patient's lesions (Schneider-Maunoury et al. 1987).

2.1.2.2 Rabbit Cell Lines. There is no report of cultures of primary skin carcinomas associated with CRPV, but two cell lines (Georges et al. 1985) and, most probably, a third one (McVay et al. 1982; Nasseri and Wettstein 1984b) have been isolated from the transplantable VX2 carcinoma of the domestic rabbit, a tumor established four decades ago from a carcinoma derived from a CRPV-induced skin wart. The VX2 carcinoma is an anaplastic carcinoma, frequently metastasizing, which may be considered as expressing to an extreme degree the malignant properties of a PV-associated carcinoma. The three cell lines contain the CRPV genome as integrated head-to-tail tandem repeats, at the same copy number and with the same integration and transcription patterns as the VX2 tumor in vivo (McVay et al. 1982; Nasseri and Wettstein 1984b; Georges et al. 1985). As in vivo, the three viral transcripts hybridize only to the transforming region of the CRPV genome. Interestingly, the two cell lines (VX2T, VX2R) which were isolated from the same tumor show different levels of keratinocyte differentiation and transplantability (Georges et al. 1985). The VX2T cells retain the capacity to produce tumors in rabbits and the low expression of epidermal keratinocyte differentiation of the VX2 tumor cells. The VX2R cells are no longer serially transplantable in the rabbit and express the differentiated properties of a keratinocyte, but since they grow as an established cell line, show anchorage-independent growth, and induce carcinomas in nude mice, VX2R cells are still transformed keratinocytes. The anaplastic characteristics and the transplantability of VX2 carcinoma cells to immunecompetent allogeneic host may thus be lost without any detectable modification of the physical state and transcription of the CRPV genome. It seems likely that culture conditions have selected a cell variant with a different chromosomal distribution resulting in the reexpression of differentiated characteristics and an antigenic make-up which renders the cells susceptible to the immune reactions of the host, precluding their development in vivo (Georges et al. 1985). Recently, transfection of VX2T, VX2R, and rabbit keratinocytes with expression vectors containing CRPV or HPV1 noncoding regions inserted upstream of the bacterial chloramphenicol acetyltransferase *(CAT)* gene showed that only the CRPV promoter was activated and was active to a greater extent in the more differentiated VX2R cells (Giri et al. 1985). VX2T and VX2R cells may provide good models for studying the cellular events conditioning the phenotypes of PV-transformed keratinocytes and could also prove useful in the study of PV gene functions by providing viral and cellular factors involved in the regulation of the expression of PV genomes.

2.1.3 Hamster Sarcoma Cell Lines

Two cell lines (HT2, HT3) have been established from the first transplant of two different hamster sarcomas induced by BPV1. HT2 cells have a fibroblastic morphology, whereas HT3 cells are epitheloid. Both cell lines are tumorigenic for hamsters (Breitburd et al. 1981). HT2 cells harbor about 25 copies per cell of BPV1 genome, predominantly as free oligomeric or catenated molecules. HT3 cells contain about 100 copies per cell, with most, if not all, integrated to the cell genome. Transcripts corresponding to the transforming region of BPV1 were found only at early passages and were no longer detected at late passages, either in cell cultures or in hamster tumors induced by grafting the cells, although the copy number and the physical state of the viral genomes remained unchanged (Jaureguiberry et al. 1983). As in the progression of BPV4-induced alimentary tract papillomas of cattle, which takes place in the absence of the viral genome (Campo et al. 1985), the maintenance of the transformed state and the tumorigenicity of HT2 and HT3 cells require little, if any, expression of the viral genome (Jaureguiberry et al. 1983). It is not known whether this reflects specific properties of the virus or of the cells.

2.2 Introduction of PV Genomes into Cultured Cells

Another approach to obtain model systems for the study of PV replication and gene expression has been to infect with viral particles, or transfect with viral DNA or recombinant viral DNA molecules, keratinocytes or fibroblasts, as primary cultures or established cell lines.

2.2.1 PV Infection and Transfection of Keratinocytes

Early in vitro experiments had shown that infection of domestic rabbit skin fragments (Coman 1946) or epidermal cell monolayers (Kreider et al. 1967) with CRPV was effective although no evidence for virus replication was obtained. When implanted back into the liver of the same rabbit, the cells gave rise to masses of proliferating squamous epithelium with a papillomatous pattern (Coman 1946), or, when transplanted to the cheek pouch of immunosuppressed hamsters, typical papillomas developed (Kreider et al. 1967). Infection at the time of plating of epidermal keratinocytes from adult domestic rabbit or cottontail rabbit growing on collagen gels showed no hyperproliferative response as compared to uninfected keratinocytes (Fig. 1; Orth et al. 1974; Taichman et al. 1984). Vegetative viral DNA synthesis could be detected by in situ molecular hybridization methods in a small number of differentiating cells of the superficial layers of cottontail rabbit keratinocyte cultures, but capsid antigens were not detected (Taichman et al. 1984). In human foreskin keratinocytes infected with HPV1 particles,

HPV1 DNA was detected as monomeric episomes at about 100 copies per cell for up to eight passages with no evidence of vegetative viral growth or cell transformation (La Porta and Taichman 1982). Viral DNA copy number was not increased in the more differentiated cells or by vitamin A-deprived media or by subcutaneous inoculation to nude mice, indicating that viral DNA replication was taking place in proliferating cells (Taichman et al. 1984). All these studies demonstrate clearly that PVs infect keratinocytes in vitro, and that the lack of virus replication results from a defect in the expression of some cellular or viral factors in vitro. The maintenance of PV genomes as stable episomes in keratinocytes provides a good model for the study of latent infection. Although the finite lifetime of the cultures is a major drawback, it proved to be useful for assaying the transforming properties of HPV16. Transfection with HPV16 DNA (Dürst et al., manuscript submitted for publication) or with a recombinant DNA containing a head-to-tail dimer of HPV16 DNA and the *neo* bacterial gene (Pirisi et al. 1987) resulted in the establishment of immortalized keratinocyte clones harboring HPV16 DNA with some evidence for the presence of transcripts containing HPV16 sequences. This could constitute a test for the genetic analysis of the HPV16 transforming region.

To obviate the finite lifetime of the cultures, immortalized cell lines originating from cutaneous (SCL1, SCC12B2) or oral (SCC9, SCC25) human squamous cell carcinomas were infected with HPV1. HPV1 DNA persists as a multicopy monomeric episome for up to 4-5 subcultures but is no longer detected thereafter in carcinoma cells (Reilly et al. 1985). Similarly, cottontail rabbit epidermal keratinocytes established in vitro after transfection with CRPV DNA maintain CRPV DNA at a low copy number only for a number of passages (Taichman et al. 1984). A possible explanation could be that infected cells are overgrown by uninfected cells. Cotransfection of PV DNA with dominant selectable markers, like recombinant plasmids containing bacterial genes for resistance to selective media, allows rapid selection of transfected cells. After transfection of SK-v cells with HPV1 DNA, only 1 of 32 clones resistant to selection was found to contain HPV1 DNA sequences, at low copy number, integrated into the cell genome, and with no evidence of transcription (Taichman et al. 1984).

Another way of establishing a keratinocyte line containing PV genomes has been to cotransfect human fetal trunk keratinocytes with cloned dimeric HPV1 DNA and cloned SV40 (origin minus) DNA and to select transfectants by their transformed phenotype. One of such clones contained 2-4 copies of integrated HPV1 DNA with trace amounts of episomal DNA and amounts of HPV1-specific transcripts too low to be characterized. Vegetative viral DNA synthesis was not detected even when the cells inoculated into nude mice formed keratinizing cysts (Burnett and Gallimore 1983). The results so far obtained by infection or transfection of keratinocytes indicate that PV genomes may be maintained as stable episomes, that conditions supporting vegetative viral replication have not been obtained in vitro, and that underreplication of viral DNA during active keratinocyte growth in vitro could account for the loss of the viral genome after repeated subcultures or clone expansion. The reasons for integration of HPV1 DNA are unknown but may relate to the transformed state of the cells.

2.2.2 PV-Induced Cell Transformation In Vitro

2.2.2.1 The BPV1 Model. The transformation of murine NIH 3T3 and C127 cells by BPV1 has served as a model for the study of the molecular biology and genetics of PVs (Howley et al. 1986). Studies from a number of laboratories, which will be fully detailed by Turek and Haugen in this volume, have demonstrated the following points. (1) The fully transformed phenotype requires the integrity of ORF E2, ORF E5, and the region containing ORFs E6 and E7. Analysis of cDNA have confirmed mutational analysis, and E6 and E5 proteins have been identified in BPV1-transformed cells using antisera against bacterially expressed putative genes. (2) The establishment and maintenance of plasmid replication, as well as the synchronization of plasmid replication with cell DNA duplication, requires an intact ORF E1. (3) An enhancer element upstream of the early region is transactivated by the product(s) of ORF E2, and this could be involved also in the expression of ORFs E6, E6/E7, and E1. (4) Transactivating factors controlling plasmid copy number are encoded in the ORF E6-E7 segment.

2.2.2.2 Stepwise Tumoral Progression in BPV1-Transformed Cells. Infection with BPV1 or transfection with viral DNA induces oncogenic transformation of rodent fibroblasts as immortalized pseudo-normal lines but not as primary cultures, suggesting that BPV1 has no immortalization function (Cuzin et al. 1985). Four distinct stages have been tentatively defined in the oncogenic transformation process induced by BPV1 in rat FR3T3 and mouse C127 cells on the basis of transformed cell phenotypes, i.e., morphology, growth properties, and tumorigenicity (Cuzin et al. 1985). Cell transfection with a plasmid construct containing the BPV1 transforming fragment and the bacterial *neo* gene for resistance to neomycin (G418) yielded resistant colonies which expressed a normal cell morphology and growth control and contained free nonrearranged copies of viral DNA (stage 1). These cells may produce transformed clones at a low frequency which increases by treatment with TPA. Stage 2 cell lines are isolated by focus formation or derive spontaneously from stage 1 cell lines. They show an "intermediate" morphological phenotype but produce highly invasive tumors in the animal. Stage 3 cell lines are stage 2 tumor cells reestablished in vitro which show increased growth rates and tumorigenicity but still require serum factors for growth. Stage 4 cell lines derive from stage 1 cell lines by tumor formation in the animal and show a highly transformed morphology and a further increase in tumorigenicity. At stage 4, viral genomes are highly rearranged or, in one case, entirely lost. Thus, this artificial experimental model reproduces the stepwise progression of viral papillomas towards malignancy observed in vivo and may provide a means to analyze the sequence of events in phenotypic transformation (Cuzin et al. 1985).

2.2.2.3 Cell Transformation by Other PVs. Transformation of rodent fibroblasts in vitro may not be restricted to those PVs inducing fibropapillomas or fibromas and thus with a capacity to transform fibroblasts in vivo (Howley et al. 1986). PVs associated with tumors progressing towards malignancy have been tested for their capacity to express their oncogenic potential by transforming immortalized fibroblast cell lines.

Cloned DNA from BPV4 associated with alimentary tract papillomas in cattle (Jarrett 1985) has been found to induce morphological transformation of NIH 3T3 mouse fibroblasts (Campo and Spandidos 1983). Transformed cells lose contact inhibition, are anchorage independent, require low serum, and are tumorigenic in nude mice. Recombinant plasmids are retained as multiple copies of nonintegrated monomers (Campo and Spandidos 1983). Similarly, morphological transformation of NIH 3T3 and C127 mouse cells was observed after infection with purified CRPV or transfection with CRPV DNA. Clones derived from transformed foci grow in soft agar, produce tumors in nude mice, and contain free monomeric forms of CRPV DNA persisting through subcultures (Watts et al. 1983). Foci of transformed cells were obtained by transfecting C127 cells with HPV5 and HPV1 recombinant plasmid DNA with the same efficiency as with BPV1 recombinant plasmid DNA (Watts et al. 1984). Multiple copies of non-rearranged, episomal, virus-specific DNA were found, and they persisted in subclones at copy numbers varying from 2 to 30 copies per cell between the different cell lines. No evidence was found of integration (Watts et al. 1984). Naturally occurring mutants of HPV5, with deletions in the late region of the genome, retained the ability to transform C127 cells and to replicate as persistent episomes. Subgenomic fragments corresponding to the early region also transformed cells but integrated into the host cell genome (Watts et al. 1985).

When compared with HPV5- and CRPV-transformed clones, HPV1-containing clones showed poor growth in soft agar and induced tumors in nude mice after a long delay. This appeared to be due to the interruption of the putative region involved in cellular transformation by the cloning site, since transfected cells containing viral DNA with another cloning site grew well in soft agar (Watts et al. 1984). Introduction of HPV1 DNA into a rat fibroblast line using a cloned herpes virus thymidine kinase gene as a selectable marker did not result in cell transformation, in spite of the presence of multiple copies of the entire HPV1 genome integrated into the cell genome (Burnett and Gallimore 1985). It has been proposed that in these cells de novo methylation of viral DNA results in low levels of expression of the viral genomes which integrate and thus have no effect on the growth properties of this cell type (Burnett and Gallimore 1985). Recent experiments using different recombinant HPV1 DNAs cloned at different sites indicate that transformation of C127 or primary rat kidney cells occurs rarely, if at all, or is irreproducible, at least under the conditions employed (Chow et al. 1986). When HPV1, HPV6b, or HPV11, inserted at the noncoding region or the late region into constructions containing a selectable marker like herpes virus thymidine kinase gene or the bacterial gene for resistance to neomycin, were transfected into rodent, monkey, or human cells, none of the selected colonies showed a transformed phenotype, and all contained integrated HPV DNA at low copy number (Chow et al. 1986). It is tempting to relate the lack of transforming capacity in vitro to the low oncogenic potential of these viruses in vivo.

HPV16 DNA cloned in a vector containing the bacterial *neo* gene was found to be maintained in transfected monkey kidney cells as an autonomously replicating episome at a copy number of 2–10 per cell. On the other hand, integration of the recombinant plasmid DNA within the host genome was observed in transfected mouse fibroblasts. Neither type of transfected cell showed phenotypical transformation (Chesters and McCance 1985). Contrasting with these results, when the recombinant HPV16 DNA contained a head-to-tail dimer of the full length HPV16 genome and the bacterial *neo* gene for selection, morphological transformation was induced, and the transformed murine cells were tumorigenic in nude mice (Yasumoto et al. 1986) while the in vitro lifespan of human fibroblasts was extended (Pirisi et al. 1987). This study showed that transformation with HPV16 DNA has unique features. A long latency period is required for induction of morphological transformation (over 4 weeks). Viral sequences are found as multimeric tandem repeats with some rearrangements, but it is not yet known whether they are free or integrated into the cell genome. Virus-specific RNAs are expressed in transformed cells but have not been characterized. Recently, a genomic DNA sample from a cervical cancer biopsy which contained HPV16 DNA sequences was found to induce malignant transformation of NIH 3T3 cells by transfection. Primary and secondary transformed cells contained integrated HPV16 DNA sequences and human-specific *Alu* sequences. Transcription of HPV16 sequences has been detected in primary transformed cells (Tsunokawa et al. 1986a).

2.2.2.4 Transcription of PVs in Cultured Cells Using Surrogate Promoters. Introduction of different HPV types (HPV1, 2, 3, 4, 9), together with herpes virus thymidine kinase gene, into thymidine kinase-negative mouse cells and selection of these cells for thymidine kinase-positive phenotype provided a set of cell lines containing different HPV genomes for comparison of their transcription pattern. Most of the HPV DNA sequences were not present as unit-length episomal viral DNA, and integration and rearrangements of viral DNA sequences were observed. Viral transcripts contained both HPV and plasmid sequences, and it was not known whether transcription was initiated by HPV promoters (Brackmann et al. 1983). Using another approach, electron microscopic analysis of heteroduplex RNA-DNA molecules has shown that transcriptional promoters of cloned HPVs transiently expressed in monkey COS cells (Gluzman 1981) are too weak, or nonfunctional, to promote transcription and that the HPV transcripts observed arise from the activity of the SV40 promoters present in the expression vectors used (Chow and Broker 1984). Owing to this inability of HPVs to generate RNA from its own promoter in rodent, monkey, or human cells, strong foreign promoters (*Drosophila* heat-shock protein, SV40 early and late promoters) were inserted into HPV1 or HPV6 DNA cloned in several eukaryotic expression vectors. Analysis of the messenger RNA species successfully obtained was done by electron microscopic studies of RNA-DNA heteroduplex molecules. Identical splice patterns were observed, irrespective of the promoters used. The ratio of the various mRNAs changed only with the strength of the promoters. In HPV1-transfected clones, mRNAs spanning the early region alone and both the early and late region were observed, which corresponded to species characterized in HPV1-induced warts and HPV1-infected keratinocytes in vitro. HPV6 mRNAs were all polyadenylated at the end of the early region by using an alternative signal for polyadenylation since a conventional one is lacking in the HPV6 genome (Chow et al. 1986). Recently, it was shown that transformation of NIH 3T3 and 3T3 A 31 cells by HPV16 DNA could be obtained by using the murine leukemia virus long terminal repeat transcriptional unit to direct the expression of the HVP16 genome, under conditions in which HPV16 DNA failed to produce morphological transformation (Matlashewski et al. 1987a). This indicates that high levels of HPV expression may be essential to induce and maintain the transformed phenotype (Matlashewski et al. 1987a). Furthermore, it was demonstrated that transformation of primary rat cells by this construct required cooperation with an activated *ras* gene and that this activity involved the E6/E7 region of the HPV16 genome (Matlashewski et al. 1987b). Such a system should prove useful in identifying the viral and cellular genes and their products involved in the transformation process by HPVs.

3 Experimental Systems for the Study of PV-Tissue Interactions

Advances in the knowledge of the mechanisms involved in the specific interaction of PVs and their susceptible tissues would require experimental systems reproducing the PV-induced changes in normal tissue under natural conditions. This is a particularly difficult issue for human PVs. However, a new approach consisting of grafting in vitro PV-infected cells to a privileged site in nude mice has been recently successfully tested and may open new possibilities for such studies.

First attempts to graft HPV-infected human tissue to immunologically deficient animals failed (Cubie 1976; Pass et al. 1973), in contrast with the successful neoplastic transformation of normal rabbit skin by CRPV grafted to nude mice (Kreider et al. 1979). By using a transplantation site underneath the renal capsule of athymic mice, human uterine cervical tissue infected with extracts of a vulvar condyloma showed morphological alterations characteristic of PV infection, i.e., hyperplastic papillomatous epithelium with koilocytic changes, together with the expression of PV capsid antigens in the more differentiated layers of the hyperplastic epithelium (Kreider et al. 1985). Upon testing the susceptibility to HPV11 infection of other skin specimens (child and adult foreskins, vulvar, abdominal, and lower leg skin), it was found that foreskins were the most sensitive, as characterized by the extensive hyperplasia of the infected tissue. However, in spite of a partial or complete lack of morphological alterations, HPV genomes and capsid antigens were readily detected in infected skin from other anatomical sites (Kreider et al. 1986). Thus, for the first time, experimentally-produced infectious HPV particles and capsid-associated antigens may be recovered for further studies (Kreider et al. 1987). This approach is most promising for the study of HPV infection of human tissue, the pathogenesis of putative HPV-induced lesions of humans, and the contribution of cofactors, such as other viruses, UV light, and chemical carcinogens, to the malignant conversion of benign HPV-induced lesions.

In conclusion, productive infection of keratinocytes by papillomaviruses has not yet been obtained, despite numerous attempts involving a variety of experimental approaches and PV types with different biological properties. However, the difficulties encountered are revealing in themselves, and, as shown here, the accumulation of dispersed information has provided new insights into the mechanisms developed by PVs to infect and transform keratinocytes. In view of the plurality of PV types, their specific biological properties, particularly in humans, and the multistep progression of tumors induced by PVs with a higher oncogenic potential, it may be that strictly in vitro studies are out of reach. However, as recently demonstrated, the association of in vitro infection and graft to nude mice may be fruitful (Kreider et al. 1986). In vivo experimental systems, such as CRPV-associated benign and malignant tumors (Kreider et al. 1985) or germ-line transmission of PV genomes in transgenic mice, as already tried with BPV1 (Lacey et al. 1986), will prove valuable in the study of the tissue specificity of PV gene expression, the activities of PV oncogenes, and the genetic requirements of PV genomes in eliciting neoplasia.

Acknowledgements. The author wishes to thank particularly Drs. G. Orth and O. Croissant for stimulating and fruitful discussions and suggestions, for continued encouragement and support, and for making available experimental data. The efficient expertise of N. Jibard in tissue culture experiments and the skillful assistance of J. Lortholary in the preparation of the manuscript are acknowledged.

References

Baker CC, Phelps WC, Lindgren V, Braun MJ, Gonda MA, Howley P (1987) Structural and transcriptional analysis of human papillomavirus type 16 sequences in cervical carcinoma cell lines. J Virol 61: 962-971

Banks L, Spence P, Androphy E, Hubbert N, Matlashewski G, Murray A, Crawford L (1987) Identification of human papillomavirus type 18 E6 polypeptide in cells derived from human cervical carcinomas. J Gen Virol 68: 1351-1359

Barrandon Y, Green H (1985) Cell size as a determinant of the clone-forming ability of human keratinocytes. Proc Natl Acad Sci USA 82: 5390-5394

Bell E, Sher S, Hull B, Merrill C, Rosen S, Chamson A, Asselineau D, Dubertret L, Coulomb B, Lapière C, Nusgens B, Neveux Y (1983) The reconstitution of living skin. J Invest Dermatol 81: 2s-10s

Boshart M, Gissmann L, Ikenberg H, Kleinheinz A, Scheurlen W, zur Hausen H (1984) A new type of papillomavirus DNA, its presence in genital cancer biopsies and in cell lines derived from cervical cancer. EMBO J 3: 1151-1157

Boyce ST, Ham R (1983) Calcium-regulated differentiation of normal human epidermal keratinocytes in chemically defined clonal culture and serum-free serial culture. J Invest Dermatol 81: 33s-40s

Brackmann KH, Green M, Wold WSH, Rankin A, Loewenstein PM, Cartas MA, Sanders PR, Olson K, Orth G, Jablonska S, Kremsdorf D, Favre M (1983) Introduction of cloned human papillomavirus genomes into mouse cells and expression at the RNA level. Virology 129: 12-24

Breitburd F, Favre M, Zoorob R, Fortin D, Orth G (1981) Detection and characterization of viral genomes and search for tumoral antigens in two hamster cell lines derived from tumors induced by bovine papillomavirus type I. Int J Cancer 127: 693-702

Breitburd F, Croissant O, Orth G (1987) Expression of human papillomavirus type 1 E4 gene products in warts. Cancer Cells 5: 115-122

Broker TR, Botchan M (1986) Papillomaviruses: retrospectives and prospectives. Cancer Cells 4: 17-36

Brown K, Quintanilla M, Ramsden M, Kerr IB, Young S, Balmain A (1986) V-ras genes from Harvey and BALB murine sarcoma viruses can act as initiators of two-stage mouse skin carcinogenesis. Cell 46: 447-456

Burnett TS, Gallimore PH (1983) Establishment of a human keratinocyte cell line carrying complete human papillomavirus type 1 genomes: lack of vegetative viral DNA synthesis upon keratinization. J Gen Virol 64: 1509-1520

Burnett TS, Gallimore PH (1985) Introduction of cloned human papillomavirus 1a DNA into rat fibroblasts: integration, de novo methylation and absence of cellular morphological transformation. J Gen Virol 66: 1063-1072

Butel JS (1972) Studies with human papillomavirus modeled after known papovavirus systems. JNCI 48: 285-299

Campo MS, Spandidos DA (1983) Molecularly cloned bovine papillomavirus DNA transforms mouse fibroblasts in vitro. J Gen Virol 64: 549-557
Campo MS, Smith KT, Jarrett WFH, Moar MH (1985) Presence and expression of bovine papillomavirus type 4 in tumours of the alimentary canal of cattle and its possible role in transformation and malignant progression. UCLA Symp Molec Cell Biol [new ser] 32: 305-326
Chesters PM, McCance DJ (1985) Human papillomavirus type 16 recombinant DNA is maintained as an autonomously replicating episome in monkey kidney cells. J Gen Virol 66: 615-620
Chow LT, Broker TR (1984) Human papillomavirus type 1 RNA transcription and processing in COS-1 cells. Prog Cancer Res Ther 30: 125-133
Chow LT, Pelletier AJ, Galli R, Brinckmann U, Chin M, Arvan D, Campanelli D, Cheng S, Broker TR (1986) Transcription of human papillomavirus types 1 and 6. Cancer Cells 4: 603-614
Ciuffo G (1907) Innesto positiveo con filtrado di verrucae volgare. G Ital Mal Venerol 48: 12-17
Coman DR (1946) Induction of neoplasia in vitro with a virus. Experiments with rabbit skin grown in tissue culture and treated with Shope papillomavirus. Cancer Res 6: 602-607
Croissant O, Breitburd F, Orth G (1985) Specificity of the cytopathic effect of cutaneous human papillomaviruses. Clin Dermatol 3(4): 43-55
Cubie HA (1976) Failure to produce warts on human skin grafts on nude mice. Br J Dermatol 94: 659-665
Cuzin F, Meneguzzi G, Binétruy B, Cerni C, Connan G, Grisoni M, de Lapeyrière O (1985) Stepwise tumoral progression in rodent fibroblasts transformed with bovine papilloma virus type 1 (BPV1) DNA. UCLA Symp Molec Cell Biol [new ser] 32: 473-486
Doran TI, Vidrich A, Sun TT (1980) Intrinsic and extrinsic regulation of the differentiation of skin, corneal, and esophageal epithelial cells. Cell 22: 17-25
Dürst M, Schwarz E, Gissmann L (1986) Integration and persistence of human papillomavirus DNA in genital tumors. Banbury Report 21: 272-280
Fuchs E, Green H (1981) Regulation of terminal differentiation of cultured human keratinocytes by vitamin A. Cell 25: 617-625
Georges E, Croissant O, Bonneaud N, Orth G (1984) Physical state and transcription of the cottontail rabbit papillomavirus genome in warts and transplantable VX2 and VX7 carcinomas of domestic rabbits. J Virol 51: 530-538
Georges E, Breitburd F, Jibard N, Orth G (1985) Two cottontail rabbit papillomavirus - associated VX2 carcinoma cell lines with different levels of keratinocyte differentiation and transplantability. J Virol 55: 246-250
Giri I, Danos O (1986) Papillomavirus genomes: from sequence data to biological properties. Trends in Genetics 2: 227-232
Giri I, Danos O, Thierry F, Georges E, Orth G, Yaniv M (1985) Characterization of transcription control regions of the cottontail rabbit papillomavirus. UCLA Symp Molec Cell Biol [new ser] 32: 379-390
Gissmann L, Schwarz E (1986) Persistence and expression of human papillomavirus DNA in genital cancer. Ciba Found Symp 120: 190-207
Gluzman Y (1981) SV40-transformed simian cells support the replication of early SV40 mutants. Cell 23: 175-181
Green H, Fuchs E, Watt F (1981) Differentiated structural components of the keratinocyte. Cold Spring Harbor Symp Quant Biol 46: 293-301
Hawley-Nelson P, Sullivan JE, Kung M, Hennings H, Yuspa SH (1980) Optimized conditions for the growth of human epidermal cells in culture. J Invest Dermatol 75: 176-182
Howley PM, Yang YC, Spalholz BA, Rabson MS (1986) Molecular aspects of papillomavirus-host cell interactions. Banbury Report 21: 261-272
Jarrett WFH (1985) Bovine papillomaviruses. Clin Dermatol 3(4): 8-19
Jaureguiberry G, Favre M, Orth G (1983) Bovine papillomavirus type 1 genome in hamster sarcoma cells in vivo and in vitro: variation in the level of transcription. J Gen Virol 64: 1199-1204
Karasek MA (1983) Culture of human keratinocytes in liquid medium. J Invest Dermatol 81: 24s-28s
Kleiner E, Dietrich W, Pfister H (1986) Differential regulation of papillomavirus early gene expression in transformed fibroblasts and carcinoma cell lines. EMBO J 5: 1945-1950
Kreider JW, Bartlett GL (1985) Shope rabbit papilloma-carcinoma complex: a model system of HPV infections. Clin Dermatol 3(4): 20-26

Kreider JW, Breedis C, Curran JS (1967) Interaction of Shope papillomavirus and rabbit skin cells in vitro: 1. Immunofluorescent localization of virus inocula. JNCI 38: 921-931
Kreider JW, Bartlett GL, Sharkey FE (1979) Primary neoplastic transformation in vivo of xenogeneic skin grafts on nude mice. Cancer Res 39: 273-276
Kreider JW, Howett MK, Wolfe SA, Bartlett GL, Zaino RJ, Sedlacek TV, Mortel R (1985) Morphological transformation in vivo of human uterine cervix with papillomavirus from Condylomata acuminata. Nature 317: 639-641
Kreider JW, Howett MK, Lill NL, Bartlett GL, Zaino RJ, Sedlacek TV, Mortel R (1986) In vivo transformation of human skin with human papillomavirus type 11 from condylomata acuminata. J Virol 59: 369-376
Kreider JW, Howett MK, Leure-Dupree AE, Zaino RJ, Weber JA (1987) Laboratory production in vivo of infectious human papillomavirus type 11. J Virol 61: 590-593
Lacey M, Alpert S, Hanahan D (1986) Bovine papillomavirus genome elicits skin tumours in transgenic mice. Nature 322: 609-612
La Porta RF, Taichman LB (1982) Human papillomavirus DNA replicates as a stable episome in cultured epidermal keratinocytes. Proc Natl Acad Sci USA 79: 3393-3397
Lavker RM, Sun TT (1983) Epidermal stem cells. J Invest Dermatol 81: 121s-127s
Leventon-Kriss S, Ben-Shoshan J, Barzilay Z, Shahar A, Leventon G (1983) Human epithelial cell line established from a child with juvenile laryngeal papillomatosis. Isr J Med Sci 19: 508-514
Matlashewski G, Osborn K, Murray A, Banks L, Crawford L (1987a) Transformation of mouse fibroblasts with HPV type 16 DNA using a heterologous promotor. Cancer Cells 5: (to be published)
Matlashewski G, Schneider J, Banks L, Jones N, Murray A, Crawford L (1987b) Human papillomavirus type 16 DNA cooperates with activated *ras* in transforming primary cells. EMBO J (to be published)
McVay P, Fretz M, Wettstein F, Stevens J, Ito Y (1982) Integrated shope virus DNA is present and transcribed in the transplantable rabbit tumour VX7. J Gen Virol 60: 271-278
Nasseri M, Wettstein FO (1984a) Differences exist between viral transcripts in cottontail rabbit papillomavirus-induced benign and malignant tumors as well as non-virus-producing and virus-producing tumors. J Virol 51: 706-712
Nasseri M, Wettstein FO (1984b) Cottontail rabbit papillomavirus-specific transcripts in transplantable tumors with integrated DNA. Virology 138: 362-367
Nasseri M, Wettsein FO (1985) The papilloma to carcinoma progression of cottontail rabbit (Shope) papilloma virus induced tumors: characterization of papilloma derived cell lines. UCLA Symp Molec Cell Biol [new ser] 32: 487-500
Niimura M, Pass F, Wooley R, Soutor CA (1975) Primary tissue culture of human wart-derived epidermal cells (keratinocytes). JNCI 54: 563-569
Orth G (1986) Epidermodysplasia verruciformis: a model for understanding the oncogenicity of human papillomaviruses. Ciba Found Symp 120: 157-174
Orth G (1987) Epidermodysplasia verruciformis. In: Salzman NP, Howley PM (eds) The papovaviridae, vol 2. The papillomaviruses. Plenum, New York, pp 199-243
Orth G, Croissant O (1968) Caractères des cultures de première explantation de cellules de papillomes provoqués par le virus de Shope chez le lapin domestique. CR Séances Acad Sci (III) 266: 1084-1087
Orth G, Favre M (1985) Human papillomaviruses: biochemical and biologic properties. Clin Dermatol 3(4): 27-42
Orth G, Breitburd F, Favre M, Paoletti C, Croissant O (1974) Étude des facteurs gouvernant le maintien et l'expression de l'état différencié des kératinocytes dans les cultures de cellules épidermiques de mammifères adultes. INSERM 36: 297-317
Orth G, Breitburd F, Favre M, Croissant O (1977) Papillomaviruses: possible role in human cancer. Cold Spring Harbor Conf Cell Proliferation 4: 1043-1068
Pass F, Niimura M, Kreider JW (1973) Prolonged survival of human skin xenografts on antithymocyte serum-treated mice: failure to produce verrucae by inoculation with extracts of human warts. J Invest Dermatol 61: 371-374
Pater MM, Pater A (1985) Human papillomavirus types 16 and 18 sequences in carcinoma cell lines of the cervix. Virology 145: 313-318
Pirisi L, Yasumoto S, Feller M, Doniger J, DiPaolo JA (1987) Transformation of human fibroblasts and keratinocytes with human papillomavirus type 16 DNA. J Virol 61: 1061-1066

Reilly SS, Albers KM, Taichman LB (1985) Replication of HPVI DNA in malignant keratinocytes in culture. UCLA Symp Molec Cell Biol [new ser] 32: 427-433
Rheinwald JG, Beckett MA (1981) Tumorigenic keratinocyte lines requiring anchorage and fibroblast support cultured from human squamous cell carcinomas. Cancer Res 41: 1657-1663
Rheinwald JG, Germain E, Beckett MA (1983) Expression of keratins and envelope proteins in normal and malignant human keratinocytes and mesothelial cells. In: Harris CC and Antrup HN (eds) Human carcinogenesis. Academic, New York, p 85
Rous P, Beard JW (1935) The progression to carcinoma of virus-induced rabbit papilloma (Shope). J Exp Med 62: 523-547
Schneider-Gädicke A, Schwarz E (1986) Different human cervical carcinoma cell lines show similar transcription patterns of human papillomavirus type 18 early genes. EMBO J 5: 2285-2292
Schneider-Maunoury S, Croissant O, Orth G (1987) Integration of HPV16 DNA sequences: a possible early event in the progression of genital tumors J Virol 61: (n° 10) in press
Schwarz E, Freese UK, Gissmann L, Mayer W, Roggenbuck B, Stremlau A, zur Hausen H (1985) Structure and transcription of human papillomavirus sequences in cervical carcinoma cells. Nature 314: 111-114
Schwarz E, Schneider-Gädicke A, Roggenbruck B, Mayer W, Gissmann L, zur Hausen H (1986) Expression of human papillomavirus DNA in cervical carcinoma cell lines. Banbury Report 21: 281-290
Seedorf K, Oltersdorf T, Krämmer G, Röwekamp W (1987) Identification of early proteins of the human papillomaviruses type 16 (HPV16) and type 18 (HPV18) in cervical carcinoma cells. EMBO J 6: 139-144
Smotkin D, Wettstein FO (1986) Transcription of human papillomavirus type 16 early genes in a cervical cancer and a cancer-derived cell line and identification of the E7 protein. Proc Natl Acad Sci USA 83: 4680-4684
Steinberg BM, Abramson AL, Meade RP (1982) Culture of human laryngeal papilloma cells in vitro. Otolaryngol Head Neck Surg 90: 728-735
Steinberg BM, Topp WC, Schneider PS, Abramson AL (1983) Laryngeal papillomavirus infection during clinical remission. N Engl J Med 308: 1261-1264
Sun TT, Eichner R, Schermer A, Cooper D, Nelson WG, Weiss RA (1984) Classification, expression and possible mechanisms of evolution of mammalian epithelial keratins: a unifying model. Cancer Cells 1: 169-176
Taichman LB, Breitburd F, Croissant O, Orth G (1984) The search for a culture system for papillomavirus. J Invest Dermatol 82: 2s-6s
Thomas M, Boiron M, Tanzer J, Levy JP, Bernard J (1964) In vitro transformation of mice cells by bovine papillomavirus. Nature 202: 709-712
Tsunokawa Y, Takebe N, Kasamatsu T, Terada M, Sugimura T (1986a) Transforming ectivity of human papillomavirus type 16 DNA sequences in a cervical cancer. Proc Natl Acad Sci USA 83: 2200-2203
Tsunokawa Y, Takebe N, Nozawa S, Kasamatsu T, Gissmann L, zur Hausen H, Terada M, Sugimura T (1986b) Presence of human papillomavirus type-16 and type-18 DNA sequences and their expression in cervical cancers and cell lines from Japanese patients. Int J Cancer 37: 499-503
Watts SL, Ostrow RS, Phelps WC, Prince JT, Faras AJ (1983) Free cottontail rabbit papillomavirus DNA persists in warts and carcinomas of infected rabbits and in cells in culture transformed with virus or viral DNA. Virology 125: 127-138
Watts SL, Phelps WC, Ostrow RS, Zachow KR, Faras AJ (1984) Cellular transformation by human papillomavirus DNA in vitro. Science 225: 634-636
Watts SL, Chow LT, Ostrow RS, Faras AJ, Broker TR (1985) Localization of HPV5 transforming functions. UCLA Symp Molec Cell Biol [new ser] 32: 501-511
Yasumoto S, Burkhardt A, Doniger J, DiPaolo J (1986) Human papillomavirus type 16 DNA-induced malignant transformation of NIH 3T3 cells. J Virol 57: 572-577
Yee C, Krishnan-Hewlett I, Baker CC, Schlegel R, Howley PM (1985) Presence and expression of human papillomavirus sequences in human cervical carcinoma cell lines. Am J Pathol 119: 361-366
Yuspa SH, Kilkenny AE, Stanley J, Lichti U (1985) Keratinocytes blocked in phorbol ester-responsive early stage of terminal differentiation by sarcoma viruses. Nature 314: 459-462

Physical State of Papillomavirus DNA in Tumors

M. Dürst

1 Introduction

Papillomaviruses are capable of inducing benign and occasionally malignant tumors in keratinizing epithelia of a wide variety of animal hosts. The basal or germinal cells are thought to be the target of infection. In skin warts these cells show an increased mitotic index (Rashad 1969). No one has demonstrated papillomavirus DNA in basal cells as yet, probably because the number of copies of viral DNA present is below the level of detection by in situ hybridisation. During transit from the basal layer to the surface, the keratinocyte undergoes a complex series of irreversible changes collectively termed keratinization. Only keratinocytes at a certain stage of differentiation provide the necessary cellular machinery which can support productive viral replication. This probably represents the main stumbling block in establishing an in vitro cell culture system for the propagation of these viruses. Viral DNA synthesis has been demonstrated in keratinizing cells (Orth et al. 1971) as early as the first suprabasal layer (Grußendorf and zur Hausen 1979), and viral capsid antigens have been detected in the spinous, granular, and cornified layers (Noyes and Mellors 1957; Orth et al. 1977).

The papillomaviruses thus differ markedly from the other members of the papovavirus family, namely SV40 and polyomavirus, in that they are capable of transforming the target cell of their *natural* host, allowing the persistence of viral DNA.

Subsequent differentiation of these cells removes the block from productive viral replication and ensures the production of progeny virus.

The aim of this chapter is to review the available data on the persistence of papillomavirus DNA in benign as well as malignant lesions.

2 Methods Generally Employed for Resolving the Physical State of Papillomavirus DNA in Biological Material

Several methods have proven useful in elucidating the physical state of viral DNA sequences present in papillomavirus lesions.

In the simplest cases, electrophoretic separation of uncleaved DNA extracted from a tissue biopsy in one-dimensional agarose gels allows a straight-forward interpretation with respect to the physical state of the viral DNA. Two bands are

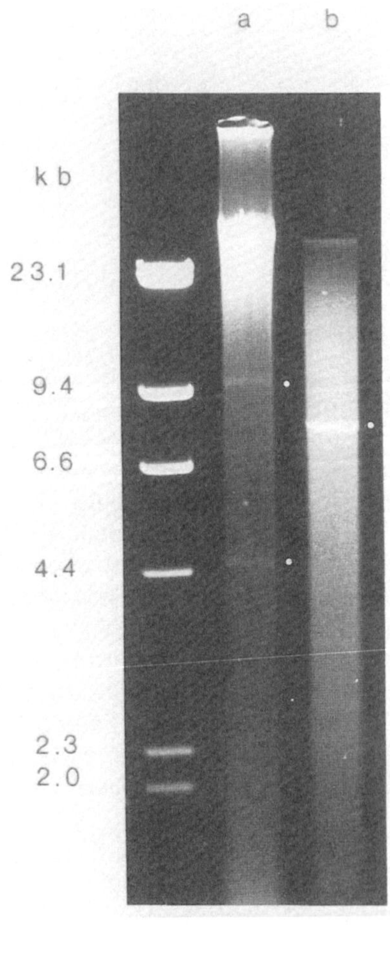

Fig. 1. Electrophoretic separation of DNA extracted from a wart in a 0.7% agarose gel. Lane *a* contains undigested DNA, lane *b* DNA digested with a single-cut restriction enzyme. After staining the gel with ethidium bromide, supercoiled and nicked circular HPV molecules (forms I and II, respectively) appear as discrete bands, indicated by *dots* in lane *a*. After digestion with a single-cut restriction enzyme the circular molecules are converted to the linear form III (lane *b*). Fragments of *Hind*III-digested lambda DNA are used as length markers (sizes are given in kilobases, kb)

usually detectable after ethidium bromide staining of the gel. The genome of the papillomavirus is composed of double-stranded, circular DNA of about 8 kilobases (kb) in length. The circular DNA is under tension and twists to form a supercoil (form I). A nick in one strand of the supercoiled DNA, which occasionally occurs because of the DNA extraction procedure, releases the tension within the molecule, giving rise to the relaxed or nicked form (form II). Form I has a greater electrophoretic mobility in agarose gels compared to form II. The linearised form of the molecule (form III), generated after shearing the DNA or digestion with a restriction endonuclease which cleaves only once within the viral genome, runs in between the two bands (Fig. 1).

If only small amounts of viral DNA are expected, the DNA from the gel is transferred onto a nitrocellulose filter (Southern 1975) and hybridised with a radioactive-labelled, homologous papillomavirus DNA probe followed by autoradiography. The sensitivity is then greatly improved and may reach 0.01 virus

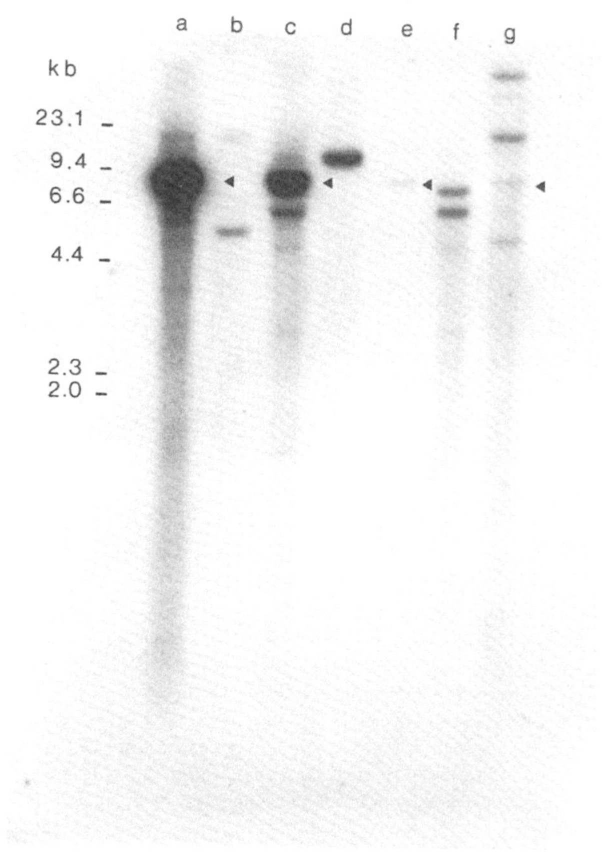

Fig. 2. Autoradiograph of a Southern blot analysis after hybridisation with ^{32}P-labelled HPV16 DNA. Ten μg of DNA from a number of cervical carcinomas *(a–g)* have been digested with a single-cut restriction enzyme (*Bam* HI). The linearised HPV16 genomes are indicated by *arrowheads*. The otherwise complex restriction fragment pattern may be due to (i) deleted HPV16 molecules, (ii) HPV16-related circular molecules of varying size which lack this particular restriction site, (iii) integrated molecules. Size markers are indicated in kilobases, (kb)

genome equivalents per diploid cell provided that 10 μg of cellular DNA is being used for electrophoretic separation.

However the fragment pattern may be more complex. In addition to the characteristic linearised form of human papillomavirus (HPV) molecules (after digestion with a single-cut enzyme), there may be additional bands (Fig. 2). These might result from circular molecules in which this particular restriction site is deleted. Furthermore, the circular DNA could be oligomerised, arranged either as catenates (analogous to the links of a chain) or in concatenated form (covalently linked genomes). Moreover, the possibility of HPV molecules, which are integrated within the host genome, must be considered.

The procedures outlined below have been found to be of use for determining the physical state of viral DNA in biological material. Essentially they can be divided into two groups: detection of free episonal DNA and detection of DNA integrated into the host cell genome.

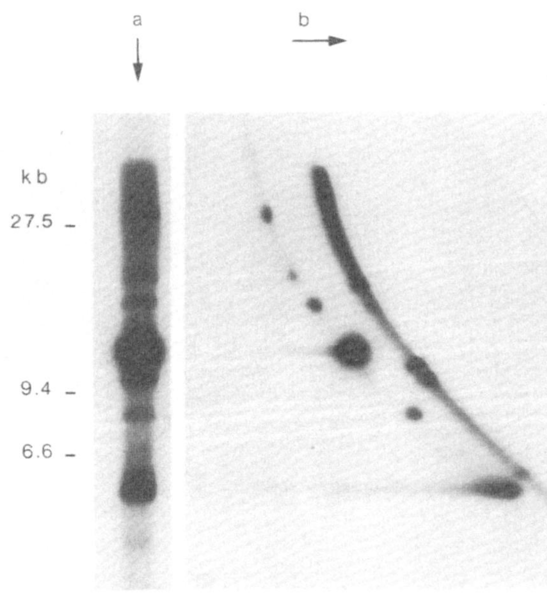

Fig. 3. Analysis of HPV16 DNA from a cervical carcinoma by two-dimensional gel electrophoresis. Twenty μg of tumor DNA was digested with *Hind* III (which has no cleavage site within the HPV16 genome) and was separated in parallel lanes (10 μg each) in a 0.4% agarose gel in the first dimension *(a)*. The DNA of one lane was transferred onto a gene screen plus filter while the DNA of the second lane was separated again in an 0.8% agarose gel in the second dimension (*b;* by rotating the track of the first dimension gel through 90° and casting a new gel). The DNA of the second dimension gel was also transferred onto a filter membrane and both filters were hybridised under stringent conditions (melting point −18 °C) with HPV16 DNA. Two tracks appear in the second dimension gel. Distinct spots representing circular DNA of varying size are seen on the left track, whereas linear DNA is seen exclusively on the right track

Detection of Free Episomal DNA

(a) With one-dimensional agarose gel electrophoresis of uncut and single cut DNA (Fig. 1), the two circular forms (forms I and II) of the papillomavirus genome are visible after electrophoresis of undigested DNA. Conversion of both circular forms to the linearised form (form III) is apparent after cleavage with a restriction endonuclease which has only one recognition site in the viral genome.

(b) Two-dimensional electrophoresis in agarose gels allows the identification of monomeric as well as variably sized oligomeric circular DNA (Fig. 3). This technique takes advantage of the fact that circular molecules possess a decreased electrophoretic mobility compared with that of linear DNA, when the gel concentration is increased (Johnson and Grossmann 1977).

(c) Caesium chloride/ethidium bromide gradient centrifugation permits the separation of supercoiled DNA from nicked circular and linear DNA (Radloff et

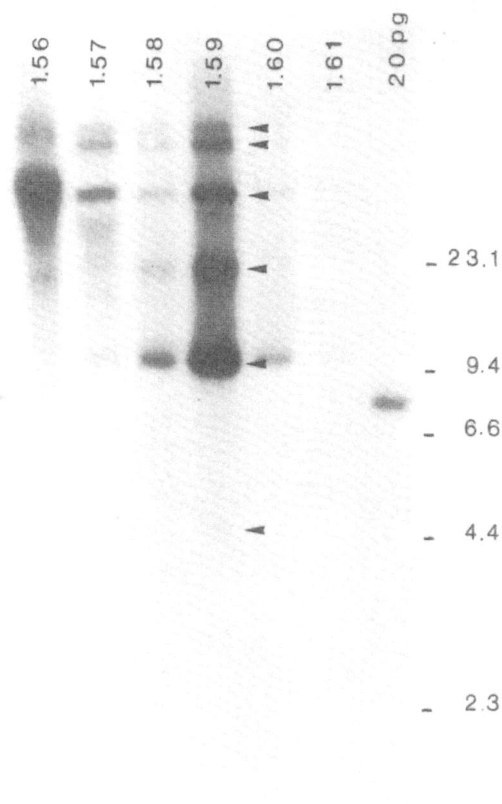

Fig. 4. Caesium chloride/ethidium bromide gradient of DNA from a cervical carcinoma. Individual fractions of the gradient with densities ranging from 1.56-1.61 g/ml were analysed. ^{32}P-labelled HPV16 DNA was used as the hybridisation probe under stringent conditions (melting point −18 °C). At a density of 1.59 g/ml, only supercoiled (form I) molecules accumulate in the gradient. The *bottom arrow* indicates supercoiled monomers of HPV-16 DNA, whereas the other *arrows* point to oligomers of increasing size. Nicked circular (form II) and linear (form III) molecules (including integrated DNA if present) accumulate at a density of 1.56 g/ml. Twenty pg of linearised cloned HPV16 DNA were used as a sensitivity marker for hybridisation. Length markers are given in kilobases (kb)

al. 1967). Fractions covering the density range from supercoiled circular DNA to nicked and linear DNA are then analysed by agarose gel electrophoresis and Southern blot analysis (Fig. 4).

Detection of Integrated Viral DNA

(a) Cleavage of tumor DNA with two or three different restriction enzymes (which have no cleavage site within the viral genome) and subsequent Southern blot analysis will only give rise to different restriction fragments if the viral DNA is covalently linked to the host genome. The electrophoretic mobility of episomal DNA will be unaffected by the digest (Fig. 5). A word of caution has to be added in this context. Cleavage with certain enzymes may lead to high molecular weight fragments (harbouring integrated viral DNA) which differ only marginally in size and thus may not be resolved satisfactorily in agarose gels.

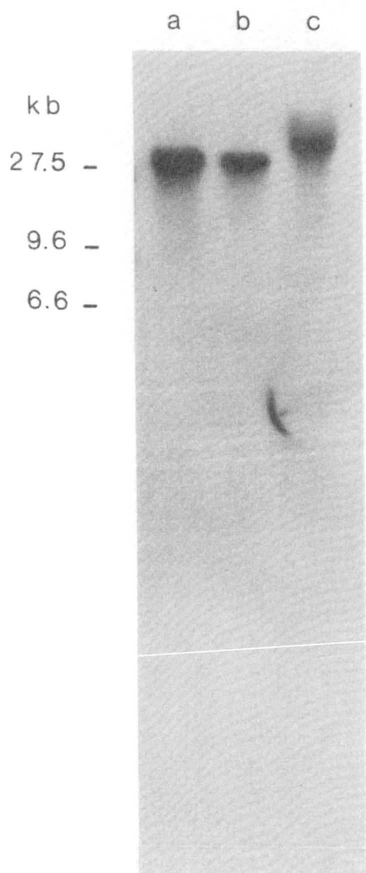

Fig. 5. Southern blot analysis of DNA from a cervical carcinoma with ^{32}P-labelled HPV16 DNA under stringent conditions (melting point − 18 °C). Lanes *a*, *b*, and *c* contain tumor DNA (10 μg each) digested with *Hind* III, *Xba* I, and *Xho* I, respectively (these enzymes have no recognition site on the HPV16 genome). The tumor DNA very likely harbours only integrated DNA as the singular bands observed for two digests have different electrophoretic mobilities

(b) Cleavage of biopsy DNA with a restriction enzyme which has only a single recognition site on the viral genome followed by electrophoresis in agarose gels and Southern blot analysis may give rise to the linearised form of papillomavirus DNA and/or some off-sized bands (Fig. 2). By using subgenomic fragments of viral DNA as a probe for hybridisation, the off-sized bands may be further characterised and assigned to the corresponding region of the papillomavirus genome. The possibility that such an off-sized band (especially those greater than genomic length) represents a virus/cell junction fragment must be considered if only part of the fragment hybridises with the viral genome.

(c) As described above, cleavage of DNA extracted from a tissue biopsy with single- or multicut enzymes may give rise to the expected regular fragment pattern with some additional off-sized bands. Double digestion of the same DNA with enzymes having no cleavage site within the viral genome may lead to the mobilisation of some of the additional bands, indicative of virus/cell junction fragments (Fig. 6).

The detection of 8 kb free viral DNA is essentially a straightforward matter. Similarly, the presence of only high molecular weight, linear, viral DNA in two-dimensional agarose gels is indicative of integrated DNA. However, in cases which disclose both large episomal as well as high molecular weight, linear viral DNA, interpretation is more difficult. Part of the linear DNA may have arisen due to the conversion of circular DNA (in particular, large oligomeric structures) to the linear form, yet some of the linear viral DNA may be covalently linked to the cellular DNA. The answer to the question of whether integration has occurred or not may be more discernible by using the above described approaches. On their own, however, they afford no absolute proof. This can only be achieved by molecularly cloning the virus/cell junction fragment and demonstrating that the cellular as well as the viral part of the cloned fragments give rise to an identical band when used as probes for the hybridisation of the biopsy-derived DNA (Dürst et al. 1985).

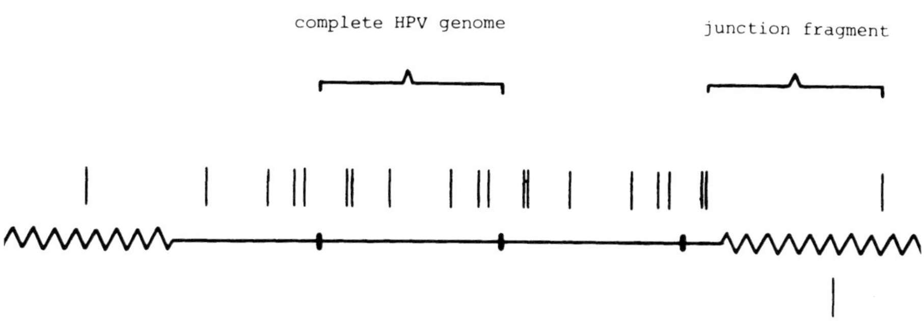

Fig. 6. Schematic drawing of HPV16 genomes, arranged as head-to-tail tandem repeats, integrated in the human genome. Cellular sequences are indicated by *zig-zag lines*. The *top vertical lines* represent the cleavage sites of an enzyme with multiple recognition sites on the HPV16 genome. The *bottom vertical line* represents a noncut enzyme for the viral DNA. The indicated junction fragment will be mobilised after digestion with both enzymes

3 Benign Lesions

The diversity of papillomavirus types is probably the result of the virus evolving in the various types of epithelia found in different body areas. This becomes evident when comparing the nucleotide sequence of a number of different papillomavirus types (Danos et al. 1984). On the basis of sequence homology, it was noted that HPV1 was more closely related to cottontail rabbit papillomavirus (CRPV), than to HPV6 and HPV11, and that bovine papillomavirus (BPV) type 1 showed the least homology with the other four viruses. HPV1 and CRPV both infect epidermal keratinocytes in their respective hosts, whereas HPV6 and HPV11 infect predominantly the genital epithelium, and BPV1 shows unique biological features such as induction of fibropapillomas.

In general the papillomas of mucosal surfaces and genital skin harbour less encapsidated viral DNA than lesions at other body sites (Grussendorf-Conen et al. 1983). Possibly the state of cellular differentiation found at these various sites and cellular factors supplied by the dermis may be responsible for this phenomenon. Despite the differences in productivity, papillomavirus DNA is found exclusively in the extrachromosomal, monomeric form in all benign lesions (Fig. 1) (Pfister 1984; Gissmann 1984). Studies by Steinberg et al. (1983) and Ferenczy et al. (1985) have shown that papillomaviruses may infect mucosal epithelium without inducing a morphologically apparent lesion. These latent or silent infections may explain the clinical phenomena of regrowth which is frequently observed after surgical removal of laryngeal papillomas or genital lesions. Indeed, the presence of papillomavirus DNA in normal epithelium at the laser excision margin (after removal of condyloma) shows a positive correlation with recurrence (Ferenczy et al. 1985). That the presence of papillomavirus DNA in tissue does not necessarily lead to clinical lesions has become particularly evident by demonstrating papillomavirus DNA in cervical scrapings of patients with no clinically apparent disease or cytological abnormality (Gissmann and Schneider 1986; E.-M. de Villiers, personal communication). At present nothing is known about the molecular basis of papillomavirus persistence or activation.

4 Malignant Lesions

The natural lesions induced by papillomavirus infection are usually benign. There is, however, ample evidence to suggest that certain papillomavirus types together with the synergistic action of physical or chemical carcinogens cause malignant growth (Orth et al. 1980; zur Hausen 1982; Gissmann 1984). Although most of the tumors still harbour papillomavirus DNA, they have become unable to produce progeny virus (Kidd and Rous 1940; Jarrett et al. 1980; Orth et al. 1980). Early reports suggested that the papillomaviruses deviate from other DNA tumor viruses, like SV40 and polyomavirus, in that they retain their extrachromosomal status in the transformed cell. This particular phenomenon was first observed in mouse C127 cells which were transformed in vitro using cloned BPV1 DNA or whole virus particles (for review see Howley 1983). However, careful analysis of the physical state of papillomavirus DNA in naturally occurring cancers has provided evidence for integrated papillomavirus genomes which are often overshadowed by a high copy number of circular molecules.

4.1 Cottontail Rabbit Carcinomas

The cottontail rabbit papillomavirus (CRPV) induces cutaneous papillomas in its natural host and under experimental conditions in domestic rabbits (Shope 1933). Conversion of papillomas into metastasising squamous cell carcinomas can be observed in about 25% of infected cottontail rabbits and in up to 75% of infected domestic rabbits (Rous and Beard 1935; Syverton 1952). Application of the carcinogen methylcholanthrene was found to have a synergistic effect, in that the papillomas became malignant earlier and at more sites (Rous and Friedewald 1944). CRPV-induced tumors therefore represent an attractive model for the analysis of the role of a virus in a multistage process leading to malignant transformation.

CRPV-induced papillomas and primary and metastatic carcinomas of domestic rabbits have been shown to harbour multiple copies (between 10–100 per diploid cell) of viral DNA (Stevens and Wettstein 1979). Further analysis has revealed that the CRPV genome in primary as well as in metastatic tumors persists extrachromosomally (Wettstein and Stevens 1982; Phelps et al. 1985). Interestingly, in carcinomas, large free episomes predominate and molecules consisting of multiple genomic equivalents of viral DNA have been resolved. In addition, Wettstein and Stevens (1982) have also provided evidence for integration of the CRPV genome within the host DNA in one of the tumors. CRPV DNA not only continues to be associated with carcinoma but is also detectable in the transplantable tumor VX2 and VX7 (Favre et al. 1982; McVay et al. 1982). These tumors were initially derived from CRPV-induced domestic rabbit carcinoma and have been serially transplanted up to the present day (Kidd and Rous 1940; Rogers et al. 1950, 1960; Smith et al. 1952). Multiple copies of the viral genome have been detected in the DNA extracted from VX2 and VX7 tumors as complexes of low electrophoretic mobility (Favre et al. 1982; McVay et al. 1982; Sugawara 1983; Georges et al. 1984, 1985) and were interpreted as head-to-tail repeats of the viral genome integrated at two or three sites within the cellular genome.

4.2 Human Genital Tumors

The human papillomaviruses types 6 and 11 are most commonly responsible for the induction of cervical, vulvar, perianal and penile condylomata acuminata (genital warts) (Gissmann et al. 1983; Kreider et al. 1985). They are benign tumors in the vast majority of cases. HPV infection is also associated with dysplastic lesions of the lower genital tract, particularly of the cervix (Meisels and Fortin 1976; Purola and Savia 1977). Dysplasias of the cervix are graded as intraepithelial neoplasia (CIN) grade I (mild) to III (severe); the latter also includes carcinoma in situ. Although most of the mild dysplasias regress spontaneously, a small percentage progress through the disease spectrum to invasive carcinoma (see Koss, this volume). HPV types 6, 11, 16 and 18 are found with the same frequency in mild dysplasias. However, the percentage of HPV16 and HPV18-positive lesions increases as the disease becomes more severe, strongly suggesting an oncogenic potential for these virus types (Dürst et al. 1983; Boshart et al. 1984; McCance et al. 1985; Gissmann and Schneider 1986). In fact only the DNA of HPV16 and HPV18 is consistently found in biopsies of invasive cervical carcinomas (Dürst et al. 1983; Boshart et al. 1984). Recently two other isolates, HPV31 and HPV33, have been added to the list of HPV types associated with genital cancer (Lörincz et al. 1986; Beaudenon et al. 1986). The physical state of HPVs in genital tumors is of particular interest as it clearly illustrates that the viral genome is not always found in an extrachromosomal state within the host cell. In each of six benign tumors (two condylomata acuminata, two low grade CIN, and two bowenoid papules) HPV16 DNA was detected exclusively as 8 kb circles (Dürst et al. 1985). However, in numerous examples of malignant lesions (including Bowen's disease) the viral DNA (HPV16 or HPV18) is integrated within the host genome. In addition, some tumors also contain oligomeric episomal molecules of viral DNA (Boshart et al. 1984; Dürst et al. 1985, 1986; Lehn et al. 1985; Matsukura et al. 1986). Multiple sites for viral integration within the host DNA, with no apparent specificity, have been observed (Dürst et al. 1986). Also the opening part on the viral genome appears to be unspecific if multiple integrations have occurred in the same cell. A selection for integrated molecules, in which preferably the distal part of the early ORF is disrupted (Fig. 7) may be postulated as the tumor progresses (see also Sect. 5).

Fig. 7. Integration pattern of HPV DNA sequences in the cellular genome of SiHa, SW756, C4-I and HeLa cells. SiHa harbours HPV16 sequences whereas the others contain HPV18 DNA. The arrangement of the open reading frames (ORF) of the HPV16 DNA (7903 base pairs in length) as determined by nucleotide sequence analysis (Seedorf et al. 1985) is indicated. *E* and *L* denote the ORFs of the putative early and late regions, respectively. The arrangement of the HPV18 ORFs is essentially similar to that of HPV16. Only E6, E7 and part of E1 can be transcribed from the early viral promotor

4.3 Epidermodysplasia Verruciformis

Epidermodysplasia verruciformis (EV) is a rare disease characterised by disseminated persistent flat warts. Unique to this disease is the genetic predisposition and plurality of the papillomavirus types involved. Of all epidermodysplasia patients about 30% develop skin cancers, mainly on sun-exposed areas (Orth 1986). HPV DNA sequences are regularly detected in EV carcinomas but in contrast to the 15 different HPV types prevalent in benign lesions, only types 5, 8, and 14 are associated with cancer.

The tumors usually harbour free HPV molecules, mostly as monomers and to a lesser extent as oligomers (Orth et al. 1980; Ostrow et al. 1982; Pfister et al. 1983; Yutsudo et al. 1985). Within some tumors the HPV genomes are deleted. Of interest also is the high average copy number (100–300 copies per cell) which may in part be ascribed to the productive replication of the virus genome occurring in a few cells, as shown by in situ hybridisation experiments (Orth et al. 1980).

5 Papillomavirus Genomes in Cell Lines Derived from Tumors

The finding that papillomavirus DNA is associated with malignant tumors led to the question of whether the DNA sequences could also be detected in cell lines derived from such tumors. In fact, six of eight human cervical carcinoma cell lines harbour either HPV16 or HPV18 DNA (Yee et al. 1985; Schwarz et al. 1985). The cell lines ME 180, SW 756, C4-I/II and HeLa contain HPV18 sequences. In Caski and SiHa cells HPV16 DNA is present, whereas cell lines C-33 A and HT-3 were found to be negative for HPV sequences. In all of the HPV-positive cell lines the viral genome persists in an integrated state (Fig. 7). Further analysis of the cell lines HeLa, C4-I, SW 756 and SiHa has shown that one break point of the integrated papillomavirus DNA is always located within a segment of the E1/E2 ORFs, thus disrupting the early region (Fig. 7) (Schwarz et al. 1985; M. Dürst, unpublished data; C. Baker and P. Howley, personal communication). Additionally the 3′-terminal part of the early region with ORFs E2, E4 and E5, together with the L2 part of the late region, is deleted in HeLa and C4-I cells. At least part of the E2 ORF of the HPV16 DNA is deleted in SiHa cells. Our data also suggest the possibility that as a consequence of integration part of the cellular DNA in the SiHa cell has been deleted at the point of integration.

The HPV DNA in HeLa, SW756 and Caski cells is amplified about 10–50-fold together with the flanking cellular sequences, whereas C4-I and SiHa contain only one copy of the respective viral DNA with no amplification in the cellular region.

More recently a cell line has been established from the vulvar lesions of a patient with multicentric pigmented Bowen's disease (S. Schneider-Maunoury, personal communication; Breitburd, this volume). After 40 passages a single copy of the HPV16 genome, amplified about ten fold together with its flanking cellular sequences, persists in the cell line. It is of particular interest to note that by using the cellular part of the virus/cell junction cloned from the cell line as a probe for hybridisation, it could be shown that integration of viral DNA had already occurred in the primary lesion which predominantly harboured free HPV16 genomes.

Georges et al. (1985) established two cell lines, VX2T and VX2R, from the transplantable VX2 carcinoma. This is a wholly anaplastic tumor induced by the Shope CRPV (Kidd and Rous 1940). The tumor contains multiple copies of the CRPV genome integrated into the cellular DNA as head-to-tail tandem repeats (McVay et al. 1982; Sugawara et al. 1983). Analysis of the two cell lines and the VX2 tumor has revealed no differences with respect to the maintenance of the viral genome and its transcriptional activity. However, one of the cell lines is no longer transplantable in the allogenic host, suggesting that this particular property of the VX2 epidermal carcinoma is reversible without any detectable modifications of the physical state or transcription of the CRPV genome (Georges et al. 1985).

6 Overview

It is generally believed that the germinal layer of the epidermis is the target for papillomavirus infection. The interaction of viral gene products with the host cell results in increased cellular proliferation giving rise to the characteristic lesions. It is only the differentiating cells of the stratum spinosum and stratum granulosum that permit productive viral replication. The synthesis of stable extrachromosomal viral molecules as detected by Southern blot analysis are part of the infectious cycle. However, clinical manifestation of the disease is not always a necessity. There is ample evidence for latent papillomavirus infection. Activation followed by productive infection may be triggered by such events as trauma, hormonal influence, genetic predisposition, or immunosuppression. There is no information concerning the physical state of the viral genome in latently infected tissue as the actual number of cells in which the virus persists is too small for molecular analysis.

The transition of benign papillomas to malignant lesions has been observed in a number of animal species, of which the cottontail rabbit system is the best studied. Also, in humans, a number of premalignant lesions is associated with papillomavirus infections, and it is tempting to speculate that these viruses are somehow involved in the process of malignant transformation. It is thus of interest to analyse the physical state of the papillomavirus genomes within carcinomas in order to disclose possible similarities which may provide some clues on the mechanisms involved in tumorigenesis.

However, when summarising the data for a variety of tumors, no general trend emerges. Integration of HPV16 and HPV18 DNA in human genital tumors as well as in derived cell lines is frequently observed. In contrast, the genomes of HPV5 and HPV8, which are associated with carcinoma in epidermodysplasia verruciformis, are invariably found in a high copy number, either as monomers or in the form of multicopy episomal molecules (Orth 1986). Similarly, large numbers of episomal viral genomes persist in CRPV-induced primary carcinoma with the exception of one tumor in which integrated papillomavirus DNA has been found (Wettstein and Stevens 1982).

The interpretation of these findings may be twofold: (a) integration of the viral genome may not be an essential step in the genesis of all papillomavirus-associated cancers; (b) integrated papillomavirus genomes simply cannot be detected in some tumors due to interference by large numbers of episomal molecules.

With this is mind, it is of interest to note that in addition to all HPV-positive cell lines established from genital tumors, the transplantable tumors VX2 and VX7 induced by CRPV also harbour only integrated papillomavirus genomes. The possibility exists that integrated papillomavirus DNA, required for the expression of the malignant phenotype, is already present in the primary carcinoma but only becomes apparent after episomal molecules are lost due to continued selection resulting from passaging either in vivo or in vitro.

The nucleotide sequence analysis of HPV16 reveals a frame shift in the ORF E1 (Seedorf et al. 1985). This particular mutation was of interest as the HPV16 DNA used for sequence analysis was cloned from a cervical carcinoma which harboured integrated as well as episomal molecules. As the E1 gene product was found to be essential for replication and episomal plasmid maintenance of BPV in transformed mouse cells (Lusky and Botchan 1984, 1985), it was thought that this particular mutation may be typical of integrated HPV16 molecules. To investigate this possibility we cloned HPV16 molecules from two condylomata acuminata (harbouring only episomal molecules of genomic length), as well as from a cervical carcinoma and the cervical carcinoma cell line SiHa, both of which harbour only integrated DNA. The sequence data (Klaus Seedorf, personal communication) revealed a continuous ORF for E1 in all isolates. The frame shift in the originally cloned HPV16 molecule arose due to a single deletion of a G at nucleotide position 1137 and appears to be untypical for HPV16.

The possibility of viral integration within a specific cellular region which may interrupt essential gene functions or conversely lead to activation of normally silent genes (analogous to the ALV promotor insertion system; Hayward et al. 1981) was also considered. For this purpose the cellular sequences flanking the integrated viral genome were cloned from one cervical carcinoma and from the cell lines SiHa, HeLa, C4-I, and SW756 and used as hybridisation probes for screening other tumor DNAs. Rearrangement of the restriction fragment pattern in a number of tumors using any of the above probes would have been indicative of viral integration within a particular region of the chromosome (Dürst et al. 1986; M. Dürst and E. Schwarz, unpublished). No common cellular site for viral integration could be found. This approach is of course restricted to the analysis of an extremely small part of the host genome. The probes are therefore assigned to their specific chromosome by hybridisation with the DNA of defined mouse-human hybrid cell lines (collaboration with K. Huebner at the Wistar Institute, Philadelphia). Evidence so far favours random integration which is in line with data obtained by in situ hybridisation on metaphase chromosomes (Mincheva et al., submitted for publication).

In most of the HPV-positive genital carcinomas, one of the opening points for viral integration is in the early region of the viral genome (E1/E2). Moreover, in cell lines derived from genital cancer it is always this region which is disrupted by integration. As a consequence only ORFs E6, E7, part of E1 and the adjacent cellular sequences can be transcribed from the early viral promotor. The analysis of cDNA clones derived from HeLa, C4-I and SW756 cells confirmed the existence of the predicted virus-cell fusion transcripts (Schneider-Gädicke and Schwarz 1986). How these transcripts (or their gene products) influence the host cell is not clear. It has been speculated that the stability of the mRNA may be increased due to the cellular 3'end. Alternatively, the disruption of the intragenomic viral regulation may lead to altered expression of viral genes.

Integration of the HPV DNA into the host genome may be a decisive factor in the development of genital cancer but it is probably not sufficient per se to explain a complex phenomenon.

Acknowledgements. I wish to thank Dr. Mark Kutcher for critical reading of the manuscript. Furthermore I am grateful to Martina Deschner for her help in preparing the manuscript and Renate Webler for assistance with the artwork.

References

Beaudenon S, Kremsdorf D, Croissant O, Jablonska S, Wain-Hobson S, Orth G (1986) A novel type of human papillomavirus associated with genital neoplasias. Nature 321: 246-249

Boshart M, Gissmann L, Ikenberg H, Kleinheinz A, Scheurlen W, zur Hausen H (1984) A new type of papillomavirus DNA, its presence in genital cancer biopsies and in cell lines derived from cervical cancer. EMBO J 3: 1151-1157

Danos O, Giri I, Thierry F, Yaniv M (1984) Papillomavirus genomes: sequences and consequences. J Invest Dermatol 83: 7-11

Dürst M, Gissmann L, Ikenberg H, zur Hausen H (1983) A papillomavirus DNA from a cervical carcinoma and its prevalence in cancer biopsy samples from different geographic regions. Proc Natl Acad Sci USA 80: 3812-3815

Dürst M, Kleinheinz A, Hotz M, Gissmann L (1985) The physical state of human papillomavirus type 16 DNA in benign and malignant tumors. J Gen Virol 66: 1514-1521

Dürst M, Schwarz E, Gissmann L (1986) Integration and persistence of human papillomavirus DNA in genital tumors. In: Peto R, zur Hausen H (eds) Viral etiology of cervical cancer. Cold Spring Harbor Laboratory, New York, pp 273-280

Favre M, Jibard N, Orth G (1982) Restriction mapping and physical characterization of the cottontail rabbit papillomavirus genome in transplantable VX2 and VX7 domestic rabbit carcinomas. Virology 119: 298-309

Ferenczy A, Mitao M, Nagai N, Silverstein SJ, Crum CP (1985) Latent papillomavirus and recurrent genital warts. N Engl J Med 313 (13): 784-788

Georges E, Croissant O, Bonneaud N, Orth G (1984) Physical state and transcription of the cottontail rabbit papillomavirus genome in warts and transplantable VX2 and VX7 carcinomas of domestic rabbits. J Virol 51: 530-538

Georges E, Breitburd F, Jibard N, Orth G (1985) Two Shope papillomavirus-associated VX2 carcinoma cell lines with different levels of keratinocyte differentiation and transplantability. J Virol 55 (1): 246-250

Gissmann L (1984) Papillomaviruses and their association with cancer in animals and in man. Cancer Surveys 3 (1): 161-181

Gissmann L, Schneider A (1986) The role of human papillomaviruses in genital cancer. In: De Palo G, Rilke F, zur Hausen H (eds) Herpes and papillomaviruses, Serono Symposia, vol 31. Raven, New York

Gissmann L, Wolnik L, Ikenberg H, Koldovsky U, Schnürch HG, zur Hausen H (1983) Human papillomavirus type 6 and 11 DNA sequences in genital and laryngeal papillomas and in some cervical cancers. Proc Natl Acad Sci USA 80: 560-563

Grussendorf EI, zur Hausen H (1979) Localization of viral DNA replication in sections of human warts by nucleic acid hybridization with complentary RNA of human papilloma virus type 1. Arch Dermatol Res 264: 55-63

Grußendorf-Conen E-I, Gissmann L, Holters J (1983) Correlation between content of viral DNA and evidence of mature virus particles in HPV 1, HPV 4 and HPV 6 induced virus acanthoma. J Invest Dermatol 81: 511-513
Hayward WS, Neel BG, Astrin SM (1981) Activation of a cellular onc gene by promotor insertion in ALV-induced lympoid leukosis. Nature 290: 475-480
Howley PM (1983) Papovaviruses: search for evidence of possible association with human cancer. In: Phillips LA (ed) Viruses associated with cancer. Marcel Dekker, New York, pp 253-305
Jarrett WFH, McNeil PE, Laird HM, O'Neil BW, Murphy J, Campo MS, Moar MH (1980) Papillomaviruses in benign and malignant tumors of cattle. In: Essex M, Todaro G, zur Hausen H (eds) Viruses in naturally occurring cancers, Cold Spring Harbor conferences in cell proliferation, vol 7. Cold Spring Harbor Laboratory, New York, pp 215-222
Johnson PH, Grossmann LI (1977) Electrophoresis of DNA in agarose gels. Optimizing separation of conformational isomers of double and single stranded DNAs. Biochemistry 16 (19): 4217-4225
Kidd JR, Rous P (1940) Cancers deriving from the virus papilloma of wild rabbits under natural conditions. J Exp Med 71: 469-485
Kreider JW, Howett MK, Wolfe SA, Bartlett GL, Zaino R, Sedlacek TV, Mortel R (1985) Morphological transformation of human uterine cervix with papillomavirus from condylomata acuminata. Nature 317: 639-640
Lehn H, Krieg P, Sauer G (1985) Papillomavirus genomes in human cervical tumors: analysis of their transcriptional activity. Proc Natl Acad Sci USA 82: 5540-5544
Lörincz AT, Lancaster WD, Kurman RJ, Jenson AB, Temple GF (1986) Characterization of human papillomavirus in cervical neoplasias and their detection in routine clinical screening. In: Peto R, zur Hausen H (eds) Viral etiology of cervical cancer, Banbury Report 21. Cold Spring Harbor Laboratory, New York, pp 225-237
Lusky M, Botchan MR (1984) Characterization of the bovine papillomavirus plasmid maintenance sequences. Cell 36: 391-401
Lusky M, Botchan MR (1985) Genetic analysis of bovine papillomavirus type 1 trans-acting replicating factors. J Virology 53: 955-965
Matsukura T, Kanda T, Furuno A, Yoshikawa H, Kawana T, Yoshiike K (1986) Cloning of monomeric human papillomavirus type 16 DNA integrated within cell DNA from a cervical carcinoma. J Virol 58: 979-982
McCance DJ, Campion MJ, Clarkson PK, Chesters PM, Jenkins D, Singer A (1985) Prevalence of human papillomavirus type 16 DNA sequences in cervical intraepithelial neoplasia and invasive carcinoma of the cervix. Br J Obstet Gynaecol 92: 1101-1105
McVay P, Fretz M, Wettstein FO, Stevens JG, Ito Y (1982) Integrated Shope virus DNA is present and transcribed in the transplantable rabbit tumor VX7. J Gen Virol 60: 271-278
Meisels A, Fortin R (1976) Condylomata lesions of the cervix and vagina: I Cytological pattern. Acta Cytol 20: 505
Noyes WF, Mellors RC (1957) Fluorescent antibody detection of the antigens of the Shope papillomavirus in papillomas of the wild and domestic rabbits. J Exp Med 106: 556-562
Orth G (1986) Epidermodysplasia verruciformis: a model for understanding the oncogenicity of human papillomaviruses. In: Howley PM (ed) Papillomaviruses (Ciba foundation symposium 120) Wiley, Chichester, pp 157-174
Orth G, Jeanteur P, Croissant O (1971) Evidence for and localization of vegetative viral DNA replication by autoradiographic detection of RNA/DNA hybrids in sections of tumors induced by Shope papilloma virus. Proc Natl Acad Sci USA 68: 1876-1880
Orth G, Breitburd F, Favre M, Croissant O (1977) Papillomaviruses: possible role in human cancers, Cold Spring Harbor Conference on Cell Proliferation, vol 4. Cold Spring Harbor Laboratory, New York, pp 1043-1063
Orth G, Favre M, Breitburd F, Croissant O, Jablonska S, Obalek S, Jarzabek-Chorzelska M, Rzesa G (1980) Epidermodysplasia verruciformis: a model for the role of papillomaviruses in human cancer. In: Essex M, Todaro G, zur Hausen H (eds) Viruses in naturally occurring cancers. Cold Spring Harbor Laboratory, New York, vol 7, pp 259-282
Ostrow RS, Bender M, Nijmura M, Kawashima M, Pass F, Faras AJ (1982) Human papillomavirus DNA in cutaneous primary and metastasized squamous cell carcinoma from patients with epidermodysplasia verrucciformis. Proc Natl Acad Sci USA 79: 1634-1638

Pfister H (1984) Biology and biochemistry of papillomaviruses. In: Adrian RH, zur Hausen H, Helnreich E, Holzer H, Jung R, Linden RJ, Miescher PA, Piiper J, Rasmussen H, Trendelenburg U, Ullrich K, Vogt W, Weber A (eds) Reviews of Physiology, Biochemistry and Pharmacology, vol 99. Springer, Berlin Heidelberg New York Tokyo, pp 111-181

Pfister H, Gassenmaier A, Nürnberger F, Stüttgen G (1983) Human papillomavirus 5 DNA in carcinoma of an epidermodysplasia verruciformis patient infected with various human papillomavirus types. Cancer Res 43: 1436-1441

Phelps WC, Leary SL, Faras AJ (1985) Shope papillomavirus transcription in benign and malignant rabbit tumors. Virology 146: 120-129

Purola E, Savia E (1977) Cytology of gynecologic condyloma acuminatum. Acta Cytol (Baltimore) 21: 26

Radloff R, Bauer W, Vinograd J (1967) A dyebuoyant-density method for the detection and isolation of closed circular duplex DNA. The closed circular DNA of HeLa cells. Proc Natl Acad Sci USA 57: 1514-1521

Rashad AL (1969) Radioautographic evidence of DNA synthesis in well differentiated cells of human skin papillomas. J Invest Dermatol 53: 356-362

Rogers S, Kidd JG, Rous P (1950) An etiological study of the cancers arising from the virus-induced papillomas of domestic rabbits. Cancer Res 10: 237

Rogers S, Kidd JG, Rous P (1960) Relationships of the Shope papilloma virus to the cancers it determines in domestic rabbits. Acta Unio Int Contra Cancrum 16: 129-130

Rous P, Beard JW (1935) The progression to carcinoma of virus-induced rabbit papillomas (Shope). J Exp Med 62: 523-548

Rous P, Friedewald WF (1944) The effect of chemical carcinogens on virus induced rabbit papillomas. J Exp Med 79: 511-537

Schneider-Gädicke A, Schwarz E (1986) Different human cervical carcinoma cell lines show similar transcription patterns of human papillomavirus type 18 early genes. EMBO J 5: 2285-2292

Schwarz E, Freese UK, Gissmann L, Mayer W, Roggenbuck B, Stremlau A, zur Hausen H (1985) Structure and transcription of human papillomavirus sequences in cervical carcinoma cells. Nature 314: 111-114

Seedorf K, Krämmer G, Dürst M, Suhai S, Röwekamp W (1985) Human papillomavirus type 16 DNA sequence. Virology 145: 181-185

Shope RE (1933) Infectious papillomatosis of rabbits (with a note on the histopathology by EW Hurst). J Exp Med 58: 607-624

Smith WE, Kidd JG, Rous P (1952) Experiments on the cause of the rabbit carcinomas derived from virus induced papillomas: I. Propagation of several of the cancers in sucklings, with etiological test. J Exp Med 95: 299-317

Southern EM (1975) Detection of specific sequences among DNA fragments separated by gel electrophoresis. J Mol Biol 98: 503-517

Steinberg BM, Topp WC, Schneider PS, Abramson AL (1983) Laryngeal papillomavirus infection during clinical remission. N Engl J Med 308 (21): 1261-1264

Stevens JG, Wettstein FO (1979) Multiple copies of Shope virus DNA are present in cells of benign and malignant non-virus producing neoplasms. J Virol 30: 891-898

Sugawara K, Fujinaga K, Yamashita T, Ito Y (1983) Integration and methylation of Shope papillomavirus DNA in the transplantable VX2 and VX7 rabbit carcinomas. Virology 131: 88-99

Syverton JT (1952) The pathogenesis of the rabbit papilloma-to-carcinoma sequence. Ann NY Acad Sci 54: 1126-1140

Wettstein FO, Stevens JG (1982) Variable sized free episomes of Shope papillomavirus DNA are present in all non-virus neoplasms and integrated episomes are detected in some. Proc Natl Acad Sci USA 79: 790-794

Yee C, Krishnan-Hewlett I, Baker CC, Schlegel R, Howley PM (1985) Presence and expression of human papillomavirus sequences in human cervical carcinoma cell lines. Am J Pathol 119: 361-366

Yutsudo M, Shimakage T, Hakura A (1985) Human papillomavirus type 17 DNA in skin carcinoma tissue of a patient with epidermodysplasia verruciformis. Virology 144: 295-298

zur Hausen H (1982) Human genital cancer: synergism between two virus infections or synergism between a virus infection and initiating events. Lancet ii: 1370-1372

Transforming and Regulatory Functions of Bovine Papillomavirus Type 1

L. P. Turek and T. H. Haugen

One might regard DNA as a big, fat, aristocratic, lazy, cigar-smoking slob of a molecule. It never does anything. It merely issues orders, never condescending to do any work itself, quite like a queen bee. How did it get such a cushy position? By ensuring the production of proteins, which do all the dirty work for it. And how does it make sure these desirable proteins get produced? Ah, that is the trick.

From Douglas Hofstadter, *Metamagical Themas* (Hofstadter, 1985)

1 Scope of This Review

Over the past several years, papillomaviruses have attracted the attention of molecular biologists, experimental oncologists, and clinicians alike. Although the existence of papillomaviruses has been known since the beginning of the century (Ciuffo 1907; Shope 1933), analysis of their genomic organization and function has progressed only since the advent of molecular cloning. The application of recombinant DNA technology to the identification and analysis of animal and human papillomaviral isolates has elucidated the sequence and organization of the papillomaviral genomes. In addition, the availability of an unlimited supply of cloned viral DNAs for hybridization probes has facilitated detection of papillomavirus DNA in clinical material. Nucleic acid homology has been used to classify human papillomavirus (HPV) isolates into more than 40 "genotypes," and to reveal a correlation between infection with specific HPV types and the location and malignant potential of the infectious lesions. These studies have shown that human papillomavirus infection may contribute to the genesis of several human malignant tumors, notably squamous cell carcinomas arising in epidermodysplasia verruciformis (Orth 1986) and malignancies involving genital squamous epithelium (zur Hausen 1985; Gissmann and Schwarz 1986).

While hybridization studies performed on clinical material clearly establish the association of HPV infection with neoplastic disease in humans, genetic analysis of those HPV functions that play a role in neoplastic cell growth has been extremely difficult. Papillomaviruses have not been successfully propagated in any tissue culture system tested so far. However, certain viral strains or their molecularly cloned DNAs can induce neoplastic transformation in rodent fibroblasts in culture. One of the most efficient transforming types is the bovine papillomavirus type 1 (BPV1). In addition to altering cellular growth characteristics, BPV1 genomes replicate in the nucleus of the transformed cell as unintegrated, circular plasmids, thus presumably mimicking the state of papillomaviral DNA in the basal epidermal layer. These features have been explored over the past 7 years to identify those BPV1 sequences and functions which are required for transformation and for plasmid replication of the viral genomes in this system. The molecular analysis of BPV1 has benefited from "reverse genetics," in that molecular mutagenesis experiments were guided in part by predictions based on available DNA sequence data. However, the functional organization of the BPV1 genome has turned out to be more complex than the initial interpretation of early experimental results had suggested. The purpose of this chapter is to summarize recent findings that have identified transforming and regulatory gene domains within the BPV genome. Similarities and differences with other papillomaviruses will be pointed out in cases where similarity in structure may reflect a similarity in function. A comprehensive discussion of papillomavirus biology can be found in several recent reviews (Lancaster and Olson 1982; Pfister 1984; Gissmann 1984), and in chapters by Breitburd, Pfister and Fuchs, and Schwarz (this volume).

2 Structure and Genomic Organization of Papillomaviruses

2.1 Virion Structure

Papillomaviruses are small DNA tumor viruses classified among the papovavirus group along with SV40 and polyoma (Melnick et al. 1974). In contrast to these, the papillomavirion particle is larger in diameter (55 v 42 nm). The eicosahedral capsid consists of 72 capsomeres and harbors a supercoiled, circular, double-stranded DNA molecule. Most papillomaviral genomes are between 7800 and 8000 n long. Over 50 independent animal and human strains have been identified and molecularly cloned; to date, the complete nucleotide sequence has been reported for 8 and the partial sequence of several other genomes is known. The general organization of all papillomaviral genomes that have been sequenced appears to be similar to that of BPV1.

2.2 Organization and Expression of the BPV 1 Genome

The complete nucleotide sequence of BPV1 DNA has been determined (Chen et al. 1982; corrected in Danos et al. 1983; and in Ahola et al. 1983). The 7945-n long circular genome contains 10 major open reading frames (ORFs), that is, stretches of DNA without termination codons that could encode peptides greater than 100 amino acids. All ORFs reside on one DNA strand, and mRNAs found in transformed mouse cells (Amtmann and Sauer 1982b; Freese et al. 1982; Heilman et al. 1982) as well as in bovine warts (Amtmann and Sauer 1982b; Engel et al. 1983) correspond to the ORF-coding strand (Fig. 1). This is in contrast to SV40 and polyoma viruses whose early and late genes are transcribed in the opposite direction (Tooze 1981).

The "early" (E) genes of BPV1 code for all functions required for fibroblast transformation (Lowy et al. 1980). These are preceded by the upstream regulatory region (URR) containing transcription control signals (Campo et al. 1983; Nakabayashi et al. 1983; Sarver et al. 1984; Spalholz et al. 1985; Haugen et al. 1987) as well as the origin of plasmid replication in transformed mouse fibroblasts (Waldeck et al. 1984). Only the E region is transcribed in BPV1 transformed mouse cells (Amtmann and Sauer 1982b; Heilman et al. 1982). The early genes are followed by a transcription termination signal, a *cis*-enhancer of transcription (Lusky et al. 1983; Spandidos and Wilkie 1983; Weiher and Botchan 1984; Sarver et al. 1985), and the "late" (L) genes, coding for viral structural proteins (Pilacinski et al. 1984). The L region is only transcribed in fibropapillomas (Amtmann and

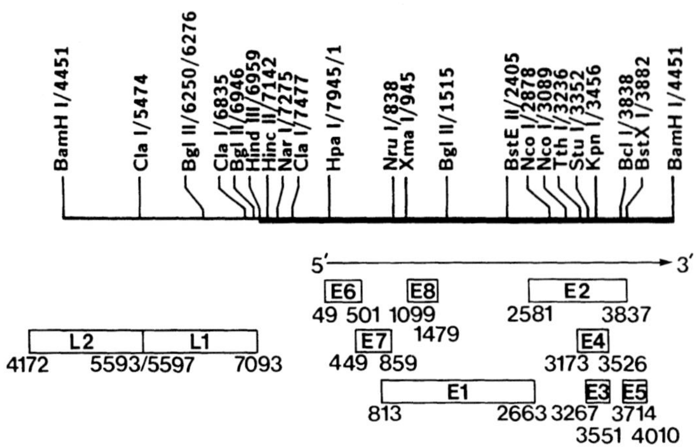

Fig. 1. Genomic organization of BPV1. As determined by DNA sequence analysis, the BPV1 genome contains ten major (> 100 amino acids) open reading frames (ORFs), eight of which are present in the 69% *Hind*III-*Bam*HI transforming fragment (thick line). These are designated "early" *(E)*. All ORFs are found on the same DNA strand, and the direction and extent of early gene transcription is indicated by the *arrow*. Immediately 5′ to E6 is the upstream regulatory region containing viral promoters, the origin of replication, and additional regulatory sequences. The downstream enhancer is located immediately upstream of the *Bam*HI site. *Numbers* indicate first and last nucleotides of each ORF.

Sauer 1982b; Engel et al. 1983). Papillomaviral transcription is discussed extensively by Schwarz elsewhere in this volume; further description in this chapter is limited to molecular dissection of viral gene regulation in transfection assays (see Sect. 5.3).

3 Papillomavirus Infection and its Outcome

3.1 Benign Epithelial Proliferations

The typical papillomavirus-induced lesion is a benign proliferation of the epidermal layer of the skin or of squamous mucosal epithelium. The great majority of these lesions show a preserved differentiation pattern. Papillomavirus infection leads to an increase in the thickness of the epithelium at the site of the lesion, either by stimulation of cell division or by disruption of normal keratinocytic differentiation (Steinberg 1986). Additional effects may include changes in the synthesis of cellular gene products involved in terminal differentiation and keratinization. While papillomavirus gene products clearly interfere with normal cellular functions, the expression of papillomaviral functions and the ultimate outcome of infection is influenced by specific factors in the host cell. This complex regulatory relationship is poorly understood. A viral particle is thought to transform an epidermal germinal cell by nonproductive infection, which leads to papillomatous proliferation. In the nuclei of germinal layer cells, viral DNA is found at low copy numbers. Synthesis of viral capsid proteins and extensive viral DNA replication is seen in the course of differentiation of the host cell, and mature capsids are finally found in the nuclei of cells of the stratum granulosum and stratum spinosum. Keratinocyte maturation and the switch to productive wart virus infection involving virion production are so intimately linked that this process has not been reproduced even in partially differentiating keratinocytes in culture (LaPorta and Taichman 1982).

3.2 Fibroepitheliomatous Lesions

Infection with several animal papillomavirus strains is associated with lesions that include a dermal component. Dermal fibroblasts underneath the warty epithelial lesion proliferate with the benign appearance of a fibroma and have been shown to contain replicating plasmid DNA genomes (see Lancaster and Olson 1982 for review). Interestingly, no human lesions that would include a dermal proliferative component have been described so far. BPV (types 1 and 2), the ovine papillomaviruses, and European elk papilloma virus (EEPV) are examples of papillomaviruses that cause large fibroepitheliomatous warts in the natural host. In addition, the deer papillomavirus (DPV) induces fibromas with minimal epidermal changes (Groff and Lancaster 1985). Members of this group of viruses readily transform established rodent fibroblasts in culture. Cell transformation by BPV1 is the best characterized genetic model used to study papillomavirus early functions in vitro. While the system has proved to be extremely useful for providing initial insights into the transforming early functions of papillomaviruses, it does not allow for the study of several important modes of virus-cell interaction (see below).

3.3 Latent Papillomavirus Infection

One aspect of papillomaviral infection that has been difficult to study in vitro is the question of viral persistence and latent infection. Regression of warts and their reappearance are well-established phenomena, and verrucous proliferations appear to be subject to immunological surveillance (reviewed in Pfister 1984). Ferenczy, Crum, and coworkers (1985) also demonstrated the presence of specific HPV types in normal mucosa at a distance from flat cervical condylomata. An even more dramatic example is viral infection in the South African multimammate rodent, *Mastomys nataliensis*. Amtmann et al. (1984) demonstrated the persistence of unintegrated *M. nataliensis* papillomaviral (MnPV) genomes in the skin, skeletal muscle, liver, colon, and even embryonic tissues. In this system, viral infection is thought to occur by horizontal transmission during the early stages of embryogenesis. Expression of the latent MnPV genomes is activated in response to carcinogens and tumor promoters (Amtmann et al. 1984), but the molecular mechanism of activation has not been analyzed in vitro.

In vitro experimental systems for the study of papillomavirus latency are still to be developed. BPV1 DNA replication in fibroblastic cell lines may occur prior to changes associated with morphological transformation, but morphologically normal cell lines containing papillomaviral genomes in a latent state usually become transformed upon passage (Law et al. 1983; Meneguzzi et al. 1984; Grisoni et al.

1984). A mechanism that has been shown to downregulate viral gene expression in other systems is gene-specific postreplicational methylation at CG dinucleotides (^{me}CG). Transfection with in vitro methylated BPV 1 DNA does not, however, appear to establish latent infection in fibroblasts (L. Turek, T. Haugen, and G. Ginder, manuscript submitted), and the ^{me}CG modification is not inherited past the initial stages of replication (Christy and Scangos 1986; Turek et al. manuscript submitted).

3.4 Progression of Benign Viral Papillomas to Malignancy

Although usually benign, some papillomavirus-induced proliferations are endowed with a low or even considerable malignant potential. Carcinomas develop especially in long-lasting papillomatous lesions, for instance, in human epidermodysplasia verruciformis, juvenile laryngeal papillomatosis, and in papillomaviral lesions of the uterine cervix (see Gissmann 1984; Pfister 1984; zur Hausen 1985, for reviews). Malignant conversion occurs frequently in connection with exposure to additional environmental factors such as chemical carcinogens, in response to chronic ingestion of bracken fern in esophageal papillomas of cattle caused by BPV4 (Jarrett et al. 1984), X-ray exposure in treated human laryngeal papillomas, or exposure to UV light which affects epidermodysplasia lesions (as reviewed by Pfister 1984). The severity and malignant potential of the papillomatous lesion is probably also influenced by genetic predisposition in the host. Furthermore, malignant conversion in human cervical neoplasia associated with HPV16/18 infection is associated with viral integration and DNA rearrangements, resulting in perturbed viral gene expression (Schwarz et al. 1985). Whereas these modified types of papillomavirus-cell interactions cannot be studied directly in the model BPV 1 transformation system, it is possible to speculate about the consequences of such rearrangements based on our current knowledge of BPV biology (see E. Schwarz, this volume).

4 Papillomavirus-Induced Neoplastic Transformation in Cell Culture

4.1 Transformation of Cultured Fibroblasts with BPV 1

Papillomaviruses have not been successfully propagated in cell culture so far. This has precluded the isolation and characterization of conditional viral mutants through classical approaches. However, Black et al. (1963), Boiron et al. (1964), and Thomas et al. (1964) observed focal neoplastic transformation with BPV isolates in conjunctival cells and fibroblasts in tissue culture. Dvoretzky, Lowy, and coworkers (1980) have developed a quantitative tissue culture focus assay in established mouse fibroblastic cell lines (NIH-3T3 and C127I). These authors have further demonstrated that transformation follows single-hit kinetics, implying that a single virion particle is sufficient for the establishment of infection. Using the calcium-phosphate DNA precipitation technique of introducing DNA into cells (transfection, Graham and van der Eb 1973), Lowy, Dvoretzky, Law, Howley, and coworkers (1980) demonstrated that neoplastic transformation can be induced by BPV1 DNA molecularly cloned in bacteria (Howley et al. 1979). Similar observations were made in other laboratories (Lancaster 1981; Moar et al. 1981) and extended to established rat (Binetruy et al. 1984), primary mouse, and hamster cells (Amtmann and Sauer 1982a). Virtually all genetic experiments with BPV1 recombinants and mutants have been performed in the mouse fibroblast cell line C127I (Lowy et al. 1978) or in NIH-3T3 cells (Jainchill et al. 1969).

The calcium phosphate coprecipitation procedure of DNA transfection into tissue culture cells (Graham and van der Eb 1973) has become a standard experimental technique. Briefly, purified DNA prepared from bacterial cloning vectors is adjusted to a concentration of 20-30 µg/ml with neutral carrier DNA in a final mix of HEPES-buffered saline with 1.5 m*M* potassium phosphate and 125 m*M* calcium chloride. The mixture is allowed to stand at room temperature for 15-20 min; a fine precipitate of calcium phosphate and calcium DNA salt gives the samples a cloudy appearance that can be monitored under the microscope. The coprecipitate is then added directly to media in culture dishes containing cells that were plated the day before, and allowed to settle on the cells for approximately 4 h. At this point, uptake of the DNA precipitate can be dramatically increased by hypertonic shock treatment with agents that influence membrane mobility, such as dimethyl sulfoxide or glycerol, or with high concentration sucrose. Residual DNA precipitate is washed off, and the cultures are incubated further. With C127I or NIH3T3 cells this procedure appears to yield the highest efficiency of introducing foreign DNA into cells in comparison to transfection of DNA in solution in the presence of agents that modify its adsorption to the membrane, such as diethyl aminoethyl (DEAE)-dextran or Polybrene. The cells are then incubated for 2-4 weeks until visible areas (foci) of transformed cells appear. Individual foci can be subcultured and expanded for analysis.

Although the transformed fibroblasts are not productively infected by the virus, cellular transformation with BPV1 DNA has provided a useful system for the study of viral functions involved in the early events in papillomavirus infection. In contrast to other transforming viral genomes, which integrate in the cellular DNA, BPV genomes replicate as supercoiled, circular DNA plasmids in the nuclei of the transformed cells (Lancaster 1981; Law et al. 1981; Moar et al. 1981). BPV-transformed cell lines maintain stable plasmid copy numbers in the range of about 50-100 genomes per cell, resembling papillomavirus replication in the dividing cells of the germinal epithelial layer. Turek, Lowy, Howley, and coworkers (1982) established that BPV-transformed mouse C127I cells which have lost the viral DNA after long-range treatment with mouse L-cell interferon revert to the non-transformed phenotype; they concluded that the maintenance of transformation requires the continuous expression of viral functions. Similarly, Sauer et al. (1984) and Amtmann et al. (1985) have determined that Syrian hamster cells lose the transformed phenotype concomitant with plasmid BPV1 DNA loss after exposure to a xanthate compound, D609. These findings opened the way to genetic dissection of the viral functions responsible for BPV transformation and plasmid replication.

4.2 Growth Properties of BPV1 Transformants

Mouse cells transformed by BPV1 DNA either extracted from purified virions or molecularly cloned in bacteria exhibit anchorage-independent growth in semisolid suspension cultures, grow to high saturation densities in low serum, and form tumors at the site of injection in athymic mice (Dvoretzky et al. 1980; Lowy et al. 1980). This is in contrast to the biological growth properties of cells established directly from bovine fibromas. These cells are contact inhibited, do not grow in soft agar, and do not form tumors in nude mice (Lancaster 1981). Whether in vitro transformed mouse fibroblasts represent tumor cells with true malignant potential remains unresolved (Pfister 1984). Cuzin and coworkers (1985) have proposed that the fully transformed phenotype in BPV1 in transfected cells is the result of a complex interaction between viral transforming gene products and as yet undefined cellular factors, whose activation is required to contribute to the neoplastic growth properties of the cell. BPV1 transformation also extends the life span of primary Syrian hamster fibroblasts in vitro; Amtmann et al. (1985) have shown that the continuous expression of viral function(s) is responsible for this "immortalization" effect, since cell clones "cured" of BPV1 sequences by D609 treatment lose both tumorigenicity and the ability to propagate indefinitely in vitro. The viral immortalization gene(s) remains to be identified.

4.3 Expression of BPV1 Genes in Transformants

Although BPV1-transformed cells have been available for some time, the analysis of viral transcripts and proteins has been very difficult due to low levels of expression of viral genes in the transformants (Amtmann and Sauer 1982b; Freese et al. 1982; Heilman et al. 1982). BPV-specific mRNA transcripts in mouse transformants represent less than 0.01% of total poly(A)-selected RNA, corresponding to about 15-30 mRNA molecules per cell. Perhaps as a consequence of this low level of expression, attempts to identify antibodies against papillomaviral early proteins in sera obtained from tumor-bearing animals have not been successful.

Recent technological advances allowed the production of antisera raised against predicted peptides (Androphy et al. 1985; Mallon et al. 1986; Schlegel et al. 1986), and the identification of BPV-specific transforming proteins (Androphy et al. 1985; Schlegel et al. 1986). In contrast to cells overexpressing the respective gene under heterologous promoter control, the proteins are found at much lower levels in cells transformed by replicating, wild-type BPV1 (see Sect. 5.1), yet wild-type BPV molecules transform established mouse cell lines almost as readily as molecular clones of viral and cellular oncogenes. These observations support the concept that a finely tuned cooperation between viral transforming genes and possibly other gene functions encoded by the BPV early region are responsible for efficient transformation.

4.4 Chromatin Structure and Function of BPV Plasmids in Transformed Cells

The replicating BPV genome in transfected mouse cells exhibits features common to cellular chromatin and other replicating viral chromosomes such as SV40. The DNA is associated with nucleosomal structures (Ostrowski et al. 1983), and there is a nucleosome-free "gap" (Waldeck et al. 1984) coincident to a DNase I-sensitive site, which maps to the upstream part of the URR (Rösl et al. 1983). This site corresponds to the origin of replication (Waldeck et al. 1984) and to some, but not all putative transcriptional control sequences in the URR (see Sect. 5.3.1). It is unclear at present which of these functions is associated with the nucleosome gap. Another DNase I-sensitive site is located at the downstream enhancer (DE); however, minor hypersensitive sites have been also detected at several other sites in the genome (Rösl et al. 1986), which may correspond to other transcription control sites.

Sarver et al. (1981) showed that a heterologous gene, the rat preproinsulin, when linked to the 69% early fragment of BPV1, becomes a stable part of the plasmid replicon, and is expressed at comparatively high levels in transformed mouse cells. DiMaio et al. (1982), Sarver et al. (1982), and Kushner et al. (1982) constructed "shuttle" vectors, in which BPV1 sequences are linked to a bacterial plasmid, such as the pBR 322-derived vector pML-1 (Lusky and Botchan 1981). The chimeric molecules frequently replicate as unrearranged, unintegrated plasmids in the transfected cell, and their integrity can be verified after "shuttling" the plasmid DNA back into bacteria. Heterologous genes inserted into BPV plasmids retain their capacity to respond to regulatory stimuli in the cell (Karin et al. 1983; Mitrani-Rosenbaum et al. 1983; Ostrowski et al. 1983; Zinn et al. 1983). Therefore, the replicating BPV plasmids respond to diffusible factors that specifically interact with *cis* elements in the plasmid sequence.

5 Analysis of Genetic Functions of BPV1 in Transfection Assays

As discussed above, the entire wild-type BPV1 genome transforms mouse C127I and NIH-3T3 cells in transfection experiments. Lowy, Howley, and coworkers (1980) digested the entire 7945-n long genome with several unique restriction endonucleases. The transforming activity was found to reside in a 5437-n long fragment extending from the *Hin*dIII to *Bam*HI restriction sites (the 69% T fragment; see also Fig. 1). Further digestion of this fragment inactivated transformation. The transforming fragment was deduced to contain "early" (E) genes of BPV1, based on analogy with other transforming viruses, such as SV40, polyoma, and adenoviruses. Abortive infection with these viral strains in nonpermissive host cells allows viral early gene expression, but not infectious virus production. Instead, expression of the E gene domain may result in cellular transformation (Tooze 1981). By analogy, BPV1 infection in fibroblasts can also be considered "abortive infection," since no infectious virus particles are produced. In contrast to transformation by SV40, polyoma, or adenoviruses, however, BPV1 does not become integrated in the cellular genome, and replicates as an unintegrated multicopy plasmid in the nucleus of the cell (Lancaster 1981; Law et al. 1981). Similar observations have been made with BPV2 (Moar et al. 1981). These early findings defined two sets of genetic functions of BPV1 amenable to molecular analysis in transfection experiments: those which alter the growth characteristics of cells in neoplastic transformation, and those involved in BPV plasmid replication. The two sets of viral functions are encoded in different gene domains and will be discussed in the following sections.

5.1 Transforming Genes of BPV1

5.1.1 Identification of Transforming Gene Regions in the Genome of BPV1

Nakabayashi et al. (1983) first created a series of BPV1 DNA deletion mutants, and correlated the location of the deletions with the capacity of the deleted viral DNAs to induce transformation in mouse cells. They found that both the upstream end, containing the URR, and the downstream end of the 69% fragment between nucleotides 2113 and 4451 were required for fibroblast transformation. These two fragments had to be physically linked together for efficient focus formation. The URR fragment could be substituted by a retroviral long terminal repeat (LTR) promoter-enhancer region, indicating that the BPV URR contains regulatory signals. Restriction enzyme analysis of the resulting transformants revealed that, in contrast to cells transformed with the entire 69% fragment, all BPV1 DNA sequences in cells transformed with smaller fragments comigrated with high molecular weight cellular DNA, and yielded integration-specific junction fragments upon further analysis (see M. Dürst, this volume). These results demonstrated that plasmid replication is not an absolute requirement for cell transformation. BPV plasmid replication in the absence of transformation was subsequently demonstrated by Lusky and Botchan (1984; see Sect. 5.2).

Surprisingly, when Schiller and Lowy (unpublished results quoted in Schiller et al. 1984) generated a series of linker insertion mutations in the entire wild-type genome, they found that many of them reduced focus-forming efficiency, but no single mutation inactivated transformation completely. Further dissection demonstrated the existence of a second, noncontiguous gene in the upstream part of the early coding region of BPV1, which can also independently transform cells (Schiller et al. 1984). Therefore, the genome of BPV1 has two, noncontiguous, transforming gene regions. Each gene transforms mouse cells alone, and their combined effect is cumulative in cotransfection experiments. However, to detect transformation with BPV E gene fragments smaller than the 69% T fragment, these have to be ligated to functional promoter-enhancer elements (Nakabayashi et al. 1983; Sarver et al. 1984; Schiller et al. 1984, 1986; Yang et al. 1985a, b). Figure 2 shows the morphology of C127I cell foci induced after transfection with wild-type BPV1 (A), and either the E6-E7 fragment (B) or the E2-E5 fragment (C) under control of the murine sarcoma virus LTR (L. P. Turek, unpublished data).

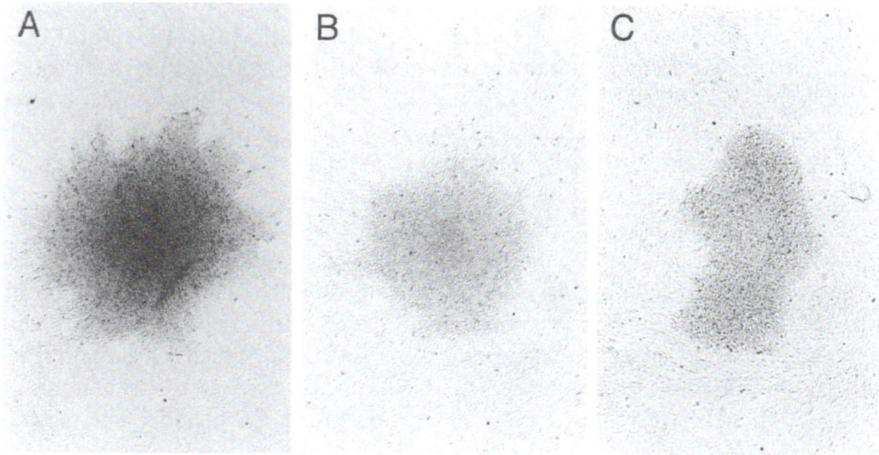

Fig. 2A-C. Transformation of C127I cells by BPV1 and its subgenomic domains. Molecular clones containing wild type BPV1/pML (**A**), the upstream (E6-E7; **B**), or the downstream (E2-E5; **C**) transforming domains of BPV1 under control of the Moloney murine sarcoma virus long terminal repeat were transfected into C127I cells by the calcium precipitation procedure. Microphotographs of transformed cell foci on flat C127I cell background were taken 3 (**A**) or 4 (**B, C**) weeks later

5.1.2 The Upstream Transforming Gene, *E6*

Genetic Mapping. The minimum size of the upstream transforming gene was initially mapped to a fragment between nucleotide 1 (*Hpa*I) and 945 (*Sma*I). The upstream transforming gene region contains the E6 and E7 ORFs. The mRNAs transcribed from this genomic segment encompass a species that could encode an E6 peptide alone as well as a spliced mRNA that would lead to translation of a fused E6/7 gene product (Stenlund et al., 1985; Yang et al., 1985a; Berg et al., 1986a). Schiller et al. (1984) demonstrated by testing a series of molecular mutants that the intact E6 ORF is required for transformation by this BPV region. Fragments carrying C-terminal mutations in the E6 ORF that do not interfere with correct E6/7 mRNA splicing do not lead to transformation (Schiller et al. 1984; Berg et al. 1986a). Yang et al. (1985a) and Berg et al. (1986a) confirmed that cDNA clones corresponding to the spliced E6/7 mRNA do not lead to C127I cell transformation, whereas the unspliced, colinear E6 cDNA clones do. The E6 transforming gene alone does not lead to visible focus formation in one of the standard recipient cell lines used to detect BPV transformation, NIH-3T3 (Schiller et al. 1984). The underlying mechanism of the differential response to E6 transformation in these immortalized cells remains to be determined.

The E6 Gene Product. A protein product of the E6 ORF has been recently identified in transformed cells. Androphy et al. (1985) have raised rabbit antisera against the E6 ORF protein obtained by molecular expression in bacteria. The immune sera precipitate a 15.5 K protein from cells transformed by wild-type BPV1. The size corresponds to the predicted gene product of the E6 ORF, initiated at the first internal methionine (translation start codon) and not subject to posttranslational modifications. Cells transformed by the E6-E7 fragment under the control of a strong retroviral promoter contain higher levels of the peptide than cells transformed by wild-type BPV. The E6 product was found only in its monomeric form even under nondenaturing conditions, suggesting that the protein is not in disulfide linkage with other E6 molecules or other peptides. Cell fractionation experiments, which were performed on lysates of cells transformed with an LTR-driven E6 ORF vector, localized the protein to both the nuclear and the membrane fraction. As these cells overexpress the E6 gene product, its high quantities may result in abnormal cellular localization. Alternatively, its dual localization may reflect its different biological functions in transformation and in the regulation of plasmid copy numbers and/or of gene expression (see Sect. 5.2.3). Which E6 protein fraction is responsible for transformation will need to be determined by the analysis of mutants with an altered intracellular localization of the E6 peptide. The mechanism of transformation by the E6 gene product remains to be explored.

5.1.3 The Downstream Transforming Gene, *E5*

Genetic Mapping. The downstream transforming function was initially mapped to a fragment between nucleotides 2113 (*Eco*RI site) and 4451 (*Bam*HI) which contains four ORFs, E2 through E5 (Chen et al. 1982; Danos et al. 1983; Ahola et al. 1983; see Fig. 1). The E2 ORF was originally considered to code for a candidate transforming gene product; however, a recent, detailed, and elegant mutational analysis performed in several laboratories unequivocally localized the transforming activity to the short E5 ORF (Yang et al. 1985b; Schiller et al. 1986; Groff and Lancaster 1986; DiMaio et al. 1986). In contrast, the E2 ORF encodes a *trans*-activator function (Spalholz et al. 1985; Yang et al. 1985b; Haugen et al. 1987), which activates the transcription of the upstream early promoter at nucleotide 89 (Haugen et al. 1987; see Sect. 5.3). An E5 ORF function is, however, also required for BPV1 plasmid replication under some conditions (Groff and Lancaster 1986; see Sect. 5.2.3).

The transforming activity of the downstream early region requires translation of the intact E5 ORF (Yang et al. 1985b; Schiller et al. 1986; Groff and Lancaster 1986; DiMaio et al. 1986; Haugen et al. 1987). DiMaio, Guralski, and Schiller (1986) have shown that the transforming activity of frameshift mutants in the E5 ORF can be restored by a second frameshift mutation which puts translation of the peptide back into the correct reading frame. As our picture of BPV transcription is still incomplete, it is unclear whether the downstream transforming gene product is translated from the E5 ORF in its entirety, or whether the E5 sequence encodes a peptide domain combined with a different terminal peptide sequence from another ORF by mRNA splicing. The latter possibility appears less likely, as Schiller et al. (1986) and Schlegel et al. (1986) were able to demonstrate that an E5 ORF BPV DNA fragment alone can transform cells when linked to a retroviral LTR promoter.

The E5 Gene Product. Schlegel and coworkers (1986) have identified the E5 transforming protein by immunoprecipitation with an antiserum against a synthetic peptide corresponding to the 20 C-terminal amino acids of the E5 ORF. The E5 polypeptide is 7 K in size and, as such, the smallest viral transforming protein that has been characterized so far. As predicted from its small size, the E5 transforming gene product could constitute an altered growth factor molecule; however, the peptide does not appear to be secreted from the cell (Schlegel et al., personal communication). If the authentic E5 methionine codon at BPV nucleotide 3879 is used for translation initiation, the resulting protein would be a short hydrophobic peptide of 44 amino acids containing a 27-amino acid alpha-helical part.

Although the predicted molecular size of the E5 ORF product is 6 K, the precipitated peptide has the electrophoretic mobility of 7 K. The reason for this apparent increase in size is unknown. The peptide does not appear to be phosphorylated (Schlegel et al. 1986) or modified by acylation (R. Schlegel, personal communication). Other posttranslational modifications of the E5 protein have not been excluded so far. As the peptide is membrane-associated, the hydrophobic alpha-helical domain may provide the transmembrane anchor portion. Complexes of approximately 14 K (possibly from disulfide bonding between C-terminal cysteines) have been identified under nonreducing conditions in the transformed cells, and it is further possible that this apparent dimeric molecule forms a complex with a cellular protein of approximately 75 K which needs to be further characterized (R. Schlegel, personal communication). Identification and characterization of the complex-forming cellular protein may well be the key to the understanding of the transforming function of the BPV E5 protein.

5.1.4 Cellular Factors Involved in Transformation

While the transforming genes of BPV1 and their peptide products have been identified, their mode of action remains unknown. Neoplastic transformation is the ultimate result of changes in cellular functions that control orderly cell division. To determine possible alterations in the pattern of cellular proteins in BPV1 transformation, R. Levenson, L. Chow, T. Broker, and D. Young have developed a large-size 2-dimensional polyacrylamide gel system which can resolve over 3000 cellular proteins. This technique revealed additional peptides in BPV-transformed C127I cells in comparison with C127I controls. Cell lines transformed by subgenomic clones of BPV1 contained "induced" peptides which could be correlated with the presence of a specific region of the BPV1 genome. This implies that the expression of BPV genes can contribute to transformation by a distinguishable pathway (Levenson et al., manuscript submitted).

Smith et al. (1986) have examined the effect of inactivation of the cellular proto-oncogene product, c-*ras*-H, on mitotic activity under conditions which allow only transformed cells to divide. These authors have found that microinjection of a neutralizing antibody against c-*ras* partially inhibits S-phase initiation in transformed ID-13 cells, in contrast to SV40 transformants. It is possible that at least one pathway of BPV transformation may involve interaction with the c-*ras* gene product. Analysis of changes in cellular gene expression at the protein or mRNA level, as well as a search for alterations in cellular factors associated with transformation, such as cellular oncogenes, growth factors, and growth factor or hormone receptors, may provide a useful approach to study the mechanism of transformation.

5.2 Genetic Analysis of BPV Plasmid Replication

5.2.1 Stages of BPV Plasmid Replication

As there is no cell culture system to propagate infectious papillomaviruses in vitro, dissection of viral replication functions has been limited to studies on plasmid replication in BPV1-transformed mouse fibroblasts, which is thought to resemble early viral replication in the epithelium in vivo. However, the comparison is based only on the similar, relatively low copy numbers of papillomavirus DNA molecules in the stratum basale and stratum spinosum cells. Formal analysis of replication in epithelial cells remains to be performed.

Papillomaviral DNA replication in vivo is thought to occur in three stages (Fig. 3 A). During the establishment ("E") stage, the infecting papillomaviral genome is amplified to about 10^2 extrachromosomal plasmids per cell nucleus. These copy numbers are maintained (maintenance, "M" stage) in the differentiating layers until as yet undefined cellular events trigger "runaway" replication ("R" stage), leading to amplification of the DNA to more than 10^4 copies per cell, and virion production.

BPV replication in fibroblastic cell lines (Fig. 3 B) is similar to the first two stages seen in warts. A single BPV genome appears to be sufficient to establish infection, based on single-hit kinetics of infection or transfection (Dvoretzky et al. 1980; L. Turek, T. Haugen, and G. Ginder, manuscript submitted) and on the presence of uniquely rearranged, amplified BPV plasmids ("plasmid clonality") in a minority of individual, transformed, cell clones (Berg et al. 1986a; Turek et al., manuscript submitted). In the E stage, the infecting BPV genome is amplified to around 10^2 copies per cell; therefore, it has to undergo a burst of rapid replication that proceeds independently of cellular DNA synthesis. After this initial stage, BPV plasmids are maintained indefinitely at stable copy numbers in transformed cells (M stage). Genetic analysis has defined several viral functions necessary for correct plasmid replication during the E and M stages, but the viral proteins involved have so far eluded detection.

5.2.2 Identification of *Cis* Elements Required for BPV Plasmid Replication

Requirement for Viral Functions. Replication of DNA viral genomes is usually controlled at the stage of its initiation (Tooze 1981). The process requires specific viral sequences present in *cis* on the viral DNA (replication origin), and at least one specific viral protein in *trans* which interacts with the origin to initiate DNA synthesis. In BPV replication, the expression of viral functions is necessary during both the E and M stages. This was first indicated by the observation that treatment with mouse L-cell interferon inhibits both establishment of BPV infection as well as plasmid maintenance in BPV transformants (Turek et al. 1982). Treatment with the antiviral xanthate compound, D609, has a similar effect (Sauer et al. 1984; Fig. 3 D). Analysis of mutants and subgenomic BPV fragments has confirmed that viral functions are directly involved in replication (see Sect. 5.2.3).

Perhaps surprisingly, mutations in all viral early genes known to encode diffusible factors can impair plasmid establishment or maintenance under specific exper-

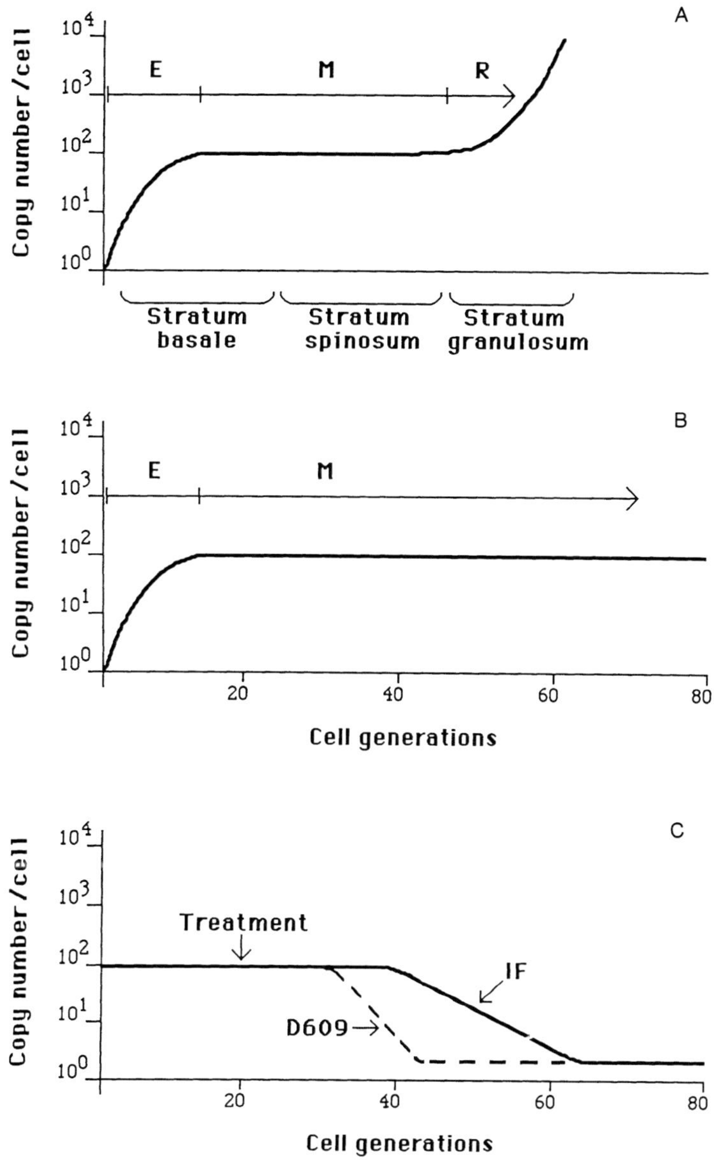

Fig. 3A–C. Papillomaviral replication. Stages of BPV replication in vivo (**A**), in transformed mouse cells (**B**), and after treatment of BPV1-transformed mouse cells with mouse L-cell interferon *(IF)* or the xanthate compound, *D 609* (**C**). *E*, establishment stage; *M*, maintenance stage; *R*, runaway replication stage

imental conditions. While some are likely to abrogate true viral replication functions, mutations in other early genes may influence replication indirectly. For instance, viral mutants in the *E2* regulatory gene (see Sect. 5.3.2) may disrupt correct transcriptional activation of other genes critically involved in replication, while mutants in the *E5* transforming gene (see Sect. 5.1.3) may prevent the production of necessary cellular factors that are usually made in dividing, but not in stationary, crowded cells. BPV plasmid replication is clearly interdependent with cellular replication at least in the M stage, and the synchronization effect is mediated by yet another viral function (see Sect. 5.2.4).

Cis Sequences in Plasmid Replication. To define functional sequences required in *cis* for the establishment and maintenance of BPV plasmid replication, Lusky and Botchan (1984) inserted different fragments of BPV1 into a plasmid containing the neomycin resistance *(neo)* gene of transposon Tn 5 under control of a retroviral LTR. Transfected mammalian cells expressing the *neo*-coded enzyme, aminoglycoside phosphotransferase (EC 2.7.1.1), grow selectively in media with the aminoglycoside antibiotic, G418 (Colbere-Garapin et al. 1981). While the parental LTR-*neo* plasmid is integrated into the cellular genome, its derivatives containing some, but not other, BPV fragments were maintained as extrachromosomal plasmids in the presence of diffusible (*trans*-acting) viral replication factors in BPV-transformed cells.

Further molecular dissection identified two *cis*-linked "plasmid maintenance sequences" (PMS1 and PMS2) that can independently support plasmid replication of the LTR-*neo* gene (Lusky and Botchan 1984). PMS1 is located in the URR, and roughly coincides with the replication origin that has been mapped by electron microscopy (Waldeck et al. 1984). The active sequences in the origin/PMS1 region (ori D1 and ori D2; see Fig. 5) have been mapped to two segments by linker insertion mutagenesis in transient transfections (Lusky and Botchan 1986a). Origin domain 1 (ori D1), located between nucleotides 6673 and 6848, is unrelated to PMS2 and overlaps with coding sequences for the capsid protein, L1. Origin domain 2 (ori D2) is found between nucleotides 7068 and 7330. There is a region of extensive sequence homology between nucleotides 7103 and 7171 in the ori D2 of PMS1 and between nucleotides 1522 and 1582 in PMS2; the homologous segments are inverted with respect to each other. Lusky and Botchan have shown that both ori domains of PMS1 are required for PMS1-mediated replication, but that ori D1 can be substituted by a strong heterologous enhancer of transcription (see also Sect. 5.3.1).

The biological function of PMS2, which overlaps with the E1 ORF in the mid-early region, remains undefined. While PMS1 may provide the replication origin during stable maintenance of plasmid replication in fibroblasts (M stage), it is possible that PMS2 is utilized during initial plasmid amplification (E stage) or in epithelial cells (e.g., the R stage associated with virion production in the upper epidermal layers in warts). Furthermore, PMS2 may substitute for PMS1 in the replication of certain deletion mutants of BPV1 (Sarver et al. 1984; M. Naruto and M. Botchan, preliminary observations quoted in Lusky and Botchan 1986a).

5.2.3 Viral *Trans*-Acting Functions Involved in Plasmid Establishment and Maintenance

The E1-R (rep) Function. Plasmid replication of BPV1 is impaired by large deletions in the early 69% fragment (Lowy et al. 1980; Nakabayashi et al. 1983; Sarver et al. 1984). Lusky and Botchan (1985) have examined the effect of smaller deletion and frameshift mutants on full-length BPV1 DNA replication in stable cell lines, and assigned them to different complementation groups. Mutations in the downstream part of the E1 ORF completely abrogate BPV1 replication. This group of mutants defines the *rep* (or E1-*R*) replication function of BPV1 (Lusky and Botchan 1985; Berg et al. 1986b). It is interesting that this part of the E1 ORF shows significant amino acid sequence homology to the replication proteins (large "T antigens") of SV40 and polyoma viruses (Clertant and Seif 1984; Seif 1984). The downstream E1 ORF sequences may therefore encode a bona fide BPV1 replication protein. The peptide product(s) of the E1 ORF or mRNA species that could encode such protein(s) still need to be identified. Peptides representing portions of the BPV1 E1 ORF have been prepared in bacterial expression vectors (E. Androphy, D. Lowy, and J. Schiller, personal communication). These peptides should prove useful for the preparation of E1-specific antisera as well as in direct studies on DNA-binding properties of the E1-*R* gene product.

The E6/7 and E6 (cop) Functions. Mutations in the E6 or E7 ORFs reduce BPV1 plasmid copy numbers to about 1-5 molecules per cell and therefore define another genetic function needed for wild-type BPV replication at about 10^2 copies per cell (the "copy number" or *cop* function; Lusky and Botchan 1985; Berg et al. 1986a, b). The E6-E7 region of BPV1 encodes at least two different *trans*-acting gene products, since mutants in the E6 ORF complement those in the E7 ORF in simultaneous cotransfection experiments. One gene is coded for by the spliced E6/7 mRNA (Lusky and Botchan 1985; Berg et al. 1986a, b). The *E6/7* gene is inactivated by mutations in the E7 ORF (Lusky and Botchan 1985) and can be complemented in cotransfections with a molecular clone expressing cDNA complementary to the E6/7 message (Berg et al. 1986a). Berg et al. (1986b) have shown that a mutant disrupting the C-terminal portion of E6, pXH775 (described by Schiller et al. 1984) also has the low copy number phenotype, *cop*. The mutation would not influence the spliced E6/7 gene product, since it is located in the intron for the E6/7 message (Yang et al. 1985a; Stenlund et al. 1985). Furthermore, the E6 and E6/7 *cop* mutants complement each other in cotransfections (Berg et al. 1986b). These findings demonstrate the existence of two separate *cop* gene products, the putative spliced E6/7 peptide and a product of the E6 ORF. Although it appears likely that the E6 *cop* gene is identical to the E6 transforming gene, a formal genetic proof remains to be given.

The predicted secondary structures of the E6 and E6/7 ORF peptides share an interesting feature with other nuclear, nucleic acid-binding proteins from different organisms (see Berg 1986 for review and references). Both ORFs as well as their putative spliced gene products contain cysteine-X-X-cysteine (C-X-X-C) motifs, spaced at regular intervals of 29-33 amino acids (Chen et al. 1982; Danos et al. 1983). The cysteines of adjacent C-X-X-C motifs could form complexes with metal ions such as zinc or be possibly linked by disulfide bonding. Such coupling would generate characteristic loops in the secondary structure of these peptides (Fig. 4). The loops ("fingers") contain numerous basic amino acids, and could form complexes with phosphates of the nucleic acid backbone. The E6 ORF protein could contain two such loops, 29 and 30 amino acids in size, respectively, at a distance of 33 amino acids. An alternative "finger" of 33 amino acids would be formed by linking the two C-X-X-C motifs in the center of the peptide. The E6/7

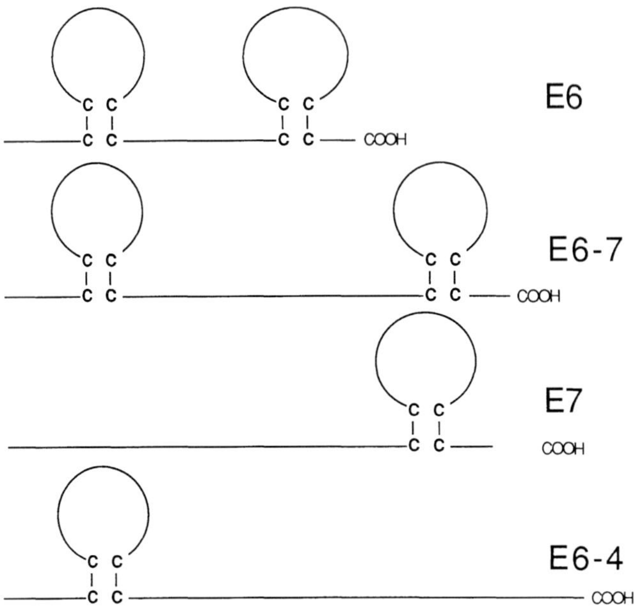

Fig. 4. E6-E7 region proteins predicted from known spliced mRNAs. Potential secondary structure is depicted (see Sect. 5.2.3)

spliced gene product would also contain two such loops, both 29 amino acids long, at a distance of 65 amino acids. A potential E7-coded peptide would share the C-terminal loop with the E6/7 spliced protein, and two additional spliced gene products, E6/4 and E6/1 (Stenlund et al. 1985; Yang et al. 1985a), would contain the N-terminal loop of the E6 peptide. Hypothetically, it is possible to speculate that these proteins consist of functional "building blocks" of identical peptide domains that retain the same function (such as binding to a specific DNA sequence or cellular protein) and that the functional differences are due to different combinations of these elements in the mature proteins encoded by differentially spliced viral mRNAs.

The functional significance of the proposed secondary structure of the putative gene products is unclear. Both the E6 and spliced E6/7 gene products are required for the maintenance of wild-type copy numbers of BPV1 (Berg et al. 1986b), but only the E6 product can transform C127I cells (Schiller et al. 1984; Yang et al. 1985a; Berg et al. 1986a). Molecular clones expressing cDNA copies of the spliced E6/4 or E6/1 mRNAs are also negative in transformation assays (Yang et al. 1985a). It remains to be determined whether the difference in primary peptide composition or in the predicted secondary structure of the E6 and E6/7 gene products are responsible for the observed difference in transforming activity. Furthermore, it is unclear whether the E6 and E6/7 products influence replication directly, perhaps as accessory replication proteins or indirectly by acting in *trans* on transcription of other viral genes. Berg et al. (1986a) have shown that BPV transcription in cell lines expressing an *E6/7⁻* mutant is low, and that it increases after supertransfection with a molecular clone expressing the E6/7 gene product. This result is consistent with either interpretation, since the E6/7 function could increase transcription indirectly via gene copy amplification of the resident genome. To resolve this problem, it will be necessary to test the *E6* and *E6/7* genes for *trans*-activation of specific BPV gene expression. So far, a *cis* element responsive to either the E6 or E6/7 functions has not been identified. One possible function of an E6/7 gene product of the E6-E7 region could be regulation of the replication gene (encoded at least in part in the E1 ORF), either directly or via viral or cellular intermediates.

The E2 Trans-Activator Function. Expression of the product(s) encoded in the E2 ORF is also necessary for plasmid replication in transformed cells (Sarver et al. 1984; Lusky and Botchan 1985; Groff and Lancaster 1986; DiMaio 1986; Kleinert et al. 1986). The E2 ORF encodes a positive *trans*-acting regulatory factor (Spalholz et al. 1985), which activates transcription of an upstream early BPV1 promoter, P89 (Haugen et al. 1987; see Fig. 5) and may therefore influence plasmid replication indirectly by activating transcription of other genes. Since the P89 promoter directs synthesis of mature, spliced, E6 and E6/7 mRNAs, indirect activation of the *cop* gene group most likely contributes to the observed effect of the *E2* gene on replication (see Sect. 5.3.2).

The E5 Transforming Function. Groff and Lancaster (1986) and Rabson, Howley, and coworkers (1986) observed that mutants in the transforming E5 ORF also interfere with BPV plasmid maintenance in stable cell lines derived from foci of transformation or biochemically selected for G418 resistance in cotransfections. Lusky and Botchan (1984), however, found that large deletion mutants in the downstream early ORFs (E2-E5) can establish plasmid replication under G418 colony selection. The E5 ORF transforming gene product may maintain or alter cell proliferation during a stage critical for plasmid establishment under conditions in which most cells in the colony become confluent and cease to divide. This possibility remains to be tested experimentally.

5.2.4 BPV Genes and Sequences Involved in Plasmid Stabilization

In comparison to other small circular viral plasmids in productive infection, as seen for example in the "runaway" replication of SV40 or polyoma virus DNA in permissive cells, the replication of BPV1 genomes in established mouse transformants has unusual features. After the initial amplification of the BPV1 plasmids (E stage, Fig. 3 B), their copy numbers are maintained at steady levels in each individual transformant (M stage). This implies that the established BPV plasmids replicate only once per cell cycle; this assumption has been confirmed experimentally in synchronized cells by Reynolds and Botchan (quoted in Berg et al. 1986b).

The E1-M (Modulator) Gene Function. In the stable replication assays discussed above, no distinction could be made between viral functions required during the E and M stages. To investigate the early phase of plasmid replication, Lusky and Botchan (1986a), and Berg et al. (1986b) tested the behavior of wild-type and mutant BPV1 in transient assays early after transfection. These experiments have shown that in uninfected cells, both wild-type BPV as well as the $E6^-$ and $E6/7^-$ mutants are rapidly amplified within 96 h. The *cop* genes are thus required only for high-copy plasmid maintenance during the M stage. In contrast, amplification of the incoming DNA is restricted in cells which contain already established BPV replicons. This restriction is due to the synthesis of a *trans*-acting factor by the incoming BPV DNA. A series of five frameshift mutants in the 5′-portion of the E1 ORF between nucleotides 945 and 1132 retain their capability to amplify even in the presence of the established BPV genome, but the amplification is suppressed in simultaneous cotransfection with wild-type or cop^- BPV DNA. Therefore, a BPV function, E1-*M*, encoded in part in the 5′-end of the E1 ORF, appears to restrict plasmid amplification during M stage, and may be responsible for synchronizing BPV plasmid replication with cellular DNA synthesis. Several frameshift mutants in the URR (between nucleotides 7613 and 7720) belong to the same complementation group, indicating that a small peptide-coding exon in the URR ("9") encodes a part of the E1-*M* gene product (Lusky and Botchan 1986b).

Negative Control of Replication. As opposed to wild-type BPV1 plasmids whose replication is synchronized with that of cellular DNA in the M stage, plasmids containing the SV40 origin undergo "runaway" (unrestricted) amplification in the presence of the SV40 replication protein (T antigen). DuBridge et al. (1985) made the observation that in C127I cells, chimeric plasmids containing BPV1 and the *cis* origin of SV40 replication (SV-*ori*) replicate like BPV1 and become amplified tenfold in the presence of the SV40 T antigen. Roberts and Weintraub (1986) analyzed the replication of a series of similar recombinant SV-*ori*-BPV constructions in monkey cells expressing the SV40 T antigen (*cos* cells). The presence of BPV1 on the chimeric replicon was found to prevent runaway replication which is typical for SV-*ori*-containing plasmids in cos cells. Instead, the SV-*ori*-BPV plasmids entered M phase, and became stably maintained at about 5×10^3 copies per cell. Genetic dissection of BPV fragments identified two BPV *cis* sequences (the negative control of replication, or NCOR, elements; see Fig. 5) and a *trans*-acting gene responsible for stable plasmid maintenance. The NCOR elements are located close to, but are not identical with the PMS1 and PMS2 elements of BPV1. The *trans*-acting function responsible for the negative control appears to be identical to the E1-*M* function, since it is also encoded at least in part in the 5′-end of the E1 ORF (Roberts and Weintraub 1986). Synchronous replication of BPV plasmids is therefore mediated by a viral function. Its activation, delayed in uninfected cells but accelerated in cells containing established BPV plasmids, is likely to be responsible for the shift between the E and M stages of replication.

5.3 Regulatory Elements and Genes of BPV1

From the genetic experiments discussed above, it is clear that the establishment and maintenance of replication and cell transformation by wild-type BPV1 are the manifestations of a highly coordinated series of events, which is ultimately mediated by correct regulation of viral genes. Transcriptional regulation may not only influence the expression of particular genes, but may also be directly involved in the initiation of replication at the viral origin (Lusky and Botchan 1986a). In turn, the degree of amplification of the viral plasmids determines the number of templates for mRNA transcription in the cell and thus influences their levels. It is therefore not surprising that a number of mutations in the coding region of BPV1 exhibit a pleiotropic effect. Additional difficulty is presented by the low levels of BPV mRNA transcription in transformed cells. The transcriptional map of BPV1 is still incomplete (Stenlund et al. 1985), as mRNAs encoding some functional gene products that have been defined genetically are yet to be found in the cell. Several *cis*-liked elements and a gene that encodes a diffusible, *trans*-acting factor have been identified in genetic experiments, but their precise function in the regulation of specific viral transcripts and in viral infection needs to be elucidated.

5.3.1 *Cis*-Acting Regulatory Elements

Cis-acting "enhancers" are short fragments of viral or cellular DNA that interact with diffusible regulatory factors in the cell; this interaction results in the activation or enhancement of transcription of the adjacent gene (Banerjee et al. 1981; Gruss et al. 1981; Moreau et al. 1981; Gorman et al. 1982; Scholer and Gruss 1984). Although enhancers are frequently located within a short distance upstream from transcriptional promoters, they can function in both orientations in different positions upstream and downstream of the gene.

The Downstream Enhancer. Lusky et al. (1983) described a short "activator" segment at the 3′-end of BPV early genes (the downstream early enhancer, DE; Lusky et al. 1983). As befits a transcriptional enhancer, the DE element was found to activate or enhance the expression of a variety of *cis*-linked heterologous genes in transfected cells: the SV40 T antigen gene used by Lusky et al. (1983), the herpes simplex virus (HSV) thymidine kinase *(tk)* gene employed by Spandidos and Wilkie (1983), or bacterial genes under control of eukaryotic promoters (T. Haugen, T. Cripe, and L. Turek, see below). The DE enhancer can be deleted from some BPV1 plasmid constructions without impairing their ability to transform or replicate in mouse cells (Schenborn et al. 1985; Howley et al. 1985); its role in natural BPV infection remains to be established.

Cis-Regulatory Elements Upstream of the Early Genes. Campo et al. (1983) tested multiple fragments of BPV1 for promoter and enhancer activity upon ligation to the HSV *tk* gene, and long-term selection of tk^{+} clones of transfected Ltk^{-} cells in HAT medium. In addition to the DE enhancer, several fragments increased resistant colony numbers; these were located in the URR and in the body of the early gene region. Several fragments could also substitute for promoter activity under these conditions but were not further characterized.

Gorman and Howard (Gorman et al. 1982) have introduced the use of the bacterial gene, *cat,* encoding chloramphenicol acetyltransferase (EC 2.3.1.28), for enhancer and promoter studies. When ligated to a suitable eukaryotic promoter and 3′ polyadenylation signals, the *cat* gene is actively transcribed in transfected mammalian cells. Expression of the gene is then measured in a convenient enzymatic assay which can be reproducibly quantitated over a 200- to 500-fold range.

Haugen and coworkers have reexamined BPV1 for *cis*-acting regulatory sequences by subcloning BPV DNA fragments into SV40 enhancer-negative, promoter-positive *cat* plasmids downstream of the chimeric gene (T. Haugen, T. Cripe, and L. Turek, manuscript in preparation). In addition to the activator element at the downstream end of BPV early genes (DE enhancer, nucleotides 3838-4451; Lusky et al. 1983; Campo et al. 1983; Spandidos and Wilkie 1983), there are other *cis*-active sequences upstream from the early genes (Fig. 5). The most active of these were found in a 248-n fragment at the downstream part of the late region (LR element, nucleotides 6697-6945), and in a 132-n fragment within the upstream regulatory region (UR element, nucleotides 7143-7275). Both these *cis* elements enhance *cat* gene expression directed by the SV40 early promoter in uninfected cells as well as in BPV-expressing fibroblasts. Although they are located in fragments near the PMS1 segment involved in extrachromosomal plasmid replication (Lusky and Botchan 1984; Waldeck et al. 1984), it is doubtful that these *cis* elements activate gene expression by increasing plasmid copy numbers, since plasmid replication and gene copy amplification of the chimeric plasmids depend on the *rep* function of BPV E1 (Lusky and Botchan 1985, 1986a), which is lacking in uninfected cells. The *cis* elements are located outside of the mRNA-coding sequences in the *cat* vectors. It is therefore probable that they enhance mRNA transcription rather than stabilize mRNA or facilitate its translation, but this assumption remains to be confirmed experimentally.

The LR Element. Although this sequence is located outside of the 69% T fragment, it may be essential for wild-type BPV plasmid replication, as it overlaps the ori D1 element of plasmid replication origin (Lusky and Botchan 1986). BPV vectors require either the authentic late region, the DE enhancer, a heterologous viral enhancer (Lusky and Botchan 1984, 1986a), or a segment of cellular DNA in this position for successful plasmid propagation (Sarver et al. 1981, 1982; DiMaio et al. 1982; Karin et al. 1983); at least one of these cellular DNA fragments contains a bona fide transcriptional enhancer (Haslinger and Karin 1985). It has not yet been formally proved that the LR element is a transcriptional enhancer, as it has only been tested in the downstream location so far. Another possibility is that this fragment contains a strong transcription termination signal that would enhance correct mRNA processing, since Engel et al. (1983) have mapped a 3′-terminus of

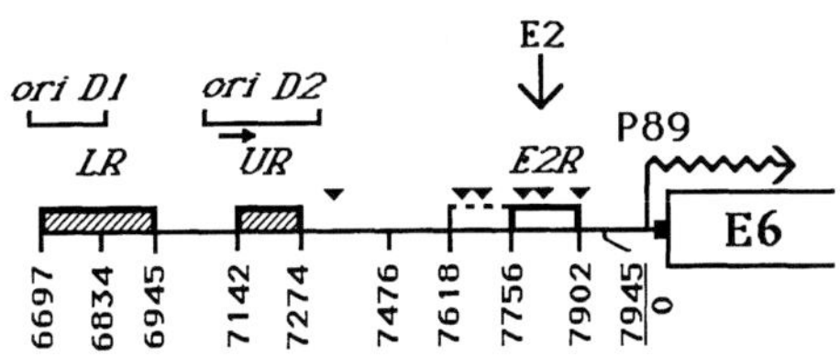

Fig. 5. Upstream regulatory region of BPV1. Nucleotide position of restriction sites used in mapping and functional *cis*-acting regulatory sequences are shown. ▼, papillomaviral palindromic sequences, $ACC(N)_6GGT$; *LR*, late region *cis* element; *UR*, upstream region enhancer; *E2R*, E2-responsive element; *ori D1*, origin of replication domain 1; *ori D2*, origin of replication domain 2 (Lusky and Botchan 1986). *Arrow* in *ori D2* designates negative control of replication (NCOR-1) element (Roberts and Weintraub 1986).

late transcripts to this region. This seems less likely as this fragment functions in both orientations.

The UR Element. This segment appears to be a true enhancer since it functions both downstream and upstream of the gene in both orientations. Furthermore, its presence results in increased mRNA levels transcribed from the E6–E7 promoter, P89 (Haugen et al., in preparation). Fragments adjacent to the LR and UR elements exhibit lower level enhancement, and may contain additional active regulatory sequences. The activity of larger fragments encompassing either the LR or UR elements is reduced. Several positive and negative *cis*-regulatory domains may be active in this region, which contains a predicted promoter (TATAA-box at nucleotide 7203), consensus enhancer core sequences, and polyadenylation signals (Danos et al. 1983; Stenlund et al. 1985; Spalholz et al. 1985).

The E2R Element. Spalholz et al. (1985) have determined that a large fragment of the URR, between nucleotides 6959 and 7945, could substitute for an enhancer of gene expression in the presence, but not in the absence, of a gene product of the E2 ORF of BPV1. Using a series of delections in the P89 promoter region, an enhancerlike element of this early BPV1 promoter that responds to E2 *trans*-activation has been identified (the E2R element; Haugen et al. 1987; see Sect. 5.3.2).

5.3.2 *Trans*-Acting Regulation of Viral Gene Expression: The *E2* Gene Function(s)

Viral gene expression can also be regulated by diffusible, *trans*-acting factors encoded by viral regulatory genes as defined in a number of DNA viruses and retroviruses. As an example, the adenovirus *E1A* gene codes for positive viral *trans*-acting regulatory factor(s), which coordinately regulate a set of five viral promoters and, in addition, stimulate the promoters of cellular *hsp*70 and beta-tubulin genes (see Nevins 1986 for review). Furthermore, an adeno E1A gene product cooperates with other transforming genes to induce neoplastic transformation in primary cells.

Trans-Activation of the Early Promoter, P89. Spalholz et al. (1985) have observed that the E2 ORF of BPV1 encodes a positive *trans*-acting factor that can activate an enhancerlike element in a large, 1000-n long URR fragment. The E2 function specifically *trans*-activates the authentic upstream early promoter with a major 5′ cap site at nucleotide 89 (the P89 promoter; Haugen et al. 1987). Fragments of the URR were linked either to the *cat* or the HSV *tk* gene, and assayed for promoter activity in cotransfections with or without an E2-expressing vector (Fig. 6). The authenticity of the P89 promoter has been verified by genetic dissection as well as by direct mapping of the 5′-end of the chimeric mRNA to the predicted correct transcription initiation site by RNase protection experiments. The E2-mediated *trans*-activation of the P89 promoter results in a more than 35-fold increase in steady state mRNA levels. Although only steady state mRNA was examined, P89 activation is likely to be due to increased transcriptional rates rather than mRNA stabilization, as the E2-mediated, specific increase in mRNA levels depends on the presence of a short *cis*-linked element, the E2R sequence, which is located

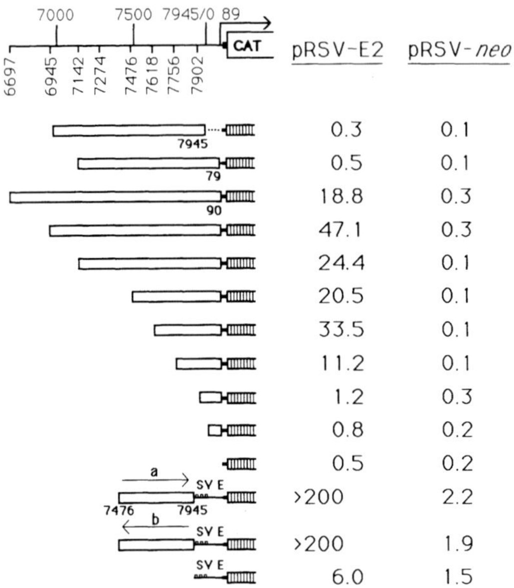

Fig. 6. Deletional mapping of the boundaries of the BPV1 P89 promoter and the E2R target element. Different BPV upstream regulatory region fragments were tested for promoter activity when linked to the *cat* gene in cotransfections with *pRSV-E2*, expressing the E2 function of BPV1, or with pRSV-*neo* as control. The E2R region was examined for enhancer activity linked to the SV40 early promoter (SV E) in the presence and absence of E2. The *cat* gene activity is given in mU × culture^{-1} × h^{-1}

outside of the transcriptional unit. The E2R element maps to nucleotides −277 to −131 upstream of the P89 cap site (BPV nucleotides 7756–7902) and can act as a strong transcriptional enhancer in either polar orientation with the authentic BPV P89 early promoter as well as with two heterologous promoters, the SV40 early promoter and the HSV *tk* promoter. The specific interaction between the *cis* element, E2R, and the diffusible E2 gene product therefore provides a well-defined model for genetic and biochemical studies on mammalian gene activation.

E. Androphy, D. Lowy, and J. Schiller (1987) have determined that a purified E2 peptide prepared by molecular manipulations in bacteria specifically binds in vitro to several restriction fragments of cloned BPV DNA. Two E2 binding sites with the strongest affinity are located within the E2R element that has been mapped by in vivo transfections (nucleotides 7758–7902), and another two are found in the 5′-adjacent URR segment which further potentiates the E2 response (nucleotides 7618–7758; Haugen et al. 1987; see Fig. 6). The core sequence for E2 peptide binding corresponds to the palindrome, A-C-C-$(N)_6$-G-G-T, which is also present in the URR of other papillomaviruses (Schwarz et al. 1983; Dartmann et al. 1986). The URR of HPV6b and HPV16 (T. Cripe, T. Haugen, and L. Turek, unpublished data) as well as the HPV8 URR (R. Seeberger, T. Haugen, L. Turek, and H. Pfister, unpublished data) also specifically respond to the BPV E2 *trans*-activator in enhancer tests. It may well be possible that the specific *trans*-acting E2 function is mediated by direct binding of the E2 ORF gene product to the E2R target element. This would be of great interest since most viral *trans*-acting proteins, such as the E1A proteins of adenoviruses, interact with cellular factors rather than bind directly to specific DNA sequences (Nevins 1986).

Trans-Activation of Heterologous Promoters. The *E2 trans*-activator gene resembles the adeno *E1A* gene in that its product is also capable of nonspecific transcriptional *trans*-activation of several heterologous genes in transient cotransfections by a factor of three- to ten fold (this adeno E1A-like function has been designated "general" *trans*-activation, or E2 function B, by Haugen et al. 1987). Unrelated promoters which appear to respond to E2 function B *trans*-activation so far include the SV40 early promoter with or without the 72-n enhancer repeat, the long terminal repeat enhancer-promoter regions of RSV and MSV, and the HSV1 *tk* promoter. It is unclear whether both specific and general *trans*-functions reside in the same peptide molecule or alternatively are carried out by proteins which result from differential mRNA splicing of the E2 region. In the latter case, general E2 *trans*-activation (function B) might act via another mechanism, such as mRNA stabilization, since it increases the expression of all the chimeric promoter clones tested so far (T. Haugen, T. Cripe, and L. Turek, manuscript in preparation).

One attractive model is that an E2 *trans*-activator protein would have at least two separate domains, one for E2R sequence binding and one for transcriptional activation. E2 function B would then represent the general capability of this E2 gene product to stimulate promoter utilization. Specific binding of the peptide to the E2R element would effectively increase the concentration of the activator domain at the selected promoter, and mediate function A. In at least two other specific *trans*-activators of eukaryotic mRNA transcription, the glucocorticoid receptor (Giguere et al. 1986) and the GCN4 protein of yeast (Hope and Struhl 1986), DNA binding and regulatory domains on the same protein molecule have been identified by site-specific mutagenesis. Since a translation termination mutant, located in the N-terminal portion of the putative E2 peptide, inactivates both specific and general E2 *trans*-activation, it will be of interest to see if these E2 functions can be separated by mutations localized further towards the carboxyl terminus of the E2 ORF. These possibilities need to be resolved in complementation tests with different E2 mutants, and by direct characterization of the physical and binding properties of the wild-type and mutant proteins.

5.3.3 Role of *Trans*-Acting Regulation in Viral Infection

Although the E2 function A appears to be required for P89 transcription and therefore may act as an immediate early gene whose expression precedes the activation of other viral genes, no information is available on the temporal sequence of transcriptional regulation in infection.

As the transcriptional map of the early region of BPV1 is still incomplete, we can only speculate about which viral genes are influenced by E2 action. Viral mRNAs with a transcription start site at nucleotide 89 are spliced to encode an intact E6 peptide, and a potential spliced E6/7 gene product. The E6 protein represents one of the two transforming gene products of BPV1 (Schiller et al. 1984; Yang et al. 1985a; Androphy et al. 1985), and both the E6 as well as the E6/7 gene products are required for positive modulation of plasmid copy numbers in stable plasmid maintenance (the *cop* function; Lusky and Botchan 1985; Berg et al. 1986a, b). This could explain why mutations in the E2 ORF inpair transformation efficiency and abolish unintegrated plasmid replication (Sarver et al. 1984; Lusky and Botchan 1985; DiMaio 1986; Groff and Lancaster 1986; Kleinert et al.

1986). In addition, P89-initiated mRNAs with coding potential for E6/4 and E6/1 proteins of unknown function have been found in BPV-transformed mouse cells (Yang et al. 1985a; Stenlund et al. 1985). Whether the P89 promoter directs the synthesis of less abundant mRNAs representing other genes remains to be determined.

It is predictable that other viral gene functions directly or indirectly restrict or suppress P89-directed transcription, since the levels of P89-initiated mRNAs are low during the M stage in BPV transformants (Stenlund et al. 1985). Furthermore, transient treatment of BPV transformants with the protein synthesis inhibitor cycloheximide results in a ten-fold BPV mRNA increase (Kleiner et al. 1986), suggesting the existence of a labile, negative-regulatory protein in the cells. A P89-specific repressor peptide(s) could share the E2R-binding interaction of the C-terminal domain of the E2 gene product(s) but not have the capability to increase transcription due to a different N-terminal part appended by differential splicing. Preliminary data of Lambert and Howley (personal communication) are compatible with this possibility.

It is also possible that the E2 gene product regulates other viral promoters. First, the E2R element can act as an enhancer at a distance from the promoter (Spalholz et al. 1985). Second, the E2 *trans*-activator gene can modulate mRNA levels of unrelated genes via its function B. Third, Androphy, Lowy, and Schiller (1986) have identified two additional E2 binding domains outside of the region upstream of P89; these may mediate specific regulation of additional viral promoters. Several biological response-modifying agents that increase or decrease BPV transformation or stable plasmid replication, such as the tumor promoter 12-O-tetradecanoyl-phorbol-13-acetate (TPA; Amtmann et al. 1982a), mouse L-cell interferon (Turek et al. 1982), or the xanthate compound D 609 (Sauer et al. 1984), appear to modulate early BPV transcription and could therefore modulate the expression of the E2 *trans*-activator itself, or alternatively, modulate P89 *trans*-activation. It is possible to investigate these alternatives using the transient in vitro transfection systems described here.

Finally, since the E2 *trans*-activator gene(s) have the capability to influence heterologous gene expression, they could potentially contribute to cell transformation by activating cellular genes. The E2 ORF region itself does not transform established mouse cell lines (Yang et al. 1985b; Schiller et al. 1986; Haugen et al. 1986). The molecular constructions defined in these studies can be used to determine whether the E2 *trans*-activator gene product(s) plays a role in cell immortalization or transformation in primary cells in culture.

5.4 Genetic Functions of BPV1 Defined in Transfection Assays: A Recapitulation

Genetic analysis of BPV1 functions in transfected tissue culture cells has recently defined papillomaviral genes and sequences required in cellular transformation, plasmid replication of viral DNA, and regulation of viral gene expression (Table 1). Two distinct early gene domains of BPV1 can transform fibroblasts in culture (Nakabayashi et al. 1983; Schiller et al. 1984). Peptide products of each of

Table 1. BPV1 functions defined in transfection experiments

E6-E7 region:	- cell transformation (E6) - efficient replication (*cop* functions; E6/7, E6)
E1 region:	- plasmid replication (*rep* function, E1-*R*; 3′ E1) - plasmid synchronization (*modulator* function, E1-*M*; 5′ E1)
E2-E5 region	- transcriptional regulation (E2) - cell transformation (E5)
URR:	- origin of plasmid replication - *cis*-regulatory transcriptional elements

these noncontiguous segments of the viral genome can transform cells independently, and their effect is additive (Yang et al. 1985a). The transforming functions have been mapped to the E6 (Schiller et al. 1984) and E5 (Yang et al. 1985b; Schiller et al. 1986; Groff and Lancaster 1986; DiMaio et al. 1986; Haugen et al. 1987) peptide ORFs by mutational analysis. The peptide products of the transforming genes of BPV1 have been identified in immunoprecipitation experiments, using antisera raised against synthetic or bacterially expressed peptide domains. An "E6" protein of the approximate predicted size (15.5K) is found both in the nuclear and cytoplasmic membrane fractions of transformed cells (Androphy et al. 1985). A small "E5" peptide corresponding roughly to the size of the E5 ORF (7 K) has features of a membrane-associated protein (Schlegel et al. 1986).

A finely tuned regulation of expression of the viral transforming genes and of the viral replication and regulatory functions is essential for efficient cell transformation. The establishment and maintenance of plasmid replication requires the E1-*R* (or *rep*) gene, encoded in part in the C-terminal portion of the E1 ORF (Lusky and Botchan 1985). The E1-*M* gene, assigned to the N-terminal half of the E1 ORF, appears to synchronize established BPV plasmid replication with the duplication of cellular DNA (Berg et al. 1986b; Lusky and Botchan 1986b; Roberts and Weintraub 1986). The search for endogenous *cis*-acting regulatory sequences in the BPV genome by molecular manipulations has led to the discovery of several elements that can substitute for transcriptional promoters and enhancers. The promoter upstream of the early gene region, P89, is activated in *trans* by peptide product(s) of the E2 ORF (Haugen et al. 1987). Another potential *trans*-acting regulatory function is encoded in the E6-E7 ORF segment. Mutations in this domain lower plasmid copy numbers (Lusky and Botchan 1985, 1986a, b; Berg et al. 1986a, b), and molecular clones overexpressing this region may *trans*-activate as yet undefined viral or cellular genes, or increase gene expression via enhanced amplification of the BPV genome (Berg et al. 1986a; T. Haugen and L. Turek, unpublished). Viral early gene expression appears to be down-regulated in stable transfectants by at least one labile, diffusible, negative protein factor of either cellular or viral origin (Kleiner et al. 1986). Regulation of viral gene expression, modulated in response to cellular *trans*-acting factors, therefore appears to control the correct temporal switching of viral gene expression in infection and cell transformation. Experiments in the BPV1 system provide a blueprint for meaningful functional studies on the transforming and regulatory capacity of human papillomaviral genomes associated with benign and malignant neoplastic lesions.

6 Genetic Analysis of Other Papillomaviruses

Other animal papillomavirus strains which induce lesions with dermal fibroblast proliferation in animal hosts, such as the ovine, European elk (Stenlund et al. 1983), and deer papillomaviruses (Groff and Lancaster 1985), also transform mouse fibroblasts in culture. Fibroblast transformation in vitro has been ascribed to the ability of these viruses to induce fibroblast proliferation in the natural lesion. However, the correlation between fibroblast transformation in in vitro transfection assays and the induction of dermal fibroblast proliferation in vivo is not absolute, since transformation has been detected with DNAs of some papillomavirus strains which cause warts without fibroblast involvement. The cottontail rabbit (Shope) papillomavirus (Watts et al. 1983), BPV4 (Campo and Spandidos 1983), and more recently HPV1a, HPV5 (Watts et al. 1984), and HPV16 (Yasumoto et al. 1986; Tsunokawa et al. 1986) transform mouse fibroblasts in culture despite the fact that they cause purely epitheliomatous lesions in the natural host. Cellular transformation with these viral strains is frequently less efficient and takes significantly longer. The general organization of these papillomaviral genomes is similar to that of BPV1; however, the nucleotide and predicted amino acid sequences of BPV and other papillomaviruses vary between homologous genomic regions. Therefore, their gene products may play the same role in viral infection as their BPV1 counterparts but may differ in their capability to alter the phenotype of the infected cell. It is also possible that their peptides are functionally equivalent to BPV1 transforming gene products but are inefficiently expressed in the target cell. Clearly, both possibilities could explain differences in the outcome of transfection with different papillomaviral DNAs in vitro. Inefficient transformation has hampered the genetic dissection of transformation by these viral types compared to the BPV1 system. It is likely that genetic tests implementing transfections with predicted functional HPV DNA gene domains or with chimeric molecules between BPV1 and other papillomavirus genomes, as well as the complementation of defined BPV1 mutants, will accelerate progress in this area.

7 Conclusions

Availability of papillomavirus DNAs amplified in bacteria has allowed the identification of papillomavirus genomes in tissues, revealed the existence of multiple viral types, and enabled studies on the structure of the viral genome. Today, the complete nucleotide sequence of three animal and five human papillomavirus types, as well as the partial sequences of several additional viral strains, are known (see H. Pfister and P. G. Fuchs, this volume). Knowledge of the organization of the viral genome has facilitated genetic studies on papillomavirus functions involved in neoplastic transformation of the cell, and those required in the establishment and maintenance of plasmid replication. Molecular manipulations revealed the presence of several *cis*-acting regulatory elements and *trans*-acting regulatory functions. Indirect evidence implicates these in the regulation of viral genes, which is further modulated by cellular, differentiation-specific factors.

Several areas of papillomavirus research are likely to receive much attention in coming years. Studies on the mechanism of viral transformation will be aided by the availability of antisera against viral early proteins which have been obtained with the aid of gene technology. Molecular dissection of viral plasmid replication may reveal features of DNA replication control in the cellular genome. Analysis of *trans*-acting regulation of papillomaviral transcription will help define interactions between *cis* elements and diffusible factors. Elucidation of the correct temporal regulation of papillomaviral genes will aid our understanding of their contribution to cell transformation and of malignant conversion of viral papillomas, both in experimental models and in clinical disease in humans. It is likely that interest in papillomaviruses will continue to expand as more information about their unusual biology and genetics is revealed.

Acknowledgments. The authors would like to thank many colleagues for discussion and for sharing their observations before publication, T. Cripe, W. Lancaster, D. Lowy, H. Pfister, and J. Schiller for critical comments on the manuscript, M. Kucera and C. Clabby for editorial help, and P. Schmid for Macintosh graphics. Experimental work in this laboratory was assisted by C. Ledet, F. Tabatabai, and J. Turk, and aided by grants from the Veterans Administration (VA), and from the University of Iowa Diabetes and Endocrinology Research Center. Both authors receive Research Career Development Awards from the VA.

References

Ahola H, Stenlund A, Moreno-Lopez J, Pettersson U (1983) Sequences of bovine papilloma virus type 1 DNA - functional and evolutionary implications. Nucleic Acids Res 11: 2639-2650

Amtmann E, Sauer G (1982a) Activation of non-expressed bovine papilloma virus genomes by tumour promoters. Nature 296: 675-677

Amtmann E, Sauer G (1982b) Bovine papilloma virus transcription: polyadenylated RNA species and assessment of the direction of transcription. J Virol 43: 59-66

Amtmann E, Volm M, Wayss K (1984) Tumour induction in the rodent *Mastomys natalensis* by activation of endogenous papilloma virus genomes. Nature 308: 291-292

Amtmann E, Muller K, Knapp A, Sauer G (1985) Reversion of bovine papillomavirus-induced transformation and immortalization by a xanthate compound. Exp Cell Res 161: 41-50

Androphy E, Schiller J, Lowy D (1985) Identification of the protein encoded by the E6 transforming gene of bovine papillomavirus. Science 230: 442-445

Androphy E, Schiller J, Lowy D (1987) A peptide encoded by the bovine papillomavirus E2 *trans*-activating gene binds to specific sites in papillomavirus DNA. Nature 324: 70-73

Banerjee J, Rusconi S, Schaffner W (1981) Expression of a beta-globin gene is enhanced by remote SV 40 DNA sequences. Cell 27: 299-308

Berg J (1986) Potential metal-binding domains in nucleic acid binding proteins. Science 232: 485-487

Berg L, Singh K, Botchan M (1986a) Complementation of a bovine papilloma virus low-copy-number mutant: evidence for a temporal requirement of the complementing gene. Mol Cell Biol 6: 859-869

Berg L, Stenlund A, Botchan M (1986b) Repression of bovine papilloma virus replication is mediated by a virally encoded *trans*-acting factor. Cell 26: 753-762

Binetruy B, Meneguzzi G, Breathnach R, Cuzin F (1984) Recombinant DNA molecules comprising BPV-1 DNA linked to plasmid DNA are maintained in a plasmidal state both in rodent fibroblasts and in bacterial cells. EMBO J 1: 621-628

Black P, Hartley J, Rowe W, Huebner R (1963) Transformation of bovine tissue culture cells by bovine papilloma virus. Nature 199: 1016-1018

Boiron M, Levy J-P, Thomas M, Friedman J, Bernard J (1964) Some properties of bovine papilloma virus. Nature 201: 423-424

Campo M, Spandidos D (1983) Molecularly cloned bovine papillomavirus DNA transforms mouse fibroblasts *in vitro*. J Gen Virol 64: 549-557

Campo M, Spandidos D, Lang J, Wilkie N (1983) Transcriptional control signals in the genome of bovine papillomavirus type 1. Nature 303: 77-80

Chen E, Howley P, Levinson A, Seeburg P (1982) The primary structure and genetic organization of the bovine papillomavirus type 1 genome. Nature 299: 529-534

Christy B, Scangos G (1986) In vitro methylation of bovine papilloma virus alters its ability to transform mouse cells. Mol Cell Biol 6: 2910-2915

Ciuffo G (1907) Innesto positivo con filtrado di verrucae volgare. Giorn Ital Mal Venerol 48: 12-17

Clertant P, Seif I (1984) A common function for polyoma virus large-T and papilloma virus E1 proteins? Nature 311: 276-279

Colbere-Garapin F, Horodniceanu F, Kourilsky P, Garapin AC (1981) A new dominant hybrid selective marker for higher eukaryotic cells. J Mol Biol 150: 1-14

Cuzin F, Meneguzzi G, Binetruy B, Cerni C, Connan G, Grisoni M, de Lapeyriere O (1985) Stepwise tumoral progression in rodent fibroblasts transformed with bovine papilloma virus type 1 (BPV1) DNA. In: Howley PM, Broker TR (eds) Papillomaviruses: Molecular and Clinical Aspects. Vol 32. Alan R Liss, New York, pp 473-486

Danos O, Engel L, Chen E, Yaniv M, Howley P (1983) A comparative analysis of the human type 1a and bovine type 1 papillomavirus genomes. J Virol 46: 557-566

Dartmann K, Schwarz E, Gissman L, zur Hausen H (1986) The nucleotide sequence of genome organization of human papilloma virus type 11. Virology 151: 124-130

DiMaio D (1986) Nonsense mutation in open reading frame E2 of bovine papillomavirus DNA. J Virol 57: 475-480

DiMaio D, Treisman R, Maniatis T (1982) Bovine papillomavirus vector that propagates as a plasmid in both mouse and bacterial cells. Proc Natl Acad Sci USA 79: 4030-4034

DiMaio D, Guralski D, Schiller J (1986) Translation of open reading frame E5 of bovine papillomavirus is required for its transforming activity. Proc Natl Acad Sci USA 83: 1797-1801

DuBridge R, Lusky M, Botchan M, Calos M (1985) Amplification of a bovine papillomavirus-simian virus 40 chimera. J Virol 56: 625-627

Dvoretzky I, Shober R, Chattopadhyay S, Lowy D (1980) A quantitative in vitro focus assay for bovine papilloma virus. Virology 103: 369-375

Engel L, Heilman C, Howley P (1983) Transcriptional organization of bovine papillomavirus type 1. J Virol 47: 516-528

Ferenczy A, Mitao M, Nagai N, Silverstein S, Crum C (1985) Latent papillomavirus and recurring genital warts. N Engl J Med 313: 784-788

Freese U, Schulte P, Pfister H (1982) Papilloma virus-induced tumors contain a virus-specific transcript. Virology 117: 257-261

Giguere V, Hollenberg S, Rosenfeld M, Evans R (1986) Functional domains of the human glucocorticoid receptor. Cell 46: 645-652

Gissmann L (1984) Papillomaviruses and their association with cancer in animals and in man. Cancer Surv 3: 161-181

Gissmann L, Schwarz E (1986) Persistence and expression of human papillomavirus DNA in genital cancer. In: Evered D, Clark S (eds) Papillomaviruses, Ciba Foundation Symposium vol 120. Wiley, New York, pp 190-207

Gorman C, Merlino G, Willingham M, Pastan I, Howard B (1982) Rous sarcoma virus long terminal repeat is a strong promoter when introduced into a variety of eukaryotic cells by DNA mediated transfection. Proc Natl Acad Sci USA 79, 6777-6781

Graham FL, van der Eb, AJ (1973) A new technique for the assay of infectivity of human adenovirus 5 DNA. Virology 52: 456-461

Grisoni M, Meneguzzi G, de Lapeyriere O, Binetruy B, Rassoulzadegan M, Cuzin F (1984) The transformed phenotype in culture and tumorigenicity of Fischer rat fibroblast cells transformed with BPV-1. Virology 135: 406-416

Groff D, Lancaster W (1985) Molecular cloning and nucleotide sequence of deer papillomavirus. J Virol 56: 85-91

Groff D, Lancaster W (1986) Genetic analysis of the 3′ early region transformation and replication functions of bovine papillomavirus type 1. Virology 150: 221-230

Gruss P, Dhar R, Khoury G (1981) Tandem repeated sequences as an element of the early promoter. Proc Natl Acad Sci USA 78: 943-947

Haslinger A, Karin M (1985) Upstream promoter element of the human metallothionein-IIA gene can act like an enhancer element. Proc Natl Acad Sci USA 82: 8572-8576

Haugen T, Cripe T, Karin M, Turek L (1987) *Trans*-activation of an upstream early gene promoter of bovine papilloma virus-1 by a product of the viral E2 gene. EMBO J 6: 145-152

Heilman C, Engel L, Lowy D, Howley P (1982) Virus-specific transcription in bovine papillomavirus-transformed mouse cells. Virology 119: 22-34

Hofstadter D (1985) Metamagical themas: questing for the essence of mind and pattern. Basic Books, New York, p 696

Hope I, Struhl K (1986) Functional dissection of a eukaryotic transcriptional activator protein, GCN4 of yeast. Cell 46: 885-894

Howley P, Law M, Heilman C, Engel L, Alonso M, Israel M, Lowy D, Lancaster W (1979) Molecular characterization of papilloma virus genomes. In: Essex M, Todaro G, zur Hausen H (eds) Viruses in naturally occurring cancers. Cold Spring Harbor, New York, pp 233-246

Howley P, Schenborn E, Lund E, Byrne J, Dahlberg J (1985) The bovine papillomavirus distal "enhancer" is not *cis* essential for transformation or plasmid maintenance. Mol Cell Biol 5: 3310-3315

Jainchill J, Aaronson S, Todaro G (1969) Murine sarcoma and leukemia viruses: assay using clonal lines of contact-inhibited mouse cells. J Virol 4: 549-553

Jarrett W, Campo M, Blaxter M, ONeil B, Laird H, Moar M, Sartirana M (1984) Alimentary fibropapilloma in cattle: a spontaneous tumor, nonpermissive for papillomavirus replication. JNCJ 73: 499-504

Karin M, Cathala G, Nguyen-Huu M (1983) Expression and regulation of a human metallothionein gene carried on an autonomously replicating shuttle vector. Proc Natl Acad Sci USA 80: 4040-4044

Kleiner E, Dietrich W, Pfister H (1986) Differential regulation of papillomavirus early gene expression in transformed fibroblasts and carcinoma cell lines. EMBO J 5: 1945-1950

Kushner P, Levinson B, Goodman H (1982) A plasmid that replicates in both mouse and *E. coli* cells. J Mol Appl Genet 1: 527-538

Lancaster WD (1981) Apparent lack of integration of bovine papillomavirus DNA in virus-induced equine and bovine tumor cells and virus-transformed mouse cells. Virology 108: 251-255

Lancaster W, Olson C (1982) Animal papillomaviruses. Microbiol Rev 46: 191-207

LaPorta R, Taichman L (1982) Human papilloma viral DNA replicates as a stable episome in cultured epidermal keratinocytes. Proc Natl Acad Sci USA 79: 3393-3397

Law M, Lowy D, Dvoretzky I, Howley P (1981) Mouse cells transformed by bovine papillomavirus contain only extrachromosomal viral DNA sequences. Proc Natl Acad Sci USA 78: 2727-2731

Law M, Byrne J, Howley P (1983) A stable bovine papillomavirus hybrid plasmid that expresses a dominant selective trait. Mol Cell Biol 3: 2110-2115

Lowy D, Rands E, Scolnick E (1978) Helper-independent transformation by unintegrated Harvey sarcoma virus DNA. J Virol 26: 291-298

Lowy D, Dvoretzky I, Shober R, Law M, Engel L, Howley P-M (1980) In vitro tumorigenic transformation by a defined sub-genomic fragment of bovine papilloma virus DNA. Nature 287: 72-74

Lusky M, Botchan M (1981) Inhibition of SV40 replication in simian cells by specific pBR322 DNA sequences. Nature 293: 79-81

Lusky M, Botchan M (1984) Characterization of the bovine papilloma virus plasmid maintenance sequences. Cell 36: 391-401

Lusky M, Botchan M (1985) Genetic analysis of bovine papillomavirus type 1 trans-acting replication factors. J Virol 53: 955-965
Lusky M, Botchan M (1986a) Transient replication of bovine papilloma virus type 1 plasmids: cis and trans requirements. Proc Natl Acad Sci 83: 3609-3613
Lusky M, Botchan M (1986b) A bovine papillomavirus type 1-encoded nodulator function is dispensable for transient viral replication but is required for establishment of the stable plasmid state. J Virol 60: 729-742
Lusky M, Berg L, Weiher H, Botchan M (1983) Bovine papilloma virus contains an activator of gene expression at the distal end of the early transcription unit. Mol Cell Biol 3: 1108-1122
Mallon R, Sisk W, Defendi V (1986) Expression of the E2 open reading frame of papilloma virus BPV-1 and HPV6b in *Escherichia coli*. Gene 42: 241-251
Melnick J, Allison A, Butel J, Eckhart W, Eddy B, Kit S, Levine A, Miles J, Paganbo J, Sachs L, Vonka V (1974) Papovaviridae. Intervirology 3: 106-120
Meneguzzi G, Binetruy B, Grisoni M, Cuzin F (1984) Plasmidial maintenance in rodent fibroblasts of a BPV1-pBR322 shuttle vector without immediately apparent oncogenic transformation of the recipient cells. EMBO J 3: 365-371
Mitrani-Rosenbaum S, Maroteaux L, Mory Y, Revel M, Howley P (1983) Inducible expression of the human interferon beta 1 gene linked to a BPV DNA vector and maintained extrachromosomally in mouse cells. Mol Cell Biol 3: 233-240
Moar M, Campo M, Laird H, Jarrett W (1981) Persistence of nonintegrated viral DNA in bovine cells transformed *in vitro* by bovine papillomavirus type 2. Nature 293: 749-751
Moreau P, Hen R, Wasylyk B, Everett R, Gaub M, Chambon P (1981) The SV40 72 base pair repeat has a striking effect on gene expression both in SV40 and other chimeric recombinants. Nucleic Acids Res 9: 6047-6068
Nakabayashi Y, Chattopadhyay S, Lowy D (1983) The transforming function of bovine papillomavirus DNA. Proc Natl Sci USA 80: 5832-5836
Nevins J (1986) Control of cellular and viral transcription during adenovirus infection. CRC Crit Rev Biochem 19: 307-322
Orth G (1986) Epidermodysplasia verruciformis: a model for understanding the oncogenicity of human papillomaviruses. In: Evered D, Clark S (eds) Papillomaviruses, Ciba Foundation Symposium vol 120. Wiley, New York, pp 157-174
Ostrowski M, Richard-Foy H, Wolford R, Berard D, Hager G (1983) Glucocorticoid regulation of transcription at an amplified, episomal promoter. Mol Cell Biol 3: 2045-2057
Pfister H (1984) Biology and biochemistry of papillomaviruses. Rev Physiol Biochem Pharmacol 99: 111-181
Pilacinski W, Glassmann R, Krzyzek P, Sadowski P, Robbins A (1984) Cloning and expression in *E. coli* of the BPV L1 and L2 open reading frames. Biotechnology 1: 356-360
Rabson M, Yee C, Yang Y-C, Howley P (1986) Bovine papillomavirus type 1 3' early region transformation and plasmid maintenance functions. J Virol 60: 626-634
Roberts J, Weintraub H (1986) Negative control of DNA replication in composite SV40-bovine papilloma virus plasmids. Cell 46: 741-752
Rösl F, Waldeck W, Sauer G (1983) Isolation of episomal bovine papillomavirus chromatin and identification of a DNase I-hypersensitive region. J Virol 46: 557-566
Rösl F, Waldeck W, Zentgraf H, Sauer G (1986) Properties of intracellular bovine papillomavirus chromatin. J Virol 58: 500-507
Sarver N, Gruss P, Law M-F, Khoury G, Howley P (1981) Bovine papilloma virus deoxyribonucleic acid: a novel eukaryotic cloning vector. Mol Cell Biol 1: 486-496
Sarver N, Byrne J, Howley P (1982) Transformation and replication in mouse cells of a bovine papillomavirus-pML2 plasmid vector that can be rescued in bacteria. Proc Natl Acad Sci USA 79: 7147-7151
Sarver N, Rabson M, Yang Y, Byrne J, Howley P (1984) Localization and analysis of bovine papillomavirus type 1 transforming functions. J Virol 52: 377-388
Sarver N, Muschel R, Byrne J, Khoury G, Howley P (1985) Enhancer-dependent expression of the rat preproinsulin gene in bovine papillomavirus type 1 vectors. Mol Cell Biol 5: 3507-3516
Sauer G, Amtmann E, Melber K, Knapp A, Muller K (1984) DNA and RNA virus species are inhibited by xanthates, a class of antiviral compounds with unique properties. Proc Natl Acad Sci USA 81: 3263-3267

Schenborn E, Lund E, Mitchen J, Dahlberg J (1985) Expression of a human U1 RNA gene introduced into mouse cells via bovine papillomavirus DNA vectors. Mol Cell Biol 5: 18-26

Schiller J, Vass W, Lowy D (1984) Identification of a second transforming region in bovine papillomavirus DNA. Proc Natl Acad Sci USA 81: 7880-7884

Schiller J, Vass W, Vousden K, Lowy D (1986) E5 open reading frame of bovine papillomavirus type 1 encodes a transforming gene. J Virol 57, 1-6

Schlegel R, Wade-Glass M, Rabson M, Yang Y-C (1986) The E5 transforming gene of bovine papillomavirus encodes a small, hydrophobic polypeptide. Science 233: 464-467

Scholer H, Gruss P (1984) Specific interaction between enhancer-containing molecules and cellular components. Cell 36: 403-411

Schwarz E, Durst M, Demankowski C, Lattermann O, Zech R, Wolfsperger E, Suhai S, zur Hausen H (1983) DNA sequence and genome organization of genital human papillomavirus type 6b. EMBO J 2: 2341-2348

Schwarz E, Freese U, Gissmann L, Mayer W, Roggenbuck B, Stremlau A, zur Hausen H (1985) Structure and transcription of human papillomavirus sequences in cervical carcinoma cells. Nature 314: 111-114

Seif I (1984) Sequence homology between the large tumor antigen of polyoma viruses and the putative E1 protein of papilloma viruses. Virology 138: 347-352

Shope R (1933) Infectious papillomatosis of rabbits; with a note on the histopathology. J Exp Med 58: 607-624

Smith M, DeGudicibus S, Stacey D (1986) Requirement for c-*ras* protein during viral oncogene transformation. Nature 320: 540-543

Spalholz B, Yang Y, Howley P (1985) Transactivation of a bovine papilloma virus transcriptional regulatory element by the E2 gene product. Cell 42: 183-191

Spandidos D, Wilkie N (1983) Host specificities of papillomavirus, Moloney murine sarcoma virus and simian virus 40 enhancer sequences. EMBO J 2: 1193-1199

Steinberg B (1986) Laryngeal papillomatosis is associated with a defect in cellular differentiation. In: Evered D, Clark S (eds) Papillomaviruses, Ciba Foundation Symposium vol 120. Wiley, New York, pp 208-220

Stenlund A, Moreno-Lopez J, Ahola H, Pettersson U (1983) European elk papillomavirus: characterization of the genome, induction of tumors in animals, and transformation in vitro. J Virol 48: 370-376

Stenlund A, Zabielski J, Ahola H, Moreno-Lopez J, Pettersson U (1985) Messenger RNAs from the transforming region of bovine papilloma virus type I. J Mol Biol 20 182: 541-554

Thomas M, Borion M, Tanzer J, Levy JP, Bernard J (1964) In vitro transformation of mouse cells by bovine papilloma virus. Nature 202: 709-710

Tooze J (1981) DNA tumor viruses. Cold Spring Harbor Laboratory, New York

Tsunokawa Y, Takebe N, Kasamatsu K, Terada M, Sugimura T (1986) Transforming activity of human papillomavirus type 16 DNA sequences in a cervical cancer. Proc Natl Acad Sci 83: 2200-2203

Turek L, Byrne J, Lowy D, Dvoretzky I, Friedman R, Howley P (1982) Interferon induces morphologic reversion with elimination of extrachromosomal viral genomes in BPV-transformed cells. Proc Natl Acad Sci USA 79: 7914-7918

Waldeck W, Rosl F, Zentgraf H (1984) Origin of replication in episomal bovine papilloma virus type 1 DNA isolated from transformed cells. EMBO J 3: 2173-2178

Watts S, Ostrow R, Phelps W, Prince J, Faras A (1983) Free cottontail rabbit papillomavirus DNA persists in warts and carcinomas of infected rabbits and in cells in culture transformed with virus or DNA. Virology 125: 127-138

Watts S, Phelps W, Ostrow R, Zachow K, Faras A (1984) Cellular transformation by human papillomavirus DNA in vitro. Science 225: 634-636

Weiher H, Botchan M (1984) An enhancer sequence from bovine papilloma virus DNA consists of two essential regions. Nucleic Acids Res 12: 2901-2916

Yang Y-C, Okayama H, Howley P (1985a) Bovine papillomavirus contains multiple transforming genes. Proc Natl Acad Sci USA 82: 1030-1034

Yang Y-C, Spalholz B, Rabson M, Howley P (1985b) Dissociation of transforming and trans-activation functions for bovine papillomavirus type 1. Nature 12: 575-577

Yasumoto S, Burkhardt A, Doniger J, DiPaolo J (1986) Human papillomavirus type 16 DNA-induced malignant transformation of NIH 3T3 cells. J Virol 57: 572-577

Zinn K, DiMaio D, Maniatis T (1983) Identification of two distinct regulatory regions adjacent to the human beta-interferon gene. Cell 34: 865–879
zur Hausen H (1985) Genital papillomavirus infections. In: Melnick JL, Ochoa S, Oro J (eds) Viruses, Oncogenes and Cancer, Progress in medical virology vol 32. Karger, Basel, pp 15–21

Transcription of Papillomavirus Genomes

E. Schwarz

Abbreviations

BPV	bovine papillomavirus
CRPV	cottontail rabbit papillomavirus
HPV	human papillomavirus
kb	kilobase pairs
kd	kilodalton
LCR	long control region
NCR	noncoding region
ORF	open reading frame
URR	upstream regulatory region

1 Introduction

Gene expression of papillomaviruses is tightly linked to the state of differentiation of the host epithelial cells. In the basal layer of the epidermis, the viral genome is thought to be present in a few copies, and only a subset of viral genes is expressed, which probably leads to the enhanced cell proliferation. As the cells undergo differentiation, the complete viral replication cycle is turned on, and mature virus particles are synthesised. Presumably due to this intimate association of viral gene expression with the differentiation programme of keratinocytes, no tissue culture system for in vitro propagation of papillomaviruses has yet been established (Taichman et al. 1984).

To understand the role of papillomaviruses in carcinogenesis, insights into the expression and function of papillomaviral genes and their regulation by viral and host cell factors are obligatory. In particular, two animal papillomaviruses have provided valuable model systems for analysis. One is the cottontail rabbit papillomavirus (CRPV) (Shope 1933) which induces papillomas and carcinomas in cottontail and domestic rabbits. The other is the bovine papillomavirus type 1 (BPV1) which belongs to a subgroup of papillomaviruses that induce fibropapillomas and are able to transform rodent cells in vitro. BPV1 has served as a prototype for studying the molecular biology of papillomaviruses. Various functions involved in cell transformation, transcription and replication have been assigned to different parts of the BPV1 genome (see Turek and Haugen, this volume). These data should be transferable at least to some extent to other papillomaviruses since sequence analysis has revealed a very similar genome organisation in all human and animal papillomaviruses analysed so far (see Pfister and Fuchs, this volume). Transcriptional analysis of human papillomavirus types associated with malignant tumors of the genital tract has been greatly facilitated by the finding that cell lines established from cervical carcinomas contain and express HPV16 or HPV 18 DNA.

2 Bovine Papillomavirus Type 1

The structural and functional organisation of the BPV1 genome has been analysed in considerable detail (for reviews see Petersson et al. 1986; Turek and Haugen, this volume). Two independent transforming functions have been assigned to open reading frames (ORFs) E5 and E6 located at the 5′- and 3′-end, respectively, of the early region (Fig. 1). A *trans*-activating factor is encoded by ORF E2 which acts on an enhancer located in the noncoding region and activates transcription from BPV1 promotors. Maintenance of the BPV1 DNA in the transformed mouse cells as an extrachromosomal multicopy plasmid requires an intact ORF E1. The copy number of BPV1 plasmids is regulated by ORF E6–E7 functions. Furthermore, two ORFs, L1 and L2, constituting the late region, code for viral structural proteins. The noncoding region located between the 3′-end of the late region and the 5′-end of the early region harbours several regulatory elements for gene expression and DNA replication. Since evidence was recently found that it may also contain

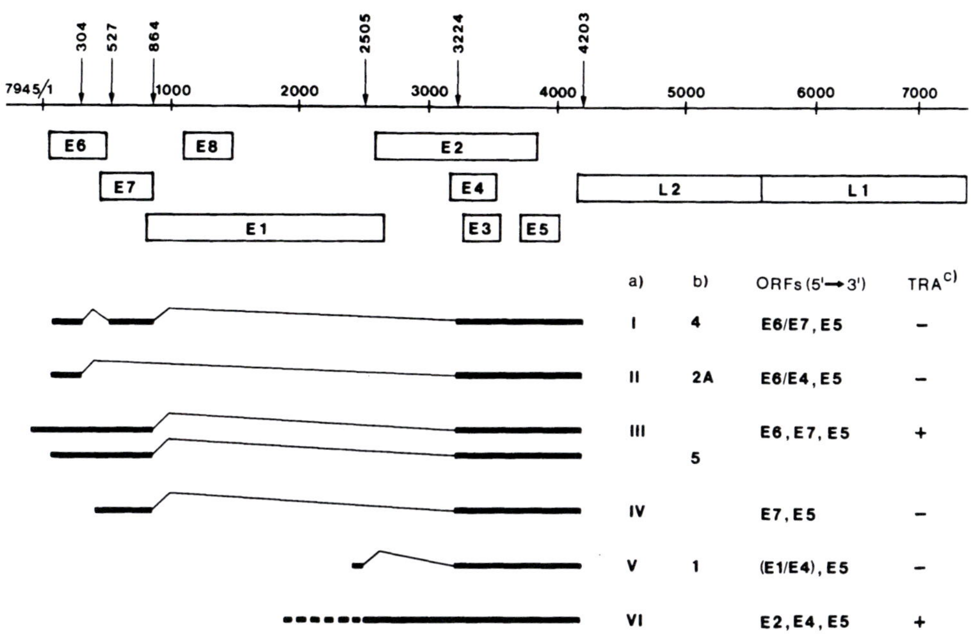

Fig. 1. Structures of BPV1 cDNA clones and mRNA species expressed in BPV1-transformed C127 mouse cells. The genome organisation of BPV1 is given *on top*. Open reading frames are indicated by *open boxes*. The positions of splice sites and of the common polyadenylation site are indicated on the *size scale*. *Bottom*, exon sequences of mRNAs and cDNAs are shown by *heavy lines;* introns by *thin slanted lines*. The heterogeneity of the 5′-ends of class VI cDNA clones (positions 1889–2534; Yang et al. 1985) is indicated by the *dotted line*. Column *a*, cDNA classes as described by Yang et al. (1985); column *b*, mRNA types according to Stenlund et al. (1985). ORFs present in the mRNAs and cDNAs are listed (ORF E3 is not taken into consideration). The transforming capacities *(TRA)* of cDNA clones from the various cDNA classes (Howley et al. 1986) are given in column *c*

protein-coding sequences, it is now often referred to as the upstream regulatory region (URR) or long control region.

The following sections will concentrate on a description of the structures of BPV1 mRNAs detected in transformed mouse cells and in productively infected cells and on their potential protein-coding properties. Genetic analyses performed to identify BPV1 functions involved in DNA replication, gene expression and cell transformation, and the BPV1 proteins already identified in the transformed cells are discussed in detail by Turek and Haugen (this volume).

2.1 BPV1 Transcription in Transformed Mouse Cells

In BPV1-transformed mouse cells, viral mRNAs are present at very low levels, comprising less than 0.01% of the total poly(A)$^+$RNA in these cells (Heilman et al. 1982). Detailed information on the structures of the BPV1 transcripts was obtained by electron microscopic heteroduplex analysis and S1 nuclease analysis (Stenlund et al. 1985) and by cDNA cloning and sequence analysis (Yang et al. 1985).

Several spliced mRNA species were identified which are all transcribed from the early region and have a common poly(A) addition site (map coordinate 53, position 4203) which is located 24 bases downstream from an A-A-T-A-A-A polyadenylation signal (position 4179). The structures of the different mRNA species and cDNA clones are presented in Fig. 1.

A major cap site was determined at nucleotide 89, located only three bases 5′ to the ATG codon of ORF E6, indicating that the T-A-T-A-A-A sequence at position 58 constitutes a component of an "early" RNA polymerase II promoter (the P89 promotor). It could be deduced from the cDNAs and mRNAs that additional promoters are also used: (a) the most common mRNA species (type 1, Stenlund et al. 1985) and the equivalent cDNA clones (class V, Yang et al. 1985; Fig. 1) have 5′-ends mapping to base 2440, indicating that they are initiated at a promoter located within the ORF E1 coding sequence. The TATA-boxlike sequence T-A-A-T-A-T-T is present upstream at positions 2414–2420. Surprisingly, the 5′-terminal nucleotide sequences of the class V DNAs deviate from the BPV1 genomic sequence (Yang et al. 1985). (b) The 5′-end of one cDNA clone is located at nucleotide 7879 and thus 165 bases upstream from the major cap site (position 89).

Several splice donor and acceptor sites were identified (Stenlund et al. 1985; Yang et al. 1985). The major splice donor sites are located at positions 304, 864, and 2505, whereas the major splice acceptor sites are found at positions 527, 557, and 3224. The different mRNA species are generated by joining these splice donor and acceptor sites in different combinations (Fig. 1).

From the structures of the mRNAs and cDNAs, predictions could be made of the proteins possibly encoded by the mRNAs. Assuming that the mRNAs are functionally monocistronic, the following six 5′-proximal ORFs are the most likely candidates for being translated into proteins (Kozak 1983).

(a) ORF E6 is the 5′-terminal ORF in class III cDNA clones (Yang et al. 1985) and in type 5 mRNAs (Stenlund et al. 1985). By mutational analysis, ORF E6 was found to encode a transforming gene that can transform C127 mouse cells but not NIH3T3 mouse cells (Schiller et al. 1984). Furthermore, by the use of specific antisera the 15.5K E6 gene product was directly identified in BPV1-transformed cells (Androphy et al. 1985; Turek and Haugen, this volume).

(b) ORF E7 is the 5′-proximal ORF in one cDNA clone (class IV) (Yang et al. 1985). No equivalent mRNA, however, was detected by heteroduplex and S1 nuclease analysis (Stenlund et al. 1985).

(c) The unspliced class VI cDNAs (with 5′-ends located between nucleotides 1889 and 2534) contain ORF E2 intact and could therefore direct synthesis of a 48K E2 protein (Yang et al. 1985). Spalholz et al. (1985) showed that ORF E2 encodes a trans-activating factor which acts on enhancer elements located in the URR region. Conversion of a TAC codon into a TAG stop codon within ORF E2 was shown to cause a significant decrease in focus-forming activity and the integration of the viral DNA (DiMaio 1986). This indicates a role for the E2 gene product both in transformation and DNA replication, probably indirectly through its *trans*-activation of transcription.

(d) Splicing together nucleotides 304 and 527 generates an E6/E7 fusion ORF that could encode a 21K protein. Berg et al. (1986a) identified a viral function

encoded by the E6/E7 gene which is necessary for high copy number replication of BPV1 DNA and for maintenance of transformation.
(e) Joining nucleotide 304 to 3224 generates a hybrid E6/E4 ORF that could encode a 20K protein.
(f) Splice 2505/3224 results in the fusion of ORF E1 and E4. The fused ORF, however, contains no AUG initiation codon in the class V cDNA clones (Yang et al. 1985).

All the different BPV1 mRNA species contain more than one ORF. Thus, translation may not be restricted to the 5′-terminal ORFs. The following potential protein-coding sequences are located in the middle or 3′-part of the mRNA-/cDNA-species (compare Fig. 1):
(a) ORF E7. The role of the complete ORF E7 (i.e. not spliced to ORF E6; see above) in BPV1 transformation and virus replication is not known. Evidence for a putative involvement of the E7 gene product in maintenance of the malignant phenotype has been obtained for human papillomaviruses HPV16 and HPV18 (see below).
(b) ORF E3. No function has yet been assigned to ORF E3 of BPV1. Since ORF E3 is not conserved in other papillomavirus genomes (Pfister and Fuchs, this volume), its functional importance is questionable.
(c) ORF E4. No function has yet been assigned to ORF E4 of BPV1. There is evidence from HPV1, however, that ORF E4-encoded proteins may play a role in virus maturation (Doorbar et al. 1986; Pfister and Fuchs, this volume).
(d) ORF E5. This is the 3′-terminal ORF of the early region of BPV1 and is present in all BPV1 mRNA species identified in transformed cells. ORF E5 was shown to encode a transforming gene (DiMaio et al. 1986; Groff and Lancaster 1986; Schiller et al. 1986) and the 6K E5 gene product which is encoded entirely within the 3′-terminal half of ORF E5 was recently identified (Schlegel et al. 1986; Turek and Haugen, this volume). It is not yet known, however, which of the different mRNA species direct the synthesis of the E5 transforming protein. A possible candidate seems to be the most abundant 1.2 kilobase (kb) mRNA.

Those mRNA or cDNA species containing a complete ORF E1 could not be detected in analyses by both Stenlund et al. (1985) and Yang et al. (1985). Genetic studies, however, clearly demonstrated that ORF E1 gene products are required in the establishment and maintenance phase of BPV1 plasmid replication; furthermore, the ORF E1 region may encode at least two gene products exerting different functions in BPV1 plasmid replication (Lusky and Botchan 1985, 1986; Berg et al. 1986b; Turek and Haugen, this volume). Thus, E1-encoding mRNAs are expected to be present in the cells but have yet to be identified.

2.2 BPV1 Transcription in Fibropapillomas

BPV1 transcription was also studied in productively infected bovine fibropapillomas. Several transcripts specific for the productive system were detected in addition to those RNA species also present in BPV1-transformed mouse cells (Amtmann and Sauer 1982; Engel et al. 1983). These wart-specific BPV1 RNAs ranged in size from 1.7 to 8.0 kb and had a 3′-terminus located at the 3′-end of the BPV1 late region (Engel et al. 1983). A BPV1-specific 2 kb RNA was described by Amtmann and Sauer (1982) which was found exclusively in the keratinised periphery of the warts, where virus replication takes place.

Detailed insight into the splicing patterns of these RNAs was obtained by construction of a cDNA library from bovine fibropapilloma mRNA and sequence analysis of BPV1-specific cDNA clones (C.C. Baker and P.M. Howley, personal communication). From six cDNA classes which could be distinguished, five had 5′-termini around nucleotide 7250 located in the URR. The upstream sequence G-G-T-A-C-A-C-A-T-C-C (positions 7212–7222) closely resembles the late promoter sequence G-G-T-A-C-C-T-A-A-C-C of the SV40 papovavirus, indicating that the BPV1 sequence element may constitute part of a "late" or "wart-specific" promoter.

A splice donor at nucleotide 7385 defines the 3′-end of the leader exon in three cDNA classes. Four cDNA classes contain sequences from the noncoding region and the early region only and are polyadenylated around position 4200, suggesting that the early poly(A) signal A-A-T-A-A-A (position 4179) is used. The leader exon is spliced to ORF E2–E4 exons with splice acceptors at positions 3225 and 3605, respectively. These mRNAs could direct synthesis of an E4 protein (103 amino acids) and a C-terminal E2 protein (80 amino acids). In two other cDNA classes, ORF E6 represents the 5′-proximal cistron.

Two cDNAs contain ORF L2 and/or L1 sequences; they are terminated using the "late" poly(A) signal A-A-T-A-A-A at nucleotide 7175. The L1-containing cDNA is composed of two leader exons (derived from the noncoding region and the E2–E5 region, respectively) and the L1-coding exon.

In summary, the BPV1 mRNAs transcribed in productive systems differ from those produced in transformed cells in at least two aspects: (a) 5′-termini. A wart-specific promoter located around nucleotide 7220 in the URR is used for initiation of transcription. This promoter seems to be inactive in transformed cells; (b) ORF L1–L2 transcription. Whereas BPV1 mRNAs in transformed mouse cells are terminated at the early poly(A) addition site (base 4203), a subset of RNAs in fibropapillomas contains the information for the L1 and/or L2 viral capsid proteins.

3 Cottontail Rabbit Papillomavirus

The cottontail rabbit papillomavirus (CRPV or Shope papillomavirus) induces benign skin papillomas in its natural host, the cottontail rabbit, and in domestic rabbits (Sundberg, this volume). Progression to malignant carcinoma occurs with a threefold higher frequency in domestic rabbits (approximately 75%) than in the natural host (approximately 25%) and can be enhanced by treatment with carcinogens. Virus production takes place in cottontail rabbit warts whereas it is low or even absent in domestic rabbit warts and completely absent in carcinomas. The Shope papilloma-carcinoma complex thus provides an interesting animal model system with which to analyse papillomavirus functions involved in tumor development and maintenance and the sequential steps in tumor progression (for review see Kreider and Bartlett 1981).

CRPV transcription has been analysed in virus-producing papillomas of cottontail rabbits, in nonvirus-producing benign tumors of domestic rabbits, in malignant tumors, and in the serially transplantable carcinomas of the domestic rabbit, VX2 and VX7, which have been passaged in domestic rabbits for decades. Two major RNA species of 1.25-1.3 kb and 2.0 kb are present in all these cells (Nasseri et al. 1982; McVay et al. 1982; Georges et al. 1984; Nasseri and Wettstein 1984a, b; Phelps et al. 1985). They are transcribed from the CRPV early region, are spliced using a splice donor site at nucleotide 1371 and an acceptor site at nucleotide 3714, and share a common 3′-exon (positions 3714-4367; ORFs E2, E4). Polyadenylation occurs 20 bases downstream from the first A-A-T-A-A-A polyadenylation signal beyond the early ORFs (position 4348) (Danos et al. 1985).

Both RNA species consist of subclasses with heterogeneous 5′-ends (Giri et al. 1985). In the case of the 2.0 kb RNA (which contains 1860 bases of CRPV-specific sequences plus poly(A) tail; Danos et al. 1985), the 5′-termini are located either upstream (position 87) or downstream (positions 157-159) of the first AUG codon (position 154) of ORF E6. Thus, two different E6 proteins (273 and 175 amino acid residues, respectively) could be translated from the 2.0 kb mRNA population. The 5′-end heterogeneity (positions 903, 904, 907, 908, 970, 975) of the 1.25 kb mRNA (1100 bases of CRPV-specific sequences plus poly(A)tail), however, does not affect the coding properties of the RNA subspecies which all could be translated into an E7 protein (first ATG codon of ORF E7 at position 1075; Danos et al. 1985). In addition to the 5′-proximal ORF E6 (2.0 kb RNA) or E7 (1.25 kb RNA), ORF E7 and/or a hybrid E1-E4 ORF are contained within the transcribed sequences (Fig. 2). The latter consists of the 5′-end of ORF E1 (10 nucleotides) fused in frame with ORF E4. It could be translated into a E1-E4 fusion protein if the mRNAs are functionally polycistronic (Danos et al. 1985).

Benign and malignant tumors differ in the relative amounts of the 2.0 kb and 1.25 kb mRNA species present. The 2.0 kb RNA is predominant in benign tumors, whereas the 1.25 kb RNA is the most abundant CRPV-specific transcript in carci-

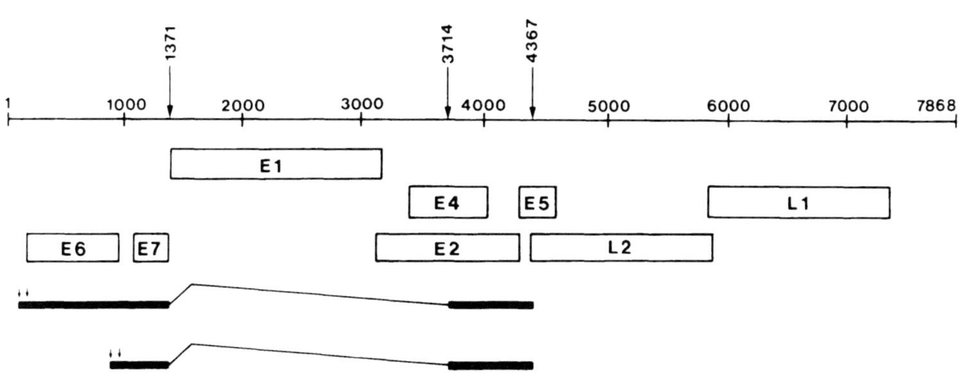

Fig. 2. Genome organisation of the cottontail rabbit papillomavirus and structures of the two major viral transcripts. Open reading frames are shown from the first ATG to the stop codon (open boxes). The positions of splice sites and of the early region polyadenylation site are indicated on the *size scale*. The *small vertical arrows* indicate the heterogeneity of the 5′-ends of the two RNA species

nomas and in the VX2 carcinoma line (Nasseri and Wettstein 1984a; Georges et al. 1984). In the VX7 carcinoma line, a third major transcript of 3.1 kb is present in addition to several minor RNA species (2.4–5.4 kb). The minor RNA species contain sequences from the late region of the CRPV genome and thus can potentially encode CRPV structural proteins (Georges et al. 1984). This could account for the CRPV-specific antibodies found in animals with VX7 carcinoma. In virus-producing papillomas of cottontail rabbits, three specific RNA species of 0.9, 2.6, and 4.8 kb have been detected in addition to the common 1,25 and 2.0 kb RNAs (Nasseri and Wettstein 1984a, b). Since the 3′-exons of the 2.6 and 4.8 kb RNAs cover the late region these two RNA classes are suspected to direct the synthesis of the CRPV capsid proteins. Transcripts of the 4.8 kb size class show a heterogeneity in the splicing pattern of the L-region sequences in that some of them have removed two very small introns located at the 5′-end of ORFs L2 and L1, respectively (Nasseri and Wettstein 1984b). The possible effects for the encoded proteins, however, are not yet known.

The continuous transcription of the integrated CRPV DNA in the two transplantable carcinomas, VX2 and VX7, which have been passaged in domestic rabbits for more than 40 years, is a strong argument for a functional role of certain CRPV genes (most likely E6 and E7) in the maintenance of the malignant state of these cells. The properties of two cell lines which were isolated from the VX2 carcinoma, however, strengthen the idea hat additional (cellular) factors are also important (Georges et al. 1985). One of them, named VX2R, has lost the capacity for serial transplantation in rabbits and has reacquired some of the differentiated functions of epidermal keratinocytes, whereas the other, VX2T, has fully retained the properties of the VX2 tumor. These differences, however, are not due to alterations in CRPV expression since the genome is maintained and transcribed in both cell lines as in the VX2 tumor.

The noncoding region of CRPV was cloned upstream of the chloramphenicol acetyl transferase (CAT) gene to test for promotor activity after transfection of various cell lines with the CRPV-*CAT* plasmids (Giri et al. 1985). Promotor activity was expressed only in the VX2T and VX2R cell lines. All other cells or cell lines tested (of murine or human origin, e.g. C127, HeLa, 293) were negative, as well as rabbit fibroblasts. Low CAT activity was observed in rabbit keratinocytes. Transfection of the VX2 cell lines with a CAT plasmid containing the noncoding region of HPV 1 did not result in any measurable CAT activity. These data indicate the requirement for trans-acting viral and/or cellular factors which activate the CRPV promotor.

4 Human Papillomaviruses

4.1 Human Papillomaviruses Associated with Genital Tract Lesions

DNA sequences of HPV16 and HPV18 have been found in about 70% of human cervical carcinomas, in vulval and penile cancer, and in precursor lesions (see zur Hausen and Schneider 1986 for review). The viral DNA is integrated into the host cell genome in the carcinomas (some of the tumors contain additional, mainly multimeric episomes), whereas it is present as extrachromosomal monomeric episomes in the precursor lesions (Dürst et al. 1985; Dürst, this volume). Two other HPV types, HPV6 and HPV11, are the most prevalent papillomaviruses in benign genital lesions (about 90%) where the HPV DNA is found to persist in an episomal state.

These findings raised several questions:

1. Why are some genital HPV types oncogenic?
2. Is integration of HPV16/18 DNA associated with malignant conversion or tumor progression?
3. Which viral genes are expressed in benign, precursor and malignant lesions?
4. What are the consequences of HPV DNA integration for viral transcription?

4.1.1 HPV6 and HPV11

Data on the transcription of HPV6 and HPV11 are just beginning to emerge. Lehn et al. (1984) have analysed HPV6/11 RNA in condylomata acuminata and in Buschke-Löwenstein's tumors by Northern blot analysis. Furthermore, they have mapped the 3′-segments of the mRNAs in condylomata acuminata by hybridisation of ^{32}P-labelled cDNA probes to subgenomic fragments of HPV6 DNA. Five different RNA species could be distinguished: three "early" transcripts of 1.4 (major RNA species), 1.85, and 2.7 kb, and two "late" transcripts (1.7 and 3.2 kb). The 3′-ends of the late and early mRNAs were mapped downstream from ORF L1 (corresponding to a polyadenylation signal A-A-T-A-A-A located at position 7407) and in the E5 region, respectively. No "classical" polyadenylation signal sequence A-A-T-A-A-A is present downstream from the early region ORFs in both HPV6 and HPV11 DNA (Schwarz et al. 1983; Dartmann et al. 1986).

More detailed insight into the RNA structures was recently gained from electron microscopic heteroduplex analysis of HPV11 mRNAs from a genital condyloma (L. Chow, M. Nasseri and T. Broker, personal communication) and HPV6 RNAs transcribed from a heat shock expression vector (Chow et al. 1986). In each case, several differentially spliced transcripts were detected. The ORF E1 region was spliced out in the majority of the transcripts by using a splice donor site located at the 5′-end of ORF E1 and splice acceptor sites at the 5′-ends of E2 or E4. Some of the RNA species were apparently derived from a putative promoter located near the 3′-end of ORF E7 whereas the 5′-ends of others were mapped at the beginning of ORF E6. The latter is in agreement with the presence of a TATA-box sequence (T-A-T-A-A-A) 36 base pairs (bp) upstream from the first ATG codon of ORF E6 in HPV6 and HPV11 DNA (Schwarz et al. 1983; Dartmann et al. 1986). All HPV6 RNAs transcribed from the heat shock expression vector in COS cells were derived from the HPV6 early region and terminated at the early polyadenylation site.

The precise structure of the major HPV11-specific condyloma transcript was established by cDNA cloning and sequence analysis: the 5′-end is located in *E7*, the first AUG codon is then provided by ORF E1, which is spliced in frame to ORF E4, and the transcript is terminated using the A-G-T-A-A-A polyadenylation site at position 4371 (M. Nasseri, L. Chow and T. Broker, submitted for publication). Two of the HPV11 mRNA species from the condyloma contained ORF L1 and ORF L1 plus L2 sequences, respectively. Thus they are possible candidates for the translation of viral capsid proteins.

4.1.2 HPV16 and HPV18

Analysis of the transcriptional activity of the cancer-associated types HPV16 and HPV18 was greatly facilitated by the finding that human cell lines established from cervical carcinomas harbour these viral DNAs integrated into the host cell genome (Boshart et al. 1984; Schwarz et al. 1985; Pater and Pater 1985; Yee et al. 1985; Schwarz and Schneider-Gädicke 1986). Integration consistently results in a disruption of the papillomavirus genome within the ORF E1-E2 segment, thereby joining the 5′-part of the early region (ORF E6-E7-E1 segment) to downstream host-cell sequences (Fig. 3) (Schwarz et al. 1985; M. Dürst, unpublished data and this volume). This integration pattern is typical not only for these cell lines, but has also been observed in cervical carcinoma biopsies (Dürst et al. 1986; Lehn et al. 1985).

4.1.2.1 HPV18. A comparative analysis of HPV18 transcription in the three cervical carcinoma cell lines, HeLa, C4-1, and SW756, was performed by Northern blot hybridisation (Schwarz et al. 1985) and in greater detail by nucleotide sequence analysis of HPV18-positive cDNA clones isolated from a HeLa, C4-1, and SW756 cDNA library, (Schneider-Gädicke and Schwarz 1986). The results can be summarised as follows:

HPV18 RNAs. Each cell line contains two or three major HPV18-positive poly $(A)^+$RNA species ranging in size from 1.2 to 6.5 kb. The RNAs exclusively hybridise to a subgenomic probe of HPV18 DNA that contains the E6-E7-E1 part of the early region.

Chimeric Structure of cDNA Clones. In all three cell lines, the HPV18-positive cDNA clones were found to be derived from chimeric viral-cellular transcripts that are composed of 5′-terminal HPV18-specific sequences from the E6-E7-E1 part of the early region and 3′-terminal host cell sequences. The viral sequences are spliced to the 3′ cotranscribed cellular sequences by using a splice donor signal located at the 5′-end of HPV18 ORF E1 in combination with cellular splice acceptor sites.

5′-Ends of mRNAs. In the three cell lines the 5′-ends of the HPV18-containing mRNAs were mapped by primer-extension analysis directly at or a few (1-8) nucleotides upstream from the ATG codon of ORF E6. The T-A-T-A-T-A-A-A sequence located 34 bases in front of the ORF E6 ATG thus most likely represents part of an early HPV18 RNA polymerase II promoter active in the cervical carcinoma cells. The primer-extension analysis provided no evidence that another T-A-T-A-T-A-A-A sequence located 81 bases upstream from the E6 ATG codon may also be used. There is, however, preliminary evidence that additional viral or cellular promoters located further upstream may also direct initiation of transcription.

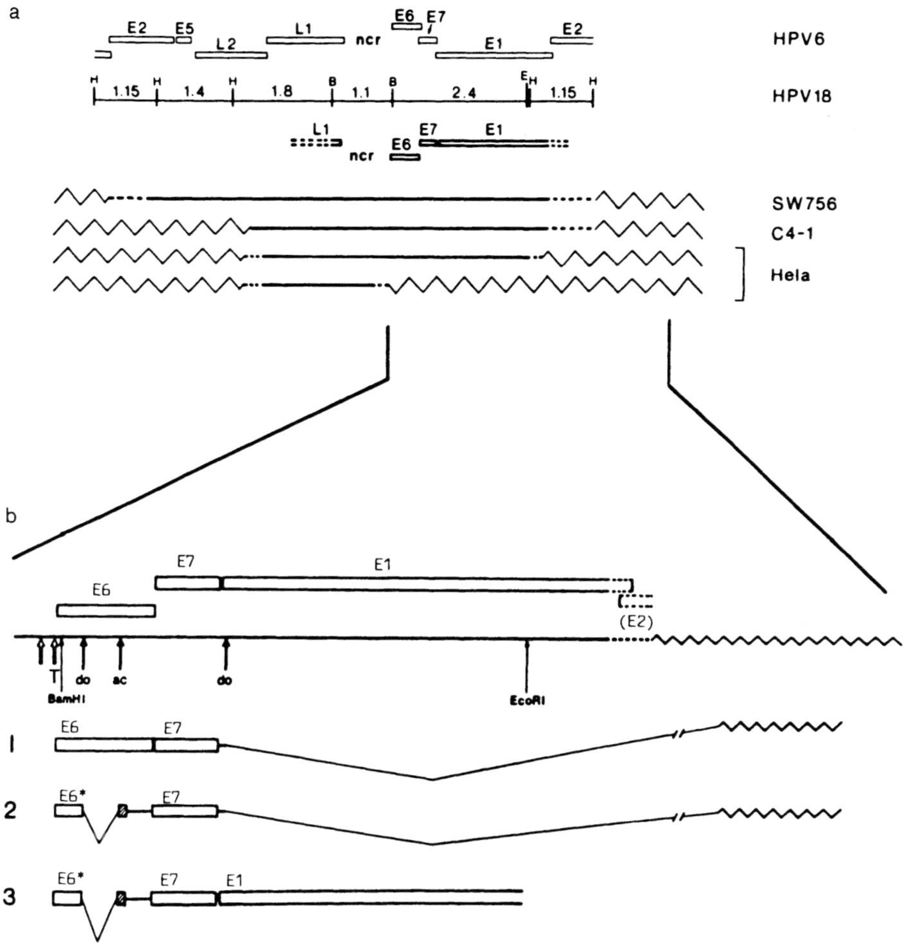

Fig. 3a, b. Integration and transcription patterns of HPV18 sequences in human cervical carcinoma cell lines HeLa, C4-1, and SW756. **a** Integrated HPV18 sequences are indicated by *solid lines* and cellular sequences by *zig-zag lines*. Viral-cellular junction sites are shown by *dotted lines*. A restriction map of the linearised prototype HPV18 DNA (Boshart et al. 1984) is shown *above* indicating the cleavage sites for *Bam*HI *(B)*, *Eco*RI *(E)*, and *Hin*cII *(H)*; sizes of restriction fragments are given in kilobases (kb). Open reading frames and the noncoding region *(ncr)* of HPV18 and HPV6 DNA are shown *below* and *above the restriction map*, respectively. **b** ORF E6-E7-E1 part of integrated HPV18 genome and cDNA structures. ORFs are shown from first ATG to stop codon; *do* and *ac* denote splice donor and acceptor sites, respectively. *Open arrows* indicate TATA-box sequences from which only one *(T)* seems to be part of an active RNA polymerase II promoter (see Sect. 4.1.2). cDNA clones are classified into three types according to the splice patterns in the 5'-terminal HPV18 sequences. The 3'-terminal part of ORF E6* is given as a *hatched box*. The 3' cotranscribed host cell sequences are drawn only schematically. Type 3 cDNAs have been identified in the HeLa cDNA library only and probably represent the 5' part of the major 3.5 kb RNA species which also contains cellular sequences in the 3'-terminal portion (Schneider-Gädicke and Schwarz 1986). This figure is reprinted with permission from Schwarz et al. (1987)

Splice Patterns. Three types of cDNAs/mRNAs can be distinguished according to the different splice patterns found in the 5′-terminal HPV18 segments (Fig. 3 B). In addition to the E1 splice donor, a splice donor and a splice acceptor site both located in the sequence covered by ORF E6 are used in the processing of the mRNAs. Removal of a 182-bp intron generates a new ORF designated E6* in the type 2 and 3 cDNAs. The three types of differentially spliced RNAs are present in rather different amounts relative to each other in the HeLa, C4-1 and SW756 cell lines. Type 3 cDNA (containing ORFs E6*, E7 and E1) is represented by the most abundant 3.5-kb RNA species found in HeLa cells but seems to be rarely, if at all, produced in C4-1 and SW756 cells. On the contrary, HeLa cells probably contain only very small amounts of type 1 RNA (harbouring ORFs E6 and E7; Fig. 3 B).

Potential Protein-Coding Properties. (a) ORFs E6 and E6* represent the 5′-proximal ORFs in the chimeric viral-cellular transcripts, suggesting that they are probably translated into protein. Since translation of the putative E6 and E6* proteins (158 and 57 amino acid residues, respectively) would start at the same AUG codon, their primary structures are identical in the N-terminal 43 amino acids, but diverge at the point where the E6* exon sequences are spliced together. Thus the two proteins would differ in their C-termini both in the number and in the sequence of the encoded amino acids. (b) ORFs E6 and E6* are then followed 3′ by ORF E7 in all three cDNA types and, in the case of the major 3.5-kb RNA species of HeLa cells, even further by ORF E1 (it is not yet known, however, whether the complete ORF E1 is present in this RNA). Therefore, synthesis of an E7 protein in each cell line and of an E1 protein in HeLa cells can be predicted if the mRNAs are functionally polycistronic. The recent identification of an E7 polypeptide in the three cell lines and of an E1 protein translated in vitro from HeLa poly(A)$^+$RNAs (Seedorf et al. 1987) renders the possibility of polycistronic mRNAs likely. (c) The presence of 3′-terminal host cell sequences raised the possibility that the chimeric transcripts may contain information for specific cellular or hybrid viral-cellular proteins. This, however, seems unlikely due to the following observations: no sequence homology has been detected in the 3′ cellular sequences of cDNA clones from the three different cell lines; cDNA clones from the same cell line also contain different 3′ cellular sequences; spliced viral-cellular ORFs (which start with the 5′-terminal 16 nucleotides of HPV18 ORF E1 and continue in frame with cellular sequences) and cellular ORFs in the cDNA sequences have coding capacities of only up to 24 amino acids and 16–87 amino acids, respectively; the cellular ORFs are always preceded by two or even three HPV18-specific ORFs.

In summary, it was concluded from the structural organisation of the chimeric viral-cellular transcripts that they most likely direct the synthesis of HPV18 early gene products E6, E6*, E7, and E1 (in HeLa) and *not* of specific cellular proteins.

E6 Splice Sites.* Sequence comparison revealed that the E6* splice sites are also present in the DNA sequences of HPV16, HPV33 (Cole and Streeck 1986), and HPV31 (A. Lörincz, personal communication). In HPV6 and HPV11, however, the corresponding nucleotide sequences do not conform to the GT-AG rule due to single nucleotide exchanges (Table 1). In agreement with these sequence data, the

Table 1. E6* splice sites in HPV18 DNA and corresponding sequences in other HPV types

Virus	Splice donor	Nucleotide positions	Splice acceptor	Nucleotide positions	Reference
HPV18	GAGGTATTT	226-234	TATTAATAAGGT	401-412	Schneider-Gädicke and Schwarz (1986)
HPV16	GAGGTATAT	224-232	TGTTAATTAGGT	399-410	Seedorf et al. (1985)
HPV31	GAGGTATTA	233-241	TGTTAATTAGGT	408-419	A. Lörincz (personal cummunication)
HPV33	GAGGTATAT	229-237	TATTAATTAGGT	404-415	Cole and Streeck (1986)
HPV6b	GAGATTTAT	225-233	TGCTAATTCGGT	400-411	Schwarz et al. (1983)
HPV11	GAGATATAT	225-233	TGTTAATTCGTT	400-411	Dartmann et al. (1986)
HPV1	GAGAAGCTG	212-220	TTGAACTCCGTT	387-398	Danos et al. (1982)

consensus sequence: - - - - A/C A G ↓ G T A/G A G T - - - - - - - consensus sequence: PyPyPyPyPyPyX C A G ↓ G/T - - - - - - -

exon ✱ intron ✱ exon

E6* splice pattern has been identified in HPV16 RNAs from the CaSki cervical carcinoma cell line (Smotkin and Wettstein 1986; see below) but is absent in HPV6 and HPV11 transcripts (see Sect. 4.1.1). Thus it is tempting to speculate that the E6* splice pattern and consequently the synthesis of a truncated E6* protein may be correlated with the oncogenic properties of genital HPV types. It remains to be determined whether the postulated E6* protein is synthesised in the cell lines. For this purpose, a type 2 cDNA clone from HeLa was used for expression of ORF E6* in *E. coli* and subsequent production of monoclonal antibodies (A. Schneider-Gädicke, unpublished data).

HPV18 DNA Integration and Chimeric Transcripts. Integration of HPV18 DNA within the E1-E2 segment disrupts the early transcription unit by removing the 3′-part (ORFs 2, E4, E5) from the upstream promotor/ORF E6-E7-E1 segment (Fig. 3 A). From RNA analyses of other papillomaviruses, however, the 3′-part is known to harbour signals for processing of the early region transcripts, e.g. splice acceptor sites and the early polyadenylation signal (see Fig. 1 for BPV1, Fig. 2 for CRPV). Production of the chimeric transcripts is thus a consequence of the integration mode of HPV18 DNA in these cell lines. Since HPV DNA integration within the E1-E2 region has also been observed in carcinoma biopsies, it is tempting to speculate that in genital tract lesions certain alterations in HPV gene expression caused by this type of integration event are required for malignant conversion and/or maintenance of the malignant state. These may include (a) abolishing expression of the 3′-ORFs of the early region and/or of the late region ORFs L1 and L2, (b) disconnecting intragenomic regulation mechanisms and/or control elements for viral gene expression, and (c) stabilising the mRNAs encoding the E6/E6*/E7 proteins by the addition of 3′-cellular sequences.

Inhibition of protein synthesis by cycloheximide had no effect on the levels of HPV18 transcripts in HeLa and C4-1 cells, indicating that HPV18 transcription is not controlled by any labile proteins in these cells (Kleiner et al. 1986). This is in contrast to results obtained with BPV1-transformed cells, where cycloheximide treatment led to a tenfold increase in the amounts of viral transcripts due to increased transcriptional activity and prolonged half-lives of the transcripts.

The noncoding region of HPV18 is completely included within a 1.05 kb *Bam*HI restriction fragment framed by the 3′-terminal 205 nucleotides of ORF L1 at the left side and the 5′-terminal 6 codons of ORF E6 at the right side. This *Bam*HI fragment was assayed by Thierry et al. (1986) for promotor activity in different cell lines using the chloramphenicol acetyl transferase (CAT) assay of Gorman et al. (1982). HPV18-CAT constructs containing the *Bam*HI fragment in sense orientation showed CAT activities depending strongly on the cell line used for transfection. Significant activities were observed in HeLa cells and even more in the SW13 cell line (derived from a human adrenocortical carcinoma). CAT activity was also found in SV40-transformed COS monkey cells, whereas a very low or even no activity was detected in HepG2 cells (derived from a human hepatocarcinoma), 293 cells (adenovirus 5-transformed human embryonic fibroblasts), CV1 monkey cells, and rodent cells including C127 and ID13 (BPV1-transformed C127 cells). It was furthermore shown that transcription from the HPV18 promotor can be activated by the SV40 T antigen, whereas it is repressed by the adenovirus E1a gene product.

4.1.2.2 HPV16. HPV16 transcription was analysed in the cervical cancer cell line CaSki and in a cervical carcinoma (Smotkin and Wettstein 1986). CaSki cells contain several hundred copies of exclusively integrated HPV16 DNA whereas the analysed tumor biopsy contained predominantly extrachromosomal dimeric HPV16 DNA in addition to some integrated HPV16 DNA (Smotkin and Wettstein 1986). In both the cell line and the tumour, the HPV16-specific transcripts were derived from the early region (major RNAs of 1.5, 2.3, and 4.5 kb in CaSki cells). A common cap site was mapped at position 97, located between the first and second ATG codon of ORF E6 (positions 83 and 104, respectively). Two different splice patterns in the E6 region were identified. One of them is identical to the E6* splice pattern of the HPV18-positive transcripts in HeLa, C4-1 and SW756 cells. The other (generated by removal of an intron of 266 nucleotides) uses different splice sites and has no equivalent in the HPV18 mRNAs. Thus the HPV16 RNAs differ in their E6-derived 5′-proximal ORFs. Translation would result in three different proteins which would be identical in their N-terminal amino acids. Similar to the structures of the HPV18 RNAs in HeLa, C4-1 and SW756 cells, the sequences of all HPV16 RNAs then include ORF E7 intact and continue without interruption up to a splice donor site located at the 5′-end of ORF E1. Translation of ORF E7 was directly demonstrated by the identification of an E7 protein in CaSki cells (Smotkin and Wettstein 1986; see Sect. 4.1.3).

The E6-E7 containing exons were found to be spliced to exons from the E2-E4 region. This is different to the situation observed in the HPV18 mRNAs where 3′-host cell sequences are spliced to the E6-E7 exons (see above). The 3′-ends of the E2-E4 exons, however, were mapped upstream of the polyadenylation site of the HPV16 early region. This indicates that also in these two cases the viral transcripts may continue into 3′-cellular sequences. Interestingly, this further suggests that in the analysed tumor the mRNAs are *not* transcribed from the extrachromosomal HPV16 copies but more likely from the small fraction of integrated HPV16 genomes (Smotkin and Wettstein 1986). Minor transcripts from the tumor, but not from the CaSki cell line, contain an intact ORF E2 which thus may code for an E2 protein.

Lehn et al. (1985) analysed the transcriptional activity of integrated HPV16 genomes in four cervical carcinoma biopsies. Since HPV16-specific RNAs could be detected in only one of the four tumors, the authors concluded that continuous expression of the viral genome is not required for the maintenance of the malignant state.

In another study, however, Northern blot analysis of RNAs isolated from HPV16 DNA-containing human cervical carcinoma biopsies revealed the presence of HPV16-positive RNA species in 47 of 56 tumor biopsies examined (84%) (G. Orth, personal communication). Furthermore, HPV18-specific transcripts were detected in four of six HPV18-positive cervical carcinomas. Thus these data support the concept that continuous expression of certain papillomavirus sequences is necessary for the maintenance of the malignant phenotype.

4.1.3 Identification of HPV16 and HPV18 Proteins in Cervical Carcinoma Cells

To identify HPV-encoded proteins in cervical carcinoma cells, various ORFs of HPV16 and HPV18 DNA were cloned into prokaryotic expression vectors, and antisera were raised against the different fusion proteins (HPV16 E7: Smotkin and Wettstein 1986; HPV16 E6, E7, E4, L1, and HPV18 E6, E7, L1: Seedorf et al. 1987).

Smotkin and Wettstein (1986) identified a 20K E7 protein by immunoprecipitation in cells of the HPV16-containing CaSki and SiHa cervical carcinoma cell lines. Using Western blot analysis, Seedorf et al. (1987) showed that the E7 protein represents the most abundant HPV protein in cell lines containing HPV16 DNA (CaSki, SiHa) or HPV18 DNA (HeLa, C4-1, SW756). Cell fractionation indicated that E7 is a cytoplasmic protein, with a size of 15K (in HPV16) and 12K (in HPV18).

Antisera directed against HPV16 E6 or HPV18 E6, however, failed to detect any specific proteins in all these cell lines despite transcription of the E6 sequences (see Sect. 4.1.2). Enrichment of viral poly(A)$^+$RNA by hybrid selection and in vitro translation of these RNAs led to the detection of an E6 protein (11K) in CaSki cells, but not in HeLa cells (Seedorf et al. 1987). The failure to detect an E6 protein in HeLa cells may in part be explained by the fact that mRNAs containing the complete ORF E6 are rarely if at all produced in this cell line (Schneider-Gädicke and Schwarz 1986). Furthermore, the synthesis of low or undetectable amounts of E6 proteins in the cell lines may be a consequence of the 5′-terminal structures of the polycistronic mRNAs: the short distances between the 5′-ends of the mRNAs and the position of the E6/E6* AUG codon (0–8 bases in HPV18 mRNAs in HeLa, C4-1, and SW756 cells; 7 bases in HPV16 mRNAs in CaSki cells) may affect initiation of translation in such a way that it is not possible at or is not restricted to the first (E6/E6*) AUG codon. Consequently, the AUG codon of ORF E7 would be recognised more efficiently by the ribosomes.

The complete ORF E6 of HPV16 DNA has a coding capacity for a 19K protein

(Seedorf et al. 1985). The considerably smaller size of the E6 protein found in CaSki cells (11K) suggests that it may be derived from spliced mRNAs.

By in vitro translation of hybrid-selected viral mRNAs, an HPV18 E1 protein (70K) could be identified in HeLa cells (Seedorf et al. 1987). This method furthermore enabled the detection of an E4 protein (10K) in CaSki cells.

4.1.4 HPV18 Transcription in Human Cell Hybrids

The similar transcription patterns of the HPV16/18 ORF E6-E7-E1 region in the cervical carcinoma cell lines HeLa, C4-1, SW756 and CaSki (Schwarz et al. 1985; Schneider-Gädicke and Schwarz 1986; Smotkin and Wettstein 1986) and the identification of some of the encoded gene products (Smotkin and Wettstein 1986; Seedorf et al. 1987) suggests a possible function of the viral proteins for the transformed and malignant phenotype of these cells. A suitable model system for further analysis is provided by human cell hybrids created by fusing HeLa cells with normal human fibroblasts or keratinocytes (Stanbridge 1976; Stanbridge et al. 1982).

In contrast to the malignant HeLa parental cell, the HeLa × normal cell hybrids have a nontumorigenic phenotype, i.e. they do not form tumors in nude mice. Rare tumorigenic segregants, however, have been isolated which are characterised by the loss of specific chromosomes from the normal parental cell (Stanbridge et al. 1981; Stanbridge cited in Hunter 1986). Despite the different growth properties in vivo, all HeLa × normal cell hybrids express transformed properties when grown in tissue culture (e.g. anchorage independence). Thus the major difference of the nonmalignant cell hybrids in comparison with their tumorigenic segregants and HeLa cells is their ability to respond to growth regulatory signals in the intact animal.

Analysis of HPV18 DNA and RNA in these human cell hybrids has revealed so far (Schwarz et al. 1987) that (a) HPV18 DNA is present in the genome of all analysed hybrid cell lines with an integration pattern identical to that of HeLa cells; (b) no differences in the steady state levels of HPV18 RNAs are observed between nontumorigenic and tumorigenic hybrids and HeLa cells when the cells are grown in tissue culture; (c) HPV18 RNAs are expressed in the tumorigenic hybrids when the cells grow in nude mice.

Based on these data, the question can now be addressed whether or not growth arrest in vivo of the nontumorigenic hybrids is associated with suppression of HPV18 gene expression, and, if so which cellular genes are involved.

4.1.5 Cellular Genes in Human Genital Cancer

Cervical carcinoma cells are characterised by (a) the presence of certain types of HPV DNA, mostly integrated into the host cell genome and (b) the transcription of viral early genes (E6, E6*, E7). Several lines of evidence, however, clearly indicate that HPV infection itself is not sufficient for malignant transformation. Usually there is a long latency period (up to 30 years) between HPV infection and cancer appearance. Only a small proportion of infected persons develop a malignant tumor and there seems to be a high frequency of inapparent HPV infections (see Gissmann et al. 1987). Furthermore, the malignant tumors are monoclonal.

Based on these observations, a model was proposed postulating that human genital cancer results from elimination of intracellular surveillance mechanisms for papillomavirus gene expression (zur Hausen 1986). In this model, unregulated expression of HPV early genes due to the loss of cellular control functions is considered to represent a key factor in determining the malignant phenotype of cervical carcinoma cells. One experimental system useful for analysing the postulated papillomavirus-host cell interactions is provided by the nontumorigenic and tumorigenic HeLa × normal cell hybrids mentioned above. Another cell system are keratinocyte lines 'immortalised' by transfection with HPV16 DNA, which are nontumorigenic in nude mice (see Sect. 4.4).

Another type of study suggests that activation of cellular oncogenes may be an additional event at least in progression of cervical carcinomas (Riou et al. 1984, 1985). Amplification of the *c-myc* and/or *c-Ha-ras* genes was observed in more than 50% of the tumors analysed, preferentially in those corresponding to stages 2, 3 and 4 of malignant progression. Furthermore, high levels of *c-myc* RNA were found in cervical carcinomas of advanced stages.

Increased steady state levels of cytoplasmic *c-myc* RNA were also found in the HeLa, C4-1 and CaSki cell lines (Marcu et al. 1983; Dürst et al. 1987). In HeLa and C4-1 cells, the integrated HPV18 DNA was mapped to chromosome 8 (A. Mincheva, L. Gissmann and H. zur Hausen, submitted for publication; Dürst et al. 1987). Furthermore, in CaSki cells which carry multiple chromosomal integration sites for HPV16 DNA, two of the HPV16-positive marker chromosomes contain part of chromosome 8 (Mincheva et al., submitted). Thus, it may be possible that at least in some genital tumors *cis* - activation of cellular oncogenes by HPV may contribute to malignant transformation (Dürst et al. 1987).

4.2 HPV1

The structures of HPV1 mRNAs were studied by electron microscopic heteroduplex analyses. The RNAs were derived either from plantar warts or from HPV1-infected human foreskin epithelial cell culture (T. Broker, L. Chow, L. Taichman and S. Reilly, unpublished data). In another set of experiments, HPV1 DNA cloned into expression vectors containing the *Drosophila* heat shock promoter were transfected into COS cells, and HPV1 transcripts produced under control of these surrogate promoters were analysed (Chow and Broker 1984; Chow et al. 1986). In each case several RNA species with two or three exons and different splice patterns were identified. They are polyadenylated either at the early or late polyadenylation site.

Major splice sites are:

1. A splice donor site in the ORF E7-E1 region which most likely constitutes the well-conserved splice donor located a few nucleotides downstream from the putative AUG start codon of ORF E1
2. A plice acceptor site located at the 5′-end of the ORF E4-E2 overlapping segment

3. A splice donor site at the 3′-end of the ORF E4–E2 overlapping region
4. A splice acceptor site located at the 5′-end of ORF L1

Some RNA species were shown to carry a 3′-terminal ORF L1 exon which, however, was always preceded by an E4–E2 exon. Furthermore, in all RNA preparations RNA species were identified which were presumably initiated at a promoter located near the 3′-end of ORF E7.

The most abundant transcripts were represented by RNAs composed of a small 5′-terminal E7–E1 exon (from 3′-end of E7 to E1 splice donor) which was spliced to a 3′-E4(E2) exon. There may be a correlation between these RNA data and the recent findings of Doorbar et al. (1986), who detected ORF E4-encoded proteins in high amounts in HPV1-induced warts. Doorbar et al. suggested that these proteins may play a role in virus maturation and consequently that ORF E4 does not encode an early viral function (see also Pfister and Fuchs, this volume).

4.3 HPV5

In carcinomas from a patient with epidermodysplasia verruciformis, HPV5-specific transcripts were detected. The two major transcripts (1.4 and 1.9 kb) hybridised to a subgenomic probe containing the E6–E7 region (G. Orth, unpublished data, cited in Orth 1986).

4.4 Viral Transcription in Cells Transformed by HPV DNA In Vitro

The extreme usefulness of BPV1 for analysing the molecular biology and transforming activities of papillomaviruses was based on the capacity of BPV1 virus or its cloned DNA to induce transformation of rodent cells in tissue culture. Numerous attempts have been undertaken to establish in vitro transformation systems also with human papillomaviruses. Watts et al. (1984) reported the successful transformation of mouse C127 cells using molecularly cloned HPV5 or HPV1 DNA. HPV5-transformed cells were able to form tumors in nude mice, whereas HPV1-transformed cells were nontumorigenic. Two subgenomic fragments of HPV5 were also able to transform cultured cells although they lacked the noncoding region and part of the E6–E7 segment, suggesting that functions in the 3′-part of the early region were responsible for transformation (Watts et al. 1985). However, no data on the transcriptional activity of HPV5 in the transformed cells have yet been reported.

Malignant transformation of NIH 3T3 mouse cells with HPV16 DNA was reported recently (Yasumoto et al. 1986). In this study head-to-tail dimers of full-length HPV16 DNA cloned into the pSV2-neo vector were used for transfection. Transformed foci were detected after 4 or more weeks, and all cell lines examined were tumorigenic in nude mice. The transformed cells contained HPV16 DNA in a multimeric and probably integrated form with some rearrangements. HPV16-positive transcripts could be detected in the transformed cell lines. It was not determined, however, which parts of the HPV16 genome were transcriptionally active.

In a different approach, total genomic DNA from a HPV16-positive cervical carcinoma biopsy was used for transfection of NIH 3T3 cells (Tsunokawa et al. 1986). Primary and secondary transformants could be isolated and were shown to contain both human-specific Alu sequences and HPV16 DNA. HPV16 transcription was analysed in the primary transformant. A major 1.8 kb poly(A)$^+$RNA together with minor species of 2.7–6.8 kb was detected using total HPV16 DNA as ^{32}P-labelled probe. Detailed analysis of the transcribed HPV16 sequences, however, is required to establish further whether the malignant phenotype of the cells can be attributed to the expression of specific HPV16 genes.

Dürst et al. (submitted for publication) used normal keratinocytes from human foreskin as target cells for introducing HPV16 DNA which was oligomerised prior to transfection. Whereas mock-transfected cells underwent differentiation and senescence typical for normal keratinocytes in culture, continuous growth was observed in the HPV16-transfected cells. These cells contain HPV16 DNA in an integrated form and transcribe viral sequences from the 5′- and 3′-part of the early region into poly(A)$^+$RNA.

Altogether, these tissue culture systems seem to provide new, promising tools to dissect the functions of cancer-associated human papillomaviruses such as HPV16 that are involved in the establishment and maintenance of malignancy and to open the way for genetic studies of these clinically important papillomaviruses.

5 Papillomavirus Transcription: Comparison and Perspectives

Until now, papillomavirus transcription has been analysed in greater detail mainly in three model systems: in BPV1-transformed mouse cells, in CRPV-induced carcinomas, and in HPV16- or HPV18-containing human cervical carcinoma cell lines. The analyses have revealed a greater complexity of viral transcription than was originally anticipated from the genome organisation. Particularly for BPV1, it became clear that the pattern of viral transcripts is still incomplete. For some viral gene functions which have clearly been characterised by mutational analysis, the encoding mRNAs have yet to be identified (e.g. E1).

Viral transcription in the three different papillomavirus systems exhibits several common features:

1. Transcription is restricted to genes of the early region
2. Transcription is initiated at a major early promoter located shortly upstream from ORF E6 (additional promoters, however, are also used and these may differ in the three systems)
3. Splicing of the RNAs mostly involves a splice donor site located at the 5′-end of ORF E1. Thus, ORF E1 sequences are usually absent from the mature transcripts

Comparison of these three papillomavirus systems, however, also revealed major differences:

1. The physical state of the viral DNA is different (episomal monomers in BPV1-transformed cells vs integrated genomes with a disrupted early region in cervical carcinoma cell lines), which has a direct impact on the structures of the viral transcripts and thus also on the viral proteins synthesised in the cells (see Sect. 4.1.2; Dürst, this volume); furthermore, it should affect regulation of viral gene expression and possibly papillomavirus-host cell interactions.
2. The splice patterns affecting ORF E6 are specific for BPV1, CRPV, and HPV16/18 transcripts. In each case, a heterogeneous set of gene products containing ORF E6-encoded amino acids can be predicted. BPV1: intact E6, E6/E7 and E6/E4 fusion proteins (see Sect. 2.1 and Fig. 1); CRPV: no splicing but different 5′-ends of mRNAs result in two E6 proteins with different N-termini (see Sect. 3 and Fig. 2); HPV16/18: intact E6 and one or two spliced variants (named E6* for HPV18) with different C-termini (see Sect. 4.1.2 and Fig. 3). Some functional dissection has already been achieved for the different E6-derived proteins of BPV1 (see Sect. 2.1; Turek and Haugen, this volume). It remains to be determined whether the different E6 proteins of HPV16/18 exert different functions and whether the E6* proteins contribute to the oncogenic properties of genital HPV types as postulated from DNA sequence comparison (see Sect. 4.1.2). In this regard, it is an interesting question whether E6* expression occurs also in the lytic cycle of HPV16/18.

Insights into the functions of viral genes and regulatory elements have mainly evolved from studies of the BPV1 transformation system. Since the genome organisation of papillomaviruses is very similar, it seems likely that mechanisms analogous to those defined in BPV1 studies may also operate for other papillomaviruses. Distinct differences between groups or types of papillomaviruses, however, have also become apparent, thus emphasising the importance of defining functions and mechanisms that are specific for a given papillomavirus-host cell system.

The BPV1 E2 gene product can *trans*-activate a noncoding region enhacer, thus stimulating transcription from the major early (P89) promoter. Work from several groups has shown that the *E2* genes of other papillomaviruses also encode *trans*-activating proteins (unpublished data). The DNA-binding site for the E2 protein of BPV1 has been identified (E. Androphy et al., submitted for publication; Turek and Haugen, this volume) and includes a 12-bp inverted repeat sequence motif A-C-C-G-(N) $_4$-C-G-G-T, which is also present in the noncoding region of other papillomaviruses (Dartmann et al. 1986). In cervical carcinoma cells, HPV DNA integration always seems to eliminate expression of the *E2* gene, yet the steady state levels of viral transcripts initiated at the early promoter are usually higher than in BPV1-transformed cells. Inhibition of protein synthesis leads to a tenfold increase of viral transcripts in BPV1-transformed cells, suggesting the existence of negatively regulating labile proteins (Kleiner et al. 1986). It is not known at present whether the E2 gene product also affects the expression of cellular genes or interacts with cellular factors in regulating viral gene expression. It seems likely that cellular genes control early viral gene expression in the nonmalignant cell, and it has been postulated that inactivation of these cellular control genes may represent an essential step in the development of human genital carcinomas (zur Hausen 1986).

The HPV16/18 protein and cDNA analyses suggest that the viral E7 gene product may be involved in the maintenance of the malignant phenotype. Direct evidence, however, is still missing. The same holds true for the putative functions of the E6/E6* proteins in the establishment and/or maintenance of malignancy.

Despite the considerable insights into viral gene expression in transformed cells, the networks of papillomavirus - host cell interactions during latent infection and in the lytic viral cycle and the perturbations necessary for malignant conversion of the infected cell are an almost complete enigma at present. Detailed comparison of papillomavirus gene expression in benign, precancerous, and malignant lesions is one necessary step to identify essential viral genes. Perhaps even more important and fascinating, however, seems to be the identification of the cellular genes involved.

Acknowledgements. I wish to thank Carl Baker, Tom Broker, Matthias Dürst, Attila Lörincz, and Gerard Orth for communicating results prior to publication. Original work described in this paper was supported by the Deutsche Forschungsgemeinschaft.

References

Amtmann E, Sauer G (1982) Bovine papillomavirus transcription: polyadenylated RNA species and assessment of the direction of transcription. J Virol 43: 59-66

Androphy EJ, Schiller JT, Lowy DR (1985) Identification of the protein encoded by the E6 transforming gene of bovine papillomavirus. Science 230: 442-445

Berg LJ, Singh K, Botchan M (1986a) Complementation of a bovine papilloma virus low-copy-number mutant: evidence for a temporal requirement of the complementing gene. Mol Cell Biol 6: 859-869

Berg LJ, Lusky M, Stenlund A, Botchan MR (1986b) Repression of bovine papilloma virus replication is mediated by a virally encoded trans-acting factor. Cell 46: 753-762

Boshart M, Gissmann L, Ikenberg H, Kleinheinz A, Scheurlen W, zur Hausen H (1984) A new type of papillomavirus DNA and its presence in genital cancer biopsies and in cell lines derived from cervical cancer. EMBO J 3: 1151-1157

Chow LT, Broker TR (1984) Human papillomavirus type 1 RNA transcription and processing in COS-1 cells. In: Pearson ML, Steinberg NL (eds) Gene transfer and cancer, progress in cancer research and therapy, vol 30. Raven New York, pp 125-134

Chow LT, Pelletier AJ, Galli R, Brinckmann U, Chin M, Arvan D, Campanelli D, Cheng S, Broker TR (1986) In: Botchan M, Grodzicker T, Sharp P (eds) DNA tumor viruses: control of gene expression and replication, Cancer Cells, vol 4. Cold Spring Harbor Laboratory, New York, pp 603-614

Cole ST, Streeck RE (1986) Genome organization and nucleotide sequence of human papillomvirus type 33, which is associated with cervical cancer. J Virol 58: 991-995

Danos O, Katinka M, Yaniv M (1982) Human papillomavirus 1a DNA sequence: a novel type of genome organization among papovaviridae. EMBO J 1: 231-236

Danos O, Georges E, Orth G, Yaniv M (1985) Fine structure of the cottontail rabbit papillomavirus mRNAs expressed in the transplantable VX2 carcinoma. J Virol 53: 735-741

Dartmann K, Schwarz E, Gissmann L, zur Hausen H (1986) The nucleotide sequence and genome organization of human papilloma virus type 11. Virology 151: 124-130

DiMaio D (1986) Nonsense mutation in open reading frame E2 of bovine papillomavirus DNA. J Virol 57: 475-480

DiMaio D, Guralski D, Schiller JT (1986) Translation of open reading frame E5 of bovine papillomavirus is required for its transforming activity. Proc Natl Acad Sci USA 83: 1797-1801

Doorbar J, Campbell D, Grand RJA, Gallimore PH (1986) Identification of the human papillomavirus 1a E4 gene product. EMBO J 5: 355-362
Dürst M, Kleinheinz A, Hotz M, Gissmann L (1985) The physical state of human papillomavirus type 16 DNA in benign and malignant genital tumors. J Gen Virol 66: 1515-1522
Dürst M, Schwarz E, Gissmann L (1986) Integration and persistence of human papillomavirus DNA in genital tumors. In: zur Hausen H, Peto R (eds) Viral etiology of cervical cancer, Banbury report 21. Cold Spring Harbor Laboratory, New York, pp 273-280
Dürst M, Croce CM, Gissmann L, Schwarz E, Huebner K (1987) Papillomavirus sequences integrate near cellular oncogenes in some cervical carcinomas. Proc Natl Acad Sci USA
Engel LW, Heilman CA, Howley PM (1983) Transcriptional organization of the bovine papillomavirus type 1. J Virol 47: 516-528
Freese UK, Schulte P, Pfister H (1982) Papilloma virus-induced tumors contain a virus-specific transcript. Virology 117: 257-261
Georges E, Croissant O, Bonneaud N, Orth G (1984) Physical state and transcription of the genome of the cottontail rabbit papillomavirus in warts and in the transplantable VX2 and VX7 carcinomas of the domestic rabbit. J Virol 51: 530-538
Georges E, Breitburd F, Jibard N, Orth G (1985) Two Shope papillomavirus-associated VX2 carcinoma cell lines with different levels of keratinocyte differentiation and transplantability. J Virol 55: 246-250
Giri I, Danos O, Thierry F, Georges E, Orth G, Yaniv M (1985) Characterization of transcription control regions of the cottontail rabbit papillomavirus. In: Howley PM, Broker TR (eds) Papillomaviruses: molecular and clinical aspects: Alan R. Liss, New York, pp 379-390
Gissmann L, Dürst M, Oltersdorf T, von Knebel-Döberitz M (1987) Human papillomaviruses and cervical cancer. In: Steinberg BM, Brandsma JL, Taichman LB (eds) Papillomaviruses, Cancer Cells, vol 5. Cold Spring Harbor Laboratory, New York (to be published)
Gorman CM, Moffat LF, Howard BH (1982) Recombinant genomes which express chloramphenical acetyltransferase in mammalian cells. Mol Cell Biol 2: 1044-1051
Groff DE, Lancaster WD (1986) Genetic analysis of the 3′ early region transformation and replication functions of bovine papillomavirus type 1. Virology 150: 221-230
Heilman CA, Engel L, Lowy DR, Howley PM (1982) Virus-specific transcription in bovine papillomavirus-transformed mouse cells. Virology 119: 22-34
Howley PM, Yang Y-C, Spalholz BA, Rabson MS (1986) Papillomavirus transforming functions. In: Evered D, Clark S (eds) Papillomaviruses, Ciba Foundation Symposium 120: 39-48
Hunter T (1986) Cell growth control mechanisms. Nature 322: 14
Kleiner E, Dietrich W, Pfister H (1986) Differential regulation of papilloma virus early gene expression in transformed fibroblasts and carcinoma cell lines. EMBO J 6: 1945-1950
Kozak M (1983) Comparison of initiation of protein synthesis in procaryotes, eucaryotes, and organelles. Microbiol Rev 47: 1-45
Kreider JW, Bartlett GL (1981) The Shope papilloma-carcinoma complex of rabbits: a model system of neoplastic progression and spontaneous regression. Adv Cancer Res 35: 81-110
Lehn H, Ernst T-M, Sauer G (1984) Transcription of episomal papillomavirus DNA in human condylomata acuminata and Buschke-Lowenstein tumors. J Gen Virol 65: 2003-2010
Lehn H, Krieg P, Sauer G (1985) Papillomavirus genomes in human cervical tumors: analysis of their transcriptional activity. Proc. Natl Acad Sci USA 82: 5540-5544
Lusky M, Botchan MR (1985) Genetic analysis of bovine papillomavirus type 1 trans-acting replication factors. J Virol 53: 955-965
Lusky M, Botchan MR (1986) Transient replication of bovine papillomavirus type 1 plasmids: cis and trans requirements. Proc Natl Acad Sci USA 83: 3609-3613
Marcu KB, Harris LJ, Stanton LW, Erikson J, Watt R, Croce CM (1983) Transcriptionally active c-myc oncogene is contained within NIARD, a DNA sequence associated with chromosome translocations in B-cell neoplasia. Proc Natl Acad Sci USA 80: 519-523
McVay P, Fretz M, Wettstein F, Stevens J, Ito Y (1982) Integrated Shope virus DNA is present and transcribed in the transplantable rabbit tumor VX7. J Gen Virol 60: 271-278
Nasseri M, Wettstein FO (1984a) Differences exist between viral transcripts in cottontail rabbit papillomavirus-induced benign and malignant tumors as well as non-virus-producing and virus-producing tumors. J Virol 51: 706-712
Nasseri M, Wettstein FO (1984b) Cottontail rabbit papillomavirus-specific transcripts in transplantable tumors with integrated DNA. Virology 138: 362-367

Nasseri M, Wettstein FO, Stevens JG (1982) Two colinear and spliced viral transcripts are present in non-virus-producing benign and malignant neoplasms induced by the Shoe (rabbit) papilloma virus. J. Virol 44: 263-268
Orth G (1986) Epidermodysplasia verruciformis: a model for understanding the oncogenicity of human papillomaviruses. In: Evered D, Clark S (eds) Papillomaviruses, Ciba Foundation Symposium 120: 157-169
Pater MM, Pater A (1985) Human papillomavirus types 16 and 18 sequences in carcinoma cell lines of the cervix. Virology 145: 313-318
Pettersson U, Ahola H, Stenlund A, Moreno-Lopez J (1986) Organization and expression of papillomavirus genomes. In: Salzman NP, Howley PM (eds) The papovaviridae: the papillomaviruses. Plenum, New York (to be published)
Phelps WC, Leary SL, Faras AJ (1985) Shope papillomavirus transcription in benign and malignant rabbit tumors. Virology 146: 120-129
Riou GF, Barrois M, Tordjman I, Dutronquay V, Orth G (1984) Présence de genomes de papillomavirus et amplification des oncogenes c-myc et c-Ha-ras dans des cancers envahissants du col de l'uterus. C R Acad Sci Paris 299: 575-580
Riou GF, Barrois M, DutronquayV, Orth G (1985) Presence of papillomavirus sequences, amplification of c-myc and c-Ha-ras oncogenes and enhanced expression of c-myc in carcinomas of the uterine cervix. In: Howley PM, Broker TR (eds) Papillomaviruses: molecular and clinical aspects. Alan R. Liss, New York, pp 47-56
Sarver N, Rabson MS, Yang Y-C, Byrne JC, Howley PM (1984) Localization and analysis of bovine papillomavirus type 1 transforming functions. J Virol 52: 377-388
Schiller JT, Vass WC, Lowy DR (1984) Identification of a second transforming region in bovine papillomavirus DNA. Proc Natl Acad Sci USA 81: 7880-7884
Schiller JT, Vass WC, Vousdan KH, Lowy DR (1986) The E5 open reading frame of bovine papillomavirus type 1 encodes a transforming gene. J Virol 57: 1-6
Schlegel R, Glass M, Rabson MS, Yang Y-C (1986) Bovine papillomavirus (type 1) directs the synthesis of a small hydrophobic transforming polypeptide. Science 233: 464-467
Schneider-Gädicke A, Schwarz E (1986) Different human cervical carcinoma cell lines show similar transcription patterns of human papillomavirus type 18 early genes. EMBO J 5: 2285-2292
Schwarz E, Schneider-Gädicke A (1986) Organization and expression of human papillomavirus DNA in cervical cancer cell lines. In: DePalo G, et al. (eds) Herpes and papillomaviruses: their role in the carcinogenesis of the lower genital tract, Serono symposia publications, vol 31. Raven, New York, pp 105-113
Schwarz E. Dürst M, Demankowski C, Lattermann O, Zech R, Wolfsperger E, Suhai S, zur Hausen H (1983) DNA sequence and genome organization of genital human papillomavirus type 6b. EMBO J 2: 2341-2348
Schwarz E, Freese UK, Gissmann L, Mayer W, Roggenbuck B, Stremlau A, zur Hausen H (1985) Structure and transcription of human papillomavirus sequences in cervical carcinoma cells. Nature 314: 111-114
Schwarz E, Schneider-Gädicke A, zur Hausen H (1987) Human papillomavirus type 18 transcription in cervical carcinoma cell lines and in human cell hybrids. In: Steinberg BM, Brandsma JL, Taichman LB (eds) Papillomaviruses, Cancer Cells, vol 5. Cold Spring Harbor Laboratory, New York (to be published)
Seedorf K, Krämmer G, Dürst M, Suhai S, Röwekamp WG (1985) Human papillomavirus type 16 DNA sequence. Virology 145: 181-185
Seedorf K, Oltersdorf T, Krämmer G, Röwekamp WG (1987) Identification of early proteins of the human papillomaviruses type 16 (HPV16) and type 18 (HPV18) in cervical carcinoma cells. EMBO J 6: 139-144
Shope RE (1933) Infectious papillomatosis of rabbits. J Exp Med 58: 607-624
Smotkin D, Wettstein FO (1986) Transcription of human papillomavirus type 16 early genes in a cervical cancer and a cancer-derived cell line and identification of the E7 protein. Proc Natl Acad Sci USA 83: 4680-4684
Spalholz BA, Yang Y-C, Howley PM (1985) Transactivation of a bovine papillomavirus transcriptional regulatory element by the E2 gene product. Cell 42: 183-191
Stanbridge EJ (1976) Suppression of malignancy in human cells. Nature 260: 17-20
Stanbridge EJ, Flandermeyer RF, Daniels DW, Nelson-Rees WA (1981) Specific chromosome

loss associated with the expression of tumorigenicity in human cell hybrids. Somatic Cell Genet 7: 699-712
Stanbridge EJ, Der CJ, Doersen C-J, Nishimi RY, Peehl DM, Weissman BE, Wilkinson JE (1982) Human cell hybrids: analysis of transformation and tumorigenicity. Science 215: 252-259
Stenlund A, Zabielski J, Ahola H, Moreno-Lopez J, Pettersson U (1985) Messenger RNAs from the transforming region of bovine papilloma virus type 1. J Mol Biol 182: 541-554
Taichman L, Breitburd F, Croissant O, Orth G (1984) The search for a culture system for papillomaviruses. J Invest Dermatol 83: 2s-6s
Thierry F, Heard JM, Dartmann K, Yaniv M (1987) Characterization of a transcriptional promoter of human papillomavirus 18 and modulation of its expression by SV40 and adenovirus early antigens. J Virol 61: 134-142
Tsunokawa Y, Takebe N, Kasamatsu T, Terada M, Sugimura T (1986) Transforming activity of human papillomavirus type 16 DNA sequences in a cervical cancer. Proc Natl Acad Sci USA 83: 2200-2203
Watts SL, Phelps WC, Ostrow RS, Zachow KR, Faras AJ (1984) Cellular transformation by human papillomavirus DNA in vitro, Science 225: 634-636
Watts SL, Chow LT, Ostrow RS, Faras AJ, Broker TR (1985) Localization of HPV-5 transforming functions. In: Howley PM, Broker TR (eds) Papillomaviruses: molecular and clinical aspects Alan R. Liss, New York, pp 501-511
Yang Y-C, Okayama H, Howley PM (1985) Bovine papillomavirus contains multiple transforming genes. Proc Natl Acad Sci USA 82: 1030-1034
Yasumoto S, Burkhardt AL, Doniger J, DiPaolo JA (1986) Human papillomavirus type 16 DNA-induced malignant transformation of NIH 3T3 cells. J Virol 57: 572-577
Yee C, Krishnan-Hewlett I, Baker C, Schlegel R, Howley P (1985) Presence and expression of human papillomavirus sequences in human cervical carcinoma cell lines. Am J Pathol 119: 361-366
zur Hausen H (1986) Intracellular surveillance of persisting viral infections. Lancet II: 489-491
zur Hausen H, Schneider A (1986) The role of papillomaviruses in human anogenital cancer. In: Howley PM, Salzman N (eds) The papovaviridae: the papillomaviruses. Plenum, New York (to be published)

Papillomavirus Infections and Cancer

K. J. Syrjänen

1 Introduction

Genital warts as a disease have a long history dating back to the Roman-Hellenistic era, when these lesions were accurately described and named condylomas or figs (Bäfverstedt 1967; Oriel 1981). Throughout the centuries, genital warts were considered to the associated with venereal diseases and were commonly thought to be transmitted by homosexual intercourse among males (Oriel 1981). It was not until 1954, however, that their venereal transmission was documented (Barrett et al. 1954). Viral particles had been observed in skin warts in 1949, but it took some 2 decades to identify them in genital warts (Dunn and Ogilvie 1968). The structure of the viral particles in these two lesions proved to be identical, and the agent is currently known as human papillomavirus (HPV) (Melnick et al. 1974).

Until the early 1970s, it was believed that all types of human warts were caused by one and the same type of HPV, their variable morphology being ascribed to their different anatomical location. This concept has had to be modified recently, following the discovery of many different types and subtypes of HPV with predilections for a particular site of infection in man (Jablonska et al. 1982; Lutzner 1983; Orth et al. 1978a; zur Hausen 1977). At the same time, the previous view of genital warts as innocuous lesions has been subjected to a complete reappraisal since it was suggested that some neoplastic intraepithelial lesions of the uterine cervix were associated with HPV (Meisels et al. 1976; Purola and Savia 1977). It has now been amply documented that HPV is frequently associated with cervical intraepithelial neoplasia (CIN), carcinoma-in-situ (CIS) and with invasive squamous cell carcinomas (zur Hausen 1977; Kaufman et al. 1983; Kirkup et al. 1982; Meisels et al. 1979, 1981; Dürst et al. 1983; Meisels and Morin 1981; Syrjänen 1979a, b, 1980a, 1983, 1984a; Gissmann 1984; Fenoglio and Ferenczy 1982; Mazur and Cloud 1984).

The role of HPV as a carcinogenic agent is further suggested by reports on the malignant transformation of cutaneous lesions of epidermodysplasia verruciformis. Circumstantial evidence in reference to the carcinogenic role of HPV pertaining to other organs will be discussed in the body of this chapter. Also, recent immunohistochemical (IP-PAP) demonstrations of HPV structural protein expression in CIN and CIS and detection by DNA hybridization techniques of HPV DNA sequences in invasive cervical carcinomas and in a variety of precancerous squamous cell lesions further supported HPV as a potentially important etiological agent in human squamous cell carcinomas, particularly of the genital tract (zur Hausen 1977; Kaufman et al. 1983; Kirkup et al. 1982; Meisels et al. 1979, 1981; Meisels and Morin 1981; Syrjänen 1979a,b, 1980a, 1983, 1984a,b, 1986; zur Hausen and Gissmann 1980). The rapid progress in the field of molecular biology has also significantly contributed to some understanding of the molecular mechanisms involved in viral replication and cell transformation by papillomaviruses (Howley 1982; see also chapters by Turek and Haugen, and Breitburd, this volume).

The purpose of this chapter is to review briefly the present state of knowledge linking papillomaviruses with human squamous cell carcinomas. The link between papillomaviruses and squamous cell cancer is further suggested by the animal models of this disease. The possible synergistic factors involved in carcinogenesis by papillomaviruses will also be discussed.

2 Animal Models of Papillomavirus-Induced Carcinogenesis

Papillomaviruses are known to induce tumors of squamous epithelium in a wide variety of animals, including many mammals such as rabbits, hamsters, *Mastomys natalensis*, sheep, goat, deer, cattle, horse, dog, and monkey, and lower vertebrates, e.g., birds (Sundberg et al. 1985; Junge et al. 1984; Moreno-Lopez et al. 1984; Amtmann et al. 1984; Lancaster and Olson 1982). These lesions have been discussed in detail by Sundberg (this volume), but they will be briefly surveyed here to emphasize the data pertinent to the understanding of the connections between HPV infections and human cancer.

2.1 Squamous Cell Papilloma-Carcinoma in Rabbits

Cottontail rabbit (Shope) papillomavirus (CRPV)-induced tumors occur as a spontaneous enzootic disease of the wild cottontail rabbits *(Sylvilagus floridanus)* living in certain areas in the United States of America (Shope 1933; Lancaster and Olson 1982). Experimentally, CRPV induces squamous cell papillomas in the domestic rabbit *(Oryctolagus cuniculus)*. CRPV-induced tumors were the first experimental model system used to study the papilloma/squamous cell carcinoma sequence (Shope 1933). In domestic rabbits experimental CRPV infection leads to characteristic squamous cell papillomas which persist in more than 90% of animals. By contrast 50% of the lesions resulting from natural infection in cottontail rabbits undergo a spontaneous regression within 6-12 months (Syverton 1952). In domestic rabbits about 75% of the squamous cell papillomas become malignant within an average of 9 months, as compared with the cottontail rabbits where only 25% of the papillomas do so (Syverton 1952). This different behavior of the CRPV-induced tumors has been attributed to differences in the host reactivity in these two species, a fact that may be applicable in explaining the clinical course of some HPV infections in humans (Syrjänen 1984a).

The role of synergistic factors, currently regarded to be of crucial importance in PV-induced carcinogenesis, was studied by Rous and Friedewald in 1944: methylcholanthrene or tar, when applied repeatedly to CRPV-induced papillomas of domestic rabbits caused the tumors to undergo malignant transformation more rapidly and at multiple sites. When applied concurrently, tar and CRPV showed the same effect (Rogers and Rous 1951). Repeated tarring in itself rarely led to carcinomas, but subsequent intravenous inoculation with CRPV yielded squamous cell carcinomas in a high percentage of animals (Rous and Kidd 1938). Thus, these early studies suggest that chemical carinogens and CRPV play a synergistic role in the genesis of squamous cell carcinomas in rabbits.

Although after the malignant conversion of the CRPV-induced papillomas viral particles cannot be visualized with electron microscopy, viral DNA seems to persist in such carcinomas indefinitely. This is exemplified by the two transplantable cell lines, VX2 and VX7, maintained in animals for over 30 years, and still valuable models for research into CRPV DNA and the mechanisms regulating viral transcription, translation, and transformation (Nasseri et al. 1982). The studies of the mechanisms responsible for the spontaneous regression of the Shope rabbit papillomas and of the cellular immune mechanisms involved in this process (Hellström et al. 1969) also have some bearing on the understanding of the biological behavior of HPV infections in humans (Syrjänen 1984a,b).

2.2 Bovine Alimentary Tract Papilloma-Carcinoma

Another possible animal model of the human disease is the alimentary tract papilloma-carcinoma sequence in cattle observed in high incidence areas of Scotland (Jarrett 1978; Campo et al. 1980). This papilloma-carcinoma sequence was demonstrated only in cattle fed a diet containing bracken fern. Recently, evidence has been presented showing the presence of bovine papillomavirus type 4 (BPV4) DNA from papillomas, but not in carcinomas developing from papillomas (Jarrett 1978; Campo et al. 1980). The component of bracken fern responsible for this synergistic action with BPV4 has not yet been identified (Jarrett 1978; Campo et al. 1980). As BPV4 DNA has not been observed in cancerous tissue, the mechanisms of malignant transformation are likely to be different from that of CRPV-induced lesions (Jarrett 1978; Campo et al. 1980; Lancaster and Olson 1982). It cannot be excluded, however, that a papillomavirus different from BPV4 and present only in minute quantities in the papillomas has escaped detection so far.

2.3 Equine Sarcoids

Sarcoids are the most frequent, spontaneously occurring tumors in horses. They are connective tissue tumors that usually do not metastasize but may be locally invasive and can recur even after a radical surgical excision. The etiological role of BPV was suggested by prior transmission experiments by Olson and Cook (1951). BPV1- or BVV2-specific DNAs have been conclusively demonstrated in naturally occurring equine sarcoids (Lancaster et al. 1979). Equine sarcoid is an example of a naturally occurring induction of a nonproductive, semimalignant (borderline) tumor in an alien, nonpermissive host.

Although the animal models discussed above differ in detail from each other, they appear to have some common features. These can be summarized as follows: (1) some papillomaviruses in animals possess a definite tumorigenic potential; (2) the biological behavior of the PV-induced tumors (regression versus progression) seems to depend on host reactivity; and (3) synergistic actions by chemical and other carcinogens may be required for the development of a malignant tumor from a preexisting benign PV lesion. As will be seen later in this discussion, similar factors may also be operating in HPV-induced lesions.

3 Epidemiological Aspects of HPV Infections and Carcinogenesis

3.1 Common Warts

The infectious nature of venereal and common warts was apparently first recognized at the beginning of the first century A.D. by Celsus who stated that "sometimes one is alone, generally several grow together, either on the palms or soles of the feet. The worst, however, are situated upon the genitals, and they bleed the most" (cited by Bäfverstedt 1967).

The epidemiology of common warts has been extensively surveyed. As these data are pertinent to the subsequent discussion on the biology of HPV infections, they are briefly reviewed here. In an extensive follow-up study of 1000 institutionalized mentally handicapped children (4–20 years of age), much was learned about the natural history of common warts (Massing and Epstein 1963). The most important findings of this study were: (1) the overall incidence of these lesions increased during the 2-year follow-up from 18% to 25%; (2) two-thirds of the warts in 168 children were shown to regress within 2 years; (3) regression occurred almost twice as often in boys as in girls; (4) regression was not related to age, mental status of the children, or the presence of warts in other children; and (5) new lesions appeared thrice as often in previously infected children as in uninfected ones. In conclusion the authors suggested that therapy should be directed at reducing the reservoir in the infected person and the environment. As will be shown later, these measures could be of importance in preventing the spread of cervical HPV infections as well.

In another study of warts in school children, it was suggested that early diagnosis and treatment of plantar, palmar, or filiform warts may lead to a decrease in the incidence of these lesions in school children (Coles 1958). Coles also emphasized the nearly total absence of genital warts in children. In a survey of 2389 school children, plantar warts were observed in 4.5% of them (Grigg and Wilhelm 1953). These authors noted that about 10% of the children with plantar warts also had warts elsewhere on the body. Most plantar warts (86.7%) occurred on pressure points, and thus trauma was thought to be a predisposing factor (Grigg and Wilhelm 1953).

With the developments of modern DNA technology, over 50 types of HPV have been identified (see Pfister and Fuchs, this volume). It was recognized that not all warts are caused by one and the same HPV type. In one of the earliest studies it was shown that HPV1, HPV2, and HPV3 induced warts predominantly in children between 5 and 15 years of age, whereas those due to HPV4 were found more often in people between the ages of 20 and 35 years old (Pfister and zur Hausen 1978). There was no evidence linking HPV1, HPV2, HPV3, and condylomata acuminata, laryngeal papillomas, or any of the malignant tumors tested (Pfister and zur Hausen 1978).

In epidemiological studies of different types of common warts, the high prevalence of hand warts in meat handlers and butchers has been repeatedly observed. Of 536 meat workers 24% had hand warts compared with 8.5% in 965 control subjects (De Peuter et al. 1977). Besides HPV1 and HPV2 the virus most frequently observed in the hand warts of meat handlers is HPV7 (Ostrow et al. 1981). Still, the exact mode of transmission of these HPV-associated lesions remains obscure (De Peuter et al. 1977; Ostrow et al. 1981).

The epidemiology of common warts can be summarized: (1) skin warts are a common disease; (2) they are mainly encountered in children; (3) they are caused by specific HPV types; (4) they have certain epidemiological features in common, but still represent different diseases; (5) their epidemiology has little in common with genital warts; (6) the clinical course is apparently modified by the host's immunological reactivity and environment; and (7) malignant transformation is a distinct rarity that was documented only a few times (zur Hausen 1977).

3.2 Genital Warts

Genital warts are seen mainly in sexually mature persons and are exceptional in children (Coles 1958). In fact, finding a characteristic venereal wart in a child should always arouse the suspicion of sexual abuse (Goldman et al. 1976).

Condylomata acuminata are sexually transmitted lesions affecting the genitalia, anal region, and less frequently urethral mucosa, bladder, and ureters (Oriel 1981; Feldman 1984; see chapter by Oriel, this volume). The occurrence of a characteristic condyloma acuminatum in the uterine cervix is more common than previously thought, as will be discussed later. The usual sites of genital warts are those most frequently injured during coitus. Furthermore, there is a high incidence of genital warts in people who have had sexual contacts with individuals suffering from this disease (Oriel 1981; Feldman 1984). The link between genital warts and those of the skin is not clearly established, although some epidemiological associations may exist between these two (Oriel 1981). It has also been suggested that an infected mother can transmit a condyloma acuminatum to her newborn baby. Indeed, the occurrence of laryngeal papillomatosis in children vaginally delivered from women with genital warts suggests a relationship between the two (Oriel 1981; Feldman 1984). As calculated from the expected prevalence figures (1%–2%) of condylomata acuminata in the United States of America, the risk of developing laryngeal papillomatosis for a child born to an infected mother is roughly estimated to be between 1:100 and 1:1000 (Kashima et al. 1985; see chapter by Kashima and Mounts, this volume).

The steadily increasing incidence of genital warts through the 1970s and early 1980s recognized in the United States of America and the United Kingdom has been attributed to increased sexual promiscuity, especially among young people (Chuang et al. 1984; Feldman 1984). This increase was equal in both sexes. Among the age groups, the rate was highest in the 20- to 24-year-old group. Female patients contracted the infection at a higher rate and at a younger age than male patients. Furthermore, more female than male patients had genital lesions at multiple locations (Chuang et al. 1984). This is pertinent to the concept of the multifocal nature of genital HPV infections, linking this infection to intraepithelial neoplasia in the cervix, vagina, and vulva, which will be discussed below.

Although the rate of clinically overt condylomata acuminata has increased, the frequency of such papillary leasions in the vagina and cervix is not known and presumably remains rather low. This is in contrast to the flat, HPV-associated, cervical and vaginal lesions which will be discussed below.

3.3 Cervical Carcinoma

The synchronous occurrence of HPV infection, intraepithelial neoplasia of varying degree, and occasionally an invasive carcinoma in the uterine cervix and vulva has been repeatedly observed (zur Hausen 1977; Kaufman et al. 1983; Kirkup et al. 1982; Meisels et al. 1979, 1981; Meisels and Morin 1981; Syrjänen 1979a, b, 1980a, 1983, 1984a, 1986; Gissmann 1984; Fenoglio and Ferenczy 1982; Mazur and Cloud 1984). There is also evidence that cervical cancer behaves not unlike a sexually transmitted disease (STD) (Oriel 1981; zur Hausen 1977; Fu et al. 1983b; Kessler 1976; King 1980; Singer et al. 1976; see chapter by Koss, this volume). There is considerable evidence linking cervical cancer with the sexual behavior of women; the role of 'a high-risk male' in transmitting the disease to his sexual partners has also been stressed (Singer et al. 1976). As recently emphasized, there are more than 250 reports indicating a significant relationship between sexual activity and cervical cancer (King 1980). As a consequence of a series of extensive epidemiological surveys completed during the last 2 decades, the factors associated with an increased risk for the development of cervical carcinoma can be summarized as follows: (a) early age at first coitus, (b) multiple sexual partners, (c) poor sexual higiene, (d) poverty, (e) use of tobacco, (f) use of oral contraceptives, (g) β-carotene deficiency, (h) immunosuppression, (i) geographical variations, and (j) genetic background (Richardson and Lyon 1981; Rotkin 1967, 1973; Kessler 1976, 1977; Singer et al. 1976; Singer 1983; Guijon et al. 1985). Some of these data point to the involvement of an infectious, venereally transmitted agent or agents in cervical carcinogenesis. Owing to the known racial and geographical differences in the occurrence of cervical cancer, however, other factors including the immunological and synergistic ones must also be considered.

The search for a sexually transmitted agent responsible for the development of cervical cancer has been conducted for several years (Singer 1983), and a vast number of potential carcinogenic agents have been examined (Fenoglio and Ferenczy 1982). One of these agents was herpes simplex virus (HSV) (Kessler 1976, 1977), but its role in the genesis of cervical cancer could not be confirmed. Recently, HPV has become the agent most intensely studied and considered in the etiology of cervical carcinoma and possibly other human cancers (zur Hausen 1977; Kaufman et al. 1983; Kirkup et al. 1982; Meisels et al. 1979, 1981; Meisels and Morin 1981; Syrjänen 1979a, b, 1980a, 1983, 1984a; Gissmann 1984; Fenoglio and Ferenczy 1982; Mazur and Cloud 1984). The epidemiological data support the concept of HPV involvement in this process; HPV infection in the uterine cervix is definitely a STD encountered in sexually active women. The women with cervical HPV infections had the same risk factors as women with CIN or invasive cervical carcinoma (Syrjänen et al. 1984a; Kessler 1976; King 1980; Singer et al. 1976).

In spite of the increasing rates of clinically overt condylomata acuminata, reliable prevalence and incidence data on flat cervical HPV infections have not been established so far in an unselected population. Prevalence figures from 1.1% to 2.5% have been reported from large series of Papanicolaou (PAP) smears (Meisels et al. 1981, 1982; Meisels and Morin 1981). Currently, a study is in progress in our laboratory to establish the frequency of cervical lesions in an unselected female population, based on data derived from the mass-screening program conducted in Finland since the early 1960s (Syrjänen et al. 1985a). Preliminary results suggest that the prevalence of cervical HPV infections in women between 25 and 60 years of age is close to 1%, whereas it is 2.7% at the age of 22 (Syrjänen et al. 1986a, b, 1987d).

Currently, there seems to be a definite shift in the first appearance of CIN lesions and even invasive carcinomas from older age groups to younger ones (Andrews et al. 1978; Snyder et al. 1976; Feldman et al. 1978). Some recent reports also emphasize the onset of cervical cancer in the upper socioeconomic groups (Bain and Crocker 1983) in patients who are unlikely to belong to the accepted risk groups. This is consonant with the recent observations shown that HPV infection may be transmitted by a single, HPV-infected, sexual partner (Guijon et al. 1985).

The information on the mode of transmission of cervical flat lesions is very recent. The results of questionnaires answered by women being followed-up for cervical HPV lesions clearly demonstrated the dramatic influence of sexual behavior on the transmission of these lesions (Syrjänen et al. 1984a). The HPV-infected patients were sexually promiscuous, had poor sexual hygiene, multiple sexual partners, and a past history of venereal diseases significantly more frequently than age-matched healthy women used as controls (Syrjänen et al. 1984a). Thus, the factors known to increase the risk of cervical cancer (Kessler 1976; King 1980; Singer et al. 1976) are also prevalent in females with cervical HPV infections (Syrjänen et al. 1984a).

Another line of evidence that may support the concept of cervical carcinoma as a sexually transmitted disease (STD) has recently been provided by epidemiological reports on the association of carcinoma of the penis and carcinoma of the cervix (Martinez 1969; Goldberg et al. 1979; Graham et al. 1979; Cocks et al. 1980; Smith et al. 1980). In a study of 889 men with penile cancer in Puerto Rico, cervical cancer was observed in 8 wives whereas none was found in women in a control series (Martinez 1969). The mortality from cervical cancer in 711 wives of men dying from penile cancer was shown to be in excess of that expected (Smith et al. 1980). Consonant with the hypothesis that cervical and penile cancer may be induced by the same agent is the discovery of HPV16 and HPV18 DNA in invasive cervical carcinomas, in penile bowenoid papulosis, and in penile cancer (Ikenberg et al. 1983; Villa and Lopes 1985). The significant new findings linking HPV16 and HPV18 DNA to the host cell genome in many of the cervical carcinomas and some penile carcinomas analyzed so far might have implications for the tumorigenicity of these viruses (Gissmann 1984; zur Hausen et al. 1984; Pfister 1984).

4 Malignant Transformation of Benign Lesions

4.1 Condylomata Acuminata and Giant Condylomas: General Remarks

The presence of condylomata acuminata has now been well-documented in the external genitalia and vagina. By 1974 there were 254 condylomata acuminata reported in the cervix, of which 13 were regarded as malignant (Qizilbash 1974). Some epidemiological data are also available on malignant transformation of condylomata acuminata of the vulva in Uganda, particularly in women aged 40 or older in whom 6 cancers developed in 36 condylomas. No cancer was observed in 210 condylomas in the younger age groups (Schmauz and Owor 1980). These authors concluded that condylomata acuminata may represent precancerous lesions showing a low risk of malignancy in young age groups, and a high risk in older age groups (zur Hausen 1977; Schmauz and Owor 1980).

In 1930, Buschke and Löwenstein described penile condylomata characterized by huge size, which both grossly and histologically resembled squamous cell carcinoma. This relationship has been repeatedly emphasized, and giant condylomas in the external genitalia of both sexes have been called carcinomalike condylomas or condylomalike carcinomas, to stress the difficulty in making a distinction between these two entities. It has also been suggested that a giant condyloma is a variant of verrucous carcinoma, characterized by slow growth, fungating appearance, and ulceration without metastases (Weed et al. 1983; Väyrynen et al. 1981). According to a recent description, the ultrastructure of giant condylomas shares features in common with squamous cell carcinoma (Hull et al. 1981). The viral (HPV) etiology of the tumor has been conclusively documented, but its clinical course is distinct from that of the usual condyloma acuminatum (zur Hausen 1985; zur Hausen et al. 1984). Eventually, the genitalia may be overgrown with luxuriant condylomatous masses, which are extremely hard to eradicate but do not produce metastases (Boxer and Skinner 1977). The literature concerning giant condylomas has expanded; by 1977, a total of 65 cases had been reported, in which a giant condyloma had been subsequently classified as a squamous cell carcinoma, although metastases had not been observed (Boxer and Skinner 1977).

4.2 Vulvar HPV Lesions

A number of reports is available on malignant transformation (or association with a malignancy) of genital warts in the vulva (Rhatigan and Saffos 1977; zur Hausen 1977; Daling et al. 1984; Shafeek et al. 1979; Schmauz and Owor 1980; Rastkar et al. 1982). Epidemiological surveys of vulvar carcinomas pointed out the high frequency of association with other neoplastic lesions in the anogenital area (Jimerson and Merrill 1970; Franklin and Rutledge 1972). This multicentricity in turn suggests a common (viral?) etiology of these tumors (Kimura et al. 1978; zur Hausen 1977, 1985). It is to be emphasized, however, that most of the reported malignant cases are characterized by long latency periods, several years passing between the appearance of genital warts and the frankly invasive squamous cell carcinoma (Rhatigan and Saffos 1977; Daling et al. 1984; Shafeek et al. 1979; Rastkar et al. 1982). The fact that these may represent separate events cannot be excluded because of the lack of reliable prospective follow-up data. A strong epidemiological association was recently described between condyloma acuminatum and vulvar squamous cell carcinoma, suggesting a causal relationship (Daling et al. 1984).

In this context, special reference should be made to an entity known as verrucous carcinoma (Weed et al. 1983; Väyrynen et al. 1981). In some series, this lesion has comprised close to 25% of the vulvar squamous cell carcinomas (Rastkar et al. 1982). The relative frequency of this particular lesion was reported to have increased fourfold between 1976 and 1980 as compared with that in the period 1951-1970 (Rastkar et al. 1982). It was emphasized that many of these lesions were originally diagnosed as condylomas, and not surprisingly, HPV particles were found in 67% of them (Rastkar et al. 1982). Quite recently, HPV6 DNA was disclosed in vulvar verrucous carcinomas (Rando et al. 1986). According to our experience using an in situ DNA hybridization technique, such lesions also contain HPV16 DNA (S. Syrjänen et al. 1986, unpublished observations).

Recently, a substantial amount of work has been done on vulvar intraepithelial neoplasia (VIN) (Crum et al. 1982a,b). The interest in these lesions has increased in parallel with the growing body of data on the HPV-CIN associations in the uterine cervix (see chapter by Koss, this volume). As in the cervical lesions, koilocytosis, a cytopathic effect of HPV (Koss and Durfee 1956), and HPV particles have been found in a substantial (70%) proportion of diploid VIN lesions (Crum et al. 1982a,b). When stained for HPV structural proteins, 50%-64% of VIN lesions proved to be positive (Pilotti et al. 1984; Crum et al. 1982a). HPV antigens and HPV DNA have been also demonstrated in vulvar in situ and invasive carcinomas (Pilotti et al. 1984; Crum et al. 1982a). Further evidence on the presence of HPV DNA in VIN lesions may develop with specific-type HPV DNA probes (HPV6, 10, 11, 16, 18, 31, 33, 35) (Gissmann et al. 1984; zur Hausen et al. 1984; zur Hausen 1985; McCance 1986).

Bowenoid papulosis, a lesion distinct from genital warts, was recently described in the vulva (Berger and Hori 1978). Morphologically this lesion closely resembles Bowen's disease, which by definition is an in situ squamous cell carcinoma (see chapter by Gross, this volume). Typical HPV particles and HPV antigens have been discovered in bowenoid papulosis lesions (Steffen 1982). Although the biological behavior of the majority of cases reported so far has been entirely benign and many of them undergo spontaneous regression, bowenoid papulosis of the vulva and penis was shown to contain DNA of the high-risk types HPV16 and HPV18 (Ikenberg et al. 1983; Gross et al. 1985a,b; Obalek et al. 1986) and may constitute an important source of viral transmission.

4.3 Vaginal HPV Lesions

It is well established that HPV lesions in the female genital tract are frequently multicentric, that is the external genitalia, cervix, and vagina are simultaneously affected (Stanbridge and Butler 1983; Walker et al. 1983). This is also consistent with our experience with the prospective follow-up of cervical HPV infections, which are accompanied by concomitant or successive HPV lesions in the vagina in a high percentage of cases (Syrjänen et al. 1985a,c,d,e, 1986a,b, 1987d).

Vaginal HPV lesions have been recently subjected to an extensive survey of their epidemiology, cytology, histology, colposcopy, electron microscopy, and clinical significance (Roy et al. 1981). According to these authors, the abnormal vaginal cytology (e.g., the cytopathic changes of HPV) shows a tendency to persist, but the development of carcinoma is uncommon (Roy et al. 1981). However, there are also a few, well-documented cases in the vagina, in which carcinoma have developed from a preexisting condyloma even in young women (Beck 1984).

It is commonly believed that squamous cell carcinomas in the vagina evolve from intraepithelial neoplasia (VAIN) and CIS (Benedet and Sanders 1984). Accordingly, vaginal in situ squamous cell carcinoma seems to be associated with identical lesions elsewhere in the genitalia in a high percentage of cases (Hummer et al. 1970). More than 50% of vaginal carcinomas occur in women with previous cervical cancer or CIS (summary in Koss 1979). As discussed previously for the other locations, verrucous carcinomas may rarely occur in the vagina, and they were recently shown to contain HPV DNA by hybridization techniques (Okagaki et al. 1984).

Evidence to support the involvement of HPV in vaginal carcinogenesis was recently provided by the analysis of VAIN by DNA hybridization techniques; of the 19 cases studied, 15 (79%) contained HPV6 DNA, and the rest, DNA of an at that time unidentified HPV type (HPV16?) (Okagaki et al. 1983). These authors also found HPV6 DNA in two vaginal verrucous carcinomas (Okagaki et al. 1984). Indeed, many of the genital HPV types have been subsequently disclosed in VAIN (McCance 1986). A double infection of HPV16 and HPV18 was recently discovered in an invasive vaginal cancer, which developed after less than 3 years of follow-up (Syrjänen et al. 1985c).

4.4 Penile HPV Lesions

The possible epidemiological connections between carcinoma of the penis and carcinoma of the cervix were discussed above (Martinez 1969; Goldberg et al. 1979; Graham et al. 1979; Cocks et al. 1980; Smith et al. 1980). Penile carcinoma is a rare disease in western countries (zur Hausen 1977) but more common in Thailand, India, China, Puerto Rico, and in certain parts of Africa, including Kenya, Tanzania, and Uganda (Schmauz et al. 1977). In general, penile condylomata acuminata are usually recorded at the age of 20-25 years in high-risk countries like Uganda (Schmauz et al. 1977). Penile carcinomas prevail in the 50-year age group in Uganda and 60-year age group in Europe (zur Hausen 1977). In areas of high risk for penile carcinoma (Uganda), however, there seems to be another peak of penile condylomas preceding the occurrence of penile carcinoma by only a few years (Schmauz et al. 1977). Such a peak might reflect the existence of the different precursors of penile carcinomas, a hypothesis that remains to be proven, however (zur Hausen 1977).

Bowenoid papulosis discussed above in reference to lesions of the vulva also occurs on the penis (see chapter by Gross, this volume). It may represent a reservoir of HPV infection in the male, which would support the concept of the high-risk male in cervical carcinogenesis (Singer et al. 1976). This view has gained substantial support from a recent discovery of the high-risk HPV16 DNA in 8 of 10 biopsies of penile bowenoid papulosis lesions, 2 of which also contained sequences of HPV18 (Ikenberg et al. 1983). In another recent study, HPV16 DNA was disclosed in all 12 cases studied with this technique (Gross et al. 1985a). The clinical course of this disease is usually benign; most seem to regress spontaneously and some persist, but there is no evidence of possible progression to malignancy so far (Berger and Hori 1978; Wade et al. 1979). It should be emphasized that no reliable data on the natural history based on prospective follow-up are available for these lesions as yet.

The attempts to detect HPV infections in the penis and urethra of male sexual partners of women with cervical HPV infections are just under way, and the role of the male as the carrier of HPV infections (whether the HPV resides deeper, etc.) needs further assessment (Levine et al. 1984). Of special interest in this respect is the recent discovery of HPV DNA in semen (Ostrow et al. 1986).

In a recent study of 114 penile warts, the gross appearance of the lesions did not bear any correlation with the type of HPV DNA disclosed (Syrjänen et al. 1987a). The detection rate of epithelial atypia was markedly different in the flat (25%), acuminatum (50%), and papular (75%) warts. A definite association of HPV16 and HPV18 with severe epithelial abnormalities was found. All of the HPV16-positive and 75% of the HPV18-positive cases had concomitant epithelial abnormalities, the corresponding figures for HPV6 and HPV11 being 40.8% and 31.2%, respectively (Syrjänen et al. 1987a).

4.5 Anal Warts

Condylomata acuminata in the anal region are generally ascribed to anal coitus (see chapter by Oriel, this volume). In a retrospective survey of 500 homosexual males, anal warts were shown to be seven times as common as penile warts (Oriel 1971). This was subsequently confirmed in an extensive study of homosexuals in New York (Carr and William 1977). Recently, anal warts from seven homosexual males were shown to be associated with intraepithelial (in situ) carcinoma within or adjacent to the condyloma tissue (Croxon et al. 1984). Four of these patients proved to have symptoms of AIDS, suggesting that perianal condylomas in male homosexuals should alert the clinician not only to the possibility of anal carcinoma but of AIDS as well (Croxon et al. 1984).

The literature dealing with cases in which anal warts have undergone malignant transformation was recently exhaustively reviewed (Lee et al. 1981), and an additional case report appeared afterwards (Ejeckam et al. 1983). In such cases, foci of transition from condyloma to epithelial dysplasia and CIS, sometimes with focal frank invasion, have been observed, confirmed by the DNA hybridization technique demonstrating that HPV DNA of various types, including the high-risk type HPV16, was present in anal lesions (Gross et al. 1985b; Krzyzek et al. 1980; Zachow et al. 1982). According to Gross et al. (1985b), the morphology of anal lesions seems to bear some relationship to the HPV types detected. In common condylomata acuminata, HPV6 DNA was found in 63%, HPV11 in 22%, and HPV16 in none of the cases. In the flat condylomalike lesions, the corresponding figures were 44%, 11%, and 33% and in the pigmented bowenoid papulosis lesions 0%, 25%, and 25%, respectively (Gross et al. 1985b). Consistent with the fact that some anal warts may undergo malignant transformation, it could prove to be of value to perform HPV typing of anal warts whenever possible to identify the high-risk HPV types.

4.6 Cervical HPV Lesions

The existence of warts in the uterine cervix has been recognized for many years and described under the name of squamous papilloma in earlier literature (Pitkin and Kent 1963; Woodruff and Peterson 1958; chapter by Koss, this volume). Formerly, these lesions were thought to be rare. A total of 254 condylomata acuminata in the uterine cervix had been described by the year 1974 (Jagella and Stegner 1974). There are few reports on the conversion of a condyloma acuminatum in the cervix into a carcinoma, although cellular atypia and even CIS in such lesions have been documented (Woodruff and Peterson 1958). In a series of 20 squamous cell papillomas, 2 were reported to have progressed to squamous cell carcinoma (Kazal and Long 1958). By 1963, a total of 13 such cases could be collected from the literature (Pitkin and Kent 1963). Three years later, only 20 cases were reported (Gilbert and Palladino 1966). Thus, reports are available showing that squamous cell papillomas or condylomas in the uterine cervix, as in the other genital sites, may sometimes progress to a malignant lesion. The subject of the association of HPV and CIN is discussed further below.

4.7 Laryngeal Papillomas

Laryngeal paillomas can be devided into juvenile- and adult-onset types (Incze et al. 1977; Arnold 1976; Kleinsasser and Oliveira e Cruz 1973). The epidemiology of these lesions is incompletely understood; an association of the juvenile-onset squamous cell papilloma with maternal HPV infections in the genital tract has been suggested (Quick et al. 1980; Mounts and Shah 1984; Kashima et al. 1985). The HPV etiology of both the juvenile- and adult-onset laryngeal papillomas has been firmly established (Resler and Snow 1967; Boyle et al. 1973; Spoendlin and Kistler 1978; Quick et al. 1980; Mounts and Shah 1984; Kashima et al. 1985; see chapter by Kashima and Mounts, this volume). HPV structural protein expression has been disclosed with variable frequency in such lesions (Lack et al. 1980; Costa et al. 1981; Braun et al. 1982), suggesting some differences in the biological behavior of HPV in these two entities.

Subsequently, HPV DNA sequences have been found in both the juvenile- and adult-onset laryngeal papillomas by the DNA hybridization technique (Mounts et al. 1982; Gissmann et al. 1982b; Steinberg et al. 1983). First, HPV DNA classified as HPV11 was detected in both juvenile- and adult-onset lesions (Mounts et al. 1982; Gissmann et al. 1982b; Steinberg et al. 1983). It is currently established that

HPV6 DNA is also frequently present in laryngeal papillomas (Gissmann et al. 1982b). Under light microscopy laryngeal squamous cell papilloma is usually a benign-appearing lesion devoid of signs of atypia or invasion (Mounts and Shah 1984). This is not always the case, however, as indicated by reports on the rare event of invasive laryngeal papillomatosis (Fechner et al. 1974) and on the occurrence of intraepithelial atypia in some of these lesions (Quick et al. 1979). Such atypical changes were shown to be most common (20%–40%) in lesions with the highest frequency of recurrences (Quick et al. 1979). Malignant conversion of a laryngeal squamous cell papilloma is an uncommon occurrence (Mounts and Shah 1984). However, there are occasional case reports of such events particularly after radiotherapy (Zehnder and Lyons 1975; Yoder and Batsakis 1980; Weiss and Kashima 1983; Mounts and Shah 1984; Shapiro et al. 1976; Kleinsasser and Glanz 1979; Galloway et al. 1960; Maier 1968). In such cases, the possible synergistic effects between HPV and radiation must be taken into account (zur Hausen 1977, 1982, 1985, 1986; discussed in the chapter by Kashima and Mounts, this volume).

Recently, HPV antigens (Syrjänen et al. 1982a) were demonstrated in a series of invasive laryngeal carcinomas, and HPV30 DNA was found in one such lesion (Kahn et al. 1986). In a systematic survey of laryngeal carcinomas, however, HPV30 DNA could not be found in a single case, but HPV16 DNA was demonstrated by in situ hybridization in some of them (Syrjänen et al. 1987b). This had been done previously for verrucous laryngeal carcinomas (Brandsma et al. 1986). Additional evidence is still needed to elucidate fully the role of different HPV types in laryngeal carcinogenesis.

4.8 Epidermodysplasia Verruciformis

Epidermodysplasia verruciformis (EV) is a rare, autosomal recessive disease, which usually begins in infancy or childhood (Lutzner 1978). EV is most commonly characterized by flat warts and by pityriasis versicolor-like skin lesions in 75% of cases (Lutzner 1978). Its viral etiology has been suspected for a long time; virus isolates from the flat warts of EV patients were used to induce skin warts in some recipients (Jablonska and Formas 1959). Subsequently HPV particles could be demonstrated by electron microscopy in skin lesions of a healthy recipient inoculated with an extract from an EV patient (Jablonska et al. 1972). This experiment documented the causative role of HPV in the verrucous lesions of EV. EV may be considered a model in the study of the role of HPV in human oncogenesis (Jablonska et al. 1972, 1978, 1979).

During a 21-year follow-up study of a family with EV, it was shown that members of one family can be infected with different HPV types (Jablonska et al. 1979). The skin lesions differed in morphology according to the type of HPV involved. It is thought that in patients with EV, cell-mediated immunity (CMI) may be impaired. No such defect was observed in patients with abortive forms of the disease (Jablonska et al. 1979).

One of the characteristic events in EV is the development of malignant skin lesions, such as Bowen's disease and squamous cell carcinoma. The lesions usually begin at a relatively early age, exclusively in Caucasians, and on the sun-exposed areas of the skin (Haustein 1982; Yabe and Koyama 1973; Ruiter 1973; Jablonska et al. 1970). In Africans the disease runs a more benign course, and cancer does not occur (Jacyk and Subbuswamy 1979). These observations suggest that hereditary factors and exposure to ultraviolet light play an important role in carcinogenesis in EV. It has been suggested that up to 30% of Caucasian patients with EV may develop malignant skin lesions (Jablonska et al. 1978, 1979; Ruiter and van Mullem 1970). Invasive carcinomas usually develop from CIS, e.g., Bowen's disease (Ruiter and van Mullem 1970; Haustein 1982). HPV particles can be disclosed in the precancerous lesions but can no longer be demonstrated after malignant conversion (Haustein 1982; Yabe and Koyama 1973; Ruiter 1973; Jablonska et al. 1970). As emphasized before, this is consonant with the animal model (Shope rabbit papilloma-carcinoma; Syverton 1952).

Within the past few years, a substantial amount of new data on HPV types associated with EV has been produced, as described in the chapter by Grußendorf-Conen (this volume). In fact, the majority of the more than 50 different, currently recognized HPV types have been isolated from EV lesions. It is known today that skin lesions associated with HPV5 and HPV8 and perhaps HPV14 tend to become malignant (Pfister et al. 1981; Claudy et al. 1982). Thus, the above three HPV types are currently regarded as potentially oncogenic in EV patients. HPV5 was also observed in immunosuppressed EV patients who were renal allograft recipients (Lutzner et al. 1980). Many additional HPV types have been disclosed in EV lesions (Lutzner et al. 1984; Pfister et al. 1983; Yutsudo et al. 1982; Kremsdorf et al. 1982, 1983). A similar situation has been observed in patients with genital tract HPV infections, in which specific HPV types (HPV16, 18, 31, 33, 35) are frequently observed in malignant lesions.

4.9 Squamous Cell Tumors at Other Sites

4.9.1 Bronchial Papillomas and Bronchogenic Carcinoma

The solitary benign squamous cell papilloma of the bronchus is a very rare lesion (Maxwell et al. 1985). So far, 14 such cases have been described (Roglic et al. 1975; Rubel and Reynolds 1979; Maxwell et al. 1985). Squamous cell papilloma of the bronchus appears to occur predominantly in adult males of the same age group as those developing bronchial cancer (Maxwell et al. 1985). The prognosis has been good in patients with a completely removed tumor, and no cases with malignant transformation were documented. In some of the lesions, epithelial atypia was found (Maxwell et al. 1985). Interestingly, the cytological changes in the exfoliated cells from the papillomas closely resembled those found in cervical smears of HPV-infected women (Roglic et al. 1975; Koss 1979; Rubel and Reynolds 1979). Indeed, cells with koilocytosis, multinucleation, and dyskeratosis were identified, suggesting a possible HPV etiology for bronchial papillomas as well (Roglic et al. 1975; Koss 1979; Rubel and Reynolds 1979).

Reports have been published in which squamous cell carcinomas were seen to develop from preexisting papillomatosis of the lower respiratory tract (e.g., laryngotracheal papillomas), usually after radiation or interferon therapy (Runckel and Kessler 1980; Schouten et al. 1983). In a series of 10 cases of adult-onset tracheobronchial papillomatosis, squamous cell cancer developed in 5 (Al-Saleem et al. 1968). A similar case with rapidly progressive metastases was recently described (Rahman and Ziment 1983).

The possibility that HPV could also be involved in bronchial squamous cell carcinomas was investigated (Syrjänen 1979c). In one case of bronchial squamous cell cancer, the epithelium adjacent to the malignant lesion contained areas identical to those previously described in cervical HPV lesions (Syrjänen 1979c). In a subsequent series of 104 bronchial squamous cell carcinomas, morphological changes suggestive of HPV lesions were found in the adjacent epithelium in 36 cases (34.6%; Syrjänen 1980b). Most lesions were flat, but some were papillary or inverted. In a subsequent series of 220 bronchial squamous cell carcinomas, morphological evidence suggestive of HPV involvement was found in 67 cases (30.4%; Syrjänen 1980c). HPV16 DNA was found in 1 of 24 lung carcinomas, in which a late metastasis of a previous cervical cancer could not be ruled out, however (Stremlau et al. 1985). In a recent systematic survey of bronchial squamous cell carcinomas, HPV DNA was found in 5 of 99 cases, when using the in situ hybridization with a mixed probe of HPV6, 11, 16, 18, and 30 DNA (Syrjänen and Syrjänen 1987).

At the moment, the role of HPV in the development of bronchogenic carcinoma is unproven. An analysis of additional cases with the DNA hybridization techniques may shed more light on this problem (Syrjänen 1980b, c; Stremlau et al. 1985; Syrjänen and Syrjänen 1987). It is generally accepted, however, that bronchial squamous cell carcinoma develops from the metaplastic squamous epithelium through different grades of intraepithelial neoplasia and CIS. The possibility of synergistic mechanisms between the virus and e.g., cigarette smoke must be considered (Syrjänen 1980b, c; Syrjänen and Syrjänen 1987).

4.9.2 Nasal Cavity/Paranasal Sinus Papillomas

The transitional cell, inverted, or schneiderian papilloma of the nose and paranasal sinuses is a well-recognized entity (Friedmann and Osborn 1982). As the lesions may contain extensive areas of squamous metaplasia, they are sometimes designated as squamous cell papillomas. The frequency of this disease has been calculated as 1/25 of that reported for common inflammatory polyps (Syrjänen et al. 1983). The transitional papillomas of the nose and sinuses have an unusual clinical behavior pattern (Friedmann and Osborn 1982; Syrjänen et al. 1983). There is a high rate (30%–62%) of recurrence even after adequate therapy. The occurrence of squamous cancer developing within the lesions in about 12% of patients was observed (summary in Syrjänen et al. 1983).

The pattern of recurrence of nasal papillomas is similar to respiratory papillomatosis in the larynx and suggests a similar infectious etiology (Syrjänen et al. 1983). Recently, morphological similarities (koilocytosis, papillomatosis, dyskeratosis) to HPV lesions elsewhere were observed in a typical nasal transitional papilloma with focal epithelial atypia. The lesion was also shown to contain HPV structural proteins (Syrjänen et al. 1983). Respler et al. (1985) isolated and characterized HPV DNA sequences from a nasal, inverted papilloma. The HPV proved to be closely related to HPV11, the type commonly present in laryngeal papillomas (Respler et al. 1985). Recently, HPV11 and HPV16 DNA was demonstrated in a series of benign nasal papillomas, and squamous cell carcinomas developed in some of them (Syrjänen et al. 1987c). Additional information is still necessary, however, to elucidate fully the role of HPV in the papilloma-carcinoma sequence in the nasal cavity and in the paranasal sinuses (Syrjänen et al. 1983, 1987c; Respler et al. 1985).

4.9.3 Esophageal Papillomas

Squamous cell papilloma of the esophagus is a rare lesion, only 20 histologically documented cases having been reported by 1980 (Colina et al. 1980). There is no evidence of malignant transformation of the esophageal squamous cell papillomas (Colina et al. 1980). It is well documented, however, that squamous cell carcinoma in the esophagus develops through the various degrees of intraepithelial neoplasia (dysplasia) and CIS (Suckow et al. 1962; Ushigome et al. 1967). A flat form (Barge et al. 1981) and a rare verrucous form of esophageal squamous cell carcinoma are known (Meyerowitz and Shea 1971; Minielly et al. 1967) The progression and prognosis of the verrucous carcinoma have been claimed to be more favorable than those of the usual forms (Meyerowitz and Shea 1971; Minielly et al. 1967).

HPV etiology of esophageal squamous cell papillomas was first suggested in 1982 by demonstrating HPV structural proteins in one such lesion (Syrjänen et al. 1982b). A prior morphological survey of 60 cases of esophageal squamous cell carcinoma disclosed 1 case with evidence implying a papillary HPV lesion, 3 cases with areas suggestive of an endophytic condyloma, and 20 cases in which the peripheral epithelium contained lesions similar to flat condyloma of the cervix (Syrjänen 1982). Following these observations, several new cases of esophageal papilloma were reported from other laboratories (Javdan and Pitman 1984; Ottenjann et al. 1984). Thus, as suggested before (Syrjänen et al. 1982b) the lesion is probably more common than previously thought because 15 such lesions were observed in 12 patients within a time period of only 9 months (Ottenjann et al. 1984). The same authors could find 26 additional lesions in a retrospective survey of a large number of patients from 1974 to 1981 (Ottenjann et al. 1984).

The findings of Syrjänen et al. (1982b) have been subsequently confirmed by other workers, who found HPV antigens in one esophageal papilloma (Lesec et al. 1985) and in 4 of 13 (31%) cases of esophageal mucosal abnormalities (Winkler et al. 1985). Thus, the involvement of HPV in some squamous cell lesions of the esophagus has been confirmed (Syrjänen et al. 1982; Lesec et al. 1985; Winkler et al. 1985; Goldsmith 1984). It is of note that Kulski et al. (1986) demonstrated HPV DNA in five invasive squamous carcinomas using a mixed HPV DNA probe. Of special interest in this context are the epidemiological data from China, where high-risk areas of human carcinomas of the esophagus and carcinomas of the gullet in chickens exist (Shu 1985). The significance of this observation is not clear and requires further scrutiny.

5 Evidence Linking HPV Infections with Cervical Intraepithelial Neoplasia

5.1 Cytology

Although the term 'koilocytotic atypia' was introduced 30 years ago (Koss and Durfee 1956), it was some 2 decades later before its full significance was established (Meisels et al. 1976; Purola and Savia 1977). The identification of koilocytes in cervical smears as evidence of HPV infection resulted in a series of reviews of the cytological patterns of cervical HPV lesions (Kirkup et al. 1982; Ludwig et al. 1981; Meisels et al. 1981; Pilotti et al. 1981; Syrjänen et al. 1981; Casas-Cordero et al. 1981; Bernstein et al. 1985).

According to our experience, cervical HPV lesions characteristically and almost constantly shed cells (either singly or in clusters) classifiable as dyskeratotic superficial cells (Syrjänen et al. 1981). The nuclei of these superficial cells may or may not show dyskaryotic changes, depending on whether concomitant CIN is present or not. The importance of evaluating the nuclear/cytoplasmic ratio in these cells is emphasized, because of its diagnostic value in assessing the grade of HPV lesions, e.g., whether without (HPV-NCIN) or with coexistent CIN (HPV-CIN). This in turn is of definite prognostic significance for the subsequent clinical course of cervical HPV infections, as established by our prospective follow-up study (Syrjänen et al. 1986a, b, 1987d).

The cell currently regarded as the most reliable sign of HPV infection is the koilocyte, the cytopathic effect of this virus (Koss and Durfee 1956; Meisels et al. 1976; Purola and Savia 1977; Kirkup et al. 1982; Ludwig et al. 1981; Meisels et al. 1981; Pilotti et al. 1981; Syrjänen et al. 1981; Casas-Cordero et al. 1981; Bernstein et al. 1985). It is to be stressed that the presence of koilocytes by no means precludes the concomitant appearance of cells with dyskaryotic changes, indicating the association of HPV infection with a coexistent CIN (Kirkup et al. 1982; Ludwig et al. 1981; Meisels et al. 1981; Pilotti et al. 1981; Syrjänen et al. 1981; Casas-Cordero et al. 1981; Bernstein et al. 1985).

5.2 Histology

Until 1976, condylomata acuminata were regarded as the only manifestation of HPV infection in the genital area. In that year it was suggested that certain flat and inverted (endophytic) lesions of the cervical epithelium were related to condylomata acuminata (Meisels et al. 1976; Purola and Savia 1977). Since then the HPV etiology of these lesions has been unequivocally confirmed first by the demonstration of HPV particles by electron microscopy (Laverty et al. 1978; Hills and Laverty 1979), HPV common antigens (Woodruff et al. 1980), and HPV DNA (Gissmann and zur Hausen 1980; Gissmann et al. 1982a). The light microscopic appearance of the flat and inverted condylomas are discussed elsewhere (see chapter by Koss, this volume). When changes consistent with CIN are encountered in the flat condyloma, the lesions have been called atypical condylomas, condylomatous atypias, condyloma with CIN, or condylomatous dysplasias (Kirkup et al. 1982; Ludwig et al. 1981; Meisels et al. 1981; Pilotti et al. 1981; Syrjänen 1979a, 1984a; Syrjänen et al. 1981; Casas-Cordero et al. 1981; Bernstein et al. 1985; Zuna 1984; Okagaki 1984; Mergui et al. 1984; Suprun et al. 1985; Crum and Levine 1984). Despite the confusing nomenclature, it is generally accepted that flat and inverted HPV lesions are found frequently associated with CIN, CIS, and occasionally with an invasive squamous cell carcinoma.

In practice, the concomitant CIN lesion can either display features of HPV infection or be situated adjacent to the characteristic condyloma. It is to be emphasized that the grading of CIN is of prognostic value in reference to the clinical outcome of the HPV lesion (Syrjänen et al. 1986a,b, 1987d), whether flat or inverted.

5.3 DNA Ploidy Patterns

Several attempts have been made recently to evaluate the clinical character of cervical HPV lesions on the basis of their DNA ploidy patterns. In such analyses of HPV-infected cells, both diploid and polyploid DNA distributions, including tetra- and octaploidy have been disclosed (Nasiell et al. 1979; Evans and Monaghan 1983; Fu et al. 1983a,b; Winkler et al. 1984; Fujii et al. 1984). In some of these data, DNA content showed a weak correlation with the degree of epithelial atypia and the number of abnormal mitotic figures, HPV-CIN lesions usually containing aneuploid DNA (Evans and Monaghan 1983; Nasiell et al. 1979; Fu et al. 1983a,b; Winkler et al. 1984; Fujii et al. 1984).

No absolute discriminants were found, however, between polyploidy and aneuploidy in most cervical HPV lesions, as evidenced by the presence of HPV antigens in 3 of 21 aneuploid lesions and abnormal mitotic figures in 5 of 17 polyploid HPV lesions (Reid et al. 1984). In addition, these investigations are impeded by inadequate follow-up, thus not permitting the correlation between DNA ploidy and the clinical behavior of these HPV lesions. The value of these DNA analyses was seriously questioned by data showing that quantitative DNA determinations in dysplastic cervical cells do not offer additional means of predicting the outcome of the lesions (Nasiell et al. 1979).

5.4 Expression of HPV Structural Proteins

In the past few years, an indirect immunoperoxidase (IP-PAP) method has been applied to paraffin sections to disclose the HPV common antigens (structural proteins) in cervical lesions. The sodium dodecyl sulphate (SDS)-disrupted virions of papillomaviruses possess common group-specific antigens, giving rise to antisera capable of reacting with all papillomavirus types known so far (Jenson et al. 1980). Such widely cross-reacting antisera have proved useful in assessing a variety of squamous cell lesions for the presence of HPV antigens, especially when commercial kits came into general use (Ferenczy et al. 1981; Kurman et al. 1983; Woodruff et al. 1980; Jenson et al. 1980).

With the IP-PAP technique the presence of HPV antigens is generally confined to the nuclei of the koilocytes and/or superficial dyskeratotic cells. In most series stained with the IP-PAP technique, some 50% of the lesions have been shown to contain HPV antigens. The same technique was recently applied to PAP smears derived from cervical condylomas, and 67% of the smears were positive for HPV antigens (Gupta et al. 1983). According to our experience, there are differences in HPV antigen expression related to lesion morphology in that the papillary warts usually show the highest and flat lesions the lowest frequency of positive results. Positive staining is inversely related to the degree of concomitant CIN in HPV lesions, e.g., the milder the degree of epithelial atypia, the higher the frequency of HPV antigen expressors (Syrjänen and Pyrhönen 1982a,b; Syrjänen 1983; Woodruff et al. 1980; Jenson et al. 1980; Ferenczy et al. 1981; Kurman et al. 1983).

In routine diagnosis this method is much less laborous and time-consuming than electron microscopy. Although of definite diagnostic value in positive cases, the applicability of IP-PAP staining in HPV research is limited by the fact that it can only disclose productive HPV infection, those in which viral structural proteins are expressed (see chapter by Schneider, this volume). In lesions with nonpermissive infection such as CIS and cancer, structural proteins are usually not expressed (Gissmann 1984; Pfister 1984). Recent results showed only insignificant differences in structural protein expression among the lesions infected by different HPV types (Syrjänen et al. 1986a,b, 1987d). As disclosed in prospectively followed-up cervical HPV lesions, no correlation exists between HPV antigen expression in the first biopsy and the subsequent clinical course, thus invalidating the use of IP-PAP test as a prognostic predictor in this disease (Syrjänen et al. 1986a,b, 1987d).

5.5 Presence of HPV DNA

Introduction of the DNA sequence-specific restriction endonucleases into general use has made possible more detailed analyses of HPV DNA, enabling the construction of physical maps of the cleavage sites in the genomes of the different HPV types (zur Hausen et al. 1974; Favre et al. 1975; Gissmann and zur Hausen 1978, 1980; Gissmann et al. 1982b; Howley 1982; Pfister 1984). Molecular cloning technology has significantly increased our knowledge of the molecular biology and biochemistry of papillomaviruses (Danos et al. 1980; De Villiers et al. 1981; Heilman et al. 1980; Gissmann et al. 1982b; see chapter by Schwarz, this volume). Based on the agreed typing criteria, more than 50 HPV types and many subtypes are currently recognized (see chapter by Pfister and Fuchs, this volume).

By assessing human genital HPV lesions with DNA hybridization, it became evident that HPV6 DNA was preferentially found in various forms of CIN (Gissmann et al. 1983), whereas HPV11 DNA was also disclosed in a few cases of cervical carcinomas. In some of these lesions, mixed HPV infections were reported (Schneider et al. 1984). Undoubtedly, one of the most important recent findings was the discovery of HPV16 and HPV18 in invasive cervical carcinomas and subsequently in bowenoid papulosis lesions (Boshart et al. 1984; Dürst et al. 1983). These findings have been confirmed by a number of authors (Ikenberg et al. 1983;

Boshart et al. 1984; Crum et al. 1984, 1985; Scholl et al. 1985; Gissmann et al. 1984; Wagner et al. 1984; Schneider et al. 1985; Fukushima et al. 1985).

It soon became apparent that the physical state of DNA in invasive carcinomas differed from that found in intraepithelial lesions (Gissmann 1984; Pfister 1984). In nearly all cases of invasive carcinomas analyzed so far, HPV16 or HPV18 DNA was integrated into the host cell genome (Dürst et al. 1985; see also chapters by Dürst and by Schwarz, this volume). Recently, this was also confirmed in six of eight cell lines derived from human cervical carcinomas, in which either HPV16 and HPV18 DNA sequences were demonstrated to be integrated in the cell genome (Boshart et al. 1984; Pater and Pater 1985; Yee et al. 1985). Integrated HPV16 and HPV18 DNA was also found in invasive cancer in one of our patients, who developed this lesion in under 3 years of follow-up (Syrjänen et al. 1985c). Thus, integration of HPV DNA into the host genome probably has implications in the tumorigenicity of these viruses (Gissmann 1984; Pfister 1984; Dürst et al. 1985; zur Hausen 1985; zur Hausen et al. 1984). Certainly there are other factors of importance, which are not defined as yet.

Using analysis with HPV16 and HPV18 DNA probes, these HPV types were present in 5.1% of condylomata acuminata, in 16.7% of flat condylomas, in 53.8% of CIS, and in 57.4% of invasive cervical cancers (Gissmann et al. 1984). Except for the fact that HPV16 and HPV18 DNA has been repeatedly found integrated in the host cell genome, its presence was shown to bear a close correlation with the presence of abnormal mitotic figures in flat warts, suggesting that such lesions represent precursors of cervical cancer (Crum et al. 1984). In 1985, three additional HPV types implicated in the etiology of cervical HPV lesions were described, classified as HPV31 (Lörincz et al. 1985a), HPV33 (Beaudenon et al. 1985), and HPV35 (Lörincz et al. 1985b). HPV31 was shown to be present in some 35% of mild dysplasias, but in less than 5% of the CIS and cervical cancer studied (Lörincz et al. 1985a). On the other hand, HPV33 and HPV35 were originally isolated and characterized from invasive cervical carcinomas. Thus, HPV16 and HPV18, and possibly HPV33 and HPV35, may represent high-risk types of HPV, more prone to induce malignant transformation than the others, e.g., HPV6, 10, 11, and 31 found in CIN or in benign genital warts (zur Hausen 1985; Gissmann 1984; Pfister 1984). This would be analogous to the situation in EV, where HPV5, 8, and 14 are the HPV types found in malignant lesions (reviewed by Lutzner et al. 1984).

5.6 Natural History of Cervical HPV Infections

According to the generally accepted concept, CIN lesions may persist a varying period of time, undergo regression, or develop into more severe forms of CIN, into CIS, and eventually into an invasive carcinoma (Koss et al. 1963; Nasiell et al. 1983). Thus, CIN is a true precancerous lesion possessing the potential to progress to CIS if left untreated. Since 1981 an extensive follow-up study has been in progress in our clinic for women with cervical HPV infection, to explore the natural history of this disease and the factors modifying it (Syrjänen et al. 1984a, 1985a). This has been entirely neglected so far, although the natural history of classical CIN has been well-characterized by a few, carefully conducted, follow-up studies (Nasiell et al. 1983).

On each attendance at the clinic (at 6-month intervals), our patients were subjected to colposcopy accompanied by PAP smears and/or punch biopsy, both being analyzed for the cytopathic changes of HPV and for concomitant CIN. In the biopsies the expression of HPV structural proteins was assessed, and HPV typing accomplished by Southern blot, spot, or in situ hybridization with the DNA probes for HPV6, 11, 16, 18, 31, and 33. Blood samples were taken for antibody determinations to HSV and cytomegalovirus (CMV) as possible cofactors. Cervical swabs are cultured for *Chlamydia* and HSV. All the patients fill in a detailed questionnaire concerning their sexual as well as smoking habits, to assess the influence of these factors in the transmittance of HPV infections (Syrjänen et al. 1984a, 1985a, 1986a).

So far, only fragmentary observations based on a short-term follow-up of a limited series of cervical HPV lesions are available, resembling the behavior of classical CIN (Walker et al. 1983). Similar observations were made in a recent retrospective follow-up of 764 cervical HPV lesions for 18 months: 26.8% regressed, 52.9% persisted, and 20.3% progressed (de Brux et al. 1983). Our preliminary data based on a 2-year follow-up of 418 females with cervical HPV-NCIN and HPV-CIN lesions showed that 24% of the HPV lesions regressed, 55% persisted, and 21% progressed, 10.6% having been coned due to progression into CIS. The clinical progression was significantly associated with the grade of HPV-associated CIN (Syrjänen et al. 1985a). In the first 103 biopsies analyzed using Southern blot and spot hybridization, HPV6 was found in 8%, HPV11 in 36%, HPV16 in 11%, and HPV18 in 8% of the lesions (Syrjänen et al. 1985d, e, f). HPV-CIN lesions were found more frequently than HPV-NCIN to be associated with HPV16 and HPV18. The progression rate was highest (45.5%) in HPV16 lesions, followed by that (27.3%) in HPV18 lesions, as contrasted with 0% and 13.3% for HPV6 and HPV11, respectively (Syrjänen et al. 1985a, e, f).

After an additional period of follow-up (mean 5 months) and study of more samples analyzed using in situ DNA hybridization, the above results have changed (Syrjänen et al. 1985a, b, 1987d): at this time the highest progression rate appears to be associated with HPV16 lesions (33.3%), which also are the least frequent ones to undergo spontaneous regression (5.6%). All the lesions that recurred after conization also contained HPV16 DNA (Syrjänen et al. 1986a, b, 1987d). This is in contrast to the lesions induced by HPV6 or HPV11, which do regress in 25.6% of cases. Noteworthy, however, is the relatively high progression rate of 25.6% currently established also for HPV6/11 lesions. Although these data support the concept of HPV16 as the high-risk HPV type in cervical carcinogenesis, it should be emphasized that HPV6, HPV11, and HPV18 also represent a potential risk for subsequent development of more severe precancerous lesions (Syrjänen et al. 1986a, b, 1987d).

Prospective follow-up of a large series of patients also permits evaluation of the eventual latent infections in cervical epithelium. This concept of latent HPV infections was first substantiated by the discovery of HPV particles (Syrjänen et al. 1985b), structural proteins, and HPV DNA sequences (Ferenczy et al. 1985) in spontaneously regressing or laser-eradicated genital HPV lesions. These data are in full agreement with our observations confirming the occurrence of HPV6, 11, 16, 18, and 31 in lesions with no morphological evidence of HPV infection under light microscopy (Syrjänen et al. 1986b). Thus, latency seems to be an established feature of the infectious cycle of these and probably other HPV types as well. Such HPV genomes present in apparently normal epithelium have been recently shown to be responsible for new lesions after a seemingly successful removal of an adjacent lesion (Ferenczy et al. 1985), a fact that should be taken into account in therapeutic considerations.

The follow-up data available on cervical HPV lesions suggest that their natural history is similar to that of the classical CIN lesions (Koss et al. 1963; Nasiell et al. 1983; Syrjänen et al. 1985d, e, f). The only substantial difference between these two entities is that HPV infections occur in women more than 10 years younger than those presenting with classical CIN. This suggests that HPV infection accelerates the development of CIN in affected women. The inherent potential of HPV16 and HPV18 lesions to progress into CIS has been established (Syrjänen et al. 1985d, e, f, 1986a, b, 1987d). These observations undoubtedly have implications in clinics for cervical HPV infections, suggesting that whenever the high-risk types are found, the lesions should be promptly eradicated, if an adequate prospective follow-up cannot be guaranteed.

6 Synergism Between Papillomaviruses and Other Carcinogens

Current evidence on the possible synergistic mechanisms in experimental and HPV-associated carcinogenesis has been extensively discussed (zur Hausen et al. 1984). As they are relevant to understanding the possible events by which papillomavirusis induce malignant growths, these data will be shortly reviewed here.

The role of synergistic factors in carcinogenesis was studied in the CRPV model by workers in the 1940s (Rous and Friedewald 1944). According to their observations, methylcholanthrene or tar, when applied repeatedly to CRPV-induced papillomas of domestic rabbits, reduced the latency period of malignant transformation, and malignancies developed at multiple sites. Thus, these early studies imply beyond any doubt that chemical carcinogens and CRPV do play a synergistic role in squamous cell carcinogenesis in rabbits.

It is now well-known that in certain parts of Scotland there are high-incidence areas of bovine alimentary tract carcinoma, shown to arise from preexisting squamous cell papillomas, which have often been demonstrated to contain BPV4 DNA (Jarrett 1978; Campo et al. 1980). Such a papilloma-carcinoma sequence was exclusively found in cattle fed on a bracken fern diet. Here again, a well-characterized system exists, in which papillomaviruses and chemical carcinogens act synergistically to produce the malignant growth from a preexisting benign one (Jarrett 1978; Campo et al. 1980).

With regard to HPV, certain HPV lesions exist, in which synergistic mechanisms analogous to those discussed above could be involved in their malignant transformation. Malignant conversion of the laryngeal squamous cell papilloma may be more common than previously thought (Mounts and Shah 1984). Malignant conversion has been frequently associated with radiation therapy of the lesions (Mounts and Shah 1984; Galloway et al. 1960; Maier 1968), but several cases have been recently reported in which malignant transformation has been found in the absence of irradiation (Mounts and Shah 1984; Yoder and Batsakis 1980; Shapiro et al. 1976; Kleinsasser and Glanz 1979). In such cases, the possible synergistic effects between HPV and carcinogens other than irradiation must be taken into account (zur Hausen et al. 1984).

As discussed previously, there is some evidence concerning HPV involvement also in bronchial squamous cell lesions (Syrjänen 1979c, 1980b,c; Stremlau et al. 1985; Syrjänen and Syrjänen 1987). In bronchial squamous cell carcinogenesis, reference should be made to possible synergistic mechanisms between the virus and chemical carcinogens present in cigarette smoke. Similarly, the implication of HPV in some squamous cell lesions of the esophagus seems to be evident (Syrjänen et al. 1982b; Goldsmith 1984; Lesec et al. 1985; Winkler et al. 1985; Kulski et al. 1986). Following the recent demonstration of HPV DNA in esophageal carcinoma, speculations can be presented on synergistic actions between HPV and chemical carcinogens (dietary factors, etc.) in esophageal carcinogenesis as well.

Another well-defined example of PVs and synergistic factors in human carcinogensis is EV, characterized by the frequent development of squamous cell carcinoma and Bowen's disease (Haustein 1982; Yabe and Koyama 1973; Ruiter 1973; Jablonska et al. 1970). Such lesions develop mostly in Caucasians and on sun-exposed areas of the skin. In black Africans, no signs of malignant conversion whatsoever can be demonstrated (Jacyk and Subbuswamy 1979). This implies that hereditary factors and physical carcinogens (UV light) play an important role in EV carcinogenesis. The HPV types most frequently found in the malignant lesions that evolve from EV are HPV5, HPV8, and occasionally HPV14 (Lutzner et al. 1984). This is consonant with the situation in genital tract HPV infections, in which specific HPV types (HPV16, 18, 33, and 35) are more common with invasive cancer.

Of special interest are the recent reports on the increased risk of cervical cancer in cigarette smokers (Clarke et al. 1982; Marshall et al. 1983; Trevethan et al. 1983; Greenberg et al. 1985). In extensive epidemiological studies heavy smokers have been shown to have a 2-3.6-fold greater risk of cervical neoplasia (Clarke et al. 1982; Marshall et al. 1983; Trevethan et al. 1983; Greenberg et al. 1985). These studies did not, unfortunately, assess the role of HPV in this process. It remains to be elucidated whether such an amplified risk can also be established in our patients prospectively followed-up for cervical HPV infections (Syrjänen et al. 1984a, 1985a).

In the discussed above examples the interaction between a specific papillomavirus and chemical or physical carcinogens is evidently an important prerequisite for malignant transformation. It was suggested that certain papillomavirus infections may functionally resemble promoters or promoted genes (zur Hausen 1986; zur Hausen et al. 1984). Based on experimental evidence that HSV2 possesses definite initiator functions in different systems, an interesting hypothesis was presented (zur Hausen 1982): possible synergistic actions between HSV and HPV might be involved in the development of cervical cancer. In such cases primary and recurrent HSV infections would lead to an increased number of initiating events, resulting in a higher risk for malignant transformation by the promoter, HPV (zur Hausen 1982). Some support for the hypothesis was provided by our recent observations in women followed-up for an established cervical HPV lesion, which showed that these two viral infections can coexist in the genital tract in some 10% of patients (Syrjänen et al. 1984b; Adam et al. 1985). Although this does not prove that these two agents act synergistically in cervical carcinogenesis, it will be of interest to discover whether concomitant infection by HSV has any influence on the natural history of cervical HPV lesions.

7 Conclusions

Current data implicating a role for HPV infections in squamous cell carcinogenesis can be summarized as follows: (1) there are animal models in which PVs induce malignant transformation; (2) HPV involvement in both benign and malignant human squamous cell tumors has been demonstrated by morphological, immunohistochemical, and DNA hybridization techniques; (3) HPV infections in the genital tract are sexually transmitted and are associated with the same risk factors as cervical carcinoma and its precursors; (4) the natural history of cervical HPV lesions is equivalent to that of CIN, e.g., they are potentially progressive to CIS and invasive carcinoma; (5) malignant transformation of PV-induced lesions probably depends on virus type and the physical state of its DNA; (6) since latent genital HPV infections have been shown to be common, malignant transformation most probably requires synergistic actions between the PVs and chemical or physical carcinogens or other infectious agents; and (7) immunological defence mechanisms of the host are probably capable of modifying the course of PV infections.

Many details of the molecular mechanisms still remain to be clarified, however. Although BPV1 is capable of transforming rodent fibroblasts, it is not yet known how papillomaviruses transform epithelial cells. Improved tissue culture systems for in vitro differentiation of keratinocytes should aid the study of the biology of papillomaviruses and their interaction with cell differentiation and transformation.

Acknowledgements. The original investigations included in this review were supported in part by research grants from the Finnish Cancer Society, from the Medical Research Council of the Academy of Finland (SA07/014), and since 1986 by PHS grant number 1 R01 CA 42010-01 awarded by the National Cancer Institute, DHHS. The skillful technical assistance of Mrs. Heli Eskelinen, Ms. Soili Finska, Ms. Helena Kemiläinen, Mrs. Maritta Lipponen, and Ms. Ritva Savolainen is gratefully acknowledged.

References

Adam E, Kaufman RH, Adler-Storthz K, Melnick JL, Dreesman G (1985) A prospective study of association of herpes simplex virus and human papillomavirus infection with cervical neoplasia in women exposed to diethylstilbestrol in utero. Int J Cancer 35: 19–26

Al-Saleem T, Peale AR, Norris CM (1968) Multiple papillomatosis of the lower respiratory tract. Clinical and pathologic study of eleven cases. Cancer 22: 1173-1184
Amtmann E, Volm M, Wayss K (1984) Tumour induction in the rodent *Mastomys natalensis* by activation of endogenous papilloma virus genomes. Nature 308: 291-292
Andrews FJ, Linehan JJ, Melcher DH (1978) Cervical cancer in younger women. Lancet ii: 776-778
Arnold W (1976) Atiologische Aspekte zur Frage der Entstehung der Larynxpapillome. Laryngol Rhinol 55: 102-111
Bäfverstedt B (1967) Condylomata acuminata-past and present. Acta Derm Venereol (Stockh) 47: 376-381
Bain RW, Crocker DW (1983) Rapid onset of cervical cancer in an upper socioeconomic group. Am J Obstet Gynecol 146: 366-371
Barge J, Molas G, Maillard JN, Fekete F, Bogomoletz WV, Potet F (1981) Superficial oesophageal carcinoma: an oesophageal counterpart of early gastric cancer. Histopathology 5: 499-510
Barrett TJ, Silbar JD, McGinley JP (1954) Genital warts - a venereal disease. JAMA 154: 333-334
Baruah MC, Sardari L, Selvaraju M, Veliath AJ (1984) Perianal condylomata acuminata in a male child. Br J Vener Dis 60: 60-61
Beaudenon S, Kremsdorf D, Jablonska S, Croissant O, Orth G (1985) Molecular cloning and characterization of a new type of human papillomavirus associated with genital neoplasia. Workshop on papilloma viruses: molecular and pathogenetic mechanisms. Kuopio, Finland, August 25-29
Beck I (1984) Vaginal carcinoma arising in vaginal condylomata. Case report. Br J Obstet Gynaecol 91: 503-505
Benedet JL, Sanders BH (1984) Carcinoma in situ of the vagina. Am J Obstet Gynecol 148: 695-700
Berger BW, Hori Y (1978) Multicentric Bowen's disease of the genitalia. Spontaneous regression of lesions. Arch Dermatol 114: 1698-1699
Bernstein SG, Voet RL, Guzick DS, Melancon JT, Ronan-Cowen L, Lifshitz S, Buchsbaum HJ (1985) Prevalence of papillomavirus infection in colposcopically directed cervical biopsy specimens in 1972 and 1982. Am J Obstet Gynecol 151: 577-581
Boshart M, Gissmann L, Ikenberg H, Kleinheinz A, Scheurlen W, zur Hausen H (1984) A new type of papillomavirus DNA, its presence in genital cancer biopsies and in cell lines derived from cervical cancer. EMBO J 3: 1151-1157
Boxer RJ, Skinner DG (1977) Condylomata acuminata and squamous cell carcinoma. Urology 9: 72-78
Boyle WF, Riggs JL, Oshiro LS, Lennette EH (1973) Electron microscopic identification of papova virus in laryngeal papilloma. Laryngoscope 83: 1102-1108
Brandsma JL, Steinberg BM, Abramson AL, Winkler B (1986) Presence of human papillomavirus type 16 related sequences in verrucous carcinoma of the larynx. Cancer Res 46: 2185-2188
Braun L, Kashima H, Eggleston J, Shah KV (1982) Demonstration of papillomavirus antigen in paraffin sections of laryngeal papillomas. Laryngoscope 92: 640-643
Buschke A, Löwenstein L (1930) Spitze Kondylome des Penis und ihre Beziehung zum Peniskarzinom. Z Haut Geschlechtskr 34: 773-774
Campo MS, Moar MH, Jarrett WFH, Laird HM (1980) A new papillomavirus associated with alimentary cancer in cattle. Nature 286: 180-182
Carr G, William DC (1977) Anal warts in a population of gay men in New York. Sex Transm Dis 4: 56-57
Casas-Cordero M, Morin C, Roy M, Fortier M, Meisels A (1981) Origin of the koilocyte in condyloma of the human cervix. Ultrastructural study. Acta Cytol 25: 383-392
Chuang T-Y, Perry HO, Kurland LT, Ilstrup DM (1984) Condyloma acuminatum in Rochester, Minn, 1950-1978: I. Epidemiology and clinical features. Arch Dermatol 120: 469-475
Clarke EA, Morgan RW, Newman AM (1982) Smoking as a risk factor in cancer of the cervix: additional evidence from a case-control study. Am J Epidemiol 115: 59-66
Claudy AL, Touraine JL, Mitanne D (1982) Epidermodysplasia verruciformis induced by a new human papillomavirus (HPV-8). Arch Dermatol Res 274: 213-219
Cocks PS, Peel KR, Cartwright RA, Adib R (1980) Carcinoma of penis and cervix. Lancet ii: 855-856

Coles RB (1958) Virus warts in school children. Public Health 71: 371-377
Colina F, Solis JA, Munoz MT (1980) Squamous papilloma of the esophagus. A report of three cases and review of the literature. Am J Gastroenterol 74: 410-414
Costa J, Howley PM, Bowling MC, Howard R, Bauer WC (1981) Presence of human papilloma viral antigens in juvenile multiple laryngeal papilloma. Am J Clin Pathol 75: 194-197
Croxon T, Chabon AB, Rorat E, Barash IM (1984) Intraepithelial carcinoma of the anus in homosexual men. Dis Colon Rectum 27: 325-330
Crum CP, Levine RU (1984) Human papillomavirus infection and cervical neoplasia: new perspectives. Int J Gynecol Pathol 3: 376-388
Crum CP, Braun LA, Shah KV, Fu YS, Levine RU, Fenoglio CM, Richart RM, Townsend DE (1982a) Vulvar intraepithelial neoplasia: correlation of nuclear DNA content and the presence of a human papilloma virus (HPV) structural antigen. Cancer 49: 468-471
Crum CP, Fu YS, Levine RU, Richart RM, Townsend DE, Fenoglio CM (1982b) Intraepithelial squamous lesions of the vulva: biologic and histologic criteria for the distinction of condylomas from vulvar intraepithelial neoplasia. Am J Obstet Gynecol 144: 77-83
Crum CP, Ikenberg H, Richart RM, Gissmann L (1984) Human papillomavirus type 16 and early cervical neoplasia. N Engl J Med 310: 880-883
Crum CP, Mitao M, Levine RU, Silverstein S (1985) Cervical papillomaviruses segregate within morphologically distinct precancerous lesions. J Virol 54: 675-681
Daling JR, Chu J, Weiss NS, Emel L, Tamini HK (1984) The association of condylomata acuminata and squamous carcinoma of the vulva. Br J Cancer 50: 533-535
Danos O, Katinka M, Yaniv M (1980) Molecular cloning, refined physical map and heterogeneity of methylation sites of papilloma virus type 1a DNA. Eur J Biochem 109: 457-461
Davies SW (1965) Giant condyloma acuminata: incidence among cases diagnosed as carcinoma of the penis. J Clin Pathol 18: 142-149
Dawson DF, Duckworth JK, Bernhardt H, Young JM (1965) Giant condyloma and verrucous carcinoma of the genital area. Arch Pathol Lab Med 79: 225-231
de Brux J, Orth G, Croissant O, Cochard B, Ionesco M (1983) Lesions condylomateuses du col uterin: evolution chez 2466 patientes. Bull Cancer (Paris) 70: 410-422
de Peuter M, de Clercq B, Minette A, Lachapelle JM (1977) An epidemiological survey of virus warts of the hands among butchers. Br J Dermatol 96: 427-431
de Villiers EM, Gissmann L, zur Hausen H (1981) Molecular cloning of viral DNA from human genital warts. J Virol 40: 932-935
Dunn AEG, Ogilvie MM (1968) Intranuclear virus particles in human genital wart tissue: observations on the ultrastructure of the epidermal layer. J Ultrastruct Res 22: 282-295
Dürst M, Gissmann L, Ikenberg H, zur Hausen H (1983) A papillomavirus DNA from a cervical carcinoma and its prevalence in cancer biopsy samples from different geographic regions. Proc Natl Acad Sci USA 80: 3812-3815
Dürst M, Kleinheinz A, Hotz M, Gissmann L (1985) The physical state of human papillomavirus type 16 DNA in benign and malignant genital tumours. J Gen Virol 66: 1515-1522
Ejeckam GC, Idikio HA, Nayak V, Gardiner JP (1983) Malignant transformation in an anal condyloma acuminatum. Can J Surg 26: 170-173
Evans AS, Monaghan JM (1983) Nuclear DNA content of normal, neoplastic and wart-affected cervical biopsies. Anal Quant Cytol 5: 112-116
Extract from the annual report of the Chief Medical Officer of the Department of Health and Social Security for the year 1983 (1985) Sexually transmitted diseases. Genitourin Med 61: 204-207
Favre M, Orth G, Croissant O, Yaniv M (1975) Human papillomavirus DNA: physical map. Proc Natl Acad Sci USA 72: 4810-4814
Fechner RE, Goepfert H, Alford RR (1974) Invasive laryngeal papillomatosis. Arch Otolaryngol 99: 147-151
Feldman MJ, Kent DR, Pennington RL (1978) Intraepithelial neoplasia of the uterine cervix in the teenager. Cancer 41: 1405-1408
Feldman YM (1984) Condylomata acuminata. Cutis 33: 118-120
Fenoglio CM, Ferenczy A (1982) Etiological factors in cervical neoplasia. Semin Oncol 9: 349-372
Ferenczy A, Braun L, Shah KV (1981) Human papillomavirus (HPV) in condylomatous lesions of

cervix. A comparative ultrastructural and immunohistochemical study. Am J Surg Pathol 5: 661-670
Ferenczy A, Mitao M, Nagai N, Silverstein SJ, Crum CP (1985) Latent papillomavirus and recurring genital warts. N Engl J Med 313: 784-788
Franklin EW III, Rutledge FD (1972) Epidemiology of epidermoid carcinoma of the vulva. Obstet Gynecol 39: 165-172
Friedmann I, Osborn DA (1982) Papillomas of the nose and sinuses. In: Friedmann I, Osborn DA (eds). Pathology of granulomas and noplasms of the nose and paranasal sinuses. Churchill Livingstone, Edinburgh
Fu YS, Braun L, Shah KV, Lawrence WD, Robboy SJ (1983a) Histologic, nuclear DNA and human papillomavirus studies of cervical condylomas. Cancer 52: 1705-1711
Fu YS, Reagan JW, Richart RM (1983b) Precursors of cervical cancer. Cancer Surv 2: 359-382
Fujii T, Crum CP, Winkler B, Fu YS, Richart RM (1984) Human papillomavirus infection and cervical intraepithelial neoplasia: histopathology and DNA content. Obstet Gynecol 63: 99-104
Fukushima M, Okagaki T, Twiggs LB, Clark BA, Zachow KR, Ostrow RS, Faras AJ (1985) Histological types of carcinoma of the uterine cervix and the detectability of human papillomavirus DNA. Cancer Res 45: 3252-3255
Galloway TC, Soper GR, Elsen J (1960) Carcinoma of the larynx after irradiation for papilloma. Arch Otolaryngol 72: 289-294
Gilbert EF, Palladino A (1966) Squamous papillomas of the uterine cervix. Review of the literature and report of a giant papillary carcinoma. Am J Clin Pathol 46: 115-121
Gissmann L (1984) Papillomaviruses and their association with cancer in animals and in man. Cancer Surv 3: 161-181
Gissmann L, zur Hausen H (1978) Physical characterization of the deoxyribonucleic acids of different human papilloma viruses (HPV). Med Microbiol Immunol 166: 3-11
Gissmann L, zur Hausen H (1980) Partial characterization of viral DNA from human genital warts (condylomata acuminata). Int J Cancer 25: 605-609
Gissmann L, de Villiers EM, zur Hausen H (1982a) Analysis of human genital warts (condylomata acuminata) and other genital tumors for human papillomavirus type 6 DNA. Int J Cancer 29: 143-146
Gissmann L, Diehl V, Schultz-Coulon HJ, zur Hausen H (1982b) Molecular cloning and characterization of human papilloma virus DNA derived from a laryngeal papilloma. J Virol 44: 393-400
Gissmann L, Wolnik L, Ikenberg H, Koldovsky U, Schnürch HG, zur Hausen H (1983) Human papillomavirus types 6 and 11 DNA sequences in genital and laryngeal papillomas and in some cervical cancers. Proc Natl Acad Sci USA 80: 560-563
Gissmann L, Boshart M, Dürst M, Ikenberg H, Wagner D, zur Hausen H (1984) Presence of human papillomavirus in genital tumors. J Invest Dermatol 83: 26s-28s
Goldberg HM, Pell-Ilderton R, Daw E, Saleh N (1979) Concurrent squamous cell carcinoma of the cervix and penis in a married couple. Br J Obstet Gynaecol 86: 585-586
Goldman L, Feldman M, Levitt S (1976) Condyloma acuminata in infants and children. Arch Dermatol 112: 1329
Goldsmith MF (1984) Papillomavirus invades esophagus, incidence seems to be increasing. JAMA 251: 2185-2187
Graham S, Priore R, Graham M, Browne R, Burnett W, West D (1979) Genital cancer in wives of penile cancer patients. Cancer 44: 1970-1974
Greenberg ER, Vessey M, McPherson K, Yates D (1985) Cigarette smoking and cancer of the uterine cervix. Br J Cancer 51: 139-141
Grigg WK, Wilhelms G (1953) Epidemiological study of plantar warts among school children. Public Health Rep 68: 985-988
Gross G, Hagedorn M, Ikenberg H, Rufli T, Dahlet C, Grosshans E, Gissmann L (1985a) Bowenoid papulosis. Presence of human papillomavirus (HPV) structural antigens and of HPV 16-related DNA sequences. Arch Dermatol 121: 858-863
Gross G, Ikenberg H, Gissmann L, Hagedorn M (1985b) Papillomavirus infection of the anogenital region: correlation between histology, clinical picture, and virus type. Proposal of a new nomenclature. J Invest Dermatol 85: 147-152

Guijon FB, Paraskevas M, Brunham R (1985) The association of sexually transmitted diseases with cervical intraepithelial neoplasia: a case-control study. Am J Obstet Gynecol 151: 185-190

Guillet GY, Braun L, Masse R, Aftimos J, Geniaux M, Texier L (1984) Bowenoid papulosis. Demonstration of human papillomavirus (HPV) with anti-HPV immune serum. Arch Dermatol 120: 514-516

Gupta JW, Gupta PK, Shah KV, Kelly DP (1983) Distribution of human papillomavirus antigen in cervicovaginal smears and cervical tissues. Int J Gynecol Pathol 2: 160-170

Haustein U-F (1982) Epidermodysplasia verruciformis Lewandowsky-Lutz mit multiplen Plattenepithel- und Bowenkarzinomen. Dermatol Monatsschr 168: 821-828

Heilman CA, Law M-F, Israel MA, Howley PM (1980) Cloning of human papilloma virus genomic DNAs and analysis of homologous polynucleotide sequences. J Virol 36: 395-407

Hellström I, Evans CA, Hellström KE (1969) Cellular immunity and its serum-mediated inhibition in Shope-virus-induced rabbit papillomas. Int J Cancer 4: 601-607

Hills E, Laverty CR (1979) Electron microscopic detection of papillomavirus particles in selected koilocytotic cells in routine cervical smears. Acta Cytol 23: 53-56

Howley P (1982) The human papillomaviruses. Arch Pathol Lab Med 106: 429-432

Howley PM (1983) The molecular biology of papillomavirus transformation. Am J Pathol 113: 414-421

Hull MT, Eble JN, Priest JB, Mulcahy JJ (1981) Ultrastructure of Buschke-Löwenstein tumor. J Urol 126: 485-489

Hummer WK, Mussey E, Decker DG, Dockerty MB (1970) Carcinoma in situ of the vagina. Am J Obstet Gynecol 108: 1109-1116

Ikenberg H, Gissmann L, Gross G, Grussendorf-Conen EI, zur Hausen H (1983) Human papillomavirus type 16-related DNA in genital Bowen's disease and in bowenoid papulosis. Int J Cancer 32: 563-565

Incze JS, Lui PS, Strong MS, Vaughan CW, Clemente MP (1977) The morphology of human papillomas of the upper respiratory tract. Cancer 39: 1634-1646

Jablonska S, Formas I (1959) Weitere positive Ergebnisse mit Auto- und Heteroinokulation bei Epidermodysplasia verruciformis Lewandowsky-Lutz. Dermatologica 118: 86-93

Jablonska S, Biczysko W, Jakubowicz K, Dabrowski H (1970) The ultrastructure of transitional states to Bowen's disease and invasive Bowen's carcinoma in epidermodysplasia verruciformis. Dermatologica 140: 186-194

Jablonska S, Dabrowski J, Jakubowicz K (1972) Epidermodysplasia verruciformis as a model in studies on the role of papovaviruses in oncogenesis. Cancer Res 32: 583-589

Jablonska S, Orth G, Jarzabek-Chorzelska M, Rzesa G, Obalek S, Glinski W, Favre M, Croissant O (1978) Immunological studies in epidermodysplasia verruciformis. Bull Cancer 65: 183-190

Jablonska S, Orth G, Jarzabek-Chorzelska M, Glinski W, Obalek S, Rzesa G, Croissant O, Favre M (1979) Twenty-one years of follow-up studies of familial epidermodysplasia verruciformis. Dermatologica 158: 309-327

Jablonska S, Orth G, Lutzner MA (1982) Immunopathology of papillomavirus-induced tumors in different tissues. Springer Semin Immunopathol 5: 33-62

Jacyk WK, Subbuswamy SG (1979) Epidermodysplasia verruciformis in Nigerians. Dermatologica 159: 256-265

Jagella HP, Stegner H-E (1974) Zur Dignität der Condylomata acuminata. Arch Gynaecol 216: 119-132

Jarrett WFH (1978) Transformation of warts to malignancy in alimentary carcinoma in cattle. Bull Cancer 65: 191-194

Javdan P, Pitman ER (1984) Squamous papilloma of esophagus. Digest Dis Science 29: 317-320

Jenson AB, Rosenthal JD, Olson C, Pass F, Lancaster WD, Shah K (1980) Immunologic relatedness of papillomaviruses from different species. JNCI 64: 495-500

Jimerson GK, Merrill JA (1970) Multicentric squamous malignancy involving both cervix and vulva. Cancer 26: 150-153

Junge RE, Sundberg JP, Lancaster WD (1984) Papillomas and squamous cell carcinomas of horses. J Am Vet Med Assoc 185: 656-659

Kahn T, Schwarz E, zur Hausen H (1986) Molecular cloning and characterization of the DNA of a new human papillomavirus (HPV 30) from a laryngeal carcinoma. Int J Cancer 37: 61-65

Kashima H, Shah K, Leventhal B (1985) Clinical and epidemiological features of recurrent respiratory papillomatosis. J Cell Biochem (Suppl) 9C: 81-82
Kaufman R, Koss LG, Kurman RJ, Meisels A, Okagaki T, Patten SF, Reid R, Richart RM, Wied GL (1983) Statement of caution in the interpretation of papillomavirus-associated lesions of the epithelium of uterine cervix. Am J Obstet Gynecol 146: 125
Kazal HL, Long JP (1958) Squamous cell papillomas of the uterine cervix. A report of 20 cases. Cancer 11: 1049-1059
Kerl H, Hödl S, Kratochvil K, Kresbach H (1980) Genitale bowenoide Papulose. Pseudomorbus Bowen der Genitalregion. Hautarzt 31: 105-107
Kessler I (1976) Human cervical cancer as a venereal disease. Cancer Res 36: 783-791
Kessler I (1977) Venereal factors in human cervical cancer. Evidence from marital clusters. Cancer 39: 1912-1919
Kienzler J-L, Laurent R, Coppey J, Favre M, Orth G, Coupez L, Agache P (1979) Epidermodysplasia verruciforme. Données ultrastructurales, virologiques et photobiologiques; à propos d'une observation. Ann Dermatol Venereol 106: 549-563
Kimura S, Hirai A, Harada R, Nagashima M (1978) So-called multicentric pigmented Bowen's disease. Report of a case and a possible etiologic role of human papilloma virus. Dermatologica 157: 229-237
King JFW (1980) Sexual activity as environmental cancer hazard. NY State J Med July: 1253-1257
Kirkup W, Evans AS, Brough AK, Davis JA, O'Loughlin T, Wilkinson G, Monaghan JM (1982) Cervical intraepithelial neoplasia and "warty" atypia: a study of colposcopic, histological and cytological characteristics. Br J Obstet Gynaecol 89: 571-577
Kleinsasser O, Glanz H (1979) Spontane Kanzerisierung nicht bestrahlter juveniler Larynxpapillome. Laryngol Rhinol Otol (Stuttg) 58: 482-489
Kleinsasser O, Oliveira e Cruz G (1973) Juvenile und adulte Kehlkopfpapillome. HNO 21: 97-106
Koss LG (1979) Diagnostic cytology and its histopathologic bases. Lippincott, Philadelphia
Koss LG, Durfee GR (1956) Unusual patterns of squamous epithelium of the uterine cervix: cytologic and pathologic study of koilocytotic atypia. Ann NY Acad Sci 63: 1245-1261
Koss LG, Stewart FW, Foote FW, Jordan MJ, Bader GM, Day E (1963) Some histological aspects of behavior of epidermoid carcinoma in situ and related lesions of the uterine cervix. A long-term prospective study. Cancer 9: 1160-1211
Kremsdorf D, Jablonska S, Favre M, Orth G (1982) Biochemical characterization of two types of human papillomaviruses associated with epidermodysplasia verruciformis. J Virol 43: 436-447
Kremsdorf D, Jablonska S, Favre M, Orth G (1983) Human papillomaviruses associated with epidermodysplasia verruciformis: II. Molecular cloning and biochemical characterization of human papillomavirus 3a, 8, 10, and 12 genomes. J Virol 48: 340-351
Krzyzek RA, Watts SL, Anderson DL, Faras AJ, Pass F (1980) Anogenital warts contain several distinct species of human papillomavirus. J Virol 36: 236-244
Kulski J, Demeter T, Sterrett GF, Shilkin KB (1986) Human papillomavirus DNA in oesophageal carcinoma. Lancet ii: 683-684
Kurman RJ, Jenson AB, Lancaster WD (1983) Papillomavirus infection of the cervix: II. Relationship to intraepithelial neoplasia based on the presence of specific viral structural proteins. Am J Surg Pathol 7: 39-52
Lack EE, Jenson AB, Smith HG, Healy GB, Pass F, Vawter GF (1980) Immunoperoxidase localization of human papillomavirus in laryngeal papillomas. Intervirology 14: 148-154
Lancaster WD, Olson C (1982) Animal papillomaviruses. Microbiol Rev 46: 191-207
Lancaster WD, Theilen GH, Olson C (1979) Hybridization of bovine papilloma virus type 1 and 2 DNA to DNA from virus-induced hamster tumors and naturally occurring equine tumors. Intervirology 11: 227-233
Laverty CR, Russell P, Hills E, Booth N (1978) The significance of non-condyloma wart virus infection of the cervical transformation zone. A review with discussion of two illustrative cases. Acta Cytol 22: 195-201
Lee SH, McGregor DH, Kuziez MN (1981) Malignant transformation of perianal condyloma acuminatum: a case report with review of the literature. Dis Colon Rectum 24: 462-467
Lesec G, Gogusev J, Fermaud H, Gorce D, Lemaitre JP, Verdier A (1985) Présence d'un antigène

viral de groupe papilloma dans un condylome oesophagien chez l'homme. Gastroenterol Clin Biol 9: 166-168
Levine RU, Crum CP, Herman E, Silvers D, Ferenczy A, Richart RM (1984) Cervical papillomavirus infection and intraepithelial neoplasia: a study of male sexual partners. Obstet Gynecol 64: 16-20
Lörincz A, Lancaster WD, Temple G (1985a) Detection and characterization of a new type of human papilloma virus. J Cell Biochem [Suppl] 9C: 75
Lörincz A, Lancaster WD, Jenson B, Kurman R, Delgado G, Sanz L, Reid R, Temple G (1985b) A study of anogenital HPV infections in the United States and South America. Workshop on papilloma viruses: molecular and pathogenetic mechanisms. August 25-29, Kuopio, Finland
Ludwig ME, Lowell DM, Livolsi VA (1981) Cervical condylomatous atypia and its relationship to cervical neoplasia. Am J Clin Pathol 76: 255-262
Lutzner MA (1978) Epidermodysplasia verruciformis. An autosomal recessive disease characterized by viral warts and skin cancer. A model for viral oncogenesis. Bull Cancer 65: 169-182
Lutzner MA (1983) The human papillomaviruses. A review. Arch Dermatol 119: 631-635
Lutzner MA, Croissant O, Ducasse MF, Kreis H, Crosnier J, Orth G (1980) A potentially oncogenic human papillomavirus (HPV-5) found in two renal allograft recipients. J Invest Dermatol 75: 353-356
Lutzner MA, Blanchet-Bardon C, Orth G (1984) Clinical observations, virologic studies, and treatment trials in patients with epidermodysplasia verruciformis, a disease induced by specific human papillomaviruses. J Invest Dermatol 83: 18s-25s
Maier I (1968) Maligne Entartung bestrahlter juveniler Larynxpapillome. Z Laryngol Rhinol 56: 862-869
Marshall JR, Graham S, Byers T, Swanson M, Brasure J (1983) Diet and smoking in the epidemiology of cancer of the cervix. JNCI 70: 847-851
Martinez I (1969) Relationship of squamous cell carcinoma of the cervix uteri to squamous cell carcinoma of the penis among Puerto Rican women married to men with penile carcinoma. Cancer 24: 777-780
Massing AM, Epstein WL (1963) Natural history of warts. A two-year study. Arch Dermatol 87: 306-310
Maxwell RJ, Gibbons JR, O'hara MD (1985) Solitary squamous papilloma of the bronchus. Thorax 40: 68-71
Mazur MT, Cloud GA (1984) The koilocyte and cervical intraepithelial neoplasia: time-trend analysis of a recent decade. Am J Obstet Gynecol 150: 354-358
McCance DJ (1986) Human papillomaviruses and cancer. Biochim Biophys Acta 823: 195-205
Meisels A, Morin C (1981) Human papillomavirus and cnacer of the uterine cervix. Gynecol Oncol 12: 111-123
Meisels A, Fortin R, Roy M (1976) Condylomatous lesions of cervix and vagina: I. Cytologic patterns. Acta Cytol 20: 505-509
Meisels A, Roy M, Fortier M, Morin C (1979) Condylomatous lesions of the cervix. Morphologic and colposcopic diagnosis. Am J Diagn Gynecol Obstet 1: 109-116
Meisels A, Roy M, Fortier M, Morin C, Casas-Cordero M, Shah KV, Turgeon H (1981) Human papillomavirus infection of the cervix. The atypical condyloma. Acta Cytol 25: 7-16
Meisels A, Morin C, Casas-Cordero M (1982) Human papillomavirus infection of the uterine cervix. Int J Gynecol Pathol 1: 75-94
Melnick JL, Allison AC, Butel JS, Eckhart W, Eddy BE, Kit S, Levine AJ, Miles JAR, Pagano JS, Vonka V (1974) Papovariridae. Intervirology 3: 106-120
Mergui JL, de Brux J, Salat-Baroux J (1984) Lesions condylomateuses: étapes de la carcinogenèse cervicale. Rev Fr Gynecol Obstet 79: 443-447
Meyerowitz BR, Shea LT (1971) The natural history of squamous verrucose carcinoma of the esophagus. J Thorac Carciovasc Surg 61: 646-649
Minielly JA, Harrison EG, Fontana RS, Payne WS (1967) Verrucous squamous cell carcinoma of the esophagus. Cancer 20: 2078-2087
Moore RL, Lattes R (1969) Papillomatosis of larynx and bronchi. Case report with a 34-year follow-up. Cancer 12: 117-126
Moreno-Lopez J, Ahola H, Stenlund A, Osterhaus A, Pettersson U (1984) Genome of an avian papillomavirus. J Virol 51: 872-875

Mounts P, Shah KV (1984) Respiratory papillomatosis: etiological relation to genital tract papillomaviruses. Prog Med Virol 29: 90-114
Mounts P, Shah KV, Kashima H (1982) Viral etiology of juvenile- and adult-onset squamous papilloma of the larynx. Proc. Natl Acad Sci USA 79: 5425-5429
Nasiell K, Auer G, Nasiell M, Zetterberg A (1979) Retrospective DNA analyses in cervical dysplasia as related to neoplastic progression or regression. Anal Quant Cytol 1: 103-106
Nasiell K, Nasiell M, Vaclavinkova V (1983) Behavior of moderate cervical dysplasia during long-term follow-up. Obstet Gynecol 61: 609-614
Nasseri M, Wettstein FO, Stevens JG (1982) Two colinear and spliced viral transcripts are present in non-virus-producing benign and malignant neoplasms induced by Shope (rabbit) papillomavirus. J Virol 44: 263-268
Obalek S, Jablonska S, Beaudenon S, Walczak L, Orth G (1986) Bowenoid papulosis of the male and female genitalia: risk of cervical neoplasia. J Am Acad Dermatol 14: 433-444
Okagaki T (1984) Female genital tumors associated with human papillomavirus infection, and the concept of genital neoplasm-papilloma syndrome (GENPS). Pathol Annu 19: 31-62
Okagaki T, Twiggs LB, Zachow KR, Clark BA, Ostrow RS, Faras AJ (1983) Identification of human papillomavirus DNA in cervical and vaginal intraepithelial neoplasia with molecularly cloned virus-specific DNA probes. Int J Gynecol Pathol 2: 153-159
Okagaki T, Clark B, Zachow KR, Twiggs LB, Ostrow RS, Pass F, Faras AJ (1984) Presence of human papillomavirus in verrucous carcinoma (Ackerman) of the vagina. Arch Pathol Lab Med 108: 567-570
Olson C, Cook RH (1951) Cutaneous sarcoma-like lesions of the horse caused by the agent of bovine papilloma. Proc Soc Exp Biol Med 77: 281-184
Oriel JD (1971) Anal warts and anal coitus. Br J Vener Dis 47: 373-376
Oriel JD (1981) Genital warts. Sex Transm Dis 8: 326-329
Orth G, Jablonska S, Breitburd F, Favre M, Croissant O (1978a) The human papillomaviruses. Bull Cancer 65: 151-164
Orth G, Jablonska S, Favre M, Croissant O, Jarzabek-Chorzelska M, Rzesa G (1978b) Characterization of two types of human papillomaviruses in lesions of epidermodysplasia verruciformis. Proc Natl Acad Sci USA 75: 1537-1541
Ostrow R, Krzyzek R, Pass F, Faras AJ (1981) Identification of a novel papilloma virus in cutaneous warts of meathandlers. Virology 108: 21-27
Ostrow RS, Zachow KR, Niimura M, Okagaki T, Muller S, Bender M, Faras AJ (1986) Detection of papillomavirus DNA in human semen. Science 231: 731-733
Ottenjann R, Kühner W, Weingart J (1984) Papillome der Speiseröhre - potentielle Präkursoren des Plattenepithelkarzinoms? Dtsch Med Wochenschr 109: 613-615
Pater MM, Pater A (1985) Human papillomavirus types 16 and 18 sequences in carcinoma cell lines of the cervix. Virology 145: 313-318
Pfister H (1984) Biology and biochemistry of papillomaviruses. Rev Physiol Biochem Pharmacol 99: 112-181
Pfister H, zur Hausen H (1978) Seroepidemiological studies of human papilloma virus (HPV-1) infections. Int J Cancer 21: 161-165
Pfister H, Nürnberger F, Gissmann L, zur Hausen H (1981) Characterization of a human papillomavirus from epidermodysplasia verruciformis lesions of a patient from Upper Volta. Int J Cancer 27: 645-650
Pfister H, Gassemaier A, Nürnberger F, Stüttgen G (1983) Human papilloma virus 5-DNA in a carcinoma of an epidermodysplasia verruciformis patient infected with various human papillomavirus types. Cancer Res 43: 1436-1441
Pilotti S, Rilke F, De Palo G, Della Torre G, Alasio L (1981) Condylomata of the uterine cervix and koilocytosis of cervical intraepithelial neoplasia. J Clin Pathol 34: 532-541
Pilotti S, Rilke F, Shah KV, Della Torre G, De Palo G (1984) Immunohistochemical and ultrastructural evidence of papilloma virus infection associated with in situ and microinvasive squamous cell carcinoma of the vulva. Am J Surg Pathol 8: 751-761
Pitkin RM, Kent TH (1963) Papillary squamous lesions of the uterine cervix. A difficult problem in diagnosis. Am J Obstet Gynecol 85: 440-44
Purola E, Savia E (1977) Cytology of gynecologic condyloma acuminatum. Acta Cytol 21: 26-31

Purola E, Halila H, Vesterinen E (1983) Condyloma and cervical epithelial atypias in young women. Gynecol Oncol 16: 34-40

Pyrhönen S (1978) Antibody response against human papilloma viruses. MD Thesis, University of Helsinki, pp 1-41

Quick CA, Foucar E, Dehner LP (1979) Frequency and significance of epithelial atypia in laryngeal papillomatosis. Laryngoscope 89: 550-560

Quick CA, Krzyzek RA, Watts SL, Faras AJ (1980) Relationship between condylomata and laryngeal papillomata. Clinical and molecular virological evidence. Ann Otol Rhinol Laryngol 89: 467-471

Qizilbash AH (1974) Papillary squamous tumors of the uterine cervix. A clinical and pathologica study of 21 cases. Am J Clin Pathol 61: 508-520

Rahman A, Ziment I (1983) Tracheobronchial papillomatosis with malignant transformation. Arch Intern Med 143: 577-578

Rando RF, Sedlacek TV, Hunt J, Jenson AB, Kurman RJ, Lancaster WD (1986) Verrucous carcinoma of the vulva associated with an unusual type 6 human papillomavirus. Obstet Gynecol 78: 70S-75S

Rastkar G, Okagaki T, Twiggs LB, Clark BA (1982) Early invasive and in situ warty carcinoma of the vulva: clinical, histologic, and electron microscopic study with particvular reference to viral association. Am J Obstet Gynecol 143: 814-820

Reid R, Fu YS, Herschman BR, Crum CP, Braun L, Shah KV, Agronow SJ, Stanhope R (1984) Genital warts and cervical cancer: VI. The relationship between aneuploid and polyploid cervical lesions. Am J Obstet Gynecol 150: 189-199

Resler DR, Snow JB (1967) Cell free filtrate transplantation of human laryngeal papilloma to dogs. Laryngoscope 77: 397-416

Respler D, Jahn A, Pater A, Pater MM (1985) Isolation and characterization of papillomavirus DNA sequences from the papillomas of the nose and vocal cord. Workshop on papilloma viruses: molecular and pathogenetic mechanisms. August 25-29, Kuopio, Finland

Rhatigan RM, Saffos RO (1977) Condyloma acuminatum and squamous carcinoma of the vulva. South Med J 70: 591-594

Richardson AC, Lyon JB (1981) The effect of condom use on squamous cell cervical intraepithelial neoplasia. Am J Obstet Gynecol 140: 909-913

Rogers S, Rous P (1951) Joint action of a chemical carcinogen and a neoplastic virus to induce cancer in rabbits. Results of exposing epidermal cells to a carcinogenic hydrocarbon at time of infection with the Shope papilloma virus. J Exp Med 93: 459-488

Roglic M, Jukic S, Damjanov I (1975) Cytology of the solitary papilloma of the bronchus. Acta Cytol 19: 11-13

Rotkin ID (1967) Adolescent coitus and cervical cancer: associations of related events with increased risk. Cancer Res 27: 603-617

Rotkin ID (1973) A comparison review of key epidemiological studies in cervical cancer related to current searches for transmissible agents. Cancer Res 33: 1353-1367

Rous P, Friedewald WF (1944) The effect of chemical carcinogens on virus-induced rabbit papillomas. J Exp Med 79: 511-537

Rous P, Kidd JG (1938) The carcinogenic effect of a papilloma virus on the tarred skin of rabbits: I. Description of the phenomenon. J Exp Med 67: 399-422

Roy M, Meisels A, Fortier M, Morin C, Casas-Cordero M, Robitaille CG (1981) Vaginal condylomata: a human papillomavirus infection. Clin Obstet Gynecol 24: 461-483

Rubel LR, Reynolds RE (1979) Cytologic description of squamous cell papilloma of the respiratory tract. Acta Cytol 23: 227-230

Ruiter M (1973) On the histomorphology and origin of malignant cutaneous changes in epidermodysplasia verruciformis. Acta Derm Venereol (Stockh) 53: 290-298

Ruiter M, van Mullem PJ (1970) Behavior of virus in malignant degeneration of skin lesions in epidermodysplasia verruciformis. J Invest Dermatol 54: 324-331

Runckel D, Kessler S (1980) Bronchogenic squamous cell carcinoma in nonirradiated juvenile laryngotracheal papillomatosis. Am J Surg Pathol 4: 293-296

Schmauz R, Owor R (1980) Epidemiology of malignant degeneration of condylomata acuminata in Uganda. Pathol Res Pract 170: 91-103

Schmauz R, Findlay M, Lalwak A, Katsumbira N, Buxton E (1977) Variation in the appearance

of giant condyloma in an Ugandan series of cases of carcinoma of the penis. Cancer 40: 1686-1696
Schneider A, Kraus H, Schuhmann R, Gissmann L (1985) Papillomavirus infection of the lower genital tract: detection of viral DNA in gynecological swabs. Int J Cancer 35: 443-448
Schneider PS, Krumholz BA, Topp WC, Steinberg BM, Abramson AL (1984) Molecular heterogeneity of female genital wart (condylomata acuminata) papilloma viruses. Int J Gynecol Pathol 2: 329-336
Scholl SM, Pillers EMK, Robinson RE, Farrell PJ (1985) Prevalence of human papillomavirus type 16 DNA in cervical carcinoma samples in East Anglia. Int J Cancer 35: 215-218
Schouten TJ, van den Broek P, Cremers CWRJ, Jongerius CM, Meyer JWR, Vooys GP (1983) Interferons and bronchogenic carcinoma in juvenile laryngeal papillomatosis. Arch Otolaryngol 109: 289-291
Shafeek MA, Osman MI, Hussein MA (1979) Carcinoma of the vulva arising in condylomata acuminata. Obstet Gynecol 54: 120-123
Shapiro RS, Marlowe FI, Butcher J (1976) Malignant degeneration of nonirradiated juvenile laryngeal papillomatosis. Ann Otol Rhinol Laryngol 85: 101-104
Shope RE (1933) Infectious papillomatosis of rabbits. J Exp Med 68: 607-624
Shu Y-I (1985) The cytopathology of esophageal carcinoma. Masson, New York
Singer A (1983) Sex and genital cancer in heterosexual women. J Reprod Med 28: 109-115
Singer A, Reid BL, Coppleson M (1976) A hypothesis: the role of a high-risk male in the etiology of cervical carcinoma. Am J Obstet Gynecol 126: 110-115
Smith PG, Kinlen LJ, White GC, Adelstein AM, Fox AJ (1980) Mortality of wives of men dying with cancer of the penis. Br J Cancer 41: 422-428
Snyder RN, Ortiz Y, Willie S, Cove JKJ (1976) Dysplasia and carcinoma in situ of the uterine cervix: prevalence in very young women (under age 22). A one-year study in a healthy plan population. Am J Obstet Gynecol 124: 751-756
Spoendlin H, Kistler G (1978) Papova-virus in human laryngeal papillomas. Arch Otorhinolaryngol 218: 289-292
Stanbridge CM, Butler EB (1983) Human papillomavirus infection of the lower female genital tract: association with multicentric neoplasia. Int J Gynecol Pathol 2: 264-274
Steffen C (1982) Concurrence of condylomata acuminata and bowenoid papulosis. Confirmation of the hypothesis that they are related conditions. Am J Dermatopathol 4: 5-8
Steinberg BM, Topp WC, Schneider PS, Abramson AL (1983) Laryngeal papillomavirus infection during clinical remission. N Engl J Med 308: 1261-1264
Stremlau A, Gissmann L, Ikenberg H, Stark M, Bannasch P, zur Hausen H (1985) Human papillomavirus type 16 related DNA in an anaplastic carcinoma of the lung. Cancer 55: 1737-1740
Suckow EE, Yokoo H, Brock DR (1962) Intraepithelial carcinoma concomitant with esophageal carcinoma. Cancer 15: 733-740
Sundberg JP, Williams ES, Hill D, Lancaster WD, Nielsen SW (1985) Detection of papillomavirus in cutaneous fibromas of white-tailed and mule deer. Am J Vet Res 46: 1145-1149
Suprun HZ, Schwartz J, Spira H (1985) Cervical intraepithelial neoplasia and associated condylomatous lesions. A preliminary report on 4764 women from Northern Israel. Acta Cytol 29: 334-340
Syrjänen KJ (1979a) Histological and cytological evidence of a condylomatous lesion in association with an invasive carcinoma of uterine cervix. Arch Geschwulstforsch 49: 436-444
Syrjänen KJ (1979b) Morphologic survey of the condylomatous lesions in dysplastic and neoplastic epithelium of the uterine cervix. Arch Gynecol 227: 153-161
Syrjänen KJ (1979c) Condylomatous changes in neoplastic bronchial epithelium. Report of a case. Respiration 38: 299-304
Syrjänen KJ (1980a) Current views on condylomatous lesions in the uterine cervix and their possible relationship to cervical squamous cell carcinoma. Obstet Gynecol Surv 35: 685-694
Syrjänen KJ (1980b) Epithelial lesions suggestive of a condylomatous origin found closely associated with invasive bronchial squamous cell carcinomas. Respiration 40: 150-160
Syrjänen KJ (1980c) Bronchial squamous cell carcinomas associated with epithelial changes identical to condylomatous lesions of the uterine cervix. Lung 158: 131-142
Syrjänen KJ (1982) Histological changes identical to those of condylomatous lesions found in esophageal squamous cell carcinomas. Arch Geschwulstforsch 52: 283-292

Syrjänen KJ (1983) Human papillomavirus (HPV) lesions in association with cervical dysplasias and neoplasias. Obstet Gynecol 62: 617-622
Syrjänen KJ (1984a) Current concepts on human papillomavirus (HPV) infections in the genital tract and their relationship to intraepithelial neoplasia and squamous cell carcinoma. Obstet Gynecol Surv 39: 252-265
Syrjänen KJ (1984b) Female genital tract infections by human papillomavirus (HPV) and their association with intraepithelial neoplasia and squamous cell carcinoma. Cervix and Lower Female Genital Tract 2: 103-126
Syrjänen KJ (1986) Human papillomavirus (HPV) infections of the female genital tract and their associations with intraepithelial neoplasia and squamous cell carcinoma. Pathol Annu 21: 53-89
Syrjänen KJ, Pyrhönen S (1982a) Demonstration of human papilloma virus (HPV) antigen in the condylomatous lesions of the uterine cervix by immunoperoxidase technique. Gynecol Obstet Invest 14: 90-96
Syrjänen KJ, Pyrhönen S (1982b) Immunoperoxidase demonstration of human papilloma virus (HPV) in dysplastic lesions of the uterine cervix. Arch Gynecol 233: 53-61
Syrjänen K, Syrjänen S (1987) Human papillomavirus DNA in bronchial squamous cell carcinomas. Lancet i: 167-168
Syrjänen KJ, Heinonen UM, Kauraniemi T (1981) Cytological evidence of the association of condylomatous lesions with the dysplastic and neoplastic changes in uterine cervix. Acta Cytol 25: 17-22
Syrjänen K, Syrjänen S, Pyrhönen S (1982a) Human papilloma virus (HPV) antigens in lesions of laryngeal squamous cell carcinomas. ORL 44: 323-334
Syrjänen K, Pyrhönen S, Aukee S, Koskela E (1982b) Squamous cell papilloma of the oesophagus: a tumor probably caused by human papilloma virus (HPV). Diagn Histopathol 5: 291-296
Syrjänen KJ, Pyrhönen S, Syrjänen SM (1983) Evidence suggesting human papillomavirus (HPV) etiology for the squamous cell papilloma of the paranasal sinus. Arch Geschwulstforsch 53: 77-82
Syrjänen K, Väyrynen M, Castren O, Yliskoski M, Mäntyjärvi R, Pyrhönen S, Saarikoski S (1984a) Sexual behaviour of the females with human papillomavirus (HPV) lesions in the uterine cervix. Br J Vener Dis 60: 243-248
Syrjänen KJ, Mäntyjärvi R, Väyrynen M, Castren O, Yliskoski M, Saarikoski M, Pyrhönen S (1984b) Herpes simplex virus infection of females with human papillomavirus (HPV) lesions in the uterine cervix. Cervix and Lower Female Genital Tract 2: 25-32
Syrjänen K, Väyrynen M, Saarikoski S, Mäntyjärvi R, Parkkinen S, Hippeläinen M, Castren O (1985a) Natural history of cervical human papillomavirus (HPV) infections based on prospective follow-up. Br J Obstet Gynaecol 92: 1086-1092
Syrjänen K, Väyrynen M, Hippeläinen M, Castren O, Saarikoski S, Mäntyjärvi R (1985b) Electron microscopic assessment of cervical punch biopsies in women followed-up for human papillomavirus (HPV) lesions. Arch Geschwulstforsch 55: 131-138
Syrjänen K, de Villiers E-M, Väyrynen M, Mäntyjärvi R, Parkkinen S, Saarikoski S, Castren O (1985c) Cervical papillomavirus infection progressing to invasive cancer in less than three years. Lancet i: 510-511
Syrjänen K, Väyrynen M, Mäntyjärvi R, Parkkinen S, Saarikoski S, Syrjänen S, Castren O (1985d) Natural history of HPV infections in uterine cervix as determined by prospective follow-up. In: Howley PM, Broker TR (eds) Papilloma viruses: molecular and clinical aspects. UCLA Symp Mol Cell Biol New Ser 32: 31-45
Syrjänen K, Mäntyjärvi R, Parkkinen S, Väyrynen M, Saarikoski S, Syrjänen S, Castren O (1985e) Prospective follow-up in assessment of the biological behaviour of cervical HPV-associated dysplastic lesions. Viral etiology of cervical cancer, Banbury report 21: 167-177
Syrjänen K, Parkkinen S, Mäntyjärvi R, Väyrynen M, Syrjänen S, Holopainen H, Saarikoski S, Castren O (1985f) Human papillomavirus (HPV) type as an important determinant of the natural history of HPV infections in uterine cervix. Eur J Epidemiol 1: 180-187
Syrjänen KJ, Mäntyjärvi R, Väyrynen M, Syrjänen S, Parkkinen S, Yliskoski M, Saarikoski S, Sarkkinen H, Nurmi T, Castren O (1986a) Assessing the biological potential of HPV infections in cervical carcinogenesis. Cold Spring Harbor Laboratory. Cancer Cells 5, Papillomaviruses (to be published)

Syrjänen K, Mäntyjärvi R, Väyrynen M, Syrjänen S, Parkkinen S, Yliskoski M, Saarikoski S, Castren O (1986b) Evolution of human papillomavirus (HPV) infections in the uterine cervix during a long-term prospective follow-up. Appl Pathol (to be published)

Syrjänen S, von Krogh G, Syrjänen K (1987a) Detection of human papillomavirus (HPV) DNA in anogenital condylomata using in situ DNA hybridization applied to paraffin-sections. Genitourin Med 63: 32-39

Syrjänen S, Syrjänen K, Mäntyjärvi R, Collan Y, Kärjä J (1987b) Human papillomavirus (HPV) DNA sequences in squamous cell carcinomas of the larynx demonstrated by in situ DNA hybridization. ORL (to be published)

Syrjänen S, Happonen R-P, Virolainen E, Siivonen L, Syrjänen K (1987c) Detection of human papillomavirus (HPV) DNA in nasal inverted papillomas and squamous cell carcinomas by in situ hybridization. Acta Otolaryngol (Stockh) (to be published)

Syrjänen K, Mäntyjärvi R, Väyrynen M, Syrjänen S, Parkkinen S, Yliskoski M, Saarikoski S, Castren O (1987d) Human papillomavirus (HPV) infections involved in the neoplastic process of the uterine cervix as established by prospective follow-up of 513 women for two years. Eur J Gynaecol Oncol 8: 5-16

Syverton JT (1952) The pathogenesis of the rabbit papilloma-to-carcinoma sequence. Ann NY Acad Sci 54: 1126-1142

Trevethan E, Layde P, Webster LA, Adams JB, Benigno BB, Ory H (1983) Cigarette smoking and dysplasia and carcinoma in situ of the uterine cervix. JAMA 250: 499-502

Ushigome S, Spjut HJ, Noon GP (1967) Extensive dysplasia and carcinoma in situ of esophageal epithelium. Cancer 20: 1023-1029

Väyrynen M, Romppanen T, Koskela E, Castren O, Syrjänen K (1981) Verrucous squamous cell carcinoma of the female genital tract. Report of three cases and survey of the literature. Int J Gynaecol Obstet 19: 351-356

Villa LL, Lopes A (1985) Presence of papillomavirus DNA in human penile carcinoma. Workshop on papilloma viruses: molecular and pathogenetic mechanisms. August 25-29, Kuopio, Finland

von Krogh G (1983) Condyloma acuminata 1983: an up-dated review. Semin Dermatol 2: 109-129

Wade TR, Kopf AW, Ackerman AB (1979) Bowenoid papulosis of the genitalia. Arch Dermatol 115: 306-308

Wagner D, Ikenberg H, Boehm N, Gissmann L (1984) Identification of human papillomavirus in cervical swabs by deoxyribonucleic acid in situ hybridization. Obstet Gynecol 64: 767-772

Walker PG, Singer A, Dyson JL, Oriel JD (1983) Natural history of cervical epithelial abnormalities in patients with vulvar warts. Br J Vener Dis 59: 327-329

Weed JC, Lozier C, Daniel SJ (1983) Human papilloma virus in multifocal, invasive female genital tract malignancy. Obstet Gynecol 62: 83S-87S

Weiss M, Kashima H (1983) Tracheal involvement in laryngeal papillomatosis. Laryngoscope 93: 45-48

Winkler B, Crum CP, Fujii T, Ferenczy A, Boon M, Braun L, Lancaster WD, Richart RM (1984) Koilocytotic lesions of the cervix. The relationship of mitotic abnormalities to the presence of papillomavirus antigens and nuclear DNA content. Cancer 53: 1081-1087

Winkler B, Capo V, Reumann W, Averill MA, La Porta R, Reilly S, Green PMR, Richart RM, Crum CP (1985) Human papillomavirus infection of the esophagus. A clinicopathologic study with demonstration of papillomavirus antigen by the immunoperoxidase technique. Cancer 55: 149-155

Woodruff JD, Peterson WF (1958) Condyloma acuminata of the cervix. Am J Obstet Gynecol 75: 1354-1362

Woodruff JD, Braun L, Cavalieri R, Gupta P, Pass F, Shah KV (1980) Immunologic identification of papillomavirus antigen in condyloma tissues from the female genital tract. Obstet Gynecol 56: 727-732

Yabe Y, Koyama H (1973) Virus and carcinogenesis in epidermodysplasia verruciformis. Gann 64: 167-172

Yee C, Krishnan-Hewlett I, Baker CC, Schlegel R, Howley PM (1985) Presence and expression of human papillomavirus sequences in human cervical carcinoma cell lines. Am J Pathol 119: 361-366

Yoder MG, Batsakis JG (1980) Squamous cell carcinoma in solitary laryngeal papilloma. Otolaryngol Head Neck Surg 88: 745-748
Yutsudo M, Tanigaki T, Tsumori T, Watanabe S, Hakura A (1982) New human papilloma virus isolated from epidermodysplasia verruciformis lesions. Cancer Res 42: 2440-2443
Zachow KR, Ostrow RS, Bender M, Watts S, Okagaki T, Pass F, Faras AJ (1982) Detection of human papillomavirus DNA in anogenital neoplasias. Nature 300: 771-773
Zehnder PR, Lyons GD (1975) Carcinoma and juvenile papillomatosis. Ann Otol Rhinol Laryngol 84: 614-618
Zuna R (1984) Association of condylomas with intraepithelial and microinvasive cervical neoplasia: histopathology of conization and hysterectomy specimens. Int J Gynecol Pathol 2: 364-372
zur Hausen H (1977) Human papillomaviruses and their possible role in squamous cell carcinomas. Curr Top Microbiol Immunol 78: 1-30
zur Hausen H (1982) Human genital cancer: synergism between two virus infections or synergism between a virus infection and initiating events? Lancet ii: 1370-1372
zur Hausen H (1985) Genital papillomavirus infections. Prog Med Virol 32: 15-21
zur Hausen H (1986) Intracellular surveillance of persisting viral infections. Human genital cancer results from deficient cellular control of papillomavirus gene expression. Lancet ii: 489-491
zur Hausen H, Gissmann L (1980) Papillomaviruses. In: Klein G (ed) Viral oncology. Raven, New York, pp 433-445
zur Hausen H, Meinhof W, Schreiber W, Bornkamm GW (1974) Attempts to detect virus-specific DNA sequences in human tumors: I. Nucleic acid hybridizations with complementary RNA of human wart virus. Int J Cancer 13: 650-656
zur Hausen H, Gissmann L, Schlehofer JR (1984) Viruses in the etiology of human genital cancer. Prog Med Virol 30: 170-186

Subject Index

SPRINGER NATURE

GPSR Compliance

The European Union's (EU) General Product Safety Regulation (GPSR) is a set of rules that requires consumer products to be safe and our obligations to ensure this.

If you have any concerns about our products, you can contact us on ProductSafety@springernature.com

In case Publisher is established outside the EU, the EU authorized representative is:

Springer Nature Customer Service Center GmbH
Europaplatz 3
69115 Heidelberg, Germany

FSC
www.fsc.org
MIX
Papier | Fördert gute Waldnutzung
FSC® C083411

Kolibri 360 GmbH
Ferdinand-Jühlke-Straße 7
99095 Erfurt, Deutschland
produktsicherheit@kolibri360.de